Lecture Notes in Civil Engineering

Volume 91

Lecture Notes in Civil Engineering (LNCE) publishes the latest developments in Civil Engineering—quickly, informally and in top quality. Though original research reported in proceedings and post-proceedings represents the core of LNCE, edited volumes of exceptionally high quality and interest may also be considered for publication. Volumes published in LNCE embrace all aspects and subfields of, as well as new challenges in, Civil Engineering. Topics in the series include:

- Construction and Structural Mechanics
- Building Materials
- Concrete, Steel and Timber Structures
- Geotechnical Engineering
- Earthquake Engineering
- Coastal Engineering
- Ocean and Offshore Engineering; Ships and Floating Structures
- Hydraulics, Hydrology and Water Resources Engineering
- Environmental Engineering and Sustainability
- Structural Health and Monitoring
- Surveying and Geographical Information Systems
- Indoor Environments
- Transportation and Traffic
- Risk Analysis
- Safety and Security

To submit a proposal or request further information, please contact the appropriate Springer Editor:

- Mr. Pierpaolo Riva at pierpaolo.riva@springer.com (Europe and Americas);
- Ms. Swati Meherishi at swati.meherishi@springer.com (Asia—except China, and Australia, New Zealand);
- Dr. Mengchu Huang at mengchu.huang@springer.com (China).

All books in the series now indexed by Scopus and EI Compendex database!

More information about this series at http://www.springer.com/series/15087

Gabriella Bolzon · Donatella Sterpi ·
Guido Mazzà · Antonella Frigerio
Editors

Numerical Analysis of Dams

Proceedings of the 15th ICOLD International Benchmark Workshop

Springer

Editors
Gabriella Bolzon
Department of Civil and Environmental Engineering (DICA)
Politecnico di Milano
Milano, Italy

Donatella Sterpi
Department of Civil and Environmental Engineering (DICA)
Politecnico di Milano
Milano, Italy

Guido Mazzà
Italian Committee on Large Dams (ITCOLD)
Roma, Italy

Antonella Frigerio
Ricerca sul Sistema Energetico—RSE SpA
Milano, Italy

ISSN 2366-2557 ISSN 2366-2565 (electronic)
Lecture Notes in Civil Engineering
ISBN 978-3-030-51084-8 ISBN 978-3-030-51085-5 (eBook)
https://doi.org/10.1007/978-3-030-51085-5

This Springer imprint is published by the registered company Springer Nature Switzerland AG
The registered company address is: Gewerbestrasse 11, 6330 Cham, Switzerland

ICOLD Technical Committee

Computational Aspects of Analysis and Design of Dams

ICOLD, the International Commission on Large Dams, appointed in 1988 the ad hoc Committee Computational Aspects of Analysis and Design of Dams that was converted into a permanent Technical Committee in 2005. The main objective assigned to the Committee was to fill the gap existing between the specialists of numerical modelling and the different professional figures involved in the dam sector, i.e. dam designers, authorities, and managers. A second objective was to contribute to the diffusion of engineering software in the field of dam engineering, introducing new approaches in a context traditionally refractory to innovations. The Committee is also strongly committed in the transfer of experience, skill and knowledge across generations.

The work done by the Committee during its long activity lead to the issuing of three Technical Bulletins (1–3).

Present and future activities of the Committee are oriented towards:

- creating a stronger link between the observed dam behaviour and the modelling process, with the aim to contribute to the preservation and maintenance of existing dams;
- making advanced tools increasingly available to the professional world, for improved design of new dams;
- promoting mathematical modelling improvements to approach safety-related problems that cannot at present be properly analysed.
- issuing guidelines to be used for educational purposes in the current practice.

The benchmarking program started in Bergamo (Italy) in 1991 and has continued over the time in different Countries, with the 15th event of the series organized in Milano (Italy) on 9–11 September 2019.

The active involvement in the benchmarking activities of technicians operating in the dam sector is visualized in Fig. 1. The examination of these data suggests the following considerations.

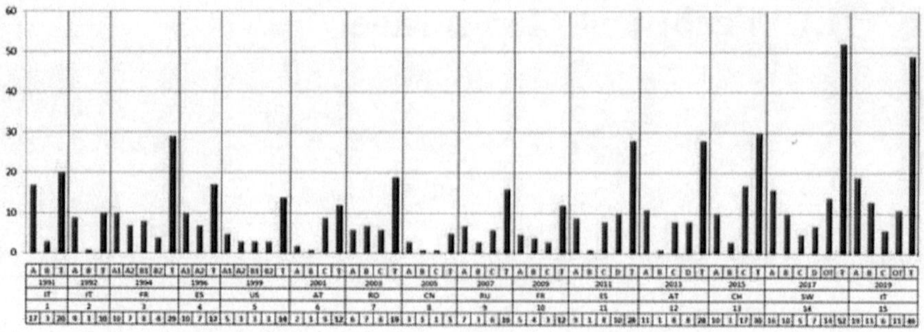

Fig. 1 Contributors to the 15 Benchmark Workshops held from 1991 to 2019 per theme and date (49 themes proposed in total); red figures and bars indicate the total number of contributors

- Until the 10th Benchmark Workshop (year 2009), the number of active participants (who provided a solution of the proposed problems) was mainly between 10 and 20; this number has been rapidly increasing in the subsequent editions.
- The total number of attendees has been always quite high, in the order of 100 people, with peaks of up to 190 participants in the last event in Milan.
- From a geographical point of view, the Benchmark Workshops took place mainly in European locations, with the exceptions of Denver (USA) in 1999 and Wuhan (China) in 2005. European centrality is partly linked to the prevalence of European members in the ICOLD Committee "Computational Aspects of Analysis and Design of Dams". However, the interest in the numerical modelling of dams has been increasing in the last few years in US and China, and these countries are prone to host these events soon.

The problem description together with the input data, and a report written by the formulator (about the individual results and their comparison) are collected by the proceedings that are available in ICOLD website.

A capitalization activity of the tremendous amount of data at disposal has been recently started with the aim to facilitate the accessibility and the critical review of the content of the Benchmark Workshops held so far. A new Bulletin—or an equivalent document—will provide a synthesis of the results, outlining the progresses made by numerical methods in dam engineering.

The benchmarking activities represent a reference for the whole dam community, and in particular for young engineers, engaged in the challenging task of ensuring dam safety.

Preface

This book collects the contributions presented at the 15th International Benchmark Workshop on Numerical Analysis of Dams held in Milano in September 2019.

These Benchmark Workshops are promoted by the Technical Committee on 'Computational Aspects of Analysis and Design of Dams' of ICOLD (International Commission of Large Dams) to provide an opportunity for engineers, researchers and operators to present and exchange their experiences and the latest developments related to the design, performance and monitoring of dams.

This series of almost biannual events was launched in Italy in 1991 and then hosted in several Countries all over the World. The topics and the approaches considered in the different editions reflect the evolution of the computer analysis tools and of the safety assessment criteria over the last decades.

The first benchmarks were focused on the validation of dedicated numerical codes, in their initial development phase. The predictive capabilities of available sophisticated computing facilities represent the challenges of the recent work. The main issues concern the selection of the most appropriate geometrical and constitutive models and the calibration of the parameters that define the conditions to be analyzed in reality.

The 15th International Benchmark Workshop on Numerical Analysis of Dams dealt with three main topics.

Theme A considers the earthquake response of a concrete gravity dam. The role of the foundation size and mass in the model, the interaction with the dam body and the non-linear material response represent the main debated aspects, with the ultimate goals of identifying the key uncertainties causing differences in results and developing the best practices in the advanced dynamic analysis of these massive structures. The benchmark problem has been formulated by: Jerzy W. Salamon and Christopher Wood (U.S. Bureau of Reclamation, USA); M. Amin Hariri-Ardebili (University of Colorado at Boulder, USA); Richard Malm (KTH Royal Institute of Technology, Sweden) and Giorgia Faggiani (Ricerca sul Sistema Energetico—RSE S.p.A., Italy).

Theme B refers to the seismic analysis of embankment dams. A case study is defined with the aim of understanding how different assumptions on the constitutive model of the rockfill material may affect the results. The stability conditions (in proximity to failure) of the dam after the assigned earthquake and, additionally, the potential damages of the bituminous facing represent problems to be specifically addressed in the proposal formulated by: Giacomo Russo (University of Napoli Federico II and University of Cassino and Southern Lazio, Italy); Manuela Cecconi, Alessia Vecchietti and Vincenzo Pane (University of Perugia, Italy); Andrea Fiorino and Sergio De Marco (So.Ri.Cal, Italy).

Theme C introduces to the pre-failure and failure behavior of a dyke on soft subsoil. The results of coupled hydro-mechanical analyses are compared with the monitoring data collected in a full-scale controlled failure test. The potential of current plane analyses to predict the failure and pre-failure response of the dyke is assessed. The improvement possibly provided by three-dimensional models taking into account the spatial variability of the soil layers is further evaluated. The benchmark formulation and the experimental information have been provided by: Cristina Jommi (Politecnico di Milano, Italy, and Delft University of Technology, the Netherlands); Tom de Gast, Elisa Ponzoni, Niccolò Valimberti and Stefano Muraro (Delft University of Technology, the Netherlands); Ludolph Wentholt and Henk van Hemert (STOWA, the Netherlands).

This volume contains both the problem definition (data concerning geometry, material properties, loading conditions, seismic input, etc.) and the throughout synthesis of the results, worked out by the formulators of each theme, together with the individual contributions by the benchmark participants.

Furthermore, the book introduces a series of topics relevant to the present practice in the computational analysis of dams and of the relevant appurtenant structures. The contributions collected in this volume focus on the study of dams with interesting features (Eberlaste, Grand Dixence, Nam Ngum 3, Ridracoli). Specific aspects of the numerical analysis of these infrastructures are discussed, with an open eye on: monitoring and identification problems; uncertainty quantification and reduction; classification models; visualisation tools and machine learning; cost risk assessment. These topics are proposed in the frame of an Open Theme session that allows to identify possible topics to be proposed in the future Benchmark Workshops.

All together, these contributions constitute a valuable reference for the whole engineering community concerned with safety, planning, design, construction, operation and maintenance of dams.

The 15th International Benchmark Workshop was organized in the main Italian financial city that hosts the headquarters of several global Companies and the seat of many International Institutions. Milano represents the cultural site and an attraction place for many professionals, including designers, fashion-operators, artists, photographers. The rich environment created by its historical heritage, together with several new quarters designed by outstanding architecture studios make Milano the Italian city of glamour with a vivid cultural and social life.

The editors acknowledge the contribution given by several Colleagues for the successful realization of the event and wish to express their deep gratitude to them. The Editors do also gratefully thank:

- the Team of Formulators for the tremendous work done to define the theme content and for providing the synthesis of the results obtained by the several participants, and Giorgia Faggiani for her assistance in the editing process;
- ITCOLD, Politecnico di Milano with the Department of Civil and Environmental Engineering, the research institute Ricerca sul Sistema Energetico-RSE, the Consiglio Nazionale degli Ingegneri for the organizational support;
- the sponsors who provided financial support and facilities for the workshop and the technical visits;
- Regione Lombardia and Comune di Milano for the patronage offered at the event.

Promoted by the ICOLD Technical Committee Computational Aspects of Analysis and Design of Dams, and organized by Politecnico di Milano, ITCOLD, Consiglio Nazionale degli Ingegneri, Ricerca sul Sistema Energetico-RSE.

ICOLD Technical Committee A
Computational Aspects of Analysis and Design of Dams

Under the patronage of Comune di Milano and Regione Lombardia.

With the sponsorship of ENEL, Mapei, Gruppo CVA, Romagna Acque, EDISON, A2A, Gelmi, and the hospitality of ENEL and Consorzio di Bonifica di Piacenza for the technical visits to Isola Serafini weir and Molato Dam.

Milano, Italy — Gabriella Bolzon
Milano, Italy — Donatella Sterpi
Roma, Italy — Guido Mazzà
Milano, Italy — Antonella Frigerio

Contents

Theme A: Seismic Analysis of Pine Flat Concrete Dam

Seismic Analysis of Pine Flat Concrete Dam: Formulation and Synthesis of Results

J. W. Salamon, C. Wood, M. A. Hariri-Ardebili, R. Malm, and G. Faggiani

Abstract ICOLD Committee on Computational Aspects of Analysis and Design of Dams organized the 15th International Benchmark Workshop in Milan, Italy, in September 2019. Theme A of the workshop is related to a seismic analysis of Pine Flat Dam. The study proposed in the Milan workshop is a continuation of investigations initiated by the United States Society on Dams Concrete Dams Committee and Earthquakes Committee during the workshop *Evaluation of Numerical Models and Input Parameters in the Analysis of Concrete Dams* held in Miami, Florida, on May 3, 2018. The purpose of Theme A investigations is to define uncertainties in numerical analyses of concrete dams in a focused, systematic, and controlled way with collaborative participation from the international dam industry and academia. The objectives of these investigations are to identify key uncertainties that may lead to differences among analysis results, to advance best practices for analyses of concrete dams, and to determine further research needs. Overall, 27 teams, representing 16 countries, submitted solutions to the formulated six study cases for Theme A, together with technical papers that documented the methods and approaches used in the analyses. The summary of the benchmark studies can serve as a reference in verification of computational models used in seismic analysis of concrete dams.

Keywords Concrete dams · Seismic analysis · Structural assessment · Verification and validation · Numerical methods · Finite element method

J. W. Salamon (✉) · C. Wood
U.S. Bureau of Reclamation, Denver, CO, USA
e-mail: jsalamon@usbr.gov

M. A. Hariri-Ardebili
University of Colorado at Boulder, Boulder, CO, USA

R. Malm
KTH Royal Institute of Technology, Stockholm, Sweden

G. Faggiani
Ricerca Sul Sistema Energetico - RSE S.p.A, Milan, Italy

G. Bolzon et al. (eds.), *Numerical Analysis of Dams*, Lecture Notes in Civil Engineering 91, https://doi.org/10.1007/978-3-030-51085-5_1

Preface to Theme A

Numerical analyses have become a primary tool in structural assessments of concrete dams over the last three decades. Expanded computing power currently available to engineers allows conducting complex structural simulations of dam behavior for static and seismic loads, which are used to describe the potential failure modes of the structure and estimate risk of potential structural failures. Technical complexity and mathematical advancement of such analyses represent a challenge to the engineering community. The primary concerns are the level of confidence in modeling and the accuracy of the results. The verification and validation (V&V) process of the computational simulations is the primary method used to build such confidence and is used for quantifying accuracy of the solutions.

The importance of the V&V process in the advanced structural analysis of dams has been emphasized and discussed by International Commission on Large Dams (ICOLD) bulletins (ICOLD 1994, ICOLD 2001, ICOLD 2013) and by the ICOLD and the United States Society on Dams (USSD) benchmark workshops. In this context, the ICOLD Committee on Computational Aspects of Analysis and Design of Dams has contributed over time in developing a verification process of numerical analysis for dams by organizing 15 benchmark workshops with the aim of offering dam engineers the opportunity to compare the computation methods used in the structural analyses of dams, as well as to work as a forum for verification and validation of the analysis results. The purpose of these benchmark workshops is to verify the results and investigate uncertainties in numerical analyses of dams in a focused, systematic, and controlled approach with collaborative participation from the international dam industry and academia.

Theme A of the ICOLD 15th International Benchmark Workshop on Numerical Analysis of Dams (2019 Benchmark Workshop), relates to a seismic analysis of Pine Flat Dam. The study proposed for this workshop is a continuation of the investigations initiated by the USSD Concrete Dams Committee and Earthquakes Committee during the workshop titled, "*Evaluation of Numerical Models and Input Parameters in the Analysis of Concrete Dams,*" organized during the 2018 USSD Annual Conference and Exhibition in Miami, Florida, on May 3, 2018.

The 2019 Benchmark Workshop is an important step towards developing a forum where:

- Analysts can verify the performance and accuracy of the modeling software before using it in seismic analyses of concrete dams.
- Software developers can verify and validate the software features and the accuracy of the software using the results of the benchmark studies.
- Researchers can further expand and investigate topics presented in the benchmark workshop.
- Dam owners can build confidence in analysis results obtained with a software that was verified with the benchmark study results.

The validity of any numerical model ultimately rests on how well the model predicts data obtained from the field (e.g., ground motion time histories from earthquakes, shaker test results, inelastic crest displacements, or mapping of cracks). Only then will there be confidence that the model can reasonably approximate real-world behavior. Ideally, a fundamentally sound computation model can be developed systematically from first principles and established empirical relations, and then independently assessed and verified for its capabilities. Unfortunately, a comprehensive framework for modeling the non-linear behavior of concrete dams and reservoirs under realistic seismic loadings has yet to be developed.

1 General

1.1 Introduction

The Theme A studies presented at the September 9–12, 2019, International Commission on Large Dams (ICOLD) 15th International Benchmark Workshop on Numerical Analysis of Dams (2019 Benchmark Workshop) continue the investigations initiated by the United States Society on Dams (USSD) Concrete Dams Committee and Earthquakes Committee during the workshop "Evaluation of Numerical Models and Input Parameters in the Analysis of Concrete Dams", organized at the USSD Annual Conference and Exhibition held in Miami, Florida, on May 3, 2018 (2018 USSD Benchmark Workshop).

This USSD initiative began with the workshop in Denver in 2016, when a benchmark analysis of a thin arch dam (Monticello Dam) was conducted. Then, during the USSD Conference in Anaheim in 2017, the seismic aspects of concrete dam analyses were discussed. In 2018 USSD Benchmark Workshop investigations, a simple linear model of Pine Flat Dam was offered. The results of that workshop, summarized in [1], become a primary reference for the Theme A formulation of the 2019 Benchmark Workshop.

The Theme A Formulation Committee (Formulators) of the 2019 Benchmark Workshop has defined new study cases for the Pine Flat Dam model (Chap. 3) based on the outcomes from the previous 2018 USSD Benchmark Workshop.

1.2 Purpose and Objectives

The purpose of the Theme A study investigations is to determine uncertainties in numerical analyses of concrete dams using a focused, systematic, and controlled approach; to determine research needs; and to advance best practices for the analyses of concrete dams with collaborative participation from the international engineering dam industry and academia.

The objective of this effort is to begin developing a framework by examining how complex problems might be divided into simpler subproblems that have tractable solutions. Development of such a framework will benefit the profession by establishing a common conceptual basis for the seismic analysis of concrete dams.

The formulated study cases are intentionally narrow in focus to clearly identify the assumptions, analysis parameters, and methods having the greatest effect. Simple model geometry and analysis input parameters are defined in the formulations to avoid overtaxing 2019 Benchmark Workshop, Theme A Contributors (Contributors), but which, in aggregate, will produce meaningful results from the work submitted by each Contributor. Little is gained if Contributors perform an analysis with input parameters and implicit assumptions that vary so widely that the key factors influencing the results cannot be clearly identified.

1.3 Deliverables

Formulation package. The Formulators prepared and distributed to Contributors a benchmark formulation package describing the scope of studies to be performed. The package included detailed description of the conceptual model of Pine Flat Dam, material parameters, selected analysis parameters, time history input load records in a spreadsheet format, as well as Excel results templates for use in documenting and submitting the analysis results from each Contributor.

Contributed results. Contributors were expected to submit complete analysis results before the 2019 Benchmark Workshop in Milan, using the spreadsheet templates provided by Formulators. However, preliminary analysis results for selected cases were solicited from those Contributors who willingly participated to highlight areas where potential computational issues might arise. The preliminary results were processed by the Formulators, and those results were made available to the participating Contributors to verify their initial analyses.

The final analysis results from all Contributors were assessed, evaluated, and summarized for a general presentation at the 2019 Benchmark Workshop.

Finally, all Contributors had an opportunity to revise their results before publication, considering the outcomes from the workshop discussion. The final study results are published in this proceeding.

Workshop presentations. During the Theme A session of the 2019 Benchmark Workshop, the Formulators described the study cases and presented a summary of analysis results submitted by Contributors. The summary of results was arranged in a logical order to highlight general findings and/or correlations within specific datasets. To avoid revealing Contributor names, each contribution was assigned an identification number varying from 11 through 34 (51 through 58 for Contributors providing results for Case B only).

During the workshop session, Contributors presented their findings, interesting aspects of the study, unexpected analysis results, and/or issues to address in the future. The presentations are available at: https://www.itcold.it/archivio-pubblicazioni.

A discussion of all the 2019 Benchmark Workshop participants followed the presentations.

Papers. Contributors prepared technical papers documenting the methods, approaches used, and the analyses results. The papers included a discussion on lessons learned, general observations regarding which model parameters had the largest impact on results, and recommendations for future studies or research on the subject. The Contributors described how they would proceed with those next steps, as well as to highlight specific issues that were encountered that should be considered in future benchmark studies.

2 Description of Contributions

2.1 Workshop Participation Statistics

Overall, 27 teams (Table 1), representing 16 countries (Fig. 1), contributed to the 2019 Benchmark Workshop Theme A studies. This included 8 Contributors representing universities and 20 Contributors representing consulting and government institutions. Some teams provided more than one solution to a case. A large dataset of results has been collected and processed by the Formulators.

2.2 Interest in the 2019 Benchmark Workshop Studies

Of the 27 teams contributing to Theme A studies, 9 teams participated once in previous ICOLD benchmark workshops, and 4 teams participated at least twice (including the RSE S.p.A. team that participated in 7 previous benchmark workshops). Twelve teams were participating for the first time in an ICOLD workshop. The Contributor's interest in the benchmark studies is summarized in Table 2.

2.3 Solution Methodology

Results from Contributors represented a wide range of software, time integration methods, boundary conditions (BC), and basic analysis approaches (Table 3). This richness in the variety of fundamental approaches used by Contributors proved to be of considerable benefit in furthering the goals of the 2019 Benchmark Workshop.

Table 1 List of 2019 benchmark workshop Contributors (in alphabetic order)

Contributor name	Organization
Cheng, H., Liu, Y., Zhang, G.	China Institute of Water Resources and Hydropower Research, China
Enzell, J., Malm, R., Abbasiverki, R., Ahmed, L.	KTH Royal Institute of Technology/SWECO, Sweden
Faggiani, G., Masarati, P., Frigerio, A.	Ricerca sul Sistema Energetico - RSE S.p.A., Italy
Farrokh, M.	K. N. Toosi University of Technology, Iran
Gallo, G., Ortega, L.H., Porras, E.	INGETEC, Colombia
Ghanaat, Y.	Quest Structures, Inc., USA
Lherbier, J.-R., Andrian, F.	ARTELIA Eau & Environnement, France
Løkke, A.	Norwegian Geotechnical Institute, Norway
Menouillard, T., Tzenkov, A.D., Schwager, M.V.	CADFEM, Switzerland
Mitovski, S., Petkovski, L., Kokalanov, G., Kokalanov, V., Panovska, F.	University Ss. Cyril and Methodius, Skopje, Macedonia
Monteiro Azevedo, N., Braga Farinha, M.L., Câmara, R., Schclar Leitão, N.	LNEC – Team DAMFA, Portugal
Mouy, V.	Tractebel, France
Naji-Mahalleh, N.	MG Consulting, Iran
Panteki, E., Goldgruber, M.	DYNARDO Austria GmbH, Austria
Peton, P., Thenint, T.	SIXENSE NECS, France
Quiroz, J.	Stantec, USA
Robbe, E., Grimal, E., Devesa, G.	EDF, France
Roberts, G.	ARQ, South Africa
Salamon, J.W., Wood, C.K., Manie, J., Geister, A.	US Bureau of Reclamation, USA & DianaFEA, Netherlands
Scolari, M., Bado, A., Gualco, D., Buraschi, L., Valsecchi, R.	RINA, Italy
Shahbazi, M., Yathon, J., Penner, O., Bergman, B.	BC Hydro, Canada
Stabile, A.F., Labbé, P., Nguyen, A.	Geodynamique et Structure, France, Brazil
Staudacher, E.J., Zenz, G.	Graz University of Technology University, Austria
Valente, S., He, Q., Capriulo, C.	Politecnico di Torino – DISEG, Italy
Vorlet, S., De Cesare, G., Manso, P.	Ecole Polytechnique Fédérale de Lausanne (EPFL), Switzerland
Wei, Su, Jin-Ting, Wang, Yan-Jie, Xu	Tsinghua University, Beijing, China
Yang, H., Wang, H., Jeremic, B., Salamon, J.W.	University of California at Davis & US Bureau of Reclamation, USA

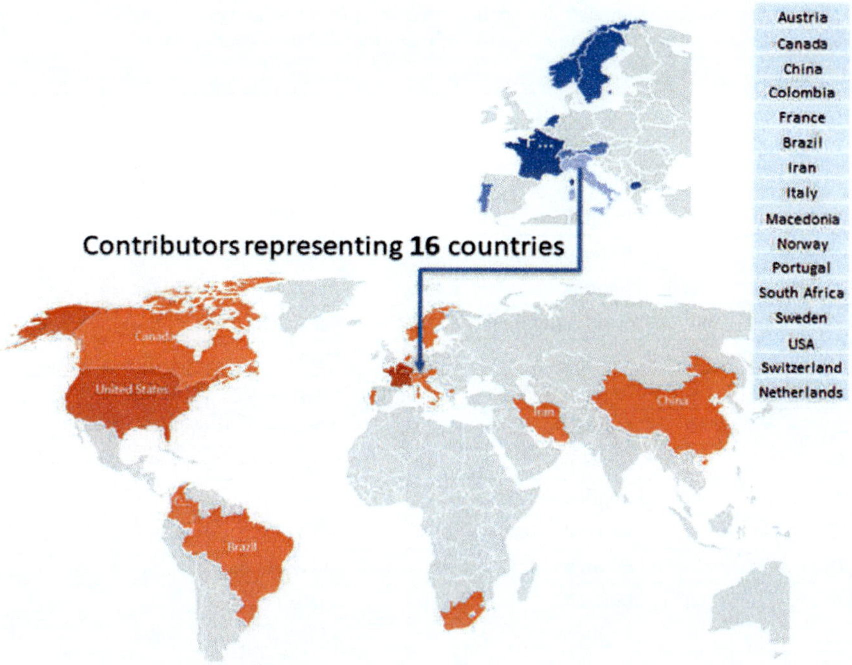

Fig. 1 Contributor representation

Table 2 Interest in the 2019 benchmark workshop

Software validation and advance modeling	Learn	Other
Verify and improve analysis accuracy	Expand and share expertise	Promote and advance proper numerical modeling for dam safety evaluations
Explore new modeling features	Gain and improve knowledge	Meet other specialists, enhance visibility, and keep up to date
Use workshop as a test case for using the software	Train their own staff	Continue our team's tradition of participating since the first edition of the benchmark workshop
Validate the code	Exchange experiences	Keep up with the best approaches
Improve numerical models	Improve computational proficiency	

Table 3 Information summary on applied method: randomly assigned contribution ID, time integration method, software name, and use of free-field boundary conditions (BCs)

Contribution ID	Time integration method	Code	Free-field BC
11	Explicit/Implicit	Real ESSI	Yes
12	Implicit	Diana	No
13	Implicit	Code_Aster	Yes
14	Implicit	ANSYS	No
15	Explicit	SAP2000	No
16	Implicit	Abaqus	Yes
17	Explicit/Implicit (A)	Parmac2D	Yes
18	Implicit	Diana	No
19	Implicit	Code_Aster	Yes
20	Implicit	Abaqus	No
21	Implicit	Code_Aster	No
22	Implicit	Diana	No
23	Implicit	ANSYS	No
24	Explicit	FLAC3D	Yes
25	Implicit	Diana	No
26	Implicit	Abaqus	Yes
27	Explicit/Implicit (A)	Parmac3D	Yes
28	Implicit	Abaqus	Yes
29	Implicit	ANSYS	No
30	Implicit	Abaqus	Yes
31	Implicit	Abaqus	No
32	Implicit	SOFiSTiK	No
33	Implicit	Abaqus	No
34	Implicit	ADINA	Yes
51	Explicit	LS-DYNA	No
52	Implicit	Abaqus	Yes
53	Explicit	LS-DYNA	Yes
54	Explicit	LS-DYNA	No
55	Explicit	LS-DYNA	Yes
56	Implicit	Diana	Yes
57	Explicit	LS-DYNA	No
58	Implicit	Abaqus	Yes

3 Formulation of the Benchmark Studies

3.1 Study Case

Pine Flat Dam, located on King's River, east of Fresno, California (Fig. 2), was constructed by the United States Army Corps of Engineers (USACE) in 1954. The dam consists of thirty-six 15.25-m-wide monoliths and one 12.2-m-wide monolith. The length of the straight gravity dam is 561 m, and the tallest non-overflow monolith is 122 m high (Fig. 3).

Pine Flat Dam was extensively studied during the 1970s and 1980s at the University of California at Berkeley [2–7, 14] and it was a subject of a seismic analysis during the 2018 USSD benchmark workshop in Miami [1]. These past investigations provide measured and calculated references for correlation and comparison to the current benchmark results.

For the Theme A analysis, the tallest non-overflow dam (Monolith No. 16) was selected (Fig. 4) because of its relatively simple geometry. The conceptual model for

Fig. 2 Downstream view of Pine Flat Dam [15]

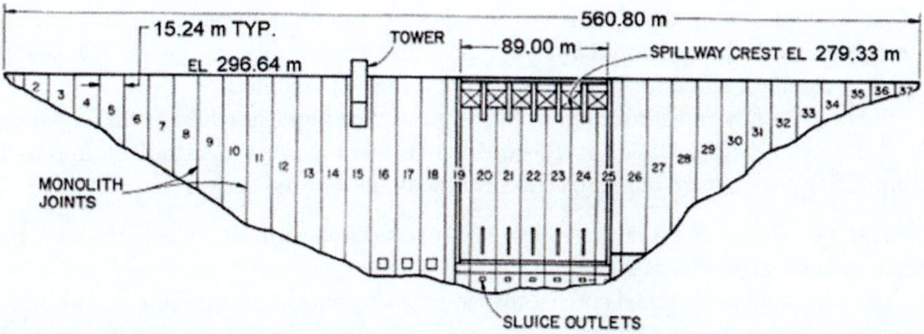

Fig. 3 Downstream view of Pine Flat Dam [14]

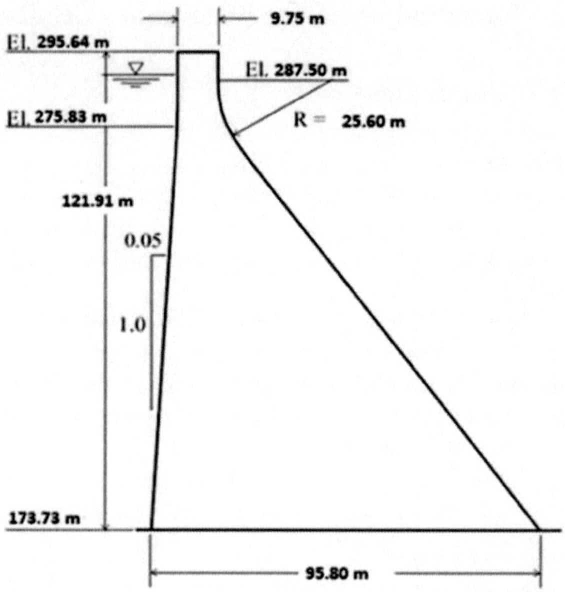

Fig. 4 Cross-section geometry of Monolith 16 [1]

Theme A was the same one used in the 2018 USSD studies with respect to the model geometry and elastic material properties for the dam, foundation, and reservoir. New seismic load records and non-linear concrete material properties are defined for the Theme A formulation.

The Formulators developed a list of study cases for Contributors to perform comparative analyses. The cases were marked as "obligatory" (to be studied by all Contributors) and "optional" (to be selected by any willing Contributor).

3.2 Conceptual Model

All input data used in the Theme A formulation, including dimensions and forces, were provided in International System (SI) units. These units are: *kg, m, MN, MPa,* and *s*. A standard value for gravity of 9.80665 m/s^2 is assumed.

For consistency and convenience, the original Pine Flat Dam dimensions and data used during the 2018 USSD workshop, which were expressed using the Imperial Unit (US) system, were converted to SI units by the Formulators.

Geometry. Monolith 16 is the tallest non-overflow section of Pine Flat Dam. Its cross–section geometry is shown in Fig. 4.

The computational model consists of the 15.24-m-wide dam monolith and a corresponding strip of the foundation and the reservoir. The origin of the axis system and key reference nodes are shown in the conceptual model cross-section (Fig. 5). The

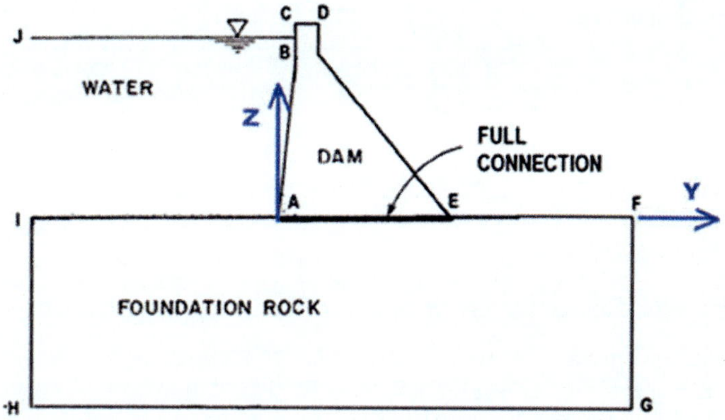

Fig. 5 Computational model cross-section

axis and reference nodes are located on the mid-width of the monolith. The axis and node system were to be used consistently for reporting results of the analysis cases.

Material properties. For consistency, the same material properties were provided for use by all Contributors; they correspond to the values (in SI units) used for the 2018 USSD studies.

Dam. Concrete is assumed to be homogeneous and isotropic throughout the entire dam, with the properties listed in Table 4.

Foundation. The foundation material is assumed to be homogeneous, isotropic, and elastic with the properties listed in Table 5.

Water. Water is considered to have a unit weight of 1,000 kg/m^3 and compression wave velocity of 1,439 m/s.

Loads. The following loads were used for the analysis cases.

Static loads. Static loads include the weight of concrete and water only (the weight of rock foundation was not to be included).

Table 4 Concrete properties

Parameter	Value
Modulus of elasticity	22,410 MPa
Density	2,483 kg/m^3
Poisson's ratio	0.20
Compressive strength	28.0 MPa
Tensile strength	2.0 MPa
Fracture energy	250 N/m
Compressive strain at peak load	0.0025
Tensile strain at peak load	0.00012

Table 5 Foundation rock properties

Parameter	Value
Modulus of elasticity	22,410 MPa
Density	2,483 kg/m^3
Poisson ratio	0.20
Shear wave velocity	1,939 m/s
Compressional wave velocity	3,167 m/s

Reservoir levels. Three reservoir water levels (RWL) were considered:

- Winter reservoir water level (**WRWL**) at El. 268.21 m.
- Summer reservoir water level (**SRWL**) at El. 278.57 m.
- Normal reservoir water level (**NRWL**) at El. 290.00 m.

Impulsive stress. This record represents an impulsive excitation (within the frequency limits of the model) and was intended to allow easy visualization of the response of the model. Its purpose was to quickly aid in the visual comparison of Contributor results. Artifacts and errors can be easily identified by visual inspection of the model output or by direct comparison to the input. Impulsive time histories representing velocity in the horizontal (Y axis) direction at the free surface were generated for the studies. As described at the end of this subsection, the free-surface velocity time–history was scaled to a shear stress to be applied at the base of the foundation. The impulsive time histories are baseline corrected to minimize drifts after the excitation has been applied.

The frequency content of the Impulsive Stress Record was determined by the depth, minimum element size, and material properties of the foundation. Characteristic frequencies and bandwidth of the impulse were selected to aid visualization and avoid inaccuracies as follows. Let H be the depth of the foundation (Fig. 5). Assuming the quarter-wavelength approximation, the lowest frequency f_0 for an elastic plane shear–wave (S-wave) is given by $f_0 = \frac{v_S}{4H}$, where v_S is the shear–wave velocity. The characteristic frequency f_N of the Impulsive Stress Record is chosen so that $N \geq \frac{1}{4}$ impulse wavelengths will fit within H, which leads to the condition $f_N = 4Nf_0$. We use values of $N = \frac{1}{4}$ and $N = 2.5$. To avoid inaccuracies, the highest frequency f_1 of the impulse is selected so that the wavelength is greater than 10 times the smallest spatial element size h, leading to the condition $f_1 \leq \frac{v_s}{10h}$. In practice, a corner frequency $f_{hi} < f_1$ is selected to low-pass filter the impulse so that its spectral amplitude at f_1 is substantially diminished from its peak value near the characteristic frequency f_N. Table 6 lists the resulting frequency limits, which are based on the foundation rock properties listed in Table 5.

Two versions of the Impulsive Stress Record were provided for the benchmark: (1) a low-frequency impulse such that the characteristic frequency corresponds to the quarter-wavelength approximation for the foundation ($N = \frac{1}{4}$); and (2) a high–frequency impulse such that several wavelengths of the impulse are contained within the foundation depth ($N = 2.5$). The time increment of the low- and high–frequency

Table 6 Foundation dimensions and impulse frequency limits

Parameter	Value
Foundation depth H	122.0 m
Maximum element size h	1.5 m
S-wave minimum frequency f_0^S	4.0 Hz
S-wave maximum frequency ..	130 Hz

time histories are 0.01 s and 0.001 s, respectively, and the positive amplitude is set to 0.01 m/s. Both records are 2 000 points long. The velocity time histories are shown in Figs. 6 and 7 (trimmed here to 10 s and 1 s, respectively), as well as their corresponding Fourier amplitude spectrum. There is a 10-point quiet period before each impulse, where the velocity is set to zero.

Incident plane S-waves can be represented as a stress time history using a non–reflecting base boundary condition initially formulated by Lysmer and Kuhlemeyer [12]. The input stress time history can be obtained by scaling the free–surface velocity time history for a medium having the same material properties as the foundation. We assume the case of a vertically-propagating plane shear wave polarized in the horizontal direction (SH-wave). Let $v_Y(t)$ be the velocity time history representing S-wave particle motion in the Y-direction at the ground surface. The velocity time history at the ground surface can be converted into a time history of shear stress $\tau_{YZ}(t)$ at the base of the foundation by using $\tau_{YZ}(t) = \rho v_S v_Y(t)$, where ρ is the foundation density, and v_S is the shear-wave velocity (e.g., Joyner and Chen [11]; Mejia and

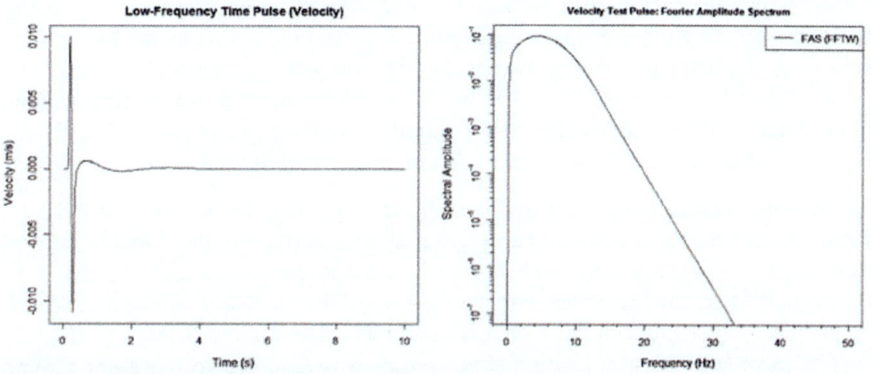

δt (s)	FNyquist (Hz)	FN (Hz)	Flo (Hz)	Fhi (Hz)	Impulse amp. (m/s)	Impulse dur. (s)	Zero-pad
0.01	50	4.0	0.5	8.0	0.01	20	10

Fig. 6 Low-frequency impulsive time history: free-surface velocity—upstream/downstream direction

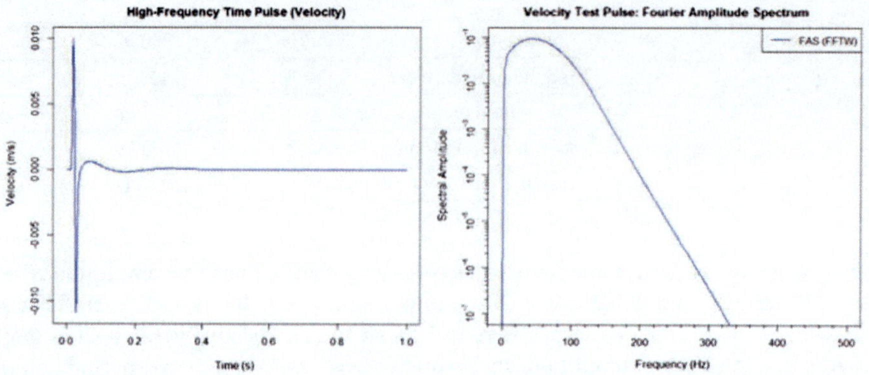

δt (s)	$F_{Nyquist}$ (Hz)	F_N (Hz)	F_{lo} (Hz)	F_{hi} (Hz)	Impulse Amp. (m/s)	Impulse Dur. (s)	Zero-pad
0.001	500	40	5	80	0.01	2.0	10

Fig. 7 High-frequency impulsive time history: free-surface velocity—upstream/downstream direction

Dawson [13]). In this benchmark, we considered only vertically–propagating SH-waves, with the shear-stress input to the base boundary using the non-reflecting BC.

The Impulsive Stress Record was to be applied in the upstream/downstream direction (Y-direction) at the base of the foundation. The Formulators provided the velocity and corresponding shear-stress time series in the Excel file, with one tab for the low-frequency impulse and another tab for the high-frequency impulse. To ensure that the excitations remain baseline-corrected, it was recommended that the full duration of the time series for the low-frequency impulse and the high-frequency impulse be input to numerical models, without truncating the record in time.

Taft Record. The M 7.3 Kern County, California, earthquake occurred on July 21, 1952. Accelerations from this earthquake were recorded at the Lincoln School (tunnel), in Taft, California. Both horizontal components were used for the 2019 Benchmark Workshop. For this workshop, we used just the baseline corrected S69E component (Fig. 8), which has a peak horizontal acceleration of 0.18 g.

The recorded Taft acceleration time history was assumed to represent ground motions at the free surface (i.e., the motions that would be recorded at the top of the foundation in the absence of the dam and reservoir).

Because free-surface motions are appropriate for input to models with a massless foundation, this acceleration record also was used for input to Case F.

In contrast, free-surface motions typically are deconvolved to base motions for input to models using a foundation with mass; therefore, the Taft free-surface acceleration record was deconvolved to the base of the foundation for the Pine Flat Dam model. The deconvolution process assumed a vertically-propagating SH-wave in a

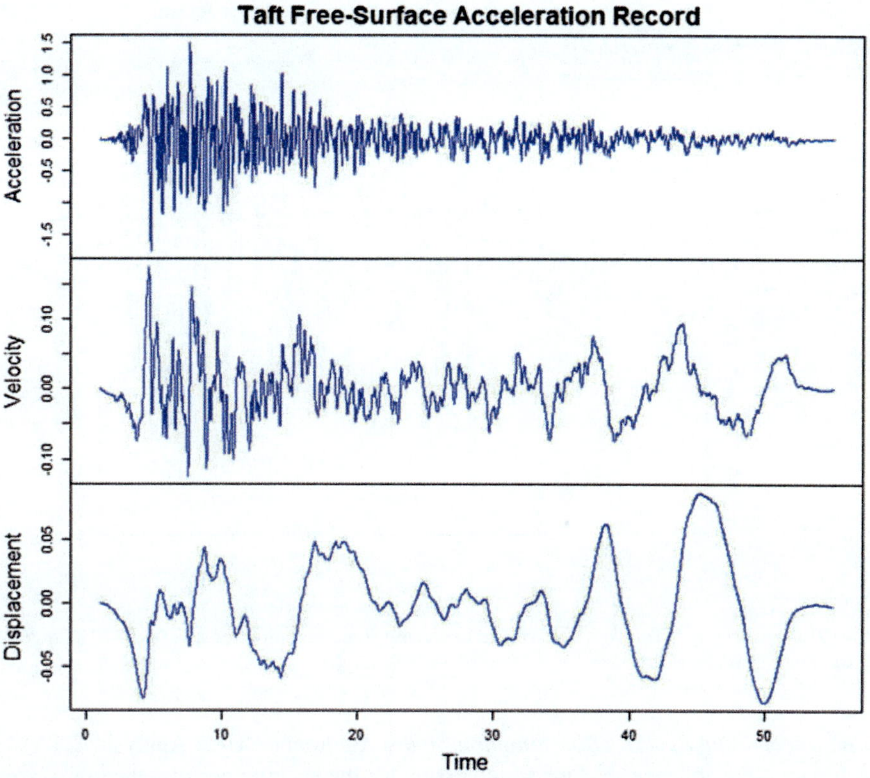

Fig. 8 Taft free-surface acceleration record—upstream/downstream direction

uniform half-space having the same material properties as the foundation (Table 5). It was computed for a depth equivalent to the base of the foundation, and it includes the Rayleigh mass and stiffness damping specified in the formulation. Figure 9 shows a comparison of the free-surface and deconvolved Taft acceleration records. The deconvolved Taft acceleration record was used for input at the base of the foundation for Cases D and E. For Contributors who preferred instead to use the stress input method for Cases D and E, an equivalent shear stress record was provided, which was to be applied at the base of the foundation. The shear stress record was computed by multiplying the free-surface Taft velocity time history (Fig. 8) by the foundation density and shear-wave velocity, as described in the formulation.

The Taft Record was to be applied in the upstream/downstream direction (Y–direction). The Formulators provided various time histories representing the Taft Record in an Excel file. The free-surface acceleration time history was intended to be used for input to Case F, while either the deconvolved acceleration time history or the shear-stress time history was intended to be used for input to Cases D and E. The free-surface velocity and displacement time histories were provided for purposes of comparison and were not meant to be used for input in this benchmark.

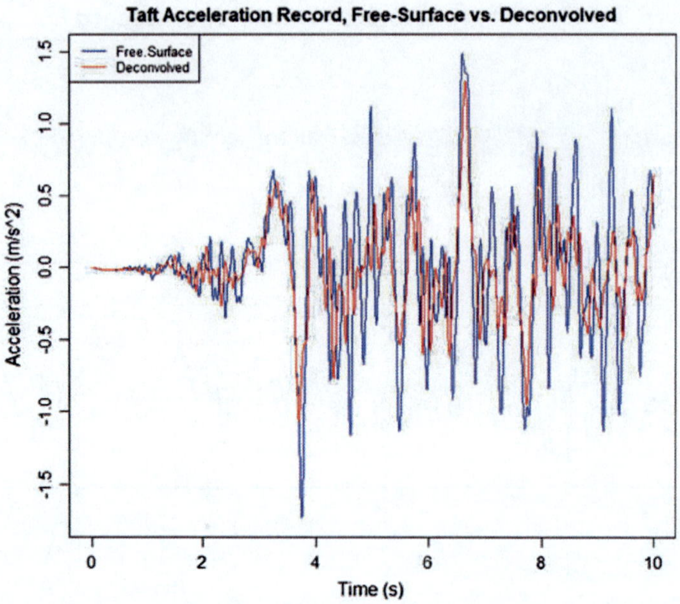

Fig. 9 Comparison of Taft free-surface and deconvolved acceleration records for 0–10 s range—upstream/downstream direction

Endurance Time Acceleration Function record. Endurance Time Analysis (ETA) is a dynamic pushover procedure which estimates the seismic performance of a dam when subjected to a pre-designed intensifying excitation. The simulated acceleration functions are aimed to shake a dam model from a low excitation level—with a response in the elastic range—to a medium excitation level—where the dam experiences some non-linearity—and finally to a high excitation level, which causes the failure. All these response variations can be observed through a single time history analysis.

The Endurance Time Acceleration Function (ETAF) is an artificially designed intensifying acceleration time history, where the response spectra of the ETAF linearly increases with time. Ideally, the profile of the acceleration time history and response spectrum increase linearly with time. Figure 10 shows a sample ETAF and its response spectra at three different times (i.e. 5, 10, and 15 s). As seen, the spectrum at $t = 5$ s is nearly one-half of the one at $t = 10$ s and one-third of the one at $t = 15$ s. In this technique, the seismic performance is determined by the duration over which the structure can endure the dynamic input.

The ETA procedure is identical to a conventional acceleration time history analysis, except that ETAFs are used as input for the numerical model instead of the real ground motions.

Figure 11 (left plot) shows the engineering demand parameter (EDP) (e.g., crest displacement) in terms of the time. Next, the absolute value of this time history is

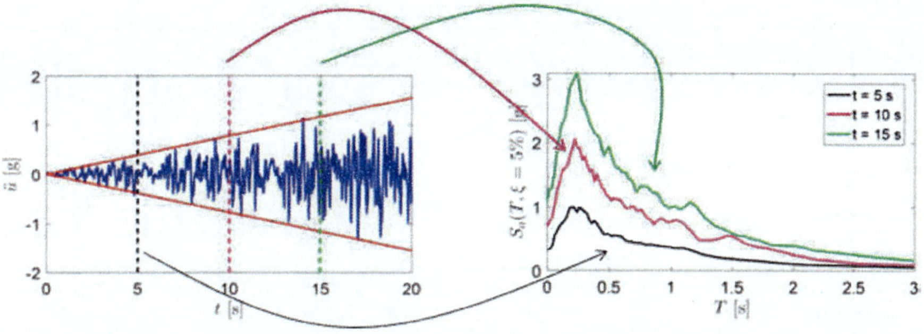

Fig. 10 Sample ETAF, its acceleration profile, and time-dependent response spectra

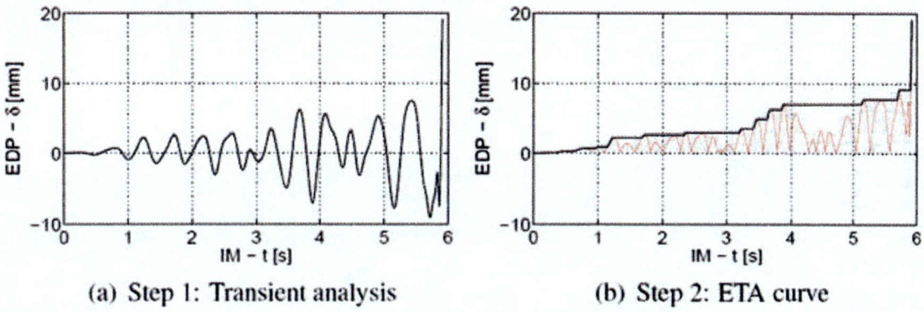

(a) Step 1: Transient analysis

(b) Step 2: ETA curve

Fig. 11 Step-by-step procedure to perform ETA and interpret the results

computed (red curve in right plot), and its cumulative absolute value is determined (black curve in right plot) versus time.

The ETAF record is shown in Fig. 12. The acceleration time history is to be applied in the upstream/downstream direction (Y-direction) at the base of the foundation. The Formulators provided the acceleration, velocity, and displacement time histories in the Excel file; however, only the acceleration time history should be used for input.

Eccentric Mass Vibration Generator (EMVG) time history. Loading of an EMVG is simulated by a harmonic-force time history, which was to be applied at the crest of the 15.24–m–wide Monolith 16 as a line load centered on the crest width. The sine wave has an amplitude of 35.4 kN and a frequency of 3.47 Hz, corresponding to the fundamental frequency of Pine Flat Dam determined during the Winter 1971 test [14]. To reduce artifacts and better emulate the mass vibration test results, the sine wave was windowed with a 25-percent Hann taper (Fig. 13), rather than a boxcar. The Formulators provided the EMVG time history record in an Excel spreadsheet.

Damping. In common practice, the term "viscous damping" is used to describe the parameter employed to stabilize a numerical model to determine the natural mode shapes and frequencies of the structure. Based on the application of viscous damping, however, the term can be characterized in several ways. Participants were instructed

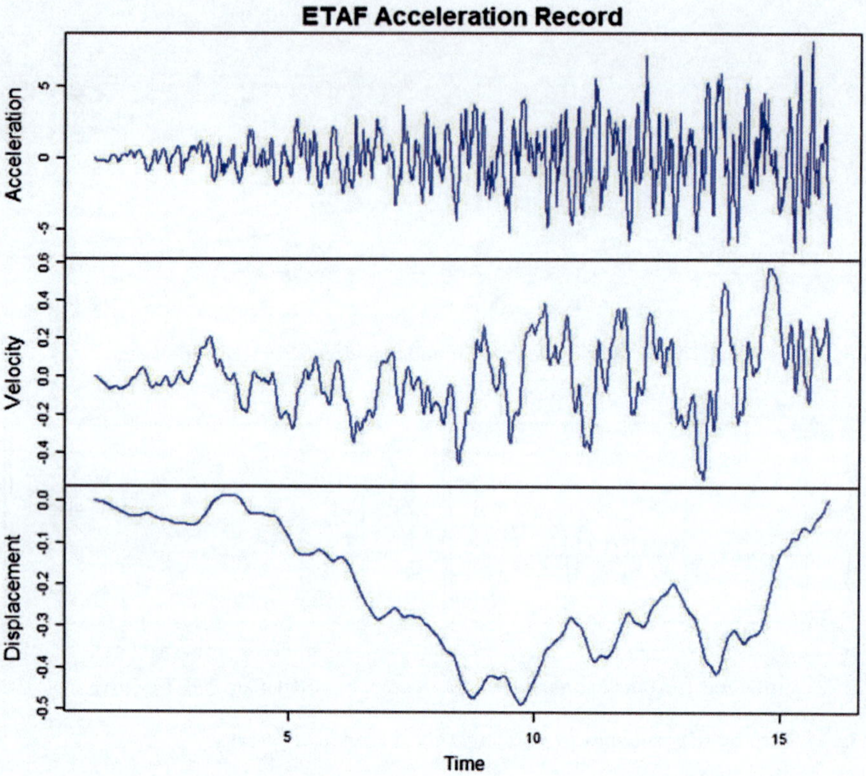

Fig. 12 ETAF record—upstream/downstream direction

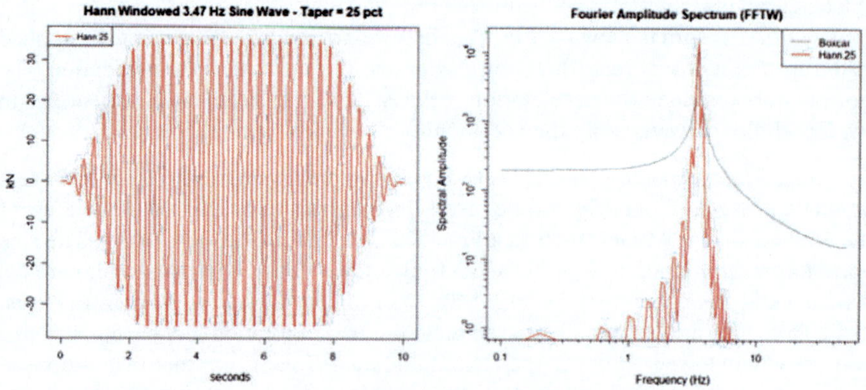

Fig. 13 EMVG time history and corresponding Fourier amplitude spectrum

to use viscous damping based on their current understanding, modeling practice, and method of analysis chosen, and to document their usage and definition in the final submitted report.

For a baseline, participants were encouraged to consider using Rayleigh viscous damping (Rayleigh damping). The Rayleigh damping matrix, by definition, is proportional to a linear combination of mass matrix (M) and stiffness matrix (K), through the constants of proportionality, denoted as α and β in the equation (Eq. 1):

$$C = \alpha M + \beta K \quad (1)$$

In reference to the values in Fig. 14 for α and β, the value of Rayleigh damping can vary, depending on the frequency and matrix constants. For comparison, in the 2019 Benchmark Workshop analysis a frequency between 4 and 10 Hz was considered, depending on the combination of dam, reservoir, and foundation. For this frequency range, the average Rayleigh damping equates to approximately 2 percent, using the input matrix constants as shown below. The constants of proportionality should be defined as $\alpha = 0.75$ 1/s and $\beta = 0.0005$ s.

Depending on the method of analysis chosen by Contributors and incorporation of viscous damping, the overall effects to dynamic response were expected to vary among the analyses. Contributors were encouraged to clearly describe how damping was applied in their analyses and to define all parameters that were used. Contributors were encouraged to verify the behavior of the structural system, according to fundamental principles, to ensure that the applied damping was reasonable.

Sign convention. For consistency among results, the sign convention for tension is "+" (positive) and the sign for compression is "–" (negative).

Base configuration. A "*base configuration*" of the model is defined as:

- Dam dimensions shown in Fig. 4.
- Foundation dimensions per Fig. 5 are:
- Length: H-G = 700 m.
- Depth: I-H = 122 m.
- Dam heel location: I-A = 305 m.
- WRWL at 268.21 m, recorded at Pine Flat Dam in winter 1971.
- Elastic properties for concrete in Table 4 and for foundation in Table 5.

3.3 Roadmap of Benchmark Studies

In the analyses defined for Theme A of the 2019 Benchmark Workshop, the same model geometry, material properties, and basic loads were intended to be used by all Contributors. To accomplish the goal of the workshop, the Formulators encouraged Contributors to study the results of the 2018 Benchmark Workshop [1] for verification purposes, conduct their analyses, and then submit their results.

The following are the 2019 Benchmark Workshop cases:

Damping parameters

Beta = 0.000500 (stiffness)

Alpha = 0.750000 (mass)

Frequency	Viscous Damping	Stiffness Damping	Mass Damping
1.00	6.13	0.16	5.97
2.00	3.30	0.31	2.98
3.00	2.46	0.47	1.99
4.00	2.12	0.63	1.49
5.00	1.98	0.79	1.19
6.00	1.94	0.94	0.99
7.00	1.95	1.10	0.85
8.00	2.00	1.26	0.75
9.00	2.08	1.41	0.66
10.00	2.17	1.57	0.60
11.00	2.27	1.73	0.54
12.00	2.38	1.88	0.50
13.00	2.50	2.04	0.46
14.00	2.63	2.20	0.43
15.00	2.75	2.36	0.40
16.00	2.89	2.51	0.37
17.00	3.02	2.67	0.35
18.00	3.16	2.83	0.33
19.00	3.30	2.98	0.31
20.00	3.44	3.14	0.30

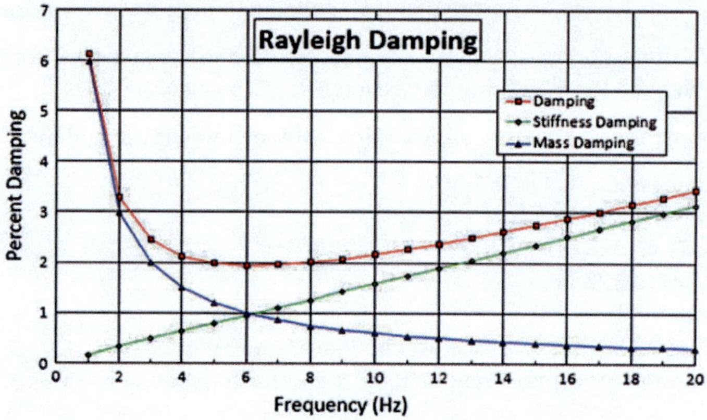

Fig. 14 Damping

- **Case A—EMVG test simulation**

  Simulation of the EMVG test performed at Pine Flat Dam in 1971 was considered for Case A.

- **Case B—Foundation analysis using impulsive loads**

  Investigate the effect of foundation size and analyze the efficiency of non–reflecting BCs in the dynamic analysis of dams. Impulsive Stress Records were to be used in the analyses.

- **Case C—Dynamic analysis using impulsive loads**

  Case B, which considers only the foundation, is extended to include the dam and reservoir. Analyses were to use the Impulsive Stress Records.

- **Case D—Dynamic analysis for various reservoir levels**

  Investigate the effect of water levels. The dam-reservoir-foundation model defined in Case C was analyzed for various reservoir water levels, using the Taft Record.

- **Case E—Non-linear dynamic analysis**

  Investigate the effect of non-linear behavior of the dam. The dam-reservoir-foundation model defined for Case D was extended to include non-linear properties for concrete. The Taft Record and the ETAF Record were to be used in the analyses.

- **Case F—Massless foundation**

  Investigate the effect of using a massless foundation. The dam-reservoir-foundation model defined for Case D was modified to use a massless foundation. The Taft Record was to be used in the analyses.

The summary of the analyses required for Theme A of the 2019 Benchmark Workshop is provided in Table 7.

Table 7 List of the cases and analysis requirements

	Analysis type			
Case	1	2	3	4
A	**Obligatory**	**Obligatory**	**Obligatory**	**Obligatory**
B	*Optional*	*Optional*	*Optional*	*Optional*
C	*Optional*	*Optional*	*Optional*	*Optional*
D	**Obligatory**	**Obligatory**	*Optional*	
E	**Obligatory**	*Optional*		
F	*Optional*	*Optional*	*Optional*	

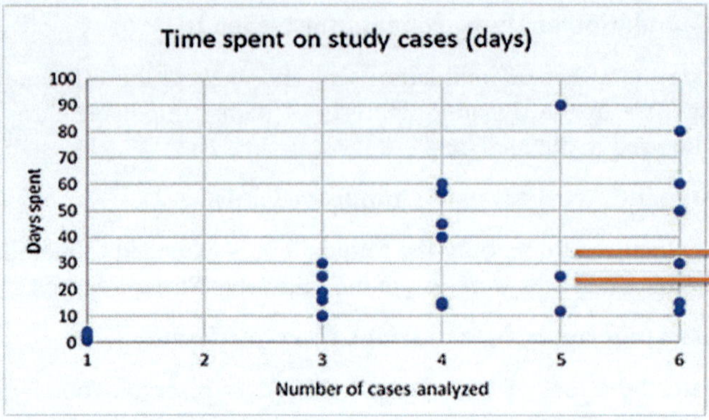

Fig. 15 Time spent on the Theme A analysis

3.4 *Estimation of Analysis Effort*

The time initially estimated by the Formulators to complete the analysis and submit the results was 10–15 days for the obligatory part, and 15–20 days for the optional part, resulting in a total of 25–35 days.

Contributors reported the time actually spent on their analyses, writing the paper, and preparing for the workshop presentation. Actual time spent ranged from 10–90 days to analyze three or more cases (Fig. 15).

3.5 *Analysis Results Identification*

Each contribution was assigned an identification number varying from 11–34 for contributions to the entire benchmark study, and 51–58 for Case B only. Names corresponding to the Contributor IDs were not provided to maintain anonymity of the results.

4 Case A—EMVG Test Simulation

4.1 *General*

The purpose of Case A was to allow Contributors to validate their models against experimental results before continuing the analysis of the other cases. For this purpose, Contributors were asked to determine the six natural frequencies and the corresponding mode shapes for the dam-foundation-reservoir system (the base model) and conduct simulation of a field test performed at Pine Flat Dam in 1971 [14].

4.2 Conceptual Model

Model configuration. The "base configuration" was considered for Case A with the EMVG excitation applied at the crest of the dam (Fig. 16).

Boundary conditions were to be defined and justified by Contributors. However, plane strain BCs at the side faces of the model were to be applied. At the bottom, upstream, and downstream faces of the foundation, each Contributor was to select appropriate conditions for a dynamic analysis (far-field/non-reflecting BCs).

Input parameters. For Case A, elastic material properties for concrete and rock were to be considered. Based on the provided information, Contributors were asked to determine the response of Monolith 16 for the specified loads. For analysis consistencies, Contributors were asked to consider the following:

- Concrete elastic properties are defined in Table 4.
- Water properties are defined in Sect. 3.2.
- Foundation elastic properties are defined in Table 5.
- The mass of the dam, reservoir, and foundation.
- Do not consider the static weight of the foundation.
- The static weight of the dam and reservoir.
- 2% viscous damping for the dam and foundation.

Loads. Loads included:

- EMVG harmonic-force time history record applied at the middle of the dam crest in the upstream/downstream direction. The load record shown in Fig. 13 was provided to Contributors in an Excel spreadsheet.
- Water levels: WRWL and SRWL.

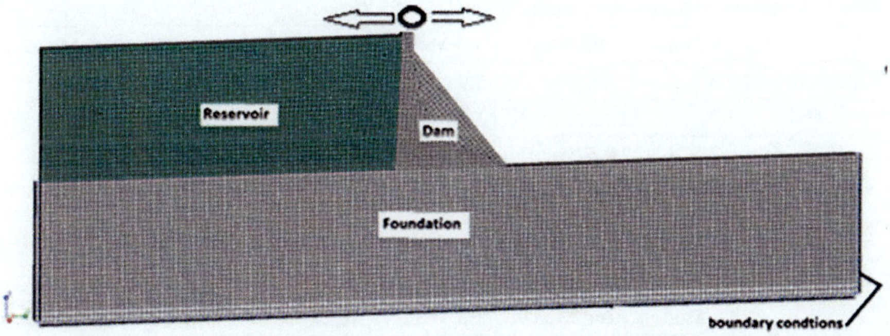

Fig. 16 Model for EMVG test simulation

4.3 Analysis

In Case A, a dynamic linear analysis of the dam-foundation-reservoir for the harmonic force record exerted by an eccentric-mass vibration generator positioned at the dam crest was considered. The analysis corresponds to the tests conducted on Monolith 16 in 1971 [14]. In this case, frequency analysis and a dynamic time history analysis were to be conducted for two different reservoir water levels, resulting in the following subcases:

- **A-1**: Natural frequencies for WRWL at El. 268.21 m.
- **A-2**: Natural frequencies for SRWL at El. 278.57 m.
- **A-3**: Dam-foundation-reservoir system with WRWL at El. 268.21 m.
- **A-4**: Dam-foundation- reservoir system with SRWL at El. 278.57 m.

4.4 Results

Based on the results provided by Contributors in the Excel sheets, a comparison was made regarding the obtained natural frequencies and mode shapes (Case A-1 and A-2) and regarding the acceleration and displacement of the dam during the harmonic excitation (Case A-3 and A-4).

Natural frequencies. Overall, 23 teams determined the first six natural frequencies and the corresponding mode shapes.

Based on the provided results, the calculated mean value of the first natural frequency was approximately 2.30 Hz for WRWL (Case A-1) and 2.20 Hz for SRWL (Case A-2). Tables 8 and 9 summarize the modes for Case A-1 and Case A-2, respec-

Table 8 Summary of the six first natural frequencies for Case A-1

	1st Mode	2nd Mode	3rd Mode	4th Mode	5th Mode	6th Mode
Median (Hz)	2.30	3.47	3.96	4.40	4.89	5.49
Mean (Hz)	2.27	3.39	4.02	4.61	5.16	5.73
SD (Hz)	0.28	0.57	0.66	0.72	0.65	0.59
COV	0.12	0.17	0.16	0.16	0.13	0.10

Table 9 Summary of the six first natural frequencies for Case A2

	1st Mode	2nd Mode	3rd Mode	4th Mode	5th Mode	6th Mode
Median (Hz)	2.19	3.20	3.83	4.36	4.87	5.44
Mean (Hz)	2.15	3.28	3.91	4.51	5.19	5.60
SD (Hz)	0.34	0.63	0.79	0.88	1.06	0.73
COV	0.16	0.19	0.20	0.20	0.20	0.13

tively. As seen in the tables, the results are, in general, more consistent for the first mode, with slightly larger scatter for the higher modes. The coefficient of variation (COV) varies between 0.1 and 0.2 for all modes with the standard deviation (SD) lowest for the first mode.

As can be seen from these tables, Contributors obtained higher natural frequencies for winter conditions (with lower reservoir) than for summer conditions (with higher reservoir). This is reasonable, since the water will contribute to more added mass that has to be excited. It also can be seen that smaller influence from variations in the water level was obtained for higher modes.

Figures 17 and 18 present the natural frequencies determined by all Contributors for Winter (WRWL, Case A-1) and Summer (SRWL, Case A-2), respectively.

In these figures, the red curves and grey histograms show the probability density of Contributor results as a function of natural frequencies. The box-plots show basic descriptive statistics for the natural frequency results, as follows:

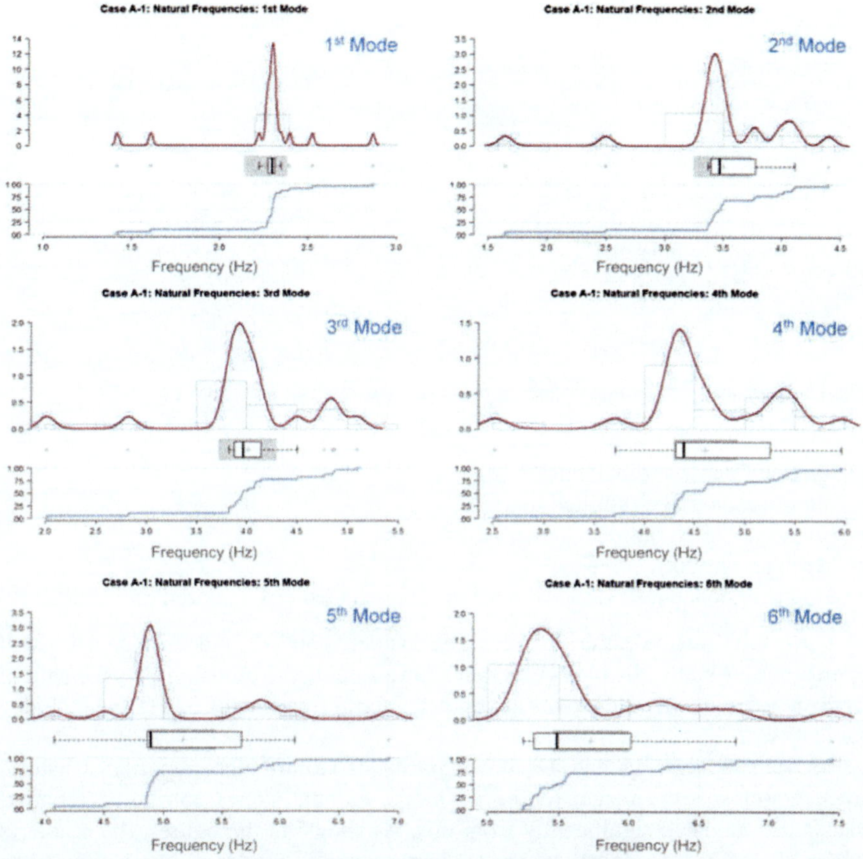

Fig. 17 Comparison of natural frequency results for the six first modes for Case A-1

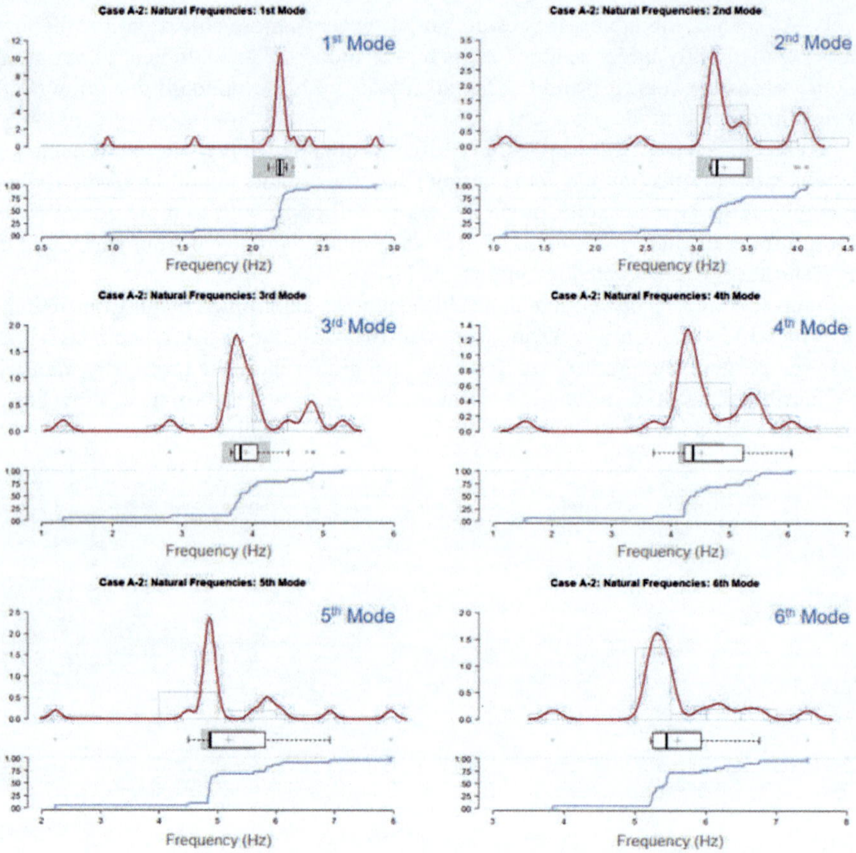

Fig. 18 Comparison of natural frequency results for the six first modes for Case A-2

- Median (heavy vertical black line).
- Interquartile range (unfilled box).
- Empirical 95% confidence interval (whiskers).
- Mean (+ symbol).
- 95% confidence interval for the mean (grey-shaded box centered on mean).

The results outside of the whiskers are considered outliers and are shown as small open circles. Finally, the blue curves show the cumulative probability of the natural frequency results. In all the plots, computed natural frequencies >8 Hz are excluded for clarity.

As seen in Figs. 17 and 18, the majority of Contributors reported a natural frequency close to the median for the first mode, although a few Contributors reported values that deviated significantly from this. As shown in the figures, the scatter is larger for the higher modes, which is clearly seen by the length of the box-plots beneath the red curves. By comparing the results in the two figures, it can also be seen that the distribution seems broader for Case A-1 than Case A-2.

Mode shapes. The mode shapes for the six first natural frequencies, obtained from all Contributors, are presented in Figs. 19 and 20. In the figures, the mode shapes are presented at the upstream face of the dam along the elevation and are normalized against the maximum displacement. The obtained mode shapes from Contributors are presented as scatter plots and illustrated with gray circles. Along with these scatter plots, a red curve is presented which represent the median normalized displacement for each elevation. As it can be seen in Figs. 19 and 20, the results are more or less similar. It can also be seen that the modal shapes for the first mode at 2.30 Hz

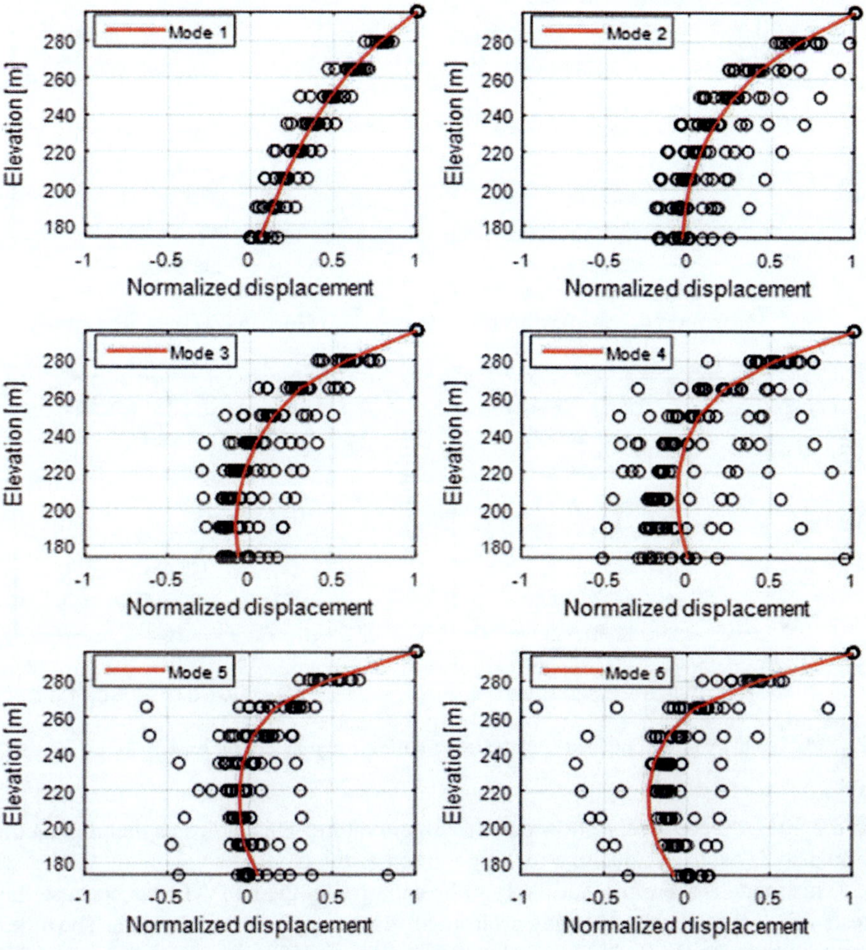

Fig. 19 Summary of the obtained mode shapes for the six first modes for Case A-1

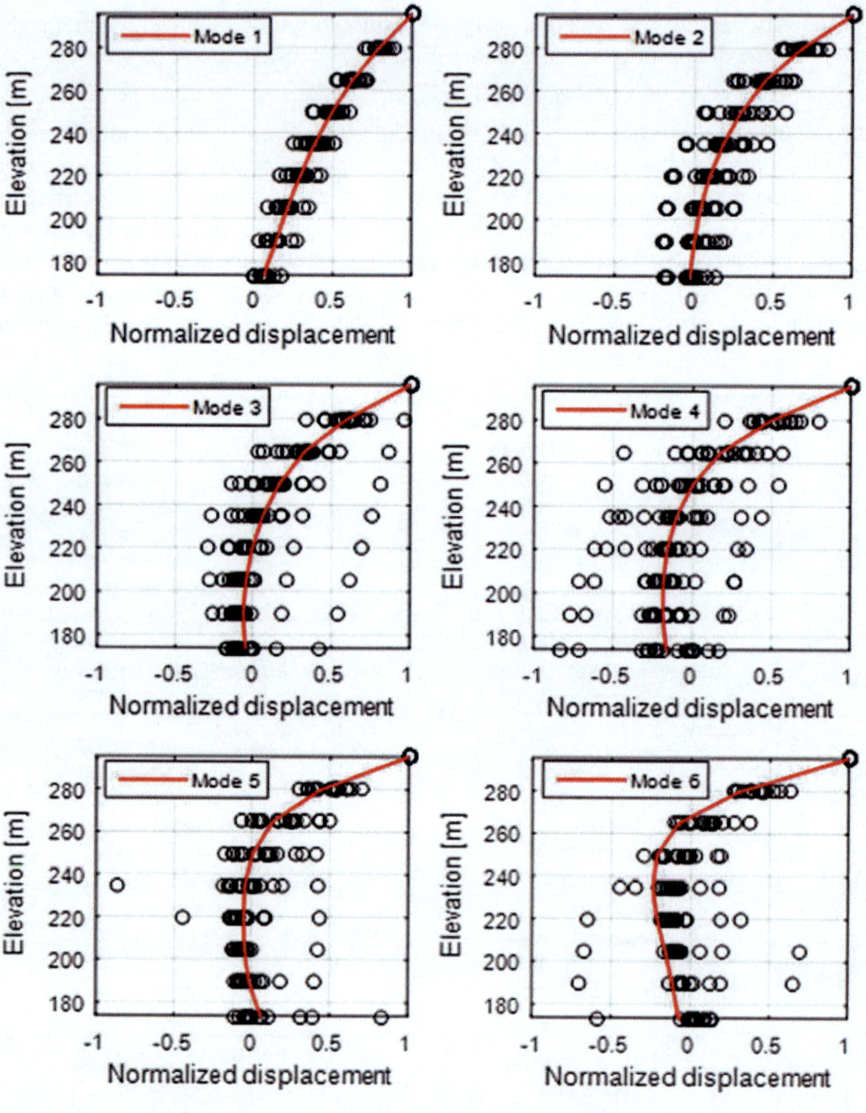

Fig. 20 Summary of the obtained mode shapes for the six first modes for Case A-2

and 2.20 Hz, respectively, show quite small curvature. This indicates that this mode primarily is occurring due to a rocking motion of the dam.

This mode is therefore not likely to be captured in the EMVG tests because the tests were conducted by applying a varying force near the crest of the dam. The other mode shapes for the 2nd to 6th frequencies, show significant curvature; however, it is not easy to interpret the global behavior of the dam based on these mode shapes.

Response during harmonic excitation. In Cases A3 and A4, Contributors were requested to perform a simulation of an EMVG test that was performed at Pine Flat Dam in 1971 [14]. The test was conducted using a harmonic excitation with a peak load of 35.4 kN and a frequency of 3.49 Hz applied at the dam crest. The excitation signal in the analysis was defined based on the applied peak load for the natural frequency measured during the test. The total duration of the signal was 10 s, where a Hann windowing function had been applied to the signal to avoid numerical artifacts.

The aim of this analysis was to compare the calculated accelerations and displacements with the corresponding values obtained from the field test. Overall, 24 teams contributed to Cases A3 and A4. In the following, only the results from Case A-3 are presented because the results for Case A-4 are quite similar. In Fig. 21, the acceleration history plots from selected contributions is shown in time domain and in frequency domain. The ideal results are that the acceleration in Point C shows the same behavior as the EMVG signal, with increasing response up to 2 s, constant between 2 and 8 s, and descending after 8 s.

The distribution of the peak acceleration in Case A-3 for the dam crest and the dam heel is shown in Fig. 22 and summarized in Table 10. The peak time history values are defined as the average between the maximum and the minimum peak.

The individual time history results for the displacements are not shown in this comparison, considering that displacements can be obtained by integration of accelerations. The distribution of the peak displacement in Case A-3 for the crest and the toe is shown in Fig. 23 and summarized in Table 11.

4.5 *Model Verification*

It could be seen from the comparison of the natural frequency results that some Contributors may have obtained non-structural modes at 1.4 and/or 1.6 Hz (see Figs. 24 and 25. These Contributors later reported an eigenfrequency of 2nd or 3rd mode to be close to 2.30 Hz. It seems reasonable to conclude that if these Contributors would have neglected the initial structural modes, even more consistent results would have been obtained. In this comparison, however, no attempts have been made to map the mode shape and corresponding natural frequencies to correct for such discrepancies.

The results from this benchmark workshop can also be compared to previous studies, conducted the 2018 USSD Benchmark Workshop [1], where eight Contributors presented the first six first natural frequencies of Pine Flat Dam, varying between 1.98 and 2.69 Hz for the normal water level elevation at 290.0 m. The results from the USSD workshop showed similar scatter as the results presented in 2019 Benchmark Workshop with a median of 2.19 Hz (in Case A-2, with El. +287.6 m), as shown in Table 9. Significantly higher discrepancy is observed in the 2018 USSD results for the higher frequencies (Fig. 24).

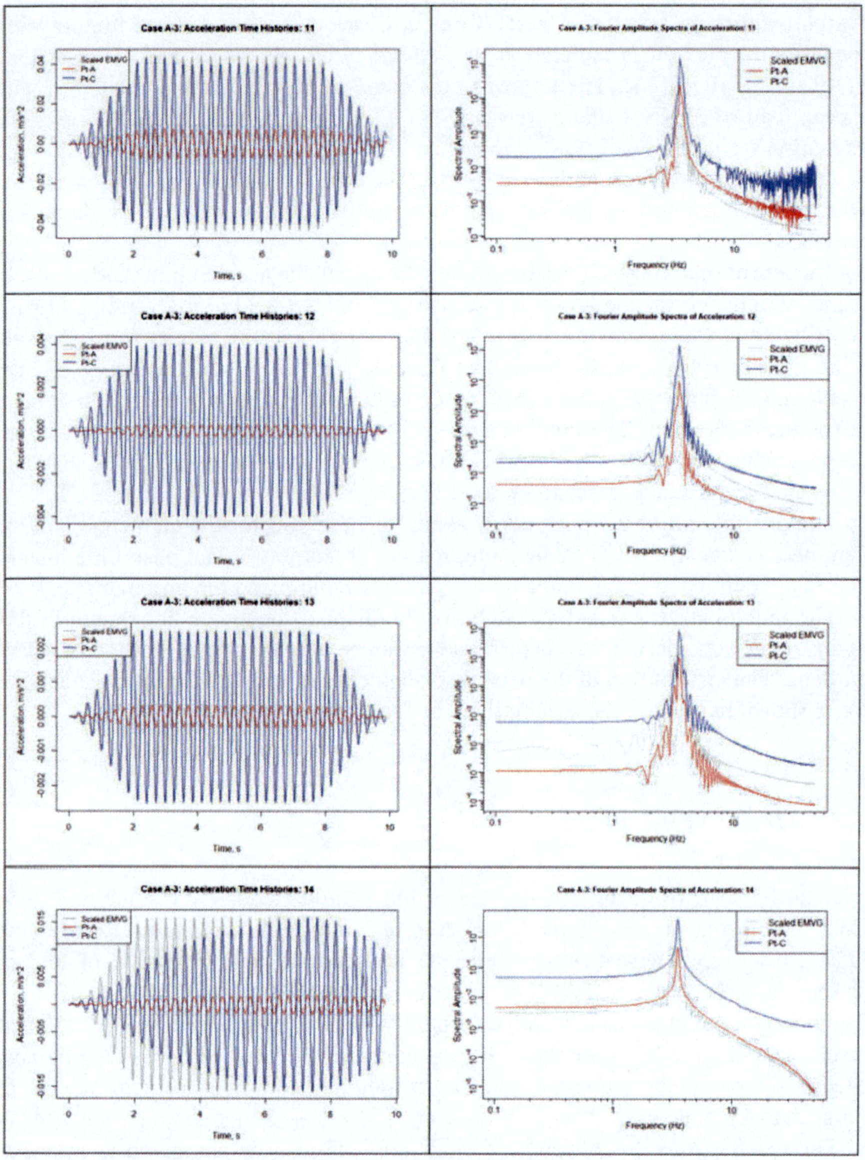

Fig. 21 Acceleration time histories (left) and Fourier responses (right) for Case A-3. The colors represent: the response at the crest (blue), the toe (red), and the scaled EMVG signal (gray)

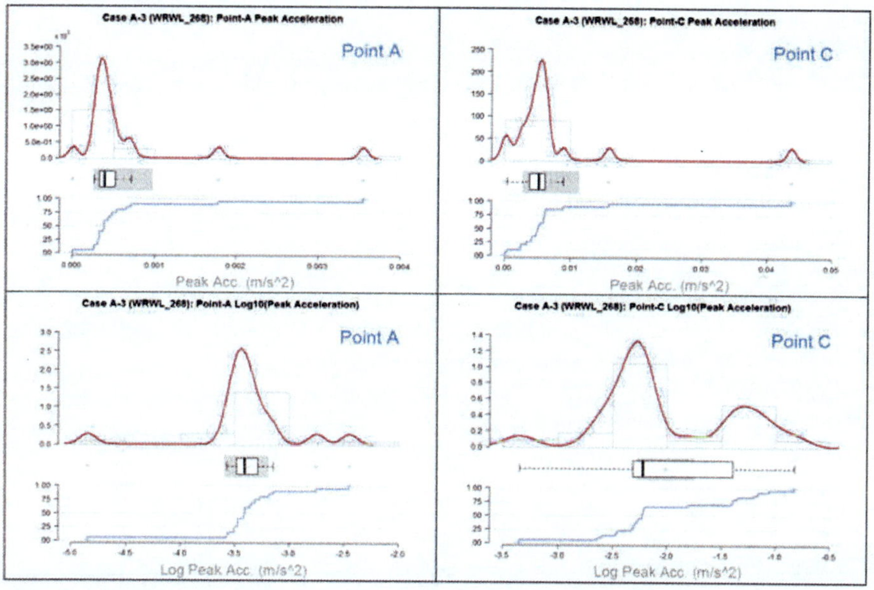

Fig. 22 Distribution of peak horizontal acceleration at Points A and C in both normal and logarithmic (base-10) units

Table 10 Summary of the characteristics of the acceleration histories for Case A-3

	Point *A*	Point *C*
Median peak acceleration (m/s^2)	3.92×10^{-4}	5.12×10^{-3}
Mean peak acceleration (m/s^2)	6.20×10^{-4}	7.27×10^{-3}
SD peak acceleration (m/s^2)	7.70×10^{-4}	9.12×10^{-3}
Median $\log_{10}$ peak acceleration (m/s^2)	−3.41	−2.29
Mean $\log_{10}$ peak acceleration (m/s^2)	−3.39	−2.30
SD $\log_{10}$ peak acceleration (m/s^2)	0.44	0.38

4.6 Model Validation

Experimental tests conducted in 1971 [14] were used to validate the numerical analysis by comparing the analysis results with the measured dam behavior.

Natural frequencies. In Table 12, the median natural frequencies from Contributors are compared with the natural frequencies obtained from the EMVG test. Not all modes that had been defined by Contributors could be identified in the field test data. In the test, only Modes 2, 3, and 6 were obtained. The numbers of the experimentally

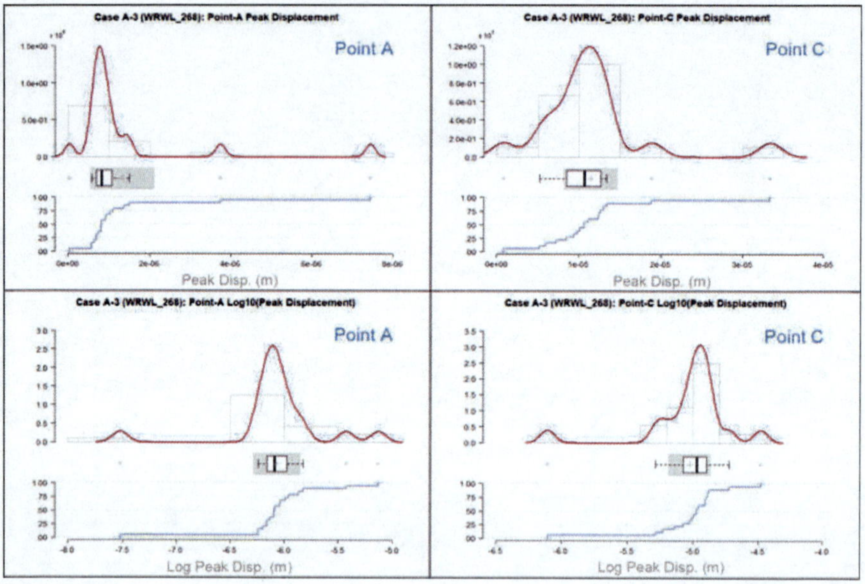

Fig. 23 Distribution of peak horizontal displacement at Points A and C in both normal and logarithmic (base-10) units

Table 11 Summary of the characteristics of the displacement histories for Case A-3

	Point *A*	Point *C*
Median peak displacement (m)	8.20×10^{-7}	1.07×10^{-5}
Mean peak displacement (m)	1.33×10^{-6}	1.14×10^{-5}
SD peak displacement (m)	1.65×10^{-6}	6.70×10^{-6}
Median $\log_{10}$ peak displacement (m)	−6.09	−4.97
Mean $\log_{10}$ peak d(m)	−6.06	−5.02
SD $\log_{10}$ peak displacement (m)	0.45	0.32

obtained modes have been renumbered here so they coincide with the numbering of the numerical results for easier comparison.

It should be noted, however, that the modeled water level is not exactly the same for the summer case, which is likely the reason why the discrepancy between numerical and experimental data is slightly larger for this case. However, the numerical models are close to the experimental data, where the difference is less than 4%. It should be noted that the natural frequency of the second mode perfectly matches the frequency from the experiment. This is promising because the frequency of 3.49 Hz has been used as input for the harmonic load in Case A-3 and A-4.

Mode shapes. The mode shapes, determined from EMVG tests [14], are presented in Fig. 25 for both field measurements and Contributors analysis results. The mode

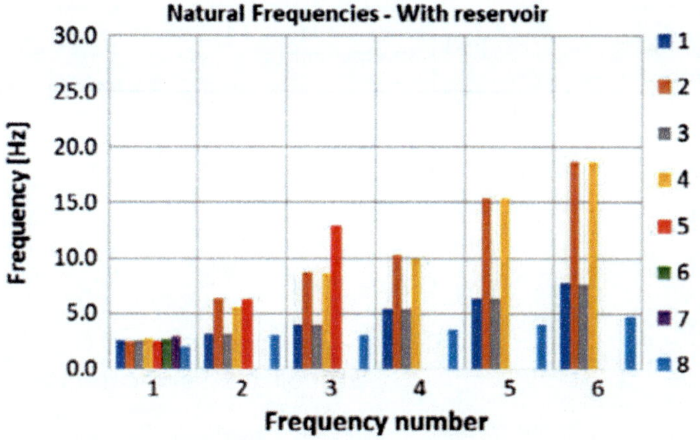

Fig. 24 Natural frequencies obtained in the 2019 benchmark workshop [1]

shapes are quite similar, especially modes 2 and 3, while mode 6 shows slightly larger curvature from the numerical models than from the field experiment. In the report by Rea and Chopra [14], the mode shapes for the whole dam were also presented, which makes it significantly easier to interpret these modes than to interpret the results of a 2D simulation. The global mode shapes are illustrated in Fig. 26, and it can be seen that the dam behaves like a monolithic structure where:

- Mode 2 can be described as the first bending mode of the whole dam,
- Mode 3 can be described as the second bending mode of the whole dam, and
- Mode 6 can be described as the fourth bending mode of the whole dam.

The third mode of the whole dam is not likely to be captured in the experiments because Monolith 16 is located closely to the expected inflection point of this mode. As seen in Figs. 25 and 26, the deformed shape of the dam is close to having a local maximum for the studied monolith (No. 16) for all these mode shapes This result clearly shows that even though a 2D model can capture the behavior of some dam monoliths with good accuracy, a 3D model that describes the whole dam is required to fully understand how the different mode shapes correspond to the global dam behavior.

Response during harmonic excitation. The results from all Contributors were previously summarized in Tables 10 and 11.

The median peak acceleration at the crest (Point *C*) was 5.12 mm/s^2, which should be compared with the maximum recorded acceleration from the EMVG test of 3.5 mm/s^2. As can be seen, the calculated acceleration is 46% higher than the experimental recorded value.

The median displacement, according to the calculations, was 0.0107 mm at the crest, which can be compared with the recorded displacement of 0.0052 mm.

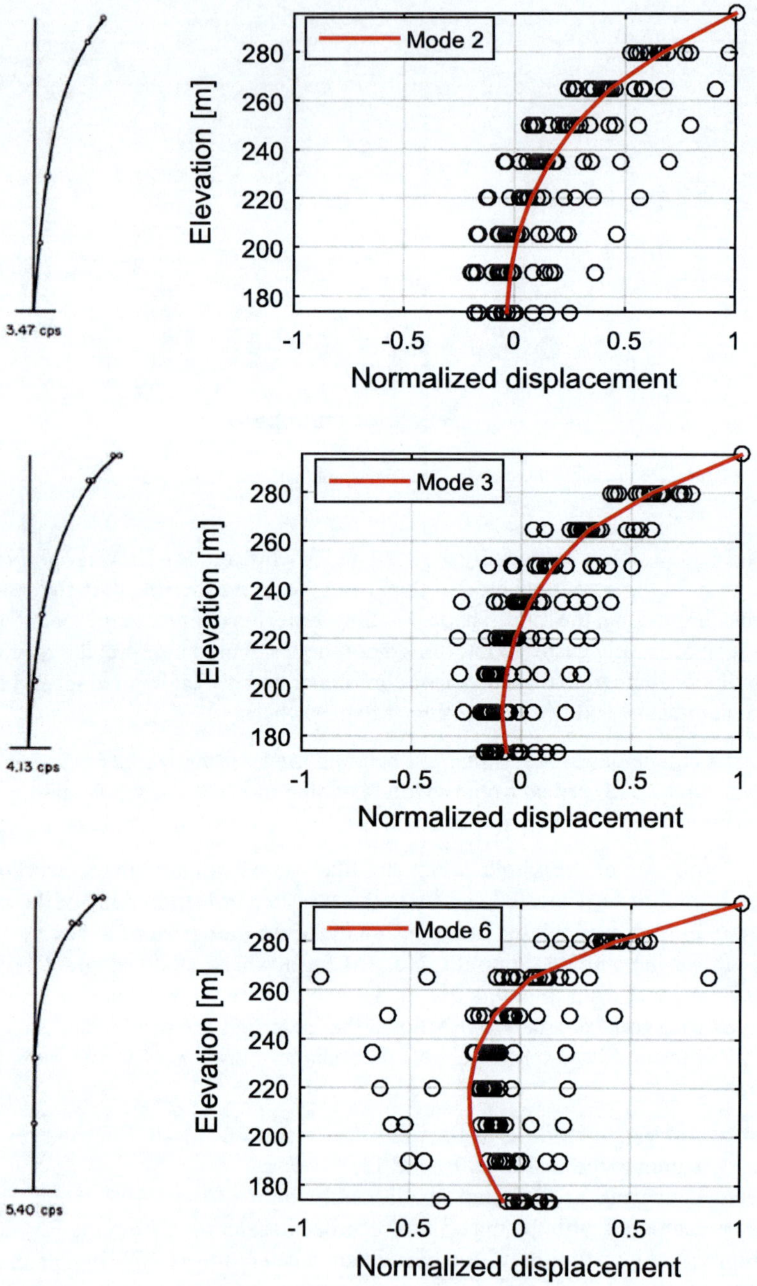

Fig. 25 Comparison of experimental and numerical mode shapes

Table 12 Comparison of natural frequencies from Contributors versus the EMVG test [14]

Analysis	Season	RWL (m)	Mode 1 (Hz)	Mode 2 (Hz)	Mode 3 (Hz)	Mode 4 (Hz)	Mode 5 (Hz)	Mode 6 (Hz)
Case A-1	W	268.2	2.30	3.47	3.96	4.40	4.89	5.49
Test [14]	W	268.2	–	3.47	4.13	–	–	5.40
Ratio			–	1.00	0.96	–	–	1.02
Case A-2	S	287.6	2.19	3.20	3.83	4.36	4.87	5.44
Test [14]	S	278.9	–	3.27	4.07	–	–	5.65
Ratio			–	0.98	0.94	–	–	0.96

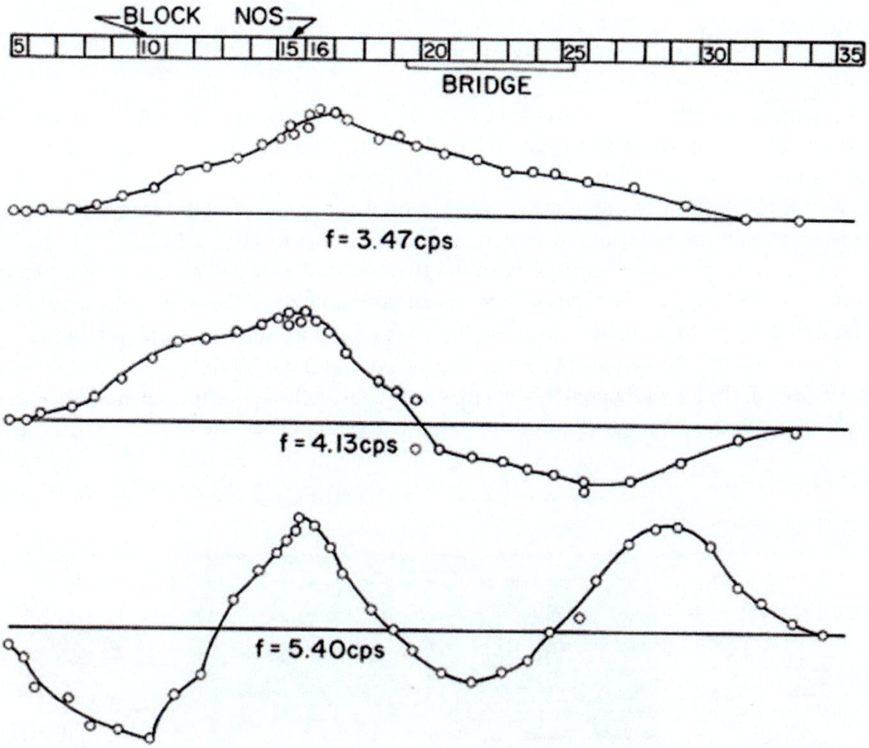

Fig. 26 Illustration of the mode shapes of the whole dam [14]

The calculated displacement is about twice as large as the recorded displacement (+105%).

One reason for the discrepancies between the experimental results and the numerical results could be the assumed damping of 2% may deviate from the actual damping of the EMVG tests. In these tests, a small excitation was obtained, which likely has not fully activated the energy absorbing effects in the concrete, such as friction in cracks, contraction joints, etc. According to Rea and Chopra [14], the damping ratio derived from the EMVG tests seems to be similar, or even slightly higher, than the damping factor used in the numerical simulations. The range of the damping factors derived from the experiments ranged between 2.0 and 3.5%.

It should also be noted that recorded displacements were very small, therefore, may have been affected by measurement accuracy.

5 Case B—Foundation Block Analysis Using Impulsive Loads

5.1 General

The purpose of the Case B study is to verify commonly used non-reflecting BCs for simulation of wave propagation in an elastic foundation block, as well as to investigate the effect of foundation size using impulsive stress records. It is assumed that a reduced-domain foundation block model must adequately represent elastic wave propagation that occurs in a semi-infinite half-space (Fig. 27 [A12]).

In general, incident energy is assumed to consist of vertically propagating plane S-waves. The upgoing S-wave motion is exactly one-half of the free-surface motion and can be applied as a shear stress at the base boundary of the foundation block.

An additional BC is needed for the reduced-domain model to keep the motions at the side boundaries equivalent to what they would be in the semi-infinite medium (i.e., equivalent to the "free-field" motion). This is the so-called "free–field" BC, and it

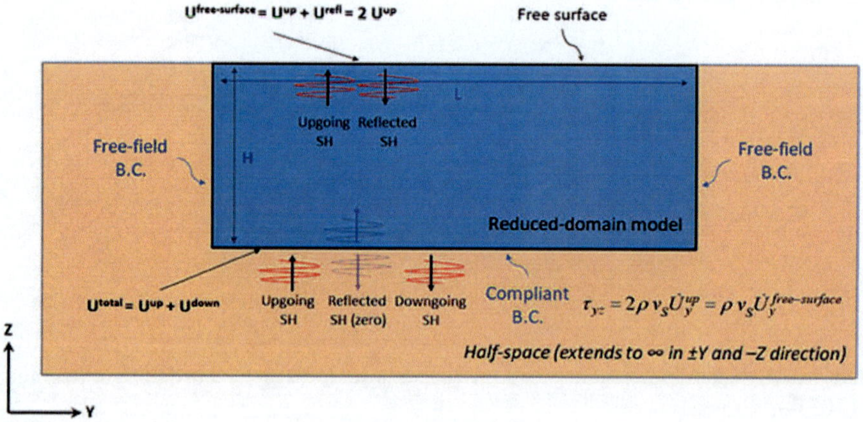

Fig. 27 Representation of a semi-infinite half-space (red) by a finite-domain foundation block (blue) [A12]

has been implemented in some FE codes used in modeling soil-structure interactions or the seismic response of dams and other structures founded on rock and soil [A12].

5.2 Conceptual Model

Model configuration. Two foundation blocks (one 700 m wide × 122 m deep, and one 3700 m wide × 122 m deep) are considered in Case B for high- and low-frequency shear stress pulses applied at the base of the block model (Fig. 28).

Contributors were asked to select the BCs at the sides and base of the model that adequately represented elastic wave propagation in a semi-infinite half-space. An absorbing BC must be applied to the sides and base of the foundation to avoid artificially trapping seismic energy within the reduced-domain model.

Input parameters. The following parameters are assumed in the analysis:

- Elastic rock material properties are defined in Table 5.
- Mass type foundation is considered.
- Static load due to weight of foundation material (gravity loads) is not considered.
- Zero viscous damping.
- The maximum element size is 1.5 m.

Loads. The time history excitation of the foundation is defined by the low- and high–frequency impulsive stress records shown in Fig. 6 and Fig. 7, respectively. These represent a vertically propagating S-wave polarized in the horizontal direction. The baseline corrected impulsive stress records are applied uniformly at the base of the foundation block, as shown in Fig. 28. The two versions of the records are: (1) a high-frequency pulse with a sample interval of 0.001 s, and (2) a low-frequency pulse with a sample interval of 0.01 s. The characteristic frequency of the high–frequency pulse was selected so that several wavelengths of the pulse would fit within the vertical extent of the foundation, while the characteristic frequency of

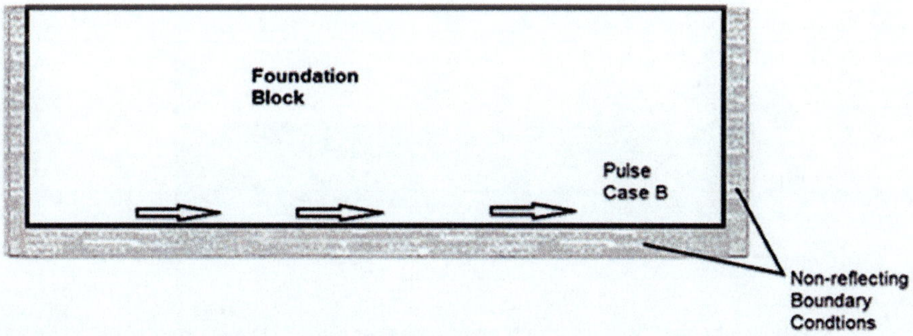

Fig. 28 Model configuration used in Case B studies

Table 13 Study cases

Case	Length L (per Fig. 29)	Impulsive signal
B-1	700 m	High frequency
B-2	700 m	Low frequency
B-3	3,700 m	High frequency
B-4	3,700 m	Low frequency

the low-frequency pulse was selected so that the vertical extent of the foundation represented approximately one-quarter wavelength of the pulse. The analysis time is 2 s for the high-frequency impulse and 20 s for the low-frequency pulse.

5.3 Analysis

Two model configurations are considered in Case B with different block lengths, but the same depth, for high- and low-frequency pulse records, as listed in Table 13.

5.4 Results

Contributors provided 22 solutions for Cases B-1 and B-2, and 14 solutions for Cases B-3 and B-4. The time histories of velocity results are compared at selected Points *a* through *j* per Fig. 29.

Screening the analysis results. For the uniformly applied stress pulse excitation, a uniform upgoing velocity pulse is expected to result at the base of the foundation block. This velocity pulse should have a maximum amplitude of 0.005 m/s (one half of the amplitude of the pulse records presented in Figs. 6 and 7).

Figure 30 shows the plots of maximum horizontal velocities provided by Contributors at the base of the foundation block at Points *b, d, f*, and *h*.

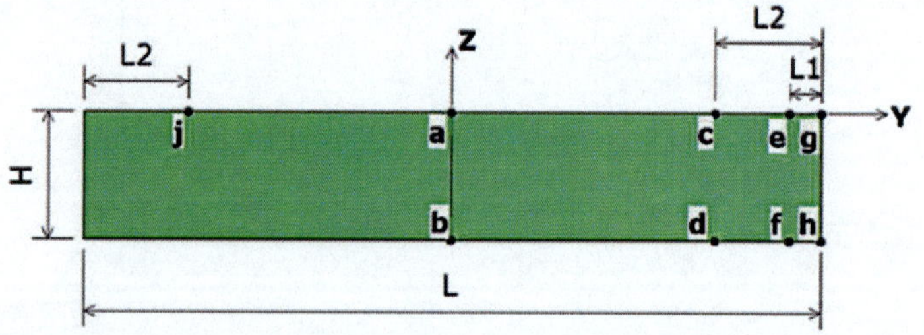

Fig. 29 Locations of points used for results comparison for Case B. Dimensions: H = 122 m; L1 = 20 m; L2= 100 m; L is provided in Table 13

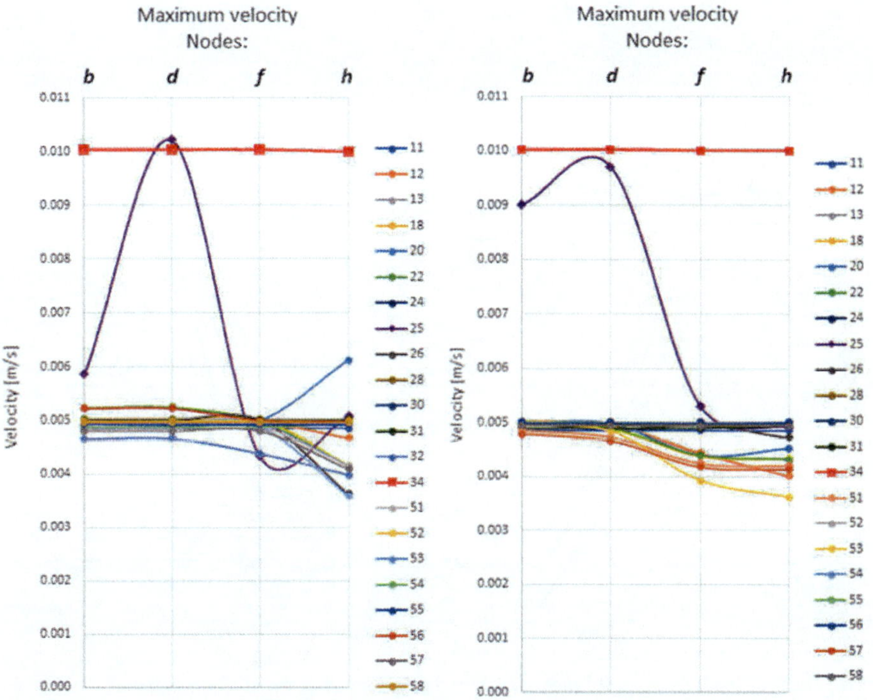

Fig. 30 Comparison of recorded peak velocities at the base of the foundation block for the high–frequency pulse (Case B-1 left) and the low-frequency pulse (Case B-2 right)

Some results required scaling or polarity reversal. For example, results from Contributor No. 32 (plotted by a red line in Fig. 30) appear to closely match the theoretical solution with a scaling factor of two; therefore, Formulators scaled down, by a factor of two, all the results provided by this Contributor for Case B. For Contributor No. 24, the excitation record appears to have been applied in the opposite direction from that specified in the formulation; therefore, Formulators reversed the sign of the provided results. It appears that Contributor No. 25 input significantly differed from the input in the formulation; therefore, the analysis results from Contributor No. 25 are included only in the overall comparison of analysis results for Case B.

Of all submitted solutions for Case B, the input record of 11 Contributors was within a 3% range of the expected theoretical value over the entire length of the foundation block. The input record for seven contributions was in good agreement at the central part of the block, with variations of around 20% at the block side edge.

Peak velocities. The ideal numerical solution for Case B would perfectly match the theoretical record for a semi-infinite half-space (i.e., a free-surface amplification factor of 2 for the upgoing wave). Figures 31, 32, 33 and 34 present the peak horizontal velocities at the top surface of the foundation block in Points *a, c, e,* and *g* submitted by Contributors. The results are compared with the analytical solution of a problem

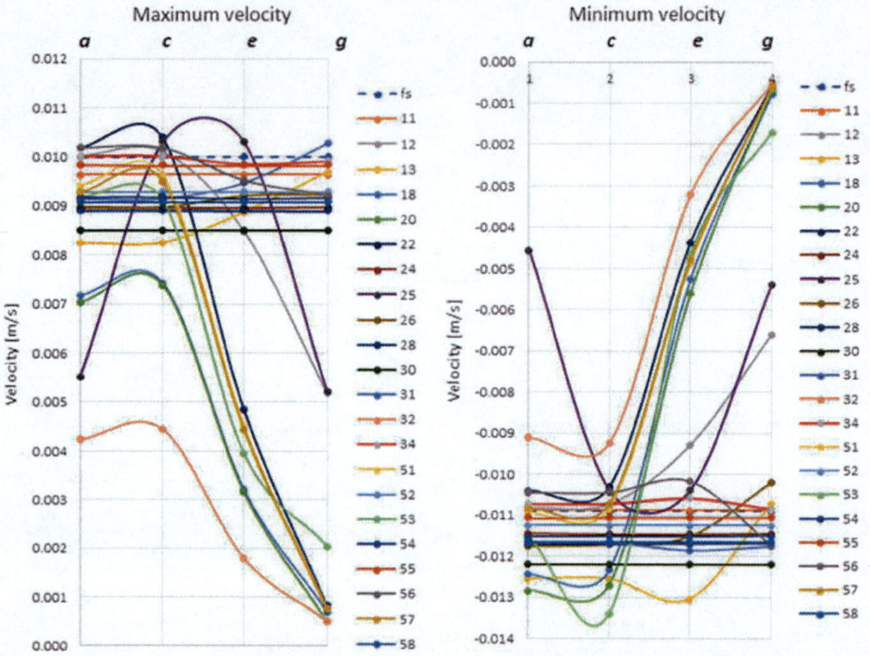

Fig. 31 Peak velocity plots at Points a, c, e, and g for Case B-1 for all contributions

for a wave propagation in a half-space, plotted in the figures as a horizontal dashed blue line (fs) at 0.010 m/s and −0.011 m/s for the maximum and minimum peak velocities, respectively.

Various BC. Significant variation in the results for Case B is observed, which is related to the type of BCs used for the analysis (Figs. 35 and 36). There is a group of results where the peak velocities are uniformly distributed along the foundation block length (Points *a, c, e,* and *g*) within a relatively narrow range. In the second group of results, the peak velocities are relatively close to the theoretical solutions at the center portion of the block (Point *a*) but they vary significantly with the distance away from the center.

Very good agreement with the theoretical solution is observed for the results when the "free-field" BCs are used (Figs. 35 and 36, left). The uniformly applied wave at the base of the foundation block remains uniform at the top surface of the block. When only the "non-reflecting" BCs are used, the relatively good agreement with the theoretical solution is observed only at the central part of the foundation block, with significant differences close to the block ends (Figs. 35 and 36, right).

Figures 37 and 38 show the typical results in time and frequency domain for the selected contributions with "free-field" and "non-reflecting" BCs, respectively.

In Figs. 39, 40, 41 and 42, the frequency distribution of the peak velocity results for all Contributors is presented.

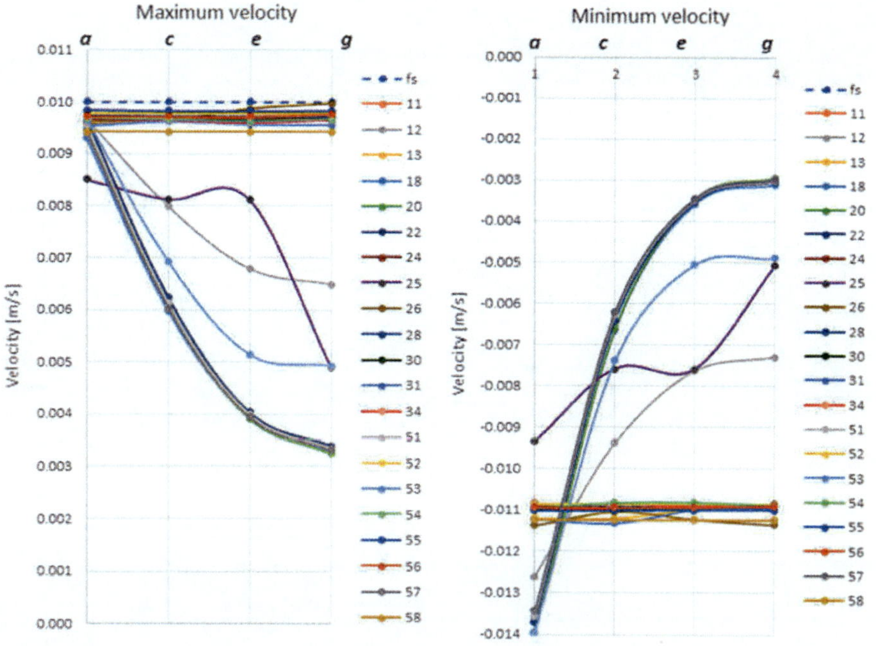

Fig. 32 Peak velocity plots in Points a, c, e, and g for Case B-2 for all contributions

Very good agreement with the theoretical solution is observed for those results provided by Contributors using the "free-field" BC. The uniformly applied wave at the block base remains uniform at the top of the block, as expected. For results based on using only the "non-reflecting" BC, however, the relatively good agreement with the theoretical solution is observed only at the center of the foundation block. There are significant differences observed away from the block center.

Results for Points *a* and *c*, which are closer to the block center, are clustered around a central value of 0.01 m/s, which is the theoretical value. In contrast, the results for Points *e* and *g* are bimodal, with one mode clustering near the theoretical value and one mode at increasingly smaller values the closer the point is to the side boundary.

Block size effect. Figures 43 and 44 compare the results for two block sizes (a small block (700 m long) and a large block (3 700 m long).

The size of the foundation block has very limited influence on the results obtained in the analysis for those Contributors using the free-field BCs. This is true for both the high- and low-frequency pulse.

As before, the results obtained from Contributors, using only the non-reflecting BCs, are relatively close to the expected theoretical solution only near the central part of the foundation block (Point *a* and *c*). For the high-frequency pulse, results differ from the theoretical solution at the locations closer to the edge of the foundation

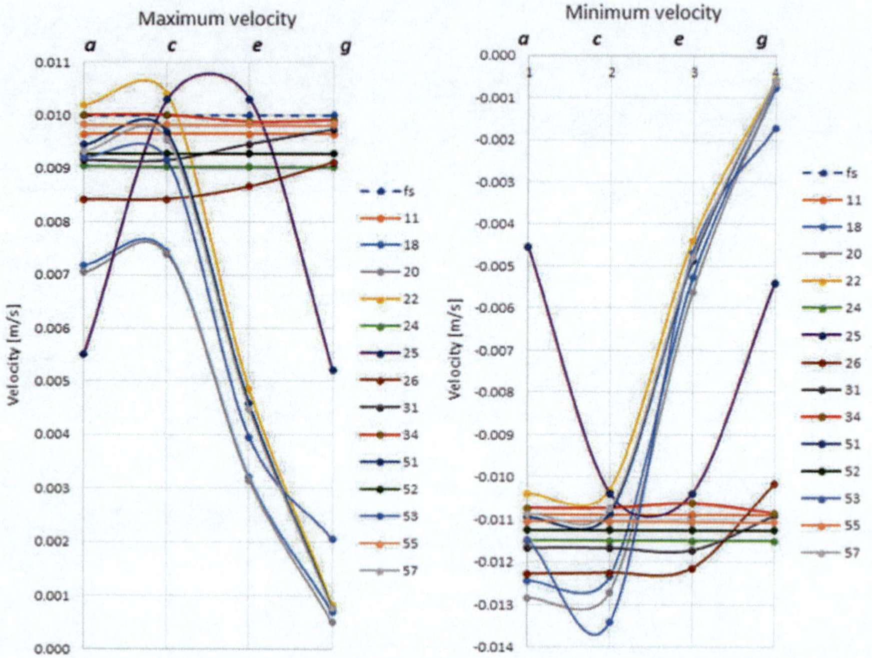

Fig. 33 Peak velocity plots in Points a, c, e, and g for Case B-3 for all contributions

block. For the low-frequency pulse, the results vary from the theoretical solution over the entire length of the foundation block.

Effect of pulse frequency. Figures 45 and 46 compare the results obtained for the high- and low-frequency pulses.

6 Case C—Dynamic Analysis Using Impulsive Loads

6.1 General

The analysis for Case C is similar to Case B; the only difference is the addition of the dam and the reservoir to the foundation block model. The purpose of this analysis is to investigate how the dam and reservoir influence wave propagation, as well as how the results compare with the free-surface motions away from the dam.

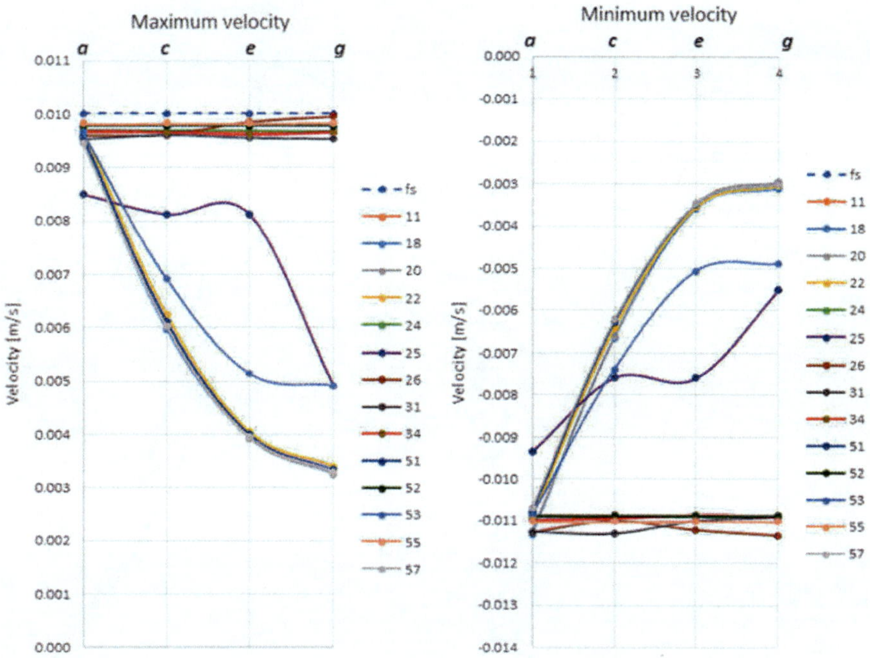

Fig. 34 Peak velocity plots in Points a, c, e, and g for Case B-4 for all contributions

6.2 *Conceptual Model*

Model configuration. In Case C, the analysis for the base model configuration is conducted using both a high- and low-frequency pulse with and without a reservoir.

Input parameters. The following parameters are assumed in the analysis:

- Concrete material properties are defined in Table 4.
- Elastic rock material properties are defined in Table 5.
- Mass type foundation is considered.
- Viscous damping is zero.
- The maximum element size is 1.5 m.

Loads. The following loads are considered:

- The excitation is defined by the horizontal input stress time histories shown in Figs. 6 and 7. The baseline corrected impulsive stress time record is applied uniformly at the base of the block in the same way it was applied in Case B.
- Static loads due to weight (gravity loads) are not considered.
- WRWS at El. 268.21 m.

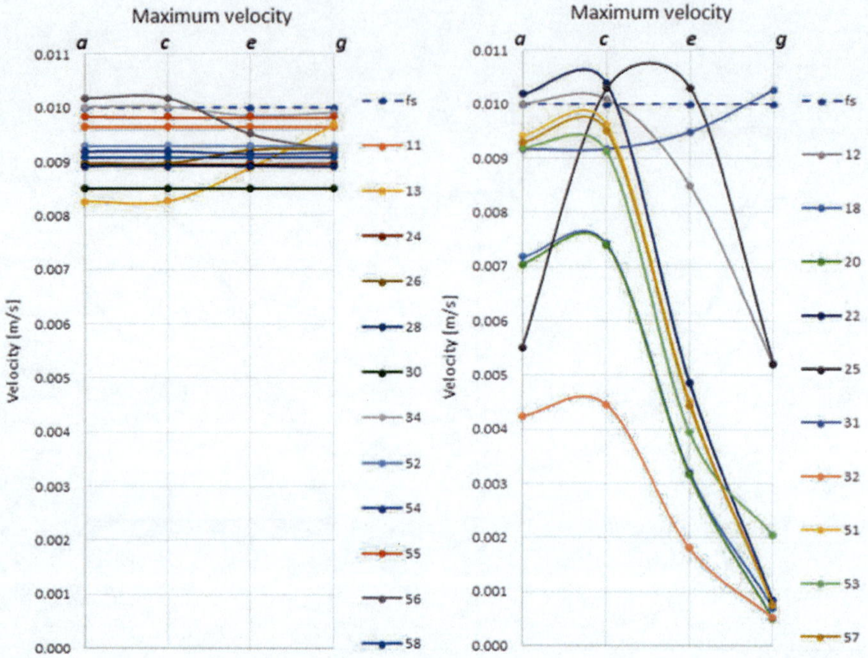

Fig. 35 Case B-1 results obtained with the "free-field" BCs (left) and with non-reflecting BCs (right)

Analysis. Model configuration and analysis type combinations are shown in Table 14. The analysis time is 2 s for the high-frequency impulse and 20 s for the low-frequency impulse. The analysis results are compared at the specified model locations, as indicated in Fig. 47.

6.3 *Results*

For Case C, 11 Contributors provided time histories velocity records at Points *a* through *j* on the foundation block and at Points *A* and *C* on the dam.

For illustrative purposes, Fig. 48 presents sequential snapshots in time of the high–frequency impulse propagating into and out of the dam-foundation model.

The high-frequency pulse (Fig. 7) was constructed for the formulation in a way that allows several wavelengths to fit within the depth of the foundation block.

The uniformly applied excitation at the base of the foundation propagates as a uniform plane wave until it splits, at the run time of about 0.008–0.09 s., into a wave that reflects from the upper face of the foundation and a wave that continues

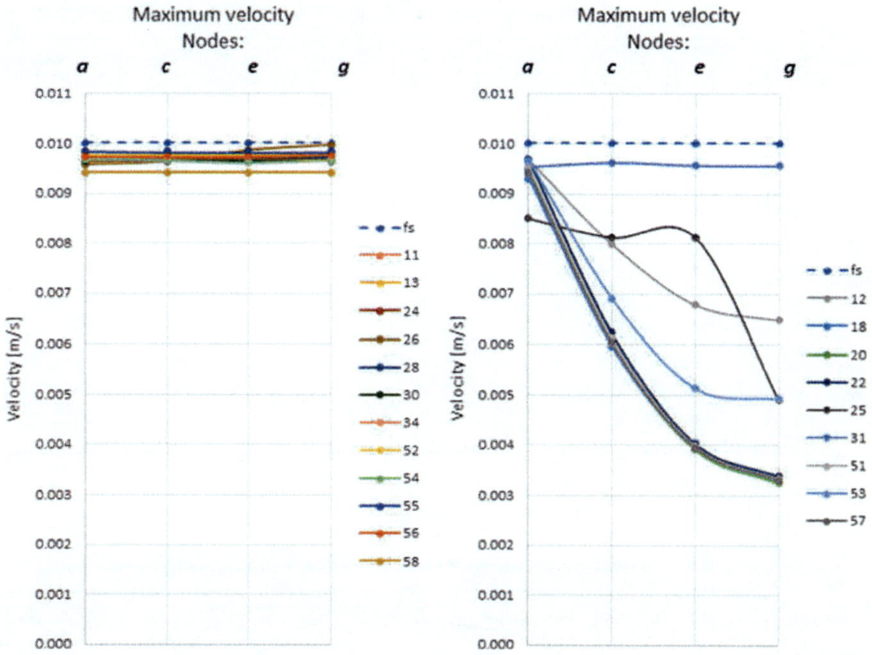

Fig. 36 Case B-2 results obtained with the "free-field" BCs (left) and with non-reflecting BCs (right)

to propagate into the dam structure, with some noise introduced at the dam heel and toe (Fig. 48) at run time of 0.094 s).

At a run time of approximately 0.015 s, the downgoing (reflected) wave in the foundation block reaches the base of the block, where it then vanishes (it is absorbed at the boundary). At about the same time, the upgoing wave within the dam reflects from the dam crest and starts propagating downwards. Propagation of the reflected wave within the body of the dam is substantially influenced by the dam geometry, with the amplitude of the wave increasing as the cross-section becomes smaller in order to preserve energy.

Results at the free surface. The ideal result would be achieved when the time histories computed at the block free surface for Case C, and far from the dam, matched the record obtained for Case B at the corresponding locations. Figures 49 and 50 compare the velocity records at Point *c* (Figs. 29 and 47) for high- and low–frequency pulse, respectively.

Effect of dam and reservoir presence. Figures 51, 52, 53 and 54 present a comparison of the maximum and minimum peak velocities for Cases C-1 and C-2 at two locations: the base of the dam (Point *a*) and the free surface (Point *c*).

Significant reduction of the velocity amplitude is observed at the dam base (Point *a*) when compared with the velocity amplitude at the free surface (Point *c*).

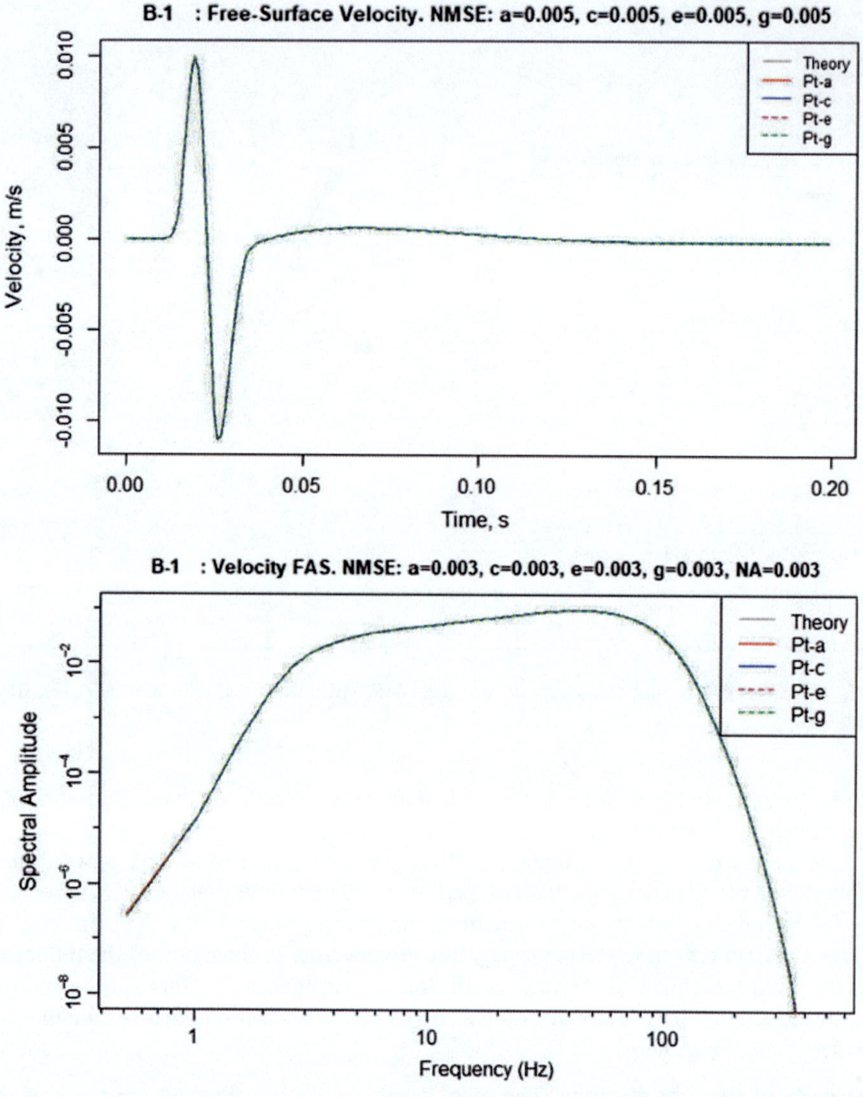

Fig. 37 The results in time and frequency domain obtained for "free-field" BCs for Case B-1

Effect of reservoir presence. Figures 55 and 56 present differences in the maximum peak velocities with and without the reservoir. In general, a smaller peak velocity for the dam is observed when the reservoir is present in the model.

The reservoir presence in the model had a limited influence on the velocity amplitude at the dam heel, but a more significant difference was observed at the dam crest.

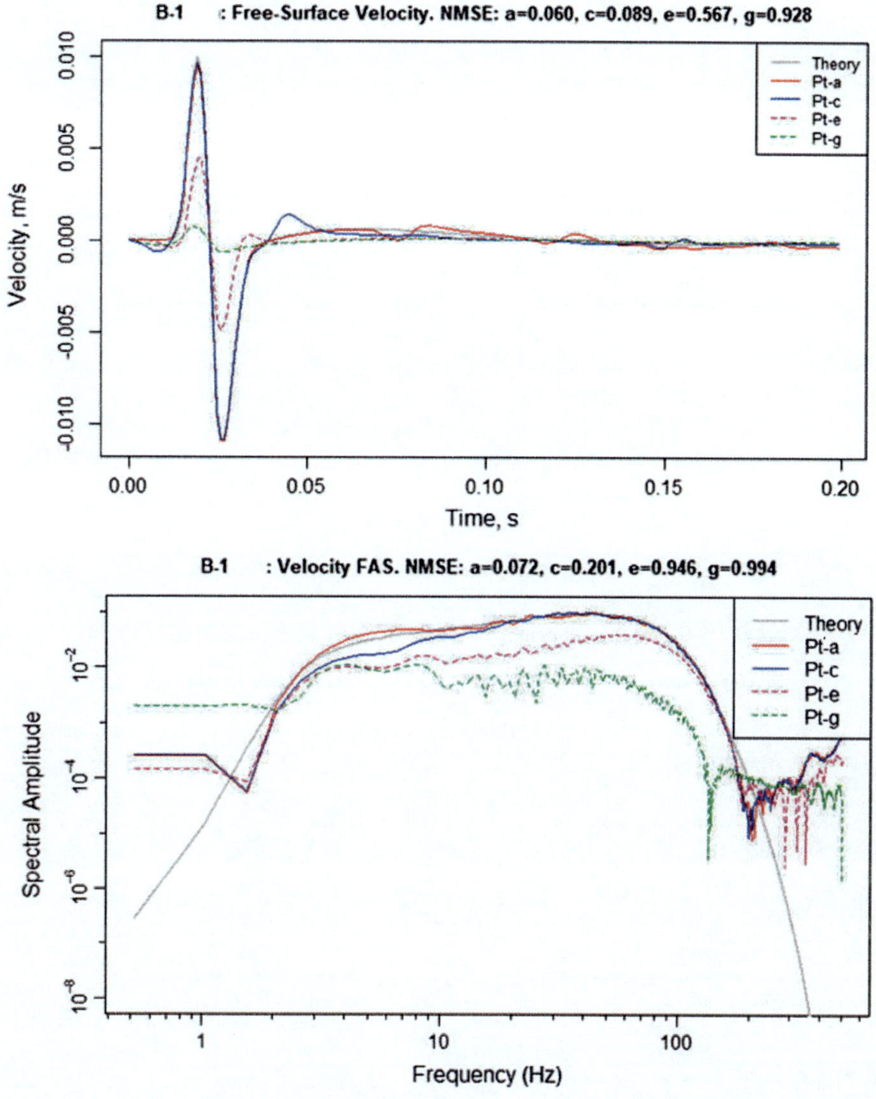

Fig. 38 The results in time and frequency domain obtained for "non-reflecting" BCs for Case B-1

Comparing the results in the time and frequency domain for Case C-1. Selected results for high frequency pulse (Case C-1) are presented in the time and frequency domain (Figs. 57 and 58) for models using free-field and absorbing BCs, respectively. The results are also summarized in Table 15.

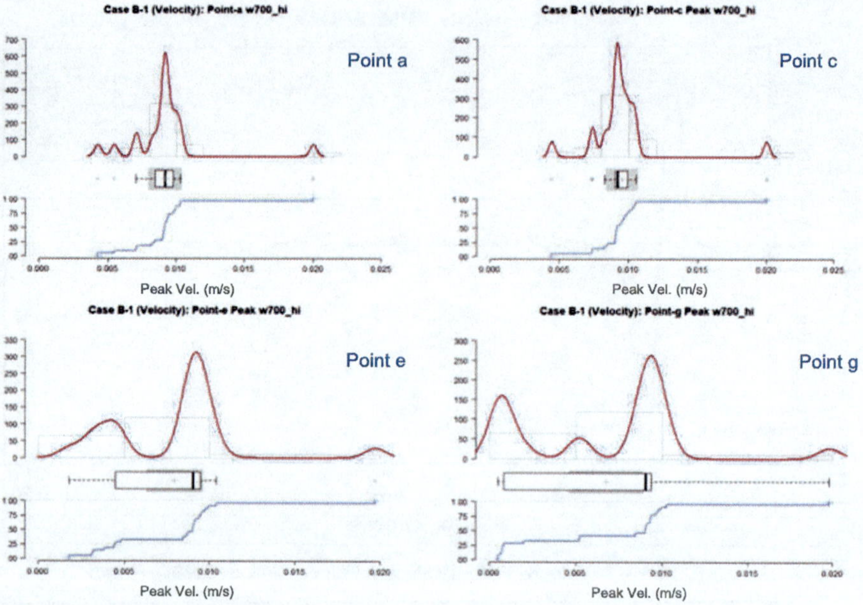

Fig. 39 Case B-1—Distribution of computed peak velocity at Points a, c, e, and g

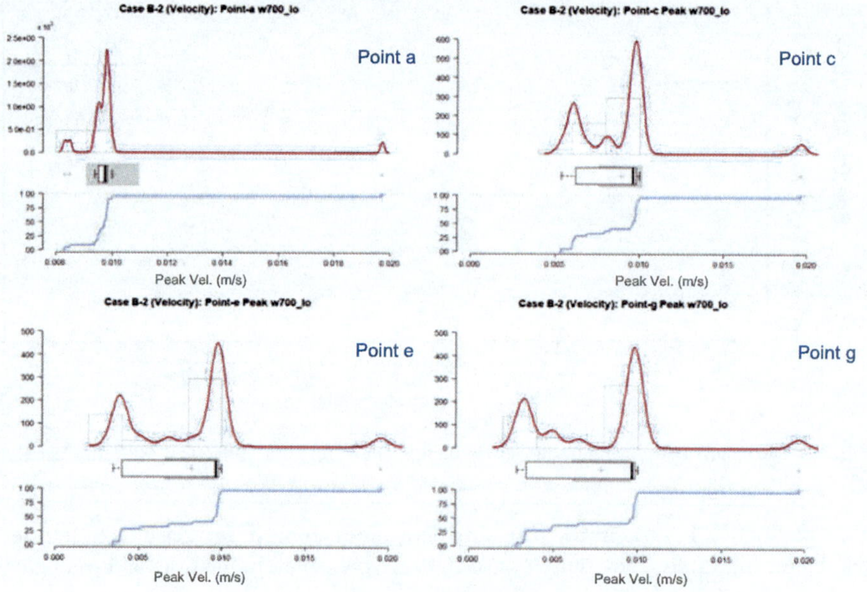

Fig. 40 Case B-2—Distribution of computed peak velocity at Points a, c, e, and g

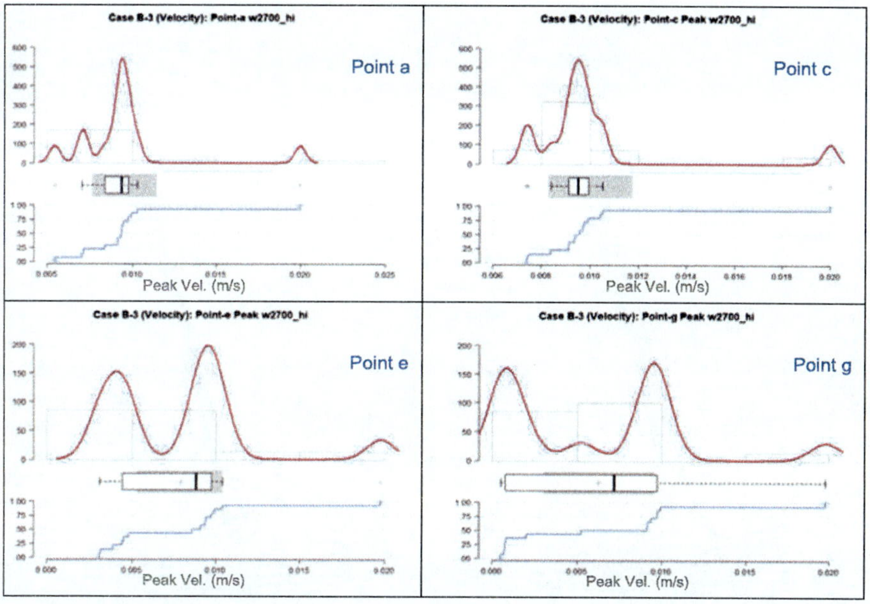

Fig. 41 Case B-3—Distribution of computed peak velocity at Points a, c, e, and g

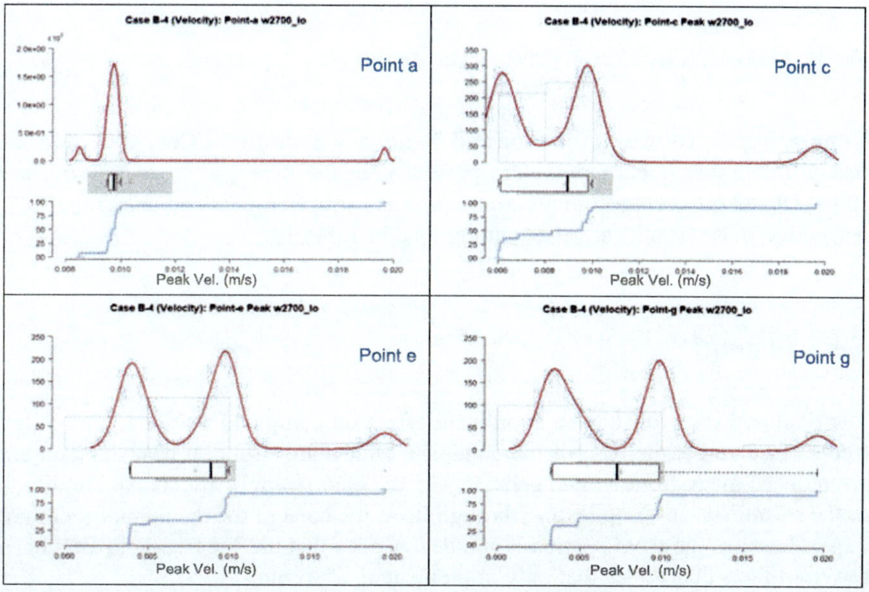

Fig. 42 Case B-4—Distribution of computed peak velocity at Points a, c, e, and g

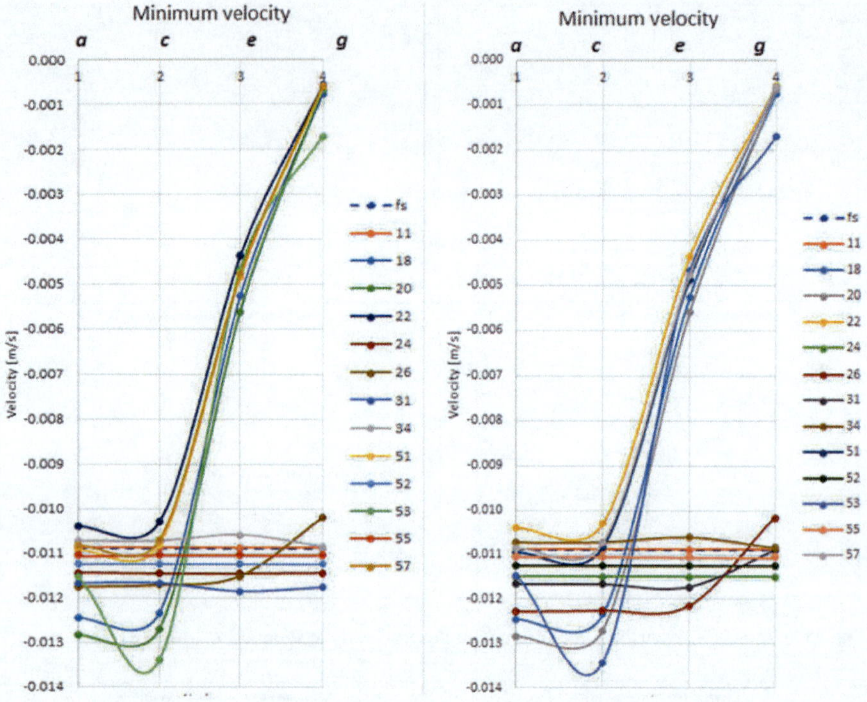

Fig. 43 Minimum peak velocities in Points a, c, e, g for Case B-1 (left) and Case B-3 (right)

Comparing the results in the time and frequency domain for Case C-2. Selected results for low-frequency pulse are presented in the time and frequency domain (Figs. 59 and 60) for the computational models using free-field and absorbing BCs, respectively. The results are also summarized in Table 16.

6.4 *Summary*

The dam and reservoir have a significant effect on computed motions, both at the free-surface and at the base of the dam. The effects are strongest near the dam but are reduced away from it. The geometry of the dam strongly affects the amplitude of the seismic waves propagating through it. At the crest of the dam, amplitudes are largest because the cross-section is smallest. Codes that use the free-field BCs have fewer artifacts than codes that only implement an absorbing BC.

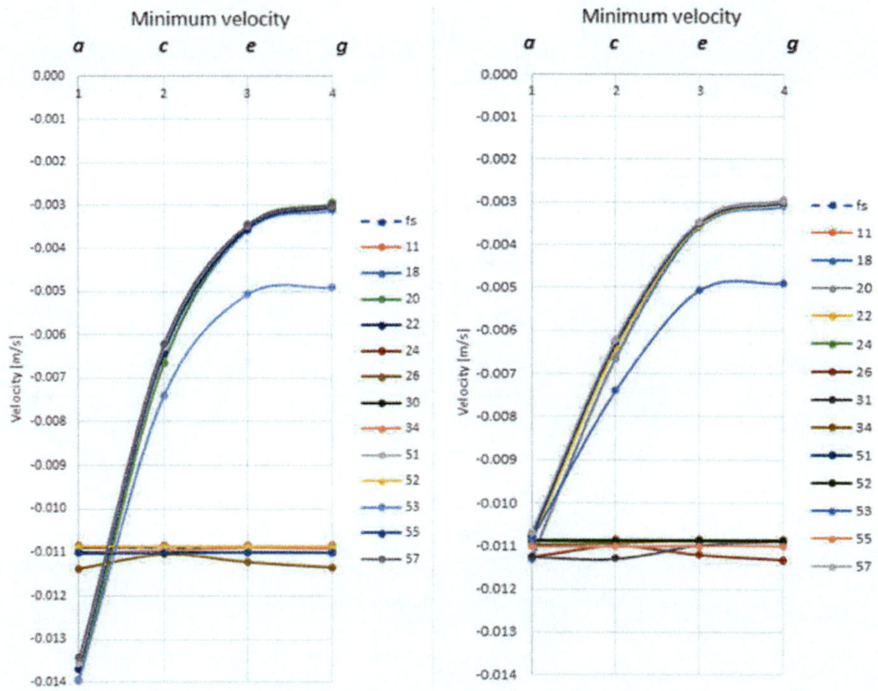

Fig. 44 Minimum peak velocities in Points a, c, e, g for Case B-2 (left) and Case B-4 (right)

7 Case D—Dynamic Analysis for Various Reservoir Levels

7.1 General

In Case D, the linear dynamic analysis of the dam-foundation-reservoir system is performed considering the elastic material properties, the Taft Record, and the reservoir water at three different levels. The intent of the analysis is to evaluate the dam response due to various reservoir levels.

7.2 Conceptual Model

Model configuration. The "base configuration" of the model, as defined in Sect. 3.2, is considered for Case D with three reservoir water levels.

Input parameters. For consistency in the analysis, the following assumptions are considered:

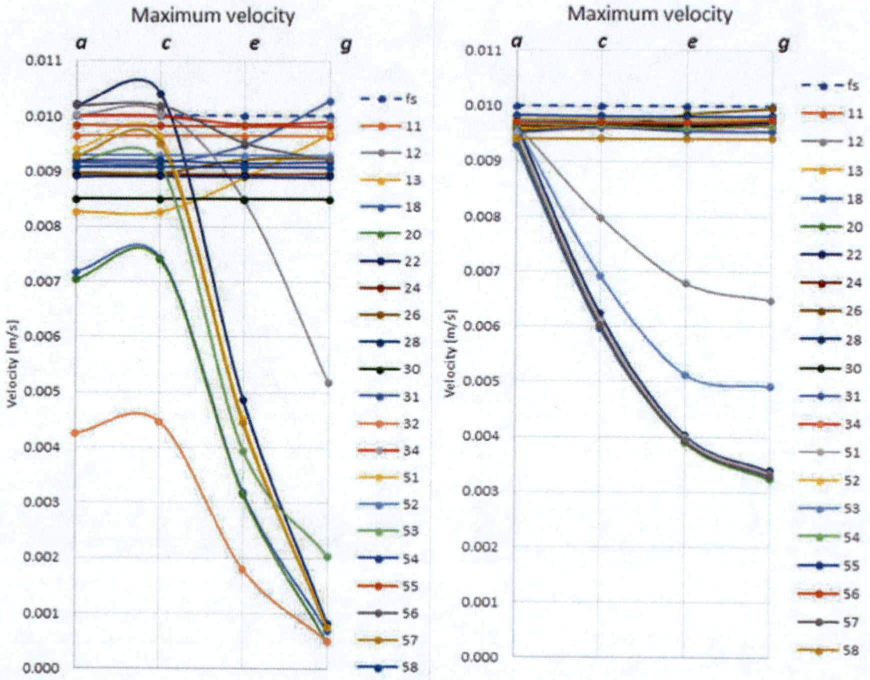

Fig. 45 Maximum velocities in Points a, c, e, and g for Case B-1 (left) and Case B-2 (right)

- Concrete elastic properties are defined in Table 4.
- Water properties are defined in Sect. 3.2.
- Foundation elastic material properties are defined in Table 5.
- The mass of the dam, reservoir, and foundation.
- The static weight of the dam and reservoir, but not the weight of the foundation.
- 2% viscous damping for the dam and foundation.

Loads. The loads include:

- The Taft Record applied in the upstream/downstream direction, using one of the following options:
    - Option 1: Taft deconvolved acceleration time history applied at the base of the foundation block for three different reservoir water levels.
    - Option 2: Taft stress time history applied at the base of the foundation block for three different reservoir water levels.
- Gravity loads applied to the dam and reservoir, but not to the foundation.

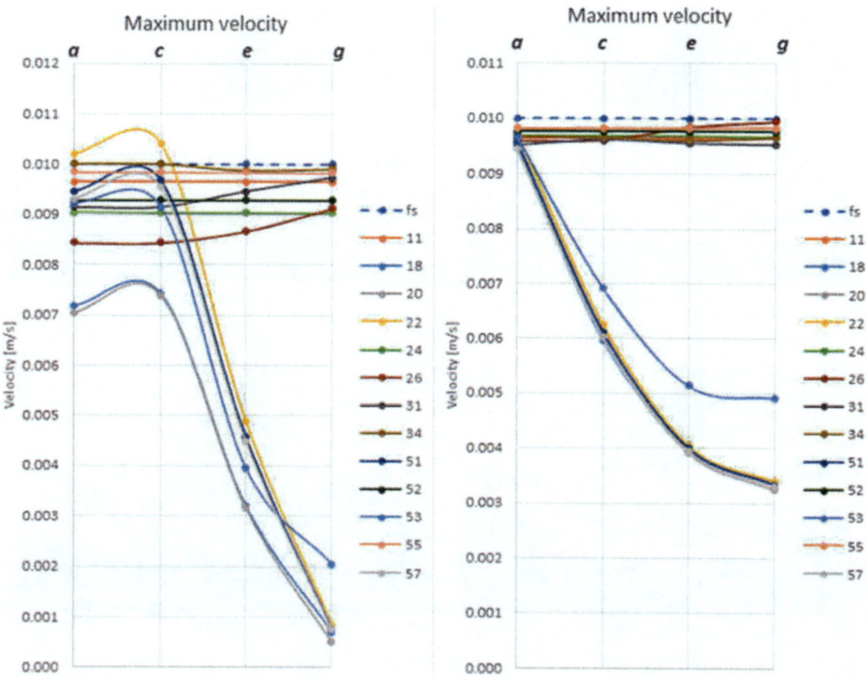

Fig. 46 Maximum velocities in Points a, c, e, and g for Case B-3 (left) and Case B-4 (right)

Table 14 Case C analysis

Case	Model configuration	Impulsive signal
C-1	Dam-reservoir-foundation	High frequency
C-2	Dam-reservoir-foundation	Low frequency
C-3	Dam-foundation	High frequency
C-4	Dam-foundation	Low frequency

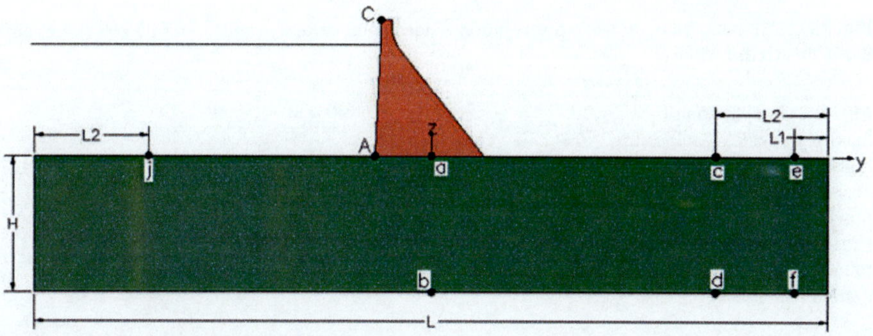

Fig. 47 Selected locations for results comparison in Case C. Dimensions: H = 122 m; L = 700 m; L1 = 20 m; L2= 100 m; H= 122 m; L = 700 m

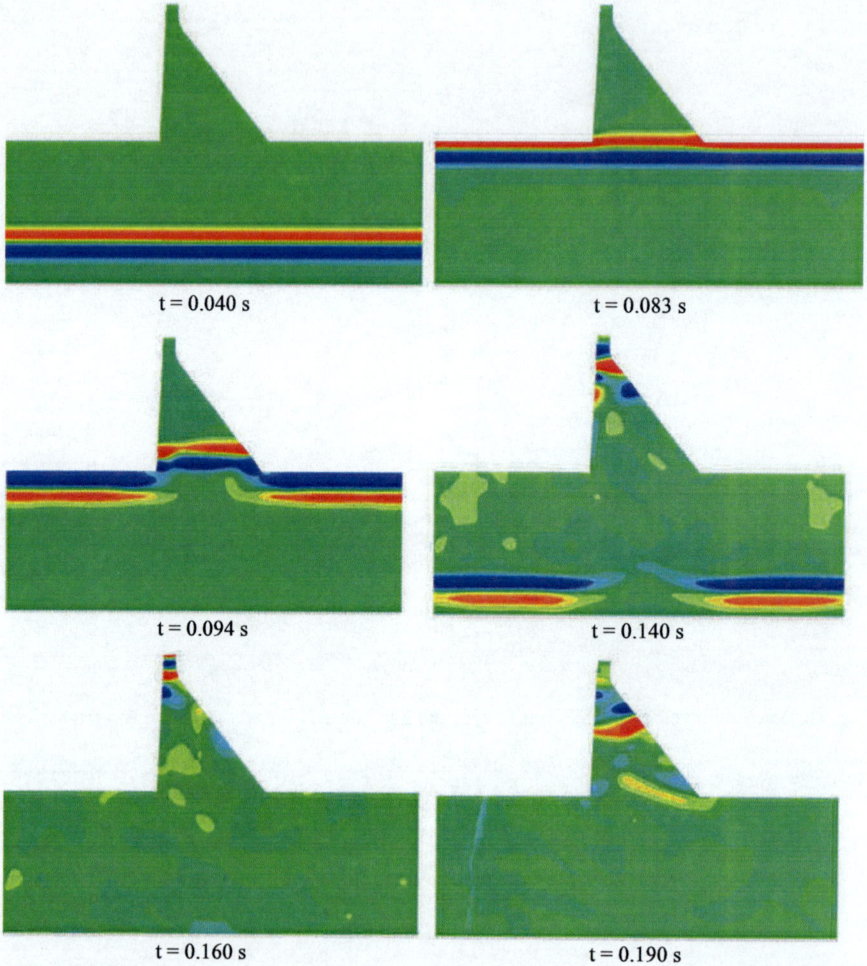

Fig. 48 Horizontal velocity wave propagation—horizontal velocity range: −0.005 m/s (blue) and 0.005 m/s (red) [A12]

Fig. 49 Comparison of maximum peak velocities for high frequency pulse excitation at Point c for foundation block model (Case B-1) and foundation-dam-reservoir system (Case C-1)

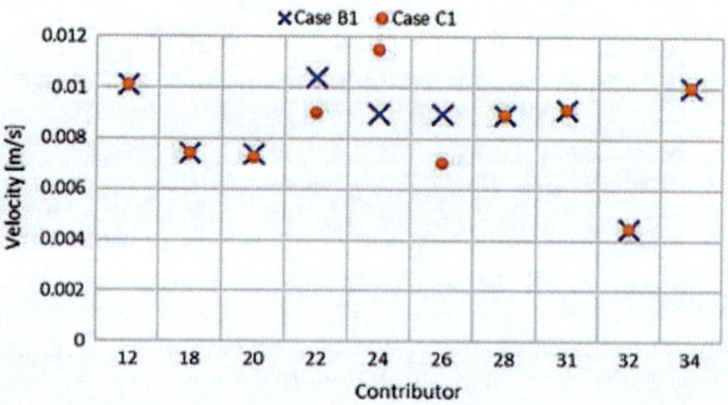

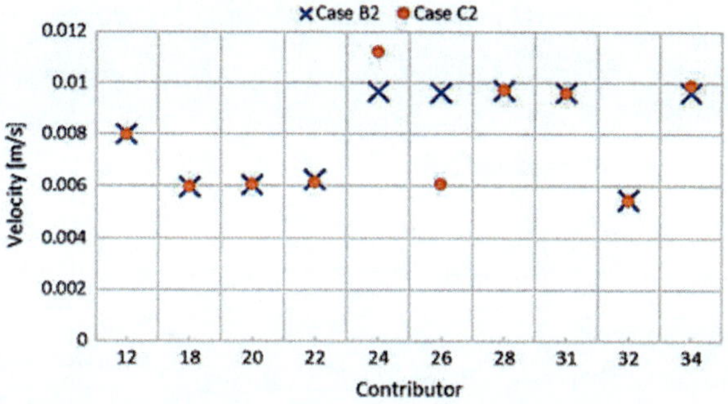

Fig. 50 Comparison of maximum peak velocities low pulse excitation at Point c for foundation block model (Case B-2) and foundation-dam-reservoir system (Case C-2)

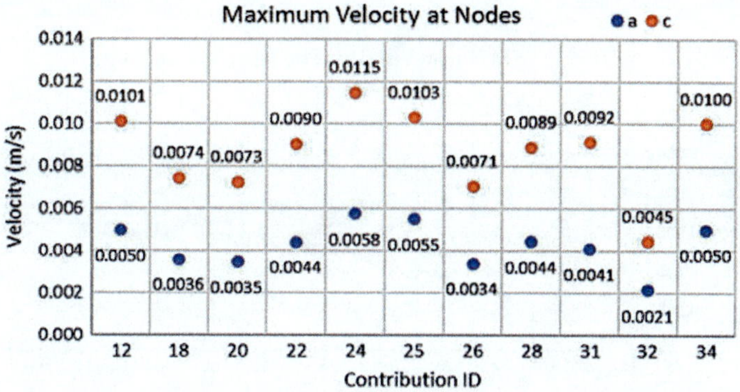

Fig. 51 Maximum peak velocity at Point a (center of the dam base) and Point c (free surface) for a high-frequency pulse, Case C-1

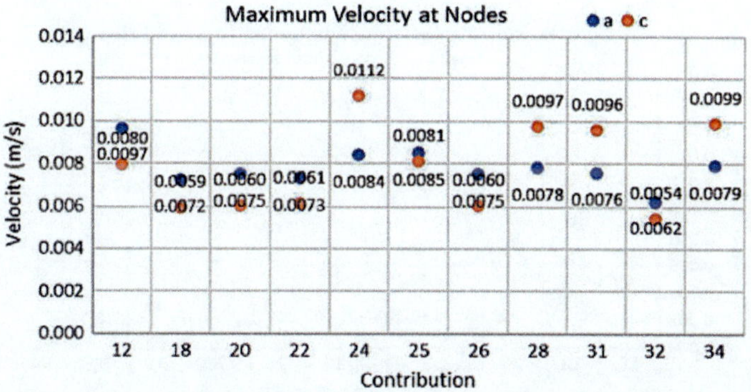

Fig. 52 Maximum peak velocity at Point a (center of the dam base) and Point c (free surface) for a low-frequency pulse, Case C-2

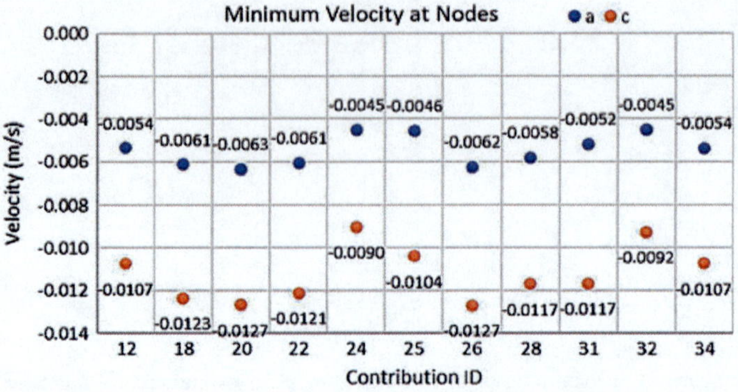

Fig. 53 Minimum peak velocity at Point a (center of the dam base) and Point c (free surface) for a low-frequency pulse, Case C-1

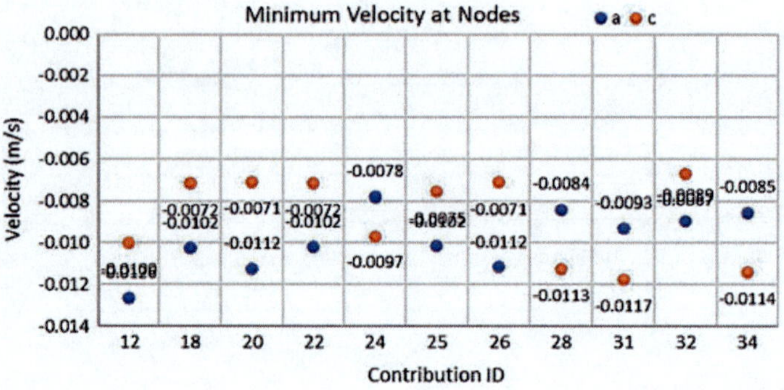

Fig. 54 Minimum peak velocity at Point a (center of the dam base) and Point c (free surface) for a low frequency pulse, Case C-2

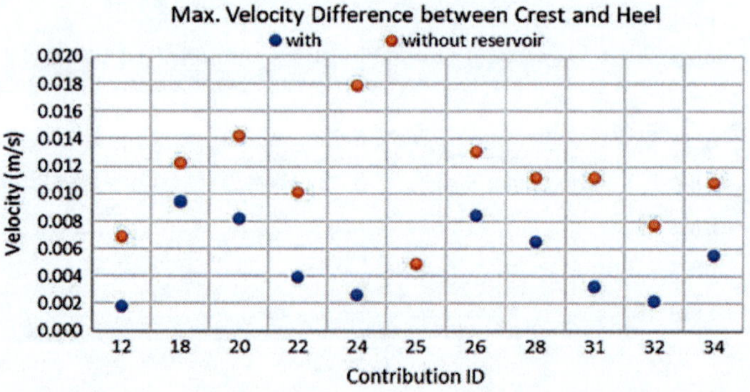

Fig. 55 Difference of maximum peak velocity between Point C (dam crest) and Point A (dam heel) for a high-frequency pulse, Case C-1

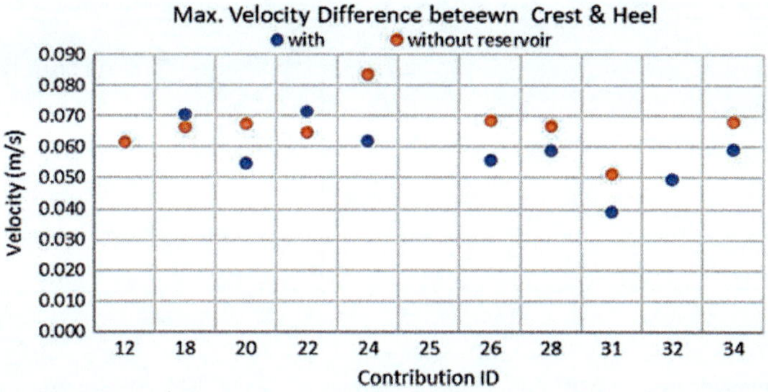

Fig. 56 Difference of maximum peak velocity between Point C (dam crest) and Point A (dam heel) for a high-frequency pulse, Case C-2

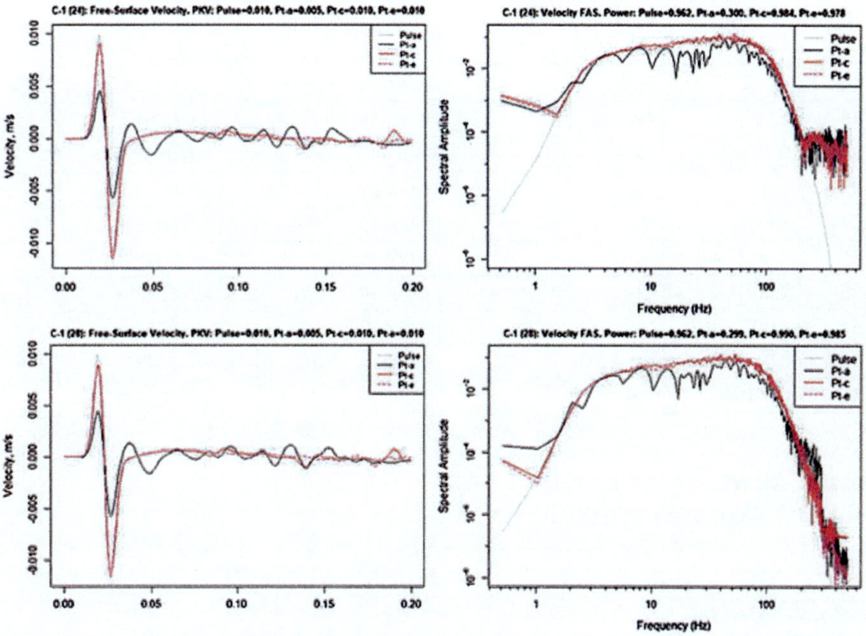

Fig. 57 Comparison of velocity time histories at Points a, c, and e, with the theoretical solution for Case C-1 for two selected contributions using "free-field" BCs

7.3 *Analysis*

The linear dynamic analysis is performed for the following water levels:

- **D-1**: WRWL at El. 268.21 m.

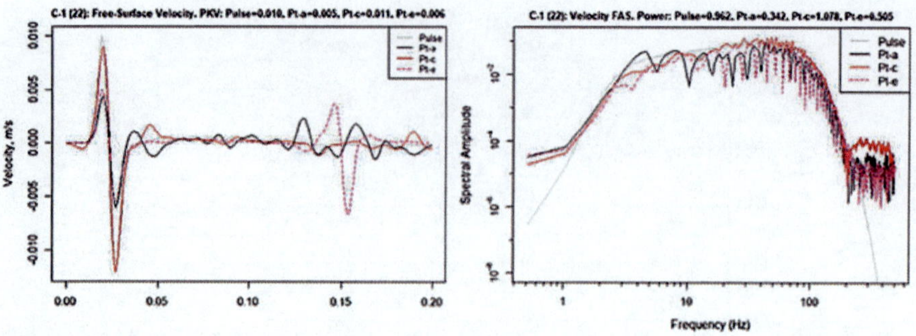

Fig. 58 Comparison of velocity time histories at Points a, c, and e, with the theoretical solution for Case C-1 for two selected contributions using “absorbing” BCs

Table 15 Summary of results for Case C-1

ID	Time step	Free-field boundary conditions	Peak positive (m/s)	Peak negative (m/s)
12	0.001	No	0.00496	−0.00538
18	0.001	No	0.00358	−0.00613
20	0.001	No	0.00346	−0.00637
22	NA	No	0.00439	−0.00597
24	3.85E−5	Yes	0.00455	−0.00569
25	2.00E−4	No	0.00545	−0.00457
28	2.00E−4	Yes	0.00448	−0.00575
31	1.00E−4	No	0.00410	−0.00520
32	0.01	No	0.00214	−0.00453
34	0.00025	Yes	0.00500	−0.00534

- **D-2**: SRWL at El. 278.57 m.
- **D-3**: NRWL at El. 290.00 m.

7.4 Results

Contributors reported the following analysis results at the points indicated in Fig. 47:

- Time history of the horizontal acceleration and horizontal displacements at dam crest (Point *C*) and dam heel (Point *A*)
- Time history of hydrodynamic pressure at the dam heel (Point *A*).

Twenty-four teams provided solutions for Cases D-1 and D-2 and 19 teams for Case D-3. Statistics for the software used in the Case D studies (Table 3) include:

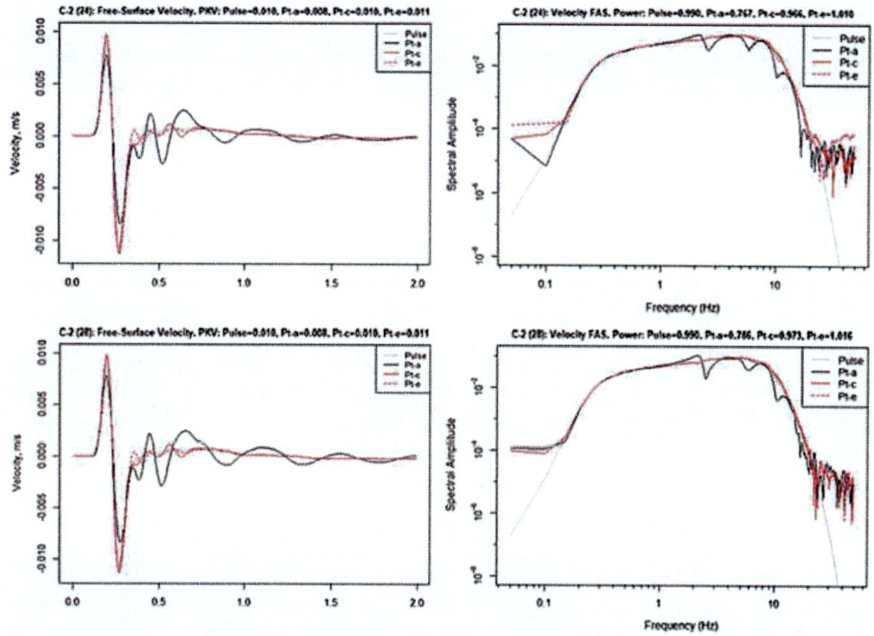

Fig. 59 Comparison of velocity time histories at Points a, c, and e, with the theoretical solution for Case C-2 for two selected contributions using "free-field" BCs

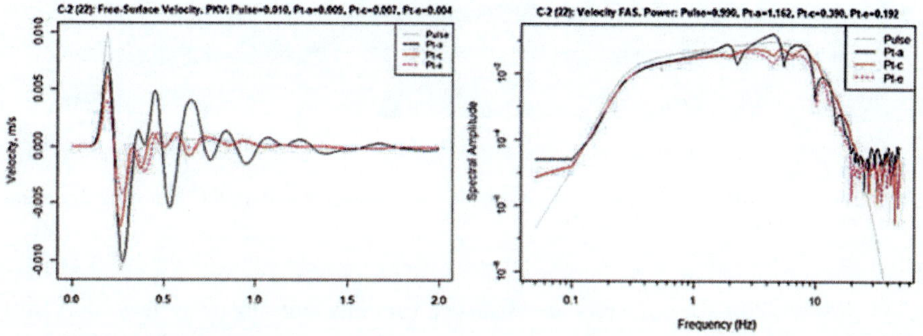

Fig. 60 Comparison of velocity time histories at Points a, c, and e, with the theoretical solution for Case C-2 for two selected contributions using "absorbing" BCs

Abaqus (7), Diana (4), Code_Aster (3), ANSYS (3). Eleven contributors used the free-field BC in their analyses.

A detailed comparison of results for Case D-1, in terms of horizontal acceleration response spectra, are reported in Figs. 61 and 62 at the dam heel (Point *A*) and in Figs. 63 and 64 at the dam crest (Point *C*) for the solutions obtained with and without free-field BCs.

Table 16 Summary of results for Case C-2

ID	Time step	Free-field BCs	Peak positive (m)	Peak negative (m)
12	0.01	No	0.00981	−0.01252
18	0.01	No	0.00729	−0.01025
20	0.01	No	0.00757	−0.01130
22	NA	No	0.00736	−0.01026
24	3.8E−5	Yes	0.00780	−0.00846
25	0.002	No	0.00850	−0.01019
28	0.002	Yes	0.00781	−0.00844
31	0.01	No	0.00758	−0.00930
32	0.01	No	0.00629	−0.00891
34	0.005	Yes	0.01581	−0.01706

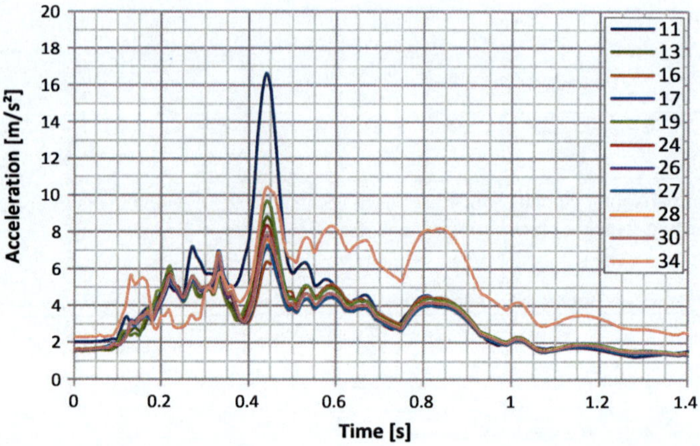

Fig. 61 Case D-1. Horizontal acceleration response spectra at the dam heel (Point A). Results are from Contributors using free-field BCs

In general, consistent results are observed for contributions using free-field BCs (Figs. 61 and 63). The results provided by other Contributors (Figs. 62 and 64) show larger scatter.

Figures 65 and 66 compare the solutions provided for Cases D-1, D-2, and D-3, in terms of peak horizontal acceleration at the dam heel (Point *A*) and dam crest (Point *C*), respectively. "Peak horizontal acceleration" ($a_{peak(average)}$) is defined as the average of the positive (a_{max}) and negative (a_{min}) peak values of the horizontal acceleration time history (Eq. 2):

$$a_{peak(average)} = 1/2(a_{max} - a_{min}) \qquad (2)$$

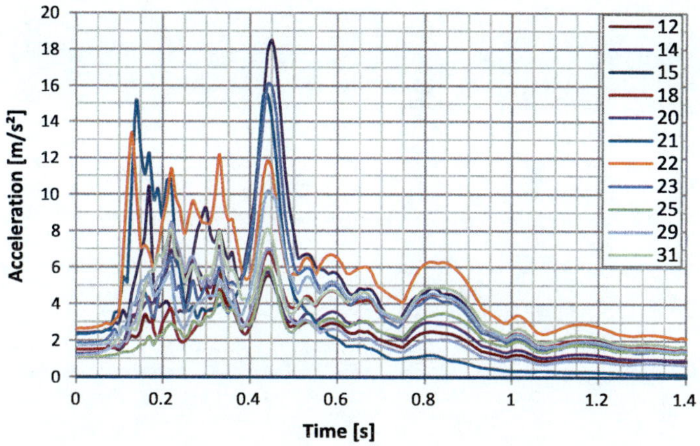

Fig. 62 Case D-1. Horizontal acceleration response spectra at the dam heel (Point A). Results are from Contributors not using free-field BCs

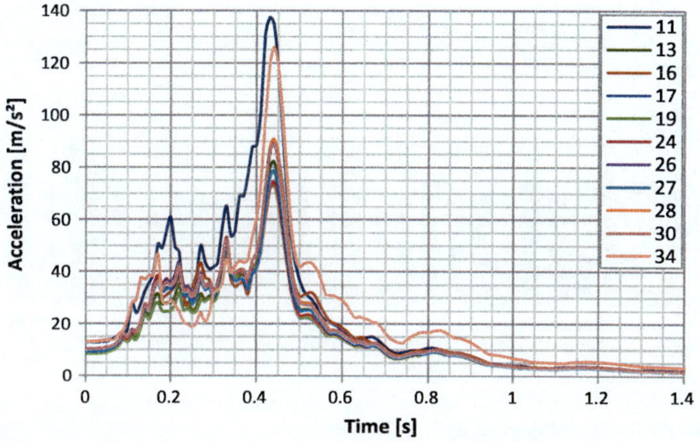

Fig. 63 Case D-1. Horizontal acceleration response spectra at the dam crest (Point C). Results are from Contributors using free-field BCs

The comparison of the peak acceleration at the dam heel shows quite similar results (also in fair agreement with the peak ground acceleration (PGA) of the Taft Record equal to 1.74 m/s^2) for most of the solutions, particularly for those solutions adopting the free-field BCs (almost all of them provide values less than the PGA of the Taft Record). In general, when comparing the three cases (D-1, D-2 and D-3), a small decrease of peak acceleration for increasing reservoir levels is observed; only in a few cases is there a greater decrease of peak acceleration or a different trend.

At the dam crest, the small differences observed among the results from the heel are amplified; however, the peak acceleration is still very similar for Contributors

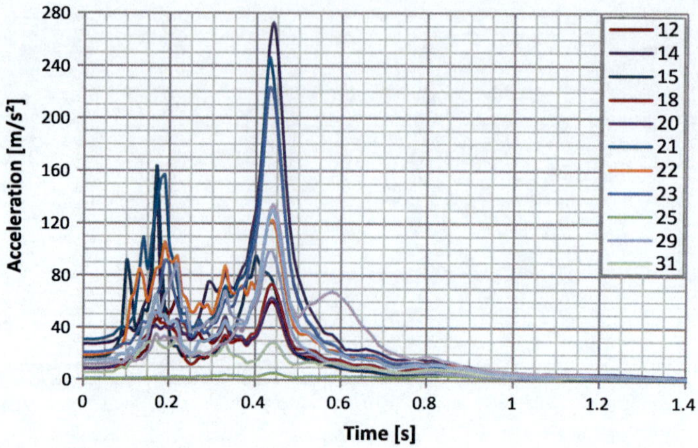

Fig. 64 Case D-1. Horizontal acceleration response spectra at the dam crest (Point C). Results are from Contributors not using free-field BCs

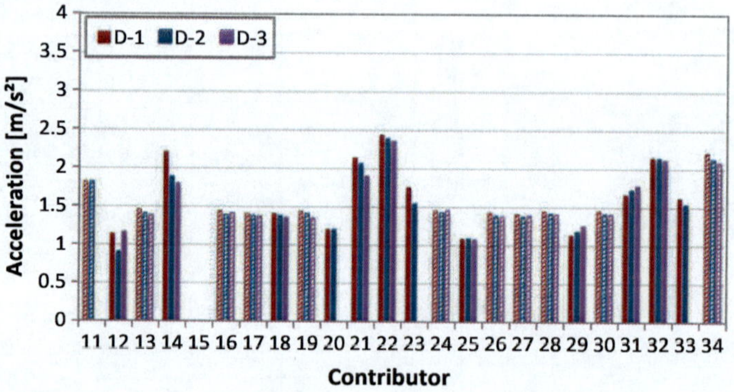

Fig. 65 Cases D-1, D-2, and D-3. Peak horizontal acceleration at the dam heel (Point A). Dashed histograms indicate Contributors using free-field BCs

using free-field BCs (about 9–10 m/s^2, 8–9 m/s^2, and 7–8 m/s^2 for Cases D-1, D-2 and D-3, respectively). Also, at the dam crest, peak acceleration decreases as the water level becomes higher, as illustrated by the results from most of Contributors.

Table 17 shows a summary of peak accelerations at the dam heel and crest, as reported for Case D, in terms of median, mean, standard deviation, and coefficient of variation over the results from all Contributors. The decreasing trend of peak acceleration with increasing water levels is confirmed, considering the median and mean values for both Point *A* and Point *C*. Results are more scattered in Point *C* than in Point *A*, as the coefficients of variation clearly highlight.

Figures 67 and 68 report the comparison of relative horizontal total displacements (between Point *C* and Point *A*) at the dam crest for maximum downstream

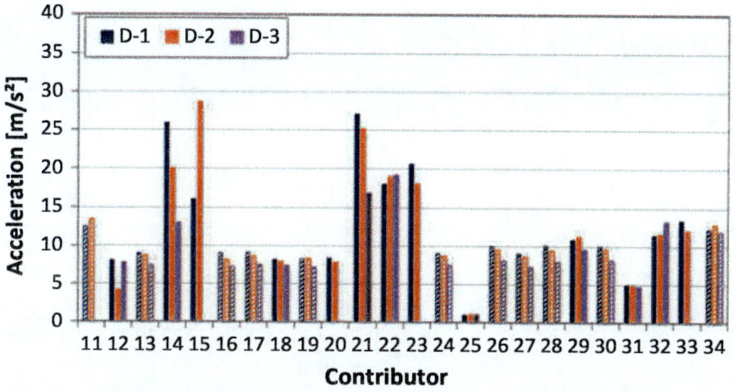

Fig. 66 Cases D-1, D-2, and D-3. Peak horizontal acceleration at the dam crest (Point C). Dashed histograms indicate Contributors using free-field BCs

Table 17 Case D. Median, mean, standard deviation and coefficient of variation of peak acceleration at the dam heel (Point A) and the dam crest (Point C)

Point	Acceleration	Units	D-1	D-2	D-3
A	Median	m/s^2	1.46	1.42	1.40
	Mean	m/s^2	1.55	1.49	1.49
	Standard deviation	m/s^2	0.50	0.48	0.48
	Coefficient of variation		0.32	0.32	0.32
C	Median	m/s^2	10.01	9.48	7.41
	Mean	m/s^2	11.76	11.53	8.69
	Standard deviation	m/s^2	6.03	6.47	4.58
	Coefficient of variation		0.51	0.56	0.53

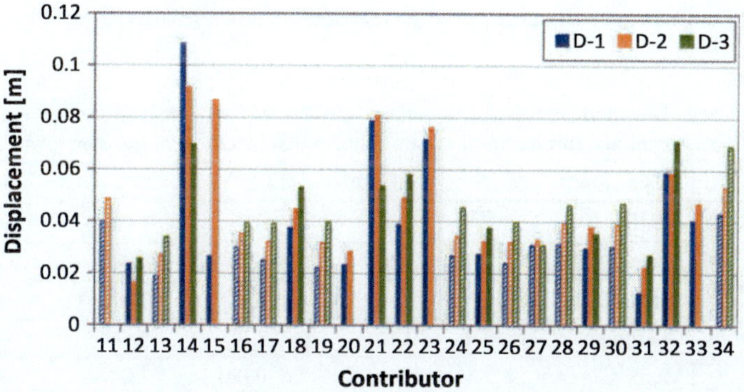

Fig. 67 Cases D-1, D-2, and D-3. Maximum downstream relative (Point C–Point A) total (static and dynamic) displacement at dam crest (Point C). Dashed histograms indicate results for free-field BCs

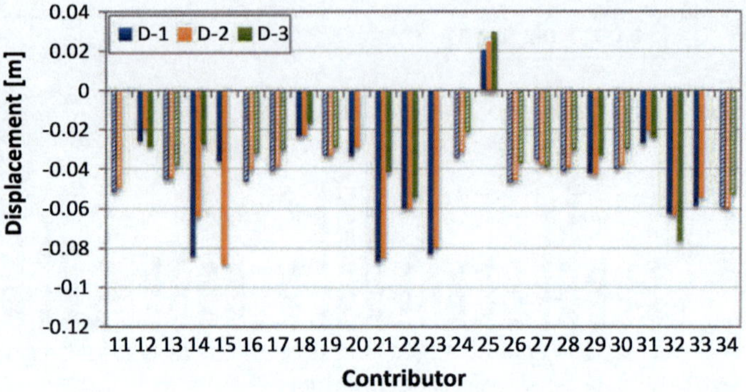

Fig. 68 Cases D-1, D-2, and D-3. Minimum upstream relative (Point C–Point A) total (static and dynamic) displacement at dam crest (Point C). Dashed histograms indicate Contributors using free-field BCs

and minimum upstream displacement, respectively (Contributor No. 12 presented dynamic displacements instead of the total displacement). Table 18 reports the summary of maximum and minimum displacements, in terms of median, mean, standard deviation, and coefficient of variation, over the results from all Contributors.

Consistent displacement results are observed for Contributors using free–field BCs; maximum downstream displacements increase with increasing water level, ranging from about 0.02–0.03 m in Case D-1, to about 0.03–0.04 m in Case D-2, and to about 0.035–0.045 m in Case D-3. Conversely, minimum upstream displacements decrease for increasing water levels. The different trend between upstream and downstream total displacement is reasonably produced by the static displacements which tends towards with increasing water level. When free-field BCs are not used, more discrepancies in the results is observed. Generally, a greater scatter is observed in the displacement results compared to the acceleration results.

Table 18 Case D—Median, mean, standard deviation, and coefficient of variation of maximum downstream and minimum upstream relative displacement at the dam crest (Point C)

	Displacement	Units	D-1	D-2	D-3
Maximum	Median	m	0.030	0.038	0.040
	Mean	m	0.038	0.045	0.045
	Standard deviation	m	0.022	0.020	0.014
	Coefficient of variation		0.579	0.444	0.311
Minimum	Median	m	−0.042	−0.041	−0.030
	Mean	m	−0.045	−0.044	−0.032
	Standard deviation	m	0.023	0.024	0.020
	Coefficient of variation		−0.511	−0.545	−0.625

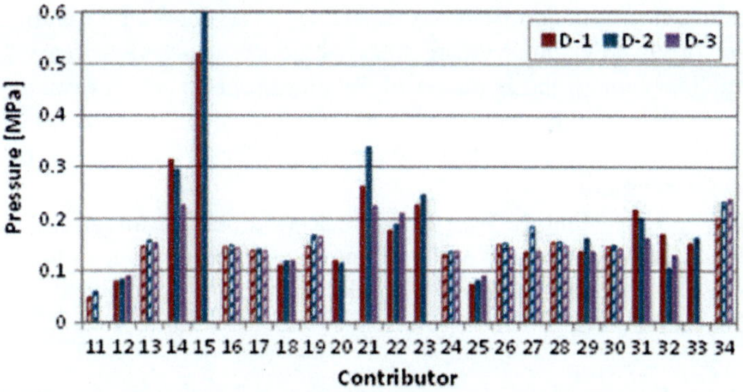

Fig. 69 Cases D-1, D-2, and D-3. Peak hydrodynamic pressure at the dam heel (Point A). Dashed histograms indicate Contributors using free-field BCs

Table 19 Case D—Median, mean, standard deviation, and coefficient of variation of peak hydrodynamic pressure at dam heel (Point A)

Point	Hydrodynamic pressure	Units	D-1	D-2	D-3
A	Median	MPa	0.147	0.158	0.144
	Mean	MPa	0.171	0.221	0.154
	Standard deviation	MPa	0.094	0.284	0.042
	Coefficient of variation		0.552	1.283	0.275

Figure 69 compares the solutions provided for Cases D-1, D-2, and D-3, in terms of peak hydrodynamic pressure at the dam heel (Point *A*). The "peak hydrodynamic pressure" is defined as the average of the positive and negative peak values of the hydrodynamic pressure time history.

In general, the results are quite consistent for Contributors using the free-field BCs. A trend of peak hydrodynamic pressure with changes in water reservoir level is not recognizable in this case.

Table 19 reports the summary of peak hydrodynamic pressure results at the dam heel, in terms of median, mean, standard deviation, and coefficient of variation, over results from all Contributors. The results are more scattered for Case D-2 (one Contributor presented peak pressures about an order of magnitude higher than the other Contributors). The maximum increase of water pressure at the dam heel induced by the Taft Record is about 15% of the hydrostatic pressure.

7.5 Summary

When comparing the various results for each different reservoir water level, consistent trends among contributions are observed, with only few outliers. The most consistent results were obtained by Contributors using free-field BCs.

The average results of contributions show that the response of the dam, in terms of acceleration (both at the dam heel and the crest), decreases with increasing reservoir water level. The same trend is observed for minimum upstream displacement; the different trend of maximum downstream displacement can be ascribed to the static initial displacements. A clear dependence on the water level is not detected for hydrodynamic pressure.

It should be noted that the Taft Record shows a peak at about 2.27 Hz, which almost coincides with the first mode of the dam-reservoir-foundation system for the WRWL, characterized, at least for accelerations and displacements, by the most severe seismic response.

8 Case E—Non-linear Dynamic Analysis

8.1 General

The intent of Case E is to perform a dynamic analysis using non-linear concrete material models for the dam.

8.2 Conceptual Model

Model configuration. Case E is an extension of Case D, with the non-linear concrete material. The "base configuration," as defined in Sect. 3.2 and used in Case D, is considered for Case E.

Input parameters. For Case E, the input parameters were defined as:

- Common parameters for the concrete material are defined Table 4. If any material model parameters were not provided in the table, Contributors were to determine them based on their engineering judgment.
- Water properties are defined in Sect. 3.2.
- Foundation elastic properties are defined in Table 5.
- The mass of the dam, reservoir, and foundation.
- 2% viscous damping for the dam and foundation.

Contributors were to select a type of a non-linear material model for concrete and define all parameters needed for the material model based on the information provided in the formulation.

Loads. Two different dynamic loads were to be applied:

- The Taft Record (time history for either deconvolved acceleration or shear stress) applied, as for Case D, at the base of the foundation block.
- The ETAF horizontal acceleration time history record applied at the base of the foundation block.

- Gravity loads are applied to the dam and reservoir, but not to the foundation, in combination with the Taft Record and ETAF Record.
- WRWL at El. 268.21 m.

8.3 Analysis

The analyses are performed for two cases:

- **E-1**: The Taft Record combined with the static load
- **E-2**: The ETAF Record combined with the static load

Contributors were asked to provide the following vectorized and scalar quantities as output parameters of their analyses:

- Crest (i.e., C) displacement, ΔC, and heel (i.e., A), ΔA displacement.
- Relative crest displacement with respect to heel, Δrelative.
- Net hydrodynamic pressure at dam heel, Phyd.
- Ratio of crest to heel amplitude acceleration spectrum.
- Damage extent: scalar value for E-1 and vector value for E-2.
- Damage index (DI): Area-based and length-based criteria.
- Failure time (reported only for Case E-2 under ETAF).

8.4 Results

In the analysis for Case E, various assumptions were made by Contributors in their models, with respect to:

- Boundary conditions for the foundation block.
- Formulation of the adapted fluid elements, and the fluid-structure interaction.
- Concrete damage model.
- Damage initiation, and propagation protocols.
- Solution technique, and time integration parameters.
- Mesh size (both in the dam and foundation).

Case E-1. Figure 70 presents the displacement time histories at the dam heel, crest, and their relative value. Blue lines show the individual contributions, while the median curve is plotted in red. As can be seen, there is a considerable difference, both in trend and values, among various Contributors.

Because it is difficult to compare displacement time histories, multiple scalar values are extracted for each response. These are:

- Initial displacement: displacement at time zero.
- Maximum normalized displacement: maximum displacement time history normalized with respect to initial displacement.

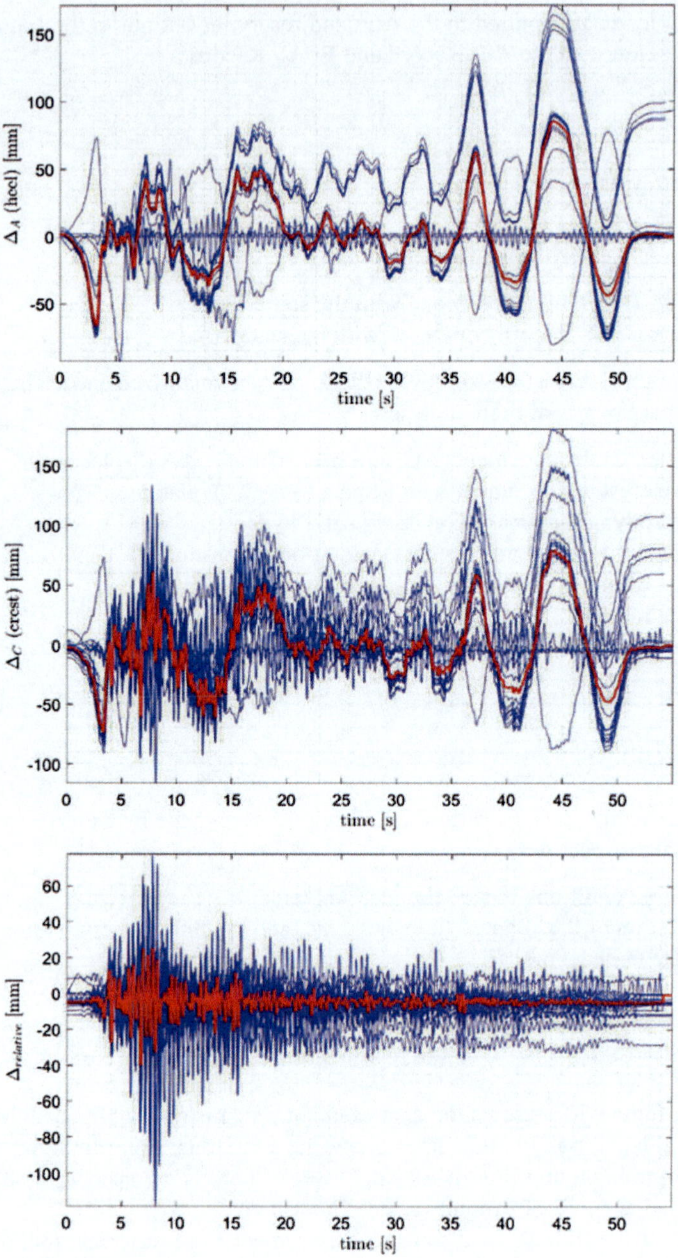

Fig. 70 Displacement time history of dam subjected to Taft Record

- Minimum normalized displacement: minimum displacement time history normalized with respect to initial displacement.

The initial displacement and the maximum and minimum normalized displacements are presented for each Contributor at the dam crest and heel and as relative displacements in Fig. 71. The solid red line in the figure represents the median of the data for each case. As shown in the figure, relatively good consistency exists among results from different Contributors. Although there are several outliers in each response quantity, most Contributors presented results close to median value. Note, however, the difference in initial (static) displacement, which is strictly independent from dynamic complexities.

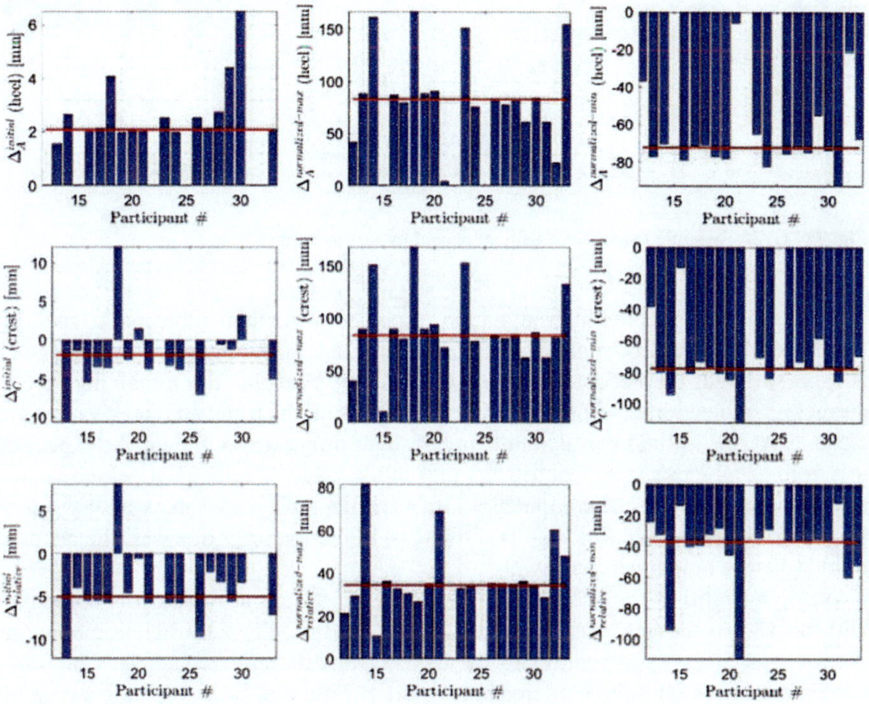

Fig. 71 Summarized static and dynamic displacement of the dam under Taft Record

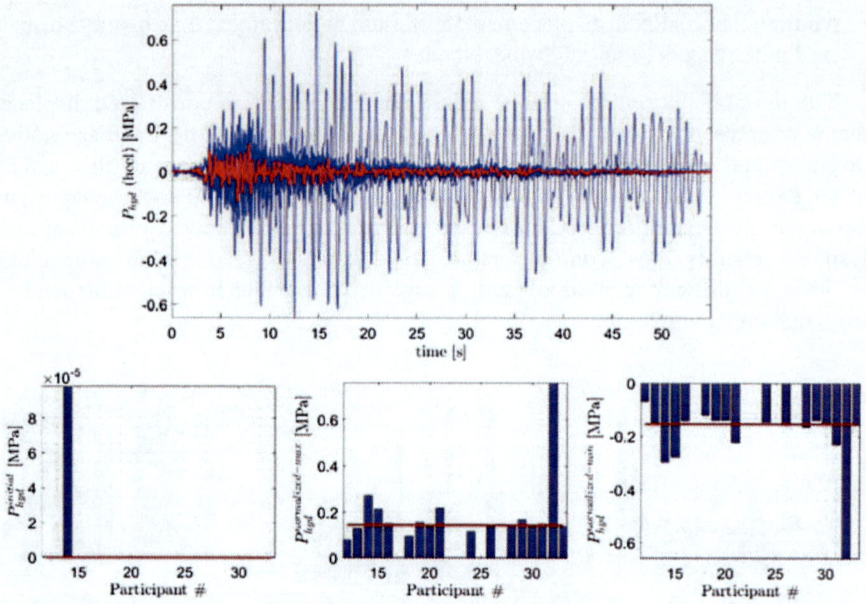

Fig. 72 Hydrodynamic response of dam subjected to Taft Record

Figure 72 presents the hydrodynamic response of the dam subjected to the Taft Record. Only the hydrodynamic pressure is presented; the initial hydrostatic pressure is ignored. With the exception of one Contributor's results, the remaining results predict a good estimate of pressure time history. Both the trend and peak values are close. Next, the initial hydrodynamic pressure (at time zero) is shown. As expected, it is zero in all cases (one Contributor presents a value of 0.01 MPa, which is also negligible). Moreover, the computed positive and negative peak values are very close (except 2–3 outliers). These results show that, in the majority of cases, the adapted fluid-structure model performs well.

Next, the ratio of the acceleration amplitude spectrum at the crest is computed with respect to the heel (see Fig. 73). This is, in fact, the ratio of the maximum acceleration at the crest relative to the base. Over different frequencies, this ratio varies widely. Around the fundamental period of Pine Flat Dam, the ratio varies by a factor of 10–20.

Damage at the dam base (i.e., dam-foundation interface) and within the body are quantified in Fig. 74. For this purpose, the concept of damage index is introduced [9, 10]. DI is the ratio of cracked length or area to the total base length or dam area:

$$DI_{base} = \frac{L_{cr}}{L_{tot}} \tag{3}$$

$$DI_{area} = \frac{A_{cr}}{A_{tot}} \tag{4}$$

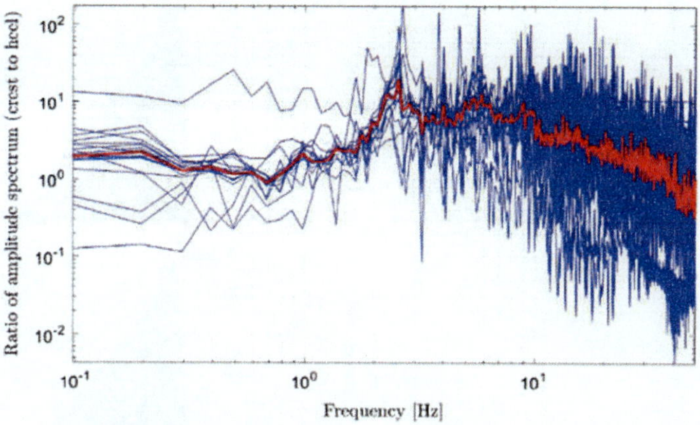

Fig. 73 Ratio of acceleration amplitude spectrum at crest to heel

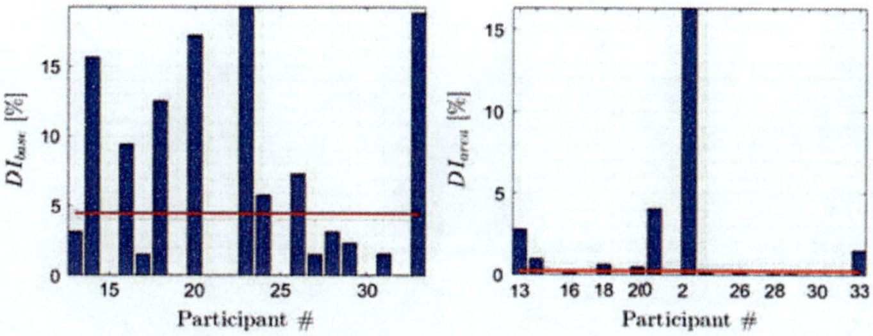

Fig. 74 Damage index comparison under Taft Record

There are multiple observations, which include:

- There is a large variation in the predicted DIs. The major sources are non-linear damage model, mesh density, and solution algorithm.
- The length-based DI is higher than the area-based DI.
- The length-based DI varies from 0 to 18%, with a median of 4.5%.
- The area-based DI varies from 0 to 16%, with a median of 0.5%.
- The potential damage zones are located at the dam heel and at the downstream face next to neck slope change.

Figure 75 shows some of the damage profiles provided by Contributors. Overall, 4 Contributors predicted no damage in the dam structure, while 11 Contributors predicted minor damage at the upstream heel, and 4 Contributors predicted moderate damage at the neck area.

Due to the complex nature of damage analysis, Contributors' major comments during non-linear seismic analysis are summarized below:

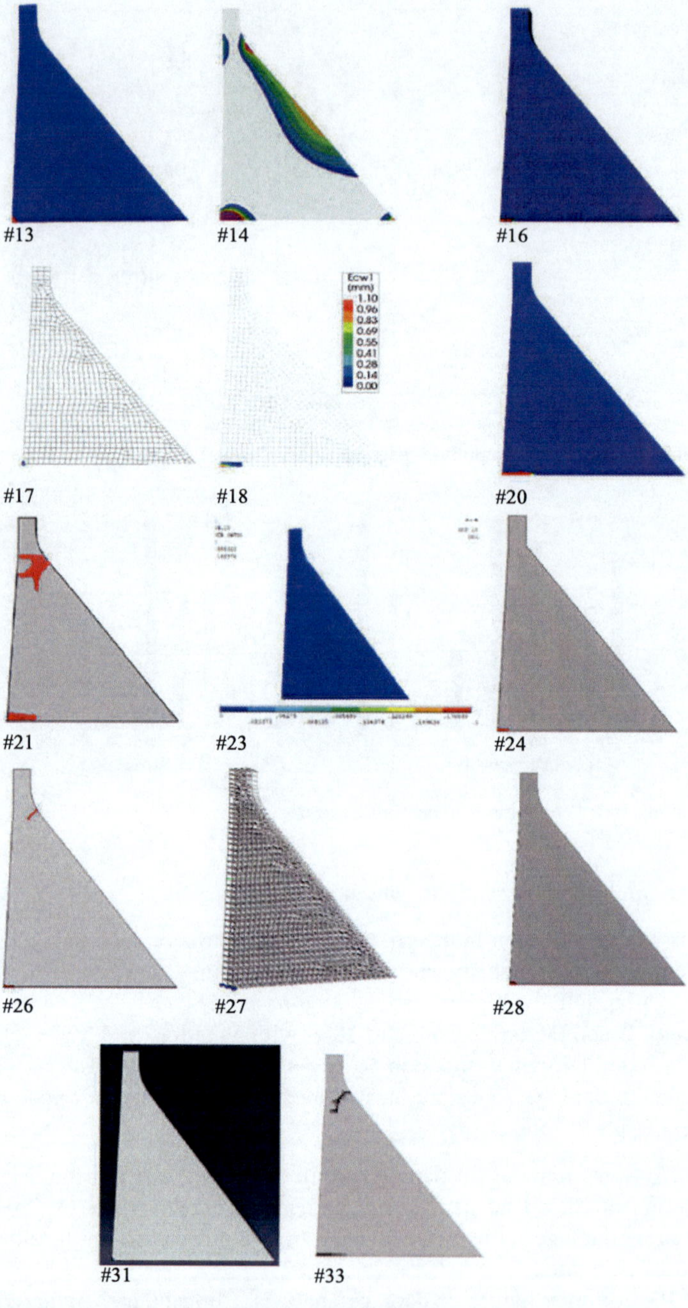

Fig. 75 Damage profile of dam subjected to Taft Record

- Contributor No. 13: The adapted damage model allows crack opening and closing. Maximum density of cracks, due to tensile stresses in the principal direction, is used with values 0 (no cracks) to 1 (fully cracks).
- Contributor No. 14: Damage (plastic strain) occurs at: (1) the upstream foot of the dam (10 m × 5 m = 50 m^2), (2) the downstream foot (1 m × 0.5 m = 0.5 m^2), and (3) the downstream face (1 m × 5 m = 5 m^2).
- Contributor No. 15: The dam does not undergo any damage.
- Contributor No. 16: Concrete damage plasticity was adapted, with fine mesh (element size 1.5 m). Viscous-spring artificial boundaries are adopted for foundation. In Case E-1, seismic input is applied by means of effective earthquake forces, both at the bottom and side boundaries of the foundation. In Case E-2, the ETAF signal is applied as motion only at the bottom of the foundation. The tensile damage variable varies from 0 to 1. If the damage variable is greater than zero in just one integration point of an element, the whole element has been assumed totally damaged.
- Contributor No. 17: A 2D plane strain seismic analysis was carried out: (1) seismic action is represented by plane waves propagating upwards from the base of the model with a non-reflecting viscous boundary; and (2) free–field conditions are forced at the lateral boundaries. The damage is based on a discrete crack modelling. Each finite element adopted in the dam body discretization interacts with the neighbouring finite elements through joints.

Interface elements. A bi-linear softening model is adopted in tension and in shear direction. Only the first interface element at the dam-foundation interface was fully cracked. All the remaining interfaces are within the elastic range.

- Contributor No. 18: Non-reflecting BCs are defined by means of particular interface elements that are characterized by a very low stiffness (both normal and tangential) and damping coefficient, depending on shear modulus and density of the foundation material. In the bottom part of the foundation, just below the dam, only one interface was modified, assigning a tangential stiffness equal to the shear modulus, so this interface could be used to transmit the input signal. Crack width was selected as the variable for damage definition. It was evaluated by multiplying the crack strain of the element and the crack bandwidth of each element. The crack bandwidth was manually set to 0.5 m in the whole dam body.
- Contributor No. 19: Damage occurs to the body of the dam when the tensile strain passes the strain corresponding to the tensile strength. For the dam–foundation interface, damage occurs when the strain is tensile ($\varepsilon > 0$). The dam is considered failed when either the body damage index or the dam–foundation interface DI is equal to 1.
- Contributor No. 20: The DI is defined as: $1 - k_{unl}/k_{first\text{-}load}$, where $k_{first\text{-}load}$ is the undamaged material stiffness, and k_{unl} is the unloading stiffness.
- Contributor No. 21: The BCs for fluid are: absorbing BC at the upstream surface, zero normal velocity on the lateral faces, and zero overpressure on the free surface. For the structure: displacement X, Y, Z restrained at the bottom of the foundation, displacement Y restrained at the upstream and the downstream surfaces of the

foundation, and displacement X fixed on the lateral faces. No absorbing BC on rock foundation is used. To capture the system non-linearity, only the tensile behaviour is modelled. Damage, d, varies from 0 to 1 (total damage). DI is the percentage of damaged material on the entity considered: $\mathrm{DI} = \Sigma\ (L_i \times d_i)/\ L$ or $\mathrm{DI} = \Sigma\ (L_i \times v_i)/\ S_{tot}$.

- Contributor No. 23: A coupled damage-plasticity model for concrete based on micro-plane formulation is used. The model has the advantage of being able to define different damage initiation and damage evolution criteria between tension and compression states, which allows to model the brittle behaviour of concrete in tension where softening starts directly after the elastic limit while hardening is observed before softening after the elastic limit in compression. In the transition from tension to compression states, the stiffness lost during tensile cracking is recovered due to crack closure. Damage is computed for each element in tension and compression, as well as a total damage, with values between 0 and 1. Damaged elements are considered to be elements where damage initiated (D>0).
- Contributor No. 26: Damage occurs if the tensile strength has been exceeded (i.e., microcracks and macrocracks are shown in red colour (d>0)).
- Contributor No. 27: Seismic action is represented by plane waves propagating upwards from the base of the model with a non-reflecting viscous boundary. Free-field conditions are forced at the lateral boundaries. The damage is based on a discrete crack modelling. Each finite element adopted in the dam body discretization interacts with the neighbouring finite elements through joint interface elements. A bi-linear softening model is adopted in tension and in shear direction. Only the first interface element at the dam-foundation interface across the model thickness was fully cracked. All the remaining interfaces are within the elastic range. In 3D, the length-based criteria should be area-based criteria and the unities should be m^2.
- Contributor No. 28: Tensile damage factor has been assumed according to the following equation: $= 1 - \sigma_t/f_t$, where σ_t is the post-cracking tensile stress, and f_t is the tensile strength. The DI and the damage profile have been respectively calculated and plotted for the region where $d_t \geq 0.2$.
- Contributor No. 29: A finite concrete volume is considered damaged when a threshold plastic strain of 0.3% is reached. The value is chosen corresponding to the tensile strain at failure.
- Contributor No. 30: Non-linear solutions are quite susceptible to input parameters of the material model. Thus, unification of model parameters in future benchmarks is recommended. No damage occurs in the dam wall for Case E-1.
- Contributor No. 31: The tensile stress in partial areas of the dam body exceeds the tensile strength of concrete materials, and concrete stiffness will degrade, resulting in a certain extent of damage and cracking.
- Contributor No. 32: No dam occurred in the dam monolith. There is no exceedance of the tension strength, 2.5 MPa, and the maximum value of the principal stresses II is 0.89 MPa.

- Contributor No. 33: The plain strain model is used with concrete damaged plasticity and Rayleigh damping. Infinite elements at the sides and base of the foundation block were implemented, and the horizontal and vertical displacements were constrained at outer nodes of the infinite elements. Finally, a non-reflecting, acoustic impedance was used at the far end of the reservoir. Damage was evaluated based on cracking (tensile failure) of the concrete. A bi-linear stress-displacement curve has been applied, where the crack width is calculated using the given fracture energy. The tensile damage parameter, d_t, relates to the defined crack widths, with a maximum value of 0.95 at maximum crack width (tensile stress = 0). Failure due to compression (crushing) has not been considered.

Case E-2. Non-linear seismic response of dams highly depends on the applied ground motion signal; therefore, it is usually recommended that a large set of ground motion records are used and scaled at different intensity levels [10]. This is very time consuming, especially when the dam-reservoir-foundation model includes many elements, and the system becomes significantly non-linear. An alternative method is to use an artificially generated, intensifying acceleration function. This is very useful because: (1) it is not a ground motion dependent input; and (2) its intensity increases over the time and, thus, covers the linear to non-linear range.

Case E-2 tends to investigate the progressive damage response of the dam subjected to an ETAF. Figure 76 presents the variation of relative crest displacement for 15 s of applied ETAF record. Some of the individual curves show results as high as 4,700 mm. Two curves seem to be outliers. The original plot is refined to show the displacement only up to 700 mm. The median (dotted block line) curve shows a total of about 400 mm displacement at t = 15 s.

Figure 77 also presents the evolution of net hydrodynamic pressure. Again, some outliers exist, and there is a large dispersion among the reported results.

Figure 78. presents the length-based and area-based DI evolution under ETAF. The participants showed great uncertainty predicting the DI. In all cases, the length–based criterion is higher than the area-based criterion. This finding in Case E–2 is validated with similar observations in Case E-1. The length-based DI appears more stable because it is less sensitive to mesh size. For very fine meshes, it is recommended that the area-based DI not be used.

Figure 79 shows the damage profile at some selective contributions and selective times. The main observations are:

- In most of the analyses, the failure occurs prior to the full duration of ETAF.
- In some cases, the analysis lasts to the end of the ETAF.
- The cracking profile is similar in the majority of cases; however, they might have different failure times.
- Increasing the intensity of the applied dynamic excitation shifts the response to non-linear phase and, subsequently, increases the uncertainty.

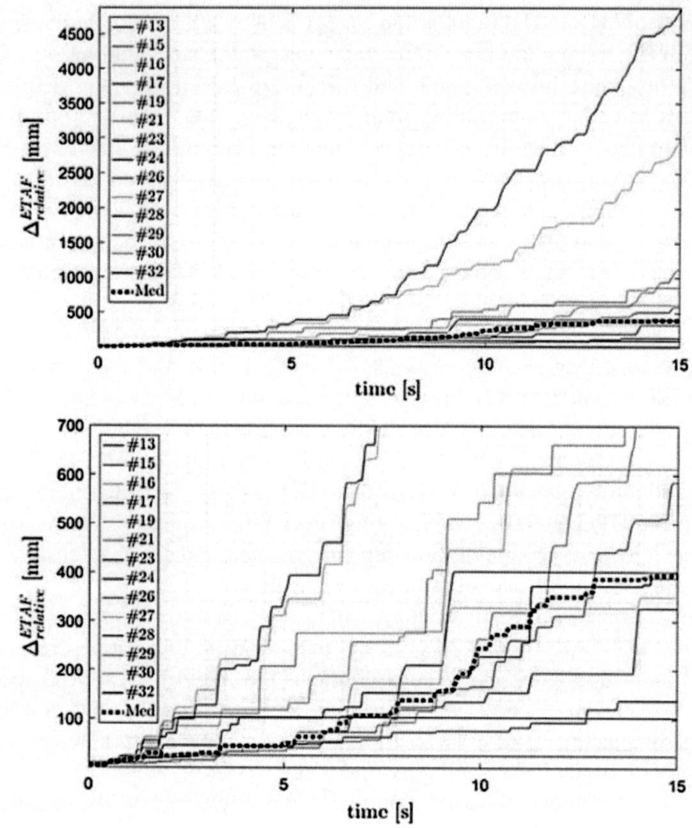

Fig. 76 Displacement evolution of the relative crest displacement under ETAF

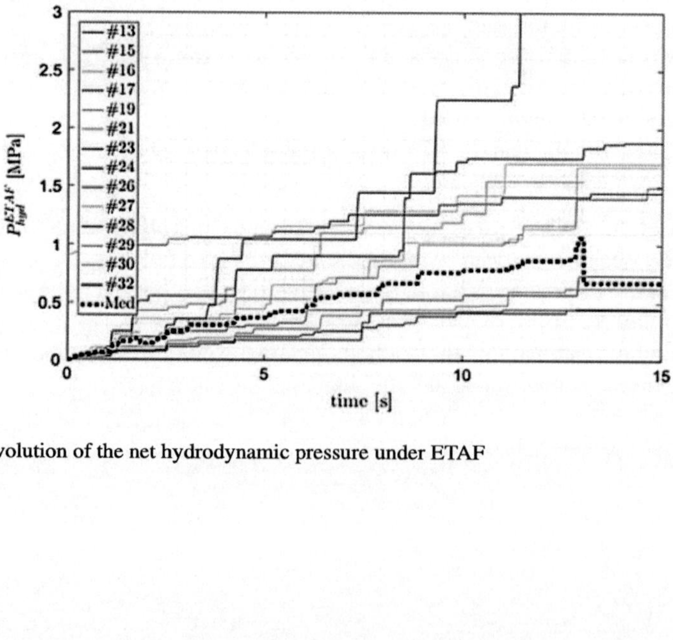

Fig. 77 Evolution of the net hydrodynamic pressure under ETAF

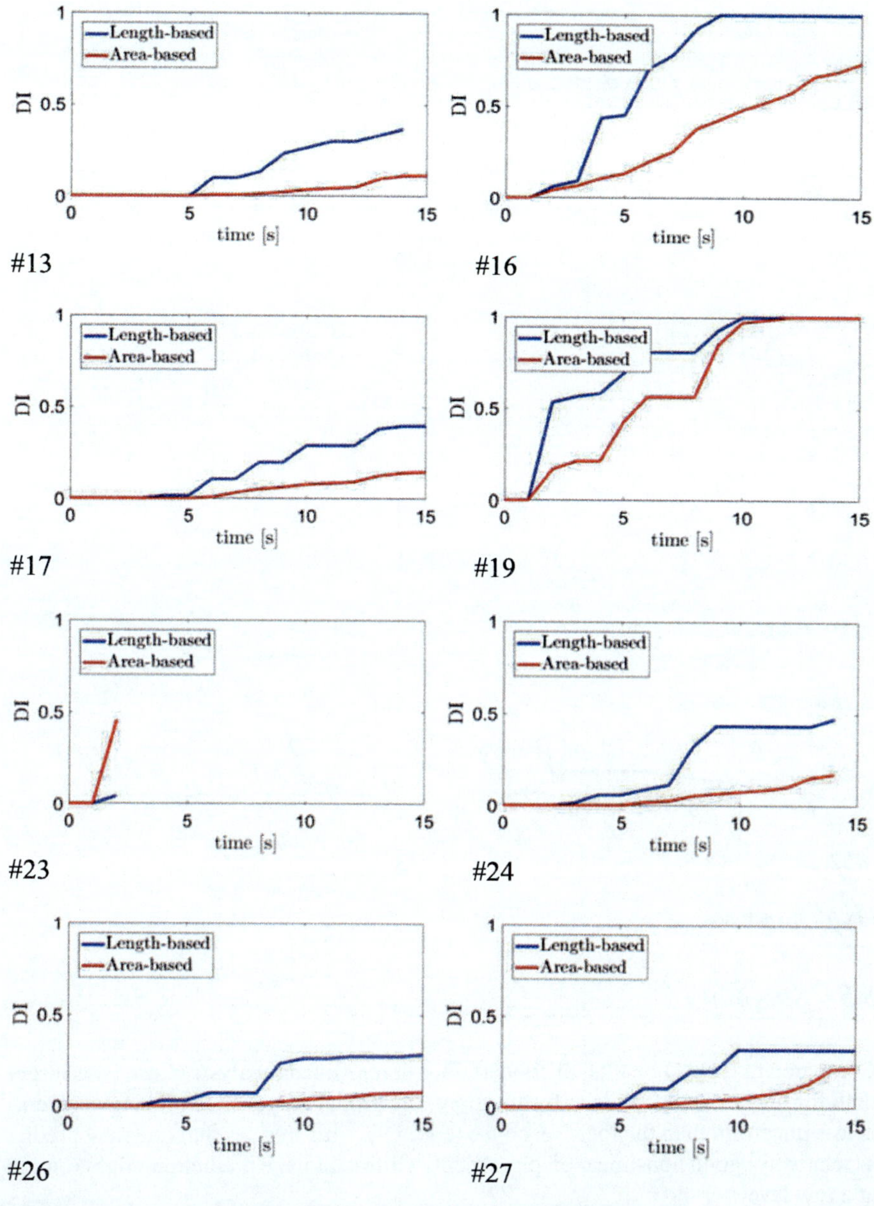

Fig. 78 Evolution length-based and area-based DIs for different Contributors

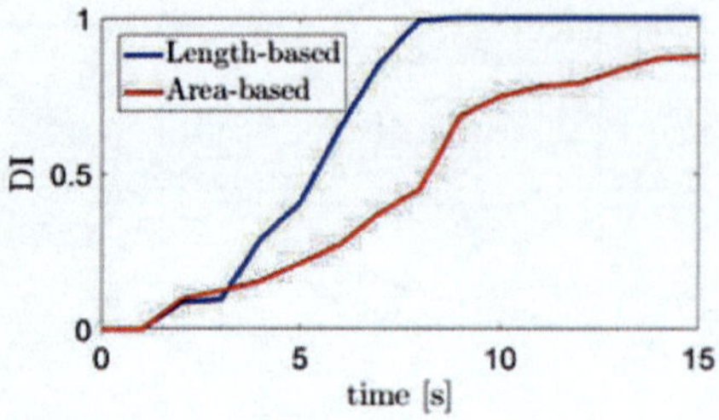

#28

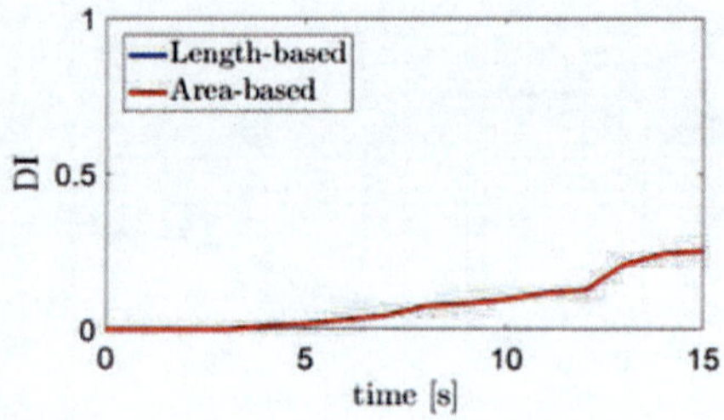

#29

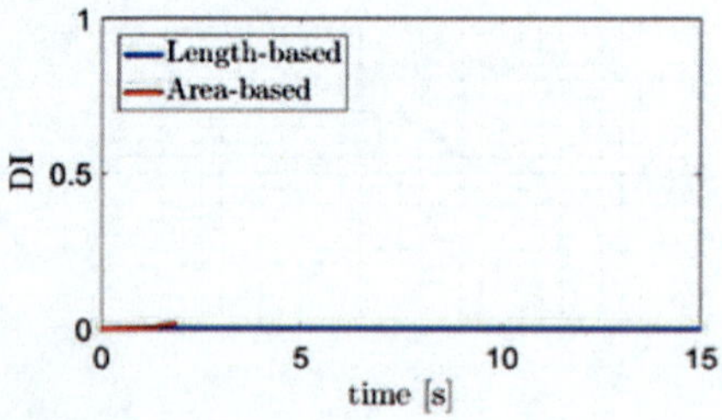

#30

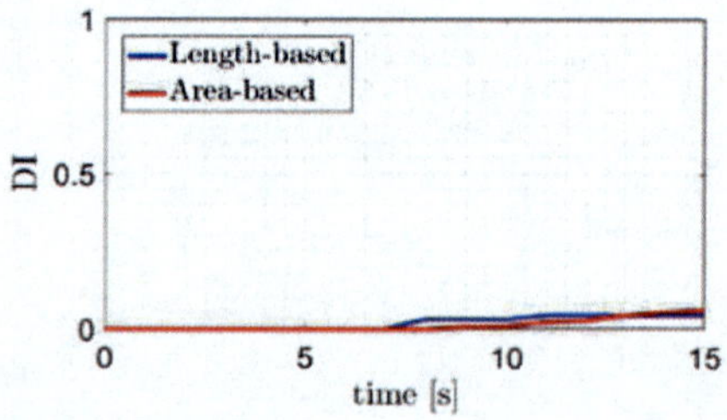

#31

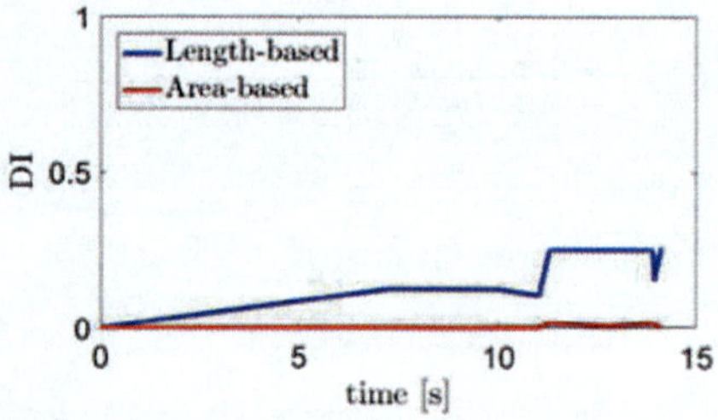

#32

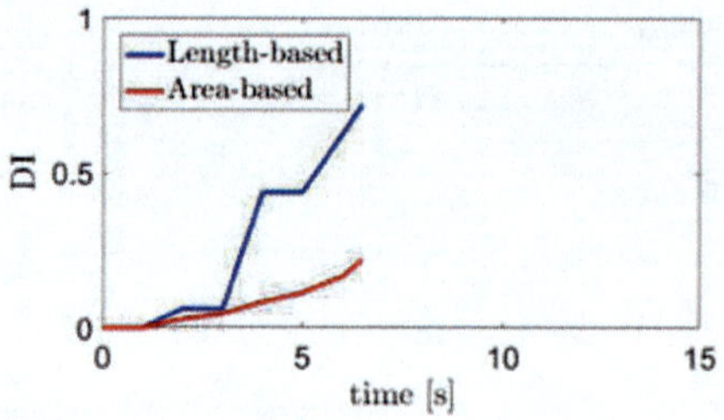

#33

Fig. 78 (continued)

8.5 *Summary*

Compared to Case D results, the uncertainty in non-linear analysis (Case E) is larger than for linear elastic models; furthermore, the global response (e.g., displacement) is less uncertain than the local response (e.g., DI). This implies that one may predict a (relatively) good non-linear displacement, while a damage prediction might remain at a low level.

Summary for Case E-1.

- Most of the participants showed no or minor cracking; hence, they obtained results similar to those obtained for Case D1.
- There already are some differences in static displacement.

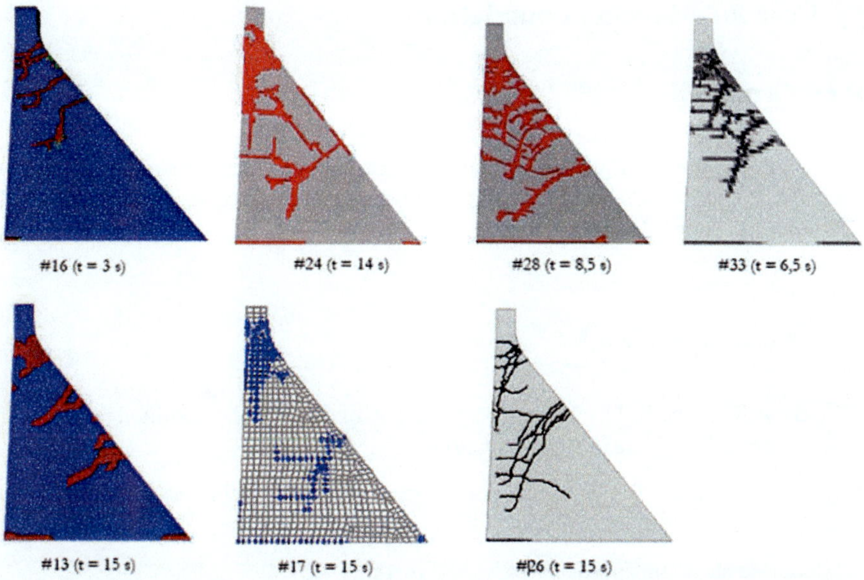

Fig. 79 Displacement damage profile of dam under ETAF for selective contributions at selective times

- Overall, there are good predictions regarding relative displacement between crest and heel, as well as the hydrodynamic pressure. Peak values are estimated better than the time histories.

Summary for Case E-2.

- There are significant differences in estimating the DI (along the concrete/rock interface), especially the DI based on damaged areas.
- It is difficult to assess the ultimate capacity from the results.

General observations. Case E can be summarized as:

- The choice of non-linear model and solution technique highly affects the non-linear analysis results.
- Uncertainty is greater in non-linear analysis than for linear cases.
- For ETAF in which the non-linear response of the system gradually increases, the uncertainty is time (or intensity) dependent and increases with the duration.
- Global response (like displacement) shows less uncertainty than local response (like DI). One may predict a (relatively) good non-linear displacement, while a damage prediction may remain at a low level.
- Finite element mesh size highly affects the DI.
- There continues to be no clear definition of a "damage" or "failure" concept in dam engineering.

9 Case F—Massless Foundation

9.1 General

In Case F, the model of the dam-foundation-reservoir system is similar to the one for Case D, except that a massless foundation subject to the free-surface acceleration Taft Record is considered.

9.2 Conceptual Model

Model configuration. The "base configuration" of the model, as defined in Sect. 3.2, is considered to be a massless foundation.

Input parameters. For analysis consistencies, the following assumptions were made:

- Concrete elastic properties are defined in Table 4.
- Water properties are defined in Sect. 3.2.
- Foundation elastic properties are defined in Table 5.
- The mass of the dam and reservoir, but massless foundation.
- 2% viscous damping.

Loads. The loads include:

- The Taft Record (free-surface acceleration time history) in the upstream/downstream direction applied to the base of the foundation block for three different reservoir water levels.
- The static weight of the dam and reservoir, but not the weight of the foundation.

9.3 Analysis

The linear dynamic analysis for the dam-reservoir-foundation system for various reservoirs is performed for the following water levels:

- **F-1**: WRWL at El. 268.21 m.
- **F-2**: SRWL at El. 278.57 m.
- **F-3**: NRWL at El. 290.00 m.

9.4 Results

Contributors were requested to report the following analysis results at the points indicated in Fig. 47:

- Time history of the horizontal acceleration and horizontal displacements at the dam crest (Point C) and dam heel (Point A).
- Time history of the hydrodynamic pressure at the dam heel (Point A).

Thirteen teams provided solutions for Cases F-1 and F-2, and 11 teams provided solutions for Case F-3. Statistics for the software used in the Case F studies (Table 1) include: Abaqus (2), Diana (2), Code_Aster (2), ANSYS (2).

A comparison of the results is reported for Case F-1, in terms of horizontal acceleration. The acceleration response spectra (2% damping) at the dam heel (Point *A*) and dam crest (Point *C*) are reported in Fig. 80 and in Fig. 81, respectively. Generally, the results are in poor agreement at both the dam heel and dam crest. Nevertheless, a group of five Contributors provide very consistent results, as highlighted in Fig. 82.

Figures 83 and 84 report a comparison between the solutions provided for Cases F-1, F-2 and F-3 by the different Contributors regarding peak upstream-downstream horizontal acceleration at the dam heel (Point *A*) and the dam crest (Point *C*). For this analysis, "peak acceleration" is defined as the average of the positive and negative peak values of the acceleration time history (Eq. 2). Reportedly, Contributor No. 21 presented peak acceleration almost null in Case F-3. As in Case D-1, at the dam heel, results from most Contributors are consistent (and also in fair agreement with the PGA of the Taft Record equal to 1.74 m/s^2), yielding a value of about 2 m/s^2 in Case F-1, 1.75 m/s^2 in Case F-2 and 1.7 m/s^2 in Case F-3; a small group of Contributors obtain peak accelerations at the dam heel that are less than

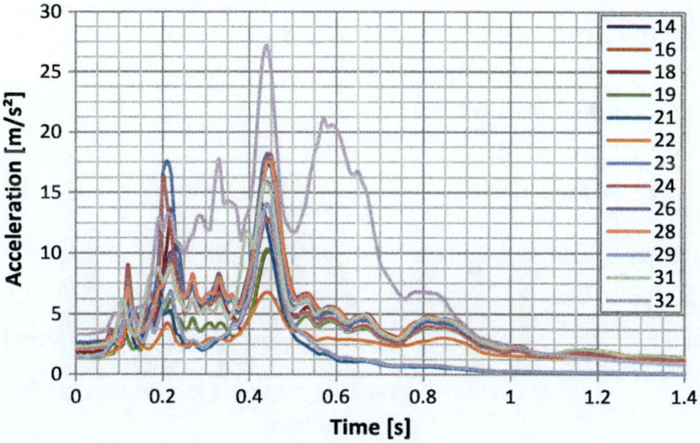

Fig. 80 Case F-1. Horizontal acceleration response spectra (2% damping) at the dam heel (Point A)

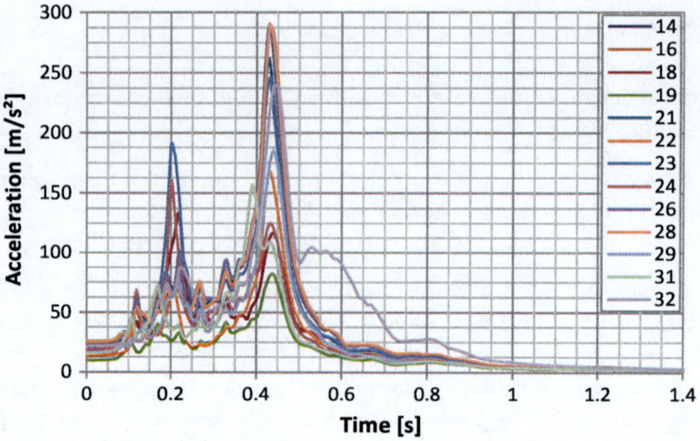

Fig. 81 Case F-1. Horizontal acceleration response spectra (2% damping) at the dam crest (Point C)

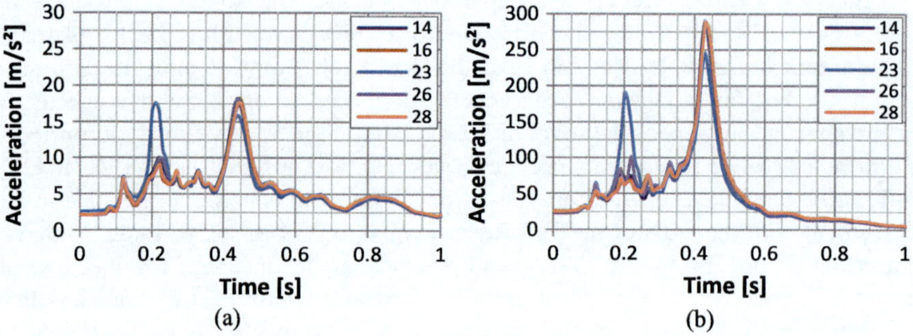

Fig. 82 Case F-1. Horizontal acceleration response spectra presented for 5 selected contributions at the dam heel (Point A) (**a**) and dam crest (Point C) (**b**)

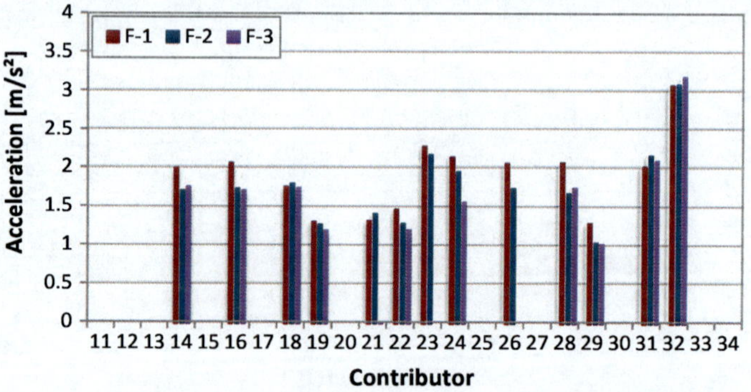

Fig. 83 Cases F-1, F-2, and F-3. Peak horizontal acceleration at the dam heel (Point A)

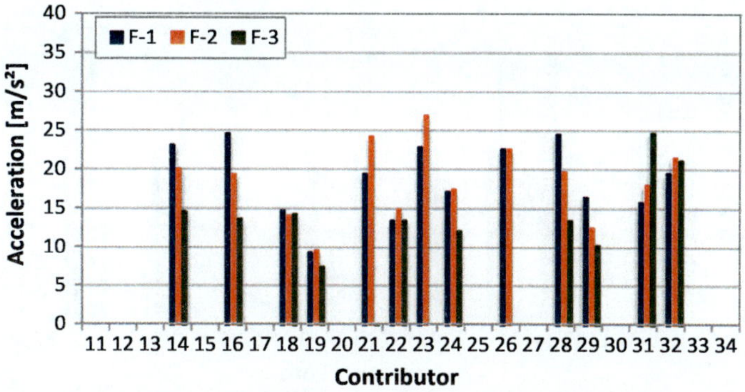

Fig. 84 Cases F-1, F-2, and F-3. Peak horizontal acceleration at the dam crest (Point C)

the PGA of the input Taft Record. The decreasing value of the peak acceleration with increased reservoir levels is observed for both Case D and Case F.

At the dam crest, peak acceleration results are greatly amplified. For seven contributors, the Point *C* peak acceleration results decrease as water levels increase. The results of the other Contributors show an opposite or indefinite trend.

Peak acceleration results at the dam heel and crest are summarized in Table 20 for Cases F-1, F-2, and F-3, respectively, for median, mean, standard deviation, and coefficient of variation over the results from all Contributors. As in Case D, the decreasing trend of peak acceleration with increasing water levels is also confirmed, considering the median and mean values for both Point *A* and Point *C*. Results are more scattered in Case F-3 than in the other two cases, as the coefficients of variation clearly highlight.

Figures 85 and 86 report the comparison of relative (Point *C*–Point *A*) total (static and dynamic) horizontal displacement at the dam crest for maximum downstream and minimum upstream displacement, respectively, for Cases F-1, F-2 and F-3.

Table 20 Case F—Median, mean, standard deviation, and coefficient of variation of peak acceleration at the dam heel (Point A) and dam crest (Point C)

Point	Acceleration	Units	F-1	F-2	F-3
A	Median	m/s^2	2.03	1.74	1.72
	Mean	m/s^2	1.92	1.78	1.57
	Standard deviation	m/s^2	0.50	0.52	0.78
	Coefficient of variation		0.26	0.29	0.49
C	Median	m/s^2	19.53	19.48	13.56
	Mean	m/s^2	18.85	18.63	13.29
	Standard deviation	m/s^2	4.73	4.84	6.45
	Coefficient of variation		0.25	0.26	0.49

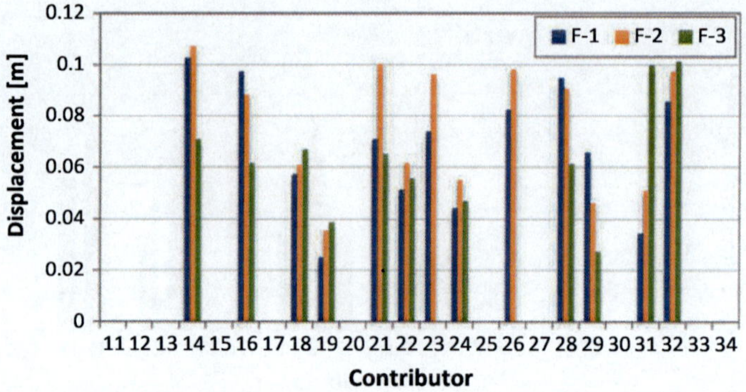

Fig. 85 Cases F-1, F-2 and F-3. Maximum downstream relative (Point C–Point A) total (static and dynamic) displacement at the dam crest (Point C)

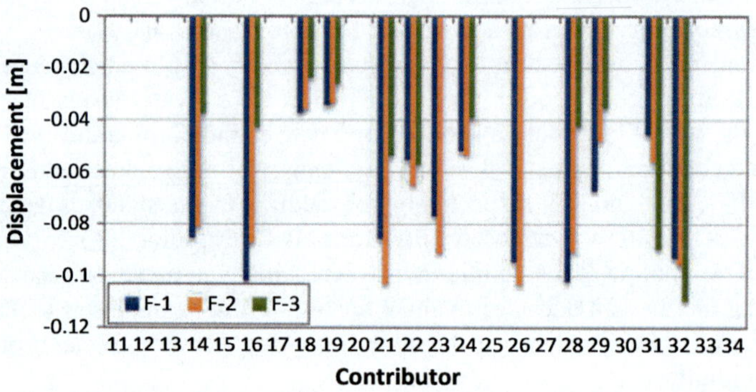

Fig. 86 Cases F-1, F-2 and F-3. Minimum upstream relative (Point C–Point A) total (static and dynamic) displacement at the dam crest (Point C)

Table 21 reports the summary of maximum and minimum displacements for median, mean, standard deviation, and coefficient of variation. Displacement results are quite scattered; relatively few cases show a recognized trend with water reservoir level because both the maximum downstream displacements and the minimum upstream displacements span a huge range of about 0.04–0.1 m (absolute values) for Cases F-1, F-2 and F3.

Figure 87 compares Contributor results for Cases F-1, F-2, and F-3 for peak hydrodynamic pressure at the dam heel (Point *A*). As for peak acceleration, "peak hydrodynamic pressure" is defined as the average of the positive and negative peak values of the hydrodynamic pressure time history.

The results from most Contributors are similar; only a small group of Contributors have results that differ significantly. In a few cases, peak hydrodynamic pressure decreases with increasing water reservoir level, but a general trend is not recognizable.

Table 21 Case F—Median, mean, standard deviation, and coefficient of variation of maximum downstream and minimum upstream relative displacement at the dam crest (Point C)

	Displacement	Units	F-1	F-2	F-3
Maximum	Median	m	0.071	0.088	0.061
	Mean	m	0.068	0.076	0.063
	Standard deviation	m	0.025	0.025	0.023
	Coefficient of variation		0.362	0.325	0.357
Minimum	Median	m	−0.077	−0.081	−0.042
	Mean	m	−0.071	−0.072	−0.050
	Standard deviation	m	0.024	0.025	0.027
	Coefficient of variation		0.343	0.348	0.528

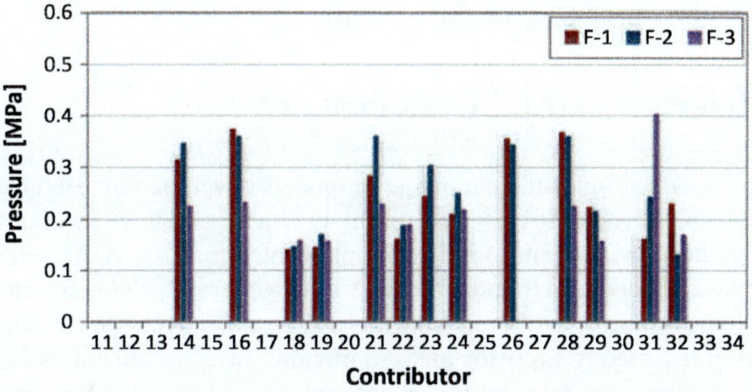

Fig. 87 Cases F-1, F-2, and F-3. Peak hydrodynamic pressure at the dam heel (Point A)

Table 22 reports the summary of peak hydrodynamic pressure results at the dam heel for Cases F-1, F-2, and F-3, respectively, for median, mean, standard deviation, and coefficient of variation over the results from all Contributors. Overall, these results present the same level of scatter as seen for the three cases with a different water reservoir level. The maximum increase of water pressure, caused by the Taft Record at the dam heel, results in about 20–25% of the hydrostatic pressure.

Table 22 Case F—Median, mean, standard deviation, and coefficient of variation of peak hydrodynamic pressure at the dam heel (Point A)

Point	Hydrodynamic Pressure	Units	F-1	F-2	F-3
A	Median	MPa	0.230	0.251	0.218
	Mean	MPa	0.246	0.263	0.215
	Standard deviation	MPa	0.085	0.087	0.070
	Coefficient of variation		0.345	0.332	0.326

9.5 Summary

Case F results, as a function of reservoir water level, show overall adeguate agreement. Compared to Case D with mass foundation, the results appear slightly less scattered.

As for Case D, peak acceleration, both at the dam heel and dam crest, decreases with increasing reservoir water level; however, a clear dependence on water level is not seen for displacements and pressure results.

At every reservoir water level, the massless foundation (Case F) produces acceleration, displacement, and hydrodynamic pressures that are larger by a factor of about 1.5–2 than the foundation with mass (Case D).

10 Overview and Result Discussion

10.1 Preliminary Model Verification Process

Preliminary verification of the computation models developed by each Contributor was found to be a critical first step for properly conducting the benchmark studies. A model verification (screening) process implemented for Case B showed that, even with a simple 2D model of a foundation block having precisely defined material properties and loads, some Contributors may misunderstand the implementation of input parameters or incorrectly apply the loads. If obvious mistakes are not identified early in the process, then the outcome of the benchmark studies will be less useful. The quality of information gained from the efforts of Contributors is improved substantially when the benchmarks include a process to recognize errors early on, and when Contributors are provided an opportunity to correct those errors before submitting final results.

10.2 Model Validation

It is a common engineering practice to validate numerical models with measured data. In the 2019 Benchmark Workshop formulation, Case A was formulated to allow Contributors to compare their calculated natural frequencies and mode shapes against the experimental tests that were conducted in 1971 by Rea and Chopra [14]. In addition, a simplified version of the EMVG test was performed to compare the calculated accelerations and displacements with those obtained in the field.

Overall, the results submitted by Contributors were in good agreement with the experimental data. For instance, the median frequency obtained from all Contributions matched the result of the first fundamental frequency of 3.47 Hz from the experiments. Regarding simulations of the EMVG test, the accuracy of the predicted accelerations was higher than the prediction of the displacements. It should be noted, however, that the displacements were small and may have been affected by the recording accuracy of the measuring system.

Even though the numerical analyses for a single monolith provide results with high accuracy, it is difficult to interpret the response of the whole dam based just on the 2D model results. In Sect. 4.6, an assessment of the mode shapes was presented and it became clear that to be able fully interpret the behavior of a gravity dam, it is thereby required that global 3D analyses are performed of the whole dam.

10.3 Boundary Conditions

In the benchmark studies, Contributors were allowed to define the BCs they considered suitable.

For the task of frequency analysis (Case A), Contributors defined BCs that constrained the deformation of the exterior surfaces of the model. The motivation for doing this was to remove non-structural modes from the foundation that otherwise would make it more difficult to assess the natural frequencies and mode shapes. Another conclusion drawn from Case A was that minimal differences exist for frequency analysis, regardless of whether foundations with or without mass are used. If a massless foundation is used in frequency analyses, then the non-structural modes from the foundation are reduced, which simplifies the evaluation of modes.

In the EMVG test of Case A, however, Contributors used a foundation with mass, and with non-reflecting boundaries, to ensure that the induced accelerations at the dam crest would not reflect back into the model as the waves reached the outer surfaces of the foundation.

For all cases where a mass foundation was used with an excitation at the base of the foundation (i.e., Cases B, C, D, and E), the results have clearly shown that the free–field BC is essential to obtain realistic ground motions and, hence, to model the excitation of the dam. A significant difference in results was observed between Contributors who used only absorbing (or non-reflecting) boundaries, compared to Contributors who used both absorbing and the so-called free–field BCs. For example, results from contributions using only absorbing boundaries exhibited large computational artifacts along the nodes of the ground surface. In contrast, Contributors using free-field BCs obtained relatively consistent results, with minimal differences in the computed velocity along the nodes of the ground surface in Case B (i.e., when only the rock mass was studied). Results obtained by Contributors using free-field BCs were also in close agreement with the theoretical solution.

Codes implementing the free-field BC do so in various ways. Some Contributors used codes that are based on a simplified approach, where the equivalent forces on the sides of the foundation block are calculated using 1D wave propagation theory. Other Contributors obtained these forces from analyses of a 1D column, according to the discrete finite element approach. From the analyses performed in this benchmark workshop, it was impossible to discern any significant differences related to the method chosen by Contributors for including free-field BCs.

Methods based on analytical 1D wave propagation theory generally neglect the influence from damping or other non-linear effects. Contributors using this method obtained results that closely agreed with Contributors using other methods that account for free-field BCs. This information shows that use of the analytical approach is a valid assumption for cases with low material damping, such as the dam considered in this benchmark study. In contrast, the availability of commercial codes that implement the free-field BC, and that are designed for modeling the dynamic behavior of non-linear soils, suggests that low material damping is not a prerequisite for use of the free-field BC.

Use of the free-field BC has been less prevalent for the analysis of concrete dams; results of this benchmark study indicate that neglecting the free-field BC results in significant computational artifacts, even for cases with low material damping. The adoption of the free–field BC should therefore be encouraged for the analysis of concrete dams.

10.4 Size of Foundation Block

The benchmark studies show that the size of the foundation block has a very limited influence on the results obtained in the seismic wave propagation analysis when the free-field BCs are used.

The results obtained with non-reflecting BCs may provide relatively good results in the central part of the foundation block, but only for times less than the travel time for P-waves from the boundaries; results also tend to vary substantially from theory as one nears the edge of the block.

10.5 Non-linear Material Model

In the analyses performed for material non-linearity, Case E, i.e. cracking of concrete was investigated. The purpose of this was to see if the consideration of non-linear material models would lead to fracture of the dam body, and how it influences the dam response.

In Case E studies, the Taft Record had been selected since it was expected by the Formulators to only result in limited damage to dam concrete. As it was presented in Sect. 8.4 for the Taft Record analysis, most Contributors obtained minor cracking at the dam heal but four Contributors predicted additional damage at the dam neck area. The factors that may have largest influence on the predicted damage include type of boundary conditions, element size, constitutive material models or definition of damage.

Despite differences in the modelling techniques and material models, most of Contributors showed minor differences in the dam responses with a linear material model (Case D1) and to non-linear material (Case E1). Thereby, Contributors verified that minor damage in the dam structure does not have any significant influence of the dam response, i.e. its accelerations, displacements etc. Some of differences may be attributed to the fact that significantly refined mesh was used in Case E1 compared to Case D1.

When more significant damage occurs, there will be larger discrepancy between analyses based on linear versus non-linear material models. Similar conclusions could be drawn from Case E2 studies, where the analysis for the ETAF Record was performed. In this case study, significant damage in the dam body led to large amplifications of the crest displacements and accelerations.

10.6 Reservoir Level

In Case A, D and F, Contributors were asked to perform analyses for various reservoir levels to highlight the influence of the reservoir on the dam response.

In Case A, where frequency analyses were conducted, results showed higher natural frequencies of the dam-reservoir-foundation system for a lower reservoir level. This result is reasonable, considering that the reservoir contributes as an added mass to the system. The results also showed that for higher modes the reservoir level effect is significantly smaller.

In Case D and Case F, three increasing reservoir levels were considered. The results obtained for both cases show that the response of the dam, in terms of acceleration both at dam heel and dam crest, decreases as the reservoir water level increases. A clear dependence on the water level was not observed for total dam displacements or for hydrodynamic pressure.

10.7 Mass Versus Massless Foundation

Amplification of dam horizontal accelerations, i.e. the ratio of peak acceleration at the crest (Point *C)* to peak acceleration at the dam heel (Point *A*), is compared between Case D and Case F. The comparison is shown for three different reservoir water levels in Figs. 88, 89, and 90.

In general, greater accelerations are predicted at the dam crest when the massless foundation approach is adopted. Computed amplifications for the massless foundation models was even 1.5–2 times larger than amplifications obtained when the mass foundation model was used.

Figure 91 provides another comparison between Case D and Case F in terms of normalized mean values of peak acceleration at the dam heel and dam crest (a C), hydrodynamic pressure at the dam heel (hp A), and maximum downstream and

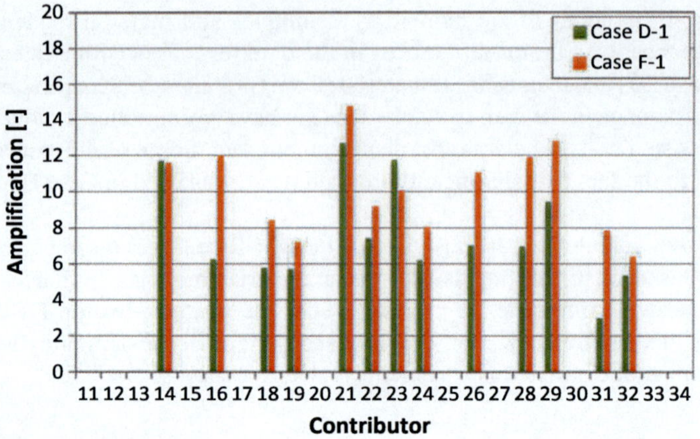

Fig. 88 Comparison between Case D-1 and Case F-1: amplification of the horizontal acceleration at the dam crest (Point C)

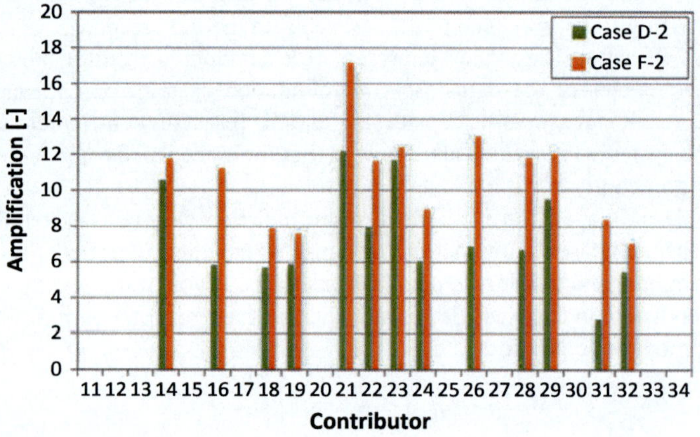

Fig. 89 Comparison between Case D-2 and Case F-2: amplification of the horizontal acceleration at the dam crest (Point C)

minimum upstream relative displacement at the dam crest (d_{max} C and d_{min} C). For the three different reservoir water levels, each quantity is normalized with respect to the results from the model with mass foundation. For horizontal acceleration at the dam crest, the response obtained from a model with the massless foundation, is about 1.5 times greater than the response from the model with mass foundation. The horizontal acceleration at the dam base is similar for both models.

The conservative use of the massless approach is widely documented in the technical literature [5, 8, 16] and has been restated in the frame of the short course: *Modern Methods for Non-linear Earthquake Response History Analysis of Concrete Dams,*

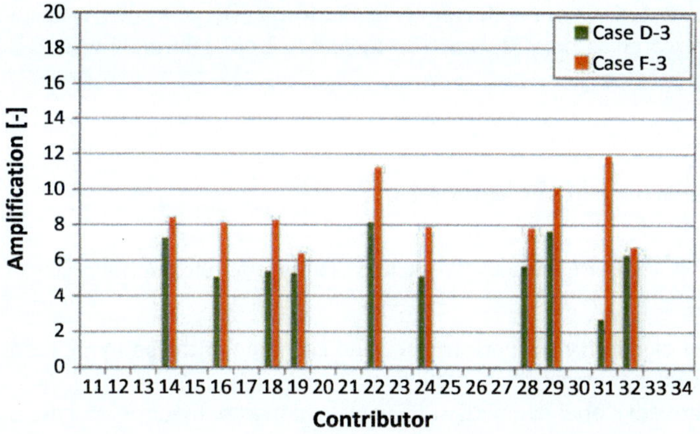

Fig. 90 Comparison between Case D-3 and Case F-3: amplification of the horizontal acceleration at the dam crest (Point C)

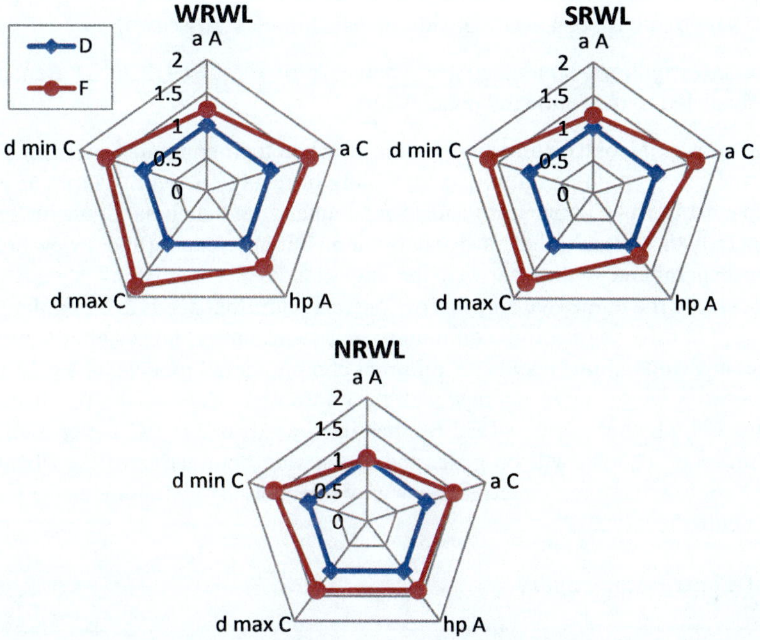

Fig. 91 Comparison between Case D and Case F. Normalized mean values: acceleration at the dam heel (a A) and dam crest (a C); hydrodynamic pressure at the dam heel (hp A); maximum and minimum relative displacement at the dam crest (d max C and d min C)

held by Prof. A.K. Chopra (University of California, Berkeley, USA) in collaboration with his research associates, N.S. Kwong and A. Lokke during the 2019 Benchmark Workshop session.

11 Summary and Conclusions

The 2019 ICOLD Benchmark Workshop is an important step towards developing a forum where:

- Analyst can verify the performance and accuracy of the software before using it in a seismic analysis of concrete dams.
- Software developer can verify and validate software features, and the accuracy of the software, using the results of the benchmark studies.
- Researcher can further expand and investigate topics presented in the benchmark workshop.
- Dam owners can build confidence in analysis results obtained using modeling software that has been verified with the benchmark study cases.

The investigations conducted for Theme A of the 2019 ICOLD Benchmark Workshops led to the following conclusions:

- The efforts of Contributors are maximized when their computational models can be verified through a preliminary screening of results to identify obvious errors. The most useful Contributor results for summarizing benchmark conclusions are generally those results that have undergone an initial screening and review process.
- The benchmark studies showed the importance of establishing a verification process for the numerical analysis of concrete dams that needs to be implemented throughout the engineering community in a coordinated and systematic way.
- In comparing various results for different computational models, one of the most significant factors affecting how well the computed results match the theoretical results is whether the free-field BC has been used. If this BC is not used, then artificial reflections will be generated at the side boundaries of the FE model, even for low-damping materials. This may substantially contaminate and bias the computed results.

List of Contributor Papers

Paper no.	Authors	Paper title
A01	Enzell, J., Malm, R., Abbasiverki, R., Ahmed, L.	Non-linear Behavior of a Concrete Gravity Dam during Seismic Excitation: a Case Study of the Pine Flat Dam
A02	Faggiani, G., Masarati, P., Frigerio, A.	Assessment of the Dynamic Response of Pine Flat Concrete Gravity Dam. FEM Simulation of Dam-Foundation Interaction

(continued)

(continued)

Paper no.	Authors	Paper title
A03	Farrokh, M.	Seismic Analysis and Damage Evaluation of Pine Flat Concrete Dam
A04	Lherbier, J.R., Andrian, F.	Numerical Seismic Analysis of Pine Flat Dam using Finite Difference Method
A05	Menouillard, T., Tzenkov, A.D., Schwager, M.V.	Dynamic Analysis of Pine Flat Concrete Dam: Acoustic Fluid-Structural Interaction with ANSYS Workbench
A06	Mitovski, S., Petkovski, L., Kokalanov, G., Kokalanov, V., Panovska, F.	Seismic Analysis of Pine Flat Concrete Dam
A07	Monteiro Azevedo, N., Braga Farinha, M.L., Câmara, R., Schclar Leitão, N.	Seismic analysis of Pine Flat Concrete Dam: Comparison of Numerical Results from 2D and 3D Analysis
A08	Naji-Mahalleh, N.	Seismic Analysis of Pine Flat Concrete Dam
A09	Panteki, E.. Goldgruber, M.	Seismic Excitation of a Concrete Dam: Analysis of the Influence of Modelling Approaches and Concrete Material Non-Linearity
A10	Peton, P., Thénint, T.	Modal, Dynamic and Seismic Analyses of the Pine Flat Concrete Gravity Dam
A11	Robbe, E., Grimal, E., Devesa, G.	Seismic analyses of Pine Flat Dam: Simplified Use of Viscous Spring Boundaries and Anisotropic Damage in the Principal Directions with the Ability of Cracks Re-closure
A12	Salamon J. W., Wood C. K., Manie J., Geister A.	Evaluation of Seismic Wave Propagation in the Analysis of Concrete Dams: Pine Flat Dam Benchmark Study
A13	Scolari M., Bado A., Gualco D., Buraschi L., Valsecchi R.	Finite Element Analyses of a Concrete Gravity Dam: Investigation on Static and Dynamic Behavior
A14	Stabile A.F., Labbé P., Nguyen A.	Analysis of Pine Flat Dam Considering Fluid-Soil-Structure Interaction and a Linear-equivalent Model
A15	Staudacher E.J., Zenz G.	2D Seismic Analysis of Pine Flat Concrete Gravity Dam Including Mass of Foundation: Approach with Infinite Elements
A16	Su W., Wang J.T., Xu Y.J.	Seismic Analysis of Pine Flat Concrete Dam
A17	Valente S., He Q., Capriulo C.	Seismic Analysis of Pine Flat Concrete Dam
A18	Vorlet S., Manso P., De Cesare G.	Seismic Behavior of Pine Flat Concrete Gravity Dam Using Microplane Damage-Plasticity Model

(continued)

(continued)

Paper no.	Authors	Paper title
A19	Yang H., Wang H., Salamon J.W., Jeremic B.	Earthquake Soil Structure Interaction Analysis of a Gravity Dam
A20	Mouy V., Molin X.	Dynamic Foundation-Fluid-Structure Interaction of a Concrete Gravity Dam: Influence of Input Signal Frequency Content and of Water Modelling on Wave Propagations—Example of Pine Flat Dam

References

1. Bureau of Reclamation (2018) Evaluation of numerical models and input parameters in the analysis of concrete dams. Result summary of the USSD Workshop, Miami, USA. DSO-19-13 report, https://www.usbr.gov/ssle/damsafety/TechDev/DSOTechDev/DSO-2019-13.pdf
2. Chavez JW, Fenves GL (1993) Earthquake analysis and response of concrete gravity dams including base sliding. Report No. UCB/EERC-93/07
3. Chavez JW, Fenves GL (1994) EAGD_SLIDE: a computer program for the earthquake analysis of concrete gravity dams including base sliding. Report No. UCB/SEMM-1994/02
4. Chopra AK, Chakrabarti P, Gupta S (1980) Earthquake response of concrete gravity dams including hydrodynamic and foundation interaction effects. Report No. UCB/EERC-80/01
5. Chopra AK (2014) Earthquake analysis of concrete dams: factors to be considered. In Proceedings of the 10th national conference in earthquake engineering, Anchorage, Alaska
6. Fenves G, Chopra AK (1980) Simplified analysis for earthquake resistance design of concrete gravity dams. Report No. UCB/EERC-85/10
7. Fenves G, Chopra AK (1984) EADG-84: a computer program for earthquake analysis of concrete gravity dams. Report No. UCB/EERC-84/11
8. Hansen KD, Nuss LK (2013) Seismic upgrades for concrete dams then and now. J Dam Saf 11(4):21–35
9. Hariri-Ardebili MA, Saouma V (2015) Quantitative failure metric for gravity dams. Earthquake engineering structural dynamics. J Int Assoc Earthq Eng 44(3)
10. Hariri-Ardebili MA, Furgani L, Meghella M, Saouma V (2016) A new class of seismic damage and performance indices for arch dams via ETA method. Eng Struct 110
11. Joyner WB, Chen ATF (1975) Calculation of non-linear ground response in earthquakes. Bull Seismol Soc Am 65(5):1315–1336
12. Lysmer J, Kuhlemeyer RL (1969) Finite dynamic model for infinite media. J Eng Mech Div 95(4):859–878
13. Mejia LH, Dawson EM (2006) Earthquake deconvolution for FLAC. In: 4th international FLAC symposium on numerical modeling in geomechanics. Itasca Consulting Group, Inc. Madrid, Spain, p 9
14. Rea D, Liaw CY, Chopra AK (1972) Dynamic properties of pine flat dam. Report No. UCB/EERC-72/7

15. Wikipedia.org, Pine Flat Dam photo (size reduce), https://commons.wikimedia.org/wiki/File:Pine_Flat_Dam.jpg
16. Zhang C, Jin F (2008) Seismic safety evaluation of high concrete dams, part I: state of the art design and research. In: 14th world conference on earthquake engineering, Beijing, China

Non-linear Behavior of a Concrete Gravity Dam During Seismic Excitation: A Case Study of the Pine Flat Dam

J. Enzell, R. Malm, R. Abbasiverki, and L. Ahmed

Abstract In this paper, seismic analyses of Pine Flat Concrete dam performed as part of theme A in the 15th benchmark workshop are presented. The results presented focuses on differences between mass and massless foundation and the influence from non-linear material behavior. The analyses performed with mass foundation using analytical free field input records and infinite boundary elements corresponded with the expected free surface results, for lower frequencies. For higher frequencies some discrepancies caused by the influence from the dam and the reservoir as expected. The corresponding analyses with massless foundation showed significantly higher accelerations but good agreement with the expected free surface displacement at the dam toe. To consider the influence from nonlinear material behavior, a dynamic push-over analysis (endurance time acceleration function, ETAF) was performed. It was possible to perform the analysis for the full duration of the record, despite significant non-linear material behavior. The initial damage occurred at the upstream toe and then showed significant induced damage as the level of excitation successively increased. In the end of the analysis, the top of the dam is cracked through which would cause an instability failure of the top of the dam.

Keywords Seismic analyses · Concrete dam · Free field boundary conditions · Non-linear analyses · Cracking

1 Introduction

The Pine Flat dam, see Fig. 1, is a 134 m high concrete gravity dam, with a length of 561 m and consists of 37 monoliths, all except one are 15.25 m wide and the other is 12.2 m wide. The dam is located in California (US) and was built in 1954 and

J. Enzell · R. Malm (✉)
SWECO, Stockholm, Sweden
e-mail: richard.malm@byv.kth.se

R. Malm · R. Abbasiverki · L. Ahmed
KTH Royal Institute of Technology, Stockholm, Sweden

G. Bolzon et al. (eds.), *Numerical Analysis of Dams*, Lecture Notes in Civil Engineering 91, https://doi.org/10.1007/978-3-030-51085-5_2

Fig. 1 Photo of the nearly completed Pine Flat Dam in 1953, from [10]

was primarily designed for flood control but also provides irrigation, groundwater recharge, recreation and since the mid-80s also hydropower generation. This dam has been extensively evaluated as a case study throughout its history for seismic, see for instance [1, 2], which makes a good choice for the comparisons of analysis methods and methodologies.

2 Case Study

The main aim of the benchmark workshop is to analyze the influence from complex phenomenon's through division into several, smaller, cases each designed to analyze a specific topic. Six different cases have been defined by the formulators of this theme where each case may include one to four analyses.

2.1 *Geometry*

The tallest, non-overflow monolith (No. 16) is studied in this benchmark, see Fig. 2, which has a height of 122 m and a thickness at the base of 95.8 and 9.8 m at the crest.

2.2 *Material Properties*

The concrete material properties were provided by the formulators as presented in [3] and are shown here in Table 1. All material properties, except the tensile strain at peak load for the concrete were used in the analyses. The tensile curve for concrete

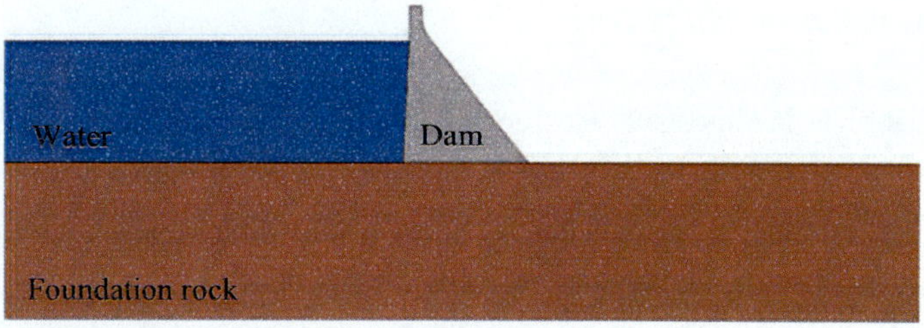

Fig. 2 Illustration of the geometry of the Pine Flat Dam

Table 1 Material properties as given by the formulators, [3]

Parameter	Unit	Concrete	Rock	Water
Elastic modulus	GPa	22.41	22.41	–
Density	kg/m^3	2483	2483	1000
Poisson's ratio	–	0.2	0.2	–
Compressive strength	MPa	28.0	–	–
Tensile strength	MPa	2.0	–	–
Fracture energy	Nm/m^2	250	–	–
Compressive strain at peak load	–	0.0025	–	–
Tensile strain at peak load	–	0.00012[a]	–	–
Shear wave velocity	m/s	–	1939	–
Compressive wave velocity	m/s	–	3167	1439

[a]In the analyses, the strain $\varepsilon_t = f_t/E = 0.00009$ was used instead

was instead simplified where it was assumed linear up to the tensile strength as seen in Table 1.

In addition to the properties specified above, material properties related to the constitutive material model *Concrete Damaged Plasticity* in Abaqus were also defined. The material properties were as follows; dilation angle $\varphi = 35°$, eccentricity $\epsilon = 0.1$ (default), ratio of biaxial and uniaxial compressive strength $f_{\text{bc}}/f_{\text{c}} = 1.16$ (default), shape of the yield surface $K_{\text{c}} = 2/3$ (default) and viscosity $\mu = 0$ (default). For further description of these material properties see, [4]. Finally, material damping has also been included in the analyses in accordance to the formulators, [3]. Rayleigh damping is defined with the following parameters, $\alpha = 0.75\ \text{s}^{-1}$ (mass) and $\beta = 0.0005\ \text{s}^{-1}$ (stiffness) which provides an average damping of 2% within the frequency interval 4–10 Hz.

2.3 Foundation Modelling

The foundation is, in almost all cases, considered as a mass-foundation, in accordance to [3]. In the models with mass-foundation, infinite boundaries are defined to prevent wave reflection of the boundaries, as seen in Fig. 3. For these, the built-in infinite elements in Abaqus are used which are based on the absorbing boundaries defined by Lysmer and Kuhlemeyer [5] for dynamic behavior and based on Zienkiewicz et al. [6] for static response. The infinite foundation elements are one-dimensional and absorbs waves travelling perpendicular to the boundary. In order to account for free field input records, the corresponding nodal forces on the nodes connected to the infinite element has been calculated according to [7]. Lateral forces, were calculated at the base of the rock, while the two vertical surface were defined with lateral and vertical forces for all the following cases with mass foundation (Case B, D, E1). For case E2 (with ETAF), see Sect. 2.6, the given input was considered as a total motion and applied only to the base of the rock in accordance to the instructions from the formulators [3].

For the frequency analysis where the natural frequencies were extracted, these infinite elements were excluded and instead displacements perpendicular to each rock boundary surface were prevented. The reason for this was that the static stiffness of these infinite elements influenced the natural frequencies obtained. In case F, where a massless foundation was studied, these infinite elements were also excluded and instead acceleration boundaries were defined on the bottom rock boundary surface, see [8].

2.4 Fluid-Structure-Interaction

The fluid-structure-interaction was considered via inclusion of compressible acoustic elements, which can describe the pressure wave propagation in water. In Abaqus, the acoustic elements do not include terms for body forces, hence it only describes the hydrodynamic pressure, [9]. Thereby, the hydrostatic pressure was defined separately, as a linear increasing pressure depending on the depth. The hydrostatic pressure at the surface of the water was defined as zero, to represent the free surface.

2.5 Mesh

In non-linear analyses, it is important that the mesh size of the concrete elements are smaller than the characteristic length of the specific concrete type to assure that the stored elastic energy that is released during cracking can be absorbed by the surrounding elements, [4]. The maximum allowed element length for a case with an exponential unloading stress-displacement curve is

$$L_{max} < \frac{E \cdot G_f}{f_t^2} \quad (1)$$

where,

E is the elastic modulus [Pa]
G_f is the fracture energy [Nm/m^2]
f_t is the tensile strength [Pa].

$$L_{max} = \frac{c}{n_{min} \cdot f_{max}} \quad (2)$$

where,

c is the shear wave velocity of the material [m/s]
n_{min} is the minimum number of points that is required to approximate a sine wave
f_{max} is the highest fundamental frequency of interest (for the structure) [Hz].

Based on the material properties given in Table 1, the element length of the rock foundation should be less than 15.8 m for compression waves and 9.7 m for shear waves if the highest fundamental frequency of interest is 20 Hz and $n_{min} = 10$. In the numerical model, the foundation has been defined to have a characteristic element length that is 6 m in most cases. The model is defined as 2D plane strain with an out-of-plane thickness of 15.24 m. In total, the model consists of about 110,000° of freedom, with the types of elements as specified in Table 2.

Table 2 List of elements used in the model

Parameter	Element type	Element name	No. elements
Concrete	3-node linear plain strain	CPE3	11,862
Foundation	4-node linear plain strain	CPE4R	2,340
Foundation (infinite)	4-node linear, one-way infinite	CINPE4	157
Water	4-node linear acoustic	AC2D4	28,858
Water (infinite)	2-node acoustic infinite	ACIN2D2	94

Fig. 3 Illustration of the foundation and the surrounding infinite foundation elements

2.6 Loads

The static loads considered in the analyses were gravity and hydrostatic pressure. The gravity load was applied to the concrete dam only and calculated directly from the volume and the given density. The hydrostatic pressure was defined to linearly increase with increasing depth on the upstream surface and as constant on the top of the foundation located upstream of the dam. In the different analyses, the dynamic loads were either;

- a harmonic force at the crest from the eccentric mass vibration test (Case A),
- nodal force input at the base and the vertical rock sides from impulse loads (Case B),
- nodal force input at the base and the vertical rock sides from the Taft earthquake (Case D, E1),
- total motion applied at the foundation base for the ETAF earthquake (Case E2),
- acceleration record for the massless foundation (Case F).

The seismic excitation was introduced as shear stresses on the bottom of the rock foundation and normal and shear stresses at the vertical sides, see Fig. 3. In Abaqus, it is not possible to define the stresses directly and thereby the stress was converted into forces based on the element length of the foundation. In Fig. 4, the Taft acceleration record (used in the massless foundation analysis) is shown along with the deconvoluted acceleration signal, which constitute the basis for the shear stress applied as excitation in the mass-foundation analyses. In the top figure, the acceleration records in time domain are shown while the bottom figure shows the response spectrum (based on 5% numerical damping) is shown. As it can be seen in the figure, there the highest excitation frequency is about 20 Hz. In Fig. 5, the corresponding acceleration record and response spectrum is shown for the ETAF signal. The response spectrum is here shown after a duration of 5, 10 and 15 s.

2.7 Solution Method

All analyses were performed with Abaqus Standard, version 2019. The static loads (gravity and hydrostatic pressure) were applied in an initial static step, while the dynamic excitation was analyzed with a dynamic implicit step with fixed time step

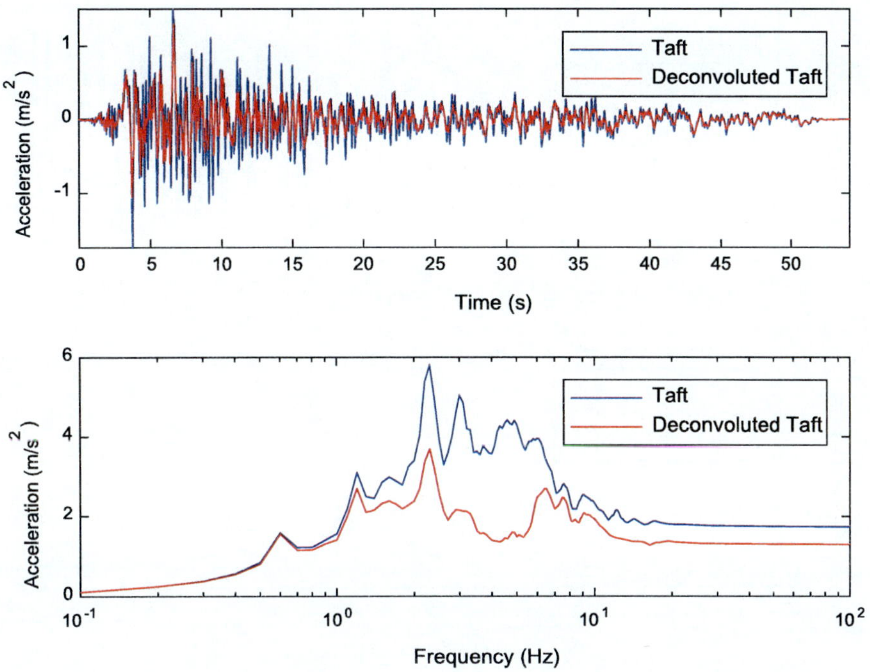

Fig. 4 Acceleration curves for Taft and the deconvoluted Taft signal, in time domain (top) and as response spectrum (bottom) with 5% damping

of at least 0.01 s. In the cases where the high frequency pulse was analyzed (case B), a fixed time step of 0.0005 s was used instead.

3 Results

A lot more results than those presented in this paper have been provided to the formulators. The authors of this paper have submitted results for 14 out of 19 cases. Due to page limit restriction, this paper focuses primarily on the seismic analyses with and without non-linear material behavior, and the influence of modelling the foundation, i.e. mass vs massless, absorbing vs free field input records at the boundaries.

3.1 Natural Frequencies

The six lowest natural frequencies was determined through a frequency analysis, for cases with different water elevations, see Table 3. These frequencies are obtained

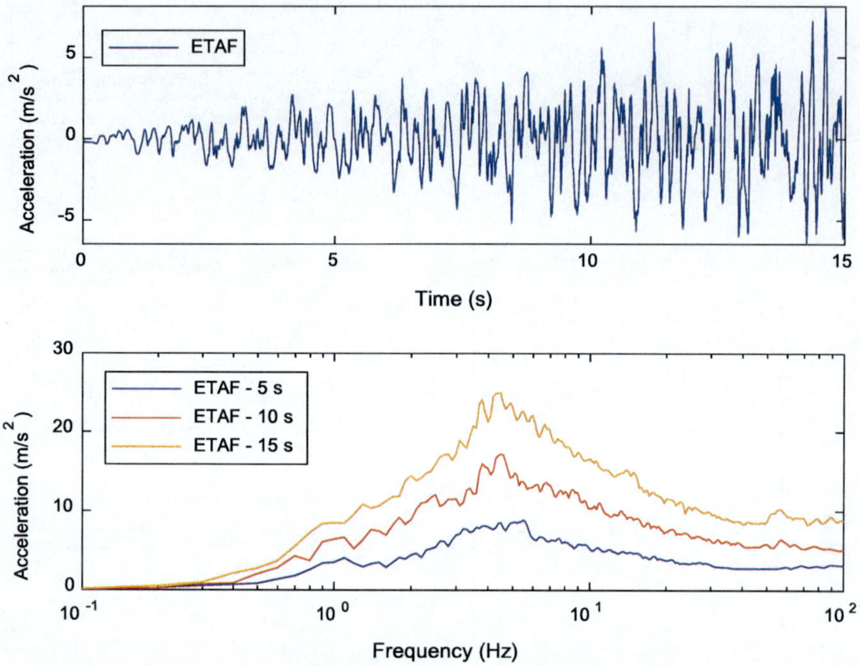

Fig. 5 Acceleration curves for ETAF signal, in time domain (top) and as response spectrum (bottom) with 5% damping

Table 3 Summary of natural frequencies, from the analyses and from the test [1]

Analysis	Season	W.L	Modes					
			1	2	3	4	5	6
Mass	W	268.2 m	2.31	3.48	3.95	4.40	4.91	5.38
Mass	S	287.6 m	2.20	3.38	4.07	4.59	4.99	5.46
Massless	W	268.2 m	2.35	3.58	4.18	–	5.07	5.70
Massless	S	287.6 m	2.24	3.47	–	4.49	5.29	5.79
Exp. [1]	W	268.2 m	–	3.47	4.13	–	–	5.40
Exp. [1]	S	278.9 m	–	3.27	4.07	–	–	5.65

with mass-foundation subjected to restrained displacements, perpendicular to each surface, instead of applying infinite elements as mentioned previously. It can be difficult to determine which of the natural frequencies that are obtained from the frequency analysis that are real structural modes. Therefore, one additional frequency analysis was also performed with massless foundation, in order to remove the non-structural modes. The results from the massless case is also presented in Table 3. As it can be seen in the table, these results are close to the mass-foundation and it is thereby shown that if only the frequencies are of interest then one suitable simplification is that massless foundation is used. All results are presented for the winter reservoir level (WRWL) and the summer reservoir level (SRWL). As it can be seen, the natural frequencies are lower for the summer conditions, which is expected since the water elevation is higher. The results from the measurements performed by [1] are also summarized in the table for winter and summer conditions respectively. The experimentally obtained mode numbers have been renumbered so that they coincide with the numbering of these numerical results for easier comparison. One thing worth noticing is that the first mode, at about 2.2–2.3 Hz, was not visible in the measurements. The second natural frequency from the model coincides well with the first mode that they obtained from the experiments, corresponding to about 3.4–3.6 Hz in the numerical analysis and 3.3–3.5 Hz from the measurements. The same is for the third mode which in the numerical is between 4.0 and 4.1 Hz and 4.1 Hz in the experiments. The modes denoted no. 4 and 5 were not observed in the measurements. The mode no. 6 is also close to the experimental results with 5.4–5.8 Hz in the numerical analysis and between 5.4 and 5.7 Hz in the experiments.

3.2 *Time History Analysis of Taft Earthquake*

In the time history analysis with linear material properties, it can be seen that the maximum principal stresses exceeds the tensile strength at a localized area near the upstream toe and that on the downstream side in the upper part of the dam (just below the section change). The Taft earthquake is however not large enough to cause significant non-linear behavior of the Pine Flat Dam. This was later confirmed in the analysis based on a nonlinear constitutive material model where cracking was initiated in these two regions, see Fig. 6a, and that the global influence from these cracks were limited since results from the linear (case D1) and nonlinear (case E1) analyses are almost identical, see Fig. 6b.

The response at point A, i.e. the upstream toe of the dam, is expected to be almost identical with the free surface record of the Taft signal for lower frequencies, see Fig. 6b. For higher frequencies there should be a difference due to the presence of dam and water. In the figure, the calculated response spectrum is shown for mass foundation (case D) with and without free field input records. The results clearly shows that neglecting free field conditions results in underestimation of the obtained surface acceleration since the infinite elements attenuates waves going both up and downwards. In case B when only the rock mass was considered for an impulse load, it could also be clearly seen that neglecting free field conditions resulted in significant difference in the velocity on the top surface of the rock mass. When using free field conditions, the velocity was more or less identical along the surface, which is a clear indication that the implementation of free field input records at the boundaries works as intended.

In Case F, similar type of analysis was performed on a massless foundation. In the model with massless foundation, no infinite elements were defined on the

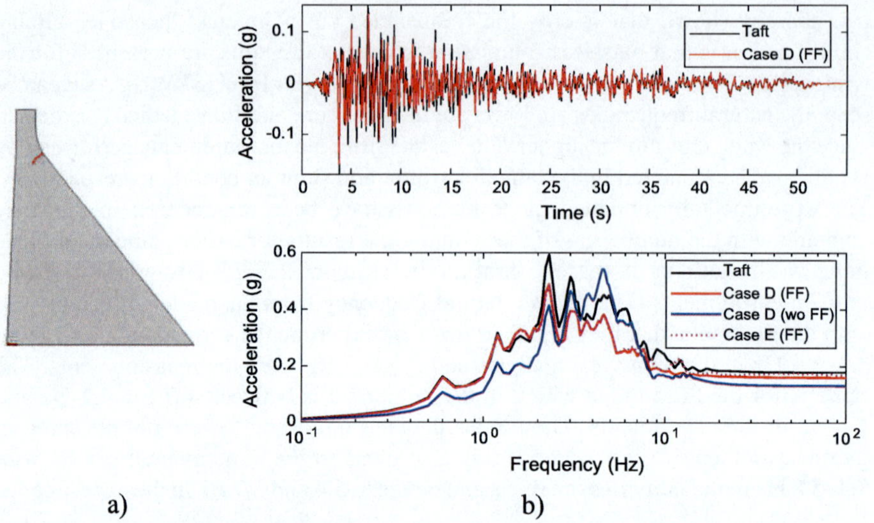

Fig. 6 **a** Cracking due to Taft earthquake (Case E1), **b** comparison of the Taft free surface acceleration and the calculated acceleration at the upstream toe (point A) with and without free field input records at the boundaries for Case D1 and E1. Top–acceleration vs time, Bottom – response spectra

rock boundaries, since these are not needed since wave reflection does not occur at the boundaries of a massless foundation. In Fig. 7, the calculated displacement for the mass foundation (Case D with and without free field conditions) and massless foundation (Case F) are shown along with the free surface displacement from the Taft record. Also from this comparison it is clear that free field boundaries are required to obtain realistic results when using a mass foundation. In the case with massless foundation and mass foundation with free field boundary conditions both are more or less identical to the free surface record from the Taft earthquake. It should however be noted that the acceleration from the massless analysis is significantly higher than the acceleration from the mass foundation and according to the free surface acceleration record. The maximum PGA at the upstream toe from the massless foundation is 2.1 m/s^2 compared to the mass foundation 1.3 m/s^2 and the Taft free surface of 1.5 m/s^2. Thereby, the analyses with massless foundation gives rise to significantly higher accelerations but gave good agreement for displacements.

3.3 *Endurance Time Acceleration Function*

The analysis with Endurance Time Acceleration Function (ETAF) is performed as a push-over analysis were the intensity of the seismic input continuously increases with time. The cracking is initiated in the model after about 4 s, were one element at the upstream toe is subjected to micro-cracks. This seems reasonable since the intensity

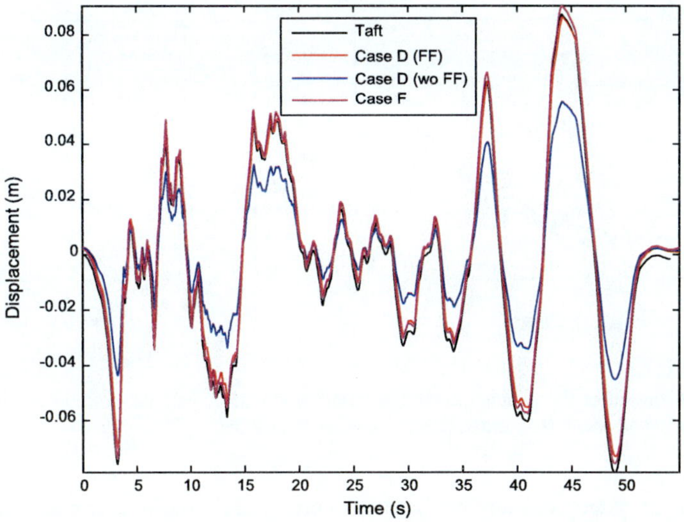

Fig. 7 Comparison of calculated displacement at the upstream toe (point A) for mass (Case D) and massless (Case F) foundations and the Taft free surface displacement

is about the same as in the Taft signal, at least for lower frequencies. As the intensity of the ETAF signal increases, the cracking propagates through the monolith. After about 5 s, an inclined crack occurs at the curved downstream part of the monolith as seen in Fig. 8a. This cracks continues to propagate, but after about 8 s a new inclined crack is initiated at the downstream side at about mid-height of the monolith, see Fig. 8b. It is possible to complete the analysis with the full 15 s duration of the ETAF signal, however as seen in Fig. 8c significant cracking has occurred in the monolith.

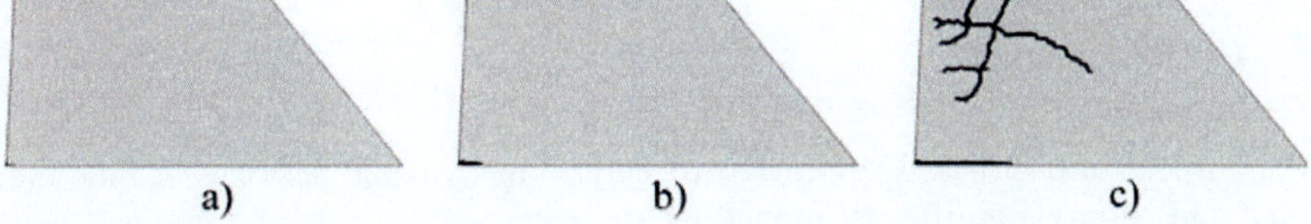

Fig. 8 Illustration of cracking from the ETAF analysis after, **a** 5 s, **b** 8 s and **c** 15 s. Induced tensile damage is shown with black color

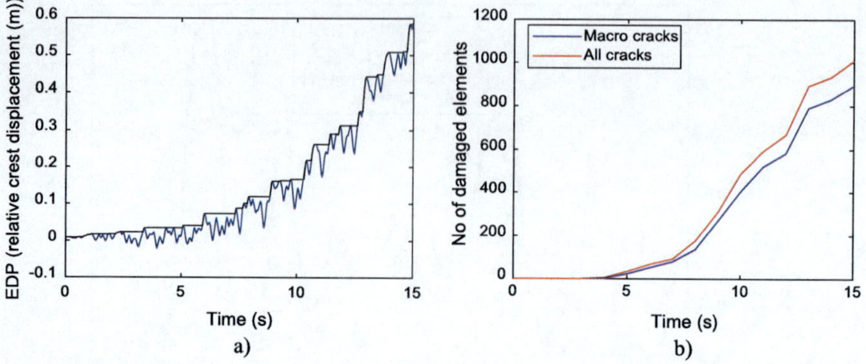

Fig. 9 Illustration of the damage evolution based on the ETAF signal, **a** Relative displacement between crest and toe, **b** No. of damage elements in the model

Based on the extensive damage that has occurred, it is likely that a collapse would occur shortly after if the intensity were increased even further.

It can also be seen from the results that the displacement of the crest does not follow the induced displacement of the ETAF signal or the displacement at the dam toe after about 8 s. The reason for this is that when the crack has propagated as shown in Fig. 8b, the top of the dam is moving in the upstream direction due to an internal failure mode of the top of the crest, similar to an internal overturning mode of the top.

In Fig. 9, the damage evolution in the model is shown. In Fig. 9a, the relative displacement between the crest and the upstream toe is shown. As it can be seen in the figure, the relative displacement is 0.6 m at the end of the simulation which shows the significant nonlinear deformation in the dam.

It can also be seen that the damage increases successively and that there is no specific time point were drastic damage evolution occurs. In Fig. 9b, the total number of damaged elements in the model is shown. The total number of elements in the dam is 11 862 in accordance with Table 2. As it can be seen, about 8% of the dam is damaged after 15 s. In the figure, two curves are shown where the first only shows macro cracks (i.e. cracks that are stress free) while the second curve shows all elements were a crack has initiated.

4 Conclusions

The purpose of this investigation was to analyze the dynamic response of Pine Flat Dam and comparing different modelling strategies. Many different analyses have been performed and submitted to the formulators. In this paper, comparisons between different modelling approaches have been compared with measured results regarding the natural frequencies of the dam. Overall, the natural modes corresponded well between both the mass foundation and the massless foundation with the experimental results for winter and summer conditions respectively. In the numerical analysis is can often be difficult to assess which are real modes and therefore, the use of massless foundation can be a good choice. Even though massless foundations, in general, are considered insufficient for time history analyses, they proved to be quite suitable for frequency analyses.

In the dynamic analyses of the Pine Flat Dam with foundation-dam-water interaction, the analyses performed with mass foundation clearly showed that it is important to include the free field boundary conditions when using absorbing boundaries such as infinite elements. Without these, the response at the base of the dam is significantly underestimated. These results also showed that the massless approach gave accurate displacements at the base of the dam but significantly overestimated the accelerations in the dam compared to the more realistic case with mass foundation and free field input records at the boundaries. Using the analytical approach presented by [7] gave good agreement with the expected results and was fairly simple to implement as nodal lateral forces at the base of the foundation and as lateral and vertical forces at the vertical sides of the foundation.

The seismic excitation from the Taft signal was large enough to initiate some damage in the elements closest to the upstream toe and at the upstream surface near the crest. However, these cracks had only minor influence of the global behavior of the dam. Therefore, an additional task was also given where a seismic pushover analysis based on ETAF was used. This analysis showed initial damage at the same location as the traditional time history analysis, and then showed significant induced damage as the level of excitation successively increased. In the end of the ETAF analysis, the top of the dam is cracked through which causes a local instability failure.

Acknowledgments The research presented was carried out as a part of Swedish Hydropower Centre—SVC. SVC has been established by the Swedish Energy Agency, Energiforsk and Svenska Kraftnät together with Luleå University of Technology, KTH Royal Institute of Technology, Chalmers University of Technology and Uppsala University. http://www.svc.nu.

References

1. Rea D, Liaw CW, Chopra AK (1972) Dynamic properties of Pine Flat Dam. Report No. EERC 72-7 University of California
2. Evaluation of Numerical Models and Input Parameters in the Analysis of Concrete Dams. Report DSO-19-13, A Summary Report of the USSD Workshop, Miami, USA (2018)
3. Salamon J, Hariri-Ardebili MA, Malm R, Wood C, Faggiani G (2019) Theme a–formulation: seismic analysis of Pine Flat Concrete Dam. In: 15th international benchmark workshop on numerical analysis of dams
4. Malm R (2009) Prediction of shear type crack initiation and growth in concrete with non-linear finite element method. PhD thesis, TRITA-BKN, Bulletin 97. KTH Royal Institute of Technology
5. Lysmer J, Kuhlemeyer RL (1969) Finite dynamic model for infinite media. J Eng Mech Div ASCE, 859–877

6. Zienkiewicz OC, Emson C, Bettess P (1983) A novel boundary infinite element. Int J Numer Methods Eng 19:393–404
7. Song Z, Wang F, Liu Y, Su C (2018) Infinite element static-dynamic unified artificial boundary. Shock Vib, 1–14. Article ID 7828267
8. Malm R (2016) Guideline for FE analyses of concrete dams. Energiforsk Rep 2016:207
9. Gasch T, Facciolo L, Eriksson D, Rydell C, Malm R (2013) Seismic analyses of nuclear facilities with interaction between structure and water–comparison between methods to account for Fluid-Structure-Interaction (FSI). Elforsk Rep 13:79
10. https://www.kingsbasinauthority.org/sixty-years-of-pine-flat-dam/

Assessment of the Dynamic Response of Pine Flat Concrete Gravity Dam. FEM Simulation of Dam-Foundation Interaction

G. Faggiani, P. Masarati, and A. Frigerio

Abstract The seismic FEM modelling of the tallest non-overflow monolith of Pine Flat Dam, a large concrete gravity dam located on King's River, in California, has been proposed as Theme A of the 15th ICOLD Benchmark Workshop on the Numerical Analysis of Dams. Different case studies have been proposed to study the response of the dam-reservoir-foundation system considering different dynamic loadings and approaches to simulate dam-foundation interaction, different behaviours for the dam concrete and different water levels: all the mandatory cases have been performed together with the non-linear case with the Endurance Time Analysis (ETA) method and the cases with massless foundation, still widely used in Italy for the seismic safety assessment of dams. The analyses have been carried out using the FEM code Abaqus, able to suitably model both the advanced dam-reservoir-foundation interaction and the non-linear behaviour of concrete. The performed set of simulations highlights the benefits of each of the studied approaches in the assessment of the seismic response of a dam-reservoir-foundation system.

Keywords Seismic FEM modelling · Concrete dams · Dam-Reservoir-Foundation interaction

1 Introduction

The seismic analyses of Pine Flat Dam, a large concrete gravity dam located on King's River, in California, has been proposed as Theme A of the 15th ICOLD Benchmark Workshop on the Numerical Analysis of Dams [1]. The theme is aimed to deepen the methods and the approaches most used in the seismic analyses of concrete dams, trying to identify the key uncertainties that may lead to differences among numerical modelling results. In particular the advanced models of structure-foundation interaction, able to reproduce the propagation of seismic waves taking into account the dissipation effects through the boundaries of the artificially truncated

G. Faggiani (✉) · P. Masarati · A. Frigerio
Ricerca Sul Sistema Energetico - RSE S.p.A, Milan, Italy
e-mail: giorgia.faggiani@rse-web.it

G. Bolzon et al. (eds.), *Numerical Analysis of Dams*, Lecture Notes in Civil Engineering 91, https://doi.org/10.1007/978-3-030-51085-5_3

foundation, are adopted. These models are supposed to provide a reliable calculation of the seismic response of the dams, as confirmed in most cases also by comparison with experimental measurements. The less complex interaction models, on the contrary, tend to significantly overestimate the seismic response of dams: nevertheless these methods are well consolidated, reliable and widespread and are still used in the common practice nowadays (at least in Italy and in other European countries). A comparison among advanced and traditional simplified methods is therefore a crucial step in the understanding of the seismic response of dams, above all when experimental measurements are not available. Different case studies have been proposed to study the response of the dam-reservoir-foundation system considering different dynamic loadings and approaches to simulate dam-foundation interaction, different behaviours for the dam concrete and different water levels: all the mandatory cases have been performed together with the non-linear case with the Endurance Time Analysis (ETA) method and the cases with massless foundation, still widely used in Italy for the safety seismic assessment of dams.

2 Geometrical and Physical Model

The various case studies are addressed by means of the FEM code Abaqus, using a 3D model realised using the provided geometry. The model (Fig. 1) includes the 15.24 m-wide dam monolith no. 16 and the relevant portions of foundation and of reservoir. For non-linear cases a more refined mesh only for the dam was realised (Fig. 2). The mesh for linear cases is made of linear elements (4 within the width): structural elements (1076 for the dam and 9360 for the foundation, for a total of 13,930 nodes) and acoustic elements (6732 for the reservoir, for a total of 8840 nodes). The different parts of the mesh (dam, foundation and reservoir) are discretized quite uniformly, with element size ranging from 3 and 6 m. The fluid domain is obtained by extruding the upstream face of the dam mesh for a length about 2.5 times the total height of

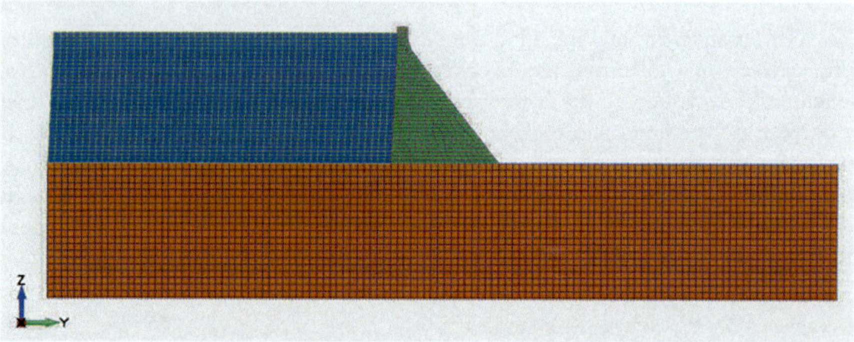

Fig. 1 FEM model of the dam monolith, the foundation and the reservoir

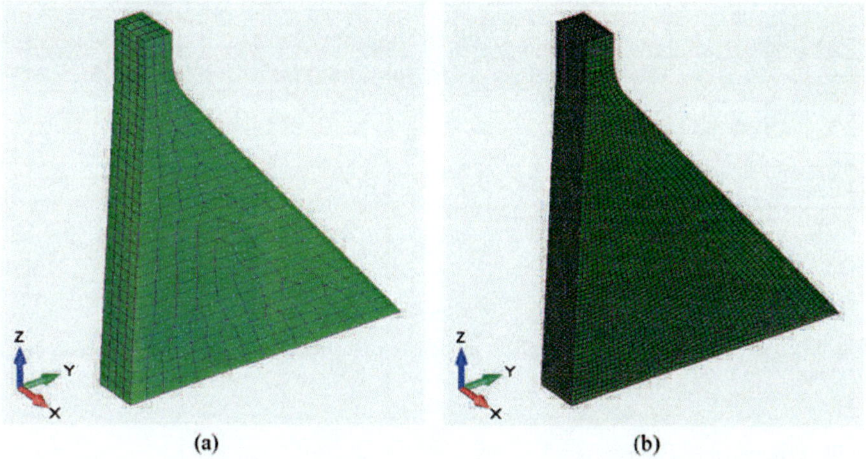

Fig. 2 FEM model of the dam used for cases A, D and F (**a**) and for case E (**b**)

the dam, reaching the upstream boundary of the foundation: the resulting mesh was only joined to the upstream dam face, not to the rock.

The dam mesh for non-linear cases consists of 25,120 elements (10 within the width) and 28,644 nodes: in this case the element size is about 1.5 m.

Foundation rock is assumed to behave linear-elastically; depending on the type of analysis (linear or non-linear) dam concrete may behave linear-elastically or according to the Concrete Damage Plasticity (CDP) model proposed by Lee and Fenves [2] and available in Abaqus [3]. According to this model, under uniaxial tensile loading the stress-strain response of concrete is linear-elastic until the tensile strength is attained; subsequent increases of strain cause the progressive damage of the material, characterized by a loss of stress for increasing deformation (softening) and by the reduction of stiffness in the unloading phase.

The loss of stress and the growth of the damage are assigned as functions of the progressive opening of the crack (Fig. 3). The damage variable can take values from zero, representing the undamaged material, to one, which represents total loss of strength.

The dynamic interaction between the dam and the reservoir is achieved through the classic structural-acoustic coupling on the upstream face of the dam, where the normal component of the dam acceleration is related to the normal gradient of the water hydrodynamic pressure [4]. No dynamic interaction is considered between the foundation and the reservoir. The upstream truncation of the reservoir is provided with non-reflecting acoustic condition. No seismic excitation is applied either to the bottom or to the upstream truncation of the reservoir.

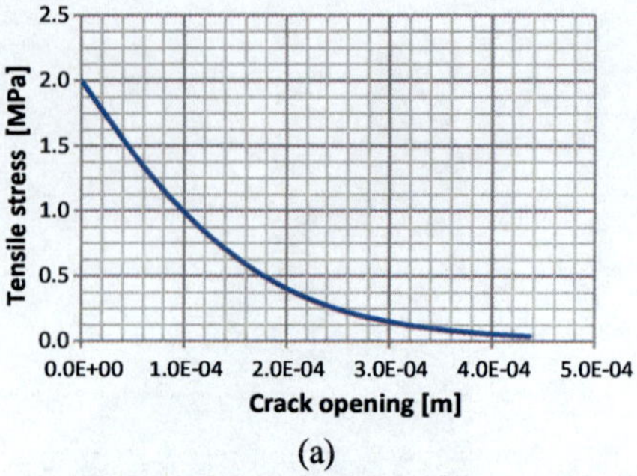

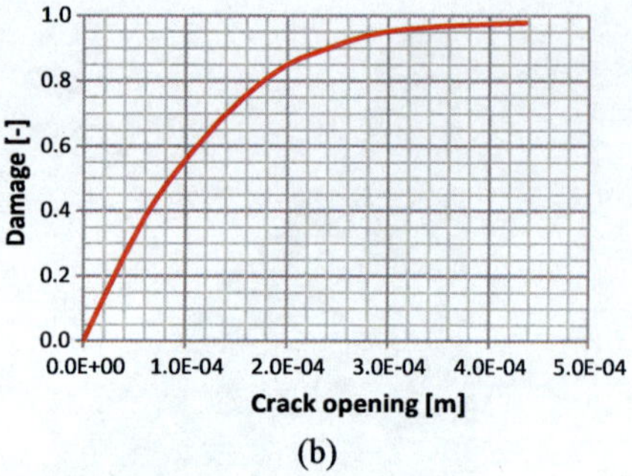

Fig. 3 Concrete Damage Plasticity (CDP) model. Tensile stress (**a**) and damage (**b**) versus crack opening

Table 1 Material properties

Parameter	Concrete	Rock	Water
Density [kg/m^3]	2483	2483 (0)	1000
Elastic modulus [MPa]	22,410	22,410	
Poisson's ratio [-]	0.2	0.2	
Compressive strength [MPa]	28.0		
Tensile strength [MPa]	2.0		
Fracture Energy [N/m]	250		
Shear wave velocity [m/s]		1939	
Compressional wave velocity [m/s]		3167	
Sound velocity [m/s]			1439

The dynamic interaction between the dam and the foundation is approached with the *viscous-spring artificial boundaries model* proposed in [5–7] based on the theoretical solution of the elastic wave problem in a half-space, except for the cases with the *massless approach* [8].

A 2% structural damping ratio is assumed as required in formulation. Table 1 summarizes the physical-mechanical parameters assumed in the model. Density of rock has been assumed null only in cases with the massless approaches.

3 Loadings

The numerical analyses have considered the effects of the following loadings/actions:

Static loads. Self-weight of the dam and the reservoir (applied as hydrostatic pressure on the upstream face of the dam and on the surface of the foundation) for three water levels:

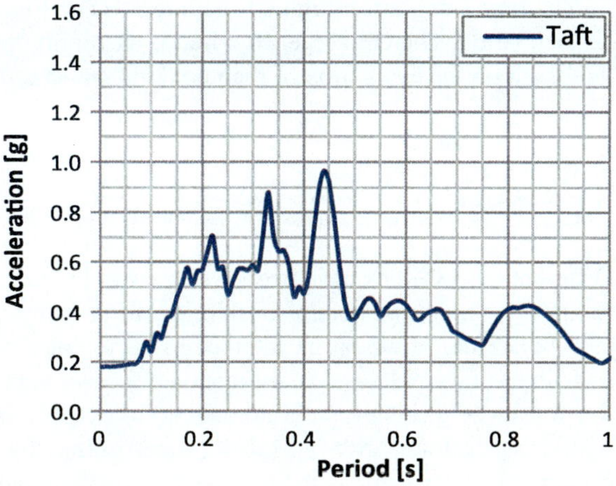

Fig. 4 Taft record. Upstream acceleration response spectrum (2% damping) at ground surface

(a) Winter reservoir water level (WRWL) at 268.21 m a.s.l.;
(b) Summer reservoir water level (SRWL) at 278.57 m a.s.l.;
(c) Normal reservoir water level (NRWL) at 290.00 m a.s.l.

Dynamic loads.

(a) Harmonic force time-history representing the load of an eccentric mass vibration generator (EMVG) located at dam crest;
(b) Taft horizontal acceleration time-history record (upstream component) of the M 7.3 Kern County (California) earthquake (peak ground acceleration of 0.18 g), considered as a free field ground motion at the surface of the foundation (Fig. 4);
(c) Endurance Time Acceleration Function (ETAF) in the upstream direction (up to peak ground acceleration of 0.83 g), applied as total motion at the base of the foundation.

The boundary conditions for the static step of the analyses consist in fixing the bottom, the upstream and downstream faces of the foundation.

In cases with massless approach the seismic motion is uniformly applied to the bottom, the upstream and downstream faces of the massless foundation.

In cases with massed foundation suitable artificial absorbing boundary conditions are set at the bottom, the upstream and downstream faces of the foundation, following the viscous-spring artificial boundaries model and using spring and dashpot elements available in Abaqus. Taft horizontal record is applied computing the effective earthquake forces following the procedure detailed in [6, 7], based on the theoretical solution of the elastic wave problem in a half-space: vertically propagating body waves are considered. The effective earthquake forces have been computed and applied both on the bottom and on the upstream and downstream faces of the foundation. ETAF is applied as total movement only at the base of the massed foundation.

Plane strain conditions are applied at side boundaries of the model.

The time integration implicit method HHT [9] is adopted, with $\alpha = -0.05$ and $\Delta t = 0.005$ s. The convergence criteria, used only in case of non-linear problems, are based on the evaluation of the nodal unbalanced forces.

4 Results

Different case studies have been proposed in [1] to study the response of the dam-reservoir-foundation system considering different dynamic loadings and approaches to simulate dam-foundation interaction, different behaviours for the dam concrete and different water levels: all the mandatory cases have been performed together with the non-linear case with the Endurance Time Analysis (ETA) method and the cases with massless foundation, still widely used in Italy for the safety seismic assessment of dams. The performed analyses are summarized in Table 2.

The following sections report and discuss the main results of the simulations.

4.1 Modal Analyses

The modal analyses provide the natural frequencies and the related mode shapes of the dam-foundation-reservoir system with winter (A-1) and summer (A-2) water level. Table 3 and Fig. 5 show the results of the analyses for the first 6 modes. The slight decrease of the natural frequencies (about 5%) in case A-2 is ascribed to the 10 m higher water level. With the only exception of the forth mode (where the effect of the massed foundation is important), in both cases mode shapes are regular.

Table 2 Summary of performed analyses

Type of analysis	Loading		Water level	Case (id)
Modal analysis	–	–	WRWL	A-1
	–	–	SRWL	A-2
EMVG test simulation	Static load	EMVG	WRWL	A-3
			SRWL	A-4
Linear dynamic analysis	Static load	Taft	WRWL	D-1
			SRWL	D-2
			NRWL	D-3
Non-linear dynamic analysis	Static load	Taft	WRWL	E-1
		ETAF	WRWL	E-2
Linear dynamic analysis with massless foundation	Static load	Taft	WRWL	F-1
			SRWL	F-2
			NRWL	F-3

Table 3 Cases A-1 and A-2. Natural frequencies

Mode [n]	Case A-1	Case A-2
	Natural frequency [Hz]	
1	2.28	2.17
2	3.75	3.44
3	4.08	3.96
4	4.60	4.41
5	4.87	4.86
6	5.56	5.48

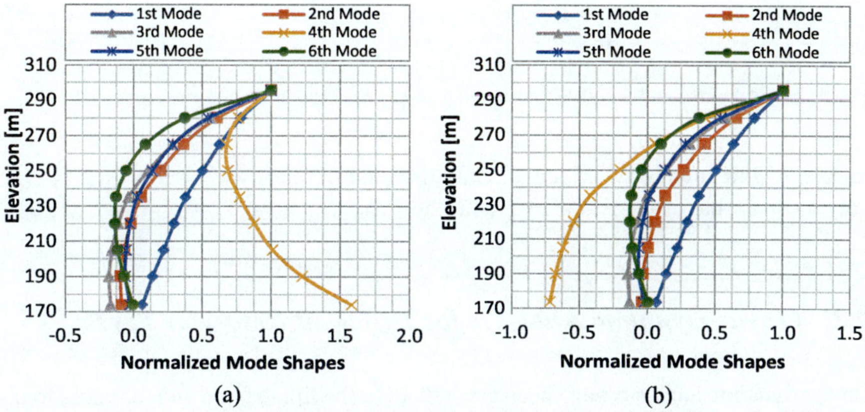

Fig. 5 Cases A-1 (**a**) and A-2 (**b**). Normalized mode shapes at upstream face

The 2-D numerical simulations reported in [10], carried out with the same material properties, had provided, for the first mode, frequencies of about 2.90 Hz and 2.85 Hz for case A-1 and A-2 respectively, higher than the values resulting from the current analyses: in these simulations however the assumption of rigid foundation was adopted. Modal analyses made with the same assumption of rigid foundation supply frequencies equal to 3.03 and 2.84 Hz, more comparable to the values in [10].

4.2 EMVG Test Simulation

The simulation consists in applying, at the crest of the 15.24-m wide monolith no. 16, the harmonic time history load of the eccentric-mass vibration generator tests conducted at Pine Flat Dam in January and June 1971 [10]. Both winter (A-3) and summer (A-4) water levels have been considered. Figure 6 reports the acceleration time history at dam heel (point A) and crest (point C) resulting from the analysis of cases A-3 (a) and A-4 (b). In case A-4 acceleration at dam crest results higher than

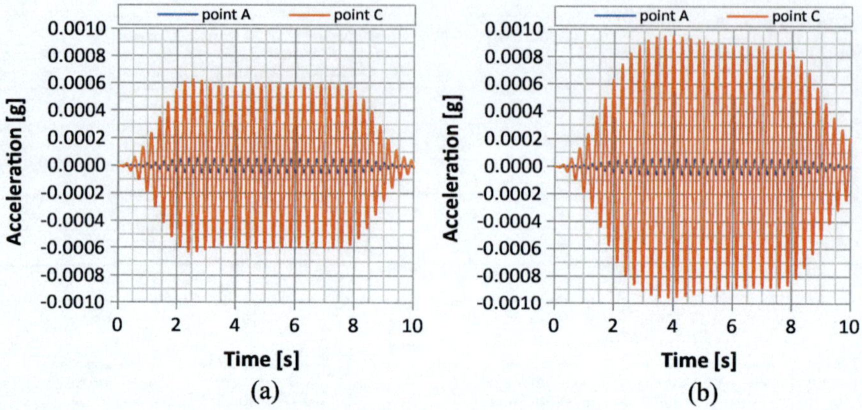

Fig. 6 Cases A-3 (**a**) and A-4 (**b**). Acceleration at dam heel (point A) and crest (point C)

in case A-3: it could be ascribed to natural second frequency of case A-2 (3.44 Hz), very close to that of the EMVG signal (3.50 Hz).

4.3 Linear Dynamic Analyses for Different Reservoir Levels

Linear dynamic analyses have been carried out both with massed and massless foundation for the three water levels specified in Sect. 3: in the first case (case D) the wave propagation problem with suitable artificial absorbing boundary conditions has been addressed, in the second (case F) the massless approach has been adopted. Figures 7 and 8 report the maximum principal stress envelope in cases D and F.

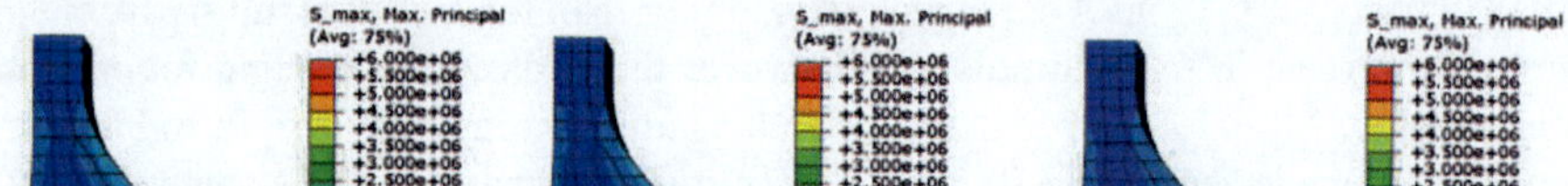

Fig. 7 Cases D-1, D-2 and D-3. Maximum principal stress envelope

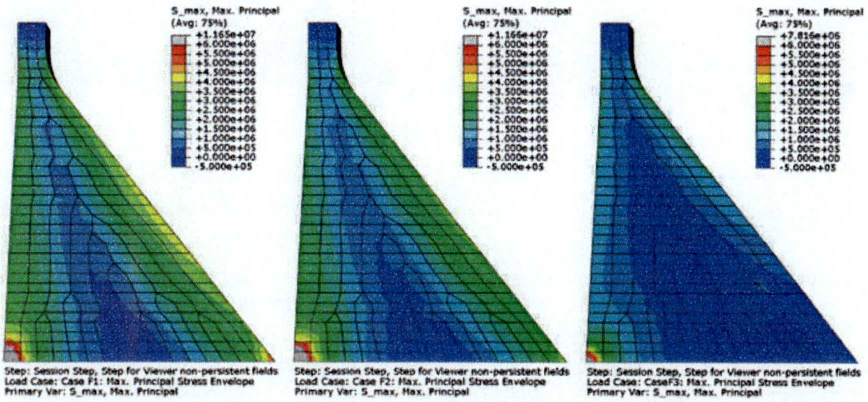

Fig. 8 Cases F-1, F-2 and F-3. Maximum principal stress envelope

In both cases the tensile stress state close to the downstream face decreases as the water level increases (moving from D-1 to D-2 and to D-3 and from F-1 to F-2 and to F-3); the trend of tensile stresses for different water levels is qualitatively similar in the area near the upstream face for case F, whereas in case D the tensile stress state becomes greater for higher water levels. In all cases stresses are directed tangential to the faces. The tensile stress state is quite more severe in cases with massless foundation (the values of tensile stress in case F are about two-three times those in case D). Tensile stress on downstream face varies from about 0.5 MPa to about 1.6 MPa in case D and from about 1.5 MPa to about 4.5 MPa in case F; also the maximum, at dam heel, grows up from about 6 MPa in case D to about 12 MPa in case F.

The comparison of the upstream acceleration response spectra (2% damping) at dam crest (point C) is reported in Fig. 9a for cases D-1, D-2 and D-3 and in Fig. 9b for cases F-1, F-2 and F-3. With respect to the seismic input (peak ground acceleration of 0.18 g), at the top of the monolith the acceleration results amplified of about $4 \div 6$ times in cases with massed foundation and absorbing boundaries condition (peak acceleration at period zero of 0.99 g, 0.94 g and 0.78 g for cases D-1, D-2 and D-3 respectively) and about $8 \div 15$ times in cases with massless foundation (peak acceleration at period zero of 2.64 g, 2.04 g and 1.46 g for cases F-1, F-2 and F-3 respectively). The whole spectra show that the use of the viscous-spring artificial boundaries model leads to a less demanding seismic response of the system with respect to that achieved with the massless approach. Nevertheless the periods of the resonant peaks resulting from the two approaches are rather comparable, as well as the dependence of the structural response on water levels since for both massed and massless cases the acceleration at dam crest reduces growing up the water level.

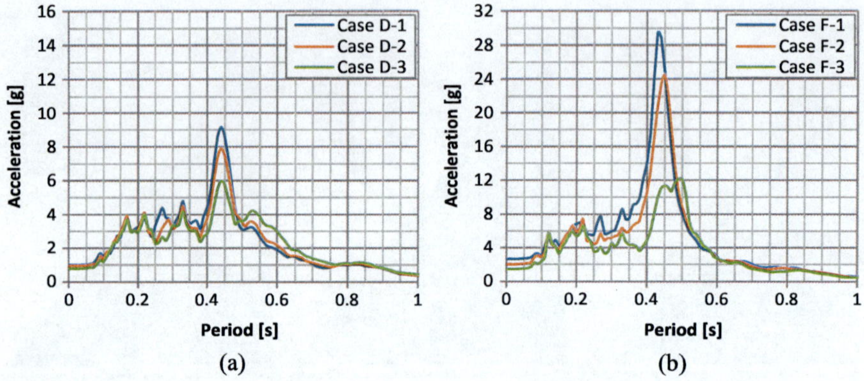

Fig. 9 Cases D-1, D-2, D-3 (**a**) and F-1, F-2, F-3 (**b**). Comparison between upstream acceleration response spectra (2% damping) at dam crest (point C)

4.4 Non-linear Dynamic Analyses for Different Time-History Signals

Non-linear dynamic analyses have been carried out with massed foundation only for the winter water level using two different seismic inputs (Taft and ETAF) and two different approaches to apply the dynamic load as specified in Sect. 3 and required in [1]: in the first case (E-1) the given input is considered as a free field ground motion at the surface, in the second case (E-2) it is applied as total motion at the base of the foundation.

The comparison between the non-linear case E-1 and the corresponding linear case D-1 in terms of upstream acceleration response spectra (2% damping) at dam crest (point C), showed in Fig. 10, highlights no difference between these two cases. When the non-linear behaviour of concrete is taken into account, a small damaged area

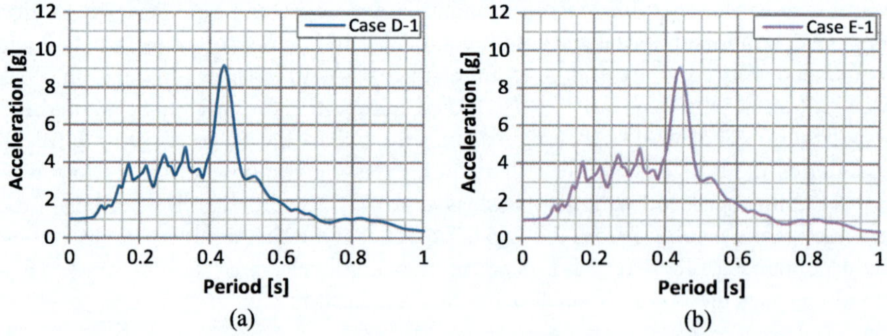

Fig. 10 Cases D-1 (**a**) and E-1 (**b**). Comparison between upstream acceleration response spectra (2% damping) at dam crest (point C)

(equal to 0.24% of the total area of the monolith) occurs at dam-foundation interface in the nearby of point A (Fig. 11b). Nevertheless, the overall structural response of the dam is essentially the same of the linear case D-1, since Taft earthquake it is not severe enough to highlight significant non-linear effects, as already noted for the acceleration and also confirmed by the trend of maximum principal stress envelope in Fig. 11a, absolutely comparable to case D-1 (Fig. 7), and of relative displacement at dam crest (Fig. 12).

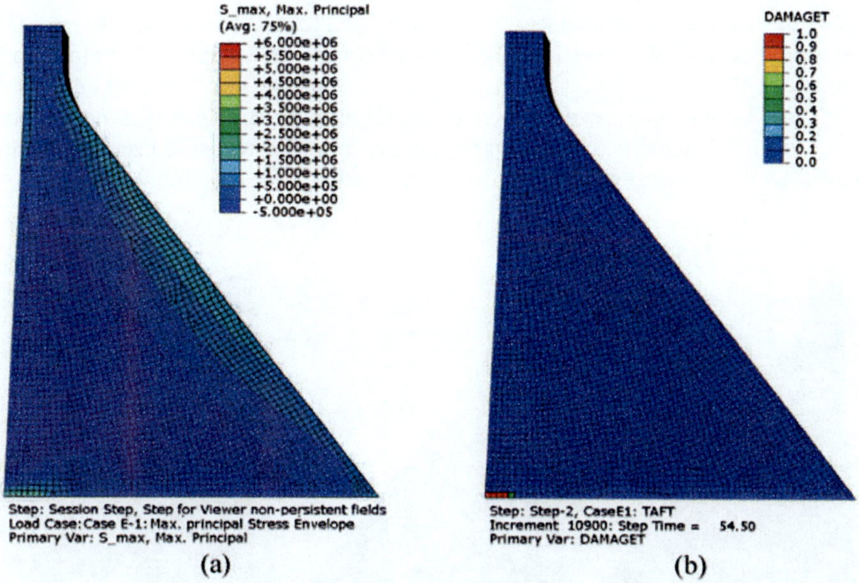

Fig. 11 Cases E-1. Maximum principal stress envelope (**a**) and damage (**b**)

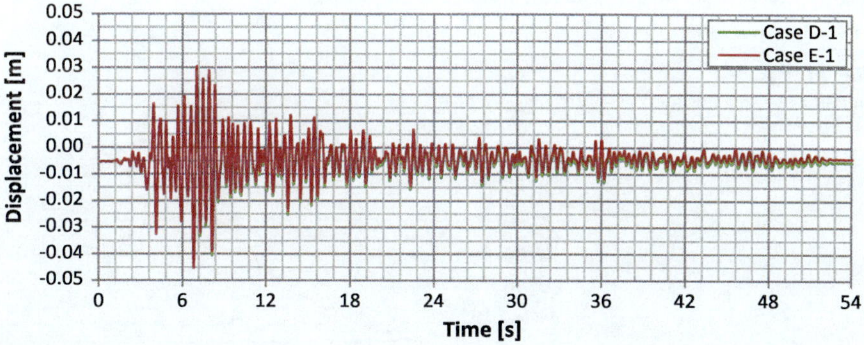

Fig. 12 Cases D-1 and E-1. Comparison between relative (respect to dam heel) displacement at dam crest (point C)

Case E-2 allows estimating the seismic performance of the dam subjected to an intensifying acceleration time history. The evolution of tensile damage relevant to case E-2 is reported in Fig. 13 for three subsequent time steps (1, 1.5 and 2 s): at 1 s no damage occurs but in the following time steps the cracking events initiate at dam-foundation interface in the nearby of point A and at the downstream face, affecting the upper part of the dam immediately below the slope change. Assuming that the failure occurs when a single crack extends from the downstream to the upstream face, the time 2 s has been identified as the time in which the dam fails. In addition, one of the failure criteria proposed for concrete gravity dams in [11] has been considered: at 2.24 s the relative displacement at dam crest is equal to 0.122 m, so the drift ratio *DR* between maximum drift and dam height results equal to 0.001, defined as a threshold value in [11].

Figure 14 shows the evolution of damaged area in case E-2: it is understood that, even if the numerical analysis goes on, the structural response provided by the analysis has a physical meaning only up to the failure time or a little later.

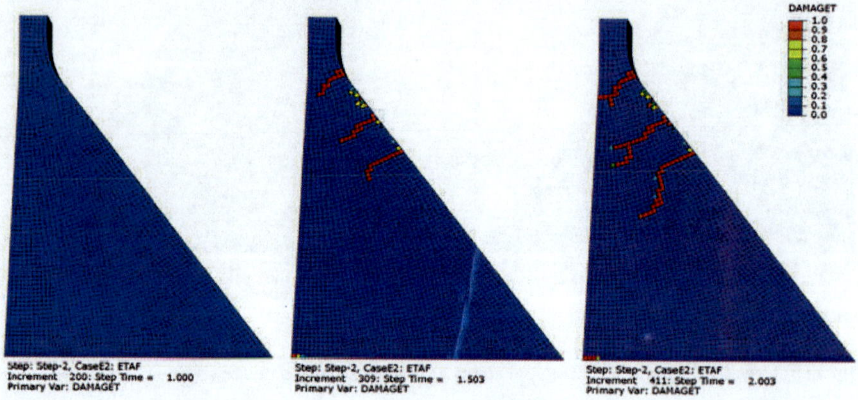

Fig. 13 Cases E-2. Damage at times 1 s, 1.5 s and 2 s (failure)

Fig. 14 Case E-2. Evolution of damaged area during ETAF

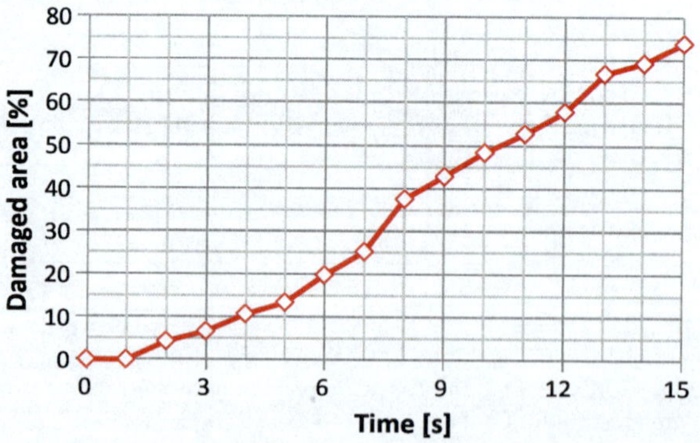

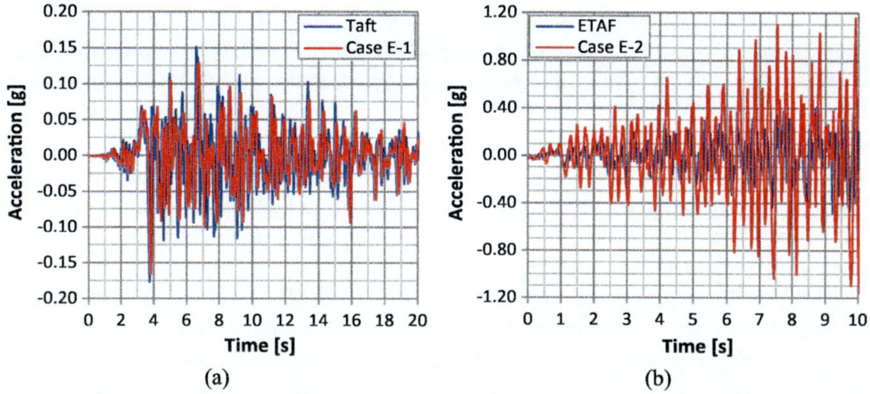

Fig. 15 Cases E-1 (**a**) and E-2 (**b**).Comparison between given seismic input and upstream acceleration at dam heel (point A)

Unlike the previous case E-1, in case E-2 important non-linear effects affect the response of the dam: as Fig. 15 clearly highlights, in case E-2 the seismic input is significantly amplified even at dam base (likely due to the different approach used to assign the seismic input) leading to a peak acceleration at 2 s about double than in case E-1.

5 Conclusions

Theme A of the 15th ICOLD Benchmark Workshop on Numerical Analysis of Dams, dealing with FEM modelling of the tallest monolith of Pine Flat Dam, has been approached by using the FEM code Abaqus, able to suitably model both the advanced dam-reservoir-foundation interaction and the non-linear behaviour of concrete. All the mandatory stages of the theme have been performed together the non-linear case with the Endurance Time Analysis (ETA) method and the cases with massless foundation.

Modal analysis represents an important step in the assessment of dams seismic behaviour since it gives some hints to better understand and verify the reliability of more complex investigations.

The results of linear dynamic analyses, carried out both with viscous-spring artificial boundaries model and massless approach for three different increasing water levels, point out that the overall structural response of the dam softens increasing the reservoir level. The models with the seismic wave propagation approach, as expected, provide a less demanding seismic response of the system with respect to that achieved with the massless approach (about the half). Nevertheless the periods of the resonant peaks resulting from the two approaches are rather comparable, as well as the dependence of the structural response on water level. For this reason massless approach, although conservative but at the same time thanks to its conservativeness, is still widely used, at least in Italy, in the seismic safety assessment of dam-reservoir-foundation system. However, in many cases it may be important to reduce this excessive conservativeness that may lead to the non-fulfilment of the performances required by the regulations: in this context, consolidating and increasing the confidence in the use of models that accurately simulate soil-structure interaction represent a crucial issue. It is worth noting that, to enhance the reliability and the confidence in using these advanced FEM models, the comparison with earthquake records on dams should be needed.

The results of non-linear dynamic analyses with Taft and ETAF time history signals show that in the first case no significant non-linear effect arises and the structural behaviour results pretty unchanged with respect to the correspondent linear case. The case carried out with the ETAF signal allows to observe the onset of increasing non-linear effects up to the failure (first the initiation of a crack along dam-foundation interface, near dam heel, then cracks developing from the downstream face, affecting the upper part of the monolith immediately below the slope change, propagating towards the upstream face). Further in-depth studies could be useful to better understand how to link ETA method and traditional time history analysis when massed foundation is assumed.

Acknowledgements This work has been financed by the Research Fund for the Italian Electrical System in compliance with the Decree of Minister of Economical Development April 16, 2018.

References

1. Salamon J, Hariri-Ardebili MA, Malm R, Wood C, Faggiani G (2019) Theme A formulation. Seismic analysis of Pine Flat Concrete Dam. 15th ICOLD International Benchmark Workshop on the Numerical Analysis of Dams, Milan, Italy
2. Lee J, Fenves GL (1998) A plastic-damage concrete model for earthquake analysis of dams. Earthq Eng Struct Dynam 27(9):937–956
3. Dassault Systèmes, Simulia (2017). Dassault Systèmes Documentation SIMULIA 2017. Abaqus
4. Zienkiewicz OC (1977) The finite element method, 3rd edn. McGraw-Hill
5. Zhang C, Pan J, Wang J (2009) Influence of seismic input mechanisms and radiation damping on arch dam response. Soil Dyn Earthq Eng 29:1282–1293
6. Chen DH, Du CB, Yuan JW, Hong YW (2012) An investigation into the Influence of Damping on the earthquake response analysis of a High Arch Dam. J Earthq Eng 16:329–349
7. Liu YS, Chen DH (2013) Earthquake response analysis of a Gravity Dam considering the radiation damping of infinite foundation. APCOM & ISCM, Singapore
8. Clough RW (1980) Non-linear mechanisms in the seismic response of arch dams. In: Proceedings of the International Research Conference on Earthquake Engineering. Skopje, Yugoslavia
9. Hilber HM, Hughes TJR, Taylor RL (1977) Improved numerical dissipation for time integration algorithms in structural dynamics. Earthq Eng Struct Dyn 5:283–292

10. Rea D, Liaw CY, Chopra AK (1972) Dynamic Properties of Pine Flat Dam. Report No. UCB/EERC-72/7
11. Hariri-Ardebili MA, Saouma V (2014) Quantitative failure metric for gravity dams. Earthq Eng Struct Dyn. Published online in Wiley Online Library (wileyonlinelibrary.com). https://doi.org/10.1002/eqe.2481

Seismic Analysis and Damage Evaluation of Pine Flat Concrete Dam

M. Farrokh

Abstract Accurate seismic analysis of concrete dams has a great impact not only on design of the new dams but also on the stability assessment of the existing ones. In this regard, one of the challenging issues is to consider the dynamic effects of radiation damping due to the mass of the foundation rock on the seismic responses. In this paper, an absorbing boundary condition, which is comprised of the viscous boundary traction and a free-field column, has been adopted in time domain. Based on the adopted methodology, a finite element model has been prepared for Pine Flat concrete dam using ABAQUS. In order to evaluate the damage level, the concrete damage plasticity model has been assumed for the concrete constitutive relation. The damage level in concrete has been evaluated under Taft earthquake. In addition, the response spectrum that Pine Flat dam can endure has been determined using an endurance time analysis.

Keywords Concrete dam · Seismic analysis · Concrete cracking · Radiation damping · Dam-reservoir-foundation interaction

1 Introduction

The main goal of this paper is to evaluate the seismic damage of Pine Flat concrete dam considering the effects of the foundation radiation damping. In this regard, damaged plasticity model has been adopted for the concrete. As the analyses are nonlinear, the frequency-domain framework is not applicable; hence, analyses have been performed in time-domain using ABAQUS software.

The challenging task in considering the effects of the foundation radiation is to adopt suitable absorbing boundary conditions. The absorbing boundary conditions (ABC) are needed when an infinite domain like the foundation of a dam is modeled using a finite portion of it. Their main role is to absorb the outgoing propagated waves without any reflection. Different kinds of these conditions have been developed so

M. Farrokh (✉)
K. N, Toosi University of Technology, Tehran, Iran
e-mail: farrokh@kntu.ac.ir

G. Bolzon et al. (eds.), *Numerical Analysis of Dams*, Lecture Notes in Civil Engineering 91, https://doi.org/10.1007/978-3-030-51085-5_4

far. There are two different kinds of ABCs: global and local conditions. Although global ABCs have high accuracy, it can be quite time consuming due to the coupling characteristics of spatial and time domains [1]. On the other hand, the local ABCs can easily be implemented even in commercial finite element software. The simplest ABC has been introduced by Lysmer and Kuhlemeyer [2] for scattering problems. Zienkiewicz and Bicanic [3] developed this simple ABC for the seismic cases where the source of the excitation is not in the truncated model. Recently, Lokke and Chopra [4] have assessed the performance of the ABC in seismic analysis of the dam-water-foundation system in comparison with the high performing perfectly matched layer (PML) [5] and they showed that it can be acceptable for practical implementation.

In this paper, seismic stability of the Pine Flat concrete gravity dam has been assessed using ABAQUS software. For modeling nonlinear behavior of concrete, the built-in damaged plasticity is adopted. In order to include the radiation damping in foundation, the built-in dashpot elements and user defined free-field column element [6, 7] are utilized. The effects of the radiation damping are evaluated under Taft record excitation and finally the seismic stability of the Pine Flat dam is assessed using an artificially generated an endurance time acceleration function record.

2 Absorbing Boundary Conditions

In the first attempt, Lysmer and Kuhlemeyer [2] have introduced viscous boundary tractions to absorb incident waves normal to the boundary. The normal f_n and tangential f_t tractions at the boundary are:

$$\begin{aligned} f_n &= -\rho c_p \frac{\partial u_n}{\partial t} \\ f_t &= -\rho c_s \frac{\partial u_t}{\partial t} \end{aligned} \tag{1}$$

where ρ is specific mass of the domain, u_n and u_t are displacement components in normal and tangential directions, respectively. P-wave velocity c_p and S-wave velocity c_s are given by:

$$\begin{aligned} c_p &= \sqrt{\frac{\lambda + 2\mu}{\rho}} \\ c_s &= \sqrt{\frac{\mu}{\rho}} \end{aligned} \tag{2}$$

where λ and μ are Lame's constants. This artificial boundary condition can be implemented by means of the dashpot elements that quite all computational software enjoys them. The boundary condition is very effective when the body waves approach the boundary at the normal incidence. For oblique angles of incidence, there is still energy absorption, but it is not perfect. As a result, it is recommended to leave relatively large margin between the boundary and the scattering origin of the waves.

Lysmer and Kuhlemeyer's viscous boundary was proposed to absorb outgoing waves. In fact, it can be used when the source of excitation is in the model. When excitation originates from outside of the model such as incoming seismic waves, theses artificial boundary conditions need to be modified so that it allows entering incoming waves and absorbs the outgoing waves and the waves due to reflection at the bottom of the model. Zienkiewicz and Bicanic [3] showed that the effects of the incoming waves can be consider by applying some tractions to the boundary which are calculated according to Eq. (1) in which the displacement components are only for incoming wave at the bottom of the model when the motion equations are in terms of the relative displacements. In framework in which the motion equations are in terms of the total displacements, the required tractions are twice of the required tractions in the previous case [8].

Tackling the issue of the absorbing boundary conditions in the lateral parts of the model is an arduous task. At each elevation of the lateral face of the foundation model the viscous boundary must absorb the movements in surplus to the free-field movements of that location. In addition, the required tractions for applying the effects of the incoming waves must be computed by the free-field movement of the considered location. Zienkiewicz and Bicanic [3] proposed a free-field column model in order to determine the free-field movements of the different elevation of foundation rock. In the free-field column model the weaves just vertically propagate. The required conditions at the base of the free-field column are same as conditions at the bottom of the main model. There is a one-way interaction between free-field column and the main model through:

$$\begin{aligned} f_n &= -\rho c_p\left(u_n^{\mathrm{m}} - u_n^{\mathrm{ff}}\right) + \sigma_n^{\mathrm{ff}} \\ f_t &= -\rho c_s\left(u_t^{\mathrm{m}} - u_t^{\mathrm{ff}}\right) + \tau_{nt}^{\mathrm{ff}} \end{aligned} \tag{3}$$

where superscript m and *ff* denote main model and free-field model, respectively. In addition, σ_n^{ff} and τ_{nt}^{ff} are respectively normal and shear stresses in the free-field column. Some software such as PLAXIS, 3DEC, and FLAC have built-in free-field column element for their standard users. ABAQUS software does not enjoy this type of element but with its user defined capability one can develop it [6, 7]. In another approach, one can decompose Eq. (3) into tractions that are calculated from the responses of a separate free-field model and dashpots that connect the foundation lateral walls to the ground. The free-field model is a single column of rock elements that has the same mesh density as the main foundation rock model adjacent to the lateral walls of the foundation with vertical and horizontal dashpots at the base. This free-field model has been utilized for simulation the free-field movements of the rock and its DOFs in the right and left faces must be tied [3]. For more information it refers to the research study conducted by Lokke and Chopra [4].

In the undisturbed free-field situation, the shear and normal stresses at the ground surface remain zeros during the movement. To satisfy these conditions, the upward and downward propagating motions at the free surface must be equal. Therefore, the outcrop motion, motion at free surface, is simply twice the upward propagating wave motion [9]. As the needed tractions at the bottom of the model is defined based on twice the upward propagating wave motion, the outcrop motion can be directly used in the calculation of the required tractions at the bottom of the both main and free-field column models. It should be noted that two series, of dashpots, horizontal and vertical dashpots, are needed at base of the models.

3 Concrete Damaged Plasticity

Nonlinear response of concrete can be represented by damage or plasticity theory. However, none of these theories alone is able to describe adequately this phenomenon. Coupling between damage and plasticity models is essential to capture the nonlinear behavior of concrete. ABAQUS enjoys a model named "concrete damaged plasticity" that its formulation is mainly based on the research study that conducted by Lee and Fenves [10]. The model is a continuum, plasticity-based, damage model for concrete. It assumes that the main two failure mechanisms are tensile cracking and compressive crushing of the concrete material. The model assumes that the uniaxial tensile and compressive response of concrete is characterized by damaged plasticity, as shown in Fig. 1. It is clear that the model follows a linear elastic relationship until the value of the tensile failure stress.

The concrete damaged plasticity model assumes that the reduction of the elastic modulus is given in terms of scalar degradation variables d_t and d_c for tension and compression, respectively. The tension stiffening will introduce unreasonable mesh sensitivity into the results while it is defined based on the post-failure stress-strain curve shown in Fig. 1a. However, it is generally accepted that Hillerborg's fracture energy proposal is adequate to allay the concern for many practical purposes. Hillerborg defines the energy required to open a unit area of crack, G_f, as a material parameter, using brittle fracture concepts. With this approach the concrete's brittle behavior is characterized by a stress-displacement response rather than a stress-strain

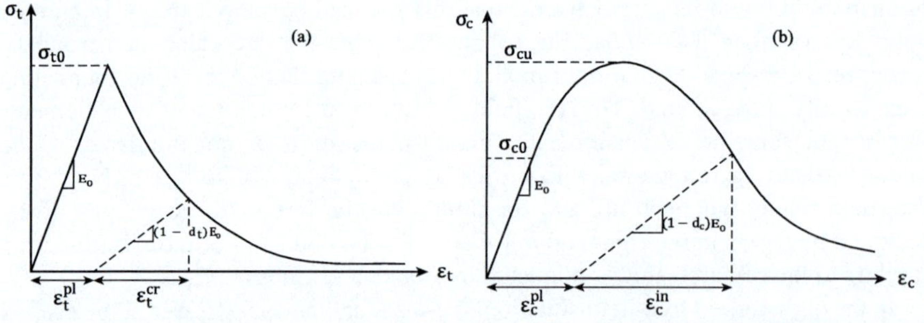

Fig. 1 Response of concrete to uniaxial loading in tension (**a**) and compression (**b**)

response. Under tension a concrete specimen will crack across some section. After it has been pulled apart sufficiently for most of the stress to be removed (so that the undamaged elastic strain is small), its length will be determined primarily by the opening at the crack. The opening does not depend on the specimen's length.

4 Pine Flat Concrete Dam

Pine Flat dam, located on King's River, east of Fresno, California, was constructed by the US Army Corps of Engineers in 1954. It consists of thirty-six 15.25 m-wide and one 12.2 m-wide monoliths. The length of the straight gravity dam is 561 m and the tallest non-overflow monolith is 122 m high. The specifications of the dam presented here have been adopted from Ref. [11].

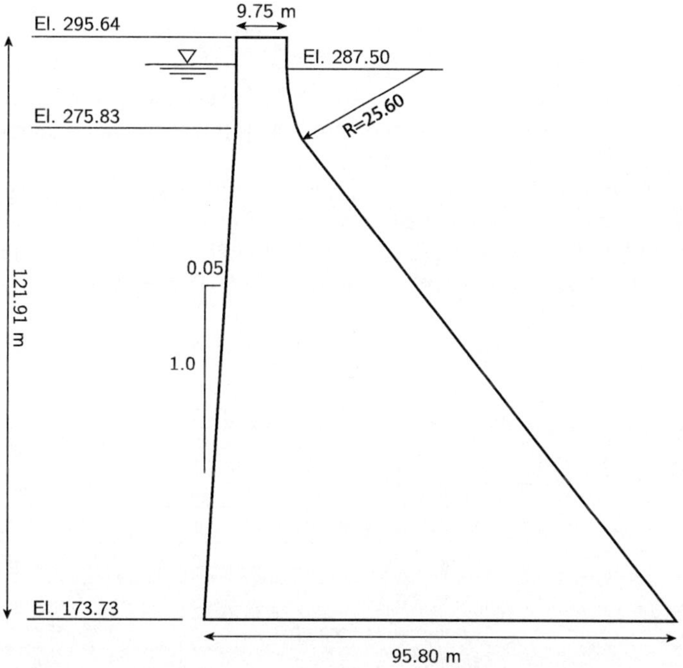

Fig. 2 Cross-section of Monolith 16 of Pine Flat concrete dam [11]

4.1 Geometry

The tallest non-overflow monolith section of the Pine flat dam shown in Fig. 2 has been chosen for modeling here.

4.2 Material Properties

Dam body. Concrete is assumed to be homogeneous and isotropic throughout the entire dam, with the properties listed in Table 1.

The nonlinear response of the concrete has been assumed to be as concrete damaged plasticity model of ABAQUS software. This model uses the uniaxial compressive and tensile behavior of concrete. In this paper, the following uniaxial compressive stress-strain relationship [12] has been adopted:

$$\sigma_c = \frac{E_0 \varepsilon_c}{1 + \left(\frac{E_0 \varepsilon_c}{\sigma_{cu}} - 2\right)\left(\frac{\varepsilon_c}{\varepsilon_{cu}}\right) + \left(\frac{\varepsilon_c}{\varepsilon_{cu}}\right)^2} \tag{4}$$

where σ_c and ε_c are, respectively, the compressive stress and strain; σ_{cu} and ε_{cu} are the experimentally determined ultimate stress and its corresponding strain, which are taken to be 28 MPa and 0.0025, respectively, according to Table 1.

As it is obvious in Fig. 1b, ABAQUS uses σ_{c0} = a limit for the linear behavior in compression that here is assumed to be half of the ultimate compressive stress σ_{cu}. It should be noted that compressive stress-strain curve after this limit must be defined in terms of the inelastic strain ε_c^{in} in ABAQUS. The inelastic strain is defined as follows:

$$\varepsilon_c^{\text{in}} = \varepsilon_c - \frac{\sigma_c}{E_0} \tag{5}$$

Table 1 Concrete properties

Parameter	Value
Modulus of elasticity	22.410 MPa
Density	2483 kg/m^3
Poisson ratio	0.2
Compressive strength	28 MPa
Tensile strength	2.0 MPa
Fracture energy	250 N/m
Compressive strain at peak load	0.0025
Tensile strain at peak load	0.00,012

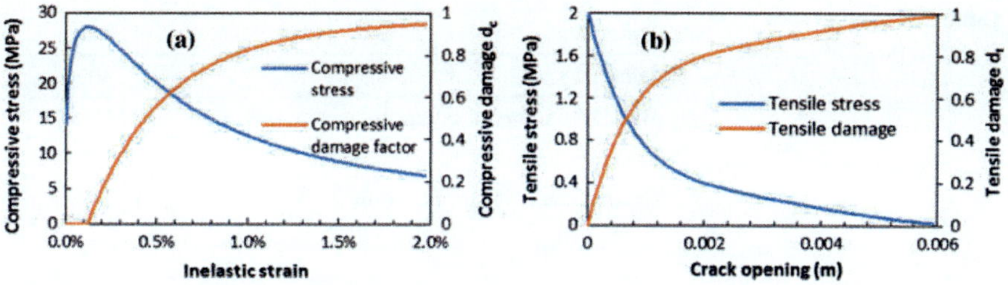

Fig. 3 Uniaxial concrete behavior in compression (**a**) and tension (**b**)

Figure 3a shows the assumed uniaxial compressive stress-strain in this paper. In damaged plasticity model the plastic strain $\varepsilon_c^{\rm pl}$ is defined in terms of compressive damage variable d_c as follows:

$$\varepsilon_c^{\rm pl} = \varepsilon_c - \frac{\sigma_c}{(1 - d_c)E_0} \tag{6}$$

Defining $b = \varepsilon_c^{\rm pl}/\varepsilon_c^{\rm in}$ and using Eqs. (5) and (6), one can get compressive damage variable as:

$$d_c = \frac{\varepsilon_c^{in}(1 - b)}{\varepsilon_c^{in}(1 - b) + \frac{\sigma_c}{E_0}} \tag{7}$$

It should be noted that damage is assumed to occur only after the concrete enters softening in both tension and compression [12]. The compressive damage variable assuming $b = 0.7$. [13] and using Eqs. (4) and (7) has been calculated and depicted in Fig. 3a.

In this paper, the stress-crack opening displacement relationship proposed by Hordijk [12, 13] is adopted:

$$\frac{\sigma_t}{f_t} = \left[1 + \left(c_1 \frac{w_t}{w_{cr}}\right)^3\right] e^{\left(-c_2 \frac{w_t}{w_{cr}}\right)} - \frac{w_t}{w_{cr}}(1 + c_1)^3 e^{-c_2} \tag{8}$$

$$w_{cr} = 5.14\frac{G_f}{f_t} \tag{9}$$

where w_t = crack opening displacement; w_{cr} = crack opening displacement at the complete loss of tensile stress; σ_t = tensile stress normal to the crack direction; f_t = concrete uniaxial tensile strength; and $c_1 = 3.0$ and $c_2 = 6.93$ are constants determined from tensile test of concrete. Using assumed concrete properties in Table 1, f_t = 2.0 MPa and $G_f = 250$ N/m, and Eqs. (8) and (9) tensile softening has been determined and depicted in Fig. 3b. In addition, the tensile damage variable d_t has been assumed to follow the following equation in this paper [14]:

Table 2 Parameters of damaged plasticity model

Parameter	Value
Dilation angle	30°
Biaxial to uniaxial compressive strength ratio	1.16
Eccentricity	0.1
Second stress invariant ration	2/3

Table 3 Foundation rock properties

Parameter	Value
Modulus of elasticity	22.410 MPa
Density	2483 kg/m^3
Poisson ratio	0.2
Shear wave velocity	1939 m/s
Compressional wave velocity	3167 m/s

$$d_t = 1 - \frac{\sigma_t}{f_t} \tag{10}$$

Equation (10) can be easily implemented and does not have intricacies of conversion of the inelastic strain into the crack opening displacement [15, 16]. The adopted tensile damage variable has been shown in Fig. 3b.

The damaged plasticity model can be used in multiaxial stress states. The corresponding parameters and their assumed values have been summarized in Table 2.

Foundation. The foundation material is assumed to be homogeneous, isotropic, and elastic with the properties listed in Table 3.

Water. Reservoir water is assumed to be a linear ideal inviscid and compressible fluid, moving with zero mean velocity. The hydrodynamic pressure in water is governed with acoustic wave pressure. Specific mass of water is 1000 kg/m^3 and compression waves move with the velocity of 1439 m/s through it.

4.3 Loads

The following loads have been used for various analysis cases.

Static loads. Static load includes weight of concrete dam and the hydrostatic reservoir pressure. Three reservoir water levels have been considered as follows:

- Winter reservoir water level (WRWL) at El. 268.21 m
- Summer reservoir water level (SRWL) at El. 278.57 m
- Normal reservoir level (NRWL) at El. 290.00 m

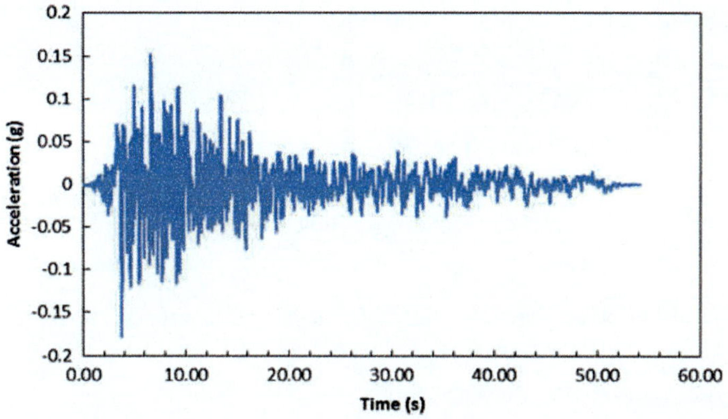

Fig. 4 Taft free-surface acceleration record—stream direction

Seismic loads. Two types of the seismic loads, a real earthquake and a synthetically generated record, have been adopted for performing seismic analysis.

Taft record. The M 7.3 Kern County, California, earthquake occurred on July 21, 1952. Accelerations from this earthquake were recorded at the Lincoln School (tunnel), in Taft, California, and have been used for this study. In particular, the baseline-corrected S69E component (Fig. 4), which has a peak horizontal acceleration of 0.18 g, has been adopted. The vertical record is not used. The recorded Taft acceleration time history is assumed to represent ground motions at the free surface, i.e., the motions that would be recorded at the top of the foundation in the absence of the dam and reservoir. An equivalent shear stress record should be provided, which also should be applied at the base of the foundation. The shear stress record has been computed by multiplying the free-surface Taft velocity time history by the foundation density and shear-wave velocity, as previously described.

Endurance Time Acceleration Function Record. Endurance Time Analysis (ETA) is a dynamic pushover procedure which estimates the seismic performance of the dam when subjected to a pre-designed intensifying excitation. The simulated acceleration functions are aimed to shake the dam from a low excitation level—with a response in the elastic range—to a medium excitation level—where the dam experiences some nonlinearity—and finally to a high excitation level, which causes the failure. All these response variations can be observed though a single time history analysis. The Endurance Time Acceleration Function (ETAF) is an artificially designed intensifying acceleration time history, where the response spectra of the ETAF linearly increases with time. Ideally, the profile of the acceleration time history and response spectrum increase linearly with time. Figure 5 shows a sample ETAF and its response spectra at three different times (i.e. 5, 10, and 15 s). As seen, the spectrum at $t \leq 5$ s is nearly one half of the one at $t \leq 10$ s and 1/3 of the one at $t \leq 15$ s. In this technique, the seismic performance is determined by the duration the structure can endure the dynamic input.

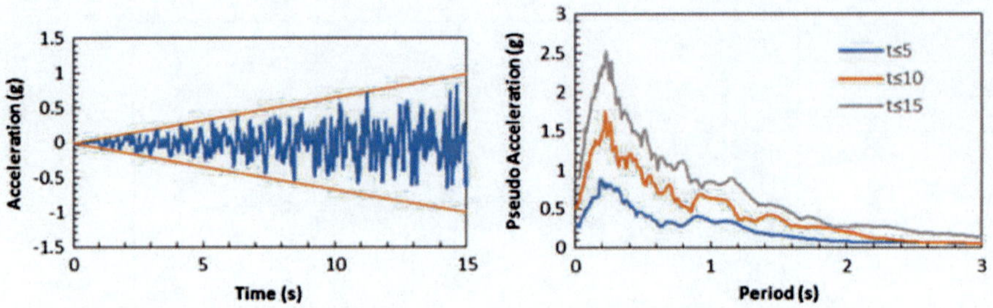

Fig. 5 Sample ETAF, its acceleration profile, and time-dependent response spectra

4.4 Damping

In this study, Rayleigh viscous damping (Rayleigh damping) has been used. Rayleigh damping matrix, by definition, is proportional to a linear combination of mass matrix (**M**) and stiffness matrix (**K**), through the constants of proportionality, denoted as α and β in Eq. (11):

$$\mathbf{C} = \alpha\mathbf{M} + \beta\mathbf{K} \tag{11}$$

Rayleigh coefficients α and β can be obtained based on the lower and upper bound frequencies of the dominant frequency interval. For the purpose this study, a frequency between 4 and 10 Hz should be considered, depending on the combination of dam, reservoir and foundation. For this frequency range, the average Rayleigh damping equates to approximately 2%. The constants of proportionality have been proposed as $\alpha = 0.75$ 1/sec and $\beta = 0.0005$ s.

5 Finite Element Model

5.1 Finite Element Mesh

A 2D finite element model for dam-foundation-reservoir system has been prepared using ABAQUS software. The dam body has been modeled according to the specifications of Fig. 2. The foundation has been modeled as a rectangle with a length of 700 and a depth of 122 m. The dam heel is located at a distance of 305 m from the upstream edge of the foundation. In addition, the reservoir has been model with a length of 305 m. It should be noted that the 4-node quadrilateral plane strain elements and the 4-node acoustic elements have been adopted for discretization of dam-foundation system and reservoir system, respectively. The maximum element dimension has been limited to a length of 1.5 m. The prepared finite element mesh has been depicted in Fig. 6.

Fig. 6 Finite element model of the dam-foundation reservoir system

5.2 Boundary Conditions

Different type of boundary conditions have been used for different analyses. For static analyses, the bottom of the foundation have been encastre while only horizontal components of the displacements have been fixed in the lateral wall of the foundation. The same boundary conditions have been used in the modal analysis for the foundation. For modal and seismic analyses, the dam-reservoir and the foundation-reservoir interaction have been activated by means of the acoustic FSI. In addition, the hydrodynamic pressure has been fixed to zero value in the free-surface of the reservoir. In the seismic analysis, a non-reflecting condition based on the Sommerfeld's condition has been applied to the truncated edge of the reservoir.

The static boundary conditions on the foundation cannot be used in the dynamic analyses. Therefore, they must be removed and to maintain the static equilibrium, the corresponding reaction forces must be applied to the foundation. In order to avoid the scattering wave reflection from the truncated edge of the foundation, bottom and the lateral walls, different treatments have been utilized. In the bottom of the foundation, the Lysmer and Kuhlemeyer's viscous boundary has been considered by means of the ABAQUS dashpot elements. The bottom of the foundation has been connected to the fixed ground by two kinds of dashpots, horizontal and vertical. The horizontal and vertical dashpot constants have been calculated according to $\rho C_s A$ and $\rho C_p A$, respectively, where A is the tributary area of the boundary node. The lateral walls of the foundation have been attached to the free-field column elements [3]. Unfortunately ABAQUS software does not enjoy the built-in free-field column element while software such as PLAXIS, 3DEC, and FLAC include it. Thanks to user subroutine to define an element (UEL) capability in ABAQUS, one can utilize this element by adopting the subroutine developed by Nielsen [6, 7]. In this study, the free-field column elements have been connected to the lateral walls of the foundation. It should be noted that the bottom of the free-field columns has been fixed to the ground by means of the dashpot elements same as the bottom of the foundation.

6 Analyses Results

6.1 *Modal Results*

The modal frequencies and damping ratios for dam-reservoir-foundation system for different reservoir water levels have been summarized in Table 4. While the reservoir water level increases, the corresponding modal frequency decreases as it is expected. In addition, the values of the modal damping ratios are in accordance with the target value of 2%.

6.2 *Seismic Analysis*

The seismic results presented in this section are for SRWL case. Figure 7 shows the crest acceleration under Taft excitation for two different linear analyses: (a) including

Table 4 Modal frequencies and damping ratios

Mode #	Frequency (Hz)			Damping ratio (%)		
	WRWL	SRWL	NRWL	WRWL	SRWL	NRWL
1	2.30	2.19	2.04	2.96	3.07	3.25
2	3.37	3.12	2.89	2.30	2.40	2.52
3	3.82	3.69	3.55	2.16	2.20	2.24
4	4.30	4.23	4.15	2.06	2.08	2.09
5	4.86	4.84	4.80	1.99	1.99	2.00
6	5.32	5.24	5.17	1.96	1.96	1.97
7	5.70	5.65	5.58	1.94	1.94	1.95
8	5.90	5.89	5.89	1.94	1.94	1.94
9	6.51	6.51	6.45	1.94	1.94	1.94
10	6.81	6.63	6.51	1.95	1.94	1.94
11	6.96	6.95	6.91	1.95	1.95	1.95
12	7.09	7.03	6.95	1.96	1.95	1.95
13	7.41	7.30	7.19	1.97	1.96	1.96
14	7.48	7.41	7.38	1.97	1.97	1.97
15	8.13	8.10	8.07	2.01	2.01	2.01
16	8.39	8.29	8.17	2.03	2.02	2.01
17	9.21	9.11	8.95	2.10	2.09	2.07
18	9.40	9.29	9.21	2.11	2.10	2.09
19	9.45	9.41	9.35	2.12	2.11	2.11
20	9.85	9.83	9.69	2.15	2.15	2.14

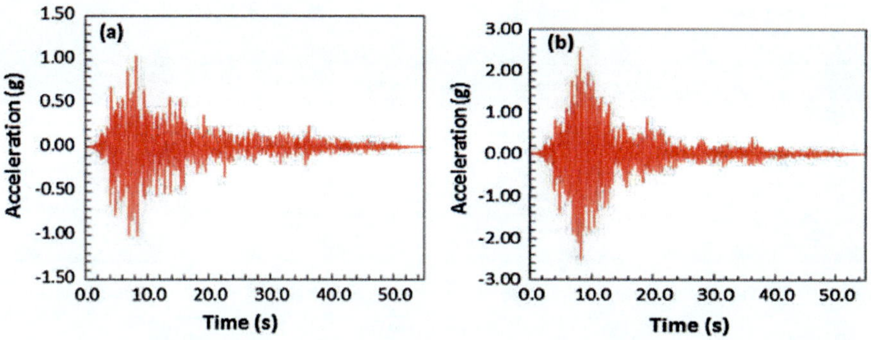

Fig. 7 Crest acceleration under Taft excitation: **a** including foundation mass, **b** excluding foundation mass

foundation mass and its attendant radiation damping and (b) conventional massless foundation. It is clear that the peak acceleration has been magnified by a factor of two while the effects of the foundation mass are neglected. In Fig. 8, the cracked regions of dam have been compared for the cases: massless foundation and with foundation mass effects under Taft excitation. As it is clear, the damage level has been reduced drastically when the effects of the foundation mass is included.

Endurance time analysis has been performed using ETAF excitation shown in Fig. 5. The sequential plot of the failure progress has been depicted in Fig. 9 under ETAF. In addition, the history of the ratio of the cracked length at dam base to the base length has been shown in Fig. 10. It shows that the dam base has been fully

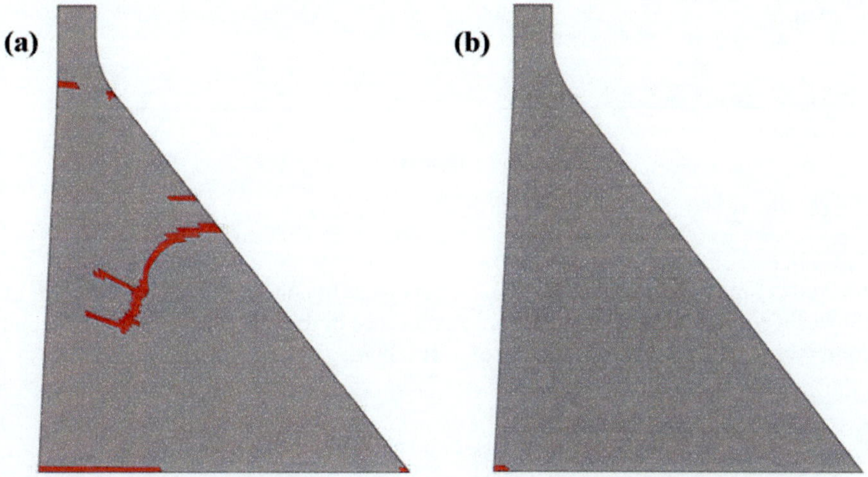

Fig. 8 Cracked region under Taft excitation: **a** massless foundation **b** including foundation mass ($d_t \geq 0.2$)

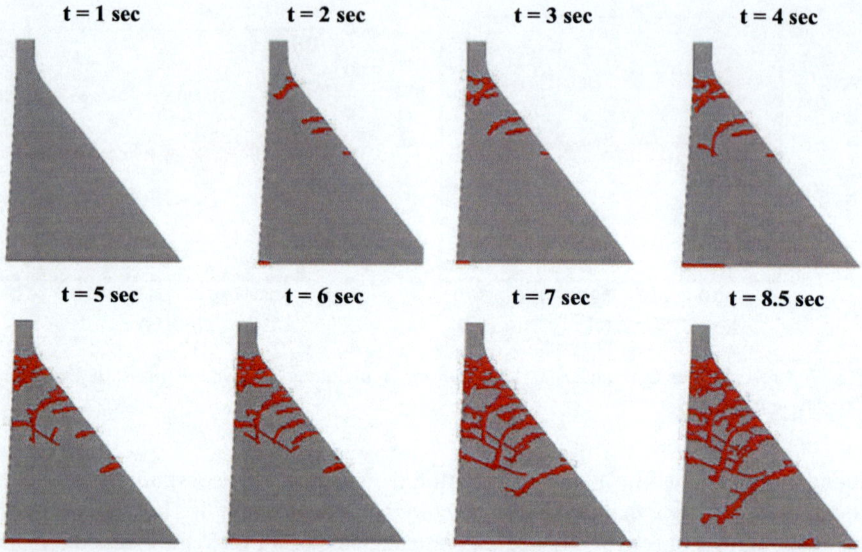

Fig. Sequential plots of failure progress in the dam under ETAF excitation ($\boldsymbol{d_t} \geq 0.2$)

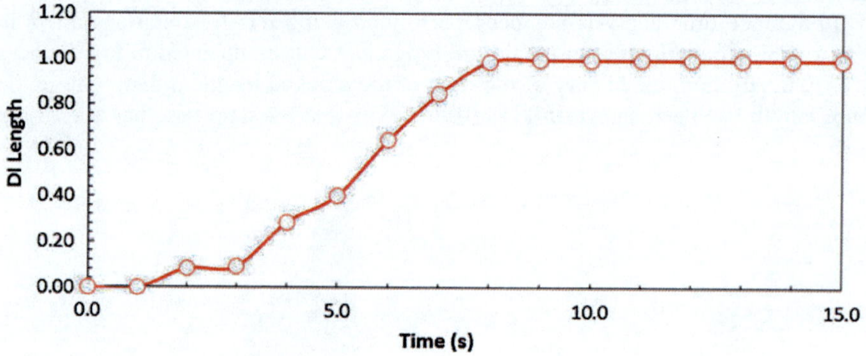

Fig. 10 History of damage index at base under ETAF excitation

cracked at time 8.5 s and after that the global instability has been occurred. Figure 11 shows the response spectrum of seismic endurance of the Pine Flat concrete dam and compares it with response spectrum of Taft record.

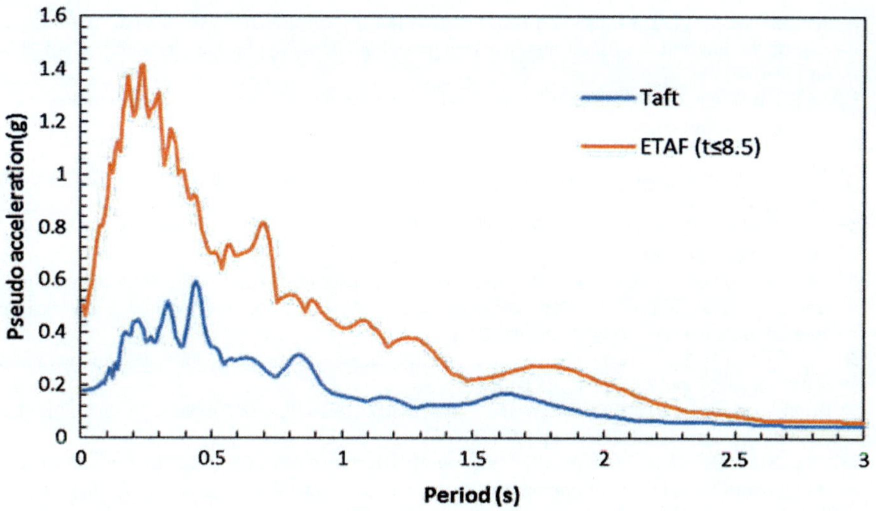

Fig. 11 Endurance response spectrum of the Pine Flat concrete dam and Taft response spectrum

7 Conclusions

In this paper, the results of seismic analyses of Pine Flat concrete dam have been presented for Taft record and ETAF excitation. The damaged-plasticity model has been adopted for concrete behavior. The results indicate that effect of the radiation damping has a great impact on the responses. The cracked regions in the dam body are substantially shrinking when the foundation radiation damping taken into account. In addition, the response spectrum that can be tolerated with the Pine Flat concrete dam has been determined using an endurance time analysis.

References

1. Liu J, Du Y, Du X, Wang Z, Wu J (2006) 3D viscous-spring artificial boundary in time domain. Earthq Eng Eng Vib 5(1):93–102
2. Lysmer J, Kuhlemeyer R (1969) Finite dynamic model for infinite media. J Eng Mech Div 95(EM4):859–876
3. Zienkiewicz OC, Bicanic N, Shen FQ (1989) Earthquake input definition and the trasmitting boundary conditions. In: Proceedings of the advances in computational nonlinear mechanics. Springer Vienna, pp 109–138
4. Løkke A, Chopra AK (2017) Direct finite element method for nonlinear analysis of semi-unbounded dam-water-foundation rock systems. Earthq Eng Struct Dyn 46(8):1267–1285
5. Basu U, Chopra AK (2004) Perfectly matched layers for transient elastodynamics of unbounded domains. Int J Numer Methods Eng 59(8):1039–1074
6. Nielsen AH (2006) Absorbing boundary conditions for seismic analysis in ABAQUS. In: ABAQUS Users' Conference, p 359–376

7. Nielsen AH (2014) Towards a complete framework for seismic analysis in Abaqus. In: Proceedings of the institution of civil engineers-engineering and computational mechanics, vol 167, no 1, pp 3–12
8. Joyner WB, Chen ATF (1975) Calculation of nonlinear ground response in earthquakes. Bull Seismol Soc Am 65(5):1315–1336
9. Mejia LH, Dawson EM (2006) Earthquake deconvolution for FLAC. In: Proceedings of the 4th international FLAC symposium, FLAC and numerical modeling in geomechanics. Madrid, Spain, p 211–219
10. Lee J, Fenves GL (1998) Plastic-damage model for cyclic loading of concrete structures. J Eng Mech 124(8):892–900
11. Salamon J, Hariri-Ardebili MA, Malm R, Wood C, Faggiani G (2009) Theme A formulation: seismic analysis of pine flat Concrete Dam. 15th International ICOLD benchmark workshop on numerical analysis of dams Milan, Italy
12. Tao Y, Chen JF (2014) Concrete damage plasticity model for modeling FRP-to-Concrete bond behavior. J Compos Constr 19(1):04014026
13. Britel V, Mark P (2006) Parameterised finite element modelling of Rc beam shear failure. In: ABAQUS Users' Conference, pp 95–108
14. Lima MM, Doh J-H, Hadi MNS, Miller D (2016) The effects of CFRP orientation on the strengthening of reinforced concrete structures. The Struct Des Tall and Spec Build 25(15):759–784
15. Alfarah B, López-Almansa F, Oller S (2017) New methodology for calculating damage variables evolution in Plastic Damage Model for RC structures. Eng Struct 132:70–86
16. Lopez-Almansa F, Alfarah B, Oller S (2015) Numerical simulation of RC frame testing with damaged plasticity model comparison with simplified models. Second European conference on earthquake engineering and seismology

Numerical Seismic Analysis of Pine Flat Dam Using Finite Difference Method

J.-R. Lherbier and F. Andrian

Abstract The seismic analysis of Pine Flat Dam, a 122 m high concrete gravity dam in California is the subject of the Theme A of the 15th ICOLD Benchmark Workshop on Numerical Analysis of Dams. For that purpose different steps of analyses have been carried out from eigenmodes determination and wave propagation calculations to linear and non-linear time history simulations. The use of free-field boundary conditions in order to model a semi-infinite medium confirms the results from previous works, such as those of the theme B of 14th ICOLD Benchmark Workshop [1]. The use of Endurance Time Acceleration Function (ETAF) record in order to assess the dynamic stability of the dam has been investigated. Finally, the hypotheses and the results of the non-linear analyses are analyzed and discussed.

Keywords Seismic analysis · Free-Field · Advanced Dam-Reservoir-Foundation interaction · Non-Linear concrete constitutive law · FLAC3D

1 Introduction

The aim of the 15th ICOLD International Benchmark Workshop on Numerical Analysis of Dams is to investigate the seismic modelling of a concrete gravity dam through extensive independent load cases in addition of those already studied during previous workshop [2]. For that purpose 3D-analyses under plane strain conditions are carried out for Pine Flat Dam, located in California and constructed by the US Army Corps of Engineers in 1954.

Eigenmodes determination, wave propagation tests in order to verify the good representation of semi-infinite domain, as well as linear and non-linear time history calculations are performed.

J.-R. Lherbier (✉) · F. Andrian
ARTELIA Eau & Environnement, Echirolles, France
e-mail: jean-remi.lherbier@arteliagroup.com

G. Bolzon et al. (eds.), *Numerical Analysis of Dams*, Lecture Notes in Civil Engineering 91, https://doi.org/10.1007/978-3-030-51085-5_5

2 General Data

2.1 Geometry

The tallest non-overflow dam monolith (no. 16) is considered in the calculation, similarly to the geometry used for previous workshop [2] (Fig. 1).

The foundation is 700 m long and 122 m deep except for case B as described hereafter.

Three water levels are considered depending on the load cases:

- Winter reservoir water level (WRWL) at El. 268.21 m;
- Summer reservoir water level (SRWL) at El. 278.57 m;
- Normal reservoir water level (NRWL) at El. 290.00 m.

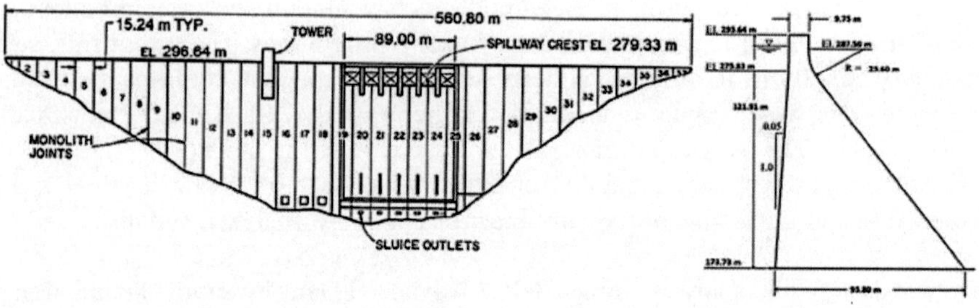

Fig. 1 Downstream view of Pine Flat Dam and cross-section of monolith no. 16

Table 1 Material mechanical properties [3]

Property	Unit	Concrete	Foundation rock	Water
Young's modulus	MPa	22,410	22,410	–
Density	kg/m3	2 483	2 483	1,000
Poisson's ratio	–	0.2	0.2	0.5
Shear wave velocity	m/s	1,939	1,939	–
Compressional wave velocity	m/s	3,167	3,167	1,439
Tensile strength	MPa	2.0	–	–
Cohesion	MPa	5.8	–	–
Internal friction angle	°	45	–	–
Fracture energy	N/m	250	–	–

2.2 Mechanical Properties

The material properties are summarized in Table 1. Since a modified Mohr Coulomb law is used for concrete in non-linear calculations, it is necessary to define friction angle as well as cohesion. Their magnitudes are defined in consistency with the specified compressive strength (28 MPa). On the other hand, the compressive strength of concrete is not explicitly used in the calculation. This is not an issue as the maximum calculated compression stress is usually lower than the compression strength for gravity dam.

2.3 Load Cases

The following analyses are performed:

- Case A: Natural frequencies determination and loading of the dam crest with an Eccentric-Mass Vibration Generator (EMVG) for several water levels;
- Case B: Analysis of the wave propagation through the foundation using impulsive loads;
- Case C: Analysis of the wave propagation considering dam+foundation or dam+reservoir+foundation systems using impulsive loads;
- Case D: Linear dynamic analysis during seismic event for various reservoir levels. The seismic signal used for this load case is corresponding to the recorded earthquake accelerations of July 21, 1952 in Taft, California with PGA = 0.18 g;
- Case E: Non-linear dynamic analysis for various levels. The seismic signal is either the same as for case D (Taft) or an Endurance Time Acceleration Function (ETAF) record (Fig. 2). The ETAF record is an artificial signal with linearly increasing acceleration time history;
- Case F: Linear dynamic analysis similar to case D but with massless foundation.

The hypotheses and the results of cases A, D, E and F are discussed in this paper.

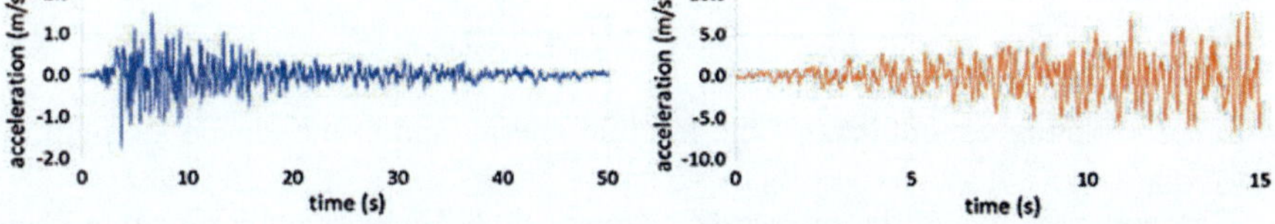

Fig. 2 Taft (left) and ETAF (right) time history accelerations

3 Numerical Model

3.1 Software

The numerical simulations are carried out using FLAC3D 6.0 developed by Itasca. It is a three-dimensional explicit finite difference program allowing non-linear dynamic analysis to be carried out. Plane strain conditions are applied on the lateral faces of the 3D model. The model width is 10 m.

3.2 Meshing Layout

The numerical models are made of hexa-dominant mixed discretization scheme elements [4]. The formulation of these elements allows the concrete, the bedrock and the reservoir to be modelled by means of the same element type. The reservoir directly interacts with the dam and the foundation.

Since the load cases require very different ranges of frequencies to be studied (up to 130 Hz for case B), the size of the element is adjusted from one load case to another. Indeed the size of each element requires refining for numerical accuracy of higher frequency wave transmission [5].

Except for dedicated analysis of very high frequencies, such as for case B, the mesh element sizes are limited so that the wave propagation equations are accurately solved up to 20 Hz. Higher frequencies are generally of lower importance in dynamic analysis of concrete gravity dams. This is confirmed by the modal analysis: the 6 first eigenfrequencies are far below 20 Hz.

Furthermore, a slightly more refined meshing of the dam for case E is considered in order to allow a better comparison with the others contributors with similar mesh elements size.

The mesh size has been calibrated according to the fracture energy approach, which is probably not necessary for this simulation since a perfectly fragile behavior law is considered by the authors. Equation (1) defines the critical length:

$$l_{cr} = \left(E.G_f\right)/f_t^2 = 1.40\,\text{m} \tag{1}$$

with E the Young's modulus, G_f the fracture energy and f_t the tensile strength of concrete.

Consequently five different meshes are created. Hereafter are illustrated the meshes used for cases A, D (linear analysis) and E (non-linear analysis) (Figs. 3 and 4).

The maximum element sizes are summarized in Table 2.

Fig. 3 Mesh for case A and D

Fig. 4 Mesh for case E

Table 2 Maximum mesh element size

Load case	Dam	Foundation	Reservoir	Number of elements
A	9.7	9.7	7.2	4 k
B	–	1.5	–	38 k/203 k
C	1.5	1.5	1.5	53 k
D	9.7	9.7	7.2	4 k
E	1.4	9.7	7.2	11 k
F	9.7	40.0	7.2	3 k

3.3 Constitutive Laws

The foundation rock, the reservoir water and the dam concrete, for linear analysis, elements are modeled by means of linear elastic constitutive laws.

For non-linear analysis (case E), the dam concrete is modeled by means of an elastic perfectly plastic constitutive law regarding shear behavior with a Mohr Coulomb failure criterion and an elastic perfectly fragile constitutive law regarding tensile behavior. This law also involves:

- The possibility of 3 mutually perpendicular cracks in each element. The first crack direction is perpendicular to the principal tensile stress, and then the second and the third crack directions are automatically imposed perpendicular to each other whenever the tensile strength is exceeded;
- A fragile behavior with regards to tension (perpendicular to crack) and shear strength (parallel to crack);
- The tracking of crack opening and closing with a partial shear strength recovery (cohesion is lost in the crack direction) when the crack closes. The tensile strength remains null once the crack has opened.

This simplified law differs from the usual concrete damage constitutive laws encountered in other commercial programs. Actually, the elastic energy is instantaneously released once the tensile strength is exceeded. Consequently, there is no residual tensile strength once a crack has opened leading to a more pessimistic constitutive law.

The ICOLD benchmark is an opportunity for the authors to compare the cracking layout and the overall results with that of the other contributors by means of a different constitutive law.

3.4 *Dynamic Features*

Dynamic viscous damping of 2% as per Rayleigh formulation was implemented in the dam and foundation according to the theme formulation when specified.

Bottom boundary conditions are modelled by means of viscous boundaries as developed by Lysmer and Kuhlemeyer [6]. It consists in series of independent dashpots in the normal and shear directions absorbing outwards waves to model semi-infinite medium hypothesis.

Lateral boundary conditions are modelled by means of free-field conditions (Fig. 5). It consists in a series of free field media connected to the boundaries of the system by means of dashpots. When the motion of the free field media differs from that of the system, the dashpots are activated so as to absorb the differential energy.

Free-field boundary conditions are systematically used by ARTELIA's teams since 2010. Without this feature the dynamical behavior of rigid dams may be overestimated by a factor up to 3 through numerical modelling [8–10].

The calculations are carried out by applying the seismic record as equivalent stresses at the foundation base. The time-step is around 8×10^{-5} and 6×10^{-6} s respectively for linear and non-linear calculations.

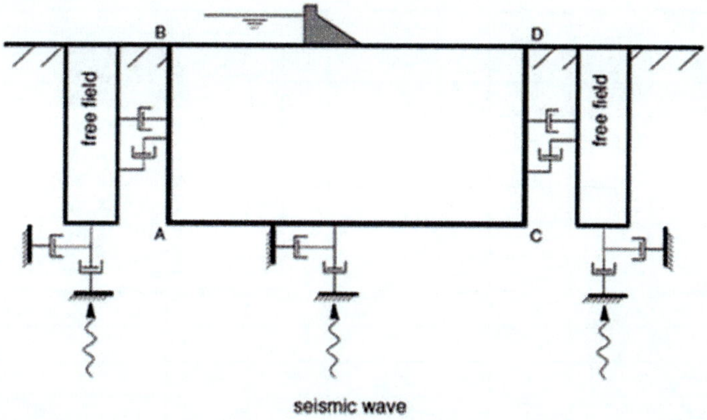

Fig. 5 Illustration of thc dynamic boundary conditions [7]

4 Results Analysis

4.1 Eigenmodes

Since FLAC3D cannot directly perform modal analyses, the eigenmodes of the system are identified by means of a different approach. A sinusoidal signal of increasing frequency is applied at the base of the foundation, whereas accelerations are monitored at several locations along the dam upstream face.

Then the amplifications of accelerations at different elevations (Fig. 6) are defined as the ratio between the spectrum of accelerations at a monitoring location and the spectrum of input accelerations. Local high values of amplification are generally corresponding to eigenfrequencies.

Even if the first eigenfrequency is clearly highlighted, here at 2.32 Hz, this method presents several drawbacks. Firstly the horizontal and vertical amplifications are independent data and it may be difficult to carry out a combined analysis of both amplifications. Then the eigenmodes identification is based on visual analysis: the identification of the first mode is quite obvious but the others may be tricky as the differences in their frequencies are low. Furthermore it cannot be used to highlight antisymmetrical eigenmodes which may be completely missed by the process.

Finally such analysis does not provide as much data as a modal analysis such as the modal masses and the mode shapes.

Nevertheless for the dynamic analysis of a concrete gravity dam, the used method is sufficient. Actually, the visual analysis of the frequencies of amplification allows the frequency ranges of the Rayleigh damping to be calibrated with acceptable accuracy.

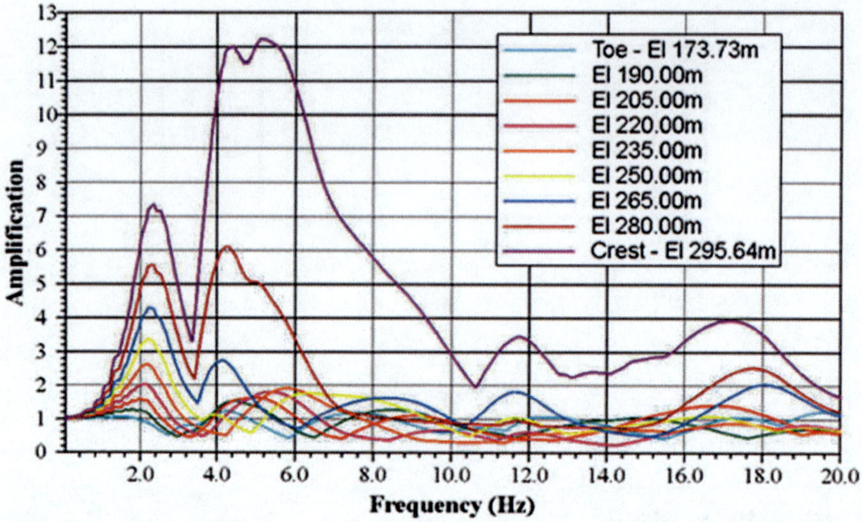

Fig. 6 Horizontal amplifications along the dam upstream face at WRWL

4.2 *Radiation Boundary Condition Effects*

The influence of radiation boundary conditions can be studied by a comparison of the results of cases D and F. A linear analysis under Taft seismic event is carried out for both cases. Only the foundation conditions vary: case D is considering massed foundation and radiation boundary conditions, whereas massless foundation with fixed boundary conditions is assumed in case F.

In order to carry out an analysis based on equivalent seismic input, a deconvolution of the signal applied at the foundation base for case D is considered.

Results in terms of acceleration and related spectra are illustrated on Figs. 7 and 8 considering WRWL.

As shown for frequencies starting from the first natural frequency (2.32 Hz) on Fig. 6, the dynamical behavior of the dam is overestimated by a factor 2 to 3 without consideration of the radiation boundary conditions which is in consistency with previous works [8, 9]. As it is now confirmed that the use of radiation boundary conditions is more realistic than the massless foundation approach [10], using the latter may lead to non-optimized design or unnecessary strengthening works. It shall however be noted that the results of previous works highlight the effect of the radiation boundary conditions may be less pronounced in case of thick dams in very narrow valley and/or if the foundation stiffness is rather high compared to that of the dam material.

Figure 9 illustrates the influence of the radiation boundary conditions for several reservoir levels in terms of the crest spectral acceleration ratios between case F and case D.

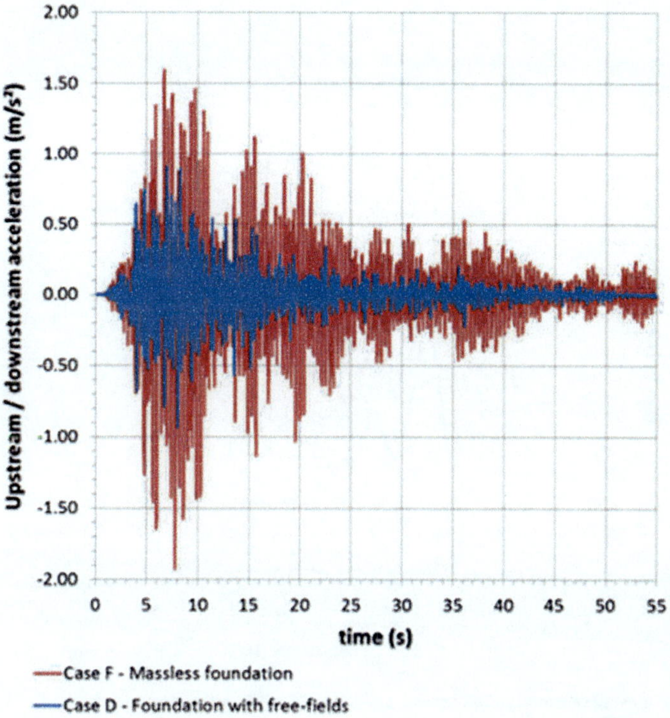

Fig. 7 Upstream/downstream accelerations at the dam crest with reservoir at WRWL

The effects of the radiation boundary conditions are slightly altered according to the reservoir levels. The reservoir level has indeed an effect on the frequency response of the mechanical system. For NRWL, the natural frequency of the reservoir (c/4H~3.09 Hz) may become close to one natural frequency of the dam/foundation system and generate resonance. Nevertheless the overestimation of the system dynamic response without consideration of the radiation boundary conditions remains of the same order of magnitude.

4.3 TAFT Time History Analysis

Both linear and non-linear analyses of the Pine flat dam under Taft signal have been carried out.

As illustrated in Fig. 10, the calculated crest spectral accelerations for both simulations are almost similar. This can be explained by the low range of yielding that the dam exhibits. Consequently the damping due to material yielding is very low. The Pine Flat seismic behavior is almost elastic except for a crack due to tensile

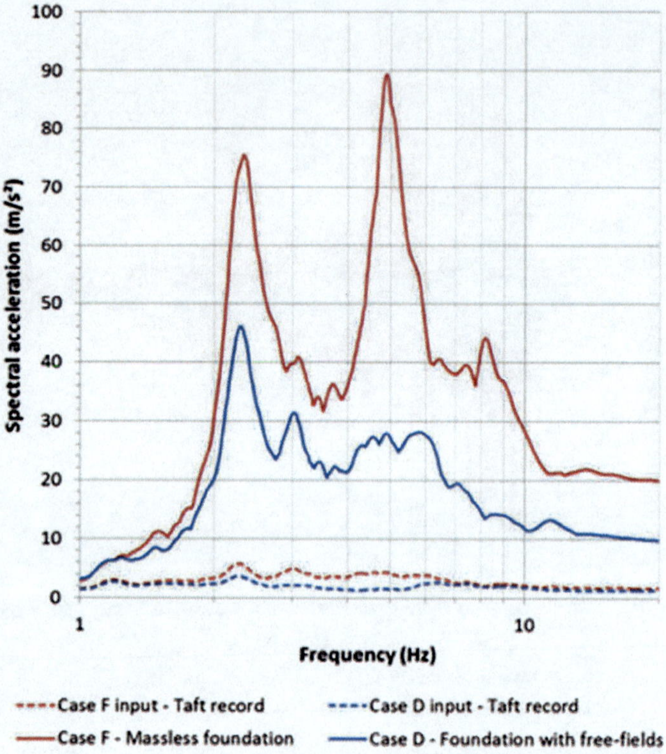

Fig. 8 Horizontal input signal and crest spectral acceleration with reservoir at WRWL

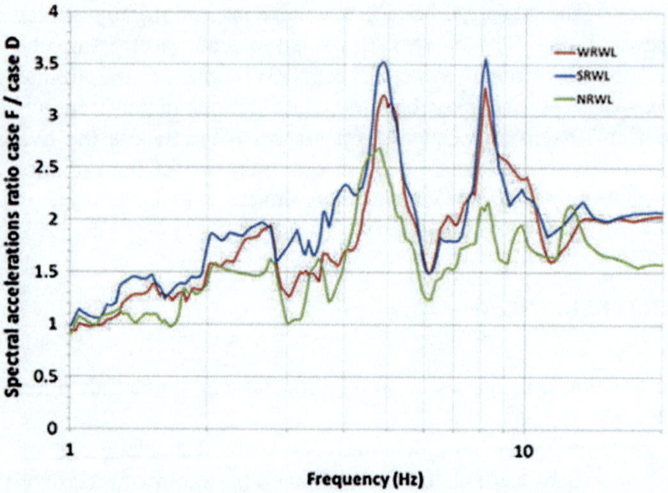

Fig. 9 Free-field boundary conditions effects at several reservoir levels

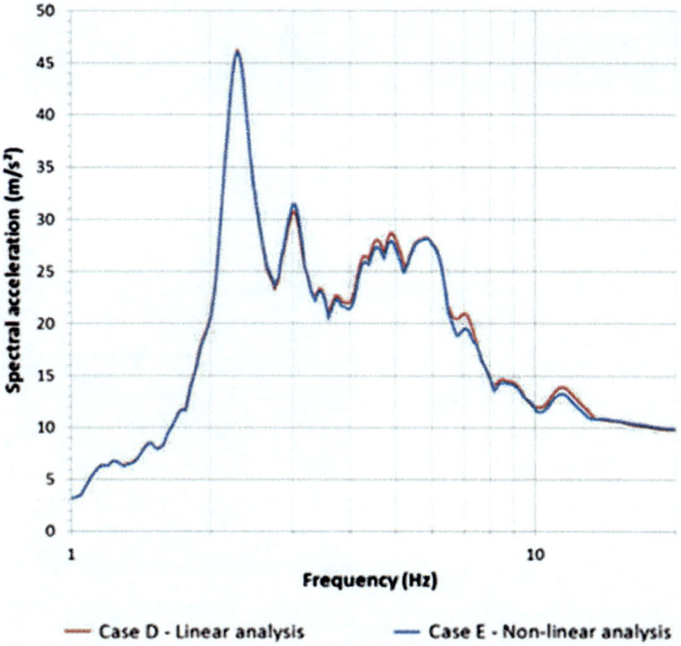

Fig. 10 Crest spectral upstream/downstream accelerations for Taft time history analysis

failure which occurs at the dam upstream toe with a depth of approximately 5.5 m as illustrated on Fig. 11 in red.

The generally admitted failure mechanisms of gravity dams are sliding along weak planes and overturning due to crack opening.

One of the main loads to be considered in stability analyses of this type of dam is uplift along the weak planes or more generally pore-pressures within the dam and the foundation materials.

In the case of Pine Flat dam, the pore pressure at the upstream toe is about 1.2 MPa for a NRWL load case. This is more than half of the value of the tensile strength of the dam concrete and leads to an apparent strength of 0.8 MPa. Moreover, it is possible that the tensile strength of the dam/foundation interface is of one or several orders of magnitude lower than that of the concrete.

Consequently, the authors believe that in order to assess the behavior of Pine Flat dam under seismic event in an appropriate manner, it is necessary to consider:

- The interface between the dam and the foundation with relevant strength parameters. This may lead to trigger irreversible displacements or at least some rearrangements at the dam/foundation interface;
- A realistic pore pressure/uplift distribution at least at the dam/foundation interface and preferably a distribution within the dam and the bedrock materials.

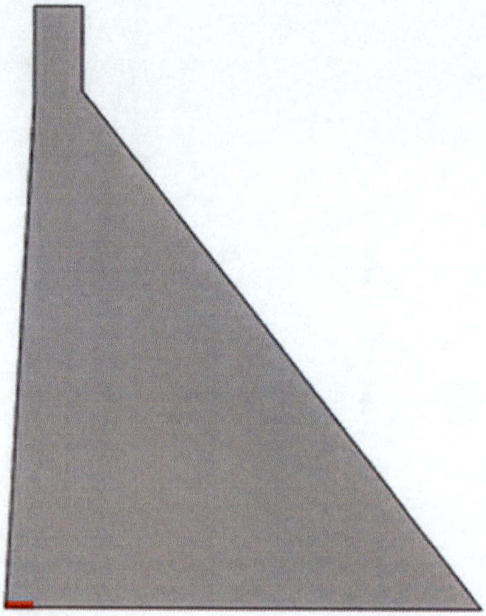

Fig. 11 Tensile failure along Pine Flat dam toe for Taft non-linear time history analysis

The consideration of the mentioned two aspects may significantly modify the currently calculated seismic response of the system and the conclusions of the analysis.

4.4 ETAF Time History Analysis

A non-linear analysis of the Pine flat dam seismic behavior under ETAF signal has been carried out. Thanks to explicit integration scheme used in FLAC3D, calculations can be performed even to model unstable mechanical states without numerical issues as long as the meshing layout is able to adequately reproduce the displacements. This probably allows a better understanding of the failure mechanisms of the dam. In this specific case, the numerical simulations have been carried out considering large strain hypothesis: the coordinates of the nodes are updated along with the simulation.

The cracking sequence as calculated is the following:

- First cracking at the dam upstream toe at around 2.5 s;
- First cracking at corner below the crest (El. 278.40 m) starting from the downstream face at around 4.0 s. This crack opening is probably amplified due to numerical effects as a simplified sharp corner is modeled between the downstream face and crest instead of the rounded shape. Still this crack is unavoidable for slightly higher PGAs;

- Downward cracking toward upstream propagating from the sharp corner below the crest between 5.0 and 6.0 s, first cracking at the dam downstream toe and superficial cracking of the downstream face;
- Diagonal cracking starting from the downstream face (El. 228.00 m) and vertical downward extension of the crest crack at around 7.5 s;
- Diagonal cracking penetrating further in the dam body and vertical upward extension of the crest cracking at around 9.0 s. At this stage, due to the extensive cracking of the crest the authors believe a collapse may be considered at around 9.5 s. Moreover, the ability of continuum mechanics approach to model such cracking is questionable. Consequently, the cracking pattern after 9.5 s is given for information only (Fig. 12).

As far as the authors understand, the ETAF record is an artificial signal with linearly increasing acceleration time history. The corresponding spectrum is also nearly linearly increasing with time. Consequently the time of each crack opening or even dam failure can be theoretically related to a specific spectrum and PGA value, and may be used to assess the seismic stability of the dam by means of a unique record.

Such approach may be relevant for linear analyses and may help engineers in saving time from running multiple earthquake records while providing with supplementary information about the "ultimate" PGA. On the other hand, its use for non-linear analyses may raise some questions.

As non-linear analyses are by essence dependent of the loading path, once a crack has opened, the corresponding softening is kept in memory. Moreover, with the linearly increasing overall intensity of the ETAF input, every crack is likely to extend further. Consequently, at higher PGAs a dam will most probably be tested under an already significantly decreased strength. More specifically, it will have gradually endured a series of lower PGAs. From a physical point of view, using ETAF record may be equivalent to a series of sub-calculations each one corresponding to a specific earthquake signal with increasing PGAs. Consequently, the approach tends to unrealistically exaggerate the non-linear effects that are triggered during calculation and may lead to alarmist conclusions about the dam safety.

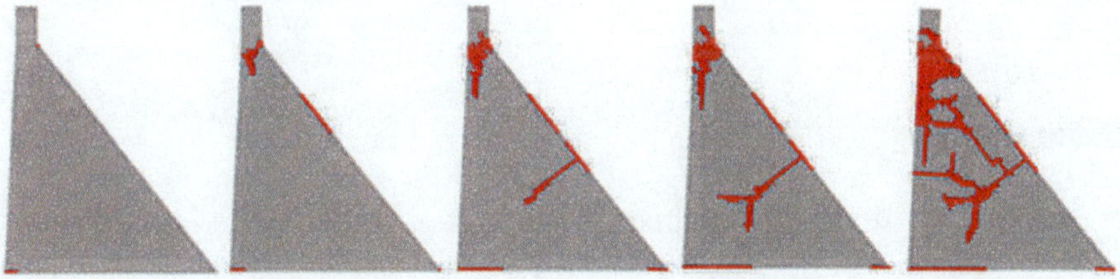

Fig. 12 Cracks through the Pine Flat dam body for ETAF non-linear time history analysis at (from left to right) 4, 6, 8, 10 and 14 s

4.5 *Miscellaneous Aspects*

Aside from the aspects discussed earlier in this paper, the authors suggest that the following points may also be considered:

- The vertical input which proved to play a major part in getting the best fitting between calculated and recorded crest accelerations for Tagokura dam in Japan [8];
- The increase in strength parameters of concrete by a factor of about 1.5 when subjected to dynamic loading.

5 Conclusions

In order to analyze the seismic behavior the Pine Flat Dam, a series of have been performed and accuracy of the model in simulating infinite half-space have been tested.

An alternative to the modal analysis has been carried out to assess the eigenmodes of the system. This approach is based on the analysis of the amplification of the input at different elevations of the dam by means of a sinusoidal record input of increasing frequencies. Despite several drawbacks and inaccuracies regarding high natural frequencies, this approach seems sufficient if the first eigenmode is of interest. Higher natural frequencies may be of lower importance in the seismic analysis of such rigid dam.

Then a comparison between linear time history simulation with and without radiation boundary conditions confirms the results of previous works [8–10]: without radiation boundary conditions, the dynamic response of the dam may be overestimated by a factor 2 to 3.

Finally, a non-linear time history analysis has been carried out considering an ETAF record, which is an artificial signal with linearly increasing accelerations. Although this approach appears to be an interesting alternative for verifying the dam safety state by means of linear analyses, its relevance for non-linear analyses may raise some questions. Actually, it is likely to lead to too much conservative results.

References

1. Andrian F, Agresti P, Mathieu G (2018) Theme B: static and seismic analysis of an arch gravity dam, Synthesis report. In: Proceedings of 14th ICOLD International Benchmark Workshop on Numerical Analysis of Dams, Stockholm, Swden, pp 240–286
2. Evaluation of Numerical Models and Input Parameters in the Analysis of Concrete Dams (2018) Report DSO-19-13, A Summary Report of the USSD Workshop, Miami, USA
3. Salamon J, Hariri-Ardebili, MA, Malm R, Wood C, Faggiani G (2019) Theme A formulation. Seismic analysis of Pine Flat concrete dam. 15th ICOLD International Benchmark Workshop on the Numerical Analysis of Dams, Milan, Italy

4. Marti J, Cundall P (1982) Mixed discretization procedure for accurate modelling of plastic collapse. Int J Num Anal Methods Geomech 6:129–139
5. Kuhlemeyer RL, Lysmer J (1973) Finite element method accuracy for wave propagation problems. ASCE J Soil Mech Found 99(SM5):421–427
6. Lysmer J, Kuhlemeyer RL (1973) Finite dynamic model for infinite media. ASCE J Eng 95(EM4):859–877
7. Itasca Consulting Group Inc. (2018)Manual FLAC3D–Fast Lagrangian Analysis of Continua in Three-Dimensions. Minneapolis
8. Chopra AK (2008) Earthquake analysis of arch dams: factors to be considered. 14th World conference on earthquake engineering. Beijing, China
9. Andrian F, Agresti P (2014) Analyses of seismic records on concrete dams in Japan, Tagokura dam, Back-analysis by means of a progressive modeling. JCOLD CFBR Collaboration, Chambéry, France
10. Andrian F, Agresti P (2018) Effects of radiative boundary conditions on seismic analysis of gravity dams. In: Fry J-J, Matsumoto N (eds.) Validation of dynamic analyses of dams and their equipment: Edited contributions to the international symposium on the qualification of dynamic analyses of dams and their equipments. CRC Press, Saint-Malo, France, 2016

Dynamic Analysis of Pine Flat Concrete Dam: Acoustic Fluid-Structural Interaction with ANSYS Workbench

T. Menouillard, A. D. Tzenkov, and M. V. Schwager

Abstract This paper presents a numerical study of the Pine Flat Dam, a 120 m high concrete gravity dam. The paper deals with the dynamic finite element analysis of the dam for different loads and water levels. The fluid structure interaction is considered through the explicit modeling of the reservoir as acoustic domain within ANSYS. The paper aims at showing the influence of the water level on the dynamic behavior of the dam. Furthermore, the differences in the dynamic behavior comparing the results of a 2D to the results of a 3D model are highlighted. The presented work is a contribution to the "15th ICOLD International Benchmark Workshop on Numerical Analysis of Dams" held in 2019.

Keywords Acoustic · ANSYS · Gravity dam · Dynamic · Concrete

1 Introduction

This study which is about modeling the dynamic behavior of a concrete dam with its reservoir is a contribution to the "15th ICOLD International Benchmark Workshop on Numerical Analysis of Dams". The presented case study deals with the Pine Flat Dam located in California U.S.A. It is a concrete gravity dam on the Kings River in the Central Valley of Fresno County, CA. The dam is 130 m high, 561 m long and consists of 36 15.25 m-wide and one 12.2 m-wide blocks. The reservoir has a capacity of 1.2 km^3 and covers 2'420 ha. Figure 1 shows a downstream view of the gravity dam. The spillway consists of six bays each controlled by a tainter gate,

T. Menouillard (✉)
CADFEM (Suisse) AG, 1020 Renens, Switzerland
e-mail: thomas.menouillard@cadfem.ch

A. D. Tzenkov
Stucky SA a Gruner Company, 1020 Renens, Switzerland

M. V. Schwager
SFOE, Swiss Federal Office of Energy, 3003 Bern, Switzerland

G. Bolzon et al. (eds.), *Numerical Analysis of Dams*, Lecture Notes in Civil Engineering 91, https://doi.org/10.1007/978-3-030-51085-5_6

Fig. 1 Downstream view of the Pine Flat Dam, California U.S.A. [ref. Wikipedia.org]

whose size is 11 m by 13 m. The construction of the Pine Flat Dam started in 1947 and was completed in 1954.

The investigated benchmark problem was selected by the ICOLD Committee A because of its relatively simple geometry and because it has been already extensively studied in the 1970's and 1980's [1–8] and during the 2018 USSD Workshop in Miami [9].

The formulation committee has defined a new problem formulation for the workshop in Milan based on the outcomes of the USSD workshop hold in Miami. The problem formulation aims among other goals at identifying key uncertainties potentially affecting numerical modeling results.

It includes several tasks: i.e. the determination of the eigenmodes and eigenfrequencies of the dam-reservoir system for different water levels (see Sect. 3.1), the calculation of the response caused by an eccentric mass vibration generator (see Sect. 3.2) and the evaluation of the dynamic response of the dam-reservoir system to an earthquake ground motion for different assumptions regarding the modeling of the foundation (see Sect. 3.3).

2 Modeling

The finite element model takes into account the geometry of the dam (see Sect. 2.1), the constitutive behavior of the structure and the foundation (see Sect. 2.2), the static (see Sect. 2.3) and inertia loads as well as the dynamic behavior of the interaction between the dam and the reservoir (see Sect. 2.4). The dynamic fluid-structure interaction is modeled by connecting the acoustic domain of the water to the structure. Thus, the dynamic pressure is calculated in the acoustic domain and transmitted to the structure.

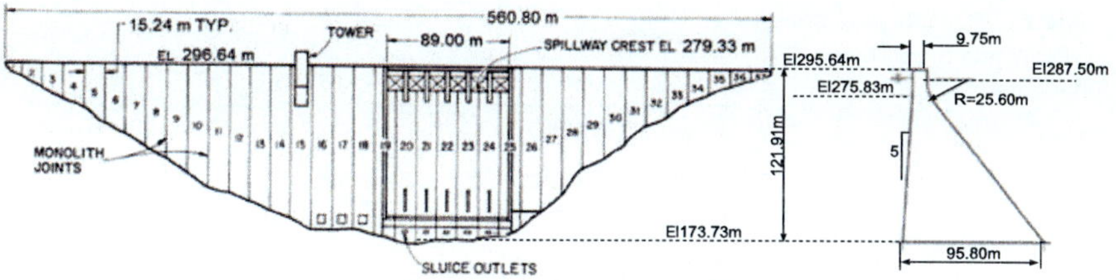

Fig. 2 Downstream view and cross-section of the Pine Flat Dam, California U.S.A

Fig. 3 View of the model including block no. 16, the foundation and the reservoir (at El. 268.21 m)

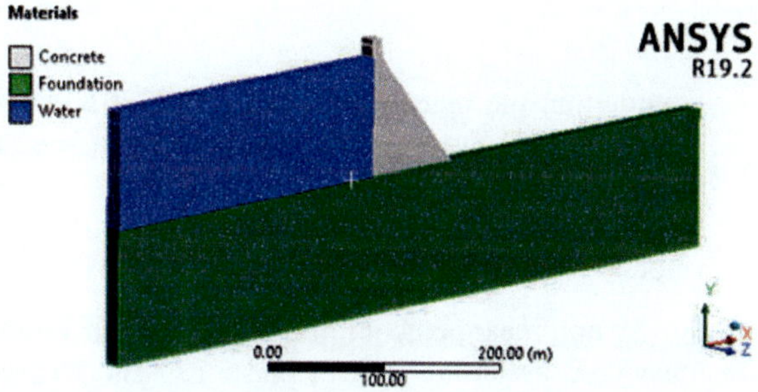

2.1 Geometry

The geometry of the dam is given in Fig. 2, which shows the downstream view of the dam and the cross-section of the tallest non-overflow block, i.e. block no. 16.

In order to simulate the acoustic fluid-structure interaction in Ansys, a 3D analysis is required. Therefore, the finite element model used consists of the entire 15.25 m-wide block no.16 and the corresponding strip of the foundation.

Three different water levels are considered:

- Winter reservoir water level (WRWL) at El. 268.21 m
- Summer reservoir water level (SRWL) at El. 278.57 m
- Normal reservoir level (NRWL) at El. 290.00 m.

Figure 3 shows the three different domains of the finite element model: i.e. block no. 16, the foundation and the reservoir.

2.2 Materials

The materials and their properties are assumed according to [2, 3] and given in Table 1.

- Concrete: the material is assumed to be homogeneous, isotropic and elastic.

Table 1 Materials properties for the concrete and the foundation

Characteristics	Concrete	Foundation
Density	2483 kg/m^3	2483 kg/m^3
Young's Modulus	22 410 MPa	22 410 MPa
Poisson's ratio	0.2	0.2
Compressive Strength	28 MPa	–
Tensile Strength	2 MPa	–
Fracture Energy	250 N/m	–
Compressive strain at peak load	0.0025	–
Tensile strain at peak load	0.00012	–

- Foundation: the material is assumed to be homogeneous, isotropic and elastic. According to [10], its properties are assumed to be identical to those of the concrete material.
- Water: the density of the water is taken to be 1'000 kg/m^3. The compression wave velocity is 1439 m/s.

For the nonlinear behavior of the concrete, the Drucker-Prager constitutive model is considered assuming perfect plasticity. The Drucker-Prager failure envelope is derived from the strength parameters given in Table 1.

2.3 *Static Loads*

For the static loads, the self-weight of the dam and the water is considered. The self-weight of the foundation material is not considered as a static load. The standard acceleration due to gravity g is considered as 9.81 m/s^2.

2.4 *Dynamic Loads*

To account for dynamic effects, different loads are applied within the analysis:

Excitation load. Excitation loads are introduced to model either the eccentric mass vibration generator or the earthquake ground motion.

Damping. Viscous Rayleigh damping has been considered. Hence, the damping matrix C is given by linear combination of the mass matrix (M) and the stiffness matrix (K): $C = \alpha M + \beta K$. The constants of proportionality considered are $\alpha = 0.75$ and $\beta = 0.0005$, corresponding to a damping ratio of around 2% within the range of relevant frequencies.

Hydrodynamic load. In addition to the hydrostatic pressures, hydrodynamic water pressures are considered because the presence of the reservoir water is changing the inertia of the dam structure and thus its dynamic behavior. The simplest way to account for hydrodynamic water pressures would be to use added masses on the wet surface of the structure according to Westergaard [11]. However, a more complex approach is chosen since the present contribution aims to explicitly model the fluid. Therefore, the fluid is considered by an acoustic domain being linked to the dam through the fluid-structure interaction.

2.5 Boundary Conditions

The bottom of the foundation is kinematically constrained. For the seismic analysis, velocities are imposed.

2.6 Equilibrium

The local equilibrium equation considers the body forces due to self-weight and inertia:

$$\mathrm{div}\sigma + \rho g = \rho \frac{\partial^2 u(x,t)}{\partial t^2} = \rho \ddot{u} \tag{1}$$

where u(x,t) is the displacement at position x and time t, ρ is the density and g is the standard acceleration due to gravity.

The space discretization equation for the fluid-structure interaction is written as follows:

$$\begin{bmatrix} M_S & 0 \\ \rho_0[R]^T & M_F \end{bmatrix} \begin{Bmatrix} \{\ddot{u}\} \\ \{\ddot{p}\} \end{Bmatrix} + \begin{bmatrix} C_S & 0 \\ 0 & C_F \end{bmatrix} \begin{Bmatrix} \{\dot{u}\} \\ \{\dot{p}\} \end{Bmatrix} + \begin{bmatrix} K_S & -[R] \\ 0 & K_F \end{bmatrix} \begin{Bmatrix} \{u\} \\ \{p\} \end{Bmatrix} = \begin{Bmatrix} \{f_S\} \\ \{f_F\} \end{Bmatrix} \tag{2}$$

where $[M_S]$, $[C_S]$, $[K_S]$ and $\{f_S\}$ denote the mass, damping and stiffness matrices and the external force vector in the structure respectively, and similarly $[M_F]$, $[C_F]$, $[K_F]$ and $\{f_F\}$ denote the mass, damping and stiffness matrices and the external force vector in the fluid. [R] is the coupled matrix and represents the coupling conditions on the interface between the acoustic domain and the structure.

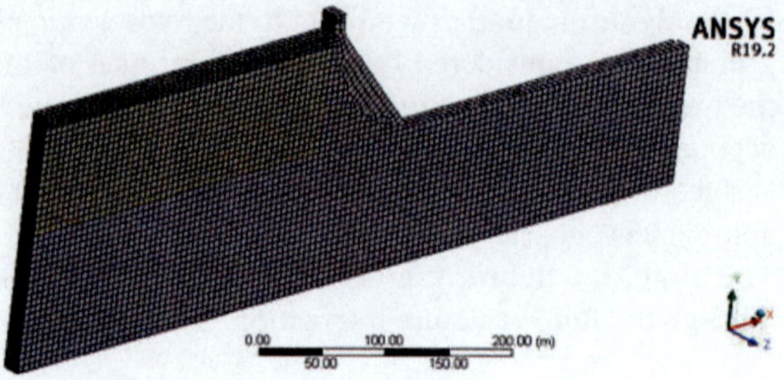

Fig. 4 Mesh of block no. 16, the foundation and the reservoir (WRWL)

2.7 *Space Integration*

The space integration is performed by means of the Finite Element Method [9, 12, 13]. Linear elements are used both for the structure and the acoustic domain. Linear shape functions are used for discretization of the displacement, velocity and acceleration fields in the structural domain as well as for the discretization of the pressure in the acoustic domain, i.e. the fluid. The meshes of the dam and the foundation share the same edges and thus their nodes. Figure 4 shows the mesh used for the linear analysis (15'236 nodes for SRWL).

2.8 *Time Integration*

The Newmark time integration algorithm allows the displacement and velocity to be updated during the transient analysis as follows:

$$\dot{u}_{n+1} = \dot{u}_n + ((1-\delta)\ddot{u}_n + \delta\ddot{u}_{n+1})\Delta t \tag{3}$$

$$u_{n+1} = u_n + \dot{u}_n\Delta t + \left(\left(\frac{1}{2}-\alpha\right)\ddot{u}_n + \alpha\ddot{u}_{n+1}\right)\Delta t^2 \tag{4}$$

The chosen parameters α and δ are 0.5 and 0.25 respectively. For the transient analysis, Eqs. (3–4) are solved.

3 Results

3.1 *Eigenvalue Analysis*

The eigenvalue analysis is performed for the two following water levels:

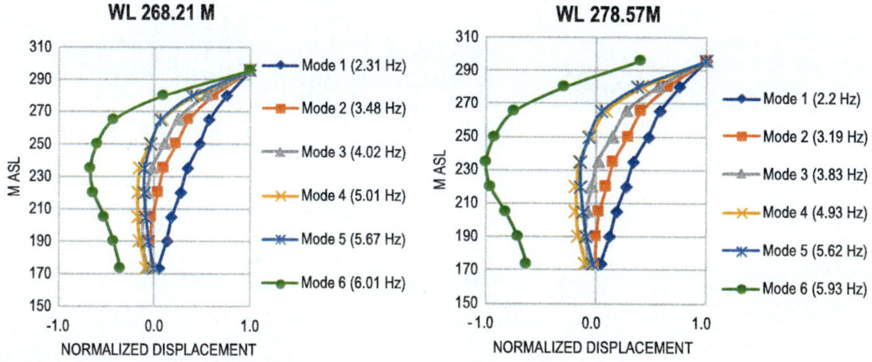

Fig. 5 Deformed shapes of the first six modes (WRWL and SRWL)

- Winter reservoir water level (WRWL) at El. 268.21 m
- Summer reservoir water level (SRWL) at El. 278.57 m.

Figure 5 shows the results in terms of the normalized deformed shape and the corresponding eigenfrequencies for the first six modes. For both water levels, the eigenmodes seem to be rather similar in terms of their shape. Furthermore, the eigenfrequencies are lower for the higher water level. The acoustic pressure in the water is mainly influencing the first mode.

3.2 *Eccentric-Mass Vibration Generator (EMVG)*

In winter 1971, block no. 16 of the Pine Flat Dam has been loaded at crest level by an eccentric-mass vibration generator. At the time, the reservoir water level was at the winter reservoir water level. The sine wave applied had an amplitude of 35.4 kN and a frequency of 3.47 Hz. To reduce artifacts and better emulate the mass vibration test results, the sine wave has been windowed with a 25% Hann taper (Fig. 6) rather than a boxcar.

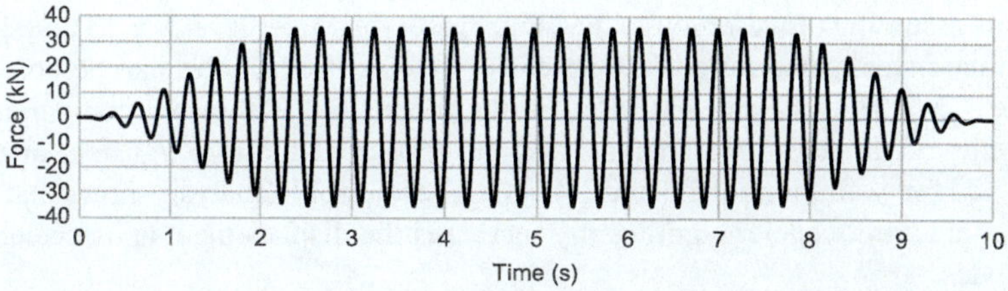

Fig. 6 Force time history generated by the eccentric-mass vibration generator (EMVG)

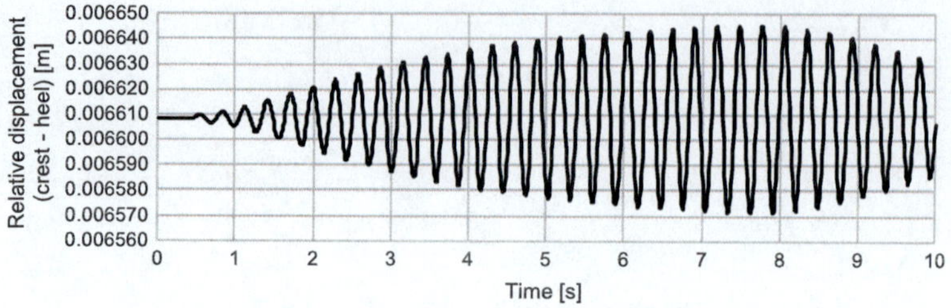

Fig. 7 Results of the transient analysis (WRWL): displacement at crest level

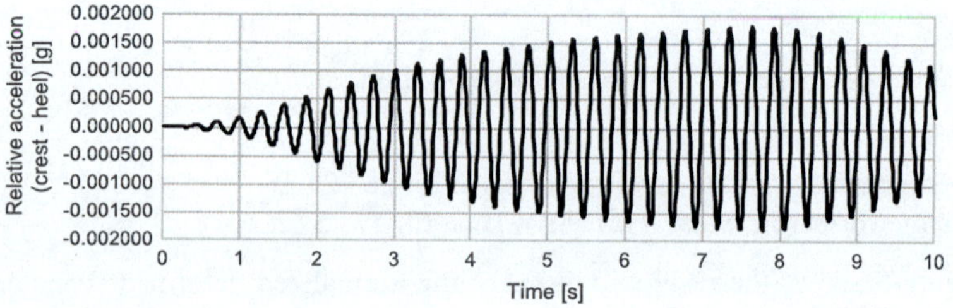

Fig. 8 Results of the transient analysis (WRWL): acceleration at crest level

Figures 7 and 8 show the results of the corresponding transient analysis in terms of displacements and accelerations at crest level respectively. Although a static force of 35.4 kN would create a downstream displacement of the crest level of 0.013 mm only, the displacement amplitude reaches 0.034 mm after 8 s (see Fig. 7) when applying the same load dynamically oscillating at 3.47 Hz. The displacement amplitude is found to increase with time because the excitation frequency is close to the second eigenfrequency (i.e. the frequency of a horizontal mode of motion). However, both the acceleration and the displacement amplitude tend to stabilize because of the damping.

Similarly, another numerical analysis has been performed considering an excitation of 35.4 kN at 2.3 Hz. It shows a displacement amplitude of more than 0.085 mm, which is about eight times larger than the static one.

Figure 9 shows the results of the harmonic analysis within the frequency range of 1 to 5 Hz assuming the reservoir at the winter reservoir water level (i.e. El 268.21 m). Within the investigated range of frequency, the eigenfrequencies can be observed. They are at 2.31, 3.47 and 4.02 Hz. Furthermore, Fig. 9 shows a comparison of the results of the harmonic analysis and the results of the two transient analyses performed for excitations at 2.3 and 3.47 Hz. The harmonic analysis shows that excitation at eigenfrequencies significantly increases the displacement and acceleration amplitudes.

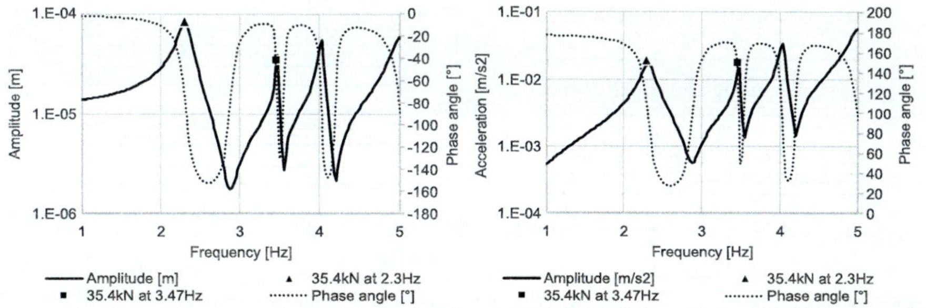

Fig. 9 Results of the harmonic analysis (WRWL): displacement and acceleration frequency responses

3.3 Earthquake

For the seismic analysis, the velocity time series is obtained from the acceleration time series shown in Fig. 10 by means of time integration. The velocity time series is applied at the base of the foundation. Only horizontal loading (acceleration or velocity) is applied, no vertical load is considered. For the reservoir level being at the winter water level, the maximum acceleration at the crown reaches 25.6 m/s^2 (see Fig. 11) whereas the maximum acceleration at the base of the foundation is equal to 1.3 m/s^2.

The hydrodynamic pressure at the upstream heel of the dam is computed within the acoustic domain and shown in Fig. 12. The maximum hydrodynamic pressure is about 300 kPa. It is compared to the hydrostatic pressure of 880 kPa at the same location. Thus, the water pressure at the heel is found to increase by around 34% for the earthquake simulated.

Assuming Drucker-Prager failure envelope allows for a preliminary assessment of the zones of the dam being potentially prone to damage. Figure 13 shows the obtained distribution of plastic strains after the earthquake. Damage is supposed to localize at the heel of the dam as well as along the downstream face below the changing point in slope. The maximum plastic strains are obtained to be 0.08%.

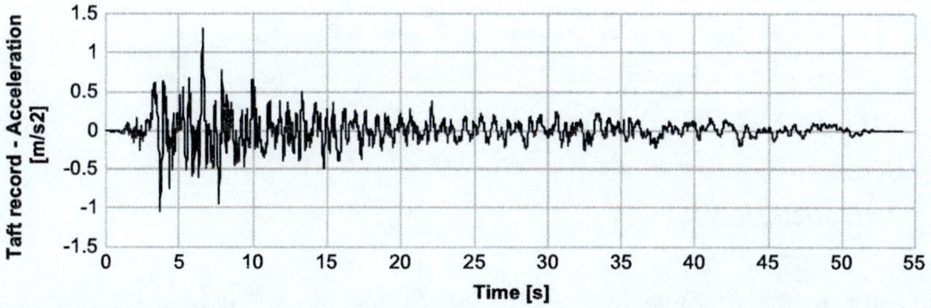

Fig. 10 Horizontal acceleration used to compute the input (velocity) at the base of the foundation

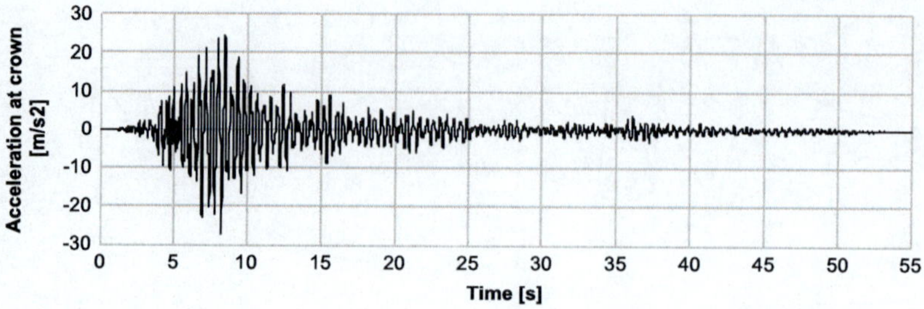

Fig. 11 Horizontal acceleration at the crown (for WRWL)

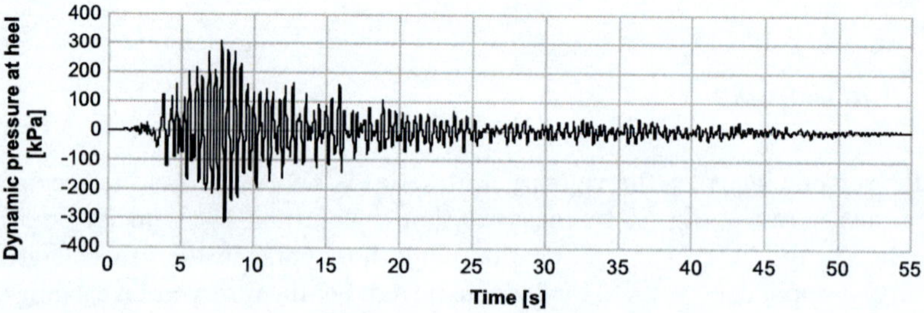

Fig. 12 Hydrodynamic pressure at the dam heel (for WRWL)

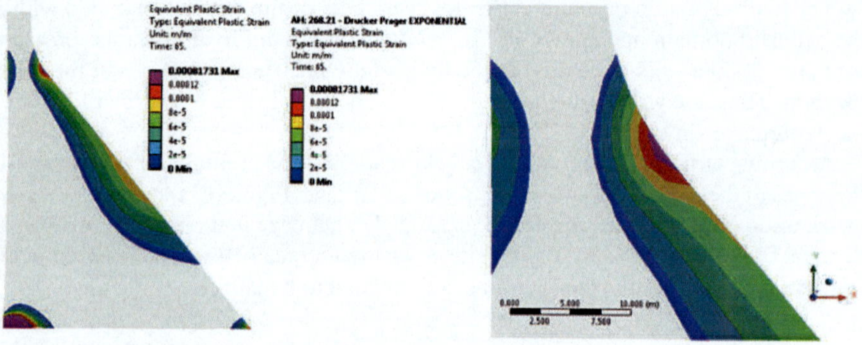

Fig. 13 Plastic strain in the dam after the earthquake (for WRWL)

4 Extended 3D Model for the Reservoir

4.1 Introduction

In addition to the problem formulation of the benchmark workshop, this section aims at showing the results of an acoustic analysis accounting for the whole Pine

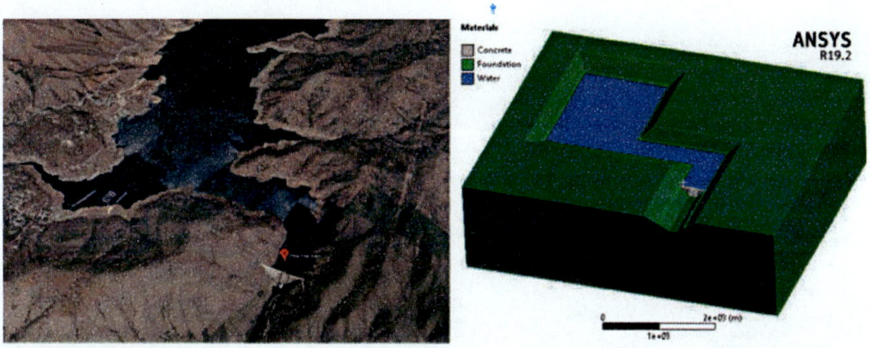

Fig. 14 Geometry of the extended 3D model

Flat reservoir and its dam. Therefore, the reservoir and the entire dam are modeled in three dimensions. For the reservoir, a simplified geometry is considered which is assumed to be appropriate. Figure 14 shows an aerial view (Google Earth) of the reservoir and the dam as well as a similar view of the extended 3D model.

4.2 *Eigenvalue Analysis*

In order to investigate the 3D-effect of the reservoir and the valley on the dynamic behavior of the dam, the eigenvalue analysis is performed again for the extended 3D model and compared to the previous results obtained for the quasi 2D model (one block). Figure 15 shows the first three eigenmodes of the dam for both WRWL and SRWL. Compared to the quasi 2D analysis, some typical 3D modes are found. Their frequencies are in the range of 3–5 Hz.

Figure 16 shows the distribution of the hydrodynamic water pressure along the boundaries of the entire reservoir for the second mode at 3.67 Hz and WRWL.

Table 2 gives a comparison of the eigenfrequencies provided by [10, 14] to the ones calculated in the presented work. For an empty reservoir, the fundamental frequencies of the quasi 2D and the 3D model are obtained to be 2.5 Hz and 3.0 Hz respectively. For a full reservoir, i.e. the water level at 290 m asl, the fundamental

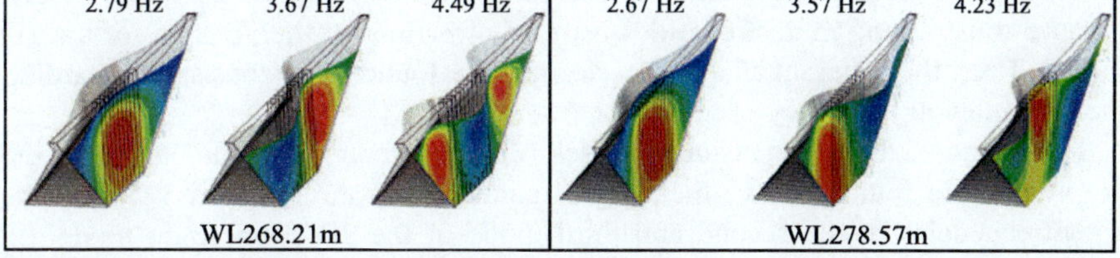

Fig. 15 Deformed shape and hydrodynamic pressure of the first 3 modes for WRWL and SRWL

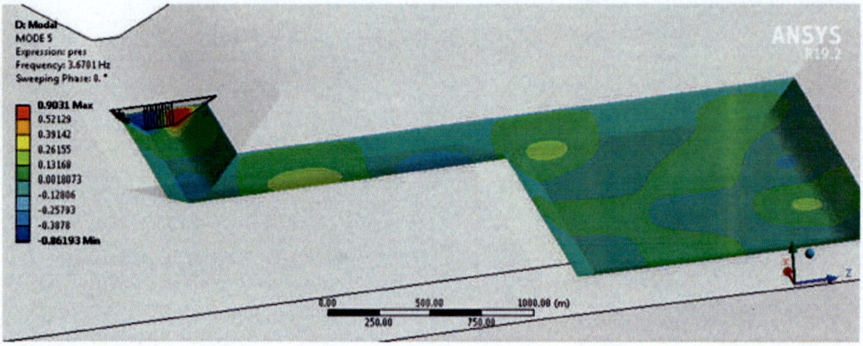

Fig. 16 Hydrodynamic water pressure along the boundaries in between the reservoir and the neighboring foundation or dam (WRWL)

Table 2 Comparison of the eigenfrequencies

Configuration	Water level	Mode 1 [Hz]	Mode 2 [Hz]	Mode 3 [Hz]	Mode 4 [Hz]
[10] foundation as dam	*Empty*	~2.50	~3*	~4*	~5*
	290.00	~2.06	~2.5*	~3*	~4*
[14] foundation as dam	*Empty*	2.56			
	290.00	2.04			
2D analysis (block no. 16)	*Empty*	2.52	5.26	5.76	8.87
	268.21	2.31	3.47	4.02	5.01
	278.57	2.20	3.19	3.82	4.93
	290.00	2.06	2.96	3.67	4.83
Full 3D analysis	*Empty*	2.99	3.82**	4.61**	5.10**
	268.21	2.79	3.67**	4.49**	
	278.57	2.67	3.57**	4.23**	
	290.00	2.48	3.35**	4.15**	4.91**

*A significant dispersion of results is observed; the value given is the average value
**Three dimensional modes

frequencies of the quasi 2D and the 3D model are 2 Hz and 2.5 Hz respectively. The 3D model is found to have a higher fundamental frequency than the 2D plain strain model. This difference is explained by the influence of the shape of the valley, which is more constraining in upstream-downstream direction for the 3D than for the 2D model. Thus, the different analyses performed are found to be consistent regarding the fundamental frequency of the dam.

Furthermore, three-dimensional modes (i.e. the second and third mode of the 3D model) are found to have their eigenfrequencies in the range of 3–5 Hz. The eigenfrequencies of the second and third mode of the 2D plain strain model do appear in the same range of frequencies. Therefore, it seems important to perform a 3D analysis in order not to miss the three-dimensional modes.

5 Conclusion

This paper investigates the dynamic behavior of the dam-foundation-reservoir system of the Pine Flat Dam using ANSYS Workbench.

A quasi 2D plane strain model is used for the harmonic and transient analyses investigating the influence of different water levels on the dynamic behavior. The additional seismic analysis shows that the downstream face as well as the heel of the dam are potentially prone to damage for the ground motion considered.

Furthermore, an extended 3D model including the shape of the reservoir is used to study 3D effects caused by the shape of the reservoir and the valley. The results obtained from the extended 3D model are compared to the results of the 2D model as well as to literature values. As to be expected, the eigenfrequencies of the 3D model are found to be higher than the ones of the 2D model. The eigenfrequencies of the three dimensional modes (i.e. the second and the third mode of the 3D model) are found to be in the same range than the eigenfrequencies of the second and third mode of the 2D model.

The 3D model used is found to allow for taking into account appropriately the effects of the three dimensional geometry of the dam, the valley and the reservoir.

References

1. Rea D, Liaw CY, Chopra AK (1972) Dynamic properties of pine flat Dam, report no UCB/EERC-72/7
2. Chopra AK, Chakrabarti P, Gupta S (1980) Earthquake response of concrete gravity dams including hydrodynamic and foundation interaction effects. Report no UCB/EERC-80/01
3. Fenves G, Chopra AK (1986) Simplified analysis for earthquake resistance design of concrete gravity Dams. Report no UCB/EERC-85/10 (1986)
4. Fenves G, Chopra AK (1984) Earthquake analysis and response of concrete gravity dams-report no UCB/EERC-84/10
5. Fenves G, Chopra AK (1984) A computer program for earthquake analysis of concrete gravity dams. Report no UCB/EERC-84/11, EAGD-84
6. Chavez JW, Fenves GL (1993) Earthquake analysis and response of concrete gravity dams including base sliding-report no UCB/EERC-93/07
7. Chavez JW, Fenves GL (1994) EAGD_SLIDE: a computer program for the earthquake analysis of concrete gravity dams including base sliding. Report no UCB/SEMM-1994/02
8. Fenves G, Chopra AK (1984) EADG-84: a computer program for earthquake analysis of concrete gravity dams, report no. UCB/EERC-84/11
9. Zienkiewicz OC, Taylor RL (2000) The finite element method, vol 1, 5th edn. The Basis, Butterworth-Heinemann, Oxford
10. Evaluation of Numerical Models and Input Parameters in the Analysis of Concrete Dams (2018) Report DSO-19-13, A Summary Report of the USSD Workshop, Miami, USA

11. Westergaard HM (1933) Water pressures on dams during earthquakes. Trans Am Soc Civ Eng 98:418–433
12. Chopra AK (1968) Earthquake behavior of reservoir-dam systems. J Eng Mech Div ASCE 94(EM6):1475–1500
13. Chopra AK (1970) Earthquake response of concrete gravity dams. J Eng Mech Div ASCE 96(EM4)
14. Løkke A, Chopra AK (2013) Response spectrum analysis of concrete gravity dams including dam-water-foundation interaction. PEER 2013/17

Seismic Analysis of Pine Flat Concrete Dam

S. Mitovski, L. Petkovski, G. Kokalanov, V. Kokalanov, and F. Panovska

Abstract Pine Flat Dam, located on King's River, east of Fresno, California, was constructed by the US Army Corps of Engineers in 1954 with height of 122 m. The dam's behavior was extensively studied in the 1970's and 1980's at the University of California at Berkeley that provide measured and calculated responses for correlation and comparison. For such purpose the dam was analyzed at action of various seismic excitations by taking in consideration linear and non-linear material properties as well and the effect of wave velocities of the rock foundation. The numerical analysis was carried out by preparation of 3D model of the dam thus applying code SOFiSTiK.

Keywords Pine flat dam · Numerical analysis · SOFiSTiK

1 Introduction

The formulation for Theme A requires case study of the seismic response of Pine Flat Dam at action of various earthquake excitations so called Cases. Namely, the formulators have developed a list of cases in order comparative studies to be performed, including linear and non-linear material models that are to be included in the numerical analysis. Total of 8 case studies were formulated, such as:

- Case A—EMVG Test Simulation (Simulation of the eccentric-mass vibration generator (EMVG) performed at Pine Flat Dam in 1971 is considered for Case A)
- Case B—Foundation Analysis using Impulsive Loads (Investigate the effect of foundation size and analyze the efficiency of non-reflecting boundary conditions in the dynamic analysis of dams. Analyses will use the Impulsive Stress Records)

S. Mitovski (✉) · L. Petkovski · G. Kokalanov · F. Panovska
Civil Engineering Faculty – Skopje, Macedonian Committee on Large Dams, University Ss. Cyril and Methodius, Skopje, North Macedonia
e-mail: smitovski@gf.ukim.edu.mk

V. Kokalanov
Faculty of Computer Science, Goce Delchev University in Shtip, Shtip, North Macedonia

G. Bolzon et al. (eds.), *Numerical Analysis of Dams*, Lecture Notes in Civil Engineering 91, https://doi.org/10.1007/978-3-030-51085-5_7

- Case C—Dynamic Analysis using Impulsive Loads (Case B, which considers only the foundation, is extended to include the dam and reservoir. Analyses will use the Impulsive Stress Records)
- Case D—Dynamic Analysis for Various Reservoir Levels (Investigate the effect of water levels. The dam-reservoir-foundation model defined in Case C is analyzed for various reservoir water levels and Taft Record)
- Case E—Non-linear Dynamic Analysis (Investigate the effect of nonlinear behavior of the dam. The dam-reservoir-foundation model defined for Case D is extended to include nonlinear properties for concrete. Analyses will use the Taft and ETAF Record)
- Case F—Massless Foundation (Investigate the effect of using a massless foundation. The dam-reservoir-foundation model defined for Case D is modified to use a massless foundation. Analyses will use the Taft Record).

Within the paper are displayed and analyzed some of the output results for Case A, Case C, Case D and Case E.

2 Dam Geometry and Modelling

Pine Flat Dam, located on King's River, east of Fresno, California, was constructed by the US Army Corps of Engineers in 1954. It consists of thirty-six 15.25 m-wide and one 12.2 m-wide monoliths. The length of the straight gravity dam is 561 m and the tallest non-overflow monolith no. 16 is 122 m high (Fig. 1), adopted for the analysis. Within the stage of numerical analysis, following steps must be undertaken, typical for this type of analysis: (1) choice of material parameters and constitutive laws (concrete. rock foundation and interface elements); (2) discretization of the dam and the rock foundation and (3) simulation of the dam behavior for the typical loading states (as required in the topic formulation).

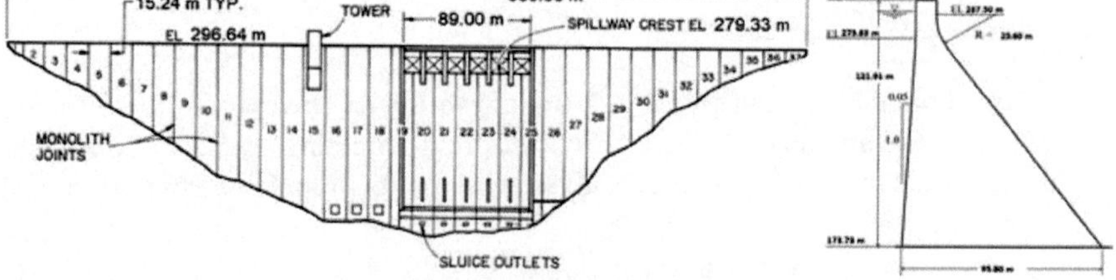

Fig. 1 Downstream view of Pine Flat dam and cross section geometry of Monolith 16

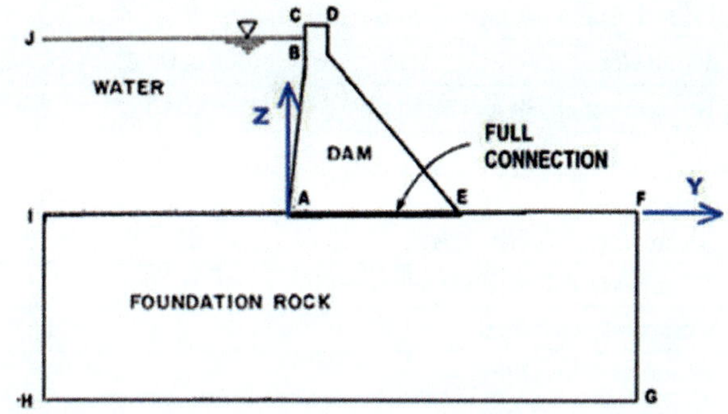

Fig. 2 Model cross section

2.1 Model Base Configuration

The model consists of the 15.24 m-wide dam monolith and a corresponding strip of the foundation. The origin of the axis system and key reference nodes are shown in Fig. 2. The axis and reference nodes are located on the mid-width of the monolith. A "base configuration" of the model is defined according to the dam dimensions (Fig. 1) and foundations dimensions' length: H-G = 700 m, depth: I-H = 122 m, dam heel location: I-A = 305 m (Fig. 2) and reservoir water level at 268.21 m, recorded at Pine Flat Dam in winter 1971.

2.2 Description of Program Package SOFiSTiK

The numerical analysis of Pine Flat dam is carried out by application of program SOFiSTiK, produced in Munich, Germany. The program is based on finite element method and has possibilities for complex modeling of the structures and simulation of their behavior. It also has possibility in the analysis to include certain specific phenomena, important for realistic simulation of dam's behavior, such as: discretization of the dam and foundation taking into account the irregular and complex geometry of the structure, simulation of stage construction, simulation of contact behavior by applying interface elements and etc. in order to assess the dam behavior and evaluate its stability. The program SOFiSTiK in its library contains and various standards and constitutive laws (linear and non-linear) for structures analysis [1].

2.3 Input Parameters and Constitutive Laws for the Materials

The choice of material parameters, as input data for the stress-deformation analysis is complex process, taking into account various factors and influences. A linear constitutive law is applied for the rock foundation and concrete (assumed to be

Table 1 Input parameters of the materials

Parameter	Rock	Concrete (linear)	Water
Modulus of elasticity [MPa]	22 410	22410	0.6
Density [kg/m^3]	2 483	2483	1000
Poisson ratio	0.20	0.20	0.499
Shear wave velocity [m/s]	1 939		
Compressional wave velocity [m/s]	3 167		1439
Compressive strength		28.0	
Shear modulus [MPa]			0.2
Bulk modulus [MPa]			2080
Tensile strength		2.0	
Fracture energy		250.0	
Compressive strain at peak load		0025	
Tensile strain at peak load		0.00012	

homogeneous and isotropic throughout the entire dam), for each Case except for Case E, whereas input data are specified in Table 1.

In Case E is applied non-linear constitutive law for concrete, according to Eurocode 2, concrete grade 30, and input parameters are specified in Chap. 6. The water load is important specific phenomena in case of dams. Rather complex is the water effect simulation in numerical models in case of seismic loads, where there are generally two approaches—by Westergaard method (added masses) or by compressible fluid (applied here below).

2.4 Discretization of Dam Body and Foundation by Finite Elements

Applied finite elements. Numerical analysis in the report are performed by spatial (3D) model, at 1 m' length in X-direction, and it can be approximated as plane analysis. Namely, the applied boundary conditions enable plane state analysis of the models. The dam body and the foundation are modeled with volume elements. A powerful and reliable finite element should be applied in case where an analysis of structure with complex geometry and behavior is required, having in consideration that the correctly calculated deformations and stresses are of primary significance for assessment of the dam stability. Generally, for discretization of the dam body and the rock foundation are applied quadrilateral finite element (as auxiliary elements, type quad, by 4 nodes), volume finite element (type brick, by 8 nodes) and interface element (dampers and water bedding) of type spring.

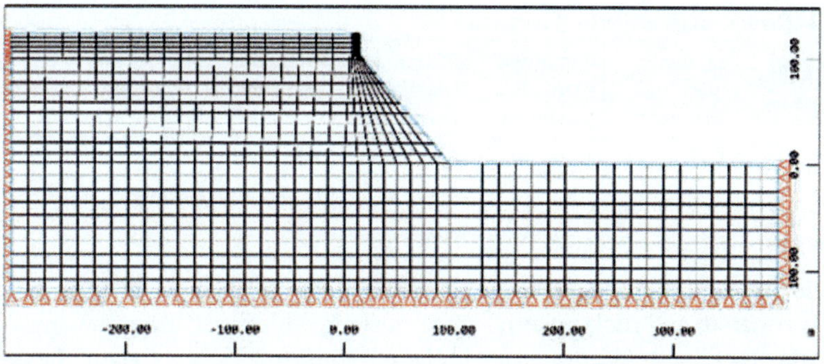

Fig. 3 Display of coarse finite element mesh (N = 816), applied for Case A, Case D, Case E and Case F

Numerical model. The model is composed of dam body, rock foundation and water fluid modeled as compressible, according to the specified geometry (Fig. 3). The model has geometrical boundaries, limited to horizontal and vertical plane whereas are defined the boundary condition of the model, varying for various Cases. The discretization is conveyed by capturing of the zones of various materials in the model—concrete and rock foundation and such model was applied for Case A, D, E and F. The model for Case C is explained in Chap. 4.

2.5 Dam Loading

Static load includes weight of concrete dam and the reservoir without weight of foundation block taken into consideration. The applied water loading includes simulation of the dam behavior for three reservoir levels: (a) winter reservoir water level at El. 268.21 m; (b) summer reservoir water level at El. 278.57 m and (c) normal reservoir level at El. 290.00 m. There are specified demands whether the static weight of the dam, the reservoir and the foundation are to be included in the analysis.

3 Case A—EMVG Test Simulation

In Case A is performed dynamic linear analysis of the dam-foundation-reservoir system for the harmonic force record exerted by an eccentric-mass vibration generator positioned at the dam crest. The boundary conditions are adopted as follows: the horizontal edge in the lowest zone of the model is adopted as non-deformable boundary condition by fixed displacements in XYZ direction, the vertical planes perpendicular on X-axis are boundary condition by applying fixed (zero) displacements in X-direction and the vertical planes perpendicular on Y-axis are boundary

Table 2 Natural frequencies for first 6 models

Water level [m]	1st mode (dam and reservoir)	2nd mode (dam and reservoir)	3rd mode (dam and reservoir)	4th mode (dam and reservoir)	5th mode (dam and reservoir)	6th mode (dam and reservoir)
268.21	1.42	1.65	2.02	2.53	4.07	5.57
278.57	0.96	1.12	1.31	1.54	2.23	3.85

condition by applying fixed (zero) displacements in Y-direction. The input parameters for concrete and rock materials are considered elastic. During the analysis the static weight of the foundation is not included, while the static weight of the dam and reservoir is taken in consideration. The dynamic load is EMVG harmonic-force time history record applied at the middle of the dam crest in the upstream/downstream direction.

The natural frequencies for water levels at 268.21 and 278.57 masl are displayed respectively in Table 2.

The displacement time history of the upstream nodes at dam crest [C] and dam heel [A] for water level at 278.57 masl are displayed on Fig. 4. The obtained value for the relative displacements point out to dominant values of displacements at point C at the dam crest compared with the low values obtained at point A. The diagrams of the time history of the displacements have expected shape of dynamic response of such dam to EMVG record. The amplitude of the relative displacements reaches value of 0.0008 m.

Very similar case is also with the obtained accelerations. Namely, the obtained accelerations at point C are higher than accelerations at point A (Fig. 5), apropos the dam response is more dominant at dam crest.

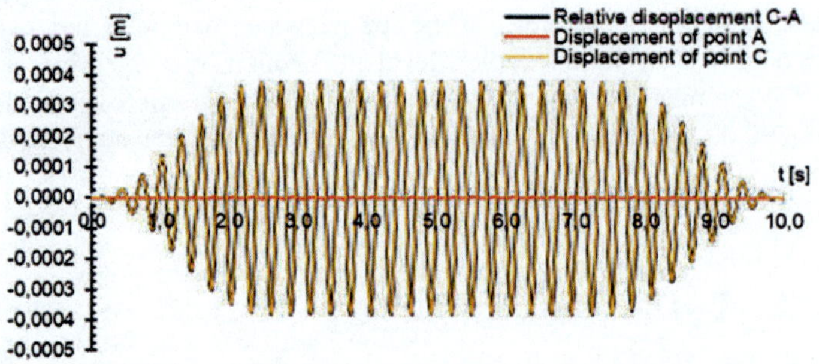

Fig. 4 The displacement time history of upstream nodes A and C and relative displacements

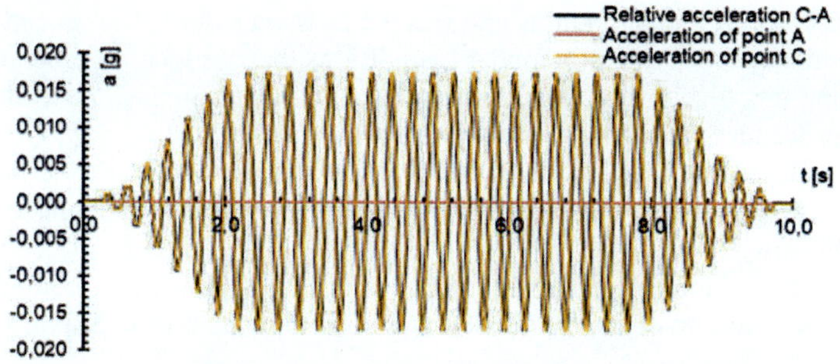

Fig. 5 The acceleration time history of the upstream nodes A and relative acceleration

4 Case C—Dynamic Analysis Using Impulsive Loads

The analysis type for Case C takes in consideration the dam and the reservoir along with the foundation model. The excitation is defined by horizontal (Y-axis) Impulsive Stress record, applied uniformly at the base of the foundation block as specified by the formulator. In this case are applied viscous boundaries that are described briefly here below. Namely, a way to eliminate waves propagating outward from the structure is to use Lysmer boundaries [2, 3]. This method consists of simply connecting dashpots to all degrees of freedom of the boundary nodes and fixing them on the other end. Lysmer boundaries are derived for an elastic wave propagation problem in a one-dimensional semi-infinite bar. The damping coefficient C_x of the dash pot equals (Eq. 1):

$$C_x = A \cdot \rho \cdot c \tag{1}$$

where A is the cross sectional area of the bar, ρ is the mass density and c the wave velocity that has to be selected according to the type of wave that has to be absorbed (shear wave velocity c_s or compressional wave velocity c_p). In two dimensions the equation takes the following form, which results in dampers C_n and C_t (Fig. 4) in the normal and tangential directions (Eq. 2):

$$C_n = A_1 \cdot \rho \cdot c_p;\ C_t C_t = A_1 \cdot \rho \cdot c_s \tag{2}$$

The shear wave velocity C_s and compressional wave velocity C_p are given by Eq. 3:

$$C_s = \sqrt{G/\rho};\ C_p = \sqrt{E(1-\upsilon)}/[(1+\upsilon)(1-2\upsilon)\rho] \tag{3}$$

where G is the shear modulus of the medium and is expressed as specified in Eq. 4:

$$G = \sqrt{E/(2(1+\upsilon))} \tag{4}$$

Here E is the Young's modulus and υ is the Poisson's ratio. The procedure leads to the frequency independent boundary conditions that are local in time and space. If the shape functions of the neighboring finite elements are used instead of crude lumping procedure, a narrow banded matrix arises, which is also easy to implement.

In our calculations, the elastic material properties are taken as specified by the formulator, while the A is the dashpot's belonging area which is calculated from the code SOFiSTiK.

The fine finite element mesh was adopted for Case C due to the requirements of the dampers at bottom of the foundation that need to be placed on distance of 1.5 m (Fig. 6).

The velocity time history for Case C1 for upstream nodes C and A for water level at 268.21 masl are displayed on Fig. 7. The maximal obtained value for the velocity are practically equal at the dam crest and dam heel. Similar case is with velocities at points a (middle point of the dam width) and b (middle point of rock foundation bottom). Namely, the obtained values are practically equal, values of 0.0045 m/s and 0.0048 m/s respectively (Fig. 8).

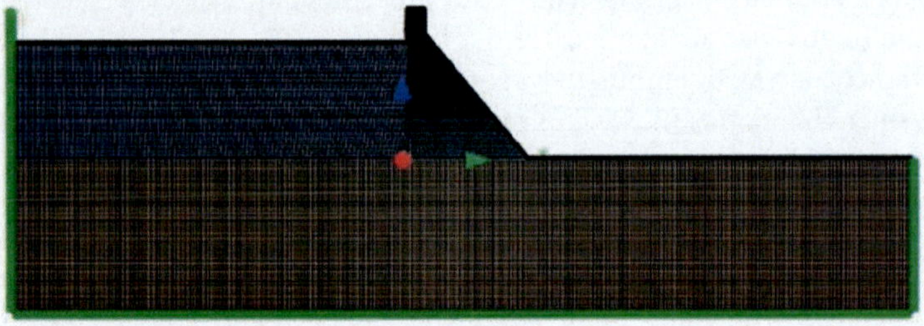

Fig. 6 Fine finite element mesh (N = 54440), applied for Case C (green lines at model edges represent the dampers)

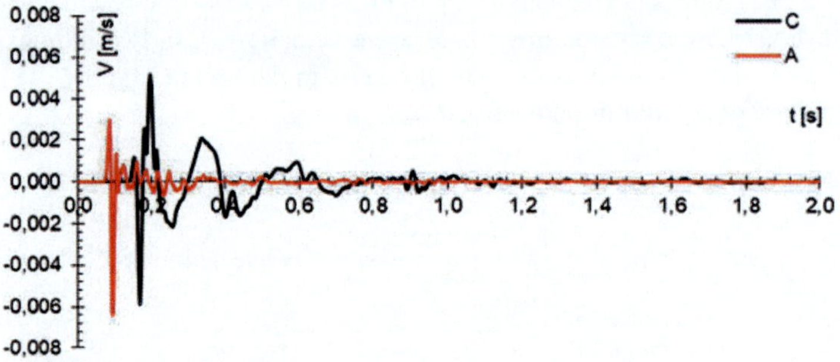

Fig. 7 The acceleration time history of upstream nodes A and C for Case C1

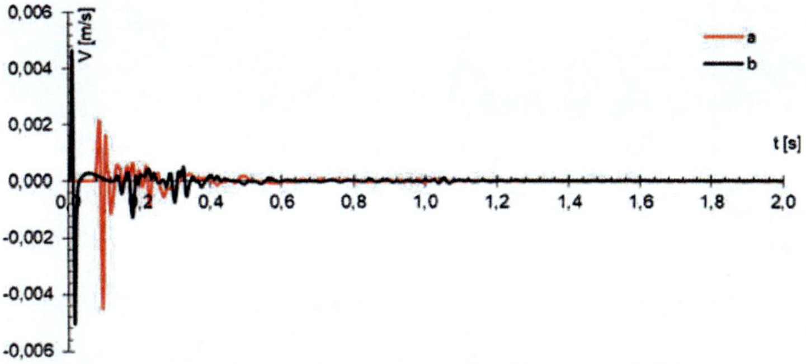

Fig. 8 The acceleration time history of nodes a (middle node of dam width) and b (middle point of foundation bottom) for Case C1

5 Case D—Dynamic Analysis for Various Reservoir Levels

In Case D, a dynamic analysis of the dam-foundation-reservoir system is performed considering the elastic material properties, the Taft earthquake record, and the reservoir water at three different elevations. The Taft record is applied as deconvolved acceleration time history at the base of the foundation block for each different reservoir water levels of 267.21 masl (D1), 278.57 masl (D2) and 290.00 m (D3) respectively. The damping parameters are adopted by Rayleigh damping calculation with input values for frequencies f1 = 4.18 Hz and f2 = 9.95 Hz.

The dam response through the various water level elevations is analyzed by the obtained values for the displacements and the hydrodynamic pressure within the excitation period for Taft record. Namely, on Fig. 9 is displayed displacements time history at point C for Case D1, D2 and D3. By the obtained values for the displacements time history there are no significant differences between analyzed water levels. Namely, the amplitude of the maximal values of the dynamic displacements ranges from 0.014 to 0.017 m for case D1 to case D3 respectively.

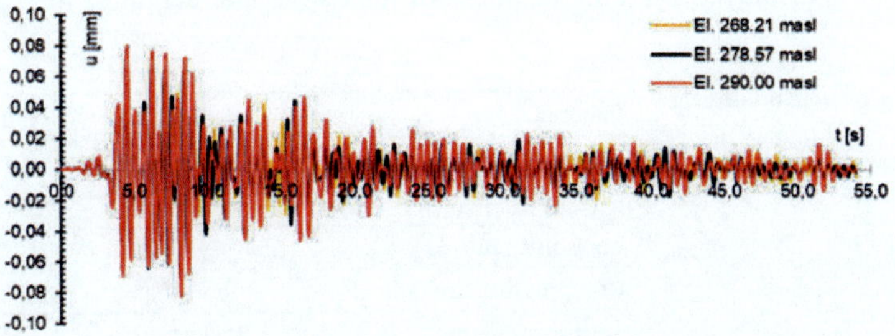

Fig. 9 The displacements time history of upstream node C for Case D1, D2 and D3

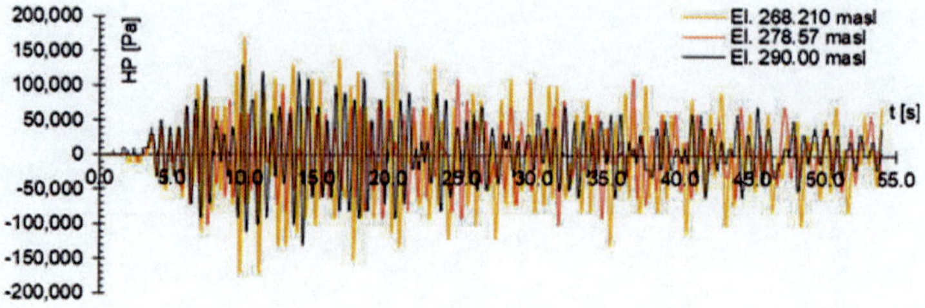

Fig. 10 The hydrodynamic pressure time history of node A for Case D1, D2 and D3

The same analogy is for the hydrodynamic pressure at the dam's heel (Fig. 10), with values ranging from of 3,40,000 Pa for case D1, 2,00,000 Pa for Case D2 and 2,60,000 for Case D3, with initial pressure set to zero.

6 Case E—Non-linear Dynamic Analysis

The analysis of Case E includes carrying out of dynamic analysis with by applying non-linear material properties for the concrete. Two different dynamic loads are applied at the base of the foundation block: The Taft (deconvolved acceleration) and ETAF Record. The non-linear material constitutive model for concrete is adopted according to Eurocode 2, EN 1992 norms [4]. For such applied non-linear law the potentially damage zones occur in case of exceedance of the tension stresses.

For Case E1 and E2 are applied Taft Record (deconvolved acceleration) and ETAF horizontal acceleration time history record at the base of the foundation block. As specified, gravity loads are applied to the dam and reservoir but not to the foundation. The analysis is performed for dam-reservoir-foundation system at water level El. 268.21 m, combined with the static load (Table 3).

The dam response for Case E1 is analyzed by the obtained values for acceleration and hydrodynamic pressure within the excitation period for Taft record. Namely, on Fig. 11 is displayed dynamic response of the dam under action of the specified

Table 3 Concrete non-linear properties

Parameter	SI unit	Value
Modulus of elasticity	MPa	31476
Density	kg/m^3	2500
Poisson's ratio		0.20
Compressive strength	MPa	33.00
Tensile strength	MPa	2.56
Fracture energy	N/m	137.0
Compressive strain at peak load		0.002

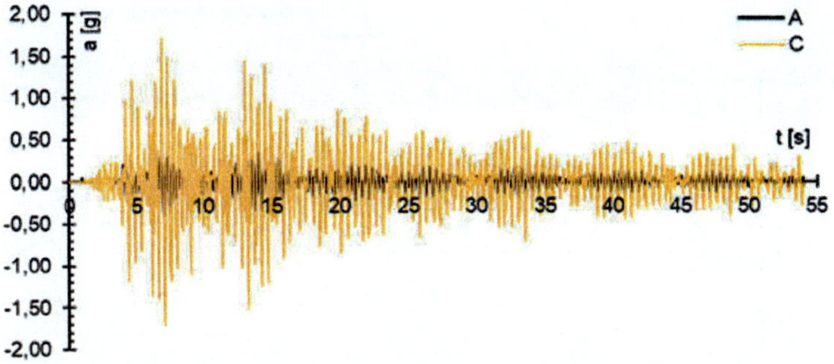

Fig. 11 The acceleration time history at node A and C for Case E1

Taft record at nodes A and C, located at dam heel and crest respectively. Maximal values for accelerations occurs at node C apropos the dam responds to Taft record amplification of the horizontal acceleration in the crest multiplied by 9.5 regarding the peak ground acceleration (dynamic amplification factor of DAF = 9.5). In case E2, on Fig. 12 is displayed dynamic response of the dam under action of ETAF record at nodes A and C. Maximal values for accelerations occurs at node C apropos the dam responds to ETAF record amplification of the horizontal acceleration in the crest multiplied by 16.2 regarding the peak ground acceleration (dynamic amplification factor of DAF = 16.2).

On Fig. 13 are displayed displacements time histories for nodes A and C and relative displacements for Case E2. The diagrams of displacements time history have expected shape of the dynamic response of the dam. The amplitude of the relative displacements has value of 0.8 m.

The hydrodynamic pressure at the dam's heel (Fig. 14), riches maximal amplitude of 9,500,000 Pa for Case E2.

On Fig. 15 is displayed potentially tension zone of the dam, for Case E1. Namely, according to the display of superimposed maximal values of principal stresses II,

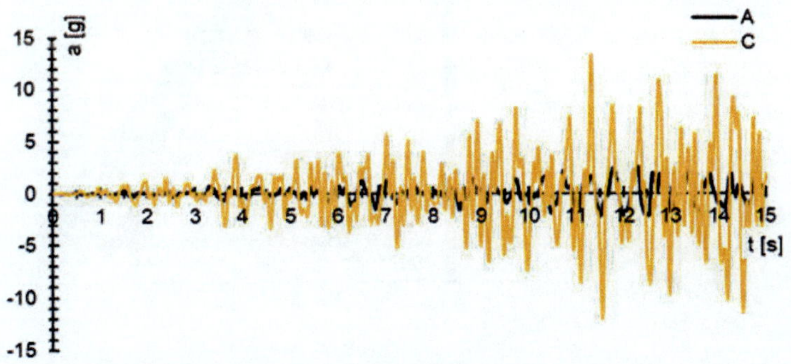

Fig. 12 The acceleration time history at node A and C for Case E2

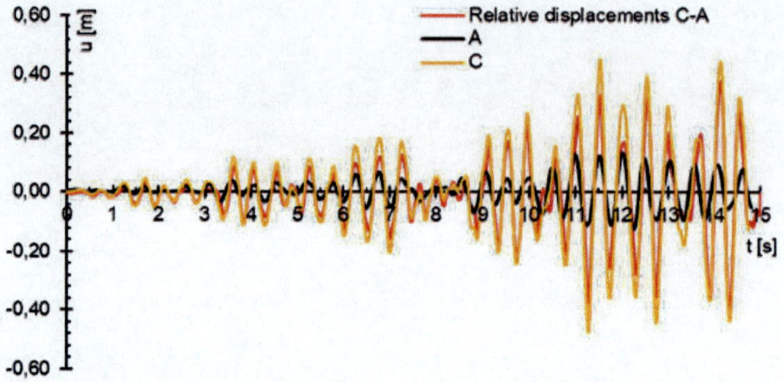

Fig. 13 The displacements time history at nodes A and C for Case E2 and the relative displacements

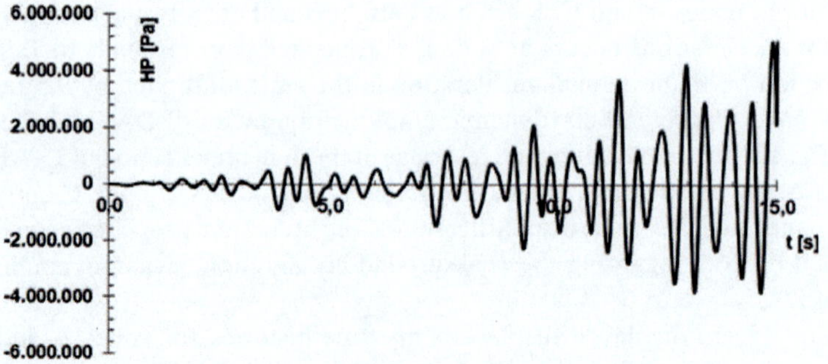

Fig. 14 The hydrodynamic pressure time history at node A for Case E2

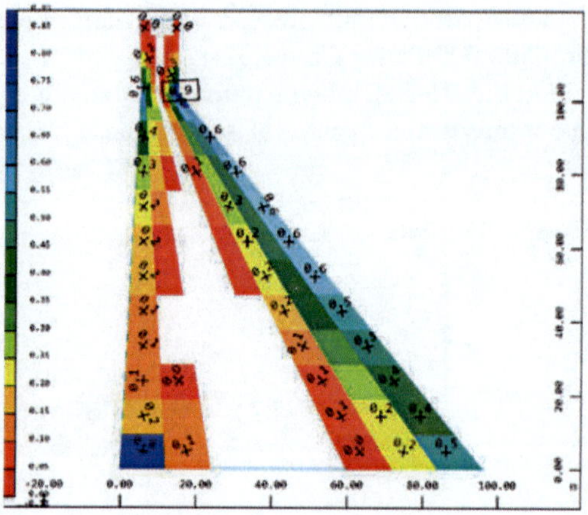

Fig. 15 Display of critical case of damaged zone for Case E1

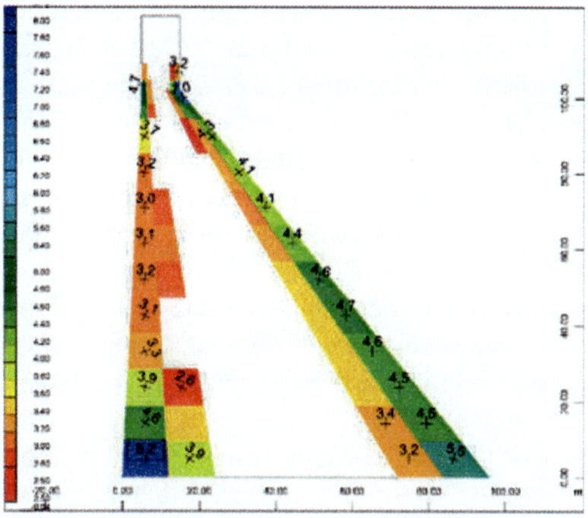

Fig. 16 Display of critical case of damaged zone for Case E2

there is no exceedance of the tension strength of 2.5 MPa, apropos the maximal value is 0.89 MPa.

On Fig. 16 is displayed potentially damaged zone of the dam, assessed as most critical, by superimposed maximal values of principal stresses II from all time steps for Case E2. Namely, within the cross section of the dam are marked zones with exceedance of the tension stress of 2.50 MPa thus assessed as potential zones for crack occurrence. Maximal value of the tension stress is 8.20 MPa, occurring at the zone of the dam's upstream heel at contact with foundation.

7 Conclusions

From the numerical experiment of simulation of the structural behavior of Pine Flat dam at various excitation records, following main conclusions and recommendations are drawn out:

- The prediction of the concrete gravity dams' behavior during construction, after construction, reservoir filling and service period by means of numerical models is of primary importance.
- The gravity dam stability includes determination of potential damaged zones, mainly occurred by exceeding the tension stress of the concrete resulting in plastification of the material. The specification of allowable tension stresses for concrete type should be done by experimental (laboratory or field) testing performed for that purpose and afterwards identification of such critical zones will be precisely determined.
- In Case A1 and A2 of the present case study are determined the first 6 fundamental frequencies of Pine Flat dam for reservoir water levels at 268.21 and 278.57 masl, ranging from 1.42–5.57 Hz and 0.96–3.85 Hz respectively, thus indicating that

the raised water level decreases the frequencies. The dam response is dominant at dam crest, as evaluated by the distribution of the relative displacements and relative accelerations during the excitation record.

- For Case C1 by the analyzed values of the velocities at upstream dam nodes A and C and nodes a (middle node of dam width) and b (middle node of foundation bottom) a modus of velocities lowering can be noticed as the nodes are more distanced then the upstream face of the dam.
- For Case D1, D2 and D3 the impact of the various water elevations results in different initial conditions due to the static load of the dam and the reservoir. The obtained values for the dynamic displacements and hydrodynamic pressure during excitation record are varying approximately for 8% for water level of 268.21 to 290.0 m.
- For Case E1 the dam responds to Taft record with dynamic amplification factor of DAF = 9.5. For case E2 the dynamic amplification factor is DAF = 16.2 for ETAF record. For Case E1 there is no occurrence of damaged zone, while for Case E2 is obtained potentially damaged zone, assessed as most critical for crack occurrence, from all time steps, with tension stress value of 8.20 MPa, occurring at the zone of the upstream heel of the dam at contact with the foundation.

From the numerical experiment of simulation of the structural behavior of Pine Flat dam at various excitation records, following recommendations for future research can be applied:

- We are on opinion that the applied non-linear constitutive law for the concrete has large impact to the obtained results regarding the damaged zones and occurrence of plastification and cracks. Namely, there are few such constitutive laws and they include various parameters for definition and determination of the damaged plasticized zones. We recommend specification of at least two non-linear constitute laws and execution of analysis for further research.
- Long term behavior of dams is extensively analyzed and researched in the recent period. The concrete properties are varying in long time period. Therefore, if possible, we believe that it will be useful to determine the some of the concrete input parameters (modulus of elasticity, compressive and tension strength) of Pine Flat dam by applying non-destructive methods and based on such parameters an innovated analysis to be conducted.
- Temperature effect is important issue in case of concrete dams. So, in the future research and analysis temperature load should be considered, in order to be obtained more comprehensive insight and assessment of the dam's behavior.

References

1. SOFiSTiK (2018) Manual
2. Lysmer J, Kuhlemeyer RL (1969) Finite dynamic model for infinite media. J Eng Mech Div ASCE 95(EM4):859–877

3. Bruman A, Maity D, Sreedep S (2010) Iterative analysis of concrete gravity dam non-linear foundation interaction. Int J Eng Sci Technol 2(4):85–99
4. Eurocode 2 (1992)Design of concrete structures, European Committee for Standardization

Seismic Analysis of Pine Flat Concrete Dam: Comparison of Numerical Results from 2D and 3D Analysis

N. Monteiro Azevedo, M. L. Braga Farinha, R. Câmara, and N. Schclar Leitão

Abstract The concrete gravity dam proposed for the 15th International Benchmark Workshop on Numerical Analysis of dams (Theme A) was numerically studied using computational finite element modules developed by the authors for dam analysis. The computational models allow non-reflecting boundary conditions, free-field boundary conditions, fluid structure interaction following a Lagrangian displacement based formulation for the fluid and nonlinear material behaviour, both continuum damage and discrete crack approaches. The seismic analysis is carried out using a central difference method, a Newmark implicit integration scheme can be adopted for other types of loading. The computational modules have been used in the assessment of the behaviour of several operating dams. The natural frequencies and force vibration analysis (Case A), the seismic analysis for various reservoir levels following a linear elastic model (Case D) and non-linear seismic type analysis including nonlinear behaviour (Case E) are assessed and discussed using both 2D and 3D modules.

Keywords Non-reflecting boundary condition · Free-field boundary condition · Fluid-structure interaction · Lagrangian models

1 Introduction

The concrete gravity dam was numerically studied using finite element modules developed by the authors for dam analysis [1–3]. These modules have been used in the safety assessment and analysis of the behaviour of several operating arch and gravity dams in both static and dynamic conditions [3–5]. The starting point was the analysis of the Miami workshop deliverables, as in this workshop the same dam was analysed by several contributors [6].

Both 2D and 3D modules use an explicit solution algorithm based on the central difference method and a dynamic relaxation algorithm for static convergence. The viscous boundary formulations developed by Lysmer and Kuhlemeyer

N. Monteiro Azevedo (✉) · M. L. B. Farinha · R. Câmara · N. Schclar Leitão
Laboratório Nacional de Engenharia Civil, Lisbon, Portugal
e-mail: nazevedo@lnec.pt

G. Bolzon et al. (eds.), *Numerical Analysis of Dams*, Lecture Notes in Civil Engineering 91, https://doi.org/10.1007/978-3-030-51085-5_8

[7] are implemented. The fluid structure interaction is carried out using a Lagrangian displacement-based fluid model with rotational penalty constraint [2, 8–11]. In 2D, a 4 node plane sub integrated fluid displacement element was adopted, and in 3D 8 node brick fluid displacement element following a mixed discretization procedure was used [2, 12]. The constraint penalty term multiplier should be as high as necessary to enforce rotational constraint, but small enough to avoid numerical ill-conditioning and overly stiff responses. In 2D a penalty term of 100 times the fluid bulk modulus was adopted and in 3D a penalty term of 0.001 of the fluid bulk modulus is required. This type of approach allows a perfect match between the solid and the fluid domains, and has, when compared with fluid pressure-based formulations, computational advantages for explicit algorithms such as the centred difference. Traditional interface edge/plane finite elements with only normal stiffness are adopted between the fluid (reservoir) and solid domains (dam and foundation).

The seismic action is represented by plane waves, propagating upwards from the base of the model with a non-reflecting character, using a viscous boundary formulation [7]. The dynamic boundary conditions of the lateral boundaries of the model, 2 in 2D and 4 in 3D, are obtained by performing independent free-field calculations, subject to the same base motion (Fig. 1).

For each side of the model, a 2D free-field mesh is created. In 3D, each 2D free field mesh has its own 1D free-field meshes at the sides [13]. This type of approach follows the general principles of the direct finite element approach proposed in [14] for earthquake analysis but follows a central difference explicit scheme.

The modelling of nonlinear behaviour can be carried out either with a two variable scalar continuum damage model (tensile induced damage and compression induced damage) [15, 16] or with a bilinear damage model developed for particle fracture studies (discrete crack approach) [17]. In this study only the latter was used.

The natural frequencies and force vibration analysis (Case A), the seismic analysis for various reservoir levels following a linear elastic model (Case D) and nonlinear seismic type analysis including nonlinear behaviour (Case E) are assessed and discussed using both 2D and 3D modules.

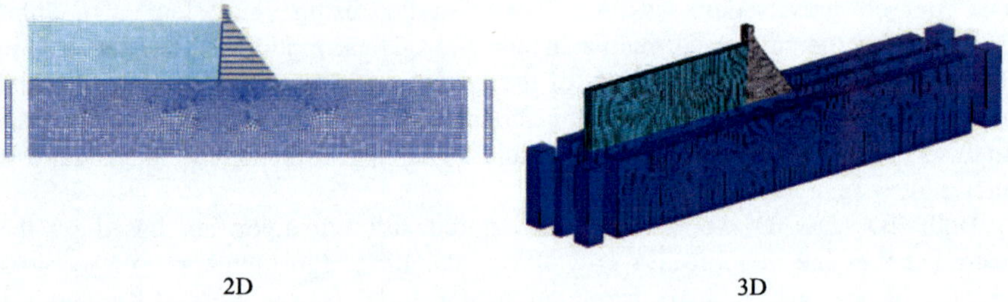

Fig. 1 Adopted numerical models—WRWL (268.21 m)

2 Numerical Models and Case Studies Analysis

2.1 Initial Considerations

Figure 1 shows the 2D and 3D developed numerical models for the winter reservoir water level (WRWL—268.21 m), see also Table 1. All adopted models follow closely the given geometry of the cross section of monolith 16 of Pine Flat dam and the corresponding vertical strip of the foundation. In the dam body discretization 4 node plane finite elements with an edge length of approximately 3.0 m were adopted. In the foundation and reservoir discretization, 4 node plane elements with a variable edge length between 3.0 to 5.0 m were adopted. The 3D mesh was obtained by extruding the 2D mesh with two layers of elements, each one 7.62 m wide, matching the total monolith width of 15.24 m.

As shown in Fig. 1 the developed model adopts interface edge/plane elements in the reservoir/dam interface, reservoir/foundation interface, dam/foundation interface and in hypothetical lift joints. In the nonlinear analyses that were carried out, all the finite elements (quadrilateral or hexahedron) adopted in the dam discretization interact with each other through interface finite elements placed at their boundaries with normal/shear stiffnesses. The nonlinear behaviour is concentrated at the interface finite elements following a bilinear softening model in the normal (tension) and shear directions [17]. This is the reason why in the nonlinear analysis the number of interfaces increases to 1405 edge elements in the 2D simulations and to 3475 plane interface elements in 3D, for the numerical model at WRWL water level (Table 1).

The material properties provided by theme A formulators were adopted, namely concrete, water and foundation properties. Table 2 presents the adopted interface material properties that need to be defined given that interface elements, edge to edge in 2D or face to face in 3D, were adopted in the dam/foundation, dam/reservoir and foundation/reservoir interactions.

Table 1 Finite element models—WRWL (268.21 m)

Model	Linear analysis		Nonlinear analysis		4 node quadrilateral/8 node hexahedron
	Nodal points	Interface elements	Nodal points	Interface elements	
2D	7865	322	4600	1405	7254
3D	74487	644	88014	3475	23862

Table 2 Interface joint elastic properties

	Dam/foundation and Dam/Dam	Reservoir/Dam and Reservoir/Foundation
Normal stiffness k_n (GPa/m)	48.4	4.0
Shear stiffness k_s (GPa/m)	19.5	0.0

2.2 Natural Frequencies and Force Vibration Analysis—Case A

Table 3 shows the natural frequencies obtained assuming that the reservoir at the WRML water level for the 2D and 3D numerical models. The same table presents the natural frequencies assuming an empty reservoir, which allows the comparison with the results obtained with the 2D and 3D models. In 2D the frequencies were defined following an eigenvalue analysis. The foundation lateral vertical and horizontal base boundaries were fixed in the global coordinates and the reservoir lateral vertical boundary was fixed in the normal direction. In 3D the frequencies were defined following a linear sweep vibration analysis. A viscous boundary condition was adopted at the reservoir lateral vertical boundary. Foundation boundaries were the same as those considered in the 2D. The different approaches and fluid boundary conditions may explain the different eigenvalues. Nevertheless, it is important to point out that both approaches predict the same fundamental frequency (2.27 Hz). As expected the predicted natural frequencies are lower when the reservoir is included in the model.

In order to further understand the differences between both numerical models, forced vibration dynamic tests were performed. Numerically, a unitary vibration force was applied at the dam crest, with frequencies ranging from 1.0 to 7.0 Hz adopting a 0.05 Hz increment. In each frequency loading a 20 s time history simulation is performed, that guarantees the stabilization of the numerical response. A Newmark integration implicit procedure with the Theme A suggested damping values was adopted. At the vertical model boundary upstream from the dam a non-reflecting

Table 3 Natural frequencies (Hz)—WRWL (268.21)

Model	Mode 1	Mode 2	Mode 3	Mode 4	Mode 5	Mode 6
2D (dam and reservoir)	2.27	3.39	3.88	4.33	4.86	5.33
3D (dam and reservoir)	2.27	3.50	4.08	4.61	4.96	5.46
2D/3D (empty reservoir)	2.46	4.18	4.83	5.63	5.91	6.74

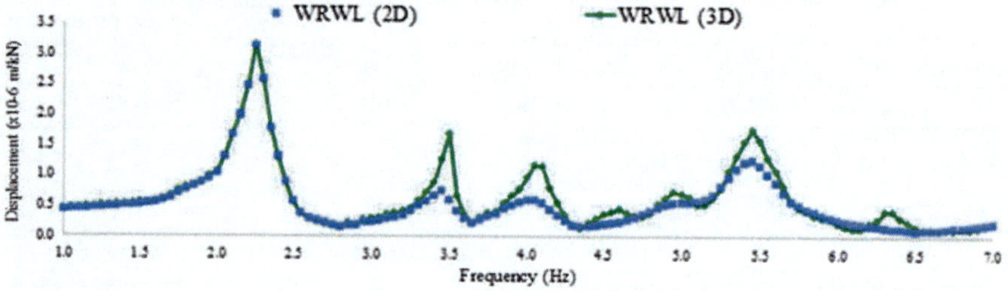

Fig. 2 Load transfer functions for the upstream/downstream displacement at the dam crest—WRWL 268.21

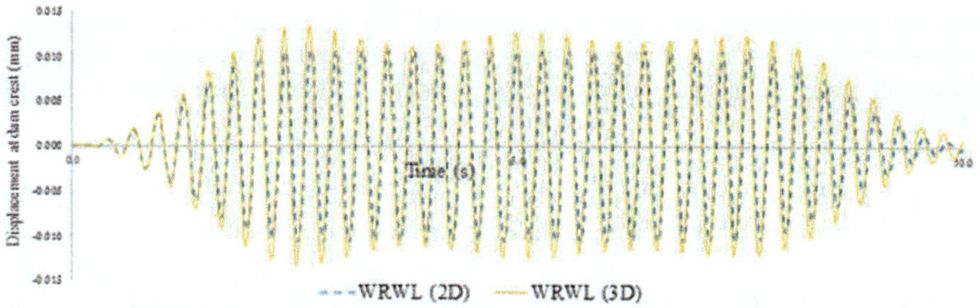

Fig. 3 Upstream/downstream displacement at the dam crest—WRWL (268.21 m)

boundary, using a viscous boundary formulation was used. As shown in Fig. 2 there is a perfect match in terms of predicted peak value and amplitude for the first peak.

Figure 2 also shows that in the 1.0–7.0 Hz range the 2D model only predicts 4 peaks (4 modes) which occur at the vicinity of peaks predicted with the 3D model. These peaks, with the exception of the 1st peak, have a lower amplitude in 2D. The two additional peak values predicted with the 3D model can be explained by the fact that with the two element discretization adopted in the 3rd dimension, the numerical model also allows some deformability in this direction, contrary to what is obtained with the 2D plane strain numerical model. Comparison of the results shown in Table 3 and in Fig. 2 for the 2D numerical model, leads to the conclusion that the adopted fluid boundary conditions influence the predicted natural frequencies.

Figure 3 shows the upstream/downstream displacement at the crest of the dam predicted by both models for the WRWL water level given an eccentric-mass vibration generator located at the crest of the dam with an amplitude of 35.4 kN and a frequency of 3.47 Hz. According to Fig. 2 in the 2D numerical model the 2nd eigenvalue occurs at 3.45 Hz and in the 3D numerical model the 2nd eigenvalue occurs at 3.50 Hz. For this reason the vibrator frequency is in close range to the mode 2 frequency in both models. A similar solution procedure to the forced vibration tests previously described was followed. The adopted numerical models predict similar responses but it is shown that in the 2D model the displacement signal has almost constant amplitude, which is not the case in the 3D response.

2.3 Seismic Analysis for Various Reservoir Levels Following a Linear Elastic Model—Case D

The seismic analysis for the Taft record with a duration of 55 s is carried out in two loading stages. Firstly, the static permanent loads are applied to the model: deadweight and the hydrostatic pressure at upstream face of the dam and at the foundation upstream interface (added mass approach). At the end of the static loading, the supports are replaced by the equivalent static reactions. Secondly, the seismic action is considered and applied as previously defined. The dynamic boundary conditions of the lateral boundaries of the model are obtained by performing independent free-field calculations, subject to the same base motion (Fig. 1). In the 3D model the displacement in the 3rd dimension is fixed in all nodal points, for this reason only the free-fields at the lateral boundaries are considered in the simulation.

The analysis was carried out using an explicit solution algorithm based on the central difference method. This is the reason why, in addition to runtime restrictions, only mass-proportional Rayleigh damping was adopted, using twice the value proposed by Theme A contributors. Figure 4 shows the fluid pressure close to the upstream face of the dam at the dam/foundation interface and Fig. 5 shows the upstream/downstream dam crest displacement. These results are obtained with the Taft record applied as a shear wave at the foundation base. It can be seen that both numerical models predict a similar response with an almost perfect match. Similar results were found for the other analysed water levels.

Table 4 presents the minimum and maximum normal and shear stresses that were recorded during the Taft record analysis for the WRWL and SRWL water levels. It can be verified that both models predict similar responses and that as the water level increases the maximum normal stress at the dam/foundation interface also

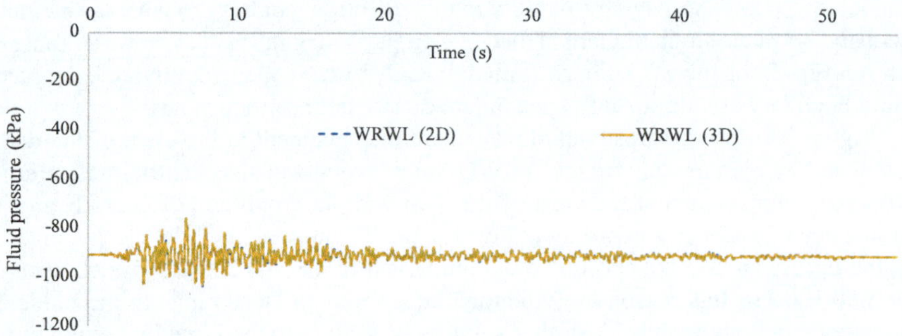

Fig. 4 Fluid pressure close to the upstream face of the dam at the dam/foundation interface—WRWL (268.21 m)

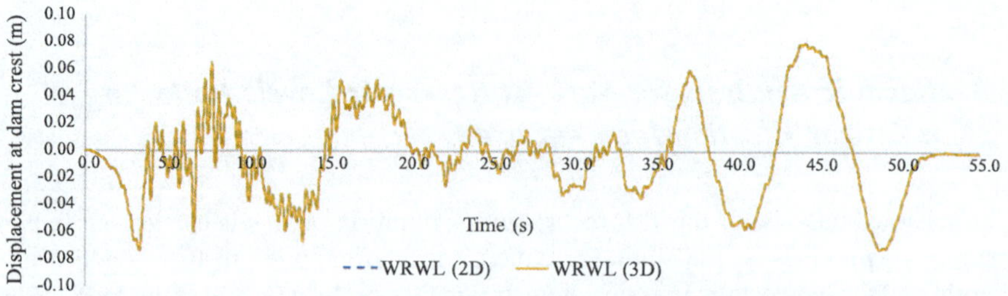

Fig. 5 Displacement of the crest of the dam in the upstream/downstream direction—WRWL (268.21 m)

Table 4 Normal and shear stresses at interface finite elements—WRWL and SRWL

Model	Dam/dam interface				Dam/foundation interface			
	σ_n (MPa)		τ (MPa)		σ_n (MPa)		τ (MPa)	
	Min	Max	Min	Max	Min	Max	Min	Max
WRWL—2D	−4.18	1.11	0.0	1.50	−7.47	2.71	0.0	3.20
WRWL—3D	−4.18	1.08	0.0	1.49	−7.49	2.69	0.0	3.20
SRWL—2D	−4.03	0.89	0.0	2.25	−6.79	4.05	0.0	3.94
SRWL—3D	−4.02	0.88	0.0	2.25	0.6.80	3.99	0.0	3.89

Table 5 Minimum and maximum recorded principal stresses at dam finite elements

Model	σ_n (MPa)	
	Min	Max
WRWL (2D)	−8.07	3.95
SRWL (3D)	−7.28	5.44
NRWL (3D)	−7.45	6.78

increases. Table 5 shows the minimum and maximum principal stresses that occur at the dam finite elements for the 2D numerical models. It can be seen that even for the lower water level the maximum principal stresses are higher than the concrete tensile strength indicated by the contributors, which means that some nonlinear behavior is expected to occur.

Figure 6 shows the fluid pressure distribution at the end of Taft record (55 s) for the SRWL water level. It can be verified that at the end of the seismic loading the fluid pressure returns to the initial static distribution.

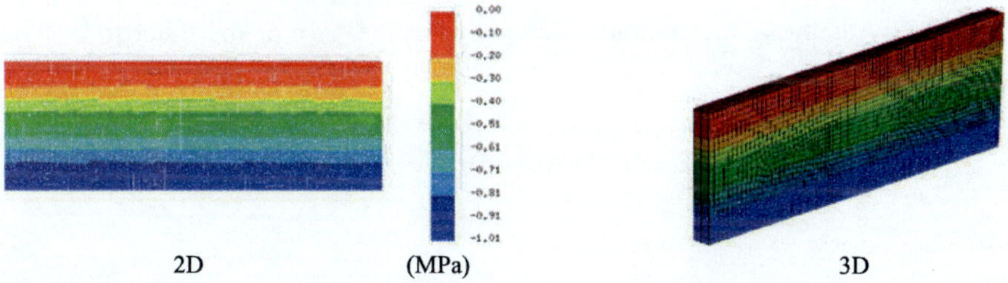

Fig. 6 Fluid pressure distribution at the end of Taft record—SRWL (278.57 m)

2.4 Non-linear Seismic Type Analysis Including Nonlinear Behavior—Case E

For the nonlinear analysis a discrete crack approach was adopted. For this reason, each finite element adopted in the dam body discretization was made to interact with the neighbouring finite elements through interface elements (edge to edge or face to face). A bilinear softening model is adopted in tension and in shear direction. In tension the tensile strength of 2.0 MPa and the fracture energy of 250 N/m were adopted as indicated by theme A formulators. In the shear direction a cohesion value of 6.0 MPa, a fracture energy of 2500 N/m and a frictional term of 1.0 were adopted, based on the experience in the application of this type of models. The same nonlinear properties were adopted for the dam/foundation interface finite elements as no information was provided for this zone.

Figures 7 and 8 show, respectively, the fluid pressure close to the upstream face of the dam at the dam/foundation interface and the upstream/downstream dam crest displacement, when the Taft record is applied as a shear wave at the foundation base. It can be seen that both numerical models predict a similar response with an almost perfect match. Given that the nonlinear behavior is much localised the response is closer to that obtained with an elastic model (Figs. 4 and 5).

Figure 9 shows the predicted crack patterns. Both numerical models predict a similar response: only the first node of the first interface element at the dam/foundation interface across the model thickness is found to be fully cracked. In 3D all the first nodes of the first interface elements in the 3rd dimensions are found to be cracked. A similar result was obtained with a two variable scalar continuum damage model (tensile induced damage and compression induced damage) [15, 16], but in this case damage only occurred at the first plane/brick finite elements close to the heel of the dam.

The velocity time history of the Endurance Time Acceleration Function (ETAF) provided by theme A formulators was applied at the base of the foundation. The

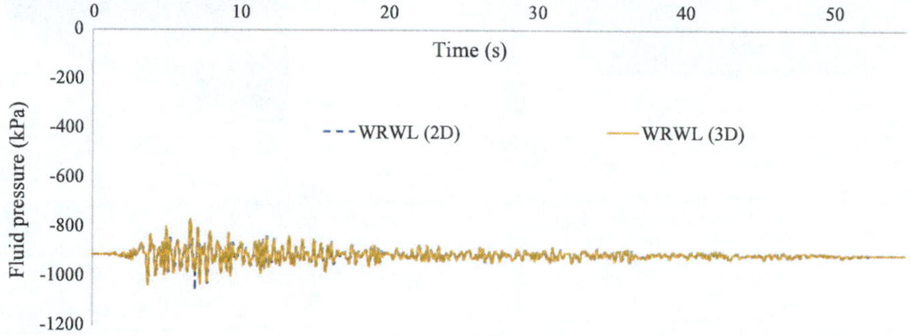

Fig. 7 Fluid pressure close to the upstream face of the dam at the dam/foundation interface WRWL—Taft record

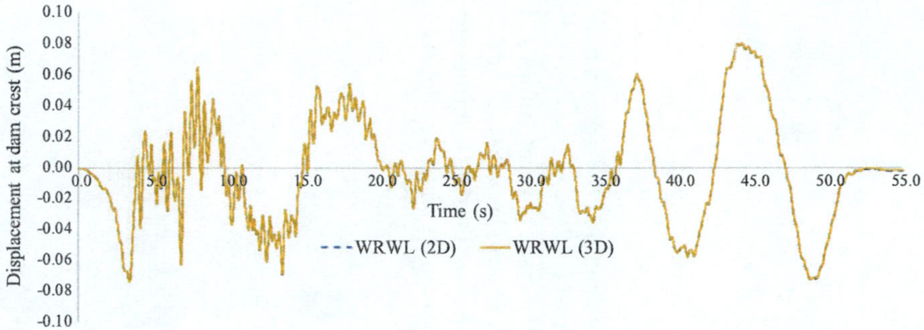

Fig. 8 Displacement of the crest of the dam in the upstream/downstream direction—WRWL—Taft record

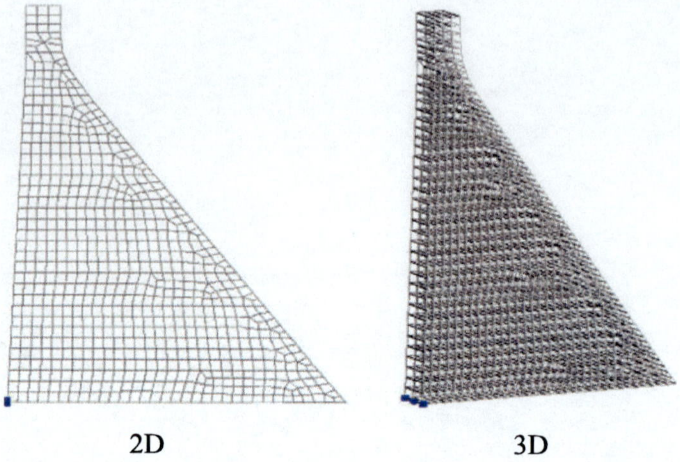

Fig. 9 Predicted crack pattern—Taft record—WRWL

ETAF record lasts 15 s and by definition the acceleration and response spectrum increase linearly within that period.

Figure 10 shows the predicted crack patterns before and after the time that was considered the dam to fail. In our point of view the criteria for failure should be chosen as the instant in which the crack crosses the dam from the upstream to the downstream face of the dam. This occurs between 6 s and 7 s in the 2D model and close to 7 s in

Fig. 10 Predicted crack pattern—ETAF record—WRWL

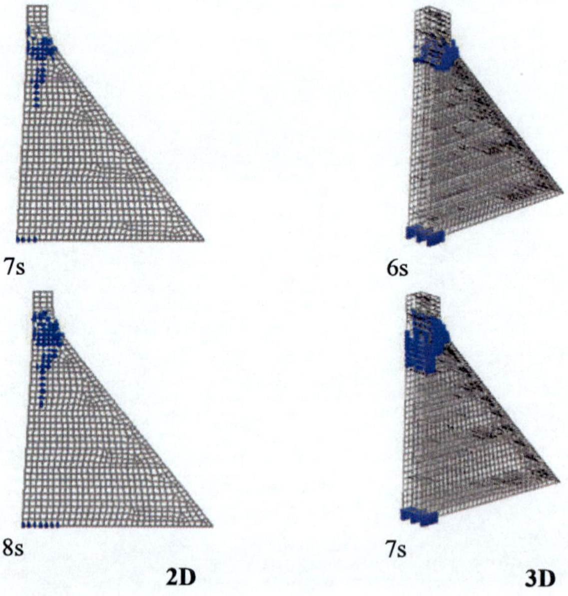

the 3D numerical model. Note that due to the explicit nature of the adopted solution algorithm it is possible to obtain for both numerical models a solution for the entire 15 s ETAF register. Nevertheless, the interface interactions may no longer be valid due to large displacements (Fig. 11) given that the adopted interface formulation is based on the hypothesis of small strains and small displacements.

Figure 12 shows the damage index evolution at the dam body for both models. Given the adopted discrete element crack approach the damage index is related to the total length/area of all the interface elements within the dam body and it does not include the interface finite elements adopted at the dam/foundation interface. The behaviour of both models has some similarities, with an evolution of both damage indexes at around 5 s. It can be seen that in 3D the damage index is slightly higher at the end of the record. Figure 13 shows the damage index for the dam/foundation interface and a similar behavior to the body index is obtained. However, at the end of the record, the damage index at the dam/foundation interface is slightly higher in the 2D model.

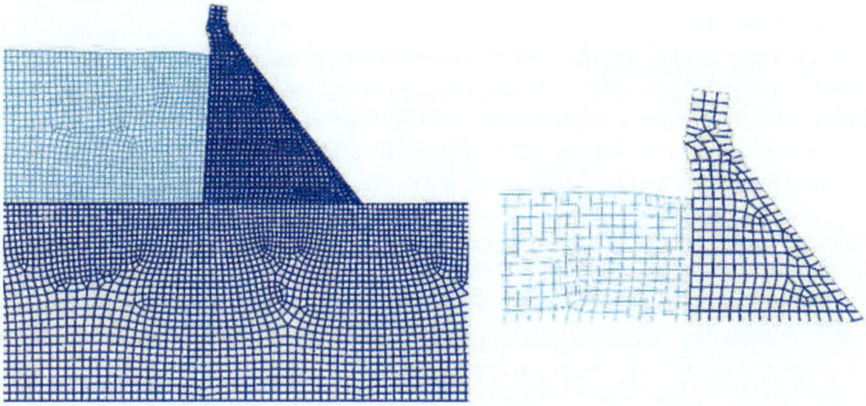

Fig. 11 Displacement field at the end of ETAF record (magnified 5 times)

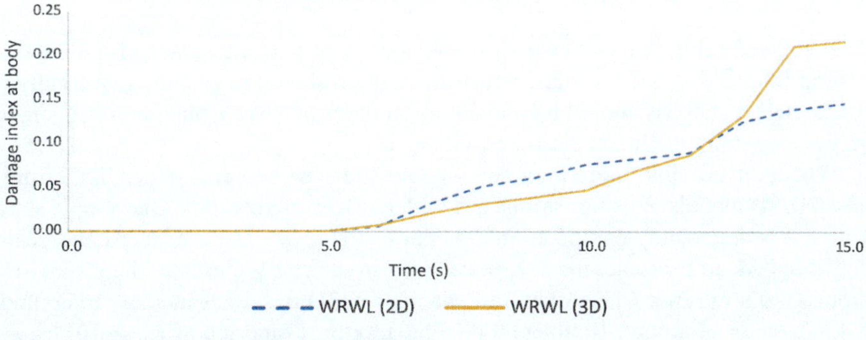

Fig. 12 Evolution of damage index at the body

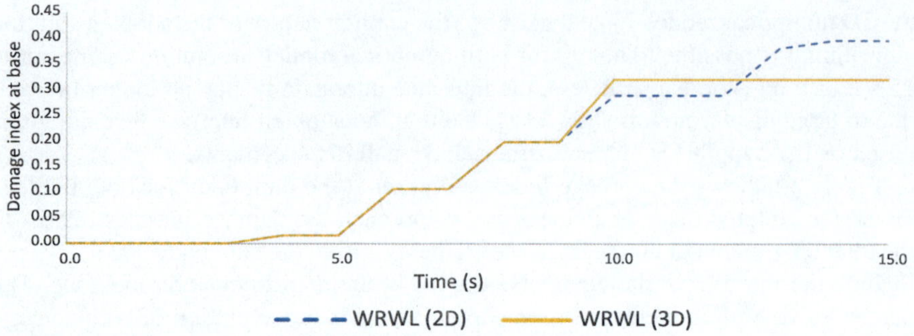

Fig. 13 Evolution of damage index at the dam/foundation interface

3 Summary of Main Results

Case A analysis showed that the eigenfrequencies for the Pine Flat dam are influenced by the adopted boundary conditions, namely the boundary condition adopted for the fluid lateral boundary. Further, the results were not presented, but if the lateral foundation boundaries were not fully fixed the predicted frequencies were significantly lower. This, in our view, explains some of the differences in the results presented in [6]. It was also shown that the 3D model predicts a slightly different response with a higher number of modes being predicted in the 1.0–7.0 Hz range.

Case D study showed that the 2D/3D response in terms of displacements and stresses at the dams are very similar. The slightly different behavior in terms of mode shapes has a less noticeable influence when seismic loading is being applied.

The nonlinear analysis that was carried out within the scope of study case E showed that both models predicted similar responses even for the ETAF record which leads to a high complex nonlinear behaviour.

4 Analysis Lessons Learned

The seismic analysis of a given dam should always be performed stepwise. First the natural frequencies and the static stress distribution should be obtained. Secondly a linear elastic analysis should be mandatory in order to asses if the predicted stress values are within the linear elastic envelope.

Whenever possible more than one model should be adopted, either like in this case 2D/3D models or using various different mesh discretizations. One should start with a less demanding computational model that allows a faster analysis and error checking. As an example, the 2D model used in the study presented here requires around 0.6 h to run a 10 s Taft record, whereas in 3D run times increase to around 5.4 h (a factor of around 10 times) for a similar critical timestep of 1.0×10^{-4} s.

5 Suggested Future Work

The analysis that was done regarding case study A showed that within the framework of a future workshop it would be interesting to carry out a dynamic characterization for a given range of frequencies, as, for example, a forced vibration analysis. This would allow the contributors to numerically define the transfer functions. Another possibility would also be to set a sine sweep providing the loading signal in order for the contributors to compare the response in terms of displacements at the dam crest.

To assess the fluid structure interaction models it would be important to assess the fluid models for different geometries, namely a gravity dam in a regular canyon, a gravity dam in a closed V shaped canyon, a perfect circular arch dam and finally an arch dam. For all the geometries the transfer functions for a given frequency range and at a given set of points and directions should be calculated. Further studies need to be carried out in order to assess whether the Lagrangian based models are able to predict correct responses for refined distorted fluid meshes.

Another important issue that needs to be addressed is 3D modelling, compared to a 2D model the discretization needs to be coarser and probably 2nd order bricks can be adopted in order to reduce the element refinement.

Finally in a future workshop related with nonlinear modelling, either static or dynamic, it would be interesting to have as an input the finite element mesh, especially for nonlinear results it is known that some nonlinear constitutive relationships are mesh dependent.

6 Conclusions

The natural frequencies and force vibration analysis (Case A), the seismic analysis for various reservoir levels following a linear elastic model (Case D) and non-linear seismic type analysis including nonlinear behaviour (Case E) were assessed and discussed using both 2D and 3D modules.

The 2D and 3D models were evaluated against each other and it was shown that the models predict, as expected, similar responses. The nonlinear behavior of both models is also similar and it would be interesting to assess the performance of both models for a seismic loading that is not as strong as the suggested ETAF loading.

Acknowledgments The study presented here is part of the research project "DAMFA: Cutting-edge solutions for sustainable assessment of concrete dam foundations" which has been supported by LNEC with the main purpose of developing a numerical multiphysic integrated tool for the sustainable assessment of concrete dam foundations, taking into account the interaction between the mechanical, hydraulic and thermal behaviours.

References

1. Câmara R (2000) A method for coupled arch dam-foundation-reservoir seismic behavior analysis. Earthq Eng Struct Dyn 29(4):441–460
2. Monteiro Azevedo N, Câmara R (2015) Dynamic analysis of concrete dams: fluid structure displacement based interaction models. Dam Eng XXV(4):113–132
3. Braga Farinha ML, Monteiro Azevedo N, Candeias M (2017) Small displacement coupled analysis of concrete gravity dam foundations: static and dynamic conditions. Rock Mech Rock Eng 50(2):439–464
4. Schclar Leitão N, Monteiro Azevedo N, Castilho E, Braga Farinha ML, Câmara R (2018) Thermal cracking computational analysis of a concrete arch dam. In: Malm R, Hassanzadeh M, Hellgren R (eds) 14th ICOLD international Benchmark workshop on numerical analysis of Dams. KTH, Civil and Architectural Engineering, Stockholm, Sweden, pp 107–117
5. Monteiro Azevedo N, Braga Farinha ML, Schclar Leitão N, Câmara R (2019) Fluid structure displacement based interaction models for the dynamic behavior of an arch dam. In: EWG2019—3rd meeting of EWG Dams and earthquakes. An International symposium, Lisbon, Portugal, pp 123–134
6. Evaluation of Numerical Models and Input Parameters in the Analysis of Concrete Dams (2018) Report DSO-19-13, a summary report of the USSD workshop, Miami, USA
7. Lysmer J, Kuhlemeyer RL (1969) Finite dynamic model for infinite media. J Eng Mech Div ASCE 95(4):859–878
8. Wilson E, Khalvati M (1983) Finite elements for the dynamic analysis of fluid-solid systems. Int J Numer Meth Eng 19(11):1657–1668
9. Calayir Y, Dumanoglu A (1993) Static and dynamic analysis of fluid and fluid-structure systems by the Lagrangian method. Comput Struct 49(4):625–632
10. Greeves E, Taylor C (1992) The use of displacement type fluid finite elements for the analysis of dam reservoir interaction. Dam engineering, III, pp 169–200
11. Akköse M, Adanur S, Bayraktar A, Dumanoglu A (2008) Elasto-plastic earthquake response of arch dams including fluid-structure interaction by the Lagrangian approach. Appl Math Model 32(11):2396–2412
12. Marti J, Cundall P (1982) Mixed discretization procedure for accurate modelling of plastic collapse. Int J Numer Anal Meth Geomech 6:129–139
13. Lemos JV (1999) Discrete element analysis of dam foundations. In: Sharma VM, Saxena KR, Woods RD (eds) Distinct element modelling in geomechanics, Balkema, pp 89–115
14. Lokke A, Chopra AK (2018) Direct finite element method for nonlinear earthquake analysis of 3-dimensional semi-unbounded dam-water foundation rock systems. Earthq Eng Struct Dynam 47(5):1309–1328
15. Oliveira S, Faria R (2006) Numerical simulation of collapse scenarios in reduced scale tests of arch dams. Eng Struct 28(10):1430–1439
16. Oliver J (1989) A consistent characteristic length for smeared cracking models. Int J Numer Meth Eng 28(2):461–474
17. Monteiro Azevedo N, Candeias M (2014) Particle modelling of failure scenarios for concrete gravity dams. In: 5as Jornadas Portuguesas de Engenharia de Estruturas, LNEC, Lisboa (in Portuguese)

Seismic Analysis of Pine Flat Concrete Dam

N. Naji-Mahalleh

Abstract Seismic safety of Pine Flat gravity dam is reevaluated by conducting a dam-reservoir-foundation interaction analysis. Contrary to conventional procedures, a foundation with mass is assumed and to avoid unrealistic detrimental effects of wave reflection at truncated boundaries of model, resort is made to non-reflecting boundary conditions. Plane strain 2D model with different element types are prepared. Solid elements are used to model dam and foundation. Compressible fluid elements is utilized in reservoir. Solid and fluid elements are tied (coupled) together on wet interfaces. To simulate elastic wave propagation in a uniform half-space and non-reflecting conditions at boundaries of truncated FE model, infinite elements along with appropriate free-field boundary conditions are incorporated. Efficiency of adopted numerical model is verified by imposing the foundation block to high and low frequency shear traction impulses applied at the base and calculating resulting velocity responses. Loading consists of self-weight, hydrostatic pressure, harmonic nodal force and base excitations due to earthquake records. Since, the FE model is a foundation with mass, thus the deconvolved acceleration records or the equivalent shear traction records are imposed at the base of foundation. Linear and nonlinear material properties are assumed for concrete. Nonlinearity is consistent with concrete damage plasticity model. In compression, uniaxial stress-strain relation as proposed by Saenz and in tension, stress-crack opening displacement relation as suggested by Hordijk are adopted. Analysis works consist of eigenvalue solution, displacement, acceleration and hydrodynamic pressure time histories for a few selected points. In addition, extent of damage suffered in dam when exposed to earthquake records of Taft and Endurance Time Acceleration Function is estimated.

Keywords Pine flat · Seismic interaction analysis · Wave reflection · Infinite elements · Truncated boundaries · Free-field boundary · Effective earthquake forces · Taft earthquake · Base shear traction · Deconvolved acceleration · Concrete damage plasticity model · Damage index

---

N. Naji-Mahalleh (✉)
Mahab Ghodss Consulting Engineering, Tehran, Iran
e-mail: nnaji1965@gmail.com

G. Bolzon et al. (eds.), *Numerical Analysis of Dams*, Lecture Notes in Civil Engineering 91, https://doi.org/10.1007/978-3-030-51085-5_9

1 Introduction

The intention of the ICOLD 15th Numerical Benchmark Workshop is to perform a seismic analysis of Pine Flat dam. The main scope of workshop is to investigate uncertainties in finite element analysis of concrete dams in a systematic and comparative approach. US Army Corps of Engineers constructed the Pine Flat dam located on King's River, California in 1954. It is a straight gravity dam consisting of thirty-six monoliths. Dam crest length is about 561 m. Its thickness at base and crest elevations are 95.8 and 9.75 m, respectively. Theme A includes an analysis of the highest non-overflow monolith No. 16 of the dam. The height of monolith is 121.91 m and its width is 15.24 m.

2 Studied Cases

Case studies introduced by Benchmark Formulators are designated as Case A to Case F. In the present study, results of a limited list of Cases are submitted.

Case A includes an eigen value analysis and a simulation of eccentric-mass vibration generator (EMVG) test performed in 1971. Case B investigates foundation size effect and efficiency of adopted non-reflecting and free-field boundary conditions in seismic events. High and low frequency shear impulses are applied at foundation base. Employed numerical model is a rectangular block with a fine mesh.

Case D investigates the effect of reservoir water level in an interaction analysis of dam-reservoir-foundation exposed to Taft earthquake. In Cases A and D, concrete is treated as a linear material.

Case E is again a seismic interaction analysis conducted for both records of Taft and ETAF (Endurance Time Acceleration Function) but with nonlinear behavior of concrete. Numerical models of Cases A, D and E are similar except in reservoir region where separate models are generated for different water levels.

3 Loads

Applied individual loads consist of self-weight, hydrostatic pressure, harmonic nodal force, impulsive shear traction and earthquake induced excitations in form of base shear tractions.

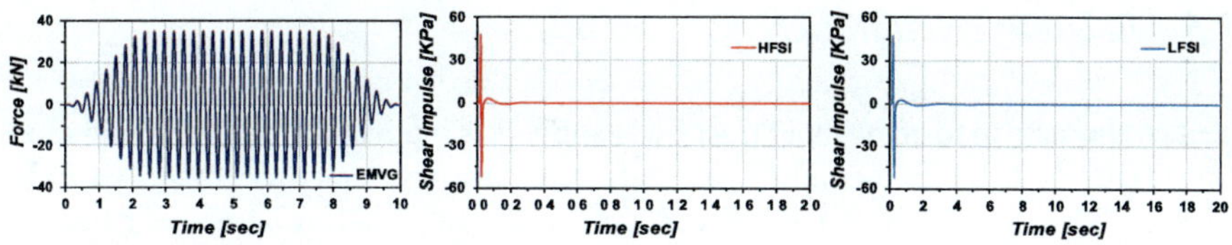

Fig. 1 EMVG harmonic force, high and low frequency shear impulses

3.1 Self-weight

It is applied only over concrete dam. No gravity effect is considered for reservoir or foundation.

3.2 Hydrostatic Load

It is applied only over wet part of the dam water face. Three reservoir levels are considered corresponding to winter at 268.21, summer at 278.57 and normal condition at 290.0 masl. No hydrostatic pressure is imposed over bottom part of reservoir at interface of rock foundation.

3.3 EMVG Harmonic Force

As is shown in Fig. 1, it is a harmonic nodal force time history applied in the middle of dam crest width and simulates an Eccentric Mass Vibration Generator (EMVG). Sine wave has a duration of 10 s, frequency of 3.47 Hz and an amplitude of 35.4 kN. In 2D plane strain case, the EMVG force is scaled by dividing it to the width (15.24 m) of the monolith No. 16.

3.4 Impulsive Shear Tractions

Efficiency of adopted numerical model in simulating non-reflecting boundary conditions is verified by imposing high and low frequency shear traction impulses at the base of foundation. Initially, it is a generated free-surface velocity (pulse) time history (v_x) where its frequency content depends on foundation model depth (H = 122 m), maximum element size (h = 1.5 m) and shear wave velocity (v_s= 1939 m/s).

As is shown in Fig. 1, velocity pulse is then scaled to shear traction by $\tau_{xy}(t) = \rho v_s v_x(t)$ and applied at foundation base in horizontal (X) direction. Both shear impulses have an amplitude of 48.15 kPa and an initial quiet period of 10 points.

3.5 Seismic Excitations

Taft earthquake in form of shear traction is applied as input at the base of foundation. Shear traction is found by multiplying Taft free-surface velocity time history (v_x) by rock foundation density (ρ) and shear wave velocity (v_s). Taft base shear traction (TBST) is shown in Fig. 2.

Endurance Time Acceleration Function (ETAF) is an artificially generated intensifying acceleration record, where its response spectra linearly increase with record duration (RDT). Seismic performance of structure is determined by the extent of time it can endure the dynamic input. The ETAF record and its response spectra is shown in Fig. 2.

4 Material Properties

Linear elastic and isotropic material properties are assumed for concrete and rock. The elastic properties are taken to be homogeneous. Reservoir water is assumed inviscid and irrotational. Linear properties are utilized in Cases A and D. Basic properties are summarized in Table 1.

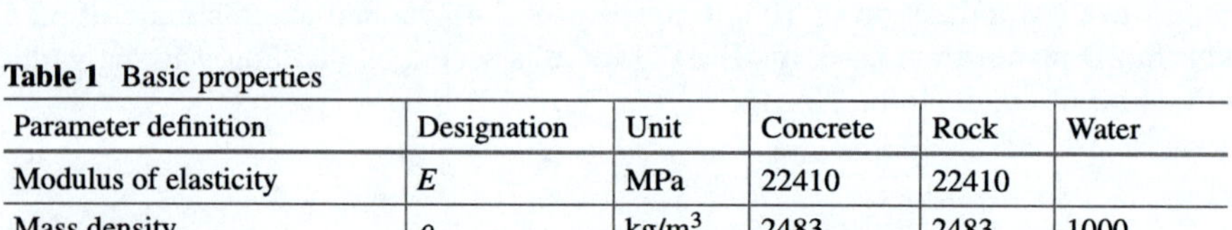

Fig. 2 Taft base shear traction, ETAF record and response spectra

Table 1 Basic properties

Parameter definition	Designation	Unit	Concrete	Rock	Water
Modulus of elasticity	E	MPa	22410	22410	
Mass density	ρ	kg/m^3	2483	2483	1000
Poisson's ratio	υ		0.2	0.2	
Compressive strength	f'_c	MPa	28.0		
Tensile strength	f'_t	MPa	2.0		
Tensile fracture energy	G_F	N/m	250		
Ultimate compressive strain	ε_c		0.0025		
Shear wave velocity	v_s	m/s		1939	
Compressive wave velocity	v_p	m/s		3167	
Bulk modulus	K	MPa			2070.721

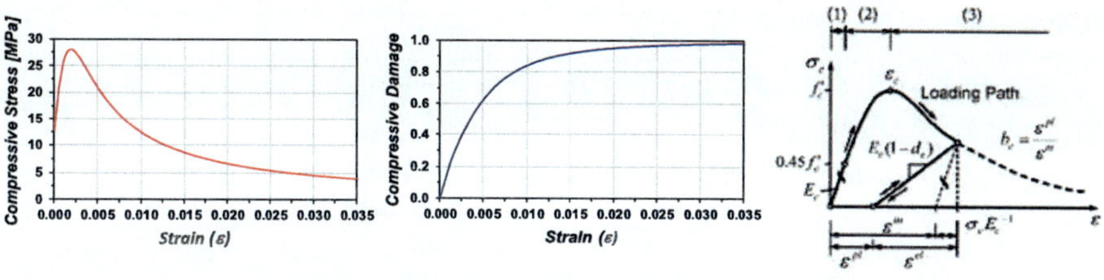

Fig. 3 Uniaxial stress-strain and damage variable in compression

In Case E, concrete in treated as a nonlinear material and its behavior is described by Concrete Damage Plasticity Model (CDPM). In uniaxial compression (σ_c), Eq. (1) defines its stress-strain relation. This relation was proposed by Saenz (1964) [1].

Initial compressive yield stress of concrete is taken equal to $f'_{c0} = 0.45 f'_c$ as per ACI 350 M-06, clause R8.5.1 [2]. In Fig. 3, variation of (σ_c) beyond initial yield stress $\left(f'_{c0}\right)$ is shown. Compressive damage variable (d_c) is defined as a function of inelastic strain (ε^{in}) by Eq. (2). The ratio of $\varepsilon^{pl}/\varepsilon^{in}$ is taken constant [3]. This ratio is found by experimental tests and in compression is assumed equal to b_c= 0.7.

$$\sigma_c = \frac{E_c \varepsilon}{1 + \left(\frac{E_c \varepsilon_c}{f'_c} - 2\right)\left(\frac{\varepsilon}{\varepsilon_c}\right) + \left(\frac{\varepsilon}{\varepsilon_c}\right)^2} \tag{1}$$

$$d_c = 1 - \frac{\sigma_c / E_c}{\varepsilon^{in}(1 - b_c) + \sigma_c / E_c}, \; b_c = \frac{\varepsilon^{pl}}{\varepsilon^{in}}, \; \varepsilon^{in} = \varepsilon - \sigma_c / E_c \tag{2}$$

The stress-strain relation (σ_t) for tensile loading consists of a linear part up to the tensile strength $\left(f'_t\right)$ and a nonlinearly descending (softening) part that depends on the specimen geometry. In the tensile softening branch, stress is defined as a function of crack opening (w) as is shown in Fig. 4.

This relation is proposed by Hordijk (1992) and is presented by Eq. (3). Critical crack opening (w_{cr}) is defined as a function of tensile fracture energy (G_F). Tensile damage variable (d_t) is defined as a function of inelastic strain (ε^{in}) as given by Eq. (4) but with $\varepsilon^{in}= w/L_t$. The average element length (L_t) is fixed to 1.5 m. Again, the ratio of $\varepsilon^{pl}/\varepsilon^{in}$ is taken constant and in tension is assumed equal to b_t= 0.1.

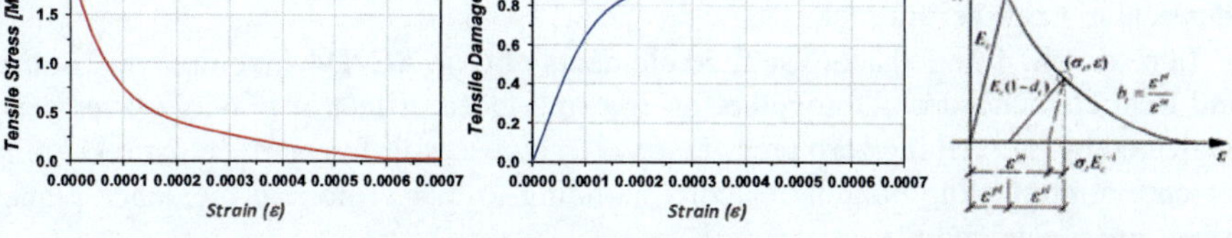

Fig. 4 Uniaxial stress-strain and damage variable in tension

Table 2 Multiaxial loading parameters

Parameter	Definition
$f'_{bc}/f'_c = 1.16$	Ratio of biaxial to uniaxial compressive strength
$K_c = 2/3$	Second stress invariant ratio
$\Psi = 35°$	Dilation angle
$\varepsilon = 0.1$	Eccentricity related to hyperbolic plastic potential
$\upsilon = 10^{-6}$	Viscosity parameter

$$\frac{\sigma_t}{f'_t} = \left[1 + \left(c_1 \frac{w}{w_{cr}}\right)^3\right] e^{-c_2 \frac{w}{w_{cr}}} - \frac{w}{w_{cr}}(1 + c_1^3)e^{-c_2},\ c_1 = 3.0,$$
$$c_2 = 6.93,\ w_{cr} = 5.14\frac{G_F}{f'_t} \tag{3}$$

$$d_t = 1 - \frac{\sigma_t / E_c}{\varepsilon^{in}(1 - b_t) + \sigma_t / E_c},\ b_t = \frac{\varepsilon^{pl}}{\varepsilon^{in}},\ \varepsilon^{in} = \frac{w}{L_t} \tag{4}$$

This value fits well with experimental results of concrete cyclic tests [3]. It is worth to mention that the crack opening (w) is equal to crack displacement (u_{ck}) as per ABAQUS designation [4]. In order to avoid numerical ill conditioning, the values of tensile stress (σ_t) is limited to $0.01 f'_t$.

In Table 2, parameters related to concrete multiaxial loading, i.e., yield surface and plastic potential function that is assumed a Drucker-Prager hyperbolic function are included.

5 Numerical Model

First order plane strain solid elements of type CPE4 are used to discretized dam and foundation. To model far-field region, infinite elements of type CINPE4 are utilized.

This type of element is capable of absorbing impinging compression and shear waves (Fig. 5). Nodes along the edge of CINPE4 elements that point in the infinite direction are positioned at about twice as far from dam as nodes on the boundary between finite and infinite elements. Material properties of infinite elements are taken identical to neighboring rock.

In reservoir, 4-noded acoustic fluid elements of type AC2D4 are employed. Solid and fluid elements are tied together on wet interfaces. Effect of gravity waves was neglected by prescribing zero pressure (p = 0) at reservoir free surface. At reservoir far end, nonreflecting boundaries corresponding to that of normal incidence plane waves are prescribed [4].

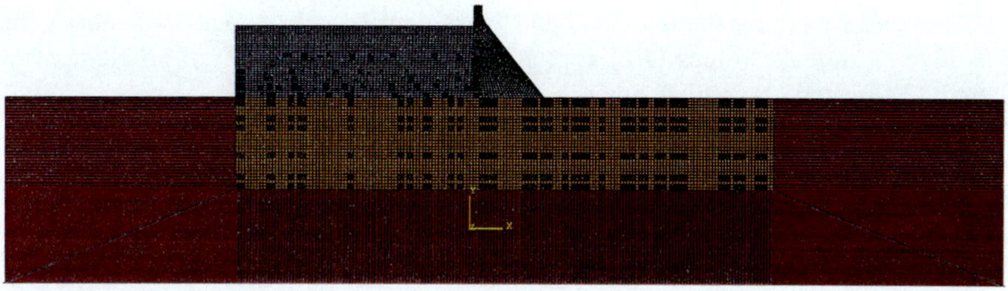

Fig. 5 Solid, fluid and infinite elements in FE model

6 Free-Field Effects

In order to find Effective Earthquake Forces (EEF) on the lateral vertical boundaries of the model, it is necessary to calculate free-field motions, i.e., displacement (u_x) and velocity ($\dot{u}_x$) responses at each nodal point located along vertical sides of the model [5]. Since, in the present study only a vertically propagating shear S-wave is taken into account, other components of the motion (u_y, $\dot{u}_y$) are equal to zero. With P-waves in seismic field, both components are involved.

As is shown in Fig. 6, analysis of this system reduces to a single column of the foundation-rock with absorbing infinite elements only at its base. The mesh density of the rock-column is taken identical to main FE model of foundation. For Cases D and E, Taft earthquake record in form of uniform Base Shear Traction (TBST) is applied at the base of rock-column. Prior to imposition of seismic load, displacement of pair of nodes positioned on both sides of rock-column model are constrained (using ABAQUS keyword: *MPC-Multi-point Constraints). Coupling (tying) is defined for both components.

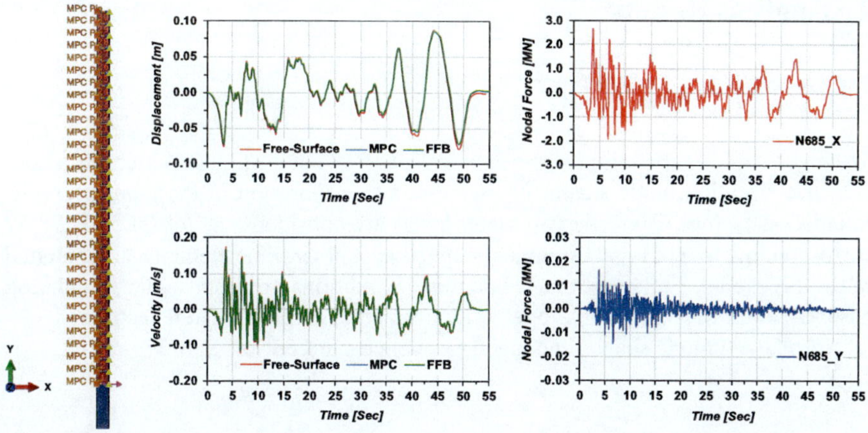

Fig. 6 Rock-column with infinite element at base and tied lateral nodes

Knowing Lamé constants (λ, G) and mass density (ρ) of rock foundation, the effective earthquake forces (P^n) applied at nodes (n and $n + 1$) of an element of length (h) located on vertical side are calculated by Eq. (5):

$$P_x^n = 0.5\left[h\rho c_p \dot{u}_x^n + l_x\lambda\left(u_y^{n+1} - u_y^n\right)\right],\ P_x^{n+1} = 0.5\left[h\rho c_p \dot{u}_x^{n+1} + l_x\lambda\left(u_y^{n+1} - u_y^n\right)\right]$$
$$P_y^n = 0.5\left[h\rho c_s \dot{u}_y^n + l_x G\left(u_x^{n+1} - u_x^n\right)\right],\ P_y^{n+1} = 0.5\left[h\rho c_s \dot{u}_y^{n+1} + l_x G\left(u_x^{n+1} - u_x^n\right)\right]$$
$$G = 0.5E/(1+\nu),\ \lambda = 2G\nu/(1-2\nu),\ c_p = \sqrt{(\lambda + 2G)/\rho},\ c_s = \sqrt{G/\rho} \tag{5}$$

where (c_p) designates P-wave and (c_s) S-wave velocities in foundation domain. For left side boundary, the coefficient ($l_x = -1$) is taken and on right side ($l_x = 1$) [6]. Nodal force calculation must be carried out for both vertical lateral boundaries, separately. With Shear Traction (TBST) imposed at the rock-column base, evidently $u_y = \dot{u}_y = 0$. Therefore, at each time increment the calculated P_x^n on both sides are identical and pointing in same direction. The same governs for P_y^n, but forces are in opposite directions. In Fig. 6, time history of resulting forces at node No. 31 (Main model 685) located on free surface (elevation 173.73) is shown.

For verification and check of results, infinite elements are added on vertical sides of the rock-column model and free-field forces are applied within a dynamic step together with TBST loading. Coefficients of Rayleigh damping are taken $\alpha = 0.75\ s^{-1}$ and $\beta = 0.0005$ s. Excellent agreement is achieved with Taft free-surface records.

In Case B, resort is made to another method known as tied lateral boundaries [7]. Procedure of tying (constraining) is quite similar to that used in rock-column model. In Case A, considering nature of its loading, no Free-Field Boundary (FFB) effect is deemed necessary to be imposed.

7 Analysis Results

7.1 *Cases A-1 and A-2*

The eigen value analysis is performed in order to find out the first six natural frequencies and vibration mode shapes of undamped free vibration of the dam-reservoir-foundation system. Two reservoir water levels are considered at 268.21 and 278.57 corresponding to winter and summer, respectively. No infinite elements are included in the FE model. Fully restrained boundary conditions are imposed at foundation base. On sides of foundation model, only horizontal displacement is restrained.

Normalized mode shapes and eigen frequencies are presented in Fig. 7.

Frequencies [Hz]			
Mode	Empty	RWL 268.21	RWL 278.57
1	2.499	2.301	2.195
2	3.818	3.368	3.117
3	4.201	3.823	3.687
4	4.492	4.304	4.230
5	4.863	4.865	4.837
6	5.679	5.321	5.248

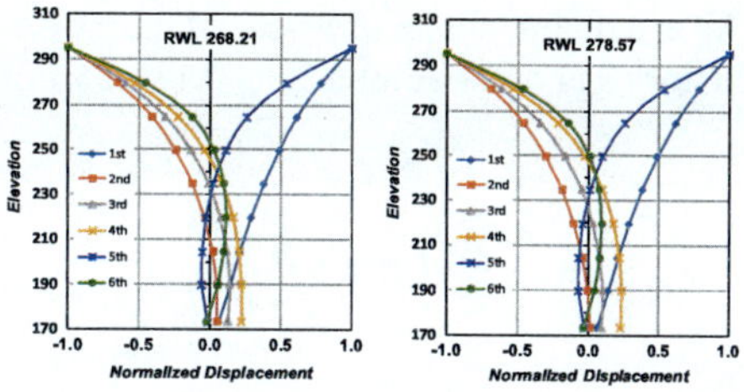

Fig. 7 Eigen frequencies and normalized mode shapes

7.2 *Cases A-3 and A-4*

A harmonic nodal force time history in the upstream-downstream d(5)irection is applied in the middle of dam crest width. It simulates an Eccentric Mass Vibration Generator (EMVG) and is equivalent to actual loading applied during a test performed in 1971. In addition, self-weight of concrete dam and hydrostatic pressure (only) on dam water face are imposed.

Rayleigh viscous damping with coefficients $\alpha = 0.75\ s^{-1}$ and $\beta = 0.0005$ s is utilized (refer to Fig. 11). This leads to a damping ratio of 3–2% in frequency range of 2–10 Hz. Integration time increment is taken equal to $\Delta t = 0.005$ s. No free-field boundary condition is considered due to nature of loading.

Relative horizontal displacement (Δu_x) results for two specific points (A, C) located on dam upstream face at heel and crest elevations are illustrated in Fig. 8. Position of points is shown in Fig. 11. Acceleration responses ($\ddot{u}_x$) are also given.

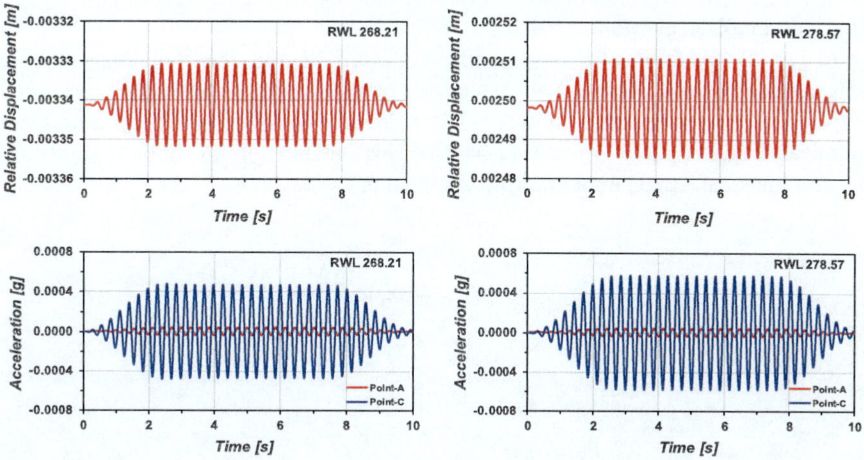

Fig. 8 Relative displacement/acceleration responses

As is evident, shape of imposed harmonic force is quite similar to acceleration time histories at crest point (C).

7.3 Cases B-1 and B-2

This is actually a wave propagation analysis in foundation domain. High and low frequency shear traction impulses are applied at the base. As is shown in Fig. 9, base shear impulses should reproduce the assumed free-surface velocity time history at the top of the model if it provides a good representation of the semi-infinite domain.

The characteristic frequency (f_N) of impulses are chosen by $f_N = 4Nf_0$ with $N_{\text{high}} = 2.5$, $N_{\text{low}}= 0.25$ and for shear S-waves $f_0 = \frac{v_s}{4H}$. Assuming a quarter-wavelength approximation and shear S-wave velocity in rock $v_s = 1939$ m/s, the lowest resolving frequency is found $f_0 = 4$ Hz. Dimensions of model are taken L = 700 by H = 122 m. Element size is confined to 1.5 m. Structural loading and viscous damping effect are excluded.

In Case B to include free-field effect, infinite non-reflecting elements located on left and right sides of the model are excluded. The infinite elements idealize a viscous boundary that could only absorb radiating waves. Instead, free-field boundary conditions are imposed to account for the effective seismic forces on vertical sides of the model. Imposition of free-field boundary conditions has a significant effect on analysis results of the nodes that are located nearby to side boundaries.

Resort is made to an approach known as tied lateral boundaries [7]. In this method, displacement of pair of nodal points positioned on lateral boundaries are constrained together. Coupling (tying) is defined for both displacement components. This leads to identical displacements of pair of nodes. For flat sites and horizontal layers, it provides an exact solution.

In Fig. 10, position of a few specific nodal points and corresponding extreme values of calculated velocities are tabulated. In addition, time history of horizontal velocity ($\dot{u}_x$) responses are shown. As is evident, responses of all top points are identical. Same behavior governs at base points. Pulse amplitude at base points are halved. Physical expectation confirms the results.

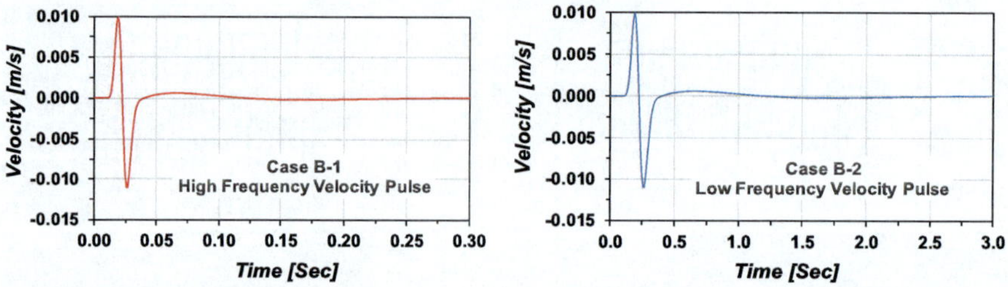

Fig. 9 Free-surface velocity pulses

Peak Velocity [m/s]										
Case		Top Points					Base Points			
		(a)	(c)	(e)	(g)	(j)	(b)	(d)	(f)	(h)
B-1	Max.	0.0085	0.0085	0.0085	0.0085	0.0085	0.0050	0.0050	0.0050	0.0050
	Min.	-0.0122	-0.0122	-0.0122	-0.0122	-0.0122	-0.0065	-0.0065	-0.0065	-0.0065
B-2	Max.	0.0096	0.0096	0.0096	0.0096	0.0096	0.0050	0.0050	0.0050	0.0050
	Min.	-0.0110	-0.0110	-0.0110	-0.0110	-0.0110	-0.0056	-0.0056	-0.0056	-0.0056

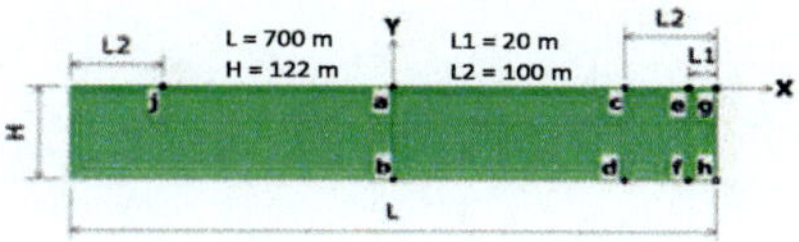

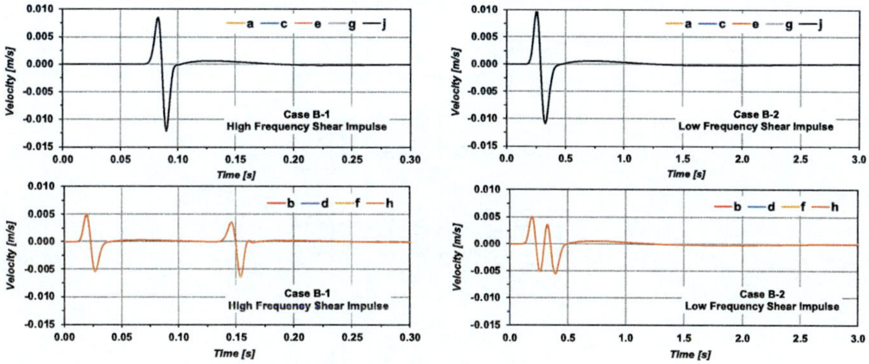

Fig. 10 Nodal velocity response histories

7.4 *Cases D*

In Case D, the effect of reservoir water level (RWL) in a dam-reservoir-foundation interaction analysis when exposed to the Taft seismic event is investigated. Reservoir water levels are given in Table 3. A separate numerical model is prepared for each level. Static loading consists of self-weight of concrete dam plus hydrostatic pressure on dam water face. Taft excitation is imposed in form of uniform shear traction (TBST) at the base of foundation (refer to Fig. 2).

Effective earthquake forces on the vertical lateral boundaries are calculated and superimposed in dynamic analysis step. These forces are extracted from a single rock-column in free-field condition.

Concrete is treated as a linear material. Rayleigh viscous damping with coefficients $\alpha = 0.75\ s^{-1}$ and $\beta = 0.0005$ s is utilized. Integration time increment is taken equal to $\Delta t = 0.01$ s. Identical time interval is used for result output. In Fig. 11, position of nodal points (A, C) and corresponding total displacement (u_x), hydrodynamic pressure (p) and acceleration ($\ddot{u}_x$) responses are shown. Extreme values are summarized in Table 3.

Following results are evident:

1. Displacement at heel point (A) almost matches free-surface displacement of the Taft record
2. Acceleration at heel point (A) nearly fits free-surface acceleration of the Taft record
3. Crest displacement is increasing with rising up of reservoir water level
4. Calculated acceleration at crest point (C) is about 10 times of heel point (A)

Table 3 Summary of results

Case	Reservoir level [mas]	Displacement (cm)				Pressure (MPa)		Acceleration (g)			
		Point A		Point B		Point A		Point A		Point C	
		Min.	Max.	Min.	Max.	Min.	Max.	Min.	Max.	Min.	Max.
D-1	286.21	−6.708	9.094	−7.399	8.842	−0.153	0.139	−0.160	0.137	−1.001	1.038
D-2	278.57	−6.569	9.232	−6.663	9.565	−0.162	0.138	−0.161	0.128	−1.029	0.932
D-3	290.00	−6.405	9.399	−5.526	10.731	−0.149	0.136	−0.163	0.124	−0.884	0.775
Taft free-surface		−7.862	8.729					−0.178	0.152		

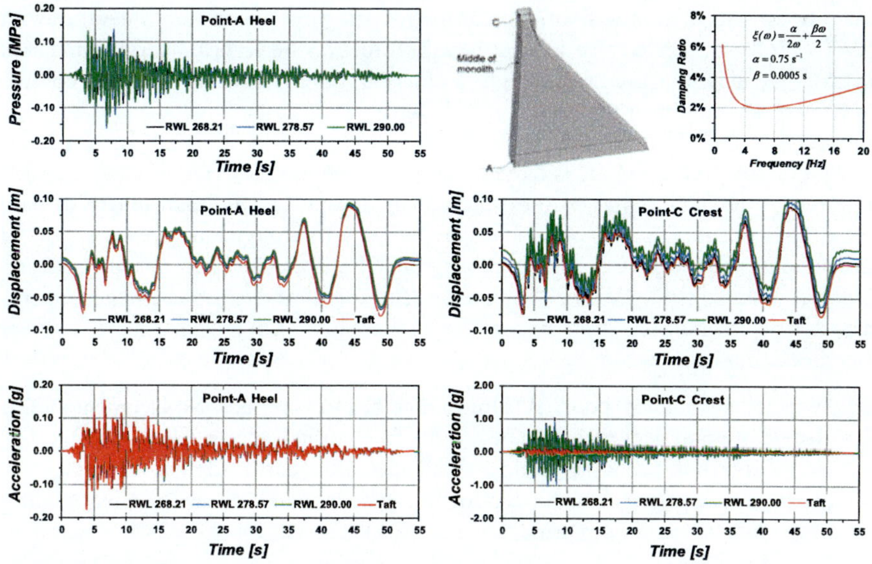

Fig. 11 Displacement, acceleration and hydrodynamic pressure responses

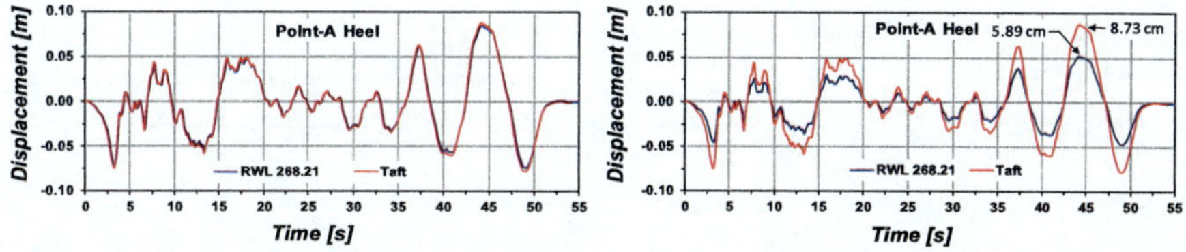

Fig. 12 EEF effects on displacement response (left: included, right: excluded)

It would be of interest to compare displacement results obtained at Point-A with and without inclusion of effective earthquake forces (EEF). In Fig. 12, the results are summarized. Without EEF, maximum response difference at A and Taft free-surface record is about 33%. Whilst, with EEF included, displacement (u_x) at heel point (A) almost fits free-surface displacement of the Taft record. Same scenario also governs for $\dot{u}_x$ and $\ddot{u}_x$ responses. It is worth to remind that, on both graphs of Fig. 12 the initial displacement due to static loading is excluded for proper comparison.

7.5 Cases E-1

In Case E-1, a nonlinear dynamic interaction analysis is performed by applying Taft excitation in form of uniform shear traction (TBST) at the base of foundation. Static loading consists of dam self-weight plus hydrostatic pressure on dam water face with RWL at 268.21. Applied effective earthquake forces on the vertical lateral boundaries is taken identical to Case D. Rayleigh viscous damping is same as Case D. Concrete is treated as a nonlinear material and its behavior is described by Concrete Damage Plasticity Model (CDPM). Automatic Time Incrementation (ATI) scheme is used.

Output interval of results is fixed at 0.01 s. Damage Index (DI) at concrete-rock interface is defined as the ratio of the damaged length to the total length of dam base (95.8 m). In Fig. 13, total displacement (u_x), hydrodynamic pressure (p) and acceleration ($\ddot{u}_x$) responses for the nodal points that are positioned on dam heel and crest (A, C) are displayed. For comparison, Taft free-surface displacement record is also added. In addition, stress history for an element at top cracking zone is given. Remarkable outcomes are:

1. Displacement at heel point (A) almost matches free-surface displacement of Taft
2. Acceleration at heel point (A) nearly fits free-surface acceleration of Taft
3. No local failure occurs within dam wall specially near its top narrow neck
4. Damage index along dam-foundation interface is found zero with no base failure
5. Magnitude of maximum principal stress (p1) is found about 1.72 MPa near dam top at t = 8.10 s.

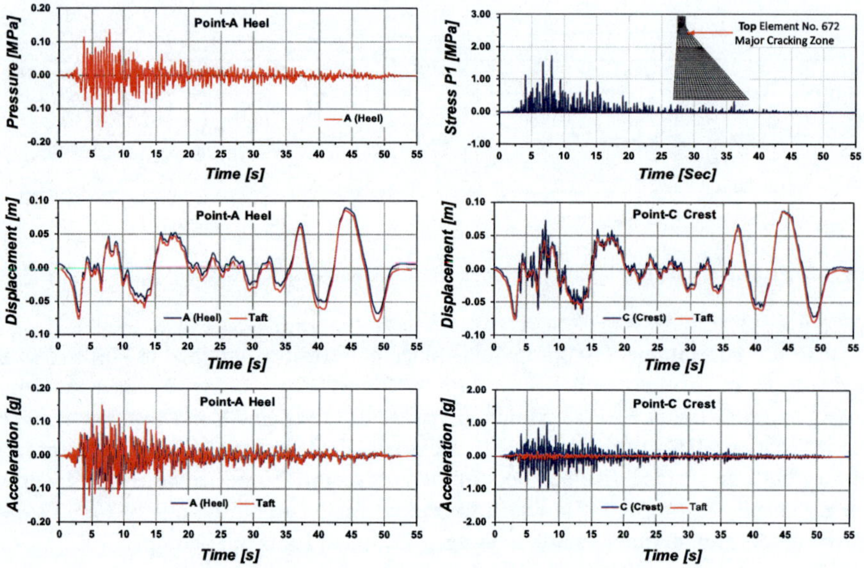

Fig. 13 Displacement, acceleration and hydrodynamic pressure responses

7.6 Cases E-2

In Case E-2, a nonlinear dynamic interaction analysis is performed by applying record of Endurance Time Acceleration Function (ETAF) at the base of foundation. It is an artificially generated intensifying acceleration record (refer to Fig. 2).

Static loadings, nonlinear properties of concrete, Rayleigh damping and integration scheme are identical to Case E-1. Seismic performance of structure is determined by the extent of time it can endure ETAF dynamic input. The damage index is defined as the ratio of sum of individual areas of local damages to total cross sectional area of dam (5710.86 m^2). At each integration time step, elements with overall value of tensile damage variable (D_t) greater than 0.25 are marked as the damaged ones.

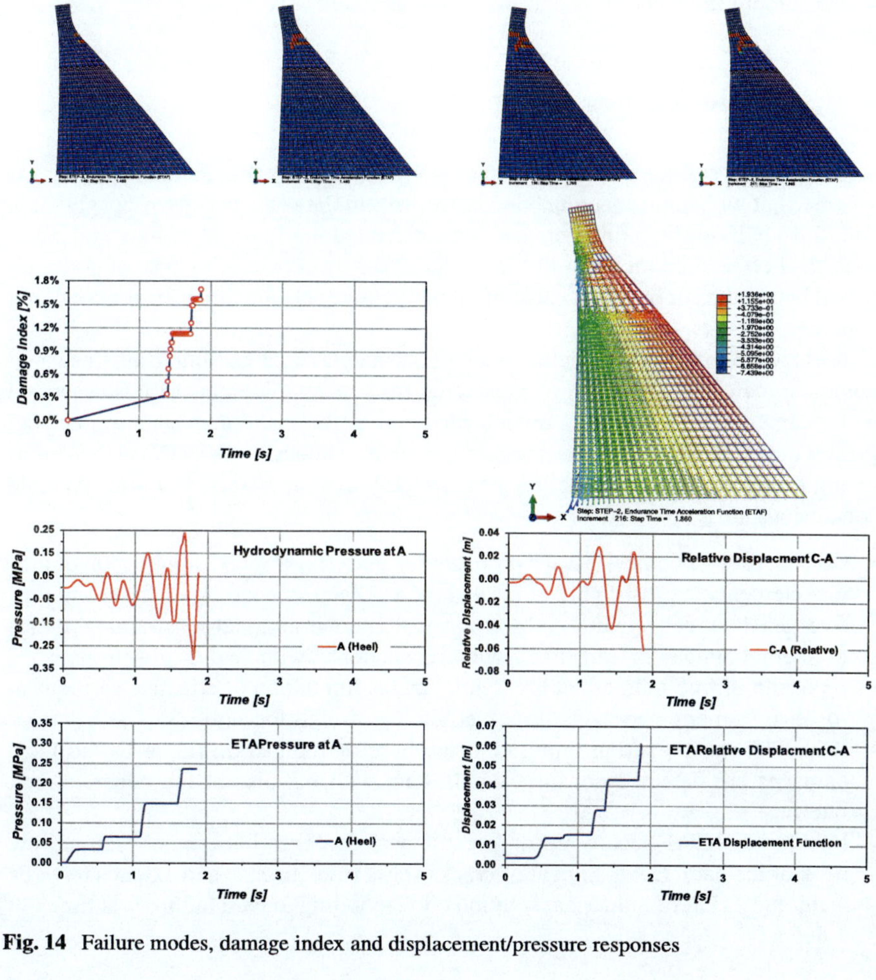

Fig. 14 Failure modes, damage index and displacement/pressure responses

In Fig. 14, percentage of damage index, pressure response at node A and relative displacement history of nodes A and C are displayed. A snapshot of in-plane principal stresses at time t = 1.86 s prior to failure is also given. In addition, four snapshots showing evolution of damage in dam wall are included. Time of the first one is at the beginning of damage (t = 1.4 s) and the last one at the end of calculation (t = 1.865 s).

Following comments are relevant:

1. Dam fails in tension as the calculated compression damage variable (D_c) is not noticeable
2. Local damages start form dam air face and propagates toward water face
3. There is a numerically ill-condition with no further convergence at failure time 1.865 s
4. Maximum (total) crest displacement at C is found −10.53 cm at time 1.865 s.

8 Conclusions

Seismic safety of Pine Flat gravity dam is reevaluated. A foundation with mass is assumed but with non-reflecting boundaries. To simulate elastic wave propagation in a uniform half-space, infinite elements are placed along foundation base and sides. Free-field effects are included in the model either by constraining pair of nodes on lateral boundaries or by direct calculation of effective seismic forces by rock-column approach.

A 2D plane strain fluid-structure interaction analysis is carried out. Both linear and nonlinear material behaviors are considered for concrete. Nonlinearity is described by concrete damage plasticity model. Elements with tensile damage variable (D_t) greater than 0.25 are marked as the damaged ones. Damage index (DI) is defined as the ratio of sum of local damaged areas to total dam cross sectional area. Notable conclusions are:

- Considering outcomes of wave propagation analysis (Case B), application of infinite elements are justifiable to idealize non-reflecting boundaries. Imposition of free-field boundary conditions by an approach known as nodal constraining (tying) of lateral boundaries leads to exact correspondence with analytical solutions
- Inclusion of free-field effect is essential and computation of effective earthquake forces by general approach of rock-column is straightforward
- With Taft shear traction imposed at the base of foundation (Case E), no local damages are detected not only within dam wall but also at the concrete-rock interface
- Results for ETAF record are different. Local damages initiate near top narrow neck of the dam. Development of cracks starts from air face and extends towards water face. Halt of numerical solution corresponding to dam failure is at time t = 1.865 s

- Results of nonlinear solutions are quite susceptible to input parameters of the material model that describes concrete nonlinear behavior.

References

1. Tao Y, Chen JF (2015) Concrete damage plasticity model for modeling FRP-to-concrete bond behavior. ASCE J Compos Constr 19(1):04014026
2. ACI 350 M-06 (2006) Code requirements for environmental engineering concrete structures. American Concrete Institute
3. Birtel V, Mark P (2006) Parameterized finite element modelling of RC beam shear failure. 2006 ABAQUS users' conference, pp 95–108
4. ABAQUS Theory guide (2016) ABAQUS Inc
5. Løkke A, Chopra AK (2017) Direct finite element method for nonlinear analysis of semi-unbounded dam-water-foundation rock systems. Earthq Eng Struct Dyn 46:1267–1285
6. Nielsen AH (2006) Absorbing boundary conditions for seismic analysis in ABAQUS. In: 2006 ABAQUS users' conference, pp 359–376
7. Zienkiewicz OC, Bicanic N, Shen FQ (1989) Earthquake input definition and the trasmitting boundary conditions. In: Proceedings of advances in computational nonlinear mechanics. Springer, Vienna, pp 109–138

Seismic Excitation of a Concrete Dam: Analysis of the Influence of Modelling Approaches and Concrete Material Non–linearity

E. Panteki and M. Goldgruber

Abstract The theme of this numerical benchmark workshop is related to the seismic analysis of Pine Flat Dam. The dam consists of 47 concrete monoliths and can thus be numerically modelled as a 2D structure. The analysis in this paper is conducted using the commercial finite element solver ANSYS. A comparison between results achieved with the typically used massless foundation and the viscous spring model is given as well as a comparison between results achieved with linear and non-linear concrete material models. In the latter case, the damaged plasticity model for concrete according to Menetrey and Willam is used. It is shown that as long as the loading is weak to moderate, the linear material model can provide a first guess of the response. However, once the loading is moderate to strong, the effects of concrete damage and cracking are significant and the models neglecting material nonlinearity are too conservative regarding expected stresses and cannot predict the real system behavior. Regarding the influence of the analysis approach, a clear tendency for the massless foundation model to reach higher peak responses that the viscous spring model is documented. The latter seems more accurate but is also more complex to set up.

Keywords Earthquake · Dam · Concrete · ANSYS · Menetrey-Willam

1 Introduction

The theme of this numerical benchmark workshop is the seismic analysis of Pine Flat Dam, see Fig. 1. The workshops purpose is to analyse the uncertainties in finite element analyses of concrete dams. Input system variables like the geometry, reference material parameters and loads are given by the formulators in [1], whereas the finite element solvers, analysis types, boundary conditions, nonlinear material models and meshes are chosen by the contributors.

E. Panteki (✉) · M. Goldgruber
DYNARDO Austria GmbH, Vienna, Austria
e-mail: kontakt@dynardo.at

G. Bolzon et al. (eds.), *Numerical Analysis of Dams*, Lecture Notes in Civil Engineering 91, https://doi.org/10.1007/978-3-030-51085-5_10

Fig. 1 Downstream view of Pine Flat Dam [2]

This paper documents and discusses the results achieved at Dynardo Austria GmbH. A comparison is given between two methodologies and special focus is given on the nonlinear earthquake simulations, since they provide the most valuable information about the risks and consequences. An outlook on open questions mostly regarding uncertain parameters and the assessment of their influence is given in the end.

2 The Finite Element Model

2.1 Approach

The finite element solver ANSYS 19.2 [3] is used for implicit transient structural analysis of the dam-foundation-reservoir under seismic loading. The fluid-structure interaction is addressed with acoustic-structural coupling using the acoustic ACT extension. In line with the specifications in [1] and since the monoliths of the structure allow for it, modelling is conducted in 2D.

In general, different methods can be used to carry out the numerical analysis of seismic excitation of dam-foundation-reservoir systems. In engineering practice, the perhaps most often used method is the massless foundation method. Here, the foundation is linear elastic and massless and the earthquake loading is applied uniformly to the structure by inertia force. Another analysis method comes with the incorporation of non-reflecting boundary conditions in which case the foundations mass is considered, and loading is applied on the truncated boundary of the domain. In this paper, both approaches are taken and compared.

For the non-linear calculations, the material model for concrete plasticity according to Menetrey-Willam [4] is used. Cracks are documented and modes of failure for a standard and a pushover earthquake loading are discussed.

2.2 Mesh

Based on the provided geometry, see Fig. 2, the mesh of the finite element model was prepared. Linear solid elements (SOLID185) were used for the dam and the foundation and acoustic fluid elements (FLUID30) were used for the water in the reservoir. The characteristic mesh size amounts to approx. 2.5 m. No contact algorithms or elements were used and adjacent bodies share common nodes instead. The non-reflecting boundary conditions were modelled with viscous spring elements (COMBIN14) and created using ANSYS Parametric Design Language (APDL). Surface elements (SURF154) were introduced internally at the interface between the reservoir and the dam and the reservoir and the foundation to account for the applied hydrostatic pressure load. Table 1 presents an overview of the number/type of elements/nodes used for the reference geometry mesh (winter reservoir water level (WRWL) of 268.21 m).

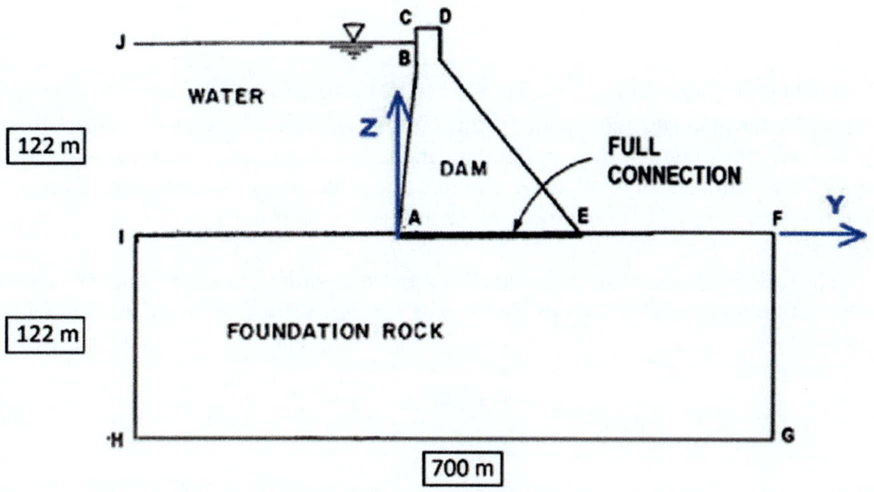

Fig. 2 Geometry of the model

Table 1 Summary of the mesh for WRWL

	Total	Solid 185	Fluid 30	Surf 154
Elements	20777	14892	4758	161
Nodes	41016			

2.3 Boundary Conditions

In order to analyze a 3D structure by means of 2D model, in the here taken 2.5D approach, solid/acoustic 8-node elements were used and their translations along one coordinate (in this case the x-coordinate) locked.

The reservoir was modelled as an acoustic body and attributed the density and wave propagation velocity of water. A constant reference acoustic pressure was defined on its free surface while an acoustic impedance boundary was introduced on its infinite side in order to simulate the non-reflecting boundary. No structure-fluid interaction conditions were required since, as stated above, no contact was defined, and the bodies share common nodes.

Rayleigh viscous damping was used for the foundation and the dam. According to the specifications of the workshop, the constants of mass and stiffness proportionality were taken as $\alpha = 0.75\ 1/s$ and $\beta = 0.0005\ s$. The damping matrix $\boldsymbol{C}$ is then computed as a linear combination of the mass $\boldsymbol{M}$ and stiffness $\boldsymbol{K}$ matrices:

$$\boldsymbol{C} = \alpha \boldsymbol{M} + \beta \boldsymbol{K} \tag{1}$$

For the viscous spring approach, COMBIN14 elements were added both in the direction normal and tangential to the bottom and the sides of the foundation. However, given that the main earthquake induced motion is horizontal and in order not to overestimate the vertical settlements due to the static weight of the dam and the reservoir, the vertical degrees of freedom at the bottom of the foundation were locked.

The values for the normal (index *BN*) and tangential (index *BT*) stiffness (*K*) and damping (*C*) coefficients of the viscous springs, were defined by the following equations:

$$K_{BN} = \frac{1}{1+\alpha}\frac{\lambda + 2\mathrm{G}}{2r_b}A, \quad C_{BN} = \beta\,\rho\,c_p A \tag{2}$$

$$K_{BT} = \frac{1}{1+\alpha}\frac{\mathrm{G}}{\alpha 2r_b}A, \quad C_{BT} = \beta\,\rho\,c_s A \tag{3}$$

where G is the shear modulus, λ is the first Lamé-constant and r_b is the distance from the wave source to the artificial boundary, taken as the accounted height and half of the length of the foundation for the vertical and the horizontal side boundaries respectively. The compressive and shear wave velocity are given by c_p and c_s respectively. The dimensionless parameters α and β are used for scaling and were taken as 0.8 and 1.1, according to [5]. *A* is lastly a mesh-dependent parameter and stands for the element area that corresponds to the node where the viscous spring is positioned.

2.4 Material Properties

Table 2 summarizes the standard provided concrete dam material properties. For the foundation, the material is always considered linearly elastic and the respective values match the ones of the concrete dam.

For the non-linear calculations, the material model for concrete plasticity according to Menetrey-Willam [4] was used. It is based on the Willam and Warnke [6] yield surface and for its use additional values were defined. They regard the concrete softening behavior in compression and tension as illustrated in Fig. 3 and are listed in Table 3.

Table 2 Standard concrete material properties

Parameter	Value	Unit
Modulus of elasticity	22410	MPa
Density	2483	kg/m^3
Poisson's ratio	0.20	–
Compressive strength	28.0	MPa
Tensile strength	2.0	MPa
Fracture energy	250.0	N/m
Compressive strain at peak load	0.0025	–
Tensile strain at peak load	0.00012	–

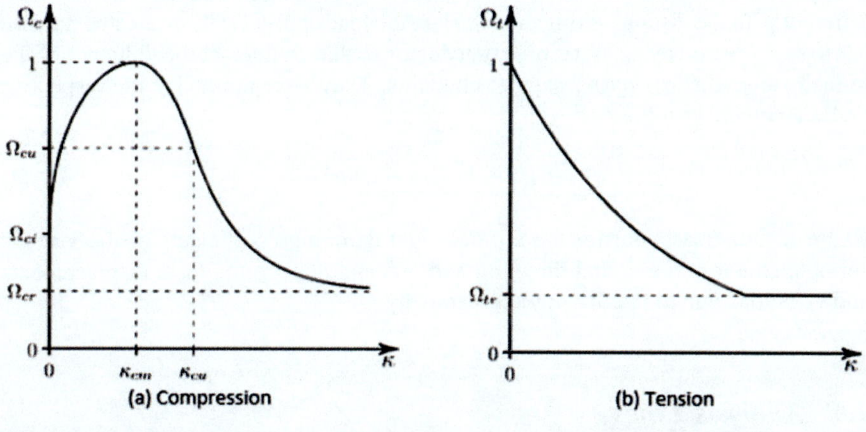

Fig. 3 Exponential softening in compression and tension from the ANSYS manual [3]

Table 3 Additionally, defined concrete material properties for use with the Menetrey-Willam concrete plasticity material model with exponential softening

Parameter	Value	Unit
Biaxial compressive strength	33.6	MPa
Dilation angle	15	°
Relative stress level Ω_{cu} at κ_u	0.85	–
Plastic strain at uni-axial compressive strength κ_{cm}	0.002411	–
Plastic strain defining start of exponential softening κ_{cu}	0.003424	–
Relative stress level at start of hardening Ω_{ci}	0.33	–
Residual relative stress level in compression Ω_{cr}	0.2	–
Residual relative stress level in tension Ω_{tr}	0.005	–

2.5 Loads

The static weight of the dam is considered by means of a static load step before the actual earthquake loading. The same applies to the hydrostatic pressure on the interface between the reservoir and the dam. The static weight of the foundation is not regarded.

The earthquake loading is given in the form of a vertically propagating SH wave, for which the acceleration and shear stress time history at the base of the foundation are known. In the case of the massless foundation approach, the acceleration signal is sufficient for the simulation of the dynamic loading. The situation is, however, different with the viscous spring model. Here the load applied at the truncated domain has to be corrected by discrete force time histories that counteract the influence of the immediately adjacent spring damper elements. They were applied at the respective nodes and were specified as:

$$F(t) = K\,u(t) + C\,\dot{u}(t) \tag{4}$$

where K and C account for the stiffness and damping coefficients of the viscous spring element in horizontal direction and u(t) and $\dot{u}$(t) are the local displacements and velocities due to the SH wave respectively.

2.6 Analysis Types

Before the actual transient structural analysis evaluating the earthquake, modal analysis of the dam-reservoir-foundation system was conducted to provide information on the characteristics of the structure. Besides the WRWL, another reservoir water level was analyzed. It differs from the first by 10.36 m of additional water which will be shown to indeed have an effect on the vibration characteristics of the structure.

Table 4 Natural frequencies (dam-foundation-reservoir system)

Mode	1	2	3	4	5	6
WRWL	2.31	3.41	3.89	4.33	4.89	5.26
SRWL	2.20	3.16	3.75	4.23	4.86	5.21

In a second step, the transient structural analysis is conducted. Two earthquake acceleration signals are applied, and their effects are documented. In order to place the excitation in relation to the natural vibration characteristics of the system, formerly identified in the modal analysis, a Fourier decomposition of the signal into its constituent frequency components is undertaken.

3 Results

3.1 Results of the Modal Analysis

Prior to the transient dynamic analysis, a modal analysis was conducted to identify the natural modes of the structure for two different reservoir levels: the winter (at El. 268.21 m) and the summer one (at El. 278.57 m). The natural mode shapes are, as expected, similar to each other with the higher water level lengthening the fundamental period of the dam-foundation-reservoir system. Table 4 summarized the results for the two models.

3.2 Results of the Fourier Analysis

The earthquake acceleration-time signal is shown in Fig. 4. Figure 5 shows its Fourier frequency decomposition along with the natural frequencies of the system for the two water levels in consideration.

3.3 Results of the Transient Structural Analysis

A comparison between the massless foundation and the viscous spring approach was set up using the linear-material model for the concrete dam. Figures 6 and 7 present the displacement and the acceleration history on the crest of the dam respectively. The results show a clear tendency for the massless foundation model to reach higher peak values.

A comparison between the linear and the non-linear material model was set up using the viscous spring approach. Figures 8 and 9 present the displacement and the

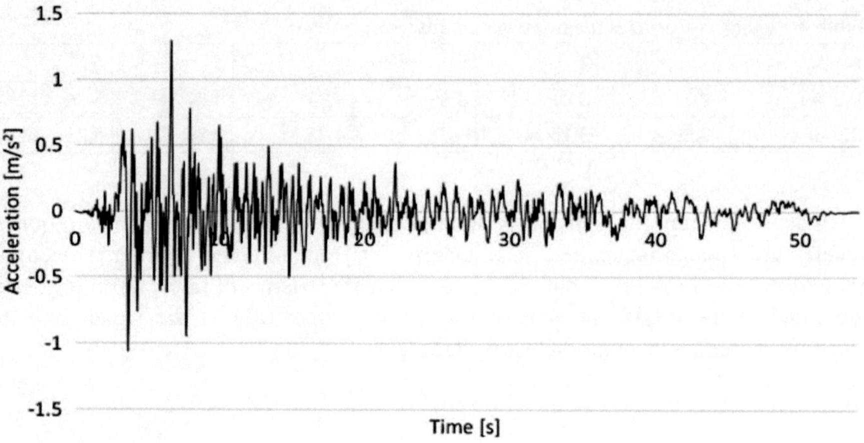

Fig. 4 Taft loading

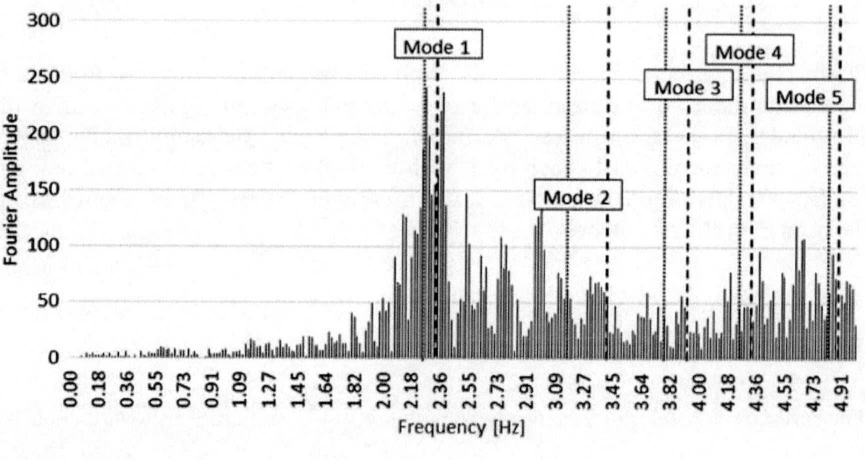

Fig. 5 Taft loading Fourier frequency decomposition and natural frequencies of the dam-reservoir-foundation system (dotted line: WRWL, dashed line: SRWL)

acceleration history on the crest of the dam respectively. The signal maxima/minima are in the same order of magnitude and in fact comparable. This finding suggests that the dam is damaged but not yet failed.

Figure 10 presents a plot of the plastic strain at the end of the Taft earthquake non-linear simulation. It shows cracks on the downstream side of the dam on water level and on the heel on the dam, however the structure is indeed still intact.

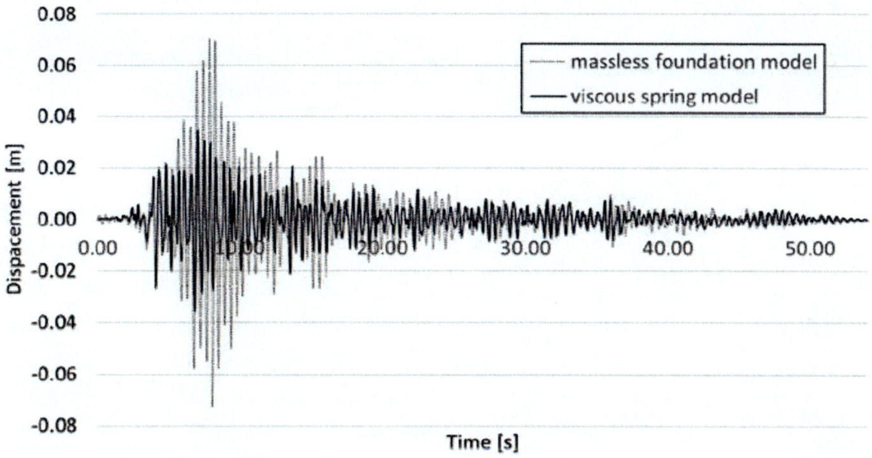

Fig. 6 Displacement histories of the crest of the dam due to the Taft earthquake and comparison of the two considered approaches

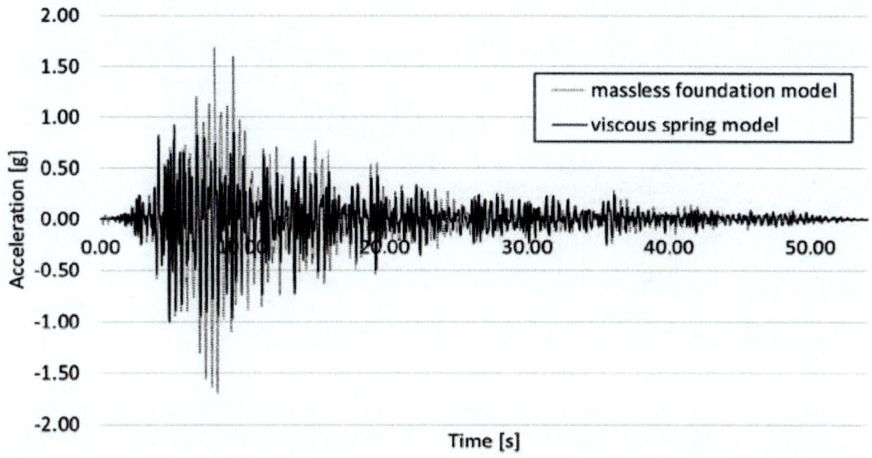

Fig. 7 Acceleration histories of the crest of the dam due to the Taft earthquake and comparison of the two considered approaches

In order to analyze the structure till failure or considerable damage is reached, an endurance time analysis (ETA) is undertaken. It shall estimate the seismic performance of the dam-reservoir-foundation system when subject to an intensifying excitation, see Fig. 11.

Plots of the plastic strain at several moments in time during the Etaft earthquake excitation are shown in Fig. 12.

Although the plots, especially towards the end of the simulation, suggest unrealistically advanced cracks, conclusion can still be drawn regarding the failure mode of

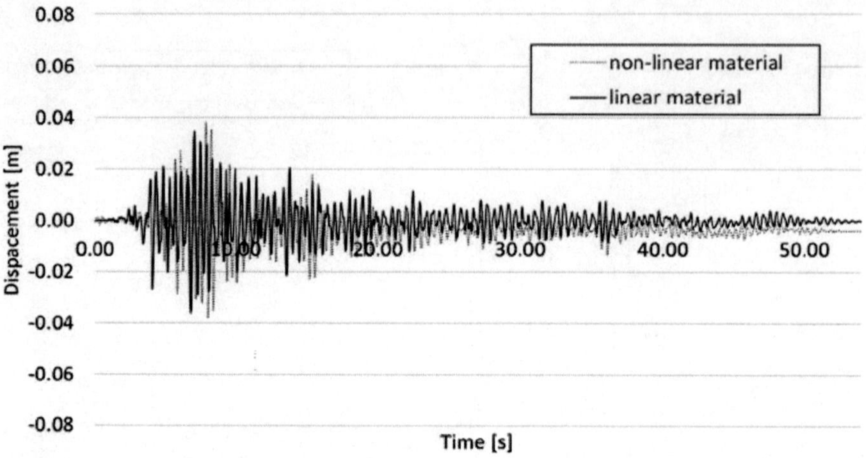

Fig. 8 Displacement histories of the crest of the dam due to the Taft earthquake and comparison of the linear and the non-linear material model

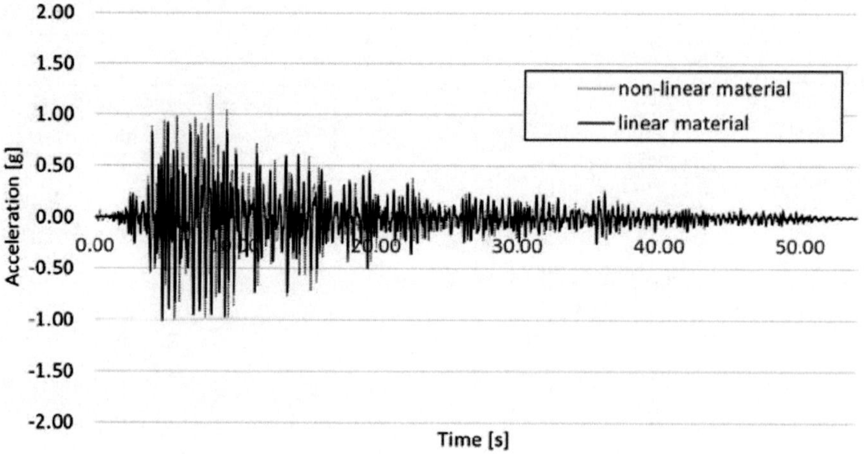

Fig. 9 Acceleration histories of the crest of the dam due to the Taft earthquake and comparison of the two considered approaches

the dam. In general, cracks form when the stress of the concrete material exceeds its tensile strength. In the case of the ETaft earthquake excitation of the dam-foundation-reservoir-systems, the dam is subject to high bending moments. The first crack initiates on the downstream side of the dam on the height of the water level and propagates over time as loading increases. On the heel point of the dam, another weak point is spotted.

Here the dam shows a tendency for separation from the foundation. The same holds for the downstream base point of the dam at later moments in time. Again, on

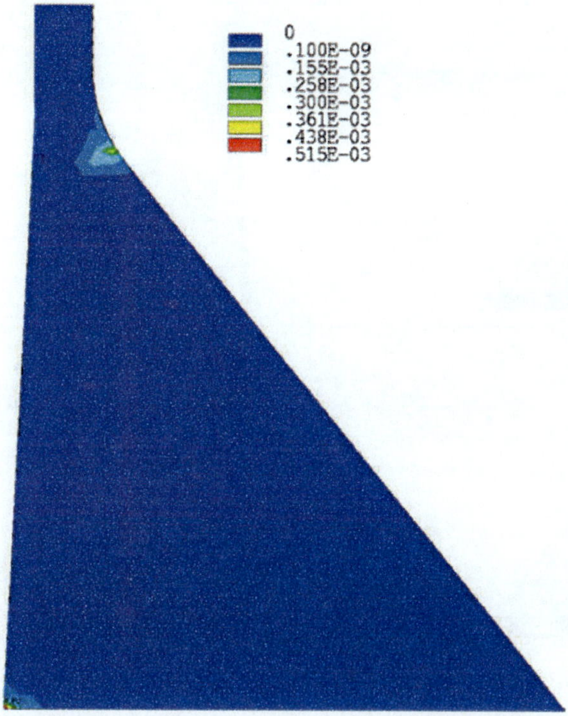

Fig. 10 Plastic strain at the end of the non-linear simulation with the Taft earthquake excitation

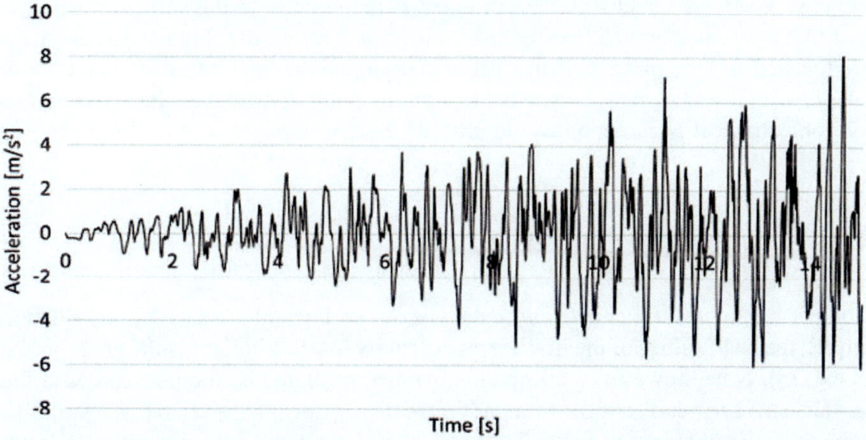

Fig. 11 Etaft loading

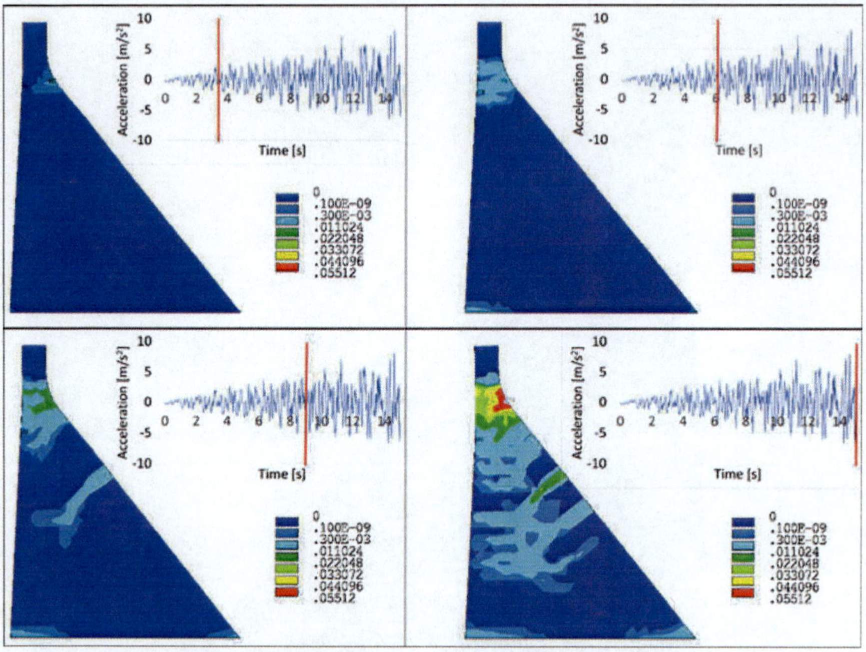

Fig. 12 Plastic strain due to the ETaft earthquake excitation

the downstream side, this time on midrise height of the dam, a secondary cracking mode is observed. Cracks that are oriented perpendicular to the surface form. The fact that they later branch, can be taken as an indication that high levels of stress are reached and the material is incapable to dissipate the total introduced amount of energy with a single crack only. Note that at this point, transient analysis of coupled hydromechanical phenomena would provide more insight.

4 Discussion and Conclusion

Comparing the results achieved with the viscous spring and the massless foundation model, the peak values of the first are significantly lower. Similar results are reported in e.g. [5]. Note, however, that the effective deconvoluted earthquake forces at the vertical side boundaries of the foundation are disregarded in the course of a simplification which might lead to errors. Note also, that shear wave reflections at boundaries like the interface between the foundation and the reservoir, the foundation and the dam and the foundation and the free field are not taken into account since forces related to these reflections are not homogeneously distributed along the length of the base of the foundation. In reality, the loading of the dam-foundation-reservoir system is not quite as homogeneous as it is accounted for. Reflections might lead to higher but also lower effective loading.

Further research questions include the influence of the applied damping. Damping is considered an important parameter in transient structural analyses but was fixed in the framework of the workshop. To vary Rayleigh damping and analyze its influence, both the applied damping ratio and the natural frequencies for its calculation can be used. Given that critical damping ratios for dam-foundation-reservoir systems are typically between 3 and 7% and assumed that for given loading measurements of e.g. deflection amplitude maxima over time at the crest of the dam are known, the Rayleigh coefficients could be varied and calibrated. This can be done with tools like ANSYS optiSLang® [7]. The same procedure can be used for the calibration of the parameters of the nonlinear material. The latter are even more crucial, but do not require large scale tests on the structure that are difficult to perform. Instead standard small-scale concrete tests can be used.

References

1. Salamon J, Hariri-Ardebili A, Malm R, Wood C, Faggiani G (2019) Thema A-Seismic analysis of Pine Flat concrete dam. In: 15th International benchmark workshop on numerical analysis of dams, Milano, Italy
2. Wikipedia: The Free Encyclopedia (2019). Accessed 3 May 29. https://en.wikipedia.org/wiki/Pine_Flat_Dam
3. ANSYS, Inc. (2019) ANSYS Documentation, ANSYS, Inc., Canonsburg, PA 15317. http://www.ansys.com
4. Menetrey P (1994) Numerical analysis of punching failure in reinforced concrete structures. Diss. Ecole Polytechnique Federale de Lausanne, Lausanne. Infoscience. Web
5. Liu YS, Chen DH (2013) Earthquake response analysis of a gravity dam considering the radiation damping of infinite foundation. APCOM & ISCM
6. Willam KJ, Warnke EP (1975) Constitutive models for the triaxial behavior of concrete. Seminar on concrete structures subjected to triaxial stresses. Int Assoc Bridge Struct Eng 19:1–30
7. Dynardo GmbH (2019) Methods for multi-disciplinary optimization and robustness analysis. www.dynardo.com

Modal, Dynamic and Seismic Analyses of the Pine Flat Concrete Gravity Dam

P. Peton and T. Thénint

Abstract This paper presents studies (models used, calculations and results) performed at Sixense Necs in the context of the 15th ICOLD International Benchmark Workshop on Numerical Analysis of Dams. This work has been proposed by United States Society on Dams (USSD) and concerns the Pine Flat Gravity Dam. The formulated case studies define several different dynamic analyses of the dam, rock foundation and reservoir system. The model consists of the 15.24 m-wide 16th dam monolith (the tallest the monolithic block) and a corresponding strip of the foundation. The tri-dimensional mesh is exclusively composed of linear hexahedron elements. Mechanical behavior of the dam and the foundation is described using 3D continuous medium elements. Displacement restrictions are implemented as dam and foundation boundary conditions, adapted for each involved face. We use the first suggested case to discuss two assumptions: first the reservoir water level modelling influence (added mass or acoustic elements) and secondly massless foundation hypothesis consequences on modal analyses. This work leads us to better describe and understand harmonic crest excitation results and highlight modal analyses dependency of single harmonic excitations. Seismic transient linear and nonlinear analyses are then computed, with seismic loading introduced as an inertial loading. Several results such as dam crest acceleration and dam heel hydrodynamic pressure are presented and compared in order to assess the importance of reservoir water levels and dam material properties. Nonlinear computations show higher crest amplifications, which indicate a loss of dam body stiffness. All numerical analyses are carried out with Code_Aster® software [1].

Keywords Gravity dam · Seism · Modal analysis · Transient analysis · Code_Aster® · Fluid-structure interaction

P. Peton (✉) · T. Thénint
SIXENSE NECS, Numerical Engineering and Consulting Services, Sceaux, France
e-mail: paul.peton@necs.fr

G. Bolzon et al. (eds.), *Numerical Analysis of Dams*, Lecture Notes in Civil Engineering 91, https://doi.org/10.1007/978-3-030-51085-5_11

1 Introduction

Formulators suggested for the workshop various case studies, all based on dynamic aspects of numerical analysis of the Pine Flat gravity dam located on King's River, east of Fresno, California. Specifically, it focuses on the 16th monolithic block, the tallest non-overflow. Dynamic properties of this block are well known thanks to eccentric mass vibration tests performed in 1971. Studies intent to analyze dynamic behavior of the dam, rock foundation and reservoir system with specified sets of parameters (dimensions, materials properties and loadings). This paper focuses on case A, E and F. Case A proposes to perform modal analyses and transient linear computations considering harmonic crest loading. Cases E and F suggest computing transient linear and nonlinear seismic analyses. All detailed specifications, including other case studies (B, C and D), and expectations of the workshop are described trough the Formulation document [2].

This paper presents the model used, test case results and additional computations performed at Sixense Necs in the context of the benchmark. First, details on the geometrical drawing, the mesh and the model are given: main dimensions, mesh sizes, modelling principles, phenomenon descriptions, boundaries conditions... Then, we use case A to discuss reservoir water level modelling influence and massless foundation hypothesis consequences on modal analyses. This work leads us to better describe and understand harmonic crest excitation results. Finally, with cases F and E, seismic transient linear and nonlinear results are presented and compared in order to assess importance of Reservoir Water Levels (RWL) and dam material properties.

2 Method and Approach

2.1 Dam and Foundation Finite Elements (F.E.) Model

The model consists of the 15.24 m-wide 16th dam monolith (Fig. 1) and a corresponding strip of the foundation. General dimensions of the mesh (foundation, dam and reservoir size and levels), presented in Fig. 2, follow the workshop specifications [2]. The axis origin is located on point A and the coordinate system is shown on every figures.

The mesh is exclusively composed of linear hexahedron elements generated from quadrangles, with a horizontal element size evolving from the dam crest ($\approx 1.5 \times 1.5$ m^2) to the foundation and reservoir upstream and downstream faces ($\approx 10 \times 32$ m^2). The monolith is discretized with four elements along its width (X direction). Foundation and dam are represented by respectively 2624 and 1824 elements.

For all linear analyses:

- Mechanical behavior of the dam and the foundation is described using 3D continuous medium elements (*3D Code_Aster*®);

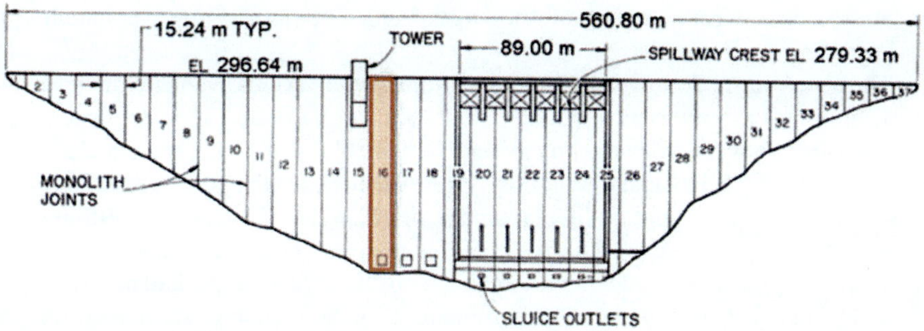

Fig. 1 Pine Flat dam—downstream view

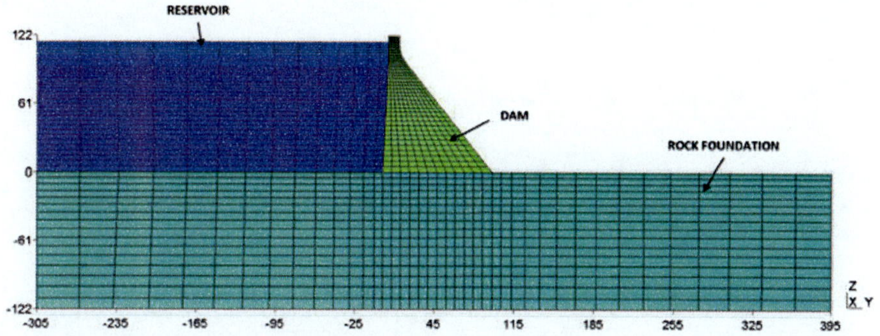

Fig. 2 Pine Flat dam mesh—general view

- Rock and concrete materials are assumed to be homogeneous, isotropic, and elastic with the following same properties: modulus of elasticity 22.41 GPa, density 2483 kg/m^3 and Poisson ratio 0.20. Rock density hypothesis will be discussed in Sect. 3.1;
- Normal displacements are restrained on dam lateral (dir. X) faces;
- Normal displacements are restrained on foundation lateral (dir. X), upstream (dir. Y), downstream (dir. Y) and underside (dir. Z) faces;
- Dam/foundation interface is assumed as a perfect bond (nodes are merged);
- Damping is introduced through Rayleigh model. Related coefficients are determined to have the desired mean damping ratio on the frequency range 4–10 Hz;
- Reservoir modelling is described in detail in the following separate section.

2.2 *Reservoir Modelling*

Empty reservoir and three RWL are studied: Winter at El. 268.21 m (WRWL), Summer at El. 278.57 m (SRWL) and Normal at El. 290.00 m (NRWL). Water is considered to have a density of 1000 kg/m^3 and a compression wave velocity of 1439 m s^{-1}.

The reservoir modelling is a challenging topic, with various available methodologies. In order to find the most suitable approach for the finite element solver used, comparisons between three tri-dimensional numerical models have been made. They all constitute of the dam and foundation system described in the previous section with a rigid foundation assumption (in practical terms, $E_{rock} >> E_{concrete}$). The differences consist in the approach of the dam/reservoir interaction. The first two models consider a classical approach with added masses on the upstream face of the dam described by Eqs. (1) and (2). This approach is recommended by the French Committee of Large Dam (CFBR) [3] to assess existing gravity dams.

$$\text{Westergaard:}\ M_{west} = \frac{7}{8}\rho_w (H_w \cdot z)^{0.5} \tag{1}$$

$$\text{Zanghar:}\ M_{zang}(z) = 0.5 H_w C_m \rho_w \left(\frac{z}{H_w}\left(2 - \frac{z}{H_w}\right) + \left(\frac{z}{H_w}\left(2 - \frac{z}{H_w}\right)\right)^{0.5} \right)$$

$$c_m = 0.735\left(\frac{2\theta}{\pi}\right)\ \theta(rad)\text{: } \textit{upstream face tilt angle}$$

with ρ_w: *water density* (kg m^3)
H_w: *reservoir depth* (m)
z: *vertical cooridinate starting from the reservoir free surface* (m) (2)

The third model takes into account the fluid of the entire reservoir with vibro-acoustic finite elements (*3D_FLUID Code_Aster®*), fluid-structure interaction elements (*FLUI_STRU Code_Aster®*) at the dam/foundation-reservoir interfaces, an absorbing boundary condition (*3D_FLUID_ABSO Code_Aster®*) at the upstream face, and free-surface (zero overpressure) condition at the reservoir-air interface. This complete fluid modelling considers a linear behavior and small perturbations around a static equilibrium. It is described with two scalar parameters: overpressure and displacement potential [4]. Comparing modal analyses performed on the NRWL with previous hypotheses and USSD Workshop in Miami results [6], the explicit fluid modelling approach is chosen.

2.3 Software

Implicit numerical resolution is used to solve finite element problems. All numerical models are made with *Code_Aster®*. Geometry and meshing of dam, foundation and reservoir models are built thanks to SALOME software 4 [7].

Code_Aster® is a free and open-source numerical simulation software, developed mainly by EDF R&D department (Major French electric utility company). ASTER means Structural Analysis and Thermomechanics for Studies and Research. It is a general software of thermo-mechanical analysis, based on the theory of continuum mechanics. It provides many finite element methods to solve different types of mechanical, thermal, acoustic, seismic and other problems. In addition to the standard functions of a thermo-mechanical simulation software, *Code_Aster®* contains many constitutive laws, finite elements and loadings. It also offers a whole range of methods of analysis and non-linear multi-physics modelling.

3 Results

3.1 Modal Analyses: Comparison Between Several Models

The first part of Case A study consists in determining natural frequencies and shapes of the full elastic model for the four reservoir filling levels defined in Sect. 2.2. Figure 3 presents the six first frequencies for the three RWL and the empty reservoir models. The evolution of the natural frequencies is consistent: for a given mode, they decrease as the water level increases. Mass contribution of the reservoir is indeed more important than its stiffness contribution.

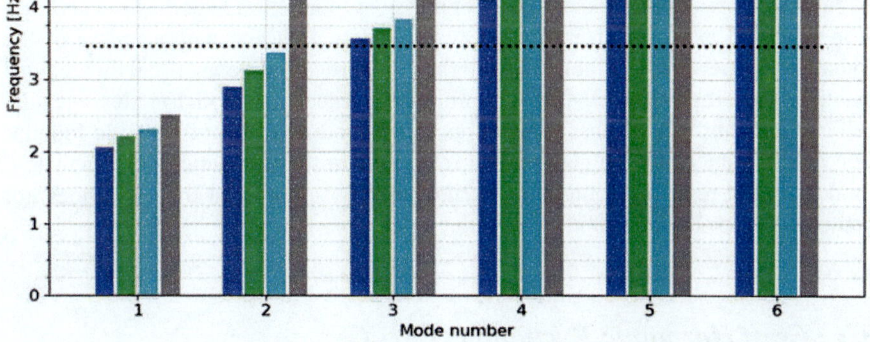

Fig. 3 Comparison of the first six eigenfrequencies (Hz) of tri-dimensional Pine Flat dam model with elastic and mass foundation for several reservoir water levels

Table 1 Comparison of the first natural mode shape (displacement and pressure) of tri-dimensional WRWL Pine Flat dam full elastic and mass model

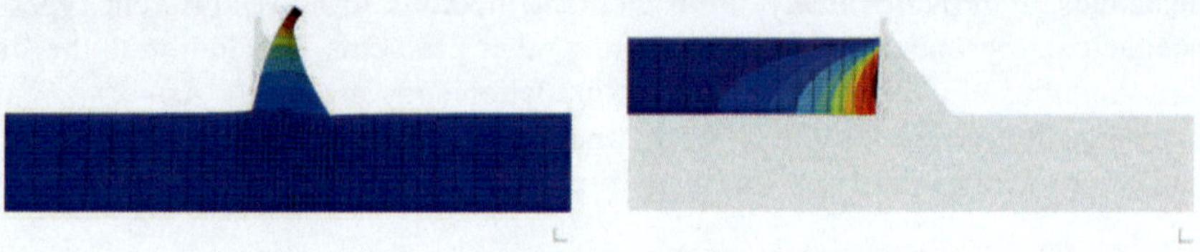

1st mode - 2.32Hz

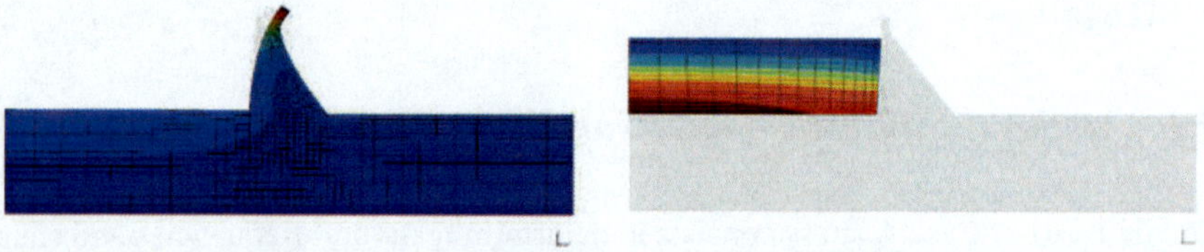

2nd mode - 3.37Hz

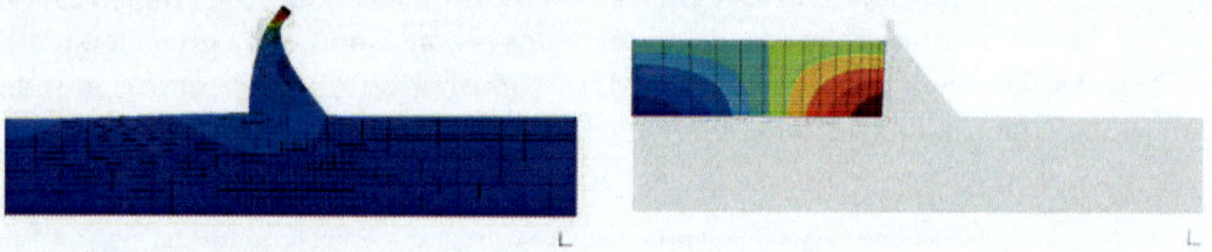

3rd mode -3.83Hz

The first three mode shapes of the WRWL model are shown in Table 1: displacement (left) and overpressure (right). Values are normalized according to maximum values.

In order to assess the consequences of massless foundation hypothesis, a comparison of modal analyses of empty reservoir models with and without mass of the foundation is performed (Table 2). Materials properties and geometrical dimensions are those specified in Sect. 2.1. The first mode excepted, modal analysis results differ significantly. The mass foundation model produces a lot of parasitic foundation modes induced by the geometrical restriction of the foundation domain. These modes are not realistic: foundation is theoretically infinite and therefore does not contribute to local modes under the dam.

3.2 *Crest Harmonic Excitation (EMVG)*

Following modal analyses, crest harmonic excitation loading is applied to the dam: an eccentric-mass vibration generator is modelled by a harmonic-force time history applied on a line load located in the middle of the dam crest. The harmonic function is windowed with a 25% Hann taper to reduce numerical artifacts. Numerical results are presented for the different RWL on Figs. 4 and 5. They all consider a linear behavior of materials and take into account the mass of the foundation (in order to respect workshop specifications) and a 2% damping hypothesis.

Single harmonic excitations are highly modal dependent. Indeed, according to the half power bandwidth method, resonance peak width is narrow ($\Delta f \approx 0.1$ Hz) for 2% damping and a 2.5 Hz eigenfrequency. The RWL that induced the maximal crest acceleration is the winter level, which is consistent with the natural frequencies

Table 2 Comparison of the first natural mode shape of tri-dimensional Pine Flat dam full elastic model with (left) and without (right) mass foundation

Full elastic and mass model	*Full elastic and massless foundation*
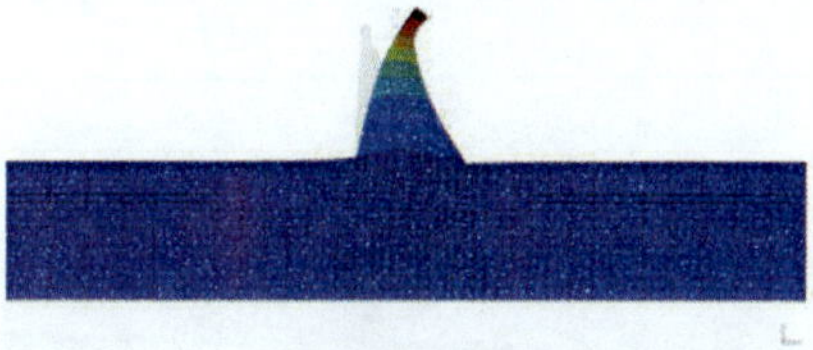 *1st mode - 2.51Hz*	 *1st mode - 2.57Hz*
 *2nd mode - 4.21Hz*	 *2nd mode - 5.34Hz*
 *3 mode - 4.87 Hz*	 *3 mode - 5.79Hz*
 *4th mode - 5.71 Hz*	

(continued)

Table 2 (continued)

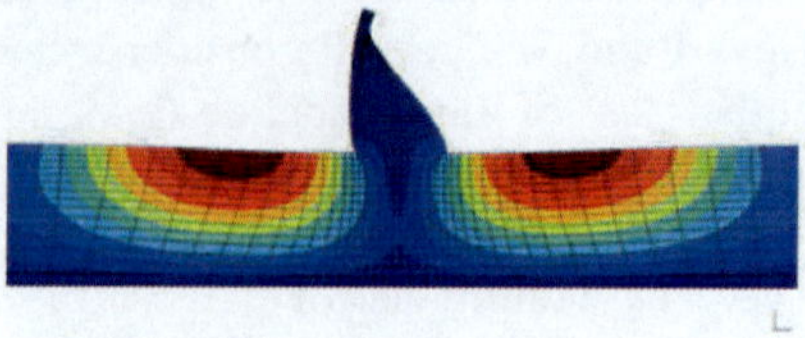 *5th mode - 5.90Hz*	*No modal contribution of foundation*
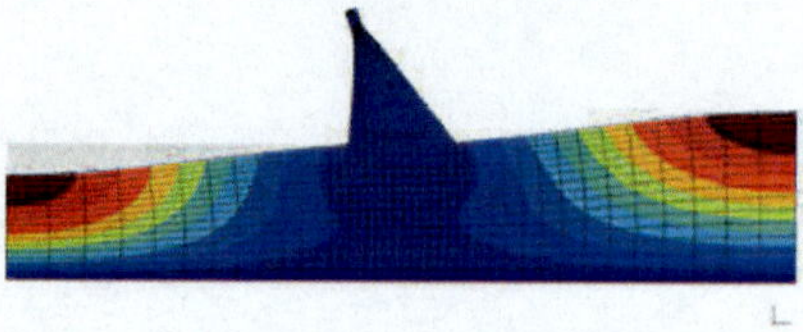 *6th mode - 6.51Hz*	
High level dam modal shapes are always coupled with foundation ones	 *4th mode - 8.93Hz*

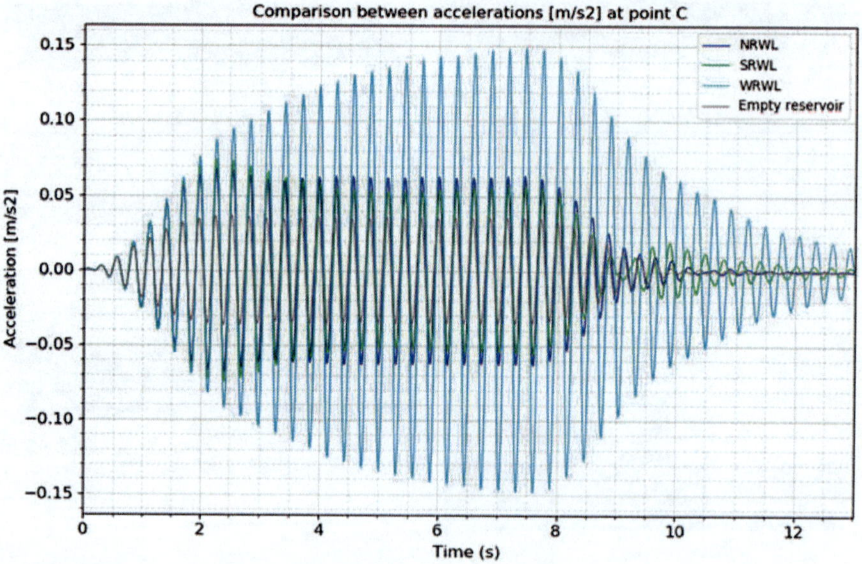

Fig. 4 EMVG—Dam crest accelerations results for several RWL

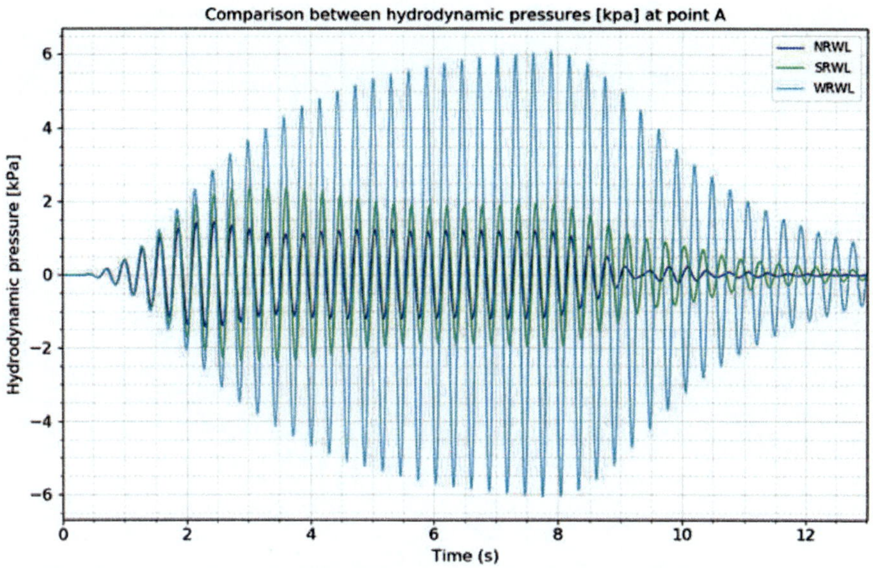

Fig. 5 EMVG—Dam heel hydrodynamic pressure results for several RWL

and the 3.47 Hz EMVG frequency (cf. Fig. 3): the frequency of its 2nd mode is the closest to EMVG frequency. A modal amplification is observed for WRWL on both acceleration and hydrodynamic pressure (Figs. 4 and 5). This implies that, even far away, the dam heel is affected by the crest loading. Although SRWL and NRWL crest acceleration shapes and magnitude are similar, hydrodynamic pressure ones are not. This point can be explained comparing pressure modal shape of 2nd and 3rd mode in Table 1. Note that variations of hydrodynamic pressure seem important but are low in front of the hydrostatic pressure (≈ 1140 kPa).

Previous calculations were performed with 2% damping ratio. Resonance peak bandwidth depends on damping: its influence is analyzed through the WRWL model calculation with 5% damping ratio. Figure 6 compares the dam crest acceleration for the two different damping ratios considered (2 and 5%). On this specific case, the maximum crest amplification decreases by 4.2% for a 3% increasing damping ratio.

A logarithmic decrement method allows us to estimate effective damping ratio of the whole numerical model. It is performed at the end of the excitation signal (>10 s) where time response can be considered as free damped oscillations. Damping ratio is determined using N pseudo-periods (cf. Fig. 6 framed) according to the Eq. (3):

$$\hat{\xi} \approx \frac{\ln\left(\frac{A_0}{A_N}\right)}{2\pi N} \tag{3}$$

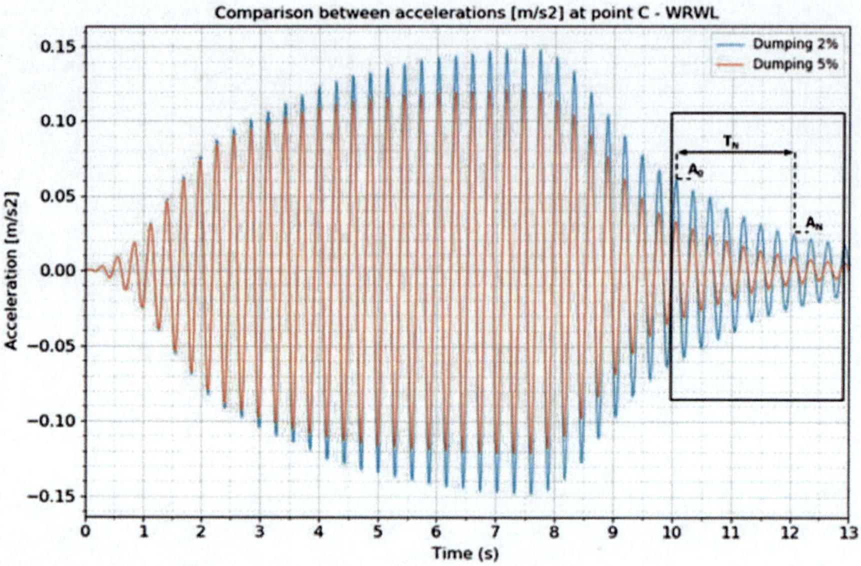

Fig. 6 EMVG—WRWL dam crest accelerations results for two damping ratios

Manual calculations give approximately $\hat{\xi}$ equal to 2 and 3% for respectively 2 and 5% concrete damping considered in the calculation models. These differences have no straightforward explanations. Further work is scheduled to tackle this point.

3.3 Seismic Analysis

Seismic transient loading is introduced as an inertial loading and relative displacement of the model is computed. This method avoids preliminary double integration of acceleration time histories and is relatively simple to set up [5]. Traditionally, this load case is computed with a free-field acceleration record and a massless foundation modelling hypothesis. This approach implies that foundation wave propagation is not considered in all computations presented in this section.

Both linear and nonlinear behavior of the dam material is considered. *Code_Aster*® ENDO_ISOT_BETON constitutive law [8] is used for dam nonlinear analysis. This modelling aims at simply describing fragile-elastic concrete behavior in tension. The material is always isotropic and its rigidity can decrease. This loss is measured by an evolving scalar from 0 (healthy material) to 1 (completely damaged material). To avoid convergence issues, a residual rigidity is kept for fully damaged elements. This constitutive law aims at only describing tensile concrete failure and supposes that the concrete compressive state remains moderate. Nevertheless, compressive strength is necessary to describe stiffness restoration due to crack

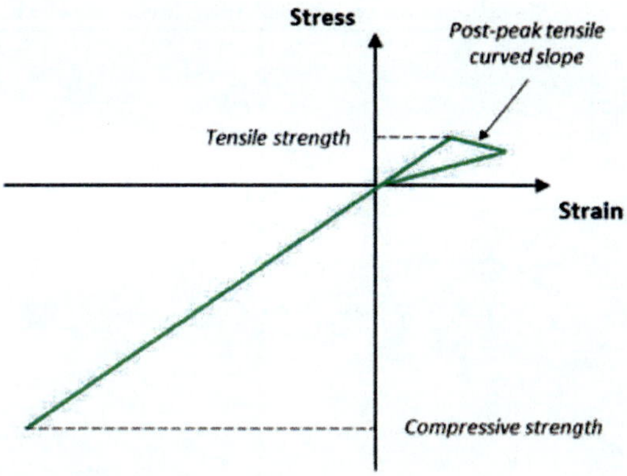

Fig. 7 Nonlinear concrete's behavior in simple uniaxial loading

Table 3 Maximum values—free field loading—linear models

Maximum values	NRWL	SRWL	WRWL	Empty reservoir
Max. crest acceleration (m/s^2)	17.6	25.9	20.0	22.3
Crest amplification: $\frac{A_{Crest}}{A_{\max\ load}}$	10.1	14.9	11.5	12.8
Max crest displacement (mm)	64	108	84	41
Heel Hydrodynamic pressure (kPa)	242	369	285	–

closing. The schematic concrete's behavior in simple uniaxial loading is shown in Fig. 7. 2 MPa tensile strength, 28 MPa compressive strength and - 4 GPa post-peak tensile slope (usual value used in studies) are considered.

Linear and non-linear analyses are performed on the several RWL model and the empty reservoir model considering the 54 s free-field loading. Following tables and figures present relevant results of the analyses for linear (Table 3 and Figs. 8, 9) and nonlinear models (Table 4 and Figs. 10, 11).

Following points can be highlighted:

- Unrealistic high crest amplification levels are due to ground damping modeling. Consciously, only ground material damping (from specifications) is considered in our computation to simplify the study. It is known that radiative damping is even more important than material damping in semi-infinite problem like dam-foundation problem.
- Crest acceleration amplification is highly RWL depends. Indeed, SRWL (medium level) presents maximum amplification values. Contrary to what one might think, the higher RWL does not seem the most critical.
- Nonlinear computations show more crest amplifications than for linear ones. Moreover, rises are not correlated with RWL. However, dam heel over pressure determined with nonlinear models decrease or stay in the same order of magnitude.

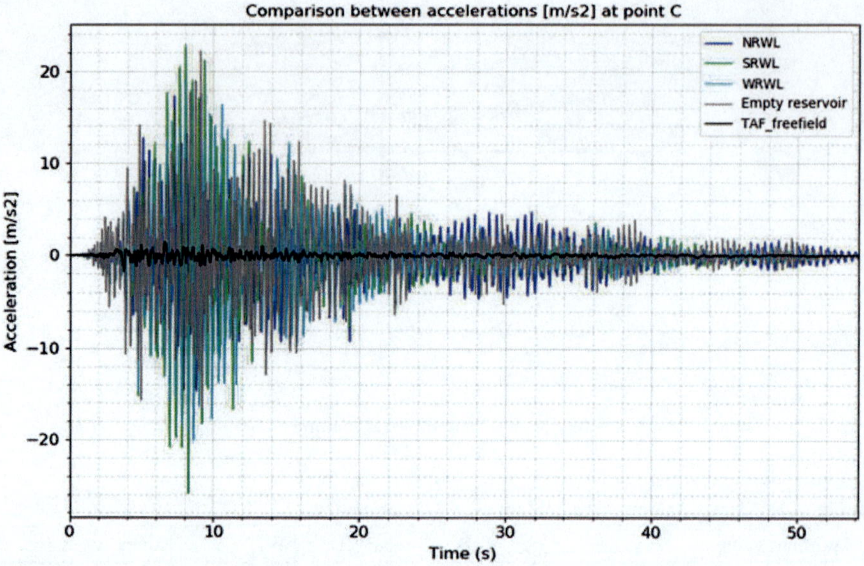

Fig. 8 Free field loading—linear models—Dam crest accelerations results for several RWL

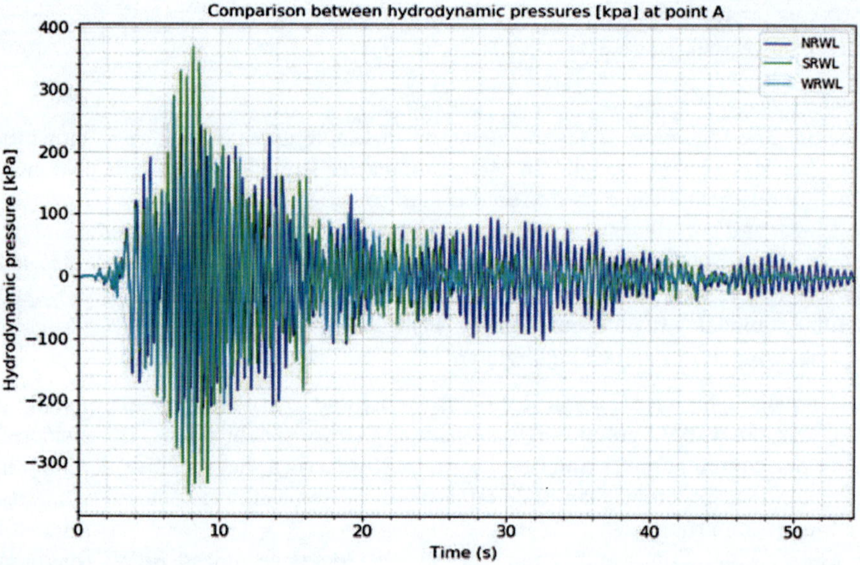

Fig. 9 Free field loading—linear models—Dam heel hydrodynamic pressure results for several RWL

Table 4 Maximum values—free field loading—nonlinear models

Maximum values	NRWL	SRWL	WRWL	Empty reservoir
Max. crest acceleration (m/s^2)	22.1	43.2	33.2	26.9
Crest amplification: $\frac{A_{Crest}}{A_{\max \text{ load}}}$	12.7	24.8	19.1	15.5
Max. crest displacement (mm)	55	104	120	42
Heel Hydrodynamic pressure (kPa)	186	252	289	–

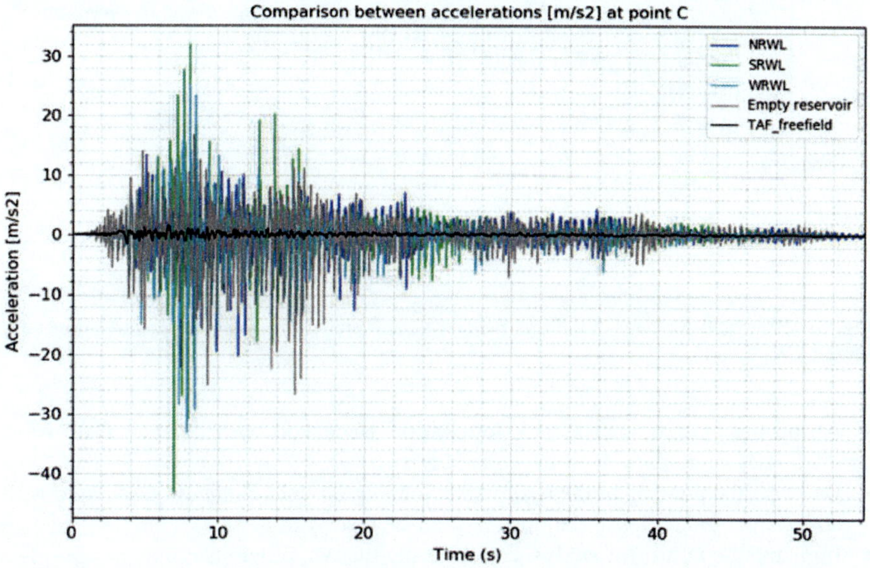

Fig. 10 Free field loading—nonlinear models—Dam crest accelerations results for several RWL

- Rigidity loss is principally located at the changing inertia area (25 m under the dam crest) and implies important consequences (in terms of amplifications) on the dynamic behavior of the dam crest.
- Variations of hydrodynamic pressure are important and the maximum history value of the SRWL model represents more than 30% of the maximum hydrostatic value.

4 Suggested Future Work

Although presented modelling gives satisfactory results, some work could be done to implement a complete nonlinear seismic analysis of a dam. In particular, we have to consider and validate all the following modelling points to reach state-of-the art practices:

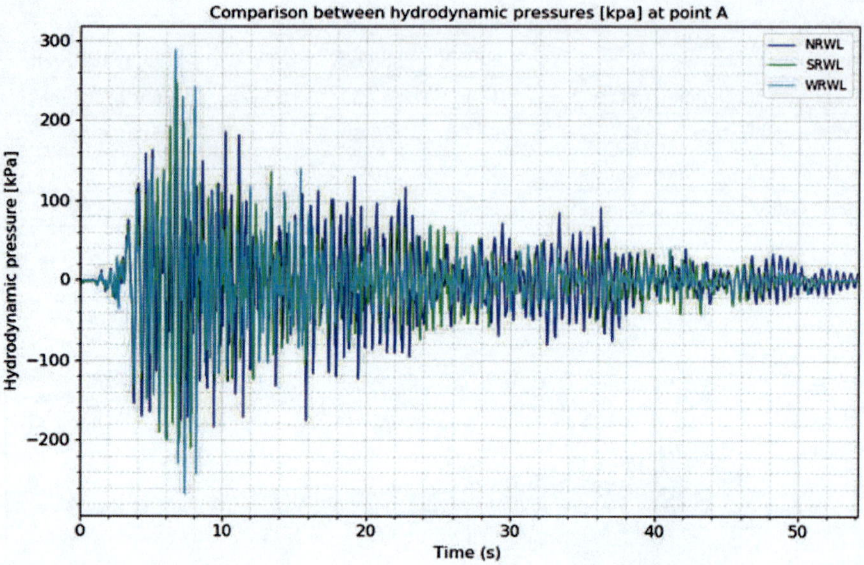

Fig. 11 Free field loading—nonlinear models—Dam heel hydrodynamic pressure results for several RWL

- Absorbing boundaries for rock foundation associated with seismic plane wave loading at the underside of the foundation;
- Describe the phasing of the hydrostatic loading on fluid elements associated with absorbing boundaries at the upstream of the reservoir;
- Implement a nonlinear analysis considering the two previous points.

Nonlinear analyses imply generally worthwhile topics and questions that could be studied in future work. So far, we highlight several points we think both interesting to develop and challenging in our current projects:

- To go further in case E, different nonlinear behaviors of concrete (for example, both in tension and compression) could be considered;
- As an extension of case E, it could be interesting to characterize the modal behavior of the dam after the seismic loading in order to evaluate the dam stiffness' degradation at the end of earthquake events. It is expected that eigenfrequencies decrease. But in which proportions? Furthermore, this information could be useful to fit Rayleigh damping on a relevant frequency range;
- Extend nonlinear material behavior on the rock foundation around the dam. Dam and foundation are highly coupled, the seismic loading can also generate degradation inside rock foundation, and maybe some local uplift;
- In all workshops' test cases, a single block of the concrete gravity dam was modelled. In reality, adjacent blocks may have an impact on the upstream/downstream behavior of the central block. Indeed, the first natural upstream/downstream frequency of the dam (WRWL in 1971) mentioned in the

formulation document do not match with different modal analyses results. Our models seem to be too flexible as if additional stiffness had been neglected. Relevant springs (determined manually or using sub-structuring methods) added on laterals faces of the dam should be taken into account.

5 Analysis Lessons Learned and Conclusions

Sixense Necs is a structural design office composed of engineers and scientists from the world of applied research. Specialized in expertise and consulting, its skills are centered on the use of digital tools for the modelling of civil engineering materials and structures behavior. One of the historic activities is about earthquake engineering of nuclear, industrial and civil structures or buildings. The diversity of technical topics mastered leads us to tackle some soil-fluid-structure interaction problems. For this reason, actions such as the participation to ICOLD benchmark are carried out to develop our understanding on those specials' topics.

The great majority of numerical analyses performed at Sixense Necs are carried out with *Code_Aster*®, a powerful open-source software dedicated to multiphysical analysis. It allows us to perform a full range of analysis with different modelling methods. This workshop served as a test case for using the *Code_Aster*® software to perform static, modal, dynamic and seismic transient linear and nonlinear (material behavior) analyses on dam structures considering fluid-structure interaction. These studies revealed that a simple geometrical problem as the dam-foundation-reservoir system raises a lot of computational and modelling problematics.

To conclude on previous results:

- Reservoir modelling provide, as expected, similar results for the first mode but significant discrepancies appear for the next ones.
- Mass foundation hypothesis create lots of artificial foundation modes due to unrealistic and arbitrary spatial restriction of the model.
- Single harmonic loadings are highly modal dependent (due to resonance peak narrowness associated with low damping ratio considered).
- Reservoir water levels and dam coupling is very complex. Critical behavior could appear for a specific intermediate level, both for harmonic crest and seismic loadings.
- Seismic analyses demonstrate that a simple nonlinear behavior of the dam concrete could induce significant acceleration amplifications (in comparison of linear dam model) at the dam crest. Stiffness degradation during seism could weaken dam sections where tensile concentration appears. Investigations should be undertaken to assess nonlinearity consequences.

References

1. Code_Aster reference and utilization documentation. http://code-aster.org/doc/v12/en/
2. Salamon J, Hariri-Ardebili MA, Malm R, Wood C, Faggiani G (2019) Theme A formulation. Seismic analysis of pine flat concrete dam. In: 15th ICOLD International benchmark workshop on the numerical analysis of dams, Milan, Italy
3. CFBR (2012) Recommendations for the justification of the stability of gravity dams
4. Code_Aster® Documentation. Vibro acoustic element—R4.02.02 (2017)
5. Code_Aster® Documentation. Seismic analysis by means of transient computation—R4.05.01 (2017)
6. Evaluation of Numerical Models and Input Parameters in the Analysis of Concrete Dams. Report DSO-19–13, A Summary Report of the USSD Workshop, Miami, USA (2018)
7. SALOME. https://www.salome-platform.org/
8. Code_Aster® Documentation. Constitutive law ENDO_ISOT_BETON—R7.05.01 (2017)

Seismic Analyses of Pine Flat Dam: Simplified Use of Viscous Spring Boundaries and Anisotropic Damage in the Principal Directions with the Ability of Cracks Re-Closure

E. Robbe, E. Grimal, and G. Devesa

Abstract This paper presents the analyses performed by EDF regarding the seismic analyses of Pine Flat dam proposed by the 2019 ICOLD Benchmark. The results will not be fully presented as they will be summarized and compared with other participants by the formulator. The paper will focus on the numerical analyses methods instead.

Keywords Dam · Earthquake · FE analyses · Damage

1 Introduction

The seismic analyses of Pine Flat dam proposed by the formulator provide an excellent opportunity to qualify numerical method to evaluate the behavior of dams under earthquake. Since several years, EDF improved its own Finite-element Software Code_Aster [1] to take into account mass of the foundation, absorbing boundaries and fluid element in order to correctly assess the safety of dams under earthquake. Comparisons with simplified analyses (massless foundation and Westergaard added masses) showed that, advanced finite-element analyses provided more accurate results compare to records on dams for example [2, 3].

The subject proposed by the formulator includes non-linear analyses taking into account concrete's damage. It was a good opportunity to evaluate one of the damage model proposed by Code_Aster and usually used in the case of swelling dams.

The following cases-studies have been performed by EDF:

E. Robbe (✉) · E. Grimal
EDF Hydro, Le Bourget-du-Lac, France
e-mail: emmanuel.robbe@edf.fr

G. Devesa
EDF R&D, Palaiseau, France

G. Bolzon et al. (eds.), *Numerical Analysis of Dams*, Lecture Notes in Civil Engineering 91, https://doi.org/10.1007/978-3-030-51085-5_12

Case A: Evaluation of the natural frequencies of Pine Flat dam taking into account Fluid Element and the mass of the foundation and response of the dam-foundation-reservoir system under an eccentric-mass vibration generator, as conducted in 1971. The 4 subcases have been performed.

Case B: Foundation analysis using impulsive loads: allow to verify the efficiency of the non-reflecting boundary conditions. Case B1 and B2 have been performed with the 700 m length foundation block.

Case C: Dynamic analyses of the dam-foundation-reservoir system under the Taft earthquake records with 3 different water levels.

Case D: Dynamic analyses of the dam-foundation-reservoir system under the Taft earthquake records and ETAF record combined with static load and taking into account concrete non-linear material properties.

As the results will be presented and compared by the formulator, this paper will focus on some lessons learned from the analyses.

2 Use of Viscous—Spring Boundaries for Dynamic Analyses of Dam-Foundation-Reservoir System

In order to take into account the mass of the foundation, absorbing boundaries have to be used in the foundation boundaries in order to avoid wave's reflection. In Code_Aster, a viscous spring boundary model is implemented as proposed and well described in [4, 5], and briefly summarized in Fig. 1. It is employed to absorb the wave energy radiating away from the dam and the foundation.

In this method, earthquake input is introduced as compression and shear waves, vertically propagating from the bottom to the top of the foundation. Side boundaries should not be neglected using free-field column providing the propagation of the wave in an unbounded foundation:

- If the foundation is homogenous, without damping, the acceleration of every point of the free-field column can be theoretically known and the acceleration at the top of the column will be twice the input at the bottom, with a delay due to propagation

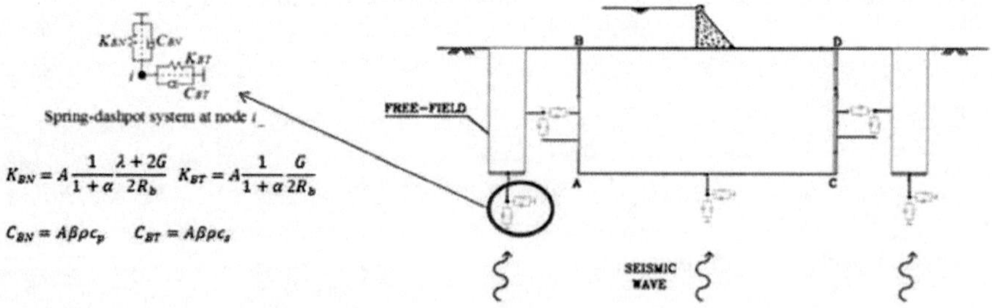

$$K_{BN} = A\frac{1}{1+\alpha}\frac{\lambda+2G}{2R_b} \quad K_{BT} = A\frac{1}{1+\alpha}\frac{G}{2R_b}$$

$$C_{BN} = A\beta\rho c_p \quad C_{BT} = A\beta\rho c_s$$

Fig. 1 Viscous spring boundary model

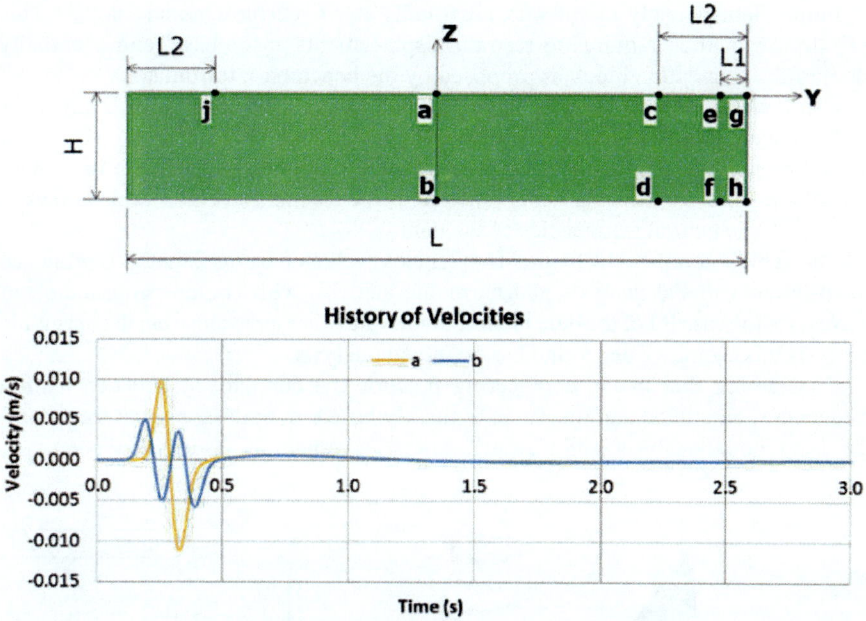

Fig. 2 Case B2 results: propagation of a shear wave in the homogenous foundation without damping

speed of the wave from the bottom to the crest. This can be seen through the results of the case B2 proposed by the formulator (Fig. 2).

- If the foundation is non homogenous and/or with damping, acceleration of every point of the free-field column has to be computed first. In addition, deconvolution should be performed to evaluate the input at the bottom to get the wanted signal at the top.

In addition the use of spring in the formulation of the viscous-spring boundaries is quite useful to provide adequate boundaries conditions for static analysis when non-linear analysis has to be performed (static and dynamic analysis can't be run independently and results solved).

3 Evaluation of Natural Frequencies (Case A)

The case A propose to evaluate the first natural frequencies and the modal shapes for the dam-reservoir system. This step is particularly useful from the engineering point of view because it allow to compare the natural frequencies of the dam that can be measured by eccentric-mass vibration generator (as done in Pine Flat in 1971) or nowadays by ambient vibration tests.

There are several ways to perform such analyses depending on the chosen numerical method.

Finite-element analyses software are usually able to perform modal analysis. This easy-to-use method requires to restrain displacements at the base (and eventually side) of the foundation block as proposed by the benchmark formulator.

Using viscous spring boundaries, modal analysis cannot be performed any more in the conventional way. It is then proposed to perform harmonic analyses with either a compression wave or a shear wave. It is then possible to visualize the imaginary part of the structure displacements function of the frequency (Fig. 3): this allows to point-pick the natural frequencies of the dam.

The Table 1 compare the natural frequencies evaluated by the 2 methods proposed (conventional modal analysis and harmonic analysis with compression and shear waves) for the case B1 of the benchmark. Some values are quite close but the harmonic analysis miss some values found by the modal analysis.

Considering that in the engineering practice the comparison between natural frequencies computed and measured is mainly based on the first natural frequency, the usual modal analysis with restrain boundaries remain the easiest way.

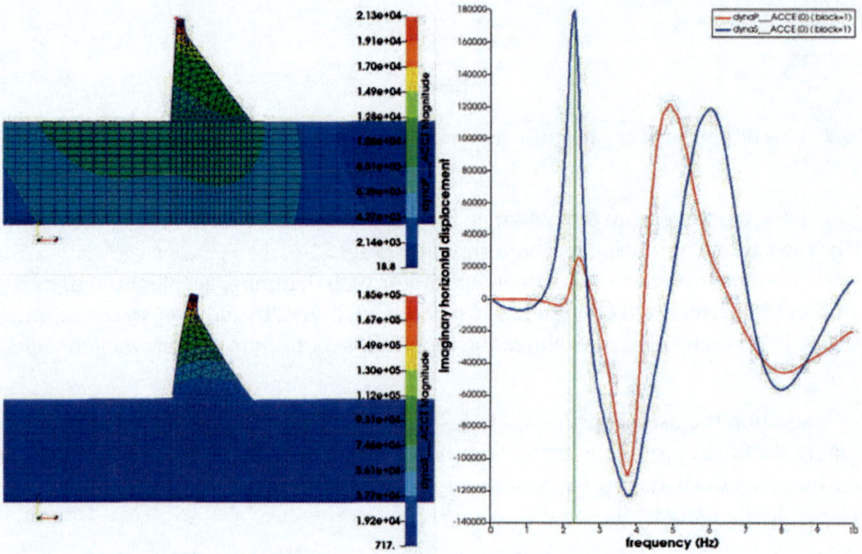

Fig. 3 Evaluation of the natural frequency with viscous-spring boundaries and harmonic analysis: top left: mode shape from the imaginary part under P wave at 2.3 Hz; bottom left: mode shape from the imaginary part under S wave at 2.3 Hz; right: imaginary part of the crest displacement (function of the frequency)

Table 1 Natural frequencies from modal analysis and harmonic analysis

Mode	Modal analysis (Hz)	Harmonic analysis (Hz)
1	2.35	2.3–2.4[a]
2	3.42	
3	3.94	3.8
4	4.38	
5	4.92	4.9
6	5.49	
7		6.0

[a]Value slightly different from the S-Wave or from the P-Wave

4 Simplified Use of Viscous-Spring Boundaries (Case B and D)

The evaluation of the dynamic response of Pine Flat dam under the Taft earthquake records has been used to assess and simplify the introduction of the seismic input in FE analyses with absorbing boundaries around the foundation.

For such analyses, there are several ways to introduce the seismic input as already described in the Sect. 2.

In a homogenous foundation without damping, the acceleration of the free field columns can be solved analytically from the wave propagation equation.

Considering that the formulators request the use of 2% damping in the foundation, the introduction of the input should be in theory more complex:

- Analysis of a 'free-field' column only with 2% Rayleigh damping.
- Deconvolution in order to get a new input at the bottom of the foundation to get the 'right' input at the surface of the foundation.
- Analysis of the 2D model considering for the free-field the results of the single column under the new input.

The second method request to perform additional analyses, time-consuming from an engineering point of view (and source of additional potential mistakes). A short sensibility study is then proposed here to evaluate the effect of several simplification considering the low value of damping in the foundation: three analyses are performed based on case D of the benchmark with the following assumptions:

- No damping for the foundation material,
- 2% Rayleigh damping in the foundation but no damping for the free field (directly evaluated from the analytical solution),
- 2% Rayleigh damping in the foundation including the free field (computed first as a soil column with input at the base).

Figure 4 shows the displacement of the crest of the dam under the TAFT earthquake for the three analyses. Results are almost the same and show that for low value of damping in the foundation, simplified seismic input procedure is suitable.

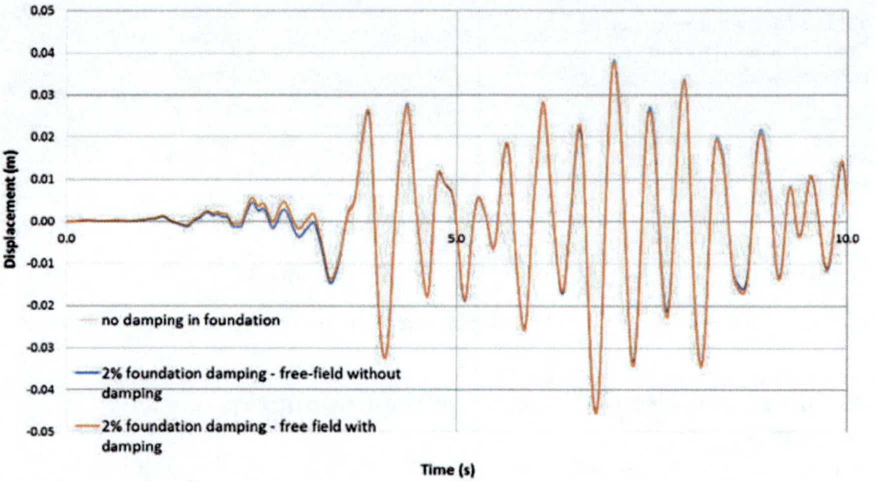

Fig. 4 Comparison of the displacement at the crest of the dam under TAFT earthquake

5 Non-linear Dynamic Analysis (Case E)

Case E of the benchmark proposes to evaluate the non-linear behavior of the dam under an increasing earthquake input (ETAF) reaching more than 5 m/s^2 at the end of the signal (Fig. 5).

The damage model 'ENDO_PORO_BETON' is embedded in the 'KIT_RGI' group of laws initially developed in CODE_ASTER for the swelling of concrete. This damage law can nevertheless be used by itself for dynamic analyses of dams. More details can be found in [6]. This law included several mechanisms:

Damage affecting tensile stresses:

- A pre-peak isotropic damage can occur if tensile peak strain is greater than the elastic strain at the peak.
- An orthotropic damage following Rankine criteria occurs during the post peak phase corresponding to the localization of tensile cracks. Localized tensile

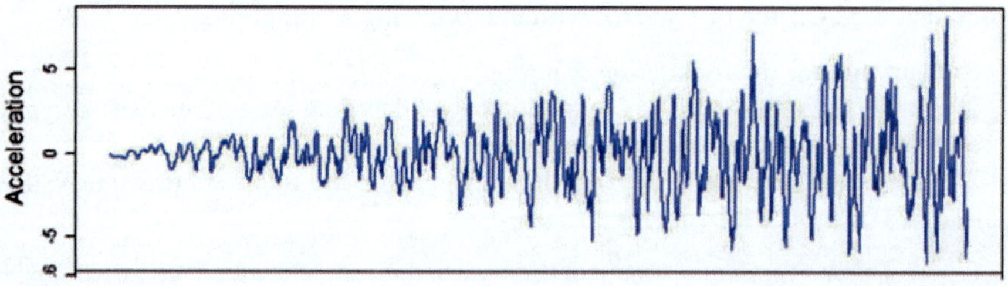

Fig. 5 ETAF acceleration record

damages are computed in the principal directions of localized crack and lead to opening of cracks. The model is able to re-close these localized cracks depending to compression stress level.
- Isotropic damage under compression stresses using Drucker-Prager criteria.

With the concrete properties proposed by the formulator (tensile strength of 2 MPa, fracture energy of 250 N/m, tensile strain at peak load of 0.00012), and the seismic methods describes in the previous chapters (viscous-spring boundaries, fluid elements), the behavior of PineFlat dam under the ETAF signal is computed (case E2 of the benchmark). Figure 6 show density of cracks in the concrete due to tensile stresses for several instants of the signal (6, 9, 12 and 15 s). This allow to visualize the progression of the damage during the earthquake.

Figure 7 compare the crest displacement computed with linear analysis (no damage) and non-linear analysis (with damage mechanism) for the ETAF signal. Crest displacement computed by the non-linear model remain in the same range than with the elastic model but starting from about 12 s, higher discrepancies start to appears. In addition, a decrease of the main frequency is observed probably due to the important damage at the top of the dam.

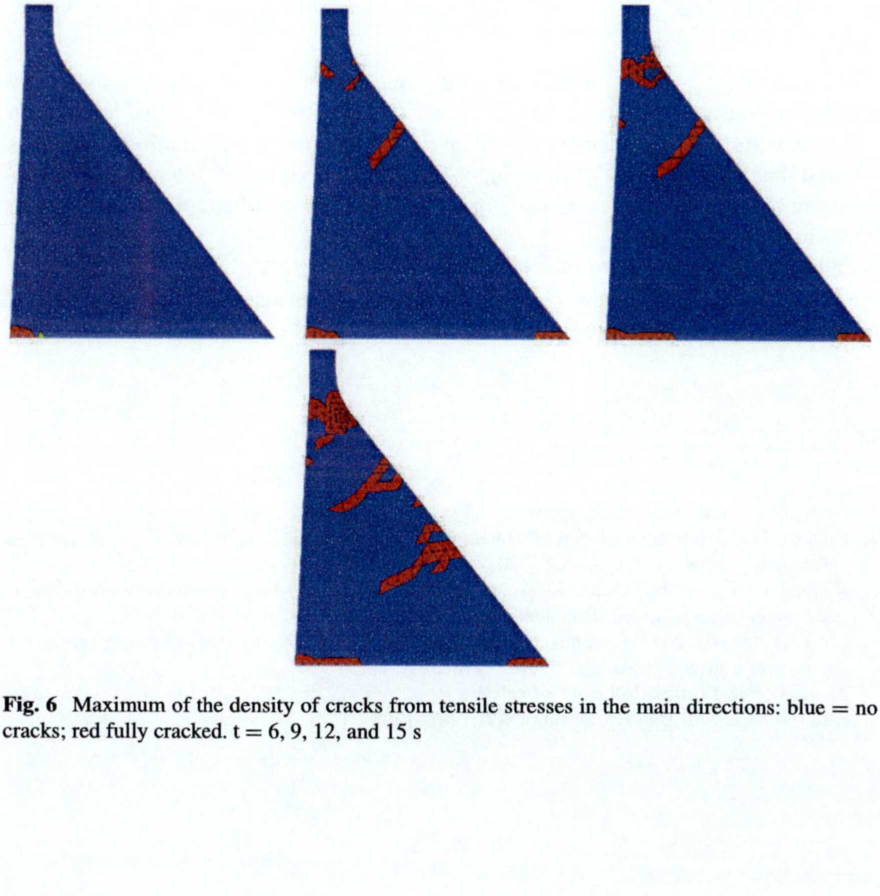

Fig. 6 Maximum of the density of cracks from tensile stresses in the main directions: blue = no cracks; red fully cracked. t = 6, 9, 12, and 15 s

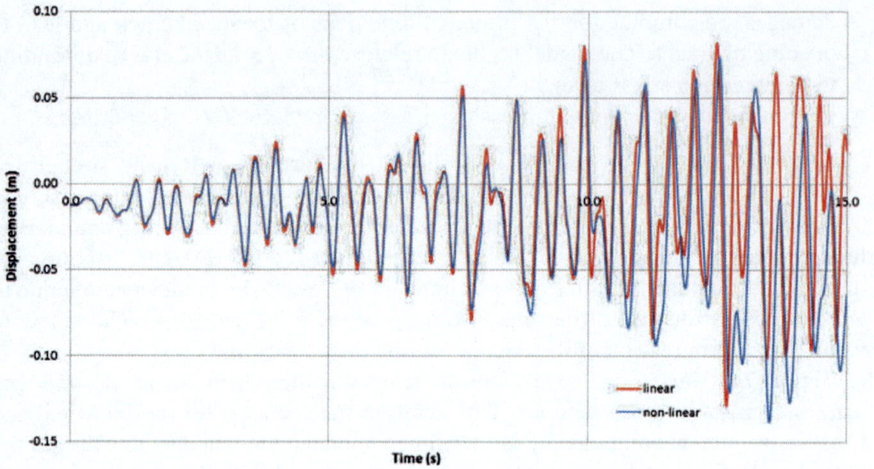

Fig. 7 Crest displacement under ETAF signal: comparison between linear and non-linear analyses

6 Conclusions

The use of viscous-spring boundaries for the analyses of concrete dams seems to be a good compromise: it allows to correctly take into account the unbounded foundation and remain quite simple to use from an engineer point of view. In addition, analyses showed than for low value of damping in the foundation, simplified analytical free-field can be used: analyses remain simple, there is no need of additional analyses to perform deconvolution.

The non-linear case of the benchmark demonstrates in addition that it can be easily combined with damage law: springs embedded in the viscous-spring boundaries allow to easily start with the static analysis, followed by dynamic analysis.

References

1. Code_Aster. http://www.code-aster.org
2. Robbe E (2017) Seismic analyses of concrete dam, comparison between finite-element analyses and seismic records. In: 16th WCEE 2017, Santiago, Chile
3. Robbe E (2017) Seismic back analysis of Monticello arch dam—blind prediction workshop and additional analyses. In: USSD annual conference, Anaheim
4. Liu YS, Chen DH (2013) Earthquake response analysis of a gravity dam considering the radiation damping of infinite foundation. In: APCOM & ISCM
5. Zhang C, Pan J (2009) Influence of seismic input mechanisms and radiation damping on arch dam response. Soil Dyn Earthq Eng 29(9):1282–1293

6. Sellier A (2018) Anisotropic damage and visco-elasto-plasticity applied to multiphasic materials. LMDC—Laboratoire Matériaux et Durabilité des Constructions de Toulouse, Toulouse; Université de Toulouse III—Paul Sabatier; INSA de Toulouse. https://hal.insa-toulouse.fr/hal-01710289

Evaluation of Seismic Wave Propagation in the Analysis of Concrete Dams: Pine Flat Dam Benchmark Study

J. W. Salamon, C. Wood, J. Manie, and A. Geister

Abstract The paper presents both a discussion of and results from a structural analysis conducted for Pine Flat Dam as formulated for Theme A of the 15th International Committee on Large Dams (ICOLD) International Benchmark Workshop on Numerical Analysis of Dams. The primary focus of this paper is to examine some basic problems that can arise when using reduced-domain models of a semi-infinite medium. We explore the use of two finite element (FE) codes to model seismic wave propagations in an elastic medium, and we evaluate their accuracy for some simple simulations where the answer is known a priori. We show that significant errors can arise due to the absence of the free-field boundary condition in FE simulations of elastic wave propagation.

Keywords Seismic wave propagation · Concrete dams · Finite element analysis

1 Introduction

This paper summarizes results of a structural analysis conducted for the tallest non–overflow monolith model at Pine Flat Dam, as defined by the Theme A Formulation Committee [1]. The concrete gravity dam—rock foundation—reservoir system is analyzed in a plane strain state using the commercial DIANA FEA [2] and LS-DYNA [3] finite element (FE) software.

Sections 2 and 3 of this paper compare the results of benchmark study Cases B and C [1], which deal with wave propagations in an elastic medium. Results for the remaining benchmark cases were submitted to the workshop organizers and are included in a workshop proceeding. This paper focuses on the simulation of seismic wave propagation in an elastic foundation, the influence of a concrete dam on

J. W. Salamon (✉) · C. Wood · A. Geister
Bureau of Reclamation, Denver, CO, USA
e-mail: jsalamon@usbr.gov

J. Manie
DIANA FEA, Delft, Netherlands

G. Bolzon et al. (eds.), *Numerical Analysis of Dams*, Lecture Notes in Civil Engineering 91, https://doi.org/10.1007/978-3-030-51085-5_13

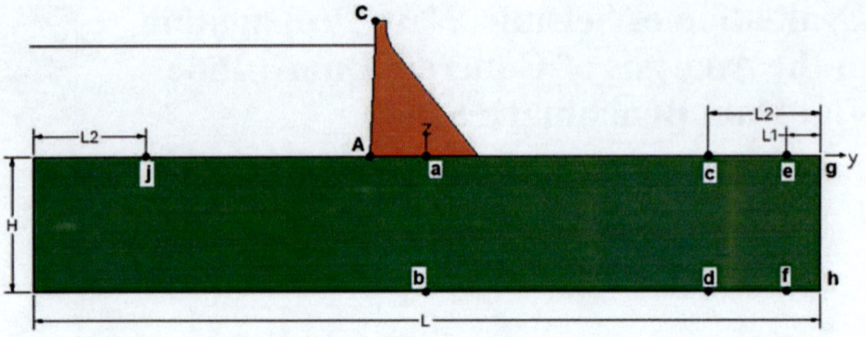

Fig. 1 Models of a reduced-domain foundation representing a homogeneous linear semi-infinite half-space for Pine Flat dam. Analysis points described in the document are shown by the lowercase letters a through j and uppercase letters A and C. See formulation document [1] for detailed model parameters

the seismic response of a model, the evaluation of various non–reflecting boundary conditions, and a comparison of the solutions obtained using the implicit versus explicit time integration method.

2 Benchmark Model and Loads

2.1 *Model Layout*

The geometry of the "base model" configuration consists of a 700 m long and 122 m deep foundation block, 121.9 m high dam with a downstream face slope at 0.7, and a reservoir depth of 94.48 m (see Fig. 1). For cases B and C, the homogeneous foundation block and the dam are assumed to be elastic and characterized by a shear-wave velocity of $v_s = 1939$ m/s. Case B of the formulation considers the response of the foundation block, while Case C considers the response of the foundation, dam, and reservoir system (Fig. 1). The detailed geometry and material properties are defined in the benchmark formulation document [1].

2.2 *Boundary Conditions*

It is assumed that the foundation block must adequately represent elastic wave propagation in a semi-infinite half-space (Fig. 2). Thus, an absorbing boundary condition must be applied to the sides and base of the foundation to avoid artificially trapping seismic energy within the finite-domain model. In the analysis, we use a variation of the boundary condition described by Lysmer and Kuhlemeyer [4].

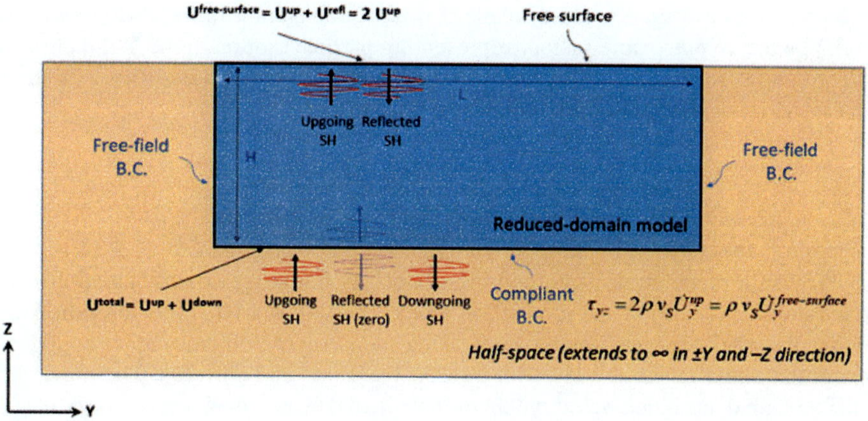

Fig. 2 Representation of a semi-infinite half-space (red) by a finite-domain foundation block (blue). In the typical Lysmer-Kuhlemeyer formulation [4], incident energy is assumed to consist of vertically-propagating plane P or S waves, and the coefficients of springs and dashpots at these boundaries are chosen to minimize reflections. In this case, the upgoing S-wave motion is exactly one-half of the free-surface motion and can be applied as a shear stress at the base boundary of the block. An additional "free-field" boundary condition must be applied on the side boundaries of the block so that the motions there are equivalent to those of the infinite-domain case

An additional boundary condition is needed for the reduced-domain model to keep the motions at the side boundaries equivalent to what they would be in the semi-infinite medium (i.e., equivalent to the "free-field" motion) [5–8]. This is the so-called free-field boundary condition, and it has been implemented in several codes used in geotechnical engineering for modeling soil-structure interactions or the seismic response of embankment dams and other structures founded on soil. Examples include the finite-difference code FLAC [9] and the FE code PLAXIS [10].

The free-field boundary condition has only had limited implementation in commercial codes typically used for dynamic analysis of concrete dams (for example, as a user–defined boundary condition written for ABAQUS [7]). As an alternative, a procedure has recently been developed that computes the equivalent loading forces for the side boundaries necessary to maintain consistency with the free-field motions [11, 12]. These forces are computed during an initial step and are then applied during a second step, along with the primary input motion. While this two-step process provides a general procedure to apply the free-field boundary condition for codes that do not directly support it, the method is somewhat complex to use in practice.

In this paper, we report on a preliminary implementation of the free-field boundary condition recently developed for DIANA FEA software and, for comparison, report the results from runs with the currently built-in, non-reflecting boundary condition.

Incident plane S-waves converted to a stress time-history can be applied to the base boundary of the foundation block by using a Lysmer-Kuhlemeyer [4] absorbing boundary condition [13, 14]. For a uniform rock foundation, the input stress time-history can be obtained through a simple scaling of the free-surface velocity time-history. Assuming plane S waves propagating in the Z-direction, let $v_Y(t)$ be the velocity time-history representing particle motion in the Y-direction at the ground surface. This time-history can be converted into a time-history of shear stress $\tau_{YZ}(t)$ at the base of the foundation block using

$$\tau_{YZ}(t) = \rho v_S v_Y(t) \tag{1}$$

where ρ is the foundation density, and v_S is the shear-wave velocity [13, 14].

We use a plane strain state and non-reflecting boundary conditions applied at the bottom and sides of the foundation block. For DIANA 10.3, the non-reflecting boundary conditions are implemented using the Lysmer-Kuhlemeyer [4] boundary condition and a far-field body with a respective mass and stiffness. The mass and stiffness are determined based on the user-defined density and dimension of the far-field body. Also, a preliminary implementation of the free-field boundary condition, to be available in the next DIANA version, is tested.

For LS-DYNA, the built-in "Boundary-Non-Reflecting" boundary condition is used; however, LS-DYNA does not currently implement the free-field boundary condition.

2.3 *Impulsive Excitation Records*

In the analysis, only vertically propagating S-waves polarized in the horizontal direction are considered. These excitations are applied as a shear stress to the base boundary of the foundation block. An impulsive stress record was developed for the formulation, and properties are defined in Table 1. This stress record is applied in the upstream/downstream direction at the base of the foundation for Cases B and C. There are two versions of the impulse record: (1) a high-frequency pulse with a sample interval of 0.001 s, and (2) a low-frequency pulse with a sample interval of 0.01 s. The characteristic frequency of the high-frequency pulse was selected so that several wavelengths of the pulse would fit within the vertical extent of the foundation, while the characteristic frequency of the low-frequency pulse was selected so that the

Table 1 Parameters of the excitation velocity pulse

Impulse	Δt (s)	FNyquist (Hz)	FN (Hz)	Flo (Hz)	Fhi (Hz)	Impulse amplitude (m/s)	Record duration (s)	Zero-pad (sample)
Low frequency	0.01	50	4.0	0.5	8.0	0.01	20	10
High frequency	0.001	500	40	5	80	0.01	2.0	10

vertical extent of the foundation represented approximately one-quarter wavelength of the pulse.

3 Case Study Results

Case B is a model of the foundation block only, whereas Case C models the foundation block together with the dam and the reservoir. Comparison of both case studies for the same selected impulsive excitation illustrates several important effects: (1) differences in computed versus theoretical motions for different boundary conditions; (2) differences in the computed motions obtained using different computer codes; (3) differences in computed motions as a function of position; and (4) differences in the deconvolution process using the foundation block versus the analysis using the foundation-dam model.

To gauge the effects of different boundary conditions, computer codes, and inputs, we compare theoretical motions with those computed using DIANA and LS-DYNA at various points. We consider four computational models: (i) DIANA with the non–reflecting boundary condition; (ii) LS-DYNA with "Boundary-Non-Reflecting" conditions and an augmented input record resampled at 1/100th of the interval of the normal input record; (iii) DIANA with a preliminary implementation of the free-field boundary condition; and (iv) LS-DYNA with "Boundary-Non-Reflecting" and a normal input record. This study was performed using DIANA 10.3 and 10.4 (beta version) with the implicit integration method for dynamic analysis. Additionally, we used LS-DYNA version 11.0, which is based on an explicit integration procedure that requires the input loading to have the same time step as the internal time step. Because the time step of the input loading typically is much greater than the internal time step, LS-DYNA interpolates the input loading to the internal time step. We wished to explore whether this interpolation was done smoothly and without generating spurious high frequencies in the results. We therefore generated an alternative input based on a Fast Fourier Transformation (FFT) augmentation procedure with the same step size as the internal time step. The FFT augmentation procedure smoothly interpolates between the original points and has the same amplitude spectrum as the original time history.

For each model, we use a misfit measure to quantify how well the computed and theoretical waveforms match. The mean square error (mse) between a computed time history v_t and theoretical time history $\hat{v}_t$ over a time window $[t_1, t_2]$ is defined as

$$mse = \frac{1}{(t_2 - t_1 + 1)} \sum_{t_1}^{t_2} \left(v_t - \hat{v}_t\right)^2 \tag{2}$$

The *mse* depends on the scale of the time history, which makes it unsuitable to compare the results between different models. We therefore use a scale-independent misfit measure, the normalized mean square error defined as:

$$nmse = \left\{ mse \Big/ \left[\left(\bar{v}_t - \bar{\hat{v}}_t \right)^2 + \sigma_v^2 + \sigma_{\hat{v}}^2 \right] \right\}^{1/2} \tag{3}$$

where $\hat{v}_t$ and $\bar{\hat{v}}_t$ are the means of the computed and theoretical time histories, and $\sigma_v^2, \sigma_{\hat{v}}^2$ are their variances [15–17].

3.1 Case B Comparisons

For Case B, the free-surface motion at the top of the foundation (points *a*, *c*, *e*, and *g*, as shown in Fig. 1) can be determined theoretically from the equivalent upgoing motion in an elastic homogeneous half-space. The incident motion for Case B is a vertically propagating plane S-wave, and the theoretical free-surface motion is simply twice the upgoing wave. The theoretical motion is the same at all points on the free surface.

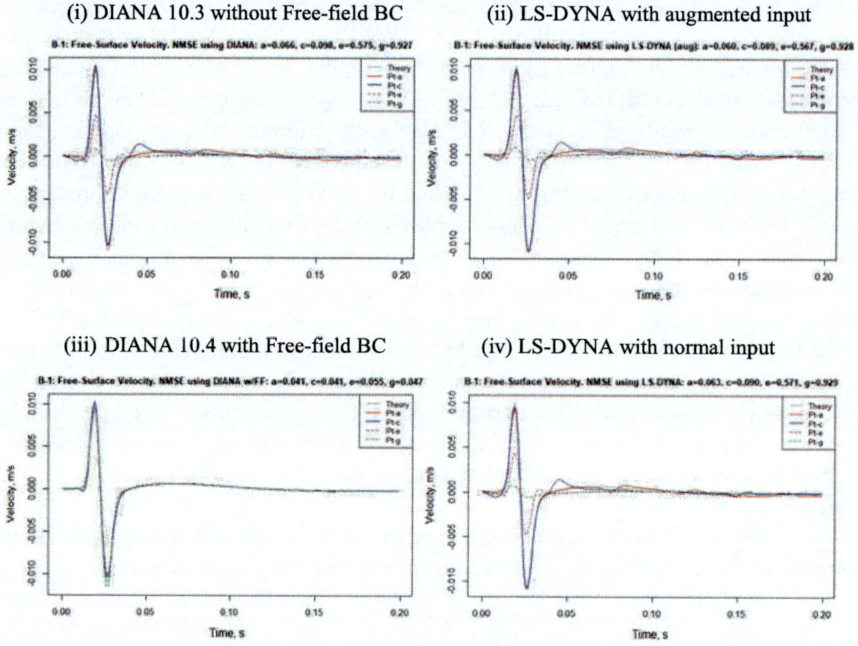

Fig. 3 Case B-1. Comparison of theoretical free-surface motions and the computed free-surface motions at points *a* (center), *c*, *e*, and *g* for the high-frequency pulse. The normalized mean square error (nmse) between the computed and theoretical motion at each point is listed in each plot. Large nmse values are seen for points near the sides of the model (*e* and *g*) unless the free-field boundary is used. See Fig. 1 for point locations

Results comparing the computed and theoretical free-surface motions at points *a*, *c*, *e*, and *g* for the high-frequency pulse are shown in Fig. 3. For those models lacking the free-field boundary condition, it is clear that the computed free-surface motions have a progressively greater misfit the closer the point is to the side boundary; nmse values range from 0.06 near the center of the foundation block to 0.93 near the edges. Motions computed near the center of the block also show reverberations, indicating contamination from reflections at the side boundaries. In contrast, the model with the free-field boundary condition (iii) has nmse values of only 0.04–0.05 at all points.

The use of an augmented input resampled at 1/100th of the original sample rate had little effect on the results for the LS-DYNA models. LS-DYNA is an explicit code and, therefore, must resample the input loading to its internal time step. Based on these results, it appears that the interpolation scheme used in LS-DYNA uses a smooth interpolation method that does not introduce spurious high-frequency artifacts in the computed motions.

We also computed the Fourier amplitude spectra of the theoretical and computed time histories for each of the four models. We computed nmse values for the full frequency bandwidth using a formula like that used for the time histories, which yielded values with similar magnitude. Results are shown in Fig. 4. For models without the free-field boundary condition, misfits are observed at all frequencies,

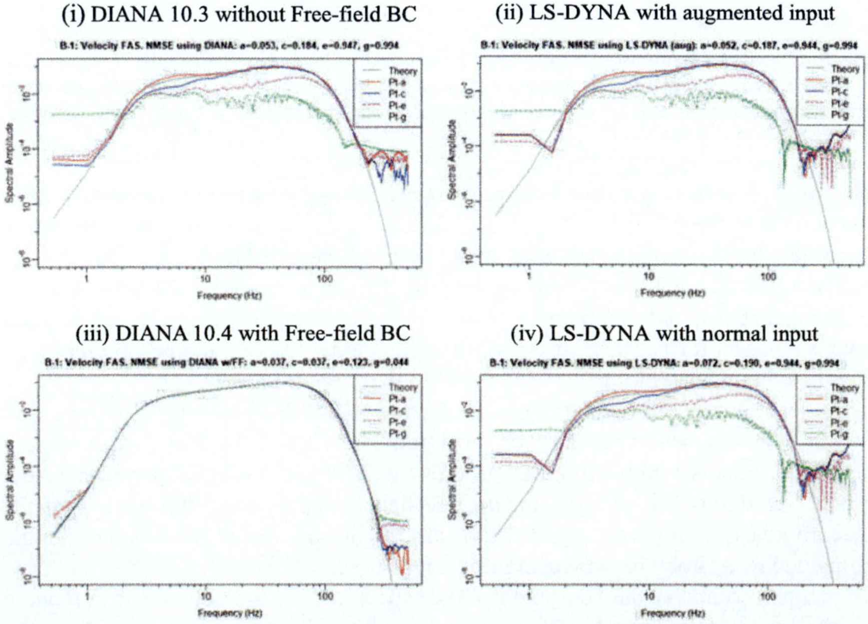

Fig. 4 Case B-1. Comparison of the Fourier amplitude spectra of the theoretical free-surface motions and the computed free-surface motions at points *a* (center), *c*, *e*, and *g* for the high-frequency pulse

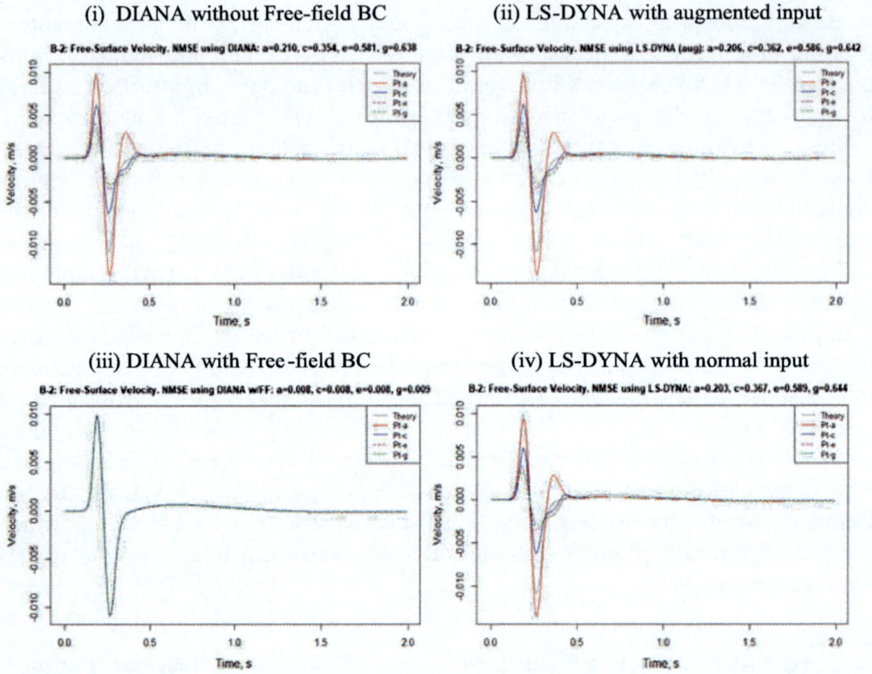

Fig. 5 Case B-2. Comparison of theoretical free-surface motions and the computed free-surface motions at points *a* (center), *c*, *e*, and *g* for the low-frequency pulse. The normalized mean square error (nmse) between the computed and theoretical motion at each point is listed above each plot

but—as was seen in the time history comparisons—the misfits increase the closer the point is to the boundary. In contrast, for model (iii) with the free-field boundary condition, there is good agreement at all but the highest and lowest frequencies.

Case B-2 uses the low-frequency impulse. Figure 5 compares the time-histories of the theoretical and computed motions for each of the four models. Large nmse values are seen for points near the sides of the model (plots i, ii, and iv) unless the free-field boundary (plot iii) is used (see Fig. 1 for point locations).

In general, the same pattern is seen for the low-frequency results as was seen for the high-frequency results. For those models without the free-field boundary condition, misfits range from 0.21 to 0.64, with larger misfits observed near the model boundaries. The model using the free-field boundary condition (iii) has misfits that are <0.01 at all points, again demonstrating the significant improvement in the computed results obtained by using this boundary condition.

Figure 6 compares the Fourier amplitude spectra of the theoretical and computed results for the four models. The spectra show the same pattern as seen from the time-history results.

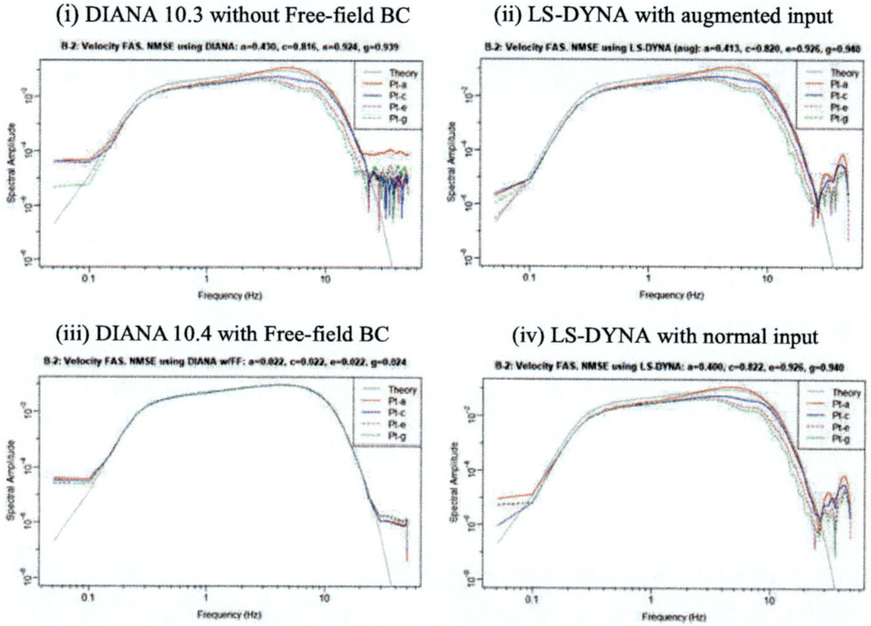

Fig. 6 Case B-2. Comparison of the Fourier amplitude spectra of the theoretical free-surface motions and the computed free-surface motions at points *a* (center), *c*, *e*, and *g* for the high-frequency pulse

The motion at the foundation-block base (points b, d, f, and h, per Fig. 1) can be determined theoretically from the equivalent motion in an elastic homogeneous half-space at the foundation depth H. That motion is simply one-half of the free-surface motion (the upgoing wave) plus an identical copy lagged by the two-way travel time (the downgoing wave).

The two-way travel time tt is given by:

$$tt = 2\,H/v_S \approx 0.12576\ \text{s} \tag{4}$$

where v_S is the shear-wave velocity. Figure 7 compares time histories for the theoretical total motion at the base of the foundation with that computed by the four numerical models. Figure 8 compares the corresponding Fourier amplitude spectra.

By not including the free-field boundary condition, large errors are generated the closer one approaches the sides of the model. For example, those models without the free-field boundary condition have an acceptable nmse of 0.07–0.09 for point a (center), but nmse values sharply increase to 0.58–0.64 at points *e* and *g*. In contrast, the model that includes the free-field boundary condition (iii) has nmse values of 0.07–0.10 at all points.

Note that in computing the theoretical motion for the high-frequency pulse, we account for the fact that the two-way travel time is not an integer multiple of the

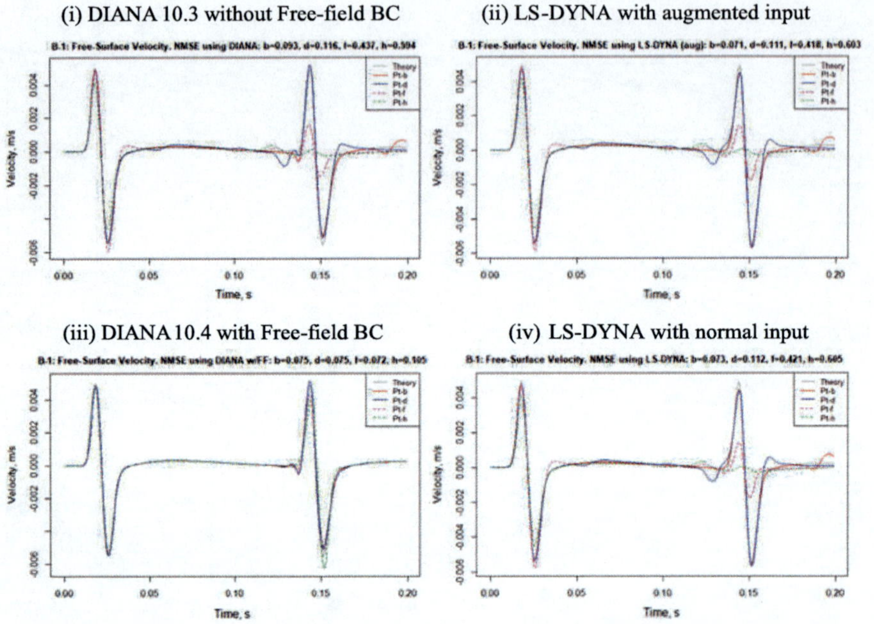

Fig. 7 Case B-1. Comparison of theoretical and computed foundation-block base motions at points *b* (center), *d*, *f*, and *h* for the high-frequency pulse. Large nmse values are seen for points near the sides of the model unless the free-field boundary is used. See Fig. 1 for point locations

sample interval of 0.001 s by using the following approximation: (1) augment the upgoing high-frequency pulse to 1/100th of the original sample interval (0.00001 s), (2) lag that pulse by 12,576 samples, (3) sum the augmented upgoing and downgoing pulses, and (4) decimate the sum back to the original sample interval of 0.001 s.

3.2 *Additional Results Comparison*

A comparison has been made to estimate the influence of the dam and the reservoir on the response of the model for wave propagation in an elastic rock foundation. Computed velocities at points *c* and *j* (Fig. 1) on the free surface, and spaced symmetrically 250 m from the block center, are listed in Tables 2 and 3 for high- and low-frequency pulse excitation, respectively. The results show the influence of the dam on the recorded horizontal wave velocities at the surface of the foundation; however, the presence of the reservoir does not have a noticeable effect. The above observations are confirmed in the graphs of the history of velocity at the center of the dam base (Point *a* in Figs. 9 and 10), for high and low-frequency pulse, respectively. Significantly lower velocities at the base of the dam are noted when compared with the free-field case study.

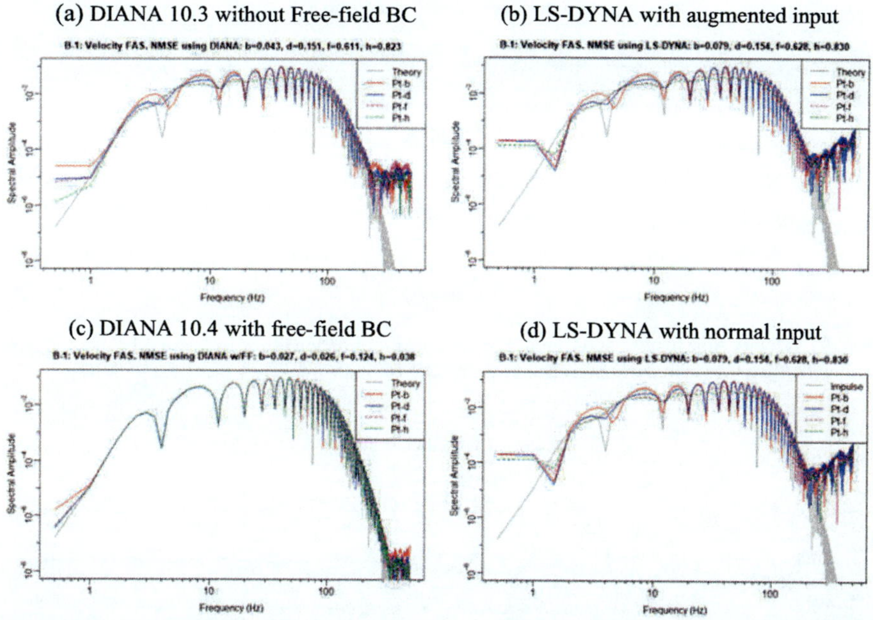

Fig. 8 Case B-1. Comparison of the Fourier amplitude spectra of the theoretical and computed motions at the base of the foundation block (points b, d, f, and h). Note that the observed scalloping shape of the spectra is simply due to the implicit Fourier transform of two delta functions lagged by the two-way travel time

Table 2 Velocity amplitudes (m/s) for high-frequency pulse at Points c and j (refer to Fig. 1)

High-frequency pulse	Point	Case B1	Case C1	Case C3
Maximum velocity	c	0.0104	0.00903	0.00903
	j	0.0104	0.00903	0.00903
Minimum velocity	c	−0.0103	−0.0121	−0.0121
	j	−0.0103	−0.0121	−0.0121

Table 3 Velocity amplitudes (m/s) for low-frequency pulse at Points c and j (refer to Fig. 1)

Low-frequency pulse	Point	Case B2	Case C2	Case C4
Maximum velocity	c	0.00623	0.00611	0.00611
	j	0.00623	0.00611	0.00611
Minimum velocity	c	−0.00641	−0.00717	−0.00708
	j	−0.00641	−0.00715	−0.00706

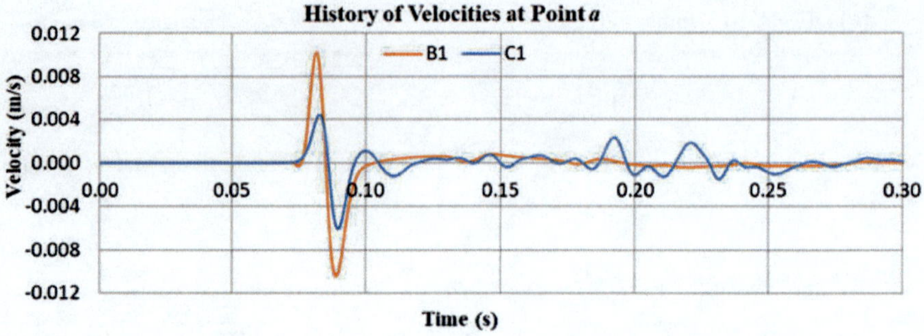

Fig. 9 Velocity time history at the center of the upper foundation block for Case B1 (without dam) and with Case C1 (with dam) presence in the model for high-frequency pulse excitation

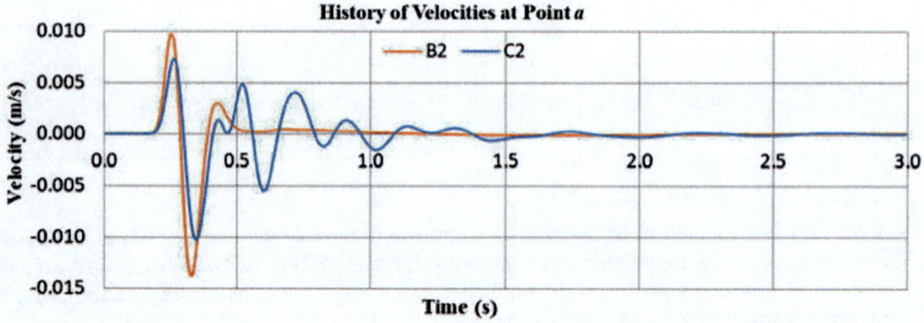

Fig. 10 Velocity time history at the center of the upper foundation block for Case B2 (without dam) and with Case C2 (with dam) presence in the model for low-frequency pulse excitation

Figure 11 presents wave propagation in a foundation-dam model. At the run time of about 0.09 s, the uniform wave splits into a part that reflects from the upper face of the foundation and a part that propagates into the dam structure until it is reflected from the dam crest; then, after the wave starts to propagate down, back to the foundation.

4 Conclusions

In comparing results for four different computational models, the most significant factor affecting the misfit between computed and theoretical results is the use of the free-field boundary condition. This effect is primarily the result of reflections generated at the side boundaries as the amplitude of the incident wave is artificially reduced to zero by the absorbing boundary condition. For codes without an implementation of the free-field boundary condition, these artifacts can be reduced somewhat by extending the dimensions of the foundation, thus delaying the onset of the artificial

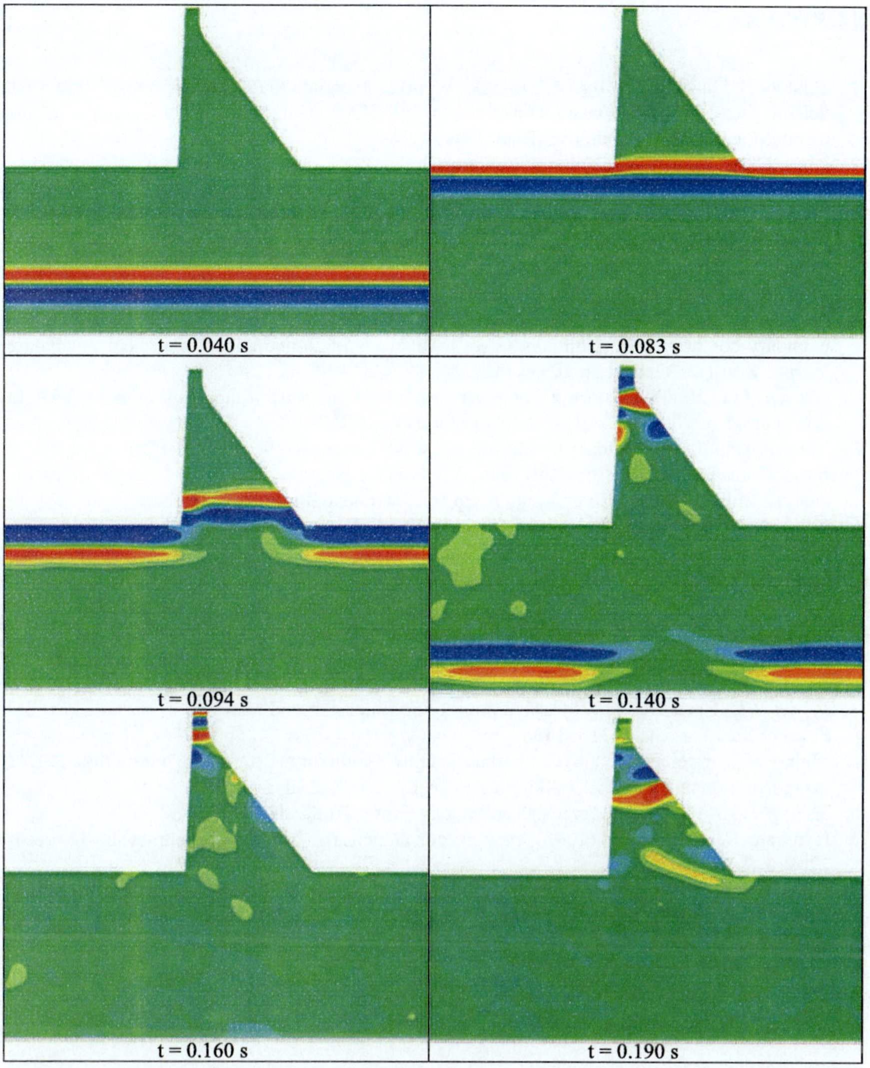

Fig. 11 Horizontal velocity wave propagation—range: 0.005 m/s (red) and −0.005 m/s (blue)

reflections to a later time. However, when modeling a large earthquake with a duration of significant shaking of tens of seconds, the dimension of the model may become prohibitively large for computation. A better solution in such cases may be to use the two-step approach of Løkke and Chopra [11, 12]. The most efficient solution for end users, however, is for commercial FE code writers to implement the free-field boundary condition in their codes. Because of the significant error introduced by neglecting this important boundary condition, this should be a high priority for the ICOLD community.

References

1. Salamon J, Hariri-Ardebili MA, Malm R, Wood C, Faggiani G (2019) Theme A formulation. Seismic analysis of Pine Flat concrete dam. In: 15th ICOLD international benchmark workshop on numerical analysis of dams, Milan, Italy
2. Diana FEA (2019) User's manual. Release 10.3
3. LSTC (2019) LS-DYNA keyword user's manual, vol 1
4. Lysmer J, Kuhlemeyer LR (1969) Finite dynamic model for infinite media. ASCE J Eng Mech Div 95(4):859–878
5. Kunar RR, Rodriguez-Ovejero L (1980) A model with non-reflecting boundaries for use in explicit soil—structure interaction analyses. Earthq Eng Struct Dyn 8(4):361–374
6. Zienkiewicz OC, Bicanic N, Shen FQ (1989) Earthquake input definition and the transmitting boundary conditions. In: Doltsinis IS (ed) Advances in computational nonlinear mechanics, Vienna. Springer Vienna, pp 109–138
7. Nielsen AH (2006) Absorbing boundary conditions for seismic analysis in ABAQUS. In: Proceedings of 2006 ABAQUS users' conference, pp 359–376
8. Nielsen AH (2009) Boundary conditions for seismic analysis. Seced Newsl 21(3):7–11
9. Itasca Consulting Group Inc. (2011) FLAC, Fast Lagrangian analysis of Continua, dynamic analysis, 5th edn. Itasca Consulting Group Inc., Minneapolis, MN
10. Galavi V, Petalas A, Brinkgreve RBJ (2013) Finite element modeling of seismic liquefaction in soils. Geotech Eng J SEAGS AGSSEA 44(3):55–64
11. Løkke A, Chopra AK (2017) Direct finite element method for nonlinear analysis of semi-unbounded dam–water–foundation rock systems. Earthq Eng Struct Dyn 46(8):1267–1285
12. Løkke A, Chopra AK (2018) Direct finite element method for nonlinear earthquake analysis of 3-dimensional semi-unbounded dam–water–foundation rock systems. Earthq Eng Struct Dyn 47(5):1309–1328
13. Joyner WB, Chen ATF (1975) Calculation of nonlinear ground response in earthquakes. Bull Seismol Soc Am 65(5):1315–1336
14. Mejia LH, Dawson EM (2006) Earthquake deconvolution for FLAC. In: 4th international FLAC symposium on numerical modeling in geomechanics, Madrid, Spain, p 9
15. Theil H (1961) Economic forecasts and policy. North Holland, Amsterdam
16. Hyndman RJ, Koehler AB (2006) Another look at measures of forecast accuracy. Int J Forecast 22(4):679–688
17. Polasek W (2013) Forecast evaluations for multiple time series: a generalized Theil decomposition. The Rimini Centre for Economic Analysis

Finite Element Analyses of a Concrete Gravity Dam: Investigation on Static and Dynamic Behavior

M. Scolari, A. Bado, D. Gualco, L. Buraschi, and R. Valsecchi

Abstract Within the "15th International Benchmark Workshop on numerical analyses of Dams", held in Milan (Italy) in September 2019 and organized by the International Commission on Large Dams (ICOLD), the static and dynamic behavior of a Concrete Gravity Dam was analyzed by means of 2D Finite Element Analyses (FEA). Aim of this paper is to investigate the advantages given by the adoption of a sophisticated FEA method with respect to a simplified analytical model.

Keywords Finite element analyses · Dam · Concrete dam · Dynamic analyses · Seismic behavior · Boundary conditions · Size effect

1 Introduction

This Paper analyzes the static and dynamic behavior of a Concrete Gravity Dam by means of 2D Finite Element Analyses (FEA). The analyzed case study, shown in Fig. 1, was Pine Flat concrete Gravity Dam. In particular, the analyses regarded the tallest non-overflow dam monolith, no. 16 in Fig. 1.

Part of the results presented in this paper was used to attend the "15th International Benchmark Workshop on numerical analyses of Dams", organized in Milan (Italy) in 2019 by the International Commission on Large Dams (ICOLD) [1]. Such case study was selected, by the workshop formulators, because of its relatively simple geometry and because it was extensively studied in the '70 s and '80 s at the University of California at Berkeley [1–7] and during the 2018 USSD Workshop in Miami [8].

The purpose of these workshops is to investigate uncertainties in FE analyses (FEA) of concrete dams in a focused, systematic and controlled approach, with collaborative participation from the international dam industry and academia.

Finite Element Analyses (FEA) involve the static and dynamic behavior of dam together with foundation and reservoir. Moreover, an investigation regarding the size and the boundary conditions effects are presented. The analyses were performed

M. Scolari (✉) · A. Bado · D. Gualco · L. Buraschi · R. Valsecchi
RINA Consulting S.P.a, Via Gran S. Bernardo, Strada 7, Palazzo R, Rozzano, MI, Italy
e-mail: matteo.scolari@rina.org

G. Bolzon et al. (eds.), *Numerical Analysis of Dams*, Lecture Notes in Civil Engineering 91, https://doi.org/10.1007/978-3-030-51085-5_14

Fig. 1 Analyzed case study: pine flat dam

using DIANA FEA BV software [9] because the complexity of the analyzed task required the adoption of a sophisticated tool.

Moreover, due to the characteristics of the analyzed case study (independent gravity dam monolith element), it has been considered useful a comparison of the FEA results with a simplified analytical calculation based on a Gravity Method (GM), implemented by means of a Rina Consulting in-house software adopted since several years for gravity dams analyses. The GM, by means of "rigid body equilibrium" and "beam theory" analyses, permits to perform stability analyses of concrete dams both in static and in seismic conditions (pseudo-static method) with relatively simple calculations.

2 FEA Model

2.1 FEA Mesh

The dam-foundation-reservoir system was modeled with variable size elements, as shown in Fig. 2. The dam, in particular, was modeled with a quite fine discretization by means of *3 m* wide elements. The foundation and the reservoir were modeled with a linear increase of the element size, from *3 to 30 m*, moving from the dam to the external boundary. Different element typologies were used in this model:

- For the dam and the foundation, plane strain elements were used. In particular, the Q8EPS quadrilateral 4 nodes element (with 4 Gauss integration point) and the T6EPS triangular 3 nodes element (with 1 Gauss integration point) were used;
- For the reservoir, the Q4HT quadrilateral acoustic 4 nodes iso-parametric element (with 4 Gauss integration point) was used;
- The interaction between dam and reservoir was taken into account in the model using a fluid-structure interface. In particular, a fluid-structure line interface was modeled using BL4S2 linear 2 + 2 nodes flow elements (with 1 Gauss integration point);

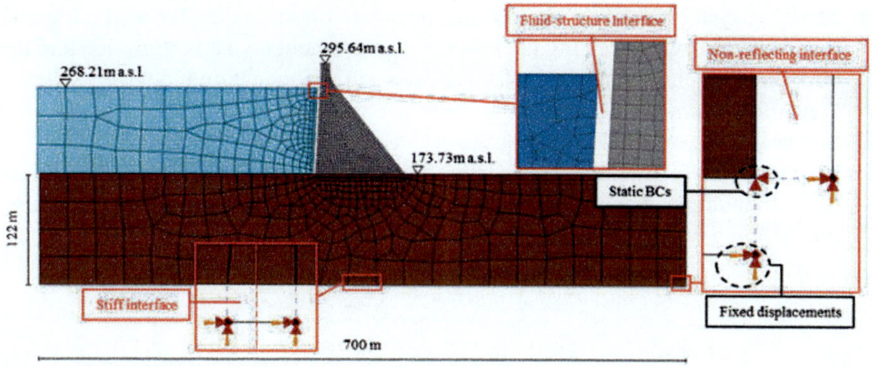

Fig. 2 Finite element model

- A non-reflecting interface along the foundation border was modeled. The non-reflecting condition was implemented by means of the L8IF 2 + 2 nodes interface, able to connect the external boundary with the foundation elements. These elements were used only in the dynamic analysis in order to have a non-reflecting behavior of the boundary conditions (BCs).

For the dynamic analyses, a 2% Rayleigh damping was assumed. The Rayleigh coefficients were evaluated considering a frequency between 4 and 10 Hz, leading to $\alpha = 0.75$ and $\beta = 0.0005$.

The Newton-Raphson convergence criterion was adopted in the analyses. The convergence norms were set to 10^{-2} for displacement and force and to 10^{-4} for energy for static analyses. On the other hand, for dynamic analyses, the convergence norms were set equal to 10^{-3} for displacement and force and to 10^{-5} for energy. Moreover, the Hilbert-Hughes-Taylor integration method was used with $\alpha = -0.3$.

2.2 *Boundary Conditions (BC)*

The reservoir was modeled by means of acoustic elements. The fundamental property of these elements is that the pressure represents their only degree of freedom. Thus, the boundary conditions must be carefully chosen:

- As already said, the interaction between the dam and the reservoir was modeled with an interface line element. This interface allows the transmission of the hydrodynamic pressure to the dam;
- Zero pressure condition was imposed at the upper and the far end face of the reservoir.

The external BCs were modeled in two different ways, depending on the analysis type:

- Static analysis: roller boundary conditions were imposed directly on the foundation elements, by restraining the vertical displacement on the bottom side and the horizontal displacement on the lateral sides of the foundation block;
- Dynamic analysis: a non-reflecting interface was used to connect the foundation borders to the external fixed boundary. To avoid reflection at the boundary, the stiffness of these interface elements must be small in comparison with the stiffness of the foundation elements ($d_n = d_s = 1$ N/m^3), as reported in [8]. At the meantime, the damping coefficients of these elements were computed as:

$$c_n = \sqrt{\rho(\lambda + 2G)}; \quad c_s = \sqrt{\rho G} \quad \lambda = \frac{\upsilon E}{(1 + \upsilon)(1 - 2\upsilon)} \tag{1}$$

where G represents the shear modulus, υ the Poisson ratio and ρ the density of the foundation.

Finally, it must be highlighted that, in order to transmit the acceleration input from the external fixed nodes to the model, a single interface, directly under the dam, was modeled with a stiffness equal to the foundation shear modulus, as shown in Fig. 2.

2.3 Loads

According to the input data provided by the Workshop formulators [1], the analyses were developed assuming a constant reservoir level of 268.21 m a.s.l., corresponding to the Winter Reservoir Water Level (WRWL), i.e. water height of 94.48 m. No uplift pressure has been considered.

In the static analysis, the following loads were considered (Fig. 3):

- Self-Weight (SW). The software directly calculates the Self-Weight based on the material density. According to the indication proposed in [10], this load affects

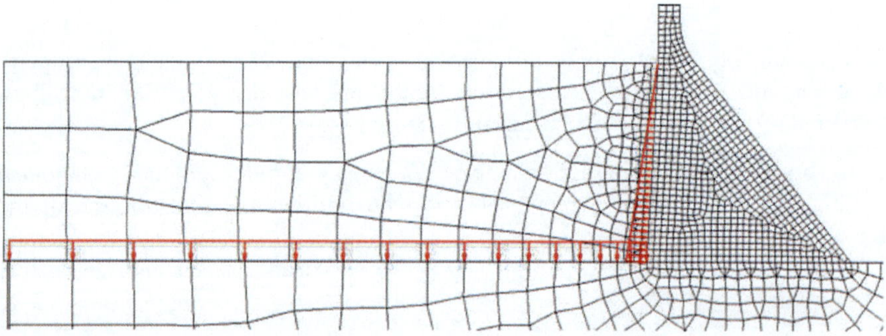

Fig. 3 FEA model, hydrostatic pressure

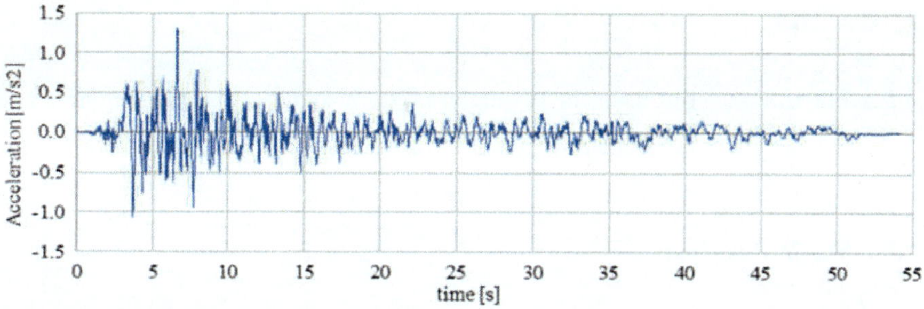

Fig. 4 Taft earthquake time history

only the state of stress because the displacement due to SW is suppressed in the analyses, in order to have an un-deformed shape after this load application;

- Hydrostatic pressure (Pr268). On the dam face, the load is applied as a linear pressure depending on the water depth and on density, from 0 MPa at the highest face elevation to 0.93 MPa at the lowest face elevation. On the reservoir bottom, a uniform pressure equal to 0.93 MPa has been applied.

In the dynamic analysis, the load is applied as a time-history acceleration record. In particular, the Taft earthquake record, reported in Fig. 4, was adopted. Such input motion was recorded during the earthquake occurred at Lincoln School in Taft, California, on July 21st, 1952.

2.4 Material Properties

The material properties adopted in the analyses, defined by the Workshop formulators [1], are resumed in Table 1.

Table 1 Linear material properties

Linear materials properties	Elastic mod. [GPa]	Shear mod. [GPa]	Density [kg/m^3]	Poisson ratio [-]
Concrete/Foundation	22.41	9.34	2483	0.20
Interface properties	Normal stiff. [N/m^3]	Shear stiff. [N/m^3]	Normal damp. [Ns/m^3]	Shear damp. [Ns/m^3]
Non-reflecting	1	1	$7.86 \cdot 10^6$	$4.82 \cdot 10^6$
Stiff	1	$9.34 \cdot 10^9$	$7.86 \cdot 10^6$	$4.82 \cdot 10^6$

3 Results

In the following, the results of the FE analyses (static and dynamic) are reported and compared with the analytical GM calculation results.

3.1 *Static Analysis*

Figure 5 reports the horizontal displacement obtained in the static FEA analyses together with a representation of the dam mesh with the sample points and faces adopted in the following to represent the stresses on the upstream and the downstream faces.

Regarding displacement, it has to be noted that the displacement associated with the Self-Weight has been suppressed in the analyses according to the hypothesis reported in Sect. 2.3.

reports the comparison between FEA and GM analytical model results in term of stress over the dam height. The stresses reported in Fig. 6 were measured oriented in the direction of the upstream and the downstream face, respectively.

The results reported in Fig. 6 highlight that FEA and GM analytical results are in good agreement. This demonstrates that analytical models (such as GM model) are able as FEA numerical analyses to reproduce, with good accuracy, the static behavior of such kind of structures.

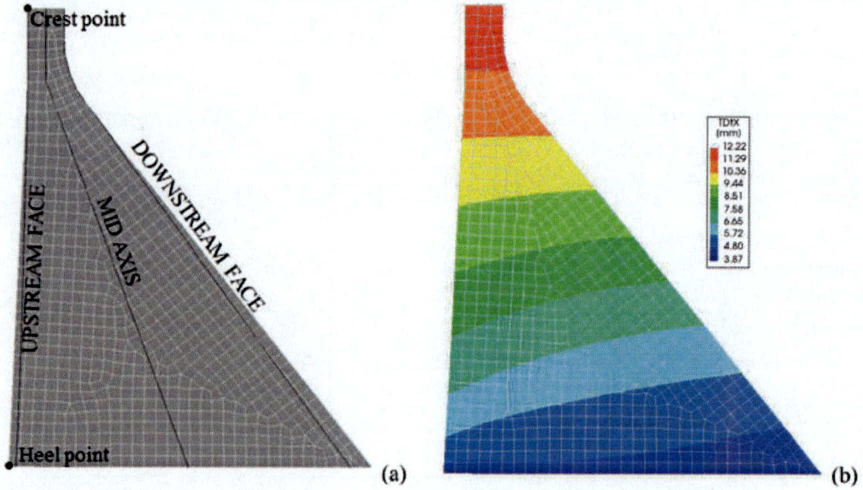

Fig. 5 **a** Dam sampling points and faces and **b** static horizontal displacements

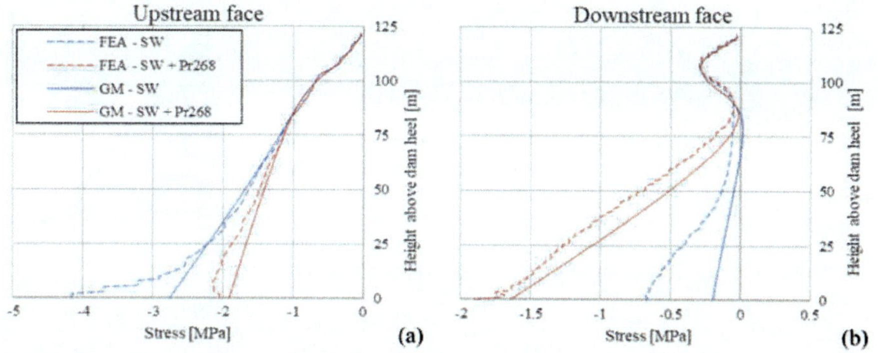

Fig. 6 Comparison between FEA and analytical GM model: stresses on **a** Upstream and **b** downstream faces (sampling faces shown in Fig. 5a)

3.2 Dynamic Analysis

Before running the time-history analysis, a modal analysis was carried out in order to highlight the principal vibration modes of the dam and their frequencies, as shown in Fig. 7.

The dynamic time-history analysis of the dam was carried out using the acceleration input reported in Fig. 4. The results of the time-history analysis are reported in Figs. 8, 9, and 10. In particular, in Fig. 8 the acceleration measured at the crest

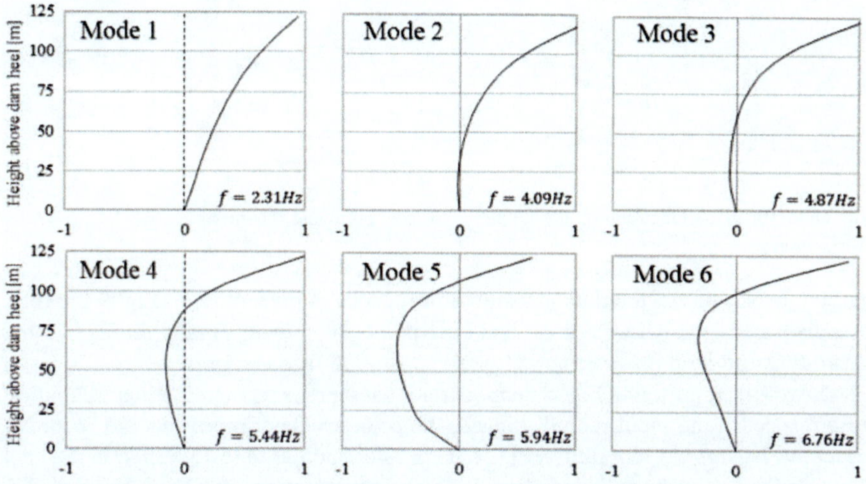

Fig. 7 Normalized modal shape and relative frequencies (first 6 modes)

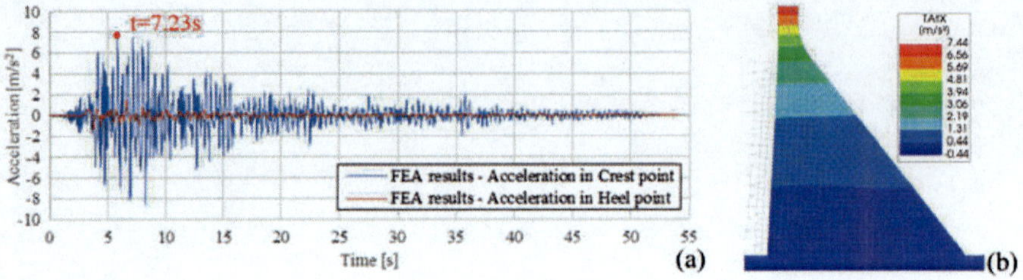

Fig. 8 Acceleration: **a** time history at crest and heel points and **b** contour at t = 7.23 s

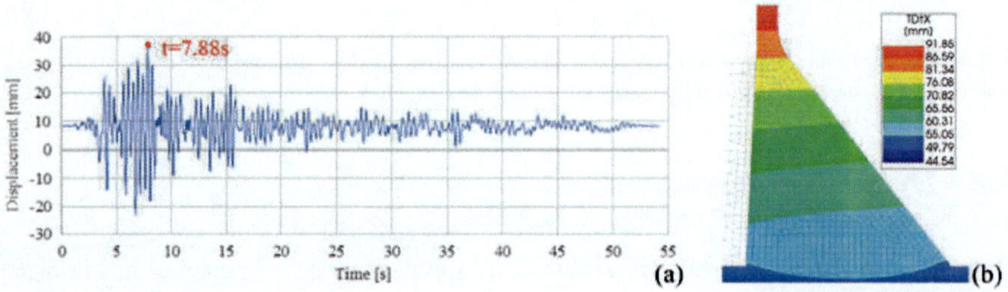

Fig. 9 Relative displacements between the crest and the heel points (including the static effect): **a** time history and **b** contour at t = 7.88 s

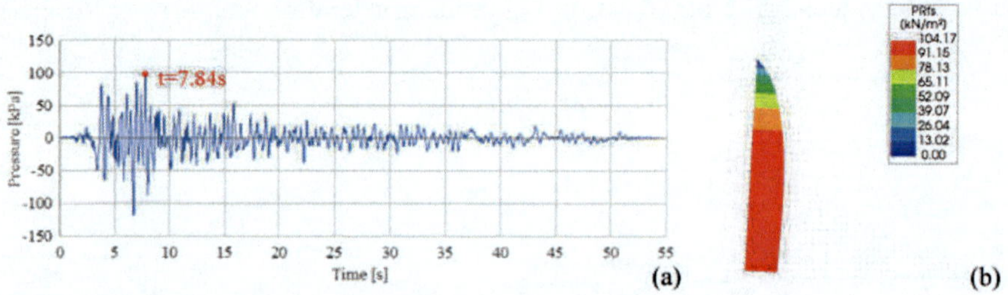

Fig. 10 Hydrodynamic pressure: **a** time history at heel point and **b** contour at t = 7.84 s

and the heel point of the dam, as shown in Fig. 5a, is presented. In Fig. 9 the relative displacements between crest and heel point are shown. Finally, in Fig. 10 the hydrodynamic fluid pressure acting on the upstream face is reported.

The results reported in Fig. 8 demonstrate that there is an increase in acceleration with respect to the input record. Indeed, the peak acceleration of the Taft record is equal to 1.30 m/s^2 as reported in Fig. 4. This value increased to 1.49 m/s^2 at the heel point and to 8.60 m/s^2 at the crest point. Such considerations can be underlined also checking the response spectrum derived from the measured acceleration, reported in Fig. 11.

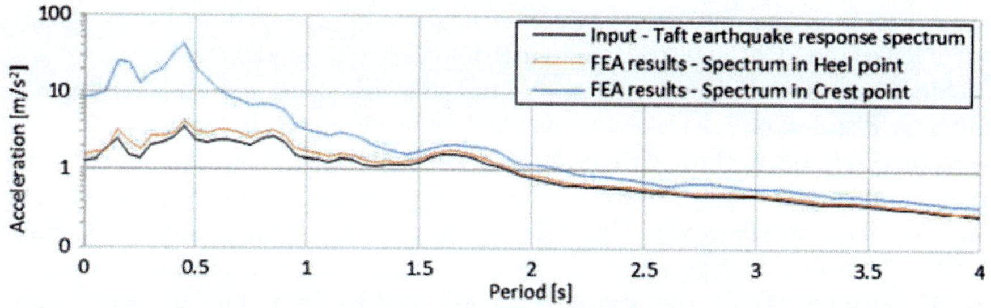

Fig. 11 Spectrum of acceleration computed for the input, Crest, and Heel point

The displacement results, represented in Fig. 9 show that in the static condition the relative displacement of the dam is equal to 8.2 mm, as shown in Fig. 5b. The further dynamic excitation subjects the dam to a maximum relative horizontal displacement of 29.24 mm in correspondence of the time step t = 7.88 s. The graphic contour of the horizontal displacements for t = 7.88 s is also reported in Fig. 9b.

Finally, in Fig. 12, the envelope of the stresses derived from FEA numerical analysis both at upstream face (Fig. 12a) and downstream face (Fig. 12b) is presented. Moreover, the FEA time-history results were compared with the results obtained with a pseudo-static method. The pseudo-static analyses were performed with both the FEA analysis (DIANA) and the analytical GM method. These analyses were based on the application of a single horizontal acceleration to the dam. Such acceleration corresponds to the acceleration derived from the Taft response spectrum (black line in Fig. 11) in correspondence of the 1st mode frequency (f = 2.31 Hz, T = 0.43 s) which leads to an acceleration of about 3.53 m/s^2.

The results reported in Fig. 12a and b highlighted that regarding the same type of analysis (pseudo-static) FEA numerical results lead to conservative values with

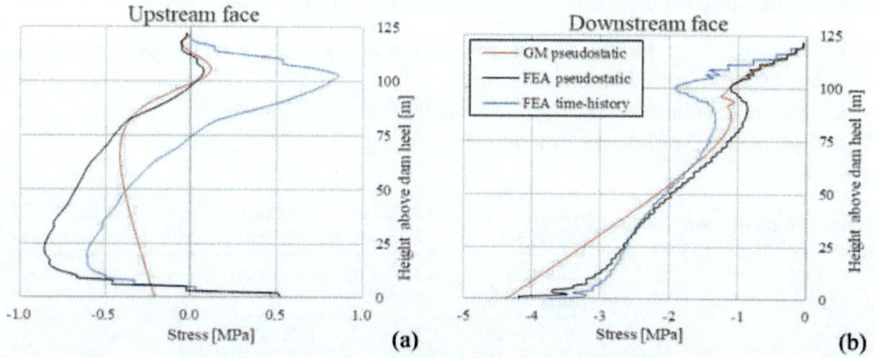

Fig. 12 Comparison between FEA and analytical GM model: Stresses at **a** Upstream and **b** Downstream faces (sampling faces shown in Fig. 5a), including the static effect

respect to analytical GM model. Such results can be expected due to the greater accuracy of the FEA model related to the simplified GM model.

Moreover, the time-history stresses compared with both pseudo-static results show different distribution over the dam height. These discrepancies are appreciable at approximately 100 m above the dam heel, in which higher stress values are reached in the time-history analysis. These results can be explained by considering the vibration modes of the dam, reported in Fig. 7. Some of the dam modes (e.g. Mode 4, 5, and 6) excite mostly the upper portion of the dam, leading to a possible increase of stresses. Such a phenomenon for sure cannot be appreciated by the simplified pseudo-static method.

4 Further Investigation of FEA Modeling

In the following, some further investigations were developed, based on the results presented in the previous chapter. Such investigations are related to three different aspects:

- Effect of element size on the results;
- Effect of reservoir level on the results;
- Effect of different Boundary Conditions on the results.

4.1 Element Size

Starting from the reference mesh, presented in Sect. 2.1 and characterized by variable element size between *3 m* (dam elements) and *30 m* (far from dam elements), two different meshes were developed. This was done in order to underline the effect of size modification on the results and on the computational time:

- Fine mesh, characterized by variable element size between *1 m* (dam elements) and *10 m* (far from dam elements).
- Coarse mesh, characterized by variable element size between *6 m* (dam elements) and *60 m* (far from dam elements).

Table 2 Comparison between different meshes

Mesh set	Fine mesh	Reference mesh	Coarse mesh
Element dimension	1÷10 m	3÷30 m	6÷60 m
N. Elements	10112	1224	340
Computational Time	70 h	1.15 h	0.35 h

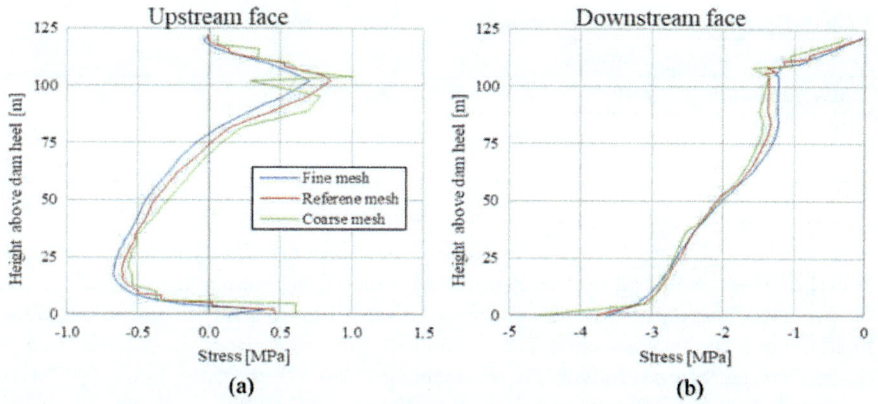

Fig. 13 Comparison between FEA and analytical GM model: Stresses at **a** Upstream and **b** Downstream faces (sampling faces shown in Fig. 5a), including the static effect

Table 2 reports the comparison between the different meshes, in terms of number of elements and computational time required for the time-history analysis.

The comparison among these meshes was extended to the evaluation of the stresses on both upstream and downstream faces, reported in Fig. 13.

The differences among the different meshes results are quite small. In particular, the finer mesh leads to very similar results in term of stresses envelope compared with reference mesh. On the other hand, the coarse mesh highlights a scatter of the stresses probably due to the lower accuracy of the model. Furthermore, the finer mesh required a computational time much higher than the other two meshes (70 h compared with approximately 1 h) due to the higher number of elements characterizing this mesh.

For this particular case study, these observations demonstrate that the reference mesh (3÷30 m) represents the perfect mix between results accuracy and computational time.

4.2 *Reservoir Level*

The effect of the reservoir level variation on the dam dynamic behavior is explained in this paragraph.

Two different reservoir levels were analyzed in addition to the WRWL (268 m a.s.l.). They are the Summer Reservoir water level (278.57 m a.s.l.) and the normal reservoir water level (290 m a.s.l.). Table 3 list the maximum relative displacements obtained in the time-history analysis for each water level. These values, namely the differences between the crest and the heel displacement, were obtained by imposing the Taft acceleration time history (Fig. 4) and neglecting the contribution of the static loads.

Table 3 Dynamic relative displacements of the dam (neglecting static effect)

Maximum relative displacement	
WRWL 268.21 m	29.24 mm
SRWL 278.57 m	31.27 mm
NRWL 290.00 m	30.42 mm

From Table 3 it can be observed the nonlinear proportionality between the horizontal displacement and the reservoir level. Indeed, differently from which it can be expected, the highest water level leads to a lower relative displacement. This phenomenon can be associated with the natural frequency variation of the dam, due to the different reservoir level. The frequencies related to the three most significant modes of the three analyzed water level are listed in Table 4.

The reservoir level influences in a considerable way the first vibration mode. Moreover, in Fig. 14a, the frequencies related to Mode 1, 2 and 4 were inserted in the Taft spectrum. In this way, the relation between the water level and the dynamic excitation of the structure can be better understood. From Fig. 14a it can be observed that the highest spectral acceleration is related to the frequency of the Mode 1, for a reservoir level of 278.57 m. Such observation can explain the fact that the reservoir level of 278.57 m leads to maximum dynamic displacement. To confirm

Table 4 Natural frequencies and excited horizontal masses of most significant modes for different reservoir level

Reservoir level	Mode 1		Mode 2		Mode 4	
	f (Hz)	M_x (%)	f (Hz)	M_x (%)	f (Hz)	M_x (%)
268.21	2.31	11.72	4.09	37.64	5.44	18.11
278.57	2.22	11.83	4.08	37.50	5.42	18.50
290.00	2.09	11.59	4.03	36.86	5.36	20.24

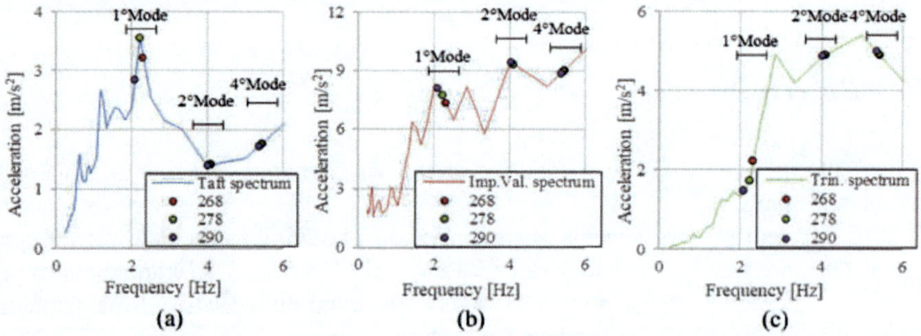

Fig. 14 Dam frequencies on different seismic spectra: **a** Taft earthquake, **b** Imperial valley earthquake and **c** Trinidad earthquake

that, two different input records were analyzed in order to better understand the displacement-reservoir relation:

- The Imperial Valley (USA) earthquake of October 15, 1979. PEER Strong Motion Database. Recorded by USGS STATION 5115. PGA = 3.09 m/s^2.
- The Trinidad (USA) earthquake of August 24, 1983. PEER Strong Motion Database. Recorded by 090 CDMG STATION 1498. PGA = 1.90 m/s^2.

Figure 14b and c report the response spectra of the two additional shaking input, together with the just described frequencies for each different reservoir level.

To easily represent the results, an equivalent acceleration was evaluated starting from the one represented in Fig. 14. It is evaluated for each different input ground motion, using the following Eq. (2):

$$a_{eq} = \frac{\sum_{i=1}^{3} a_i \cdot m_i}{\sum_{i=1}^{3} m_i} \quad (2)$$

where i represents the vibration mode, a_i represents the spectral acceleration associated with the i-th mode and m_i the excited mass of the i-th mode.

In Fig. 15a, the equivalent acceleration obtained for each input record related to different reservoir level is plotted. These accelerations were normalized to the maximum value, for each acceleration time history. Furthermore, Fig. 15b also shows the results in terms of horizontal displacements, normalized with respect to the maximum of each time history. All the displacements are referred to the dynamic behavior and to the relative values, considering only the effect on the dam body.

Based on the results plotted in Fig. 15 it can be seen that the maximum reservoir level does not always lead to the worst dynamic effect on the dam, i.e. for the Trinidad earthquake the maximum horizontal displacement was obtained for 268 m a.s.l. Moreover, it can be observed that there is a strong correlation between the evolution of the equivalent acceleration over the reservoir level (Fig. 15a) and the

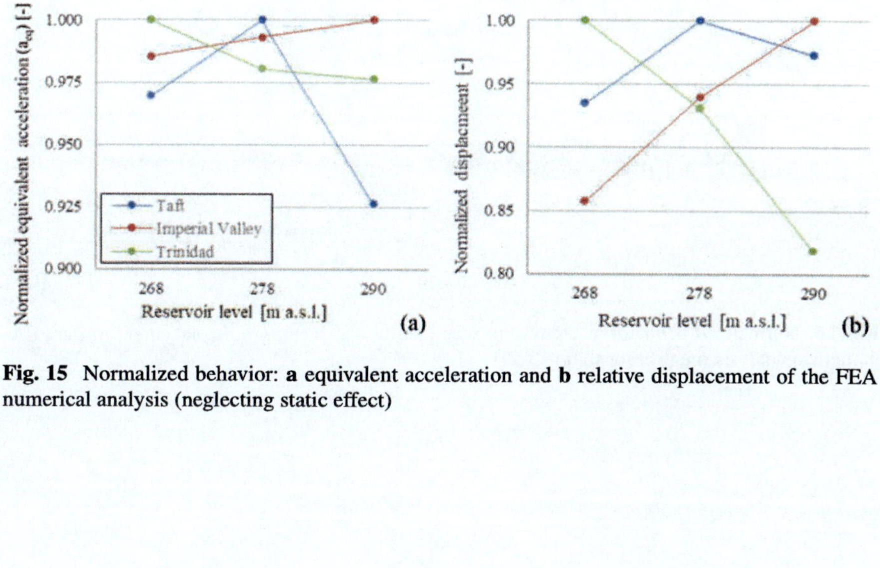

Fig. 15 Normalized behavior: **a** equivalent acceleration and **b** relative displacement of the FEA numerical analysis (neglecting static effect)

evolution of the displacement over the reservoir level (Fig. 15b). Such observation can demonstrate the importance of the variation of the dam frequencies with the reservoir level in the definition of dynamic behavior. Finally, from Fig. 15 it can be assumed that, depending on the time-history input and on the structural vibration modes, a "critical" reservoir level can be identified.

4.3 Selection of BCs

As reported in Sect. 2.2, the reference model presents a non-reflecting boundary condition at the edge of the foundation with the properties already described (Reference BCs).

In the present paragraph, a different boundary condition was analyzed to underline the effect on the dynamic behavior: a roller boundary condition (Roller BCs) were imposed directly on the foundation elements, by restraining the vertical displacement on the bottom side and the horizontal displacement on the lateral side of the foundation block. This case can be usually preferred due to its simplicity.

In Fig. 16 are plotted the acceleration derived from the FEA numerical model in two different points of the dam, crest and heel points (see Fig. 5a for the positions).

The results plotted in Fig. 16 highlighted that the simplified roller BC leads to a strong increase of the acceleration on the model. This fact can be clarified looking at the results reported in Table 5, where the maximum accelerations are presented.

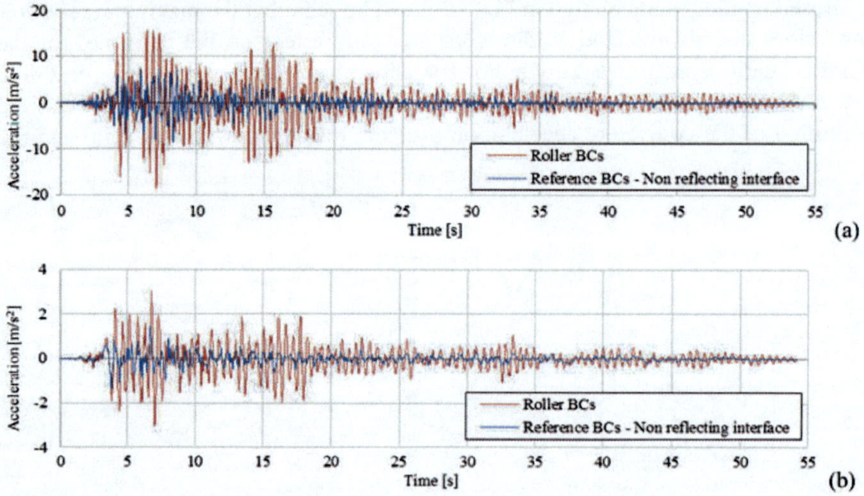

Fig. 16 Normalized behavior: **a** equivalent acceleration and **b** relative displacement of the FEA numerical analysis (neglecting static effect)

Table 5 Peaks acceleration comparison between the two types of BCs

Input	Heel		Crest	
	Reference BCs	Roller BCs	Reference BCs	Roller BCs
1.30 m/s^2	1.49 m/s^2	3.67 m/s^2	8.61 m/s^2	27.76 m/s^2

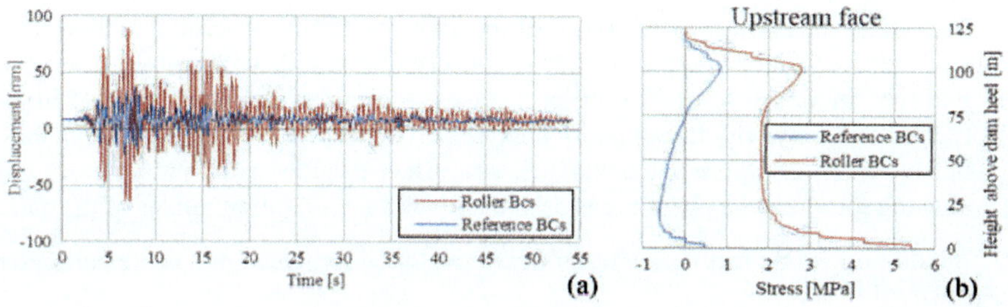

Fig. 17 Comparison between different BCs: **a** relative displacements time history between crest and heel points and **b** stresses at Upstream face (including the static effect)

From Table 5 it can be noted that in the reference BCs, characterized as already said by the presence of a non-reflecting interface at the edge of the foundation block, the maximum acceleration at the heel of the dam is approximatively equal to the maximum input acceleration. On the other hand, the Roller BCs lead to an increase of the input signal also in the foundation block. This increment of acceleration is much more evident looking at the acceleration at the crest point of the dam. The consequence of these increased accelerations is the deformation and stresses growth as it can be noted looking at Fig. 17, in which the relative displacement between crest point and heel point of the dam associated to the dynamic behavior (including static effect) is plotted.

The results plotted in Fig. 17a showed that moving from Reference BCs (non-reflecting interface) to Roller BCs, the maximum horizontal displacement strongly increased from 29.24 mm to 80.52 mm. The increment of the horizontal displacements leads to a much higher level of stresses on the structure, as can be appreciated in Fig. 17b. Considering as reliable the results referred to the "Reference BCs", as already stated in Chapter 3, it can be assumed that simplified modeling of the boundary condition (e.g. Roller BCs) could lead to strongly unrealistic and overestimated results.

5 Conclusion

This paper deals with the static and dynamic behavior of *Pine Flat* concrete gravity Dam by means of Finite Element Analyses (FEA). The numerical results are also compared with a simplified analytical calculation based on a Gravity Method (GM) model.

The main conclusions derived from the presented results can be resumed in the following points:

- Analyzing the **Static behavior** of the dam, the simplified analytical GM model leads to similar results compared with the more refined FEA model, demonstrating that, for these particular loading conditions, the adoption of a sophisticated method could not be necessary;
- Analyzing the **Dynamic behavior**, of the dam, the results presented in this paper demonstrated that FEA model leads to conservative results with respect to simplified GM model in term of state of stresses. Moreover, the analyses also highlighted that only with time-history analyses all the aspects related to the dynamic behavior can be appreciated. Indeed, simplified pseudo-static analyses are not able to reproduce the phenomena associated with the vibration modes of the dam.

In addition, some further analyses were developed to investigate some particular aspects:

- The **Boundary Conditions** play a very important role in the prediction of the dynamic behavior of the dam. Fixed boundary condition leads to an amplification of the input signal in the foundation. For this reason, the adoption of non-reflecting Boundary Condition is strongly recommended to avoid this problem;
- The investigation on different **reservoir levels** underlined that due to the variation of the vibration modes, the highest water level not always leads to the worst stresses condition on the dam;
- The **size-effect** assumes an important role in order to join the computational advantages of a coarser mesh with the benefit in term of stresses evaluation given by a finer mesh. This paper showed that the better results are obtained using an element size for the dam equal to 2.5% the height of the dam.

The results presented in this paper finally demonstrate that it could be useful to use more a refined calculation model, such as Finite Element Analyses, especially for the assessment of existing dams, where a more realistic prediction of the structural behavior is essential.

Acknowledgements This research was part of the "15th International Benchmark Workshop on numerical analyses of Dams", organized in Milan (Italy) in 2019 by the International Commission on Large Dams (ICOLD). The authors would like to thank the ICOLD organization for the possibility to be part of this project and share the relevant amount of information with the other contributors. In particular, the authors would express very great appreciation to the effort made by Jerzy Salomon as the formulator of the "Concrete Gravity Dam" topic within the Benchmark.

The authors would also like to thank DIANA FEA BV for the precious support given for this research and during years of fruitful cooperation.

References

1. Salamon J, Hariri-Ardebili MA, Malm R, Wood C, Faggiani G (2019) Theme A formulation. Seismic analysis of pine flat concrete dam. In: 15th ICOLD international benchmark workshop on the numerical analysis of dams. Milan, Italy
2. Rea D, Liaw CY, Chopra AK (1972) Dynamic properties of pine flat dam. Report no UCB/EERC-72/7
3. Chopra AK, Chakrabarti P, Gupta S (1980) Earthquake response of concrete gravity dams including hydrodynamic and foundation interaction effects. Report no UCB/EERC-80/01
4. Fenves G, Chopra AK (1986) Simplified analysis for earthquake resistance design of concrete gravity dams. Report no UCB/EERC-85/10
5. Chavez JW, Fenves GL (1993) Earthquake analysis and response of concrete gravity dams including base sliding. Report no UCB/EERC-93/07
6. Chavez JW, Fenves GL (1994) EAGD_SLIDE: A computer program for the earthquake analysis of concrete gravity dams including base sliding. Report no UCB/SEMM-1994/02 (1994)
7. Fenves G, Chopra AK (1984) EADG-84: A computer program for earthquake analysis of concrete gravity dams. Report no UCB/EERC-84/11
8. Evaluation of Numerical Models and Input Parameters in the Analysis of Concrete Dams. Report DSO-19-13. A summary report of the USSD workshop. Miami, USA (2018)
9. DIANA FEA BV DIANA User's Manual - Release 10.3 (2019)
10. Malm R (2016) Guideline for FE analyses of concrete dams. Energiforsk report 2016:270

Analysis of Pine Flat Dam Considering Fluid-Soil-Structure Interaction and a Linear-Equivalent Model

A. F. Stabile, P. Labbé, and A. Nguyen

Abstract A 2D plane strain model is adopted. After some sensitivity studies an average mesh size of 15 m is selected, resulting in about only 700 nodes. Two methods are considered for fluid-structure interaction modelling: full finite element and Westergaard's added mass approach, leading to non-significant differences in the results, at least for relatively low input motion. The dam nonlinear response is analyzed through an equivalent linearization technique, based on conventional damage model. Damage is tensile strain controlled and results in a lowered effective concrete Young Modulus. Implementation of the method requires an iterative procedure, which converges in a few iterations. Damage development in dam can be measured by the calculated effective frequency. An output is that the Taft input motion does not generate damage either in the dam body or at the dam-foundation interface. Under ETAF, with the adopted definitions of damage, the dam fails at about 9 s.

Keywords Linear-equivalent model · Seismic analysis · Concrete gravity dam · Fluid-soil-structure interaction · Concrete damage

1 Analysis and Model Description

1.1 Meshing

For all the study cases, 2D Finite Element analyses using Code_Aster, a General Public License software developed by EDF, were performed, under the assumption

A. F. Stabile (✉)
Universidade Federal do Rio Grande do Norte, Natal, Brazil
e-mail: andrestabile1@hotmail.com

A. F. Stabile · P. Labbé
École Spéciale des Travaux Publiques, du Bâtiment et de l'industrie, Cachan, France

A. Nguyen
Geodynamique et Structures, Montrouge, France

G. Bolzon et al. (eds.), *Numerical Analysis of Dams*, Lecture Notes in Civil Engineering 91, https://doi.org/10.1007/978-3-030-51085-5_15

of Plane Strain behavior, as suggested by Chopra et al. [1]. For modeling of the Foundation-Dam-Reservoir system, 3 and 4-node linear elements with maximum length of 15 m were used. To determine the size of the mesh, a sensibility study was performed, using more refined meshes. In view of the small differences between the results, it was concluded that this mesh size is adequate. For the boundaries of the foundation domain, 2-node linear paraxial absorbing boundary elements were chosen to represent the infinite domain condition. As for the fluid domain, two different models were considered, one taking into account 4-node acoustic fluid elements with absorbing paraxial elements in the upstream direction (Full FEM model) and a second one considering an added fluid mass, in accordance with Westergaard's formulation [2]. Fluid-Structure interaction was taken into account in both the Dam-Reservoir and Foundation-Reservoir interfaces. Figure 1 shows a picture of the mesh for the Full FEM model and Table 1 details its number of elements and nodes.

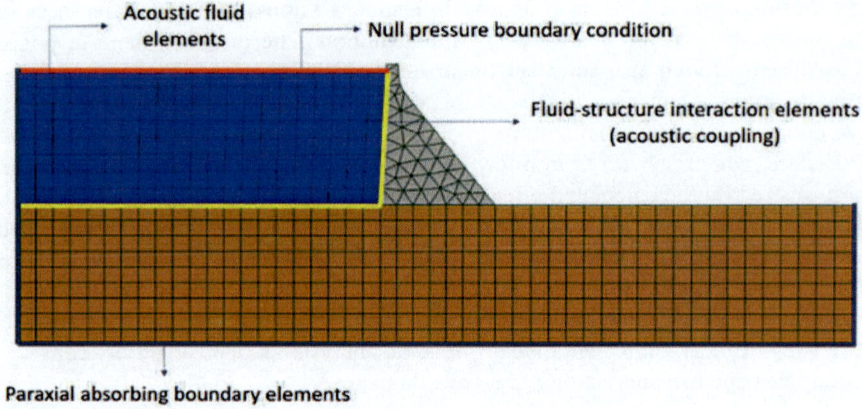

Fig. 1 Full FEM mesh

Table 1 Nodes and elements for the full FEM mesh

Part of the mesh	Element type	Number elements	Number nodes
Dam and foundation	Plane strain	497	530
Foundation boundary	Paraxial absorbing boundary	66	67
Reservoir	Acoustic elements	168	198
Upstream reservoir boundary	Paraxial fluid absorbing boundary	8	9
Dam-reservoir and foundation-reservoir interfaces	Fluid-structure interaction elements (acoustic coupling)	29	30
Total	–	768	698

1.2 Loads

For the cases including the weight of the reservoir and the foundation, the system response was determined by performing a static analysis, and was later on superposed with the dynamic results for all other analysis. The reservoir weight was imposed as a hydrostatic pressure on the upstream face of the dam and on the concerned part of the foundation.

As for the dynamic cases, the input signals were imposed as plane waves in the bottom as well as in the downstream and upstream foundation boundaries. Since Code_Aster uses the incoming wave as input, a procedure to find such a signal was adopted and is described in the Appendix. For the EMVG analysis, a horizontal force was applied on the upstream node of the dam crest.

1.3 Boundary Conditions

For both the static and modal analysis, the displacements in the foundation boundary were restrained. For models with acoustic fluid elements, null pressure as well as null displacement potential were considered on the free surface.

For the dynamic analysis, as specified earlier, paraxial absorbing elements were applied, and for their correct behavior no kind of restraint was imposed on the domain boundaries.

1.4 Damping

As suggested by the formulation document [3], Rayleigh damping was chosen to model damping due to material properties. The adopted proportionality coefficients were $\alpha = 0.75$ 1/s and $\beta = 0.0005$ s, which translates into an equivalent 2% modal damping in frequencies ranging between 4 and 10 Hz.

It is important to note, however, that those values refer only to material damping. When the absorbing boundaries are taken into account radiation damping also takes place. A free vibration analysis using the decrement-log method was conducted, and the conclusion is that radiation damping alone corresponds to a value of approximately 8.8%, which is already substantially more elevated than the original 2% material damping.

1.5 Analysis

For the non-linear study cases, a linear-equivalent, strain-based approach is proposed, in the wake of research recently carried out in ESTP by Nguyen [4]. A damage criterion was chosen in the spirit of damage model introduced by Mazars [5], based on the non-linear concrete properties given on the formulation document:

- Up until the strain equivalent to the material's tensile strength no damage is produced.
- From this point onward, the material's Young's module is reduced following the law described by Eq. (1) and depicted in Fig. 2. Such relationship was derived considering that the material's Young's module is given by the average between compressive Young's module (always equal to 22410 MPa) and an equivalent tensile Young's module. The equation describing the curve is then determined based on two points: the first corresponding to the material's tensile strength and the second to the tensile strain at peak load.
- Moreover, the condition that the curve has an asymptote corresponding to a 50% reduction in Young's module as ε goes to infinity is taken into account.

The linear-equivalent analysis is conduced, then, in an iterative manner. For each iteration, a linear analysis is performed, and the principal strains in the dam are determined. Then, for each element, the maximum principal strain throughout the time history of the calculation is considered to evaluate for damage, and a new Young's module is calculated. The convergence criterion taken into account is inspired upon Nguyen [4] and considers that the model converges when a difference no greater than 1% in the first vibration frequency is noticed from on iteration to the next one.

$$\frac{E(\varepsilon)}{E_0} = \begin{cases} 1 & for\ \varepsilon \leq 0.00009 \\ 0.5 + 0.5e^{-9629,27\varepsilon+0.8594} & for\ \varepsilon \geq 0.00009 \end{cases} \tag{1}$$

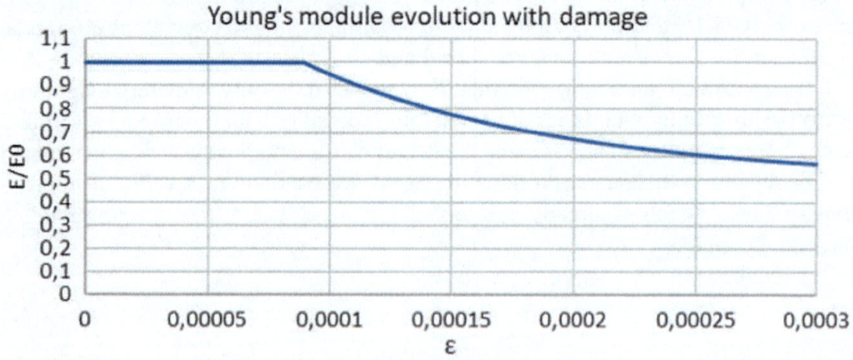

Fig. 2 Young's module evolution as a function of tensile strain

For the numerical resolution of the dynamic equilibrium equations, Newmark's integration scheme was the preferred one, with usual values of $\beta = 0.25$ and $\gamma = 0.5$. The chosen time step is of one tenth of the input signal record time, corresponding to $\Delta t = 0.001$ s for most analysis cases.

2 Results and Discussion

2.1 *Case a*

In case A, the natural frequencies of the system for two different reservoir water levels, alongside with linear dynamic analysis for a vibratory mass excitation in crest were demanded. To perform such analysis both the Full FEM and the added mass models were considered. The most relevant results from such analysis are presented in Table 2 and Fig. 3.

Table 2 First six dam natural frequencies (Hz)

Frequency number	WRWL full FEM (Hz)	WRWL added mass (Hz)	SRWL full FEM (Hz)	SRWL added mass (Hz)
1	2.40	2.34	2.29	2.21
2	3.43	4.15	3.18	4.10
3	3.96	4.77	3.81	4.73
4	4.40	5.74	4.32	5.66
5	4.94	6.30	4.90	6.18
6	5.55	7.61	5.45	7.56

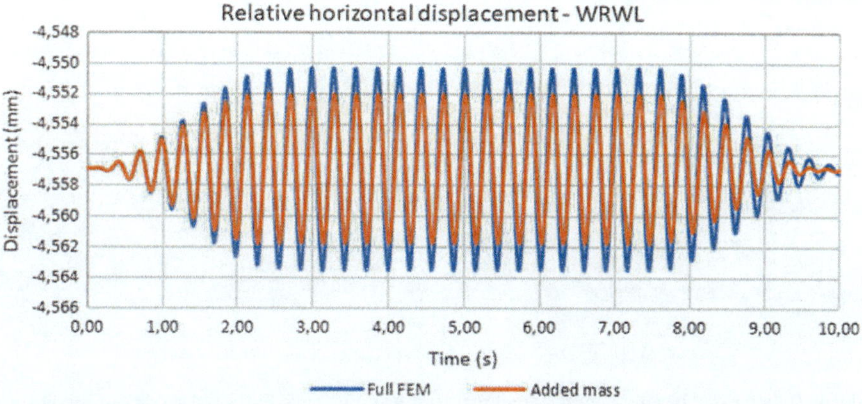

Fig. 3 Relative horizontal displacement for both models—WRWL—Case A

Considering the obtained results, it can be concluded that the added mass model, even though less sophisticate, can predict results with the same magnitude order of the Full FEM one. The interest in comparing these results arise from the much easier implementation of the first model to the detriment of the second one, which is of interest to usual engineering practice, in the measure that it doesn't demand refined analysis to be made in projects that don't have such requirements.

2.2 *Case D*

The purpose of case D was to study the dynamic behavior of the Dam-Foundation-Reservoir system, taking into account different reservoir water levels. Once again, both models (Full FEM and added mass) were used in the analysis. However, since the added mass model does not have a way of measuring hydrodynamic pressure, these results will be given only for the Full FEM model. The analysis performed took into account the three different reservoir water levels proposed. Figures 4 and 5 show a comparison between relative horizontal displacement and hydrostatic pressure, respectively, for different reservoir water levels.

From the plots, one can note that the reservoir water level has two main influences. The first is due to the change in hydrostatic pressure with the water level, represented by the displacement at time $t = 0$. The second one, mostly pronounced only for the displacements, is an amplitude amplification, caused by the reduction of inertial effects produced by the water's mass, directly proportional to the decrease of the reservoir's water level.

Figure 6 shows a comparison of horizontal relative displacement considering the SRWL for the two different models. Again, as in the case of the EMVG excitation,

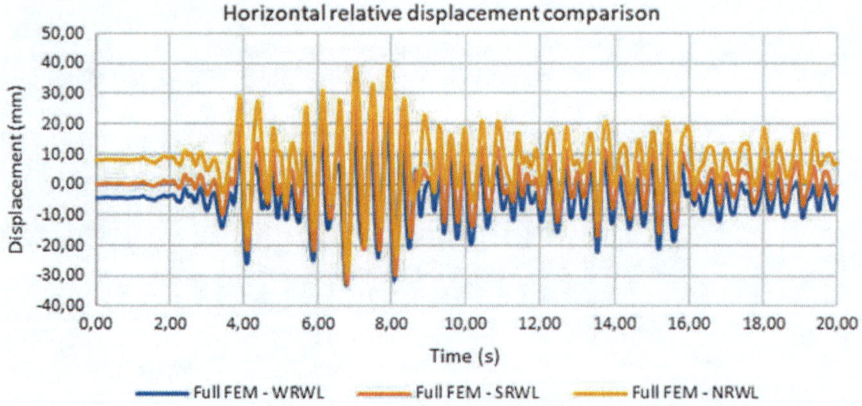

Fig. 4 Relative displacement for different reservoir water levels—Full FEM model—Case D

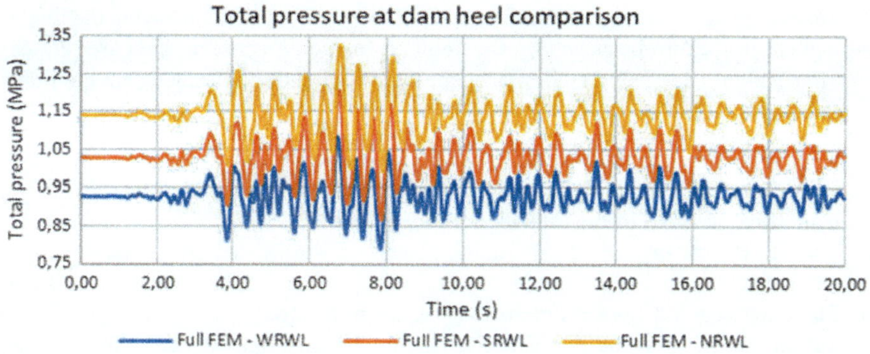

Fig. 5 Total pressure for different reservoir water levels—Full FEM model—Case D

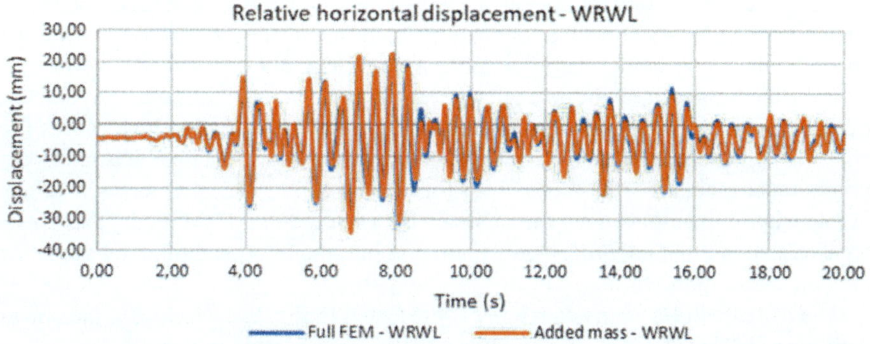

Fig. 6 Comparison of the horizontal relative displacement for the two different models—Case D

we remark that Westergaard's added mass formulation is in close accordance with the results obtained using the Full FEM model.

2.3 *Case E*

This case's focus was to study the non-linear behavior of the dam submitted to Taft's earthquake acceleration time-history (Case E-1) and to ETAF, an artificially generated time signal, intended to perform a dynamic pushover analysis (Case E-2).

The damage index (DI) in the dam body is calculated as the ratio between the number of damaged elements and the total number of elements (66). An element is regarded as damaged if the concrete tensile strain capacity is exceeded.

Regarding the dam-foundation interface, a part of it is regarded as damaged if the vertical stress is a tensile stress. The damage index is calculated as per the formulation document [3].

For case E-1, no damage in the dam body (DI = 0) was found following the linear-equivalent procedure proposed by the authors. In such a way that the case reduces to a linear dynamic analysis. Similarly, the dam-foundation interface damage index remains zero.

As for case E-2, the following procedure was adopted:

- ETAF's signal was divided into 15 parts, the i-eth of them (called ETAF i) going from time 0 to i/15 of the total signal time;
- For ETAF 1 the procedure presented in Sect. 1.5 is followed until convergence is reached;
- Then, damage indices are determined for the first signal;
- Subsequently, analysis for ETAF 2 is proceeded, starting with the damage calculated in ETAF 1's analysis;
- The analysis continues in a similar manner for the other ETAF signals.

This procedure allows taking into account damage evolution with time, and, as an advantage in comparison with a conventional non-linear analysis, to have knowledge about the dam's frequency response. Selected results will be presented hereafter.

From the figures, it can be noted that, due to the great amount of damage generated by ETAF's signal, the dam's first vibration frequency diminishes, which causes the peak values of response in Fig. 7 to be placed further away in the time axis relatively to the linear model. Furthermore, from Fig. 8, one can remark that for high values of intensity of the input signal the accuracy between the Full FEM model and Westergaard's added mass approach gets poorer, being much too amplified by the latter in comparison to the former.

The damage index in the dam body is presented in Fig. 9, according to the above definition, while the dam-foundation damage index is plotted in Fig. 10.

Adopting as failure definition that either one or the other of damage indexes becomes equal to 1, and extrapolating the dam-foundation damage index as presented in Fig. 10, we are led to the conclusion that the failure occurs at about 9.6 s.

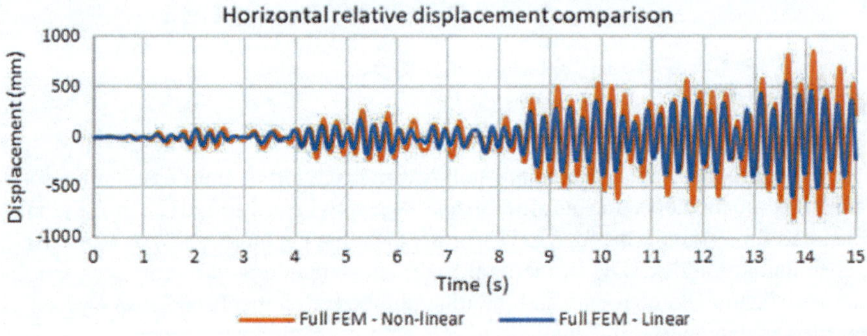

Fig. 7 Relative horizontal displacement for linear and non-linear Full FEM—Case E-2

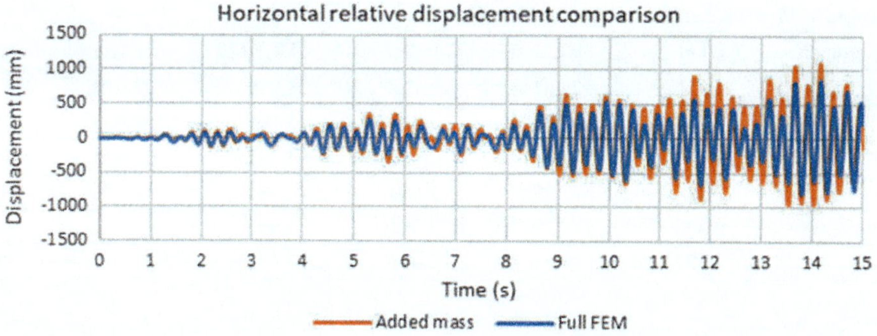

Fig. 8 Relative horizontal displacement for non-linear added mass and Full FEM—Case E-2

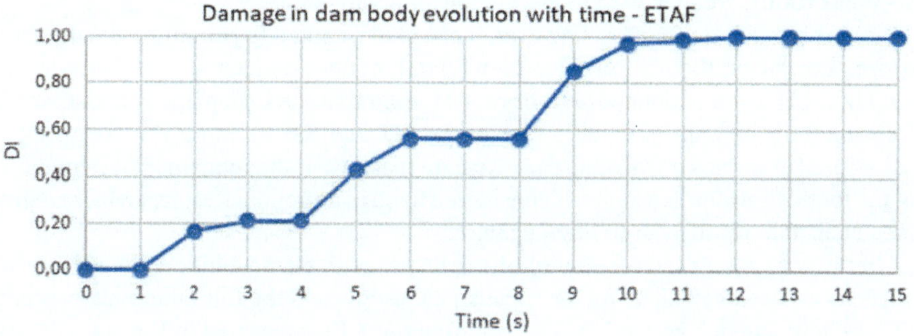

Fig. 9 Evolution of damage index in the dam body versus time—Full FEM—Case E-2

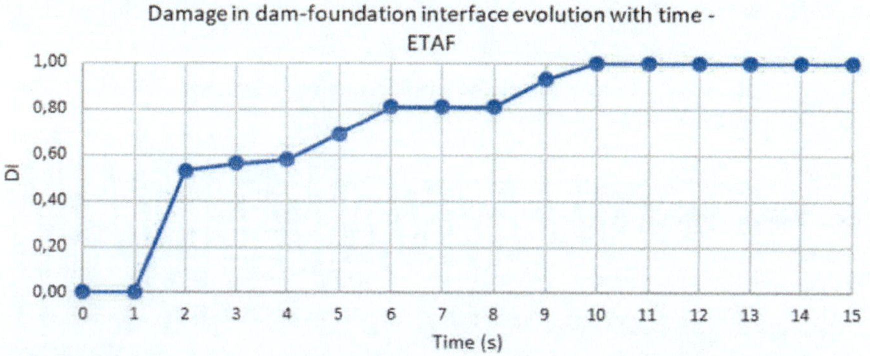

Fig. 10 Dam-foundation interface damage index versus time—Full FEM model—Case E-2

Table 3 First six dam natural frequencies (Hz)

Iteration	ETAF 3	ETAF 6	ETAF 9	ETAF 12	ETAF 15
0	2,389	2,251	2,177	2,009	1,969
1	2,388	2,250	2,177	2,008	1,969
2	2,354	2,184	2,100	1,987	1,972
3	2,344	2,179	2,088	1,984	–
4	2,338	–	2,083	1,983	–
5	2,334	–	2,081	–	–

As mentioned before, one of the advantages of a linear-equivalent analysis is that the frequency response of the structure can be observed as damage occurs. Table 3 accounts for the frequency reduction in the dam with damage.

Returning to Figs. 7 and 8, however, it can also be noted (by counting the number of zero crossings) that the response signal predominant frequency is of the order of 4,0 Hz. This result is unexpected, having in mind that the undamaged fundamental frequency of the dam is of the order of 2,4 Hz and the structure can roughly be considered as mono-modal. One can also note that 4 Hz is the fundamental frequency of the foundation domain alone, if considered fixed in its boundaries, fact which might be responsible for this surprising behavior.

In order to assess the sensitivity of the response to deconvolution, an additional analysis was conducted, using the input signal as given in the formulation document [3], without any deconvolution. The resulting output is presented in Fig. 11.

From Fig. 11 it can be noted that, without the deconvolution procedure described in the Appendix, the response signal predominant frequency is, indeed, of the order of 2,5 Hz, as expected. It is yet to be determined, however, whether the change on the

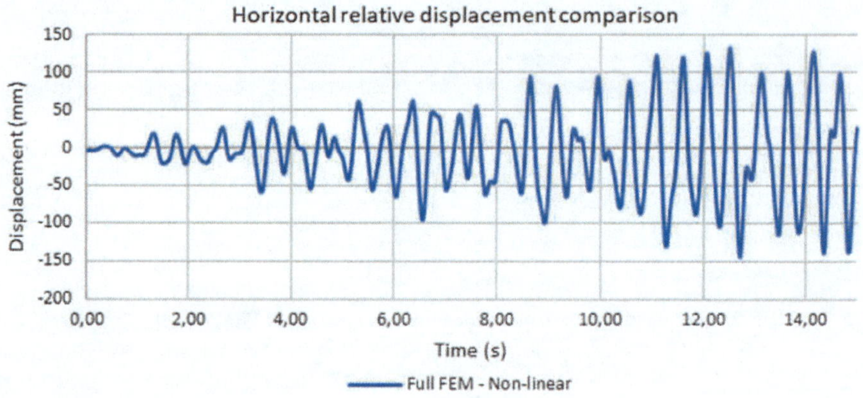

Fig. 11 Relative horizontal displacement for non-linear Full FEM—Case E-2—signal without deconvolution

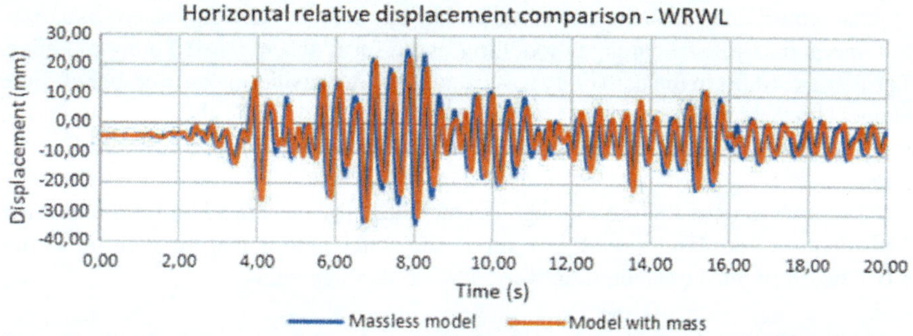

Fig. 12 Relative horizontal displacement for the models with and without foundation mass—Full FEM model

response's frequency is due to the presented deconvolution procedure or is a natural phenomenon that happened to arise in this particular case.

2.4 *Case F*

For the last analysis case, the behavior of the dam-foundation-reservoir system submitted to Taft's signal is analyzed when a massless foundation is taken into account, a common practice in dam engineering. In Fig. 12, one of the results from such analysis is presented.

It can be noted that, even though there is no physical sense in not taking into account foundation mass, the agreement between the two models is surprisingly good. A small phase difference between the signals is observed. Such a difference is due to the fact that, in a massless foundation, the wave propagation is instantaneous, whereas in a foundation with mass the wave takes a small time (of approximately 0.06 s) to propagate from bottom to top.

3 Conclusions and Lessons Learned

After a sensitivity analysis on the subject, a model with a limited number of finite elements (768) was selected for running the analyses with the software Code_Aster developed by EDF. Two options were used for fluid-structure interaction modelling, fluid elements and Westergaard's added mass. They provide very similar outputs to the extent the input motion is not too strong, for instance for the Taft input motion. The difference trends to increase with the input level. Absorbing boundary conditions were used, which result in a radiative damping of the order of 9% in the considered case.

Dam non-linear response was simulated through an iterative equivalent linearization method, based on damage modelling in the form of a reduced Young modulus. The damage index in the dam body was calculated by considering the number of finite elements in which the concrete tensile stain is exceeded, while the dam-foundation interface damage index was calculated by considering vertical tensile stresses at the interface.

Main outputs are that no damage is calculated for the Pine Flat dam under the Taft input motion. Under the ETAF input, the conclusion is that the dam fails after 9.6 s, based on the considered dam-foundation damage index.

Acknowledgements The authors would like to express their gratitude to the "Institut de Recherche en Constructibilité - IRC" for proportionating the means by which this work was achieved, and also to "CAPES Foundation – Ministry of Education of Brazil" for the concession of BRAFITEC's program scholarship to Mr. André de Figueiredo Stabile (Process n° 88887.194550/2018-00). They would also like to thank Mr. Emmanuel Robbe, engineer at EDF, for the profitable discussions and help in the usage of Code_Aster.

Appendix

The following appendix describes the procedure used to find an incoming wave at a depth z parting from the deconvolved signal at the same depth. The foundation damping is neglected throughout the whole procedure. Let:

- $\gamma(z, t)$ be the deconvolved signal at a depth z and time t;
- $\gamma(0, t)$ be the signal at the free-surface at time t;
- $\gamma_i(z, t)$ be the incoming signal at a depth z and time t;
- $\gamma_r(z, t)$ be the reflected signal at a depth z and time t;
- T be the wave-propagation time in the foundation.

We can write, by definition:

$$\gamma(z, t) = \gamma_i(z, t) + \gamma_r(z, t) \tag{2}$$

Since there is no damping in the foundation:

$$\begin{cases} \gamma_i(z, t) = \frac{1}{2}\gamma(0, t + T) \\ \gamma_r(z, t) = \frac{1}{2}\gamma(0, t - T) \end{cases} \Rightarrow \gamma_r(z, t) = \gamma_i(z, t - 2T) \tag{3}$$

Inserting (3) into (2) gives:

$$\gamma(z, t) = \gamma_i(z, t) + \gamma_i(z, t - 2T) \tag{4}$$

Since, for $t < 0$, $\gamma_i(z, t) = 0$ we can write:

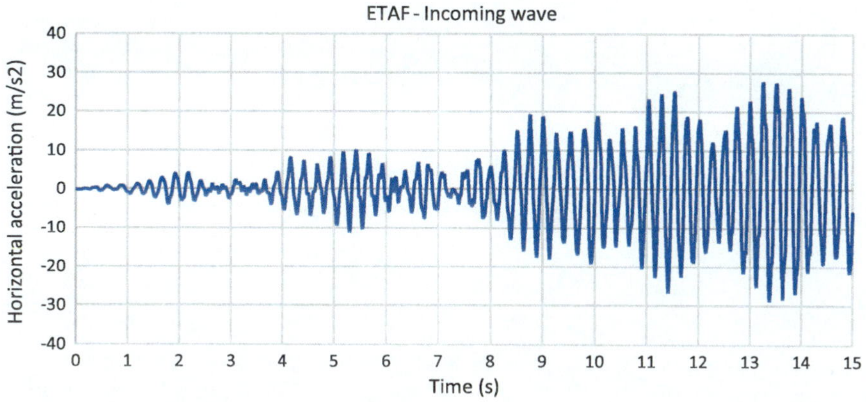

Fig. 13 ETAF's incoming signal

$$\gamma_i(z, t) = \gamma(z, t) fort < 2T \tag{5}$$

For a time $t \geq 2T$, the incoming motion $\gamma_i(z, t - 2T)$ has already been determined by means of (5), hence, we can find $\gamma_i(z, t)$ using:

$$\gamma_i(z, t) = \gamma(z, t) - \gamma_i(z, t - 2T) fort \geq 2T \tag{6}$$

The above procedure was tested for Taft's deconvolved acceleration and the incoming wave found was the same as half the free-surface acceleration provided by the Contributors. ETAF's incoming signal is illustrated on Fig. 13.

References

1. Chopra AK, Chakrabarti P, Gupta S (1980) Earthquake response of concrete gravity dams including hydrodynamic and foundation interaction effects. berkeley. Earthquake Engineering Research Center, p 201, Report no UCB/EERC-80/01. January 1980
2. Westergaard HM (1933) Water pressure on dams during earthquakes. Trans Am Soc Civil Eng 98:418–472
3. Salamon J, Hariri-Ardebili MA, Malm R, Wood C, Faggiani G (2019) Theme A formulation. Seismic analysis of Pine Flat concrete dam. In: 15th ICOLD international benchmark workshop on the numerical analysis of dams. Milan, Italy
4. Thuong AN (2017) Analyse systématique du concept de comportement linéaire équivalent en ingénierie sismique. Ph.d. [dissertation]. Paris: Université Paris-Est. Available from: TEL–Serveur de Thèses en Ligne
5. Mazars J (1986) A description of micro-and macroscale damage of concrete structures. Eng Frac Mech 25(5–6):729–737

2D Seismic Analysis of Pine Flat Concrete Gravity Dam Including Mass of Foundation: Approach with Infinite Elements

E. J. Staudacher and G. Zenz

Abstract Since 1991 the Committee on Computational Aspects of Analysis and Design of Dams within the International Commission on Large Dams (ICOLD) organizes the bi-annual held Benchmark Workshops. The 15th Benchmark Workshop, held in Milan (Italy) is dedicated to three formulated themes as well as an open theme. Numerical studies on the seismic response of the highest monolith of Pine Flat Dam are defined in Theme A, which the present paper is focused on. Several numerical studies are part of Theme A, like modal analyses and transient seismic analyses in which the mass of the foundation is respected. Generally, 2D finite element models with plain strain formulation are used. To simulate a far field solution of the given problem, infinite elements are used for static and dynamic load cases. Furthermore, linear elastic material behavior is assumed together with the so-called concrete damaged plasticity model, a non-linear material model, which is applied to the dam body. To account for hydrodynamic effects due to seismic loading acoustic elements are coupled to the solid structural parts to enable the fluid-structure-interaction.

Keywords Dam engineering · Fluid-structure-interaction · Concrete gravity dams · Structural dynamics · Hydraulic engineering

1 Introduction

The present paper summarizes the contributor's modeling assumptions and results of the numerical investigations related to Theme A of the 15th International ICOLD Benchmark Workshop on Numerical Analyses on Dams. Theme A is subdivided into different numerical studies that are obligatory or optional. All the obligatory studies are conducted on the tallest non-overflow monolith of Pine Flat Dam (California), like the determination of fundamental natural frequencies, up to dynamic analyses considering non-linear material models for the dam body. Another common feature

E. J. Staudacher (✉) · G. Zenz
Institute of Hydraulic Engineering and Water Resources Management, Graz University of Technology, Graz, Austria
e-mail: edwin.staudacher@tugraz.at

G. Bolzon et al. (eds.), *Numerical Analysis of Dams*, Lecture Notes in Civil Engineering 91, https://doi.org/10.1007/978-3-030-51085-5_16

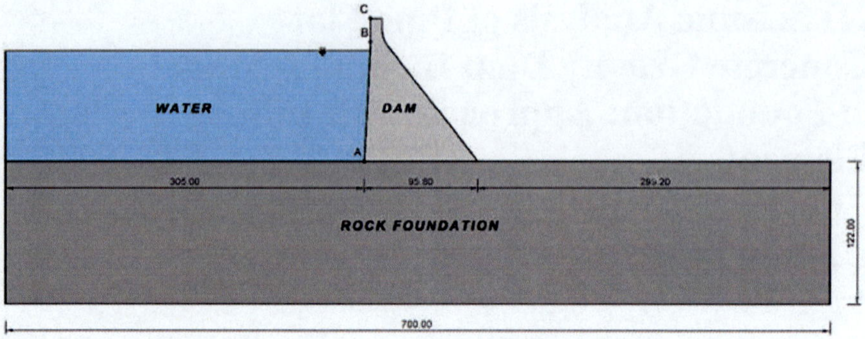

Fig. 1 Basic geometry of Pine Flat dam including foundation and reservoir

of these analyses is that the mass of the foundation block has to be considered in the analyses. This leads to the application of additional boundary conditions (BCs) at the side and bottom faces of the foundation block, like absorbing boundary conditions. The studies are conducted by discretizing the structural and fluid domains with the finite element (FE) method, using the commercial FE software Abaqus/CAE. Basically, the geometry provided by the formulators [1] consists of three parts: dam body, rock foundation and reservoir (incl. main dimensions) as depicted in Fig. 1.

The monolith in this study is about 122 m high with a base length of 95.80 m and a crest width of 9.75 m. The dimensions of the foundation block are 700 m in length and 122 m in height.

2 Finite Element Model

The formulators left the choice of using a 2D or 3D model of the 15.24 m wide Monolith 16 to the contributors. In the present studies 2D plane strain models are used based on the assumption that there is no load transfer to the rock foundation in cross-stream wise direction between adjacent blocks of the dam, which is suitable for this concrete gravity dam with a straight dam axis.

2.1 General Setup of Numerical Models

The parts of the solid domain like dam and foundation blocks share the same nodes in the dam-foundation interface, hence no differential displacements nor separation of these parts is possible. Hydrodynamic effects of the reservoir in case of dynamic analyses are accounted by acoustic elements that are coupled with the structural domain.

Table 1 Properties of considered materials

Material property	Unit	Concrete	Rock	Water
Modulus of elasticity	[MPa]	22410	22410	–
Density	[kg/m^3]	2483	2483	1000
Poisson's ratio	[-]	0.20	0.20	–
Tensile strength	[MPa]	2.0	–	–
Fracture energy	[N/m]	250	–	–
Compressive strain at peak load	[-]	0.0025	–	–
Tensile strain at peak load	[-]	0.00012	–	–
Shear wave velocity	[m/s]	–	1939	–
Compression wave velocity	[m/s]	–	3167	1439

2.2 *Material Properties*

Table 1 shows the complete set of material properties that are considered in the case studies of Theme A, provided by the formulators [1].

2.3 *Constitutive Material Model for Dam Concrete*

In the analyses for Case E "Non-Linear Dynamic Analysis" the Concrete Damaged Plasticity (CDP) model is used to account for the non-linear behavior of the dam concrete.

This material model can be used for reinforced concrete structures, but it is also suitable for plain concrete as generally used in concrete dams. Its mechanical behavior assumes to major failure mechanisms like compressive crushing and tensile cracking. The degradation (damage) of the material is defined by two variables, one represents tensile damage (d_t) and the other one compressive damage (d_c). Values for both variables are in the range from 0 (no damage) to 1 (complete loss of strength). Finally, the initial (undamaged) elastic stiffness (E_0) is reduced based on the grade of damage based on plastic strains, as shown below in the stress-strain relations (see Fig. 2a and b):

$$\sigma_t = (1 - d_t)E_0\left(\varepsilon_t - \tilde{\varepsilon}_t^{pl}\right) \quad (1)$$

$$\sigma_c = (1 - d_c)E_0\left(\varepsilon_c - \tilde{\varepsilon}_c^{pl}\right) \quad (2)$$

where σ are the stresses, ε represent the strains, $\tilde{\varepsilon}^{pl}$ the equivalent plastic strains and the indices t and c relate to tensile and compressive parts of the stress-strain curves, respectively. Further details can be found in [2].

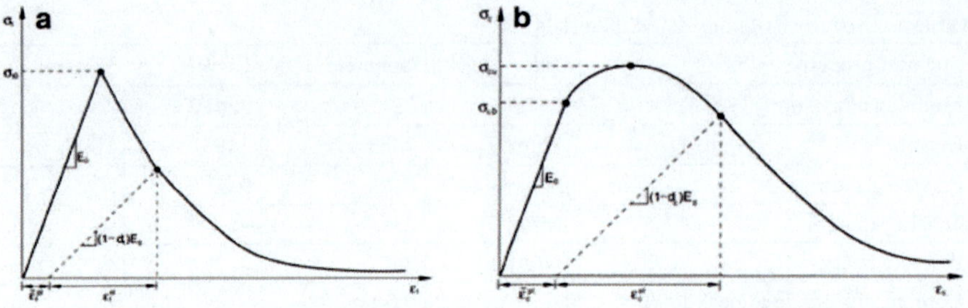

Fig. 2 **a** Uniaxial material behavior in tension. **b** Uniaxial material behavior in compression [2]

Uniaxial compression stress behavior. The compressive part of the CDP model is defined based on the specific values that are listed in Table 1, in combination with the definition of the stress–strain curve of Eurocode 2 (EC 2) [3]. Figure 3a shows the stress–strain relationship. It is assumed that compressive failure mechanism (crushing) is not decisive in the conducted studies.

Uniaxial tensile stress behavior. Generally, there are three different shapes of the tension softening curve like linear, bi-linear and exponential that may be used in analyses like these. In the presented studies, a bi-linear curve is used, as shown in Fig. 3b, which is defined by the tensile yield stress of 2.0 MPa and a value of 250 N/m for the fracture energy, as provided by the formulators (see Table 1).

Up to the tensile yield stress a linear elastic material behavior is considered, an exceedance of this strength represents the failure (or loss of strength) of the concrete due to a initiation of notional micro cracks in the concrete. At the point where the tensile strength is close to zero, the maximum crack displacement occurs. The area under this curve represents the fracture energy, the vertices that define the curve are calculated based on the definition of Hillerborg [4].

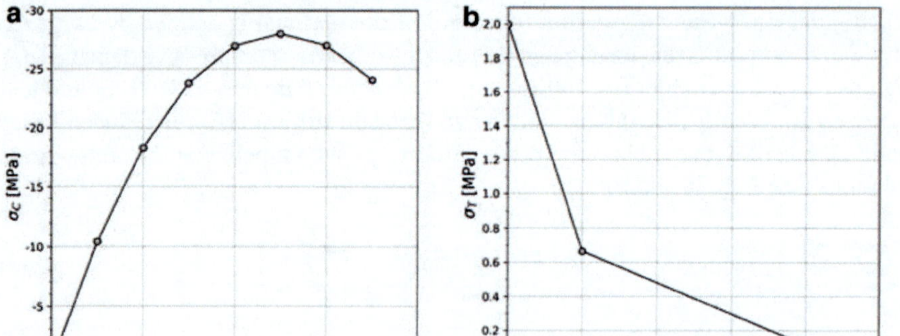

Fig. 3 **a** Non-linear compressive stress–strain curve. **b** Bi-linear tension softening curve

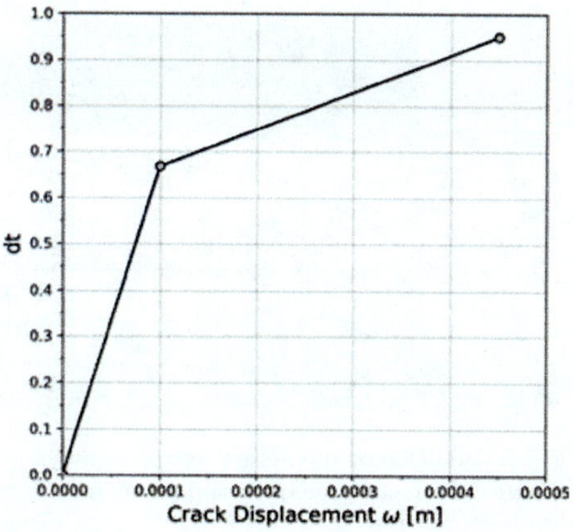

Fig. 4 Relation of tensile damage variable (d_t) to notional crack displacement

Like the tension softening curve (Fig. 3b), the tensile damage variable is defined based on crack displacement as a bi-linear curve, where at the point of maximum crack displacement relates to a value of $d_t = 0.95$, as depicted in Fig. 4.

2.4 *Discretization*

The solid parts of the provided geometry are meshed with quadrilateral, higher order elements (CPE8R) of plane strain formulation and unit thickness. For the reservoir part higher order quadrilateral acoustic elements (AC2D8) are used, with same thickness as the solid parts. Abaqus/CAE requires that the z-coordinate for plane strain elements is zero. Based on this requirement, the models are setup in the x-y plane but the results are submitted in the y-z plane, as asked by the formulators [1].

Due to the specification by the formulators that the mass of the rock foundation has to be accounted in the numerical analyses. Hence, infinite elements are used as absorbing boundaries to reduce the scattering of seismic wave in the foundation block together with finite elements. The infinite elements are applied at the vertical and bottom faces of the foundation block. In the Abaqus element library [2] the 5-noded plane strain infinite elements are named as CINPE5R and are defined based on [5] and [6] for static and dynamic loading, respectively.

Three nodes on face 1 of the infinite element share the connection to the finite element, the last 2 nodes extend away from the finite element domain. A combination of finite and infinite elements leads to a far-field solution of the problem. In dynamic analyses these elements represent a viscous damped boundary that absorbs impinging waves. Waves that impact perpendicular to the boundary are fully absorbed, the more the impact angle differs from this ideal state, the more the degree of absorption decreases [2].

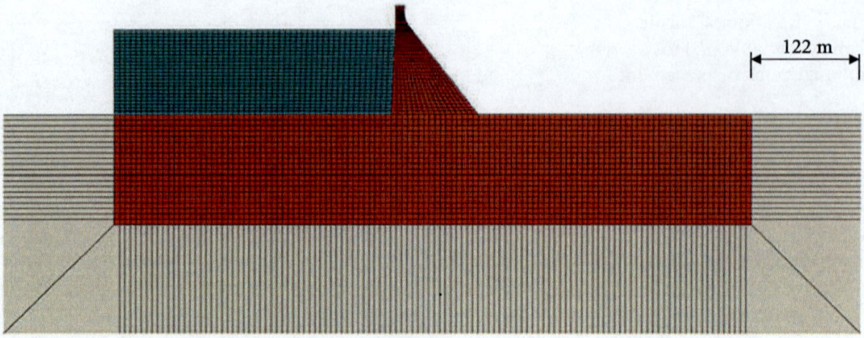

Fig. 5 FE discretization of the system, showing different element types like solid elements (red), acoustic elements (green) and infinite elements (cream)

According to Abaqus Documentation [2] the in plane thickness of the infinite layer should be at least equal to the height of the connected solid domain, in this case 122 m, which is the vertical extend of the foundation block (see Fig. 5).

2.5 *Boundary Conditions*

In the case studies A1 and A2 (modal analyses) the layer of infinite elements is supressed, since infinite elements are not contributing to natural frequencies of structural systems [2]. In modal analyses displacements degrees of freedom (horizontal and vertical) are constrained at the vertical and bottom surfaces of the model.

Acceleration time histories in the related numerical stus are applied in up and downstream direction at crest level of the dam or at the bottom nodes of the FE part of the foundation block, as shown in Fig. 6.

2.6 *Loading*

Based on the formulation document [1] two static loads are applied in the studies, which are:

- Dead load of the dam structure
- Hydrostatic water pressure at upstream surface of the dam and the reservoir-rock interface

Dynamic loading is applied accordingly for each case study, as specified in the formulation document.

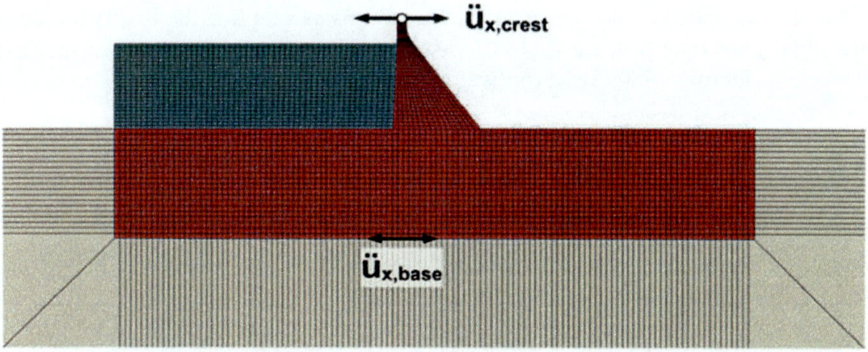

Fig. 6 Definition of structural BCs for dynamic analyses

2.7 Modelling of Dam-Foundation-Reservoir System

To account for the hydrodynamic effects during the dynamic analyses the fluid-structure-interaction (FSI) is enabled by coupling the acoustic elements to the solid domain. In the studies this connection is applied along the reservoir bottom as well as on the upstream face of Pine Flat dam. Besides the coupling of the displacement degrees of freedom of fluid and structural domain, further BCs are defined at the reservoir like:

- zero pressure condition at the free surface of the fluid and
- non-reflecting surface at the vertical back end of the reservoir.

The fluid's properties are defined by the density ρ_w and the speed of sound c_w These parameters are used to calculate the bulk modulus K_w, which is needed as input parameter in Abaqus, with this relation: $c_w = \sqrt{K_w/\rho_w}$.

Generally, the governing equation of wave propagation in the acoustic domain can be written as:

$$\frac{\partial^2 \boldsymbol{p}}{\partial t^2} - \frac{K_w}{\rho_w} \nabla^2 \cdot \boldsymbol{p} = 0 \tag{3}$$

Additionally, it is assumed that the interface of reservoir bottom and foundation is non-absorptive, which means there is a total reflection of impinging pressure waves at this interface.

2.8 Damping

Structural damping in the studies is considered by Rayleigh damping. Therefore, mass- and stiffness proportional factors of $\alpha = 0.75\ 1/sec$ and $\beta = 0.0005\ sec$ are defined as specified by the formulators. Due to this assumption the first two modes are slightly overdamped, e.g. $\xi = 3\%$ instead of the desired 2% in case of the first mode, considering winter reservoir water level.

3 Presentation of Results

Although the mandatory cases including Case E-2 as obligatory part have been studied, results are shown exemplarily in this section due to the limited number of pages. A detailed summary can be found in the synthesis report of the formulator.

3.1 *Case A*

Natural frequencies and mode shapes. Modal analyses have been conducted to identify the natural frequencies and mode shapes (normalized along the U/S face of the dam)—see Figs. 7, 8a and b. Infinite elements are excluded in these studies as described in Sect. 2.5. In addition, structural damping and the non-reflecting boundary at the vertical face at the end of the reservoir have been neglected since there is no significant influence of damping on the natural frequencies.

The displacement degrees of freedom (DOF) need to be constrained in vertical and horizontal direction at the side and bottom faces of the foundation block. If the displacements are constrained only normal to the mentioned faces, the resulting

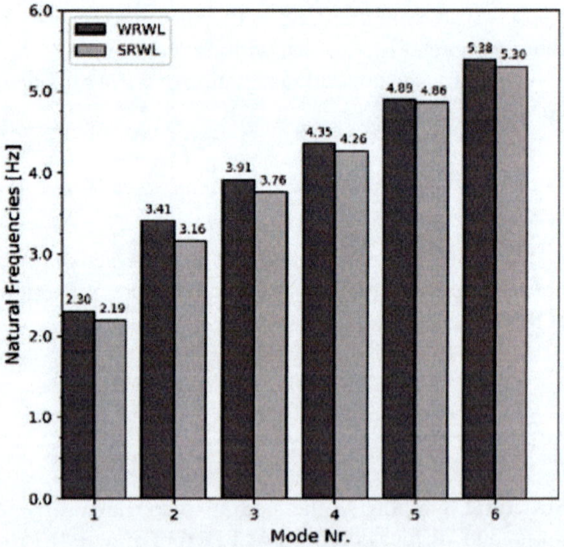

Fig. 7 Results of Case A natural frequencies for summer (SRWL) and winter reservoir level (WRWL)

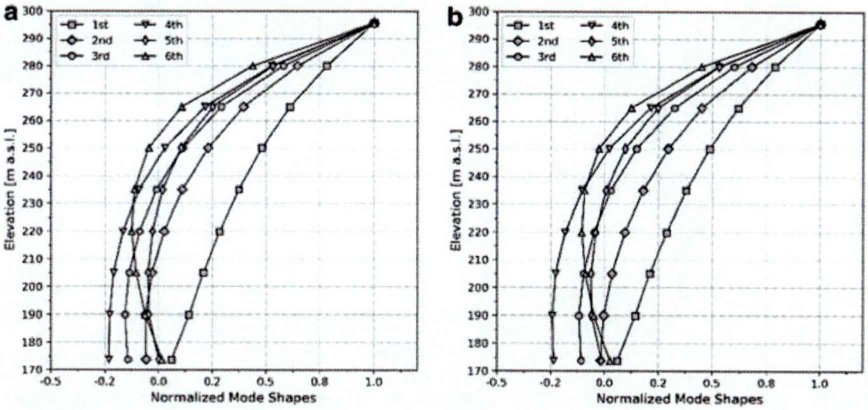

Fig. 8 **a** Normalized mode shapes along U/S face for WRWL. **b** Normalized mode shapes along U/S face for SRWL

natural frequencies are significantly lower than in the case of the fully constrained model. This behavior may be explained that the outer faces of the foundation block can deform in parallel direction, which is not possible at the prototype because of the extension of the rock of the earth's crust.

Additionally, a modal analysis with massless foundation and winter reservoir water level (WRWL) has been carried out. The resulting natural frequencies, especially the first three, are close to the ones where the mass is considered, e.g. 2.30 Hz (with mass) and 2.34 Hz (massless) in case of the first natural frequency.

3.2 *Case D*

The formulators left the choice to apply either the deconvolved TAFT acceleration time history or the TAFT stress time history at the bottom of the foundation block. In this studies option 1 (acceleration time history) is chosen. The acceleration is applied at the bottom of the FE part of the foundation block at the nodes of the interface to the infinite elements as depicted in Fig. 6. Some selected results of this studies are shown in Figs. 9, 10, 11 for the two different water levels WRWL and SRWL.

3.3 *Case E*

In previous case studies a linear elastic material behavior is assumed, both for dam and rock foundation. Case E is devoted to non-linear material behavior of the concrete dam, therefore, the so-called concrete damaged plasticity material model is used in these studies. Section 2.3 gives a brief overview about the model and its chosen

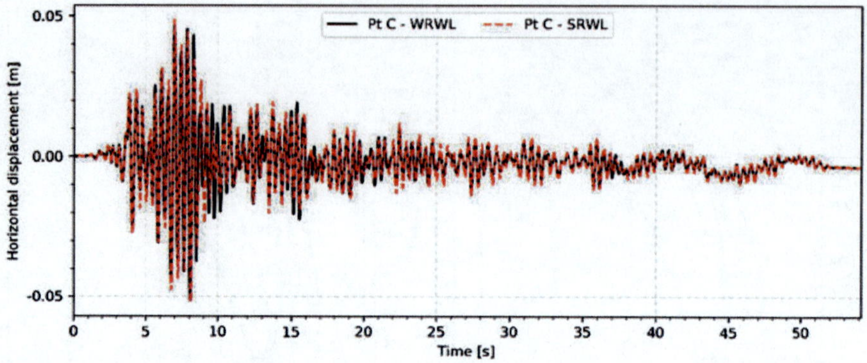

Fig. 9 Case D—relative displacements (without static deformation) at crest level (point C) for WRWL and SRWL

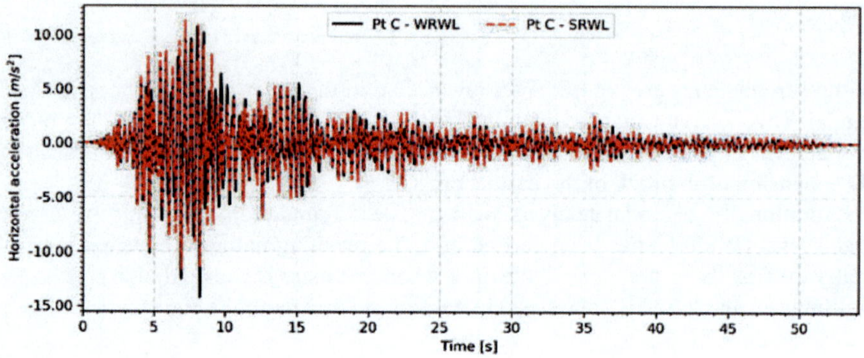

Fig. 10 Case D—relative accelerations at crest level (point C) for WRWL and SRWL

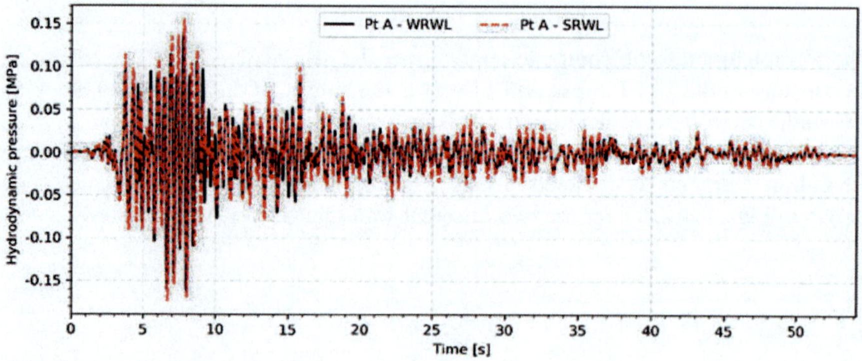

Fig. 11 Case D—hydrodynamic pressure at U/S heel (point A) for WRWL and SRWL

parameters. The acceleration time histories, TAFT earthquake and Endurance Time Acceleration Function (ETAF) are applied at the bottom nodes of the foundation block (see Fig. 6).

The FE model in Case E-1 is basically the same as Case D-1 except the mesh size at the dam body, to reduce the characteristic length (l_{ch}), to reduce convergence issues. The characteristic length depends on the concrete material parameters and represents the distance between integration points which subsequently influences the element (mesh) size [2]. According to Hillerborg [7] the characteristic length can be computed based on Young's modulus (E), tensile strength (f_t) and fracture energy (G_f) using the relationship as shown in Eq. (4).

$$l_{ch} = \frac{E \cdot G_f}{f_t^2} \tag{4}$$

Therefore, the characteristic length in these studies calculates to 1.40 m. The refined mesh comply to this criterion in the upper half of the dam body. This is a compromise that has been taken to reduce the number of elements and therefore computation time.

Case E-1—TAFT record. Figure 12 shows the areas in which the tensile damage occurred at the dam body after the application of the TAFT earthquake record. Damaged (cracked) areas are represented by the darker parts of the figure, based on the value of the tensile damage variable d_t which is plotted in the legend of Fig. 13. The crack length at the base joint is computed of about 18 m, which is about 19% of the base length of 95.80 m at the end (t=54.15 s) of TAFT ground motion.

Fig. 12 Case E-1 crack pattern, snapshot at end of TAFT record (t = 54.15 s)

Case E-2—ETAF record. In addition to the TAFT record the artificially designed ETAF acceleration time history is applied to the model. After a duration of 6.49 s the solution does not converge any more based on force equilibrium. Hence, this is can be interpreted as point or time of failure. In Fig. 13 the damaged (cracked) areas are illustrated, the evolution of the Damage Index is shown in Fig. 14a and Fig. 14b

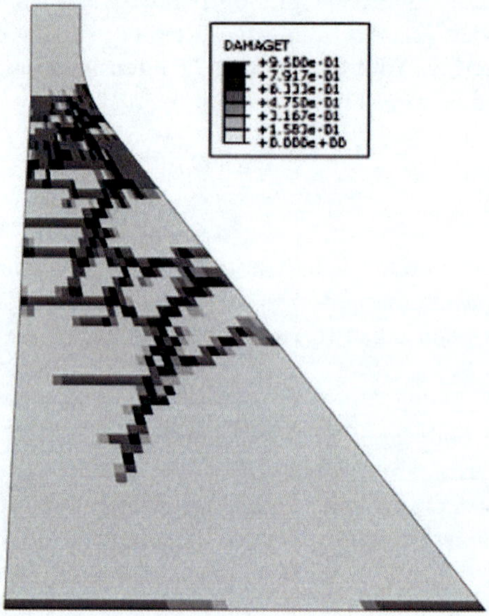

Fig. 13 Case E-2 crack pattern, snapshot at time of failure (t = 6.49 s)

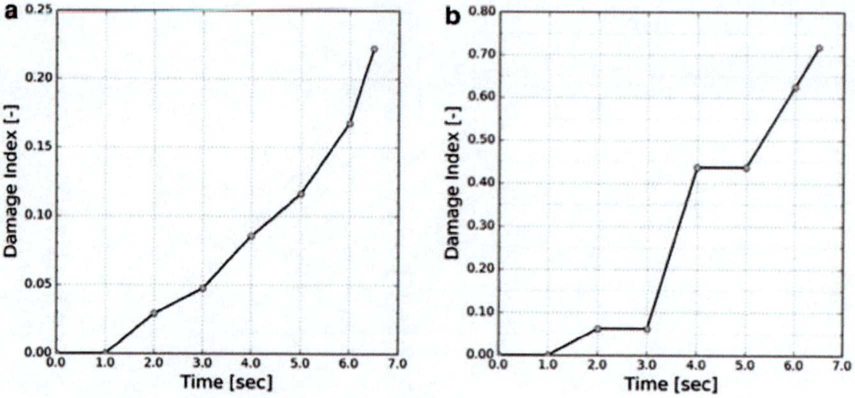

Fig. 14 **a** Evolution of cross-sectional area-based damage index. **b** Evolution of base length-based damage index

based on the cross-sectional area and base joint, respectively. The Damage Index is the ratio of damaged areas or damaged lengths to the total, undamaged quantities.

It can be seen that the majority of the dam-rock interface is cracked (~71%) and around 22% of the cross section is damaged, based on the tensile softening curve and (Fig. 3a) as well as the damage criterion (Fig. 4) that is applied to the concrete of the dam body.

4 Lessons Learned

Based on the conducted numerical studies using the FE software Abaqus knowledge has been gained as listed below.

1. In which way the boundary conditions have to be applied in modal analyses when the mass of the foundation material is considered. In the case of a 2D model, the vertical and horizontal degrees of freedom need to be constrained at the side and bottom faces of the foundation block, otherwise the natural frequencies are significantly lower because of an underconstrained model. Furthermore, infinite elements do not contribute to the modal solution.
2. The first three natural frequencies of the structure in which the mass of foundation is neglected show a good agreement with first three natural frequencies of the model including the mass of rock.
3. The application of infinite elements that are connected to the foundation block to simulate a far field solution. To ensure that the desired face (face 1) of these elements meet the faces of the finite elements in the interface a swept mesh with stacking direction has to be created.
4. The application of the concrete damaged plasticity material model, which is a non-linear material model for concrete (plain and reinforced) and calculation of the parameters and data needed as input. Regarding model assumptions it is observed that if the dam is tied to the foundation block stress concentrations (tensile stresses) appear at the U/S heel of the dam and therefore the yield stress in the elements at to the dam-rock interface is exceeded and cracking occurs. The study in which the ETAF ground motion is applied the solution do not converge at time step $t = 6.49$ s. A improvement of the mesh quality and size might improve this issue.

5 Suggested Future Work

There is a good ratio of creating the FE models to knowledge output in the studies of this workshop. In other words, it is a fairly simple FE model where basic and more sophisticated studies can be carried out. Therefore, future studies could be conducted on this dam where following points are considered:

1. Structural non-linearity (contact modeling), e.g. at the base joint or lift joint with linear and non-linear material models
2. Influence of absorbing effects at reservoir bottom on the response of the structure to account for different levels of sedimentation.
3. 3D model of the dam including block joints at boundaries of the blocks.

References

1. Salamon J, Hariri-Ardebili MA, Malm R, Wood C, Faggiani G (2019) Theme A formulation. Seismic analysis of Pine Flat concrete dam. In: 15th ICOLD international benchmark workshop on the numerical analysis of dams. Milan, Italy
2. Dassault Systèmes. Abaqus 6.14 Documentation (2014)
3. EN 1992-1-1:2004: Eurocode 2: Design of concrete structures-Part 1-1: General rules and rules for buildings (2004)
4. Hillerborg A (1985) The theoretical basis of a method to determine the fracture energy GF of concrete. Mater Struct 18(4):291–296
5. Zienkiewicz OC Emson C, Bettess P (1983) A novel boundary infinite element. International. J Numer Methods Eng 19(3), 393–404
6. Lysmer J, Kuhlemeyer RL (1969) Finite dynamic model for infinite media. ASCE J Eng Mech Div 95(4):859–878
7. Hillerborg A (1988) Application of fracture mechanics to concrete: summary of a series of lectures 1988 (Report TVBM; vol 3030). Division of Building Materials, LTH, Lund University, Sweden

Seismic Analysis of Pine Flat Concrete Dam

W. Su, J.-T. Wang, and Y.-J. Xu

Abstract This paper focuses on the Theme A of the 15th International Benchmark Workshop on Numerical Analysis of Dams. The seismic analysis of Pine Flat Concrete Dam proposed by formulators is performed by the Finite Element Method. And several two-dimensional FE models are established for the case studies. All the obligatory and optional cases have been completed, including Case A for EMVG test simulation, Case B for foundation analysis using impulsive loads, Case C for dynamic analysis using impulsive loads, Case D for dynamic analysis using Taft records, Case E for non-linear dynamic analysis, and Case F for dynamic analysis using massless foundation. Results show that the effect of reservoir level is notable in EMVG Test Simulation, and the exist of dam can influence the dynamic response on the top of foundation. Besides, the weak parts of the gravity dam body against seismic are dam heel, dam toe, dam neck, and the upper part of downstream dam body, and dynamic responses computed in model with massless foundation are overestimated.

Keywords Pine flat dam · Two-dimensional FM model · EMVG test simulation · Seismic analysis · Non-linear dynamic analysis · Massless foundation

1 Introduction

Theme A of the workshop is related to a seismic analysis of Pine Flat dam. The goals of this study are to identify key uncertainties that may significantly affect numerical modeling results, determine the need for future workshop investigations, determine research needs, and develop best practices in the advanced analysis of concrete dams. The case studies are intentionally narrow in focus so that the assumptions, factors and methods having the greatest effect can be identified. This paper will report the main analysis results using ABAQUS based on the model information, the parameters of property and the load data provided by Formulation.

W. Su · J.-T. Wang (✉) · Y.-J. Xu
Department of Hydraulic Engineering, Tsinghua University, Beijing 100084, China
e-mail: wangjt@tsinghua.edu.cn

G. Bolzon et al. (eds.), *Numerical Analysis of Dams*, Lecture Notes in Civil Engineering 91, https://doi.org/10.1007/978-3-030-51085-5_17

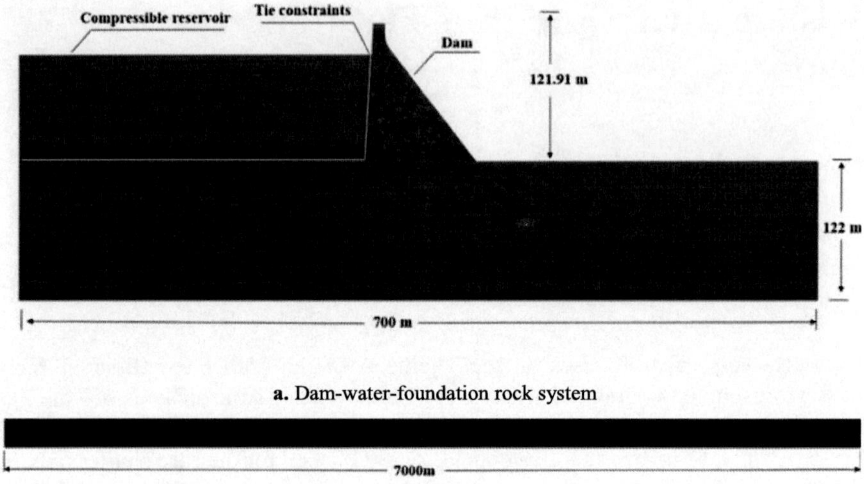

a. Dam-water-foundation rock system

b. Foundation rock with a length of 3700 m

Fig. 1 Two-dimensional FE model for case studies

2 Analysis Model

2.1 FE Mesh

The tallest non-overflow dam monolith, No.16, is selected as the analysis object for the case studies. All the cases are simulated by a two-dimensional finite element model shown in Fig. 1a. The finite element mesh consists of 3074 quadrilateral plane stress elements for the dam, 37746 quadrilateral elements for the foundation rock. The impounded water is simulated by the quadrilateral compressible acoustic fluid elements. The number of the fluid elements are 12915 for the 268.21 m water level, 14145 for the 278.57 m water levels, and 15658 for the 290.00 m water level, respectively. A tie constraint is applied for acoustic-structure coupling at the dam-water interface and foundation-water interface.

It should be noted that additional configuration is considered in some analysis cases for the foundation rock, as shown in Fig. 1b. The block length is 3700 m. The finite element mesh consists of 199827 quadrilateral plane strain elements.

2.2 Boundary Conditions

For the natural frequency and mode shape analysis, displacements in the normal and tangential directions at the base and sides of foundation are all fixed.

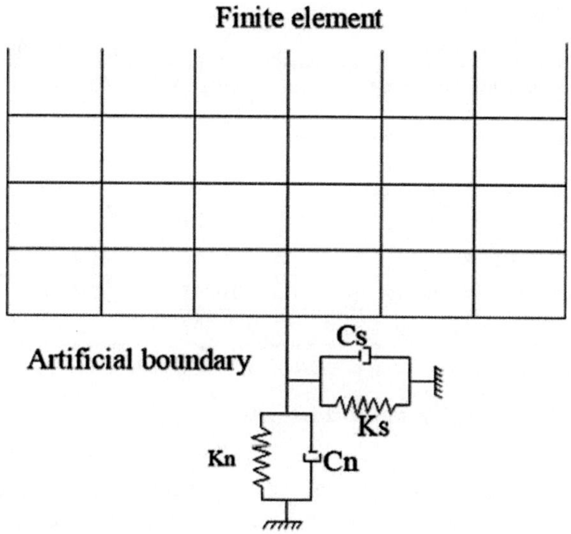

Fig. 2 Schematic of the viscous-spring artificial boundary

For static analysis, displacements are restrained in the normal direction at the base and sides of foundation.

During the earthquake response of structure, vibration energy will be dissipated by radiating outward through an infinite foundation, which is named as radiation damping effect. The case analyses simulate the radiation damping effect of infinite foundation by the two-dimensional viscous-spring artificial boundaries, as shown in Fig. 2. The essence of the artificial boundary is that the boundary can efficiently absorb the energy of the scattering waves, from the generalized structure to the infinite media [1].

The parameters of the spring-damping elements are determined by Eqs. (1) and (2) based on the theory of the viscous-spring artificial boundary [2]:

$$K_n = \alpha \frac{\lambda + 2G}{r_b}; \; C_n = \beta \rho c_p \tag{1}$$

$$K_s = \alpha \frac{G}{r_b}; \; C_s = \beta \rho c_s \tag{2}$$

where the subscripts n and s refer to the normal and tangential directions of the artificial boundary surfaces; K is the elastic stiffness of the spring; C is the viscous damping; r_b is the coordinate of artificial in the polar coordinate system; c_p and c_s are the compressional and shear wave velocities, respectively; λ and G are the Lame's constants; ρ is the mass density of foundation rock; and of α and β are modification coefficients, which may be determined from parameter analysis.

Table 1 Material properties

Concrete Properties		Foundation Rock Properties	
Modulus of elasticity	22410 MPa	Modulus of elasticity	22410 MPa
Density	2486 kg/m^3	Density	2486 kg/m^3
Poisson's ratio	0.20	Poisson's ratio	0.20
Compressive strength	28.0 MPa	Shear velocity	1939 m/s
Tensile strength	2.0 MPa	Compressional wave velocity	3167 m/s
Fracture energy	250 N/m	***Water***	
Compressive strain at peak load	0.0025	Density	1000 kg/m^3
Tensile strain at peak load	0.00012	Compressional wave velocity	1439 m/s

2.3 Analysis Conditions

Three reservoir water levels are considered in the case studies:

Winter reservoir water level (WRWL) at EI.268.21 m.

Summer reservoir water level (SRWL) at EI 278.57 m.

Normal reservoir water level (NRWL) at EI 290.00 m.

The dam and foundation are both assumed to be homogeneous, isotropic and elastic, and the reservoir water is assumed to be compressible. Their material properties are listed in Table 1.

According to the specifications for seismic design of hydraulic structure in China [3], the dynamic modulus of elasticity of concrete is set to be 33.62 GPa which is 1.5 times of static modulus, while the dynamic tensile strength is 2.4 MPa that is 120% of static tensile strength.

3 Analysis of Results

This section will report the main results of the analyzed cases.

3.1 Case A: EMVG Test Simulation

The natural frequency and mode shape analyses for winter reservoir water level (WRWL) and summer reservoir water level (SRWL) have been performed in Case A1 and Case A2. Meanwhile, the dynamic linear analyses of the dam-foundation-reservoir with two different reservoir levels have been implemented in Case A3 and Case A4, where the harmonic force record exerted by an eccentric-mass vibration generator positioned at the dam crest is considered.

Table 2 First 6 natural frequencies of the model (Hz)

Water level	1st	2nd	3rd	4th	5th	6th
EI.268.21 m	2.53	3.50	4.14	5.34	6.12	6.23
EI 278.57 m	2.40	3.22	3.93	5.23	6.09	6.14

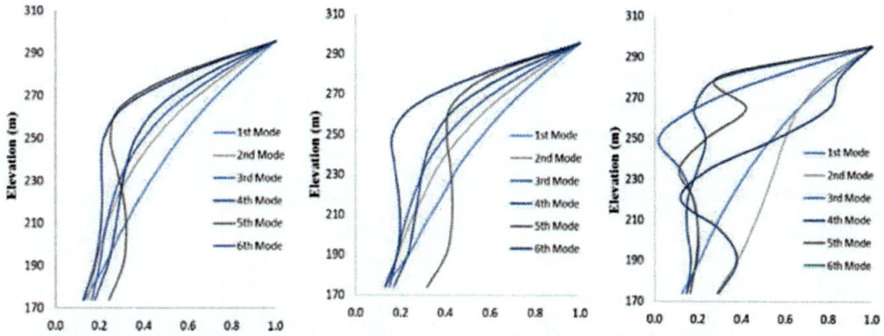

a. Winter reservoir water level **b.** Summer reservoir water level **c.** No reservoir water

Fig. 3 Normalized mode shapes of model

With the displacements in the normal and tangential directions fixed at the base and sides of foundation, the first 6 natural frequencies of the model with massless foundation are shown in Table 2, and their normalized mode shapes are plotted in Fig. 3, respectively.

It could be learned that the fundamental frequency of the system with WRWL is 2.53 Hz, while the fundamental frequency of the system with SRWL is 2.40 Hz. In general, the first 6 natural frequencies of the dam with WRWL are obviously higher than those of the system with SRWL. This basically conforms to the natural vibration characters of gravity dams.

In addition, the results show that the normalized mode shapes of two models with different water levels are similar. But the mode shapes of the model without the reservoir are quite different from those of the model with reservoir except for the 1st mode.

As for Case A3 and Case A4, the typical acceleration and displacement time histories at dam crest (Point C) and dam heel (Point A) are shown in Figs. 4, 5 and 6. It can be seen that the computed accelerations at Point A and Point C and relative displacements (Point A-Point C) have similar shapes with the scaled input force.

In Case A3, the magnitude of peak acceleration at Point C exceeds 0.3 cm/s^2, while the peak acceleration at Point A only reaches 0.028 cm/s^2. It could be concluded that the acceleration amplitude of Point C is about 10 times that of Point A. This phenomenon also could be observed in Case A4, but its acceleration amplitudes at Point A and Point C are smaller than those of Case A3.

By comparing the accelerations in Case A3 and Case A4, it can be found that the acceleration in Case A3 (reservoir water level at 268.21 m) is larger than that in Case

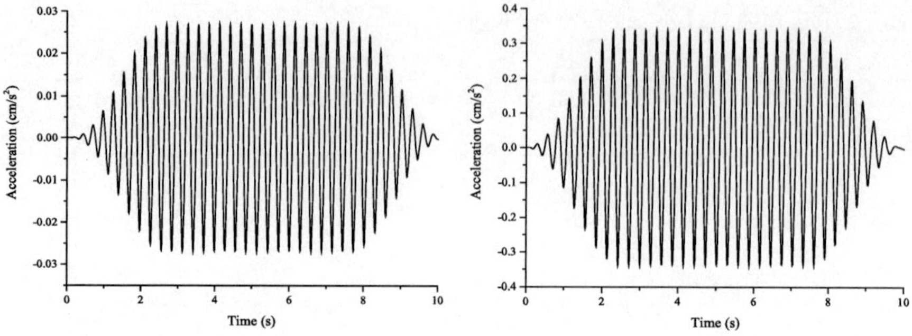

a. Point A **b.** Point C

Fig. 4 Acceleration time histories at point A and point C in case A3

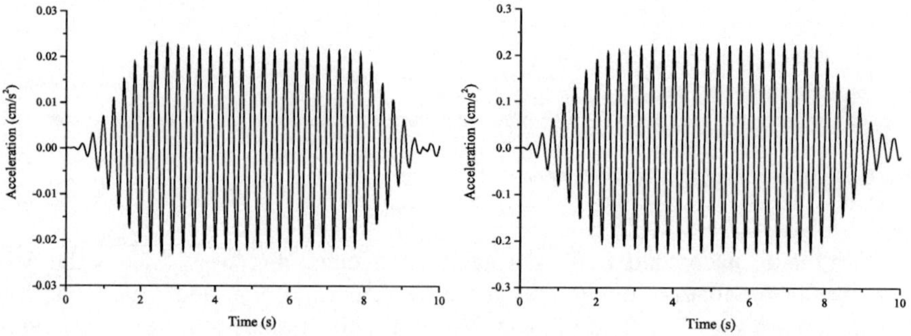

a. Point A **b**.Point C

Fig. 5 Acceleration time histories at point A and point C in case A4

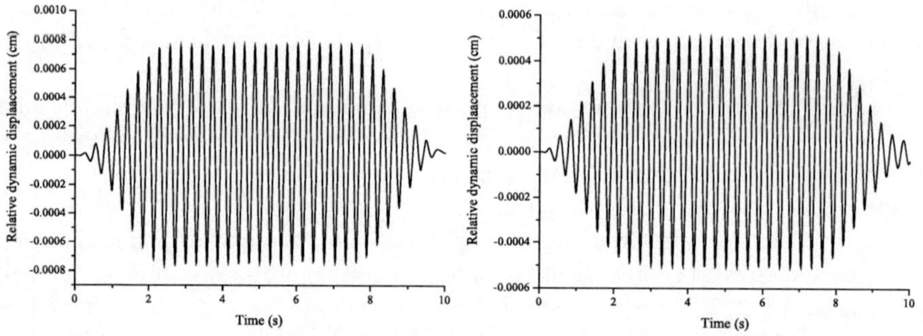

a. Case A3 **b**.Case A4

Fig. 6 Relative displacements of point A and point C

A4 (reservoir water level at 278.57 m). This demonstrates that the effect of reservoir level is notable in EMVG test simulation.

3.2 Case B: Foundation Analysis Using Impulsive Loads

Case B (Table 3) aims to explore the response due to a vertically-propagating SH wave by using an impulsive signal to investigate the effect of foundation size and to verify the efficiency of the non-reflecting boundary conditions in the dynamic analysis of dam foundations. The results are shown in Figs. 7 and 8.

The results suggest that the velocity response histories are nearly identical everywhere on the free surface. Moreover, we can obtain that the velocity responses at the top of block are amplified two times compared with the velocity time histories at the base of foundation, and the reflected waves from the free surface have propagated back to the base of foundation. It is obvious that the reflected waves are efficiently absorbed by the viscous-spring artificial boundaries. Besides, it can be seen that the velocity time histories of the points in Case B1 and Case B3 as well as Case B2 and Case B4 are nearly consistent and independent of the foundation length. Therefore, it can be concluded that the artificial boundaries applied on the foundation boundary have good accuracy.

Table 3 Case B Information

Case	Foundation length (m)	Impulse signal
B1	700	High frequency
B2	3700	Low frequency
B3	700	High frequency
B4	3700	Low frequency

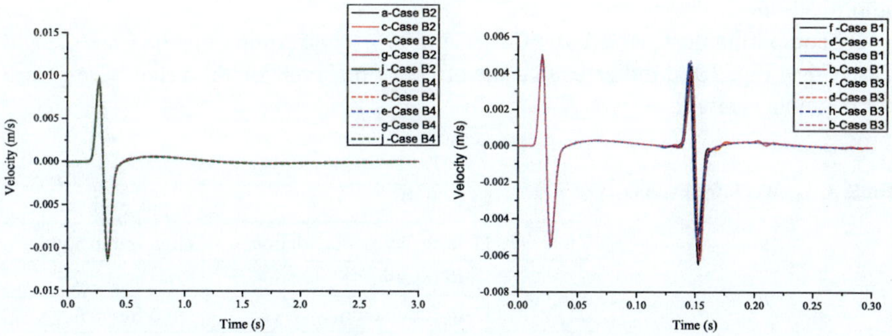

a. Points at free surface **b.** Points at foundation base

Fig. 7 Velocity time histories at the selected points in cases B1 and B3

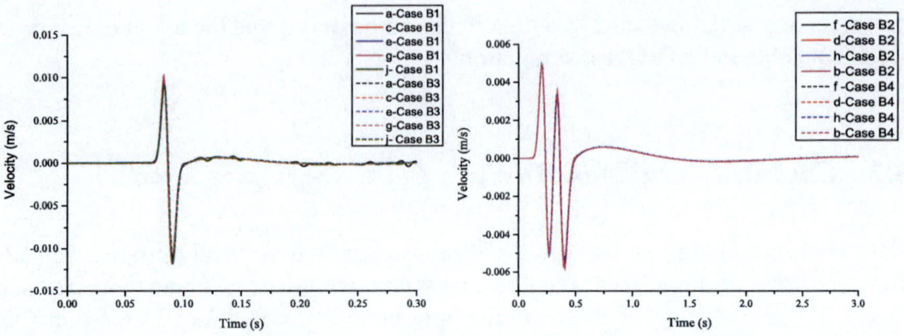

a. Points at free surface **b.** Points at foundation base

Fig. 8 Velocity time histories at the selected points in case B2 and B4

3.3 Case C: Dynamic Analysis Using Impulsive Loads

Dynamic analyses of dam-reservoir-foundation system and dam-foundation system using impulsive loads have been performed in Case C (Table 4). This section will report the velocity response histories at several specific nodes.

From Figs. 9 and 10, we can see that the velocity at the points of free surface (Point c, e, j) are identical and nearly match the impulse before the scattered waves from the dam propagate back to the points (see the velocity histories at Points c, e, j in Fig. 9b and Fig. 10b). It can be found that the amplitudes of velocity at Point a, Point c Point e and Point A are different because of the exist of dam and reservoir.

Moreover, it can be seen that an amplification of velocity occurs at the dam crest (Point C), and the amplification magnitude depends on the frequency of impulse loads. The low frequency loads (Case C2 and Case C4) have a larger amplification magnitude than the high frequency loads. This is because that the low frequency is close to the fundamental frequency of the dam. The resonance results in the larger amplification.

By comparing the Case C1 and Case C3 in Fig. 9 and comparing the Case C2 and Case C4 in Fig. 10, it indicates that the effect of reservoir on the velocity response at the points is slight.

Table 4 Case C information

Case	Model	Impulse signal
C1	Dam-reservoir-foundation	High frequency
C2	Dam-reservoir-foundation	Low frequency
C3	Dam-foundation	High frequency
C4	Dam-foundation	Low frequency

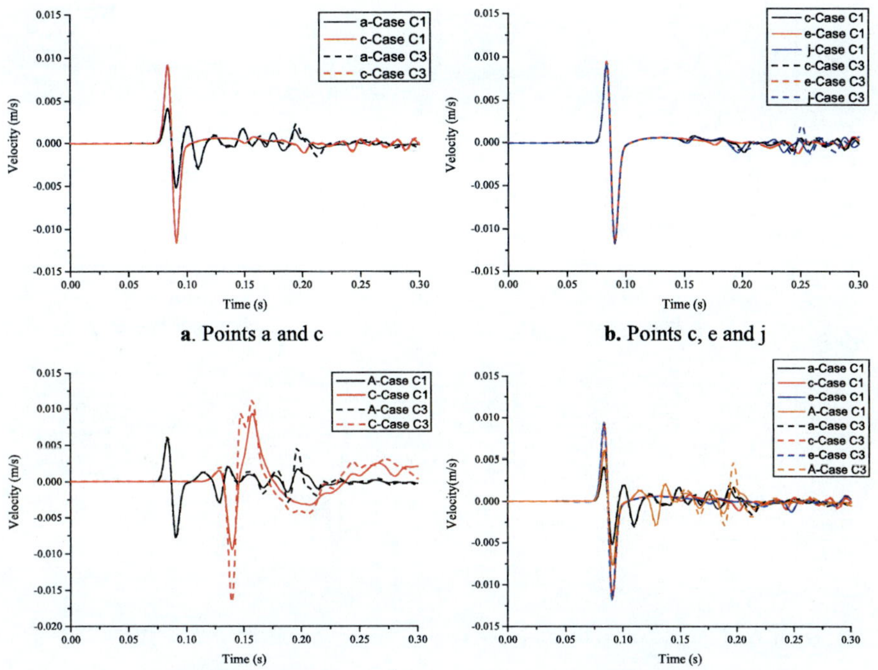

c. Points A and C **d.** Points a, c, e, and A

Fig. 9 Velocity time histories at the selected points in cases C1 and C3

3.4 Case D: Dynamic Analysis for Various Reservoir Levels

In Case D, a dynamic analysis of the dam-reservoir–foundation system is performed considering the elastic material properties, the Taft earthquake record, and reservoir water at three different elevations. The analysis results are presented in Figs. 11, 12, and 13.

It can be observed that the relative displacement time histories in three cases have similar shapes, but their magnitudes increase with the rise of reservoir water levels. The acceleration time history computed at Point A approximately matches the free-surface Taft record. Moreover, the time history computed at Point C is amplified by 2 ~ 3 times compared with that at Point A, and the peak acceleration reaches 5.36 m/s^2. In addition, the hydrodynamic pressures at the dam heel are compared in Fig. 14 for three cases. It is clear that the hydrodynamic pressure decreases with the rise of reservoir water level.

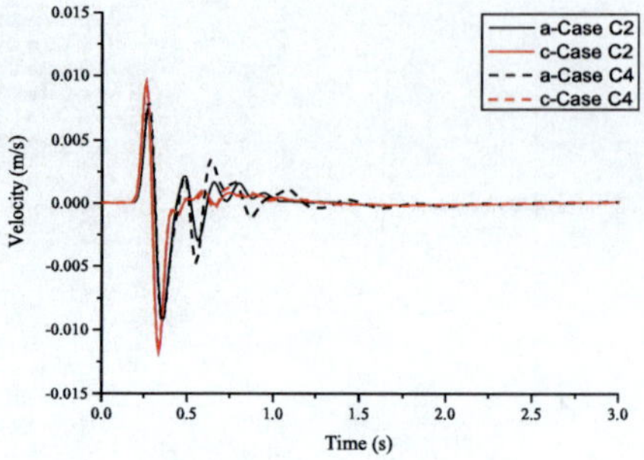

a. Points a and c

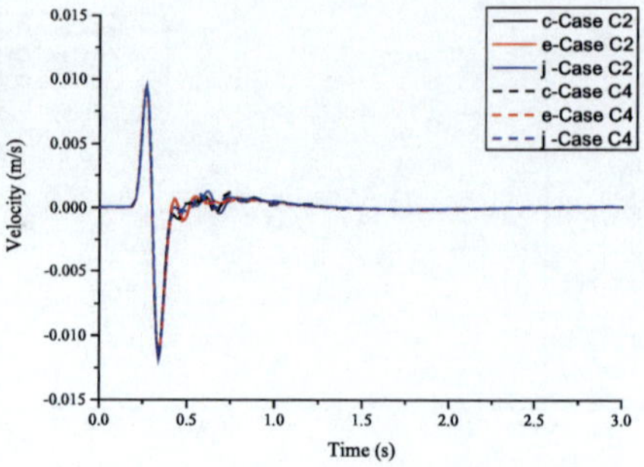

b. Points c, e and j

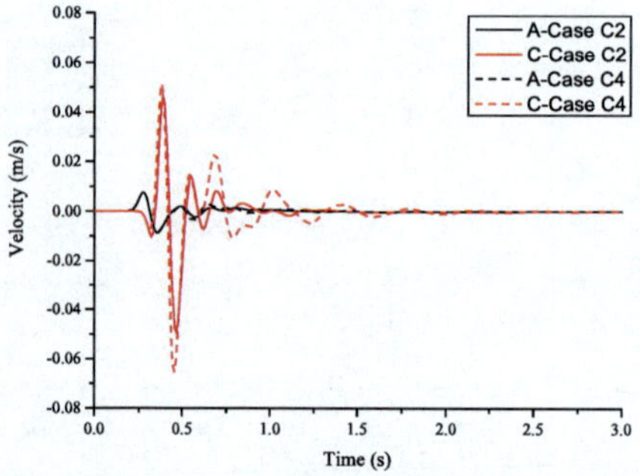

c. Points A and C

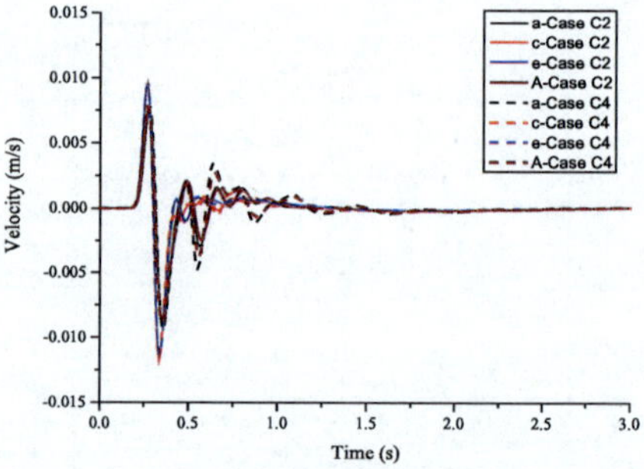

d. Points a, c, e, and A

Fig. 10 Velocity time histories at the selected points in case C2 and Case C4

3.5 *Case E: Non-linear Dynamic Analysis*

In Case E, dynamic analyses are performed considering non-linear material properties of concrete. Fracture energy and tensile strain at peak load have been provided in the formulation, and the concrete damaged plasticity model proposed by Lee and Fenves [4] is used in the analysis. When the tensile stress exceeds the tensile strength of concrete materials, the concrete stiffness will degrade, resulting in damage and cracking to a certain extent, which is regards as damage.

In Case E1 (Fig. 14), it can be seen a slight damage occurs at the dam heel under the load of Taft record, which has a peak horizontal acceleration of 0.18 g.

In Case E2, damage occurs primarily at dam heel at 7.51 s, and then damages in the middle and neck of the downstream face can be observed as the acceleration of load increases. Damages on the downstream dam surface develop diagonally downward, while damages at dam heel develop towards dam toe according to Fig. 15. It can be concluded that the weak parts of the gravity dam body against seismic are dam heel, dam toe, dam neck, and the upper part of downstream dam body. These parts should be paid more attention during seismic design.

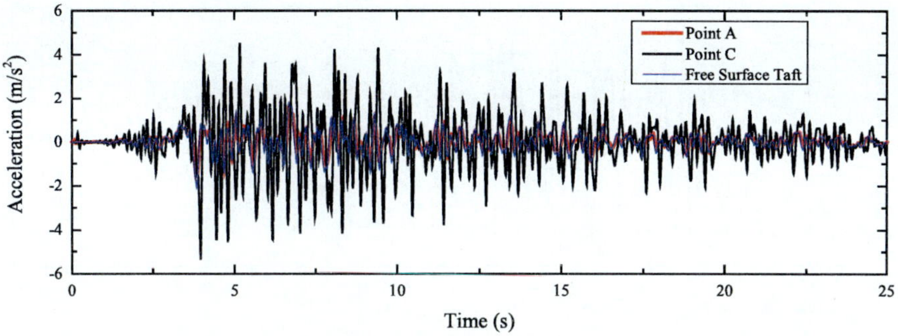

a. Case D1, water level 268.21 m

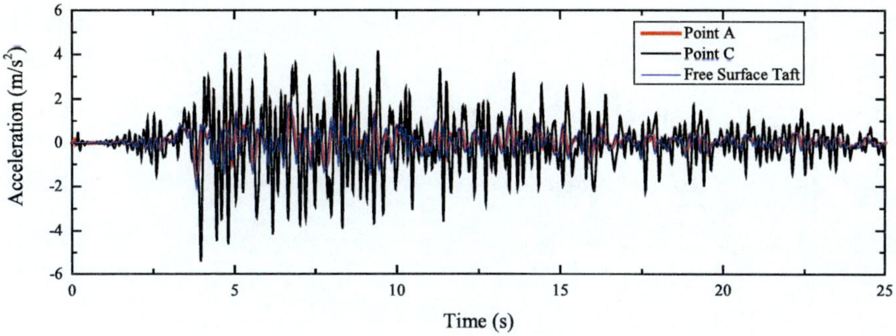

b. Case D2, water level 278.57 m

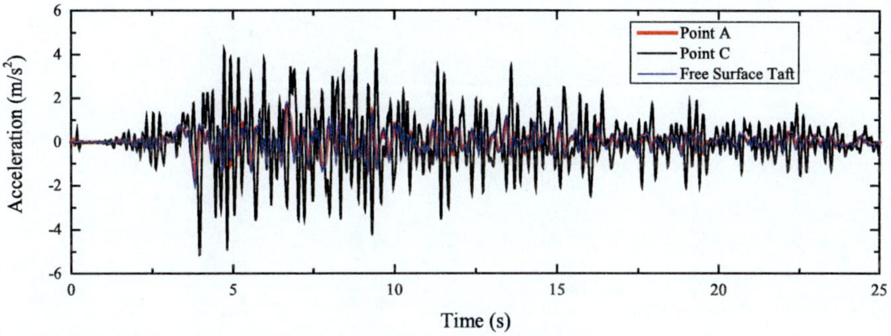

c. Case D3, water level 290.00 m

Fig. 11 Comparison of acceleration time histories at Point A and Point C

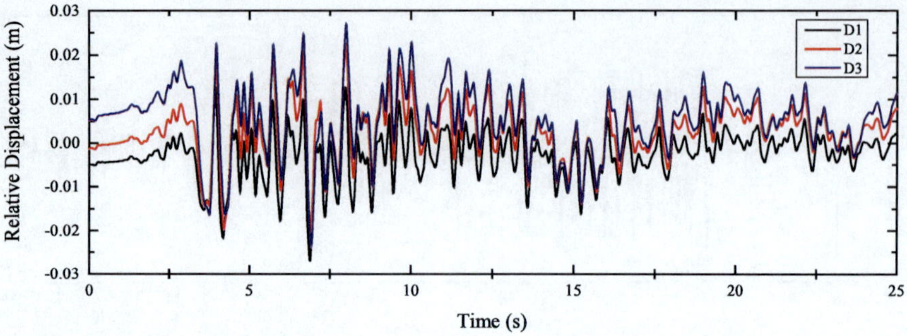

Fig. 12 Comparison of relative displacements at point A and point C in cases F1, F2, F3

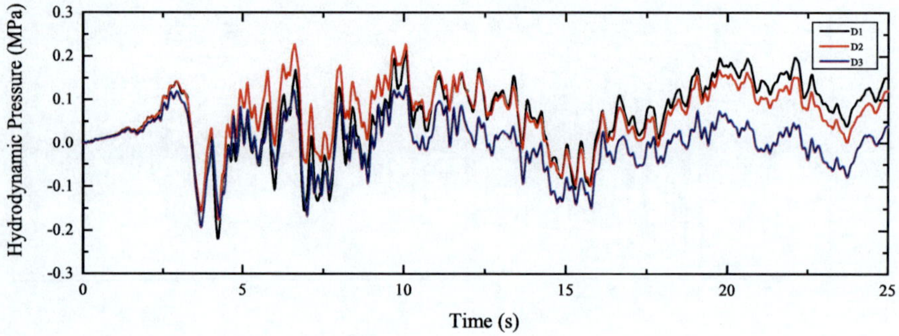

Fig. 13 Comparison of the hydrodynamic pressures at three water levels

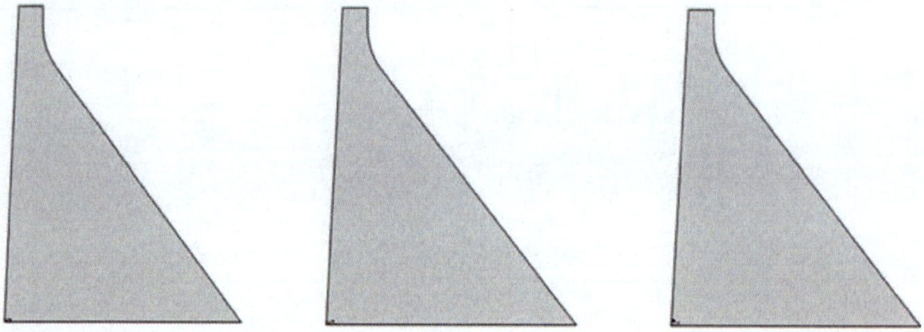

a. 10 s **b.** 30 s **c.** 54.15s

Fig. 14 Damage evolution of dam in case E1

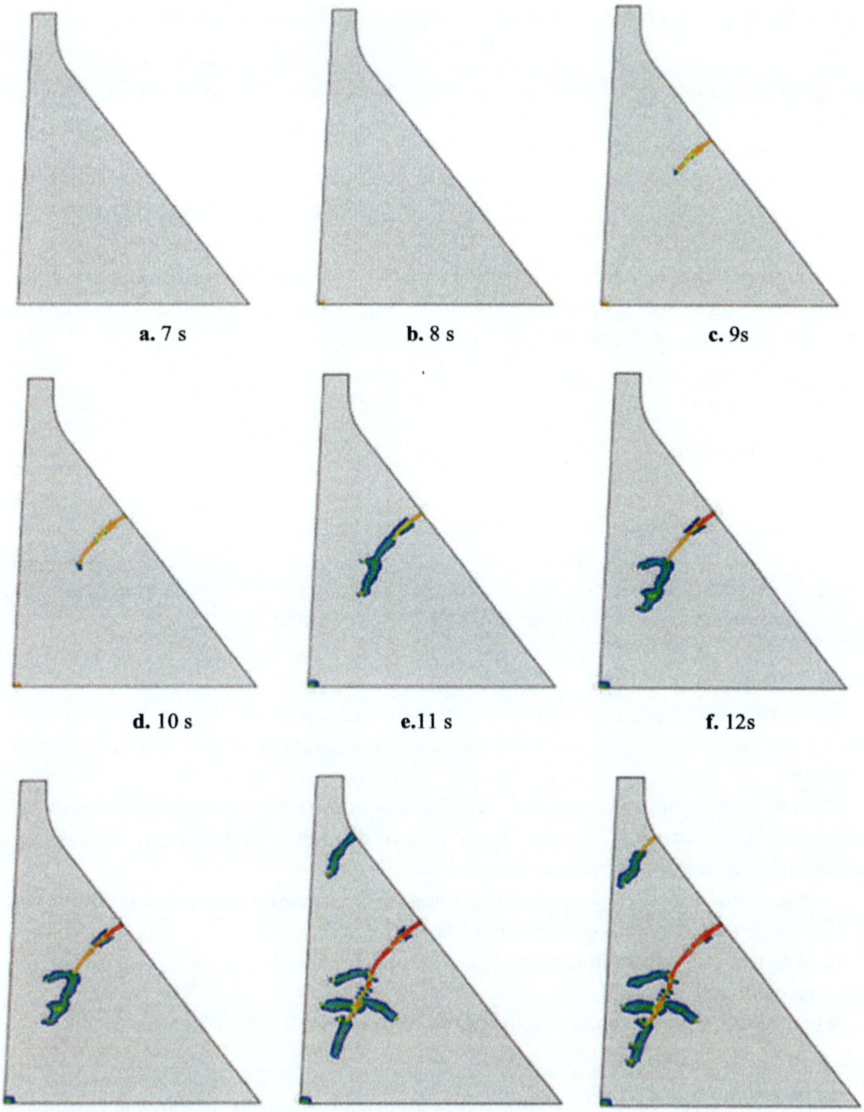

a. 7 s **b.** 8 s **c.** 9s

d. 10 s **e.**11 s **f.** 12s

g. 13 s **h.** 14 s **i.** 15s

Fig. 15 Damage evolution of dam in case E2

3.6 *Case F: Dynamic Analysis with Massless Foundation*

In Case F, the model of dam-foundation-reservoir system is similar to Case D except a massless foundation subjected to the Taft record ground motion input. This section will present the results in Case F.

The relative displacement time histories in Case F are quite different from those in Case D (Figs. 16 and 17). It can be seen that a notable amplification of acceleration occurs at the dam crest (Point C). The peak acceleration at Point C is about 7–8 times larger than the one at dam heel (Point A). The comparison of hydrodynamic pressures at three water levels presented in Fig. 18 shows that the hydrodynamic pressure increases as the water level rising.

4 Conclusions

Several conclusions are drawn from the analysis results.

In EMVG test simulation, the acceleration in Case A3 (lower reservoir water level) is larger than that in Case A4 (higher reservoir water level). Therefore, the effect of reservoir level is notable in EMVG Test Simulation.

It can be learned that the viscous-spring artificial boundary is able to absorb the reflected wave efficiently.

The exist of dam can influence the dynamic responses on the top of foundation. However, the effect of reservoir on the seismic responses of the foundation is not notable.

We can learn that the weak parts of the gravity dam body against seismic are dam heel, dam toe, dam neck, and the upper part of downstream dam body. These parts should be paid more attention during seismic design.

The amplification effect on the dam crest of the model with massless foundation is clearly larger than that of the model with the viscous-spring artificial boundary. That is to say, the dynamic responses computed in model with massless foundation are overestimated.

Meanwhile, four issues are proposed for future research.

- When compressible reservoir is considered by an acoustic-structure coupling model, we have found that the mode shapes of vibration of dam are different from those with the empty reservoir model. The further investigation is needed to explore the vibration modes with compressible reservoir and incompressible reservoir.
- In this study, the non-linear material properties of concrete are considered, while the foundation is assumed to be homogeneous, isotropic and elastic. The non-linear material properties of foundation rock should be considered for further work.

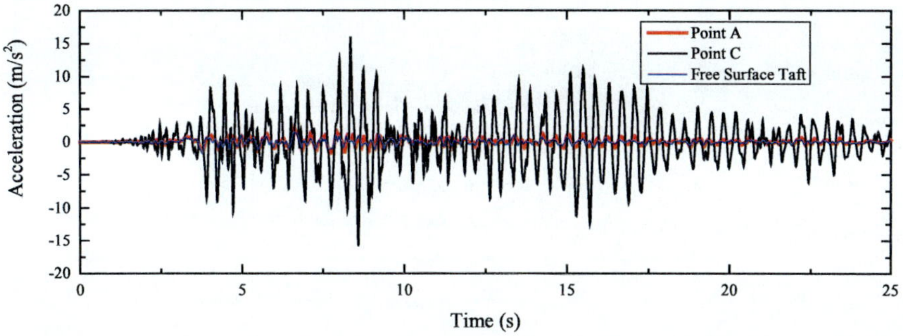

a. Case F1

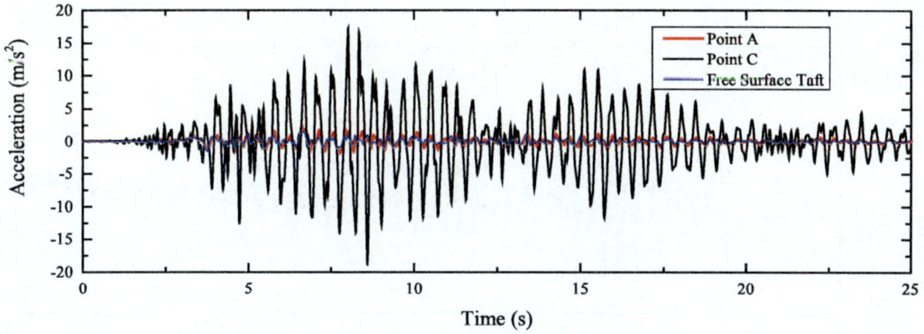

b. Case F2

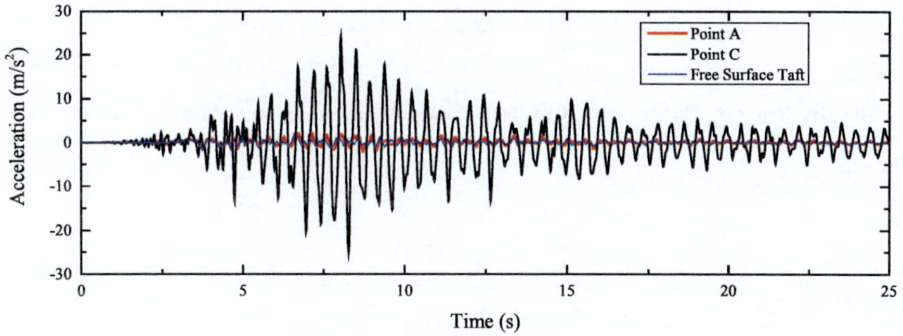

c. Case F3

Fig. 16 Comparison of acceleration time histories at point A and point C

- In non-linear dynamic analysis using ETA, the analysis results may depend on the increase rate of intensity. Therefore, the rationality of ETA should be justified in future research.
- Spatial variations in ground motion are typically ignored in engineering practice. The spatial variations in ground motion should be taken into consideration during seismic analysis.

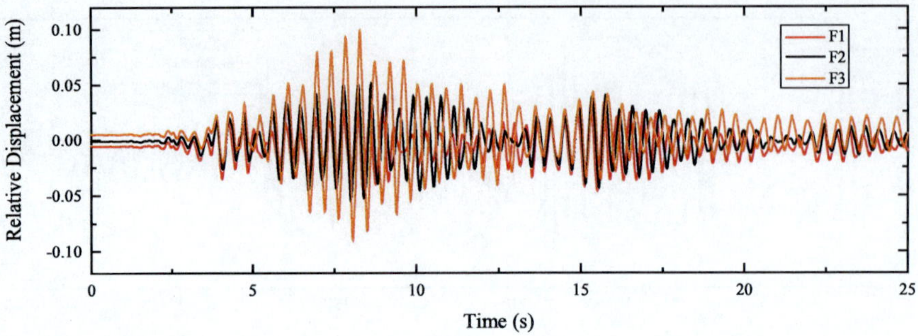

Fig. 17 Comparison of relative displacements at point A and point C in cases F1, F2, F3

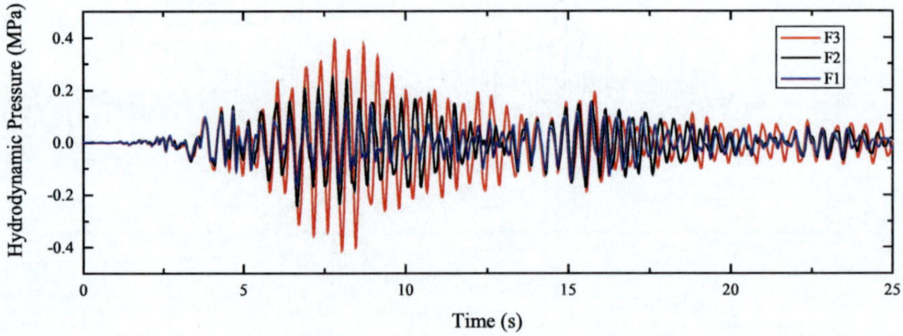

Fig. 18 Comparison of the hydrodynamic pressures at three water levels

Acknowledgements This research is financially supported by the National Natural Science Foundation of China (Nos. 51725901 and 51639006). The authors express their sincerest gratitude for the support.

References

1. Liu J, Du Y, Du X, Wang Z, Wu J (2006) 3D viscous-spring artificial boundary in time domain. Earthq Eng Eng Vib 5(1):93–102
2. Du X, Zhao M, Wang J (2006) A stress artificial boundary in FEA for near-field wave problem. Acta Mech Sin 38(1), 49–56 (in Chinese)
3. GB51247-2018: Standard for seismic design of hydraulic structures. China Planning Press, Beijing (2018)
4. Lee J, Fenves GL (1998) A plastic-damage concrete model for earthquake analysis of dams. Earthq Eng Struct Dyn 27(9):937–956

Seismic Analysis of Pine Flat Concrete Dam

S. Valente, Q. He, and C. Capriulo

Abstract Investigating the dynamic response of a concrete gravity dam (Pine Flat) is the objective of this study. The acoustic-structure coupling interaction is applied to simulate the reservoir-dam system. The effect of foundation size and the efficiency of the infinite element available in the ABAQUS code to model boundaries are studied. From the comparison of displacement, dynamic pressure and acceleration time history, the evaluation of the dam response to the rising water level is performed. The concrete damaged plasticity model is adopted to evaluate the damage and cracking of the dam caused by Taft earthquake.

Keywords ABAQUS · Acoustic-structure coupling interaction · Concrete damaged plasticity model · Concrete gravity dam · Pine flat · Reservoir-dam system

1 Introduction

This document has to be read together with the formulation document of Theme A of the 15th ICOLD International Benchmark Workshop on Numerical Analysis of Dams [1], containing more details on geometrical and mechanical data, and together with the final document, prepared by the formulators and containing the full set of results obtained by all participants to the benchmark workshop.

Pine Flat Dam, located on King's River, east of Fresno, California (Fig. 1), was constructed by the US Army Corps of Engineers in 1954. It consists of thirty-six 15.25 m-wide and one 12.2 m-wide monoliths. The length of the straight gravity dam is 561 m and the tallest non-overflow monolith is 122 m high (Fig. 2).

S. Valente (✉) · C. Capriulo
Department of Structural, Geotechnical and Building Engineering, Politecnico di Torino, Turin, Italy
e-mail: silvio.valente@polito.it

Q. He
College of Water Conservancy and Hydropower Engineering, Hohai University, Nanjing, China

G. Bolzon et al. (eds.), *Numerical Analysis of Dams*, Lecture Notes in Civil Engineering 91, https://doi.org/10.1007/978-3-030-51085-5_18

Fig. 1 Downstream view of pine flat dam

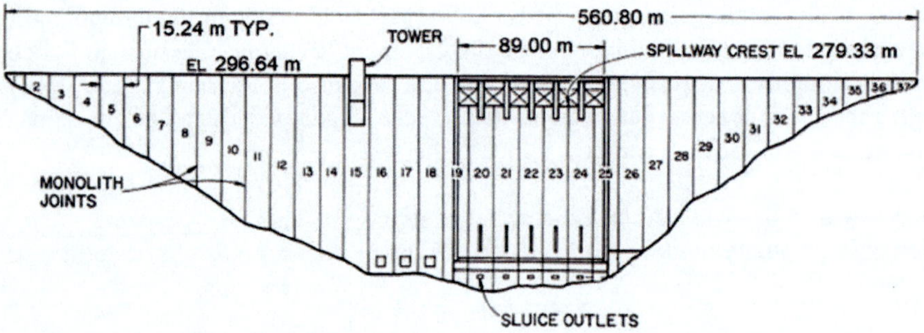

Fig. 2 Downstream view of pine flat dam [1]

This study is aimed at identifying key uncertainties that may significantly affect numerical modelling results, at determining the need for future investigations and at developing the best practices in the advanced analysis of concrete dams. This case includes an analysis of the tallest non-overflow dam monolith, no. 16 (Fig. 3).

2 Method and Approach

All the analyses were carried out by the Finite Element Method (FEM) using the ABAQUS code [2]. The models include the dam monolith 16, a relevant portion of foundation and of reservoir (complete model), with exception of model B. The mesh, shown in Fig. 4, is made of continuous solid, acoustic, and interface elements. The mesh is finer for case B (1×1 m) and case C (1×1.5 m). In case B also a larger foundation length (3700 m) is analyzed.

The reservoir was modelled by acoustic elements, considering the fluid as incompressible. The pressure in the fluid elements accounts for the hydro-dynamic effects

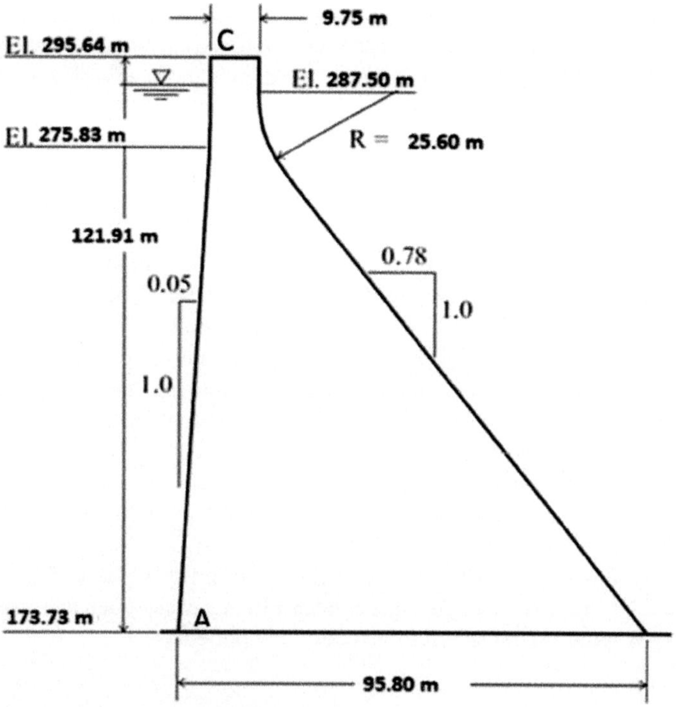

Fig. 3 Cross section geometry of Monolith 16

Fig. 4 Finite-infinite model of the monolith, the foundation and the reservoir (element size 1.48 × 1.48 m)

only, while hydrostatic pressures were directly applied at the faces where the reservoir interacts with the dam and the foundation. A boundary condition of zero pressure was applied to the upper surface of the reservoir model.

The dynamic cases were analyzed using an implicit direct integration scheme. The interaction between the dam and the reservoir is achieved through the classic structural-acoustic coupling on the upstream face of the dam, where the normal component of the dam acceleration is proportional to the normal gradient of the water hydrodynamic pressure [3]. The upstream truncation of the reservoir is provided with a non-reflecting acoustic condition.

The only considered static loads are the weights of the dam and of the water (the weight of the foundation is not included in the analysis). The weight of water is applied as hydrostatic pressure on the upstream face of the dam and on the surface of the foundation.

In all cases (A to E) a wave propagation problem in unbounded domain has to be solved. In order to truncate properly the domain analyzed, an artificial boundary condition has to be applied. Such condition is designed so that the outgoing plane waves can correctly cross the external borders of the truncated foundation. In these conditions the normal and tangential stresses are equal to the mass density multiplied by the normal and tangential component of the velocity, respectively. As a consequence, in order to prevent large rigid body motions, a self-equilibrated loading condition has to be applied. Since the self-weight of the dam and the water is not a self-equilibrated condition, in a first step the foundation boundary is assumed as fixed. In a second step, the reaction forces obtained in this way, together with the self-weight, are applied to a model with an artificial viscous boundary. This procedure can reduce the rigid body motions of the model. In any case, when the displacement time history is required, a base line correction in applied.

3 Results

3.1 Case A

Modal analysis was performed both for WRWL (winter reservoir water level) and SRWL (summer reservoir water level). Results are shown in Figs. 5, 6 and 7. The rise of reservoir water level induces a decrease of the natural frequencies (maximum

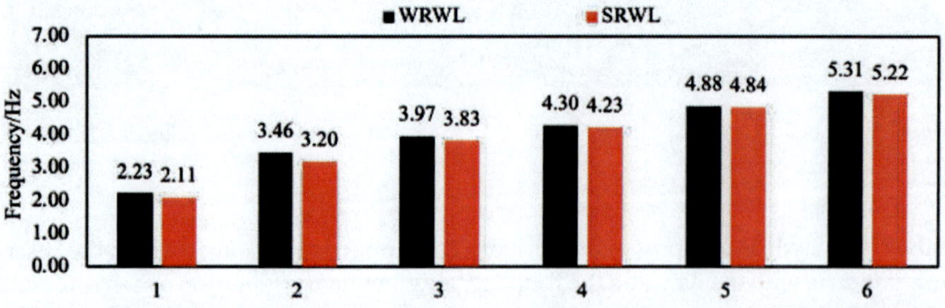

Fig. 5 Natural frequency for different reservoir water levels

7.4% for the second mode, minimum 0.7% for the fifth mode). For this analysis the boundary nodes of the foundation are assumed as fixed.

This model was also excited by an eccentric-mass vibrating generator (EMVG) applied at the dam crest. In this case, in order to allow the propagation of elastic waves to an infinite domain, a viscous artificial boundary is applied at the foundation nodes. It is true that the external force is not self-equilibrated, but it is very small. As a consequence, also the rigid body motion is very small. It is eliminated by a base line correction of the displacement histories. A viscous dumping of 2% was applied to the dam and foundation. Since in this case a concentrated force is applied to the near-field-domain (see [4]), the infinite elements provided by the Abaqus code have one function only: not reflecting the outgoing waves. This function is performed properly.

The four time-histories obtained by the authors (displacement and acceleration at the dam heel and crest) are shown in the complete report prepared by the formulators and are not repeated here.

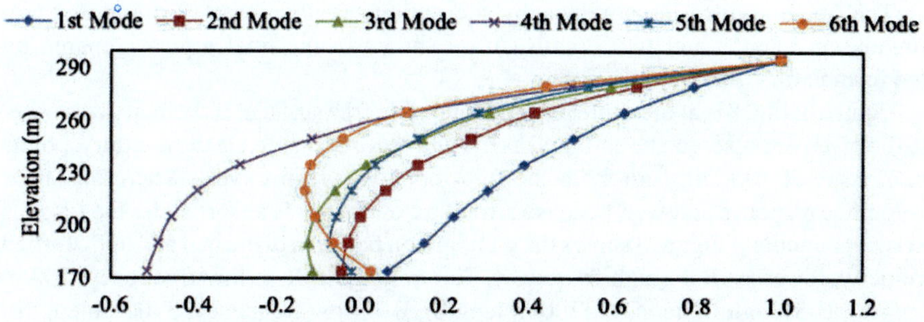

Fig. 6 Normalized mode shapes at line (A-C) for WRWL

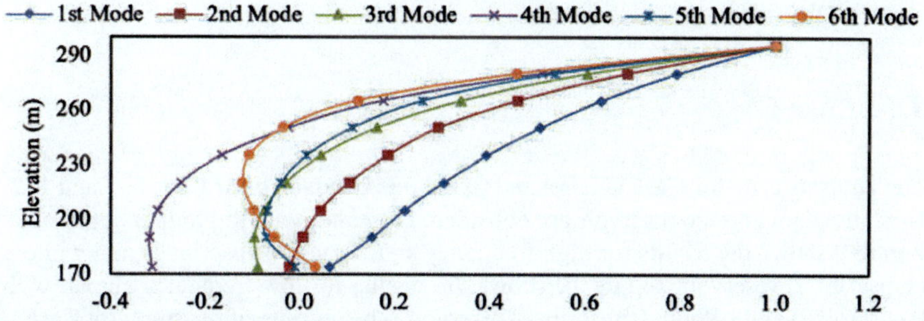

Fig. 7 Normalized mode shapes at line (A-C) for SRWL

3.2 *Case B*

Case B deals with the foundation block only, which is a reduced-domain for a uniform half-space. The loading condition is a vertically-propagating SH wave, induced by a shear-stress time-history provided for input to the base of the foundation. As a consequence, each element column behaves exactly as the adjacent one (uniaxial problem) and the time histories at the surface have to be the same. Analyzing a uniaxial problem is not the goal of this case. The goal of this case is to analyze the disturbance applied by the same boundary condition used in case C, D, E to the above-mentioned solution. No damping was applied.

Since in this case the energy arrives from a far-field source through a vertically propagating SH wave (see [4]), the infinite elements have two functions: (a) allowing the incoming waves to penetrate in the near-field-domain, (b) not reflecting the outgoing waves generated by the scattered motion induced by the dam and reservoir (see [4]). In a first analysis we didn't apply any external force on the vertical viscous boundary.

The 36 time-histories obtained by the authors (velocities at 9 points for two foundation lengths and two impulses) are shown in the final report prepared by the formulators.

Point a in Fig. 8 is at the center of the foundation free surface. Its velocity response, to the given impulse, can be assumed as the undisturbed result. On the contrary, points c, e, g are at 100, 20, 0 m from the right boundary respectively. Therefore, their velocities response is lower, because a fraction of energy is absorbed by the viscous vertical boundary. Figure 8 shows the comparison between disturbed and undisturbed velocity, for cases B-1 (high frequency, 700 m length), B-2 (low frequency, 700 m length), B-3 (high frequency, 3700 m length), B-4 (low frequency, 3700 m length). The difference observed is due to the fact that we had not enough time to implement a boundary condition on the vertical sides which is compliant with the free-field solution.

3.3 *Case C*

The analysis type for Case C is related to the one conducted for Case B-1 and B-2. Here, the dam and the reservoir are considered together with the foundation model. Figure 9 shows the results for high frequency signal, with (subscript 1) and without (subscript 3) reservoir. Figure 10 shows the results for low frequency signal, with (subscript 2) and without (subscript 4) reservoir. The purpose of the study for Case C is to investigate the effect of the dam and reservoir presence on the wave propagation in the foundation and to compare the analysis results with the free field motions studied in Case B. No damping was applied.

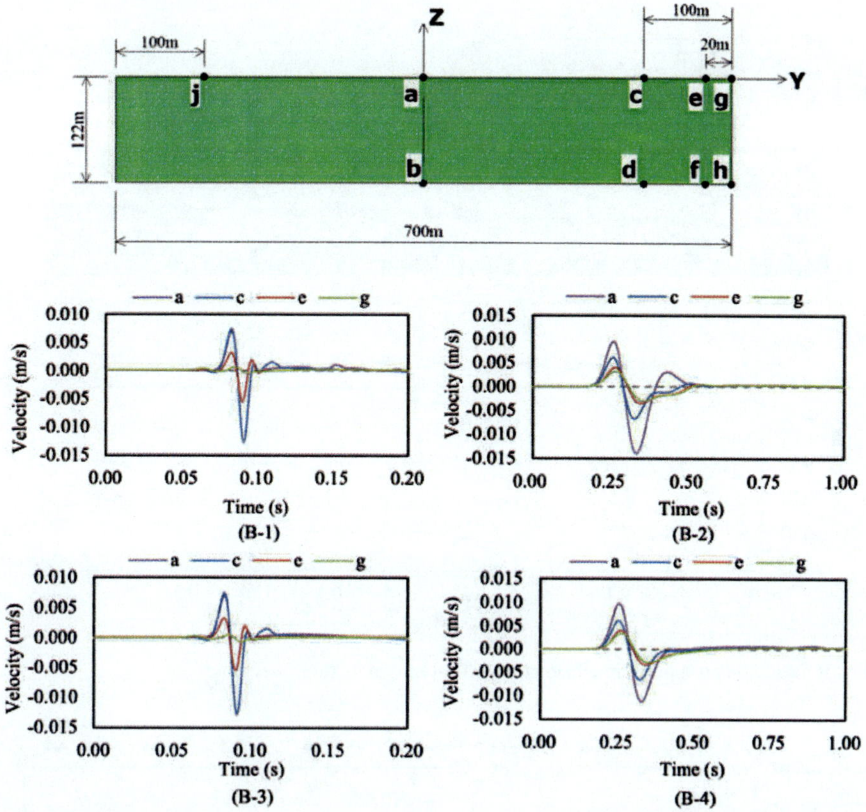

Fig. 8 Velocities response time histories at point (a), point (c), point (e), and point (g)

Points a, c, e are in the same position as in the previous case B. At point c, reverberations after the main pulse are produced primarily by the interaction of the dam and reservoir.

3.4 Case D

In Case D, a dynamic analysis of the dam-foundation-reservoir system is performed considering the elastic material properties, the Taft earthquake record, and the reservoir water at two different elevations. More precisely, the deconvolved stress time history was applied at the base of the foundation block. The intent is to evaluate the dam response due to various reservoir levels.

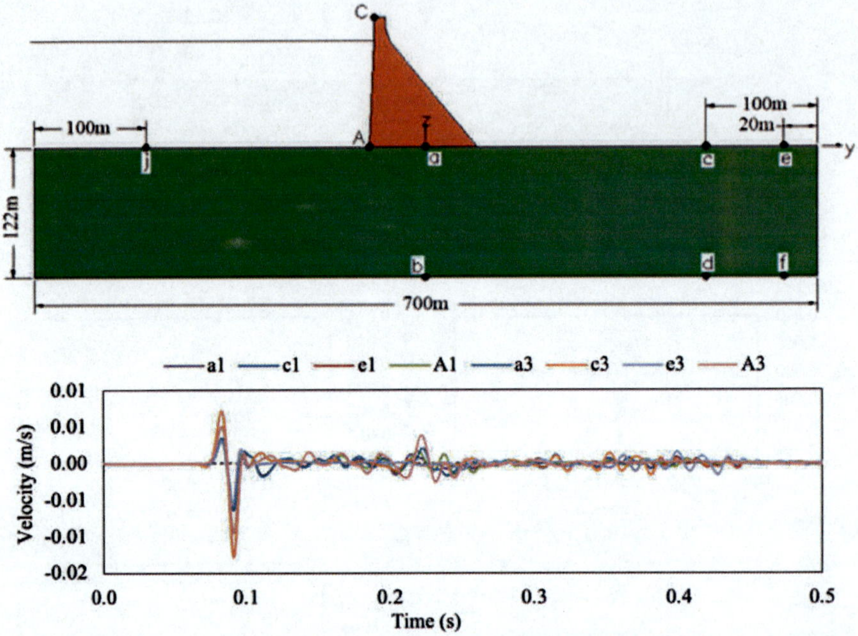

Fig. 9 Velocities response time histories at different points (high frequency)

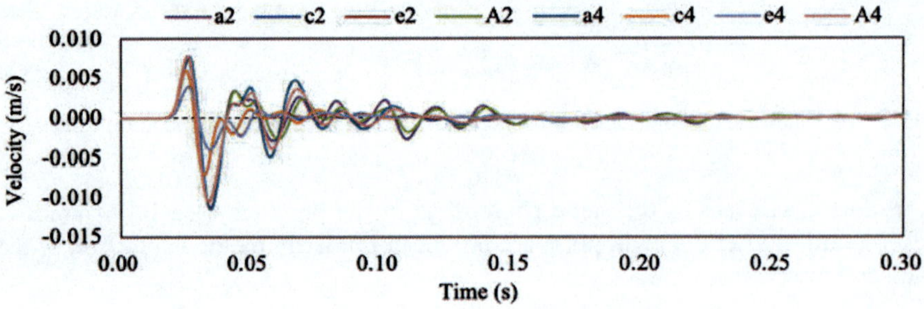

Fig. 10 Velocities response time histories at different points (low frequency)

Figure 11 shows the history of hydrodynamic pressure at heel [point A]. Figures 12 and 13 show the history of acceleration at heel [point A] and crest [point C]. The differences between the two diagrams are very small.

The acceleration time history computed at point A approximately match the free-surface Taft record, and the time history computed at point C is substantially larger.

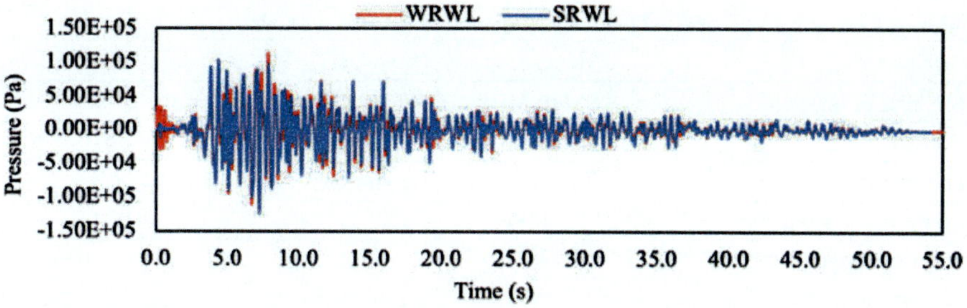

Fig. 11 History of hydrodynamic pressure at heel [point A] for WRWL and SRWL

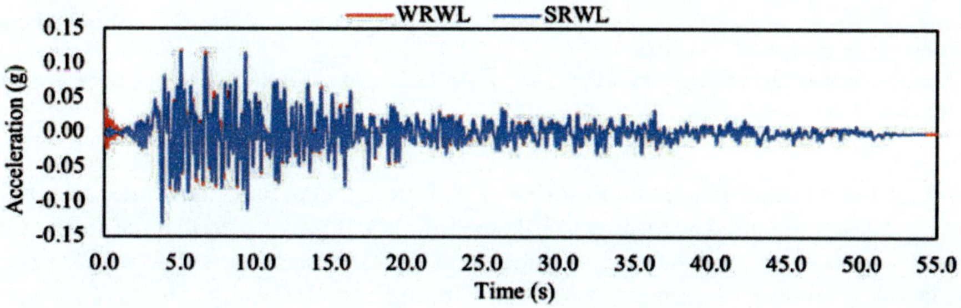

Fig. 12 History of acceleration at heel [point A] for WRWL and SRWL

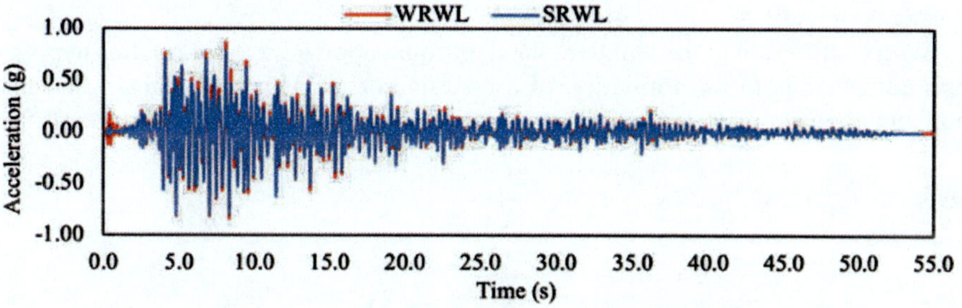

Fig. 13 History of acceleration at crest [point C] for WRWL and SRWL

3.5 Case E

The intent of Case E is to perform a dynamic analysis with concrete non-linear material properties. The dam-foundation-reservoir system used in Case D will be analyzed except the non-linear material properties of concrete are considered. The concrete damaged plasticity model is used.

During the non-linear analysis, the compressive strength of 28 MPa (see Table 1) is never achieved. On the contrary, the tensile strength of 2 MPa is reached at point

Table 1 Concrete properties and parameters applied in the Abaqus code

Parameters	Values	Parameters	Values
Modulus of elasticity	22410 MPa	Tensile strain at peak load	0.00012
Density	2483 kg/m^3	Dilation angle	30°
Poisson's ratio	0.20	Eccentricity	0.1
Compressive strength	28.0 MPa	f_{b0}/f_{c0}	1.16
Tensile strength	2.0 MPa	*K*	0.6666
Fracture energy	250 N/m	Viscosity parameter	0.0
Compressive strain at peak load	0.0025		

Note [1] f_{b0}/f_{c0} is the ratio of initial equibiaxial compressive yield stress to initial uniaxial compressive yield stress
[2] The eccentricity is a small positive number that defines the rate at which the hyperbolic flow potential approaches its asymptote
[3] *K* is the ratio of the second stress invariant on the tensile meridian to that on the compressive meridian

A (re-entrant corner at upstream side). The softening behavior of concrete, after the peak value, is assumed according to [5] (see Fig. 14). It is assumed that the softening process occurs in a narrow band of thickness equal to one element (crack-band = 1.48 m), while the elements outside this band unload elastically. This assumption allow us to transform the critical value of displacement discontinuity w_c, computed according to [5] and the fracture energy given by the formulators, into the critical value of the cracking strain $(\varepsilon_c)_c = (w_c/\text{crack-band})$ required by the concrete damaged plasticity model.

With comparison to the length of the dam-foundation joint (95.8 m), this hypothesis corresponds to the formation of a cohesive crack. It is well known that the softening behavior can induce convergence problems, due to the bifurcation of the

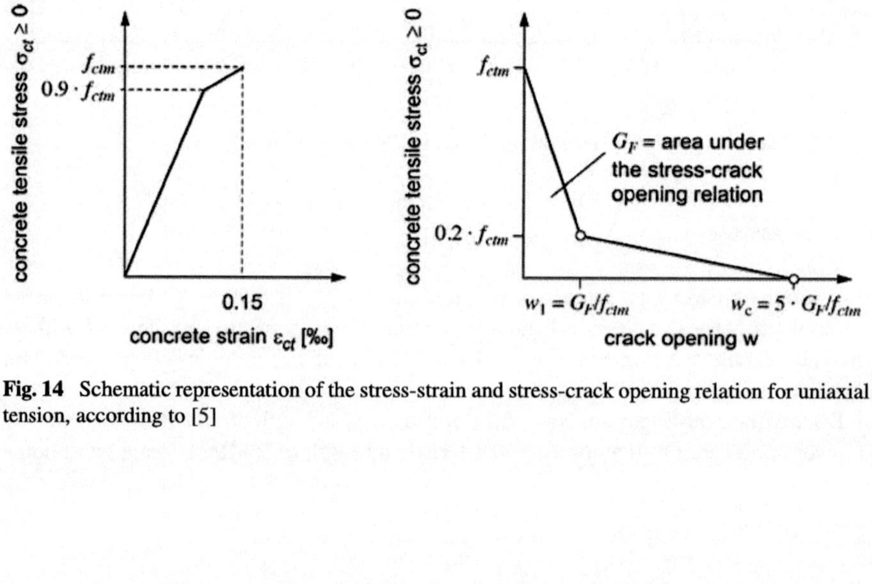

Fig. 14 Schematic representation of the stress-strain and stress-crack opening relation for uniaxial tension, according to [5]

equilibrium path in the incremental solution process. These problems occur in quasi-static analysis (see [6, 7]), as well as in this dynamic implicit analysis. In order to overcome this problem: (a) the time history of the deconvolved acceleration was used as load; (b) a set of parameters, suggested by the Abaqus manual as 'application = moderate dissipation' was used.

The material behavior in unloading condition is characterized by a stiffness k_{unl} which depends on the damage index as follow: $d = 1.0 - k_{unl}/k_{\text{first\_loading}}$. The value of d is assumed as a piecewise linear function of ϵ_c. During the first loading phase, it is assumed $d = 0$; just after the peak load, when the cracking strain ϵ_c starts to grow it is assumed $d = 0.25$; at the knee point it is assumed $d = 0.78$; when ϵ_c reach the critical value it is assumed $d = 1$ and no stress transfer occurs.

In Case E, the tensile strength is achieved at the above-mentioned re-entrant corner, for $T_1 = 3.785$ s after the excitation beginning. This is the first damage occurring in the model. Afterwards the cohesive crack starts to propagate along the dam-foundation joint and the last damage occurs for $T_2 = 7.031$ s, at a distance of 16.5 m from the re-entrant corner.

Since an automatic time-step dt is assumed, dt drops from 1.e-2 to 1.e-6 s every time the peak stress is reached in a new, undamaged element.

According to the formulation document [1], the Rayleigh damping matrix is assumed as $C = \alpha M + \beta K$, where M is the mass matrix, K is the stiffness matrix, $\alpha = 0.75$ 1/s and $\beta = 0.0005$ s. Since the Abaqus code does not propagate to the C matrix the reduction of stiffness induced by damage, on the safety side, for the first 15 concrete elements along the dam foundation joint, it was assumed $\beta = 0$.

For case E only, in order to simulate the free-field motion on both vertical sides of the foundation, a two-step-procedure was used.

In a first model B analysis, without dam and reservoir, the condition U2 = 0 was applied on both vertical sides. In this model all points in the foundation surface move in the same way. Both the histories of the related reaction force RF2 and the histories of the displacement U1 were recorded.

In a second model, the condition U2 = 0 was removed and the histories RF2 were applied. Furthermore, the viscous boundary elements were added on the vertical sides and the necessary forces to impose the previously recorded histories U1 were imposed, as well.

A specific python code was written in order to manage the above-mentioned histories. Since the application of this procedure to case B was successful, it was repeated in the case E.

According to the formulation document: (a) viscous damping of 2% was applied to the dam and foundation; (b) the uplift pressure induced by the water penetrating the crack was not taken into account (see [8]).

Figures 15 and 16 show the acceleration response histories at heel [point A] and crest [point C] for WRWL respectively. Figure 17 shows the history of hydrodynamic pressure at heel [point A].

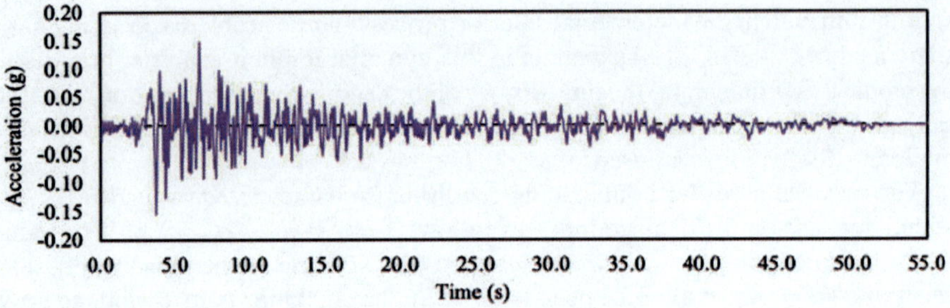

Fig. 15 History of acceleration at heel [point A] for WRWL

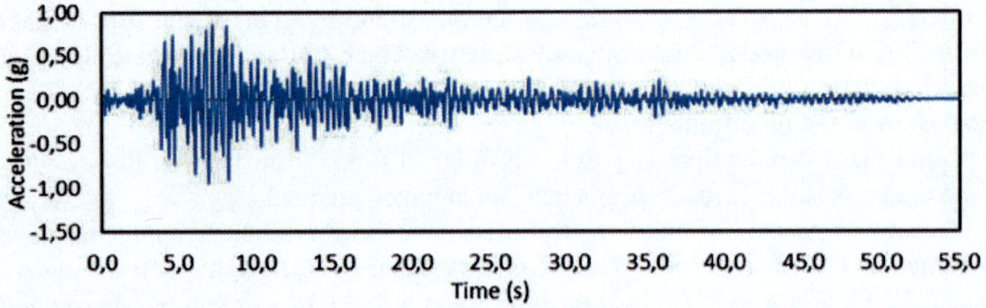

Fig. 16 History of acceleration at crest [point C] for WRWL

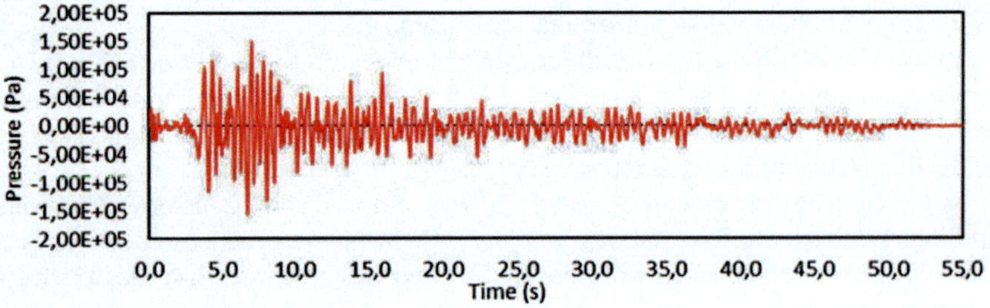

Fig. 17 History of hydrodynamic pressure at heel [point A] for WRWL

4 Conclusions

- In case of a dynamic force applied to the near field, the viscous-boundary available in Abaqus code performs properly. No specific code is necessary.
- In case of vertical propagating SH-waves, a specific python code was necessary to model the free field motion on the vertical boundaries properly. This procedure was applied to case E only.

- The Taft records induces damage on the first 16.5 m of dam-foundation joint, over a total length of 98.5 m.
- The uplift pressure induced by the water penetrating the crack was not taken into account.

References

1. Salamon J, Hariri-Ardebili MA, Malm R, Wood C, Faggiani G (2019) Theme A formulation. Seismic analysis of Pine Flat concrete dam. In: 15th ICOLD international benchmark workshop on the numerical analysis of dams. Milan, Italy
2. Dassault Systems Simulia Corp. (2018) Abaqus 6.18 documentation. Johnston, Rhode Island, USA
3. Zienkiewicz OC, Taylor RL (2000) Finite element method: volume 1, 5th edn. Butterworth-Heinemann, Oxford
4. Wolf JP (1985) Dynamic soil-structure interaction, 1st edn. Prentice Hall, Englewood Cliffs, NJ
5. International Federation for Structural Concrete (2013) fib model code for concrete structure, 1st edn. Ernst & Sohn, Berlin, Germany
6. Valente S (1992) Bifurcation phenomena in cohesive crack propagation. Comput Struct 44(1–2):55–62
7. Barpi F, Valente S, Cravero M, Iabichino G, Fidelibus C (2012) Fracture mechanics characterization of an anisotropic geomaterial. Eng Fract Mech 84:111–122
8. Barpi F, Valente S (2008) Modeling water penetration at dam-foundation joint. Eng Fract Mech 75(3–4):629–642

Seismic Behavior of Pine Flat Concrete Gravity Dam Using Microplane Damage-Plasticity Model

S. Vorlet, P. Manso, and G. De Cesare

Abstract The response of concrete gravity dams under seismic loads is a major concern of dam safety assessment in earthquake-prone areas. This study focuses on the seismic analysis of the tallest non-overflow monolith of Pine Flat Dam. The paper presents results of numerical modelling considering the effects of fluid-structure interaction for both linear and nonlinear analysis using accelerations records of the historical Taft earthquake and of an artificially designed Endurance Time Acceleration Function (ETAF). Linear analyses establish the dynamic properties of the reservoir-dam-foundation system and its behavior considering non-zero mass foundation as compared with massless foundation, for two typical reservoir water levels. Whilst the former is physically more complete, it leads to an effective horizontal displacement after the dynamic loading whereas the massless foundation model shows no residual horizontal displacement of the dam base. A damage-plasticity model based on microplane formulation is used to model the extent of damage of the dam body during nonlinear dynamic analysis. Results show the ability of the model to represent cyclic loading conditions, with recovery of the stiffness lost during cracking in the transition from tension to compression state, and subsequent failure of the dam body.

Keywords Seismic analysis · Concrete gravity dam · Microplane · Non-linear analysis · Damage-plasticity model

1 Introduction

The response of concrete gravity dams under seismic loads is a major concern of dam safety assessment in earthquake-prone areas. Due to the risks related to dam failure, it is required to understand the dynamic behavior of mass concrete during

S. Vorlet (✉) · P. Manso · G. De Cesare
Platform of Hydraulic Constructions (PL-LCH), Ecole Polytechnique Fédérale de Lausanne (EPFL), Lausanne, Switzerland
e-mail: samuel.vorlet@epfl.ch

G. Bolzon et al. (eds.), *Numerical Analysis of Dams*, Lecture Notes in Civil Engineering 91, https://doi.org/10.1007/978-3-030-51085-5_19

such events. The dynamic response of the dam body depends to some extent on the binding foundation conditions as well as on the interaction with the reservoir.

During earthquakes, gravity dams are subject to strong horizontal and vertical motions inducing stresses with peaks that may be greater than the maximal strength and consequently lead to damage of the dam body, mainly in tension state. Currently, most dam safety evaluations with finite elements (FE) analysis of reservoir-dam-foundation systems consider a linear elastic model for mass concrete with failure criteria based on maximal tensile strength [1, 2] in particular for the non-extreme load combinations.

First assessments of extreme load combinations using linear analyses allow preliminary estimates of the location and extent of tensile stress peaks greater than the maximal concrete tensile strength but cannot inform on stress and stiffness redistribution during an earthquake (damage time evolution). Nonlinear analyses with a damage model are therefore mandatory to assess the dam safety under seismic loads resulting with stress peaks leading to damage of the dam body.

This paper presents the seismic analysis of Pine Flat Dam for the 15th International Benchmark Workshop on Numerical Analysis of Dams to be held in Milan in September 2019. It focuses on the tallest non-overflow monolith of Pine Flat concrete gravity dam located on King's River, east of Fresno, California (USA).

2 Methods

The benchmark case study analysis presented herein is performed using the finite element code ANSYS (M-APDL) to discretize the governing equations on the computational domain. The seismic analysis focuses on the reservoir-dam-foundation system and is performed in two steps. Firstly, the finite element model is verified by assessing the dynamic characteristics of the system (natural frequencies and mode shapes) in the frequency domain. Secondly, a dynamic analysis in the time domain is performed using various load time histories in order to study the reproducibility of dynamic behavior of such systems by means of computational modelling including linear and nonlinear analysis.

2.1 *Governing Equations*

For typical elastic linear structural dynamics problems involving multiple degrees-of-freedom (MDoF systems), a semi-discrete equation of motion can be derived based on spatial discretization of virtual work process of a continuous structure with the finite elements method [3],

$$[M]\{\ddot{u}\} + [C]\{\dot{u}\} + [K]\{u\} = \left[F^a\right] \quad (1)$$

where $[M]$, $[C]$ and $[K]$ are the mass, damping and stiffness matrices respectively (applied to the acceleration, velocity and displacement terms), and $\{F^a\}$ is the vector of applied loads. For elastic linear systems, the internal load is linearly proportional to the nodal displacement, and the structural stiffness matrix $[K]$ remains constant. In nonlinear systems, the resisting force is an implicit function of deformations, meaning that the stiffness matrix is dependent on the current displacements [4–6]. For such systems, the equation of motion is

$$[M]\{\ddot{u}\} + [C]\{\dot{u}\} + \left\{F^i\right\} = \left\{F^a\right\} \tag{2}$$

where $\left\{F^i\right\}$ is the vector of internal loads. The modal analysis with nonzero damping is a quadratic eigenvalue problem with complex solutions [3], where λ_i is the eigenvalue and $\{\phi_i\}$ the eigenvector and $\bar{\lambda}_i = j\sqrt{\lambda_i}$

$$[K]\{\phi_i\} + \bar{\lambda}_i[C]\{\phi_i\} = -\bar{\lambda}_i^2[M]\{\phi_i\} \tag{3}$$

2.2 Finite Element's Model

A two-dimensional finite elements (FE) model of the reservoir-dam-foundation system is considered. The dam is composed of 37 monoliths and its crest is 561 m long. The tallest non-overflow monolith is 121.91 m high with a 95.8 m base length. The model mesh for linear analysis comprises 8800 nodes and 8551 elements.

The dam body and rock foundation (700 m × 122 m) are modeled with quadrilateral structural plane strain elements with two degrees of freedom per node (translational in the x and y directions). The reservoir is modeled with various reservoir water levels (RWL) using acoustic elements with inviscid and compressible behavior with a pressure degree of freedom, which accounts for hydrodynamic effects only. However, translational degrees of freedom in the nodal x and y directions are also active for acoustic elements at interfaces with the dam, where fluid-structure interactions (FSI) occur. Figure 1 shows the finite element mesh with corresponding dimensions for the linear analysis.

The dam is considered to be fully tied to the foundation. A tie constraint is also applied at the reservoir-dam and reservoir-foundation interfaces in order to couple accelerations with hydrodynamic pressures. Fluid-structure interactions are also considered at the reservoir-dam interface, in which translational DOF of the structural elements and acoustic elements at the FSI boundary are included to the governing equation. No seismic waves absorption by the reservoir bottom sediment is considered in the model [5–7]. Damping for dam body and rock foundation is modeled using Rayleigh viscous damping, defined by Rayleigh damping constants α and β. For the frequency range of interest in this study, the average Rayleigh damping is approximately 2%. The damping matrix $[C]$ is defined as a linear combination of the mass matrix $[M]$ and the stiffness matrix $[K]$.

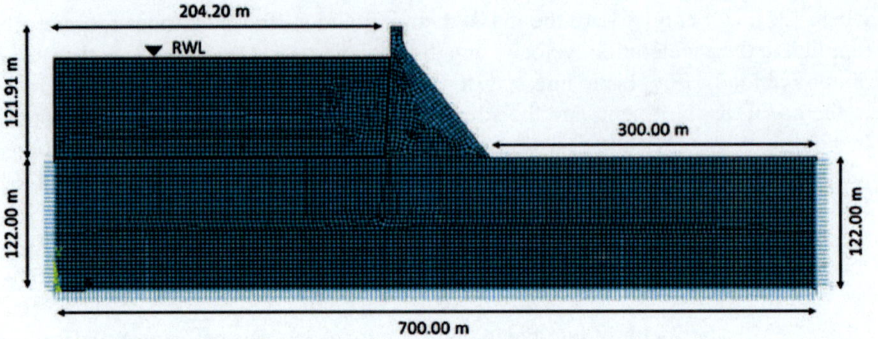

Fig. 1 Two-dimensional view of the reservoir-dam-foundation finite elements system with corresponding mesh, dimensions and reservoir water level (RWL) for the linear analysis

$$[C] = \alpha[M] + \beta[K] \tag{4}$$

At the upstream far-end of the reservoir, absorbing boundary condition (ABC) is used to truncate the unbounded domain, in order for the pressure wave to satisfy the Sommerfeld radiation condition [3, 7]. Pressure is set to zero at the free surface of the reservoir. For modal analysis only, displacements are restrained in the normal and tangential directions ($U_x = 0$; $U_y = 0$) at the sides of foundation. Finally, at the bottom, upstream and downstream faces of the rock foundation, viscous-spring boundary condition is used to model the semi-infinite medium. The spring-dashpot system absorbs the wave energy radiating away from the dam [8]. The normal and tangential spring coefficients K_B and damping coefficients C_B of the viscous-spring boundary condition are defined as follow,

$$K_{BN} = \frac{1}{1+\alpha} \cdot \frac{(\lambda + 2G)}{2 \cdot r_b}; C_{BN} = \beta\rho c_p \tag{5}$$

$$K_{BT} = \frac{1}{1+\alpha} \cdot \frac{G}{2 \cdot r_b}; C_{BT} = \beta\rho c_s \tag{6}$$

where λ and G are the medium's Lamé's first parameter and shear modulus respectively, defined by the theory of elastic moduli for homogeneous and isotropic materials, ρ is the density of the medium, c_p and c_s are the compressional wave and shear wave velocity of the medium, r_b is the distance from the wave source to the artificial boundary point, $\alpha = 0.8$ and $\beta = 1.1$ are dimensionless coefficients [8]. Figure 2 shows a schematic representation of the model's boundary conditions and interfaces for modal and dynamic analyses. For the linear analysis test cases, the dam body and rock foundation are considered homogeneous, isotropic mediums with linear elastic material properties. The elastic material properties for the reservoir-dam-foundation system are summarized in Table 1.

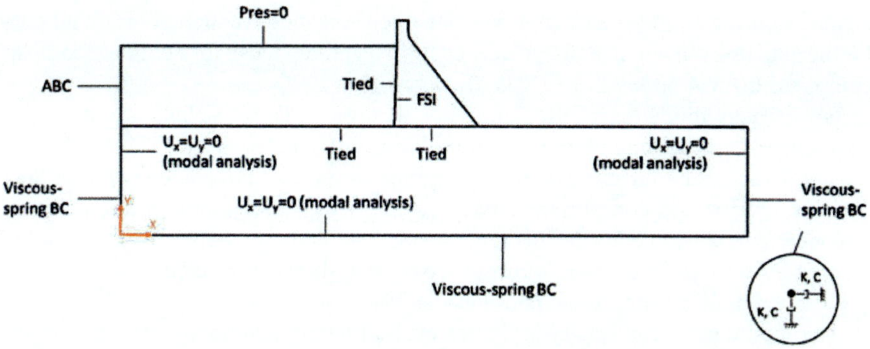

Fig. 2 FE model boundary conditions and interfaces for modal and dynamic analyses

Table 1 Model linear elastic material properties

Properties		Unit	Concrete	Rock	Water
Modulus of elasticity	E	MPa	22410	22410	–
Density	ρ	kg/m^3	2483	2483	1000
Poisson's ratio	ν	–	0.20	0.20	–
Shear wave velocity	Sw	m/s	–	1939	–
Compressional wave velocity	Pw	m/s	–	3167	1439

2.3 *Loads*

Inertia load (gravity) is only applied to the dam body and the reservoir. Static loads include hydrostatic pressure at reservoir-dam interface and hydrostatic pressure at reservoir-foundation interface. Two water levels are considered in the study:

- Winter Reservoir Water Level (WRWL) at elevation 288.21 m a.s.l.
- Summer Reservoir Water Level (SRWL) at elevation 278.57 m a.s.l.

Dynamic loads include the deconvolved Taft Record and the artificially designed Endurance Time Acceleration Function (ETAF) applied at the base of the foundation block. Thermal loads are neglected in this study. For the time integration method, the implicit dynamic $HHT - \alpha$ algorithm is used [3].

2.4 *Nonlinear Analysis*

A nonlinear analysis is conducted on the reservoir-dam-foundation system in order to analyze the extent of damage in dam body. Concrete nonlinear material properties are considered for the nonlinear analysis. Only WRWL is considered. In order to capture nonlinear effects and improve convergence, a mesh refinement is mandatory for the dam body in which damage may occur. The model mesh for nonlinear analysis comprises 80'624 nodes and 32'641 elements.

For damage modeling of the dam body, a coupled damage-plasticity model for concrete [9] based on microplane formulation [10, 11] is used. The model has the ability to define different damage initiation criteria and damage evolution laws between tension and compression states. It allows representing the brittle behavior of concrete in tension considering that softening starts directly after the elastic limit, defined by setting the damage threshold to $\gamma_t = 0$. It also considers that hardening is observed before softening in compression [9].

The model can additionally represent cyclic loading conditions, where stiffness lost during cracking is recovered due to crack closure during transition from tension to compression state, while damage sustained under compression remains upon transition to tension state [9]. The plastic behavior of concrete is defined with the Drucker-Prager yield function with hardening for compression.

The damage of concrete is defined by a damage initiation criterion and a damage evolution law. The damage parameter DMG is computed for each element of the dam body, with $DMG = 0$ representing an undamaged element and $DMG = 1$ a completely damaged element, herein assumed as local dam failure criterion. Failure of the dam body corresponds to subsequent large deformations. Figure 3 shows the one-dimensional cyclic compression stress-strain curve for the coupled plasticity-damage model. Table 2 summarizes the coupled damage-plasticity model parameters for the nonlinear analysis. It should be noted that the non-local parameters have a great influence on the spread of damage in the dam body.

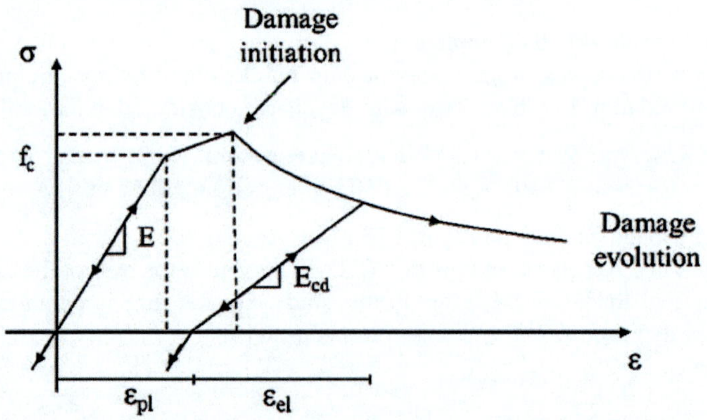

Fig. 3 One-dimensional cyclic compression stress-strain curve for the damage-plasticity model; E: modulus of elasticity, E_{cd}: modulus of elasticity for damaged element, f_c: elastic limit for concrete, ϵ_{pl}: plastic deformation, ϵ_{el}: elastic deformation; reference [9]

Table 2 Coupled damage-plasticity model parameters for nonlinear analysis

Parameter		Unit	Value
Uniaxial compression strength	f_{uc}	MPa	28
Biaxial compression strength	f_{bc}	MPa	32.2
Uniaxial tensile strength	f_{ut}	MPa	2
Tension cap hardening factor	R_t	–	1
Hardening parameter	D	GPa	40
Compression cap location	σ_v^c	MPa	−21.4
Compression cap shape	R	–	2.5
Threshold for tension damage	γ_{t0}	–	0
Threshold for compression damage	γ_{c0}	–	2E-5
Tension damage parameter	β_t	–	7500
Compression damage parameter	β_c	–	2500
Nonlocal interaction range parameter	c	m^2	0.25
Over nonlocal parameter	m	–	2.5

3 Results

Main results of numerical simulations are presented in this section for linear and nonlinear analyses. Results focus on horizontal displacements and accelerations at crest and heel of the dam body, with additional results regarding damage extent of the dam body for nonlinear analyses.

3.1 Linear Analysis

Dynamic properties of the reservoir-dam-foundation system. Dynamic properties of the reservoir-dam-foundation system under free vibration (modal analysis) are presented here. The six first natural frequencies of the system for WRWL and SRWL are shown in Table 3. As expected, a higher water level results in lower natural frequencies for each mode. Corresponding mode shapes for both water reservoir levels are shown in Fig. 4.

Table 3 Computed natural frequencies for the reservoir-dam-foundation system

Mode		I	II	III	IV	V	VI
Frequency (Hz)	WRWL	2.28	3.77	4.09	4.62	4.86	5.53
	SRWL	2.17	3.47	3.97	4.42	4.85	5.46

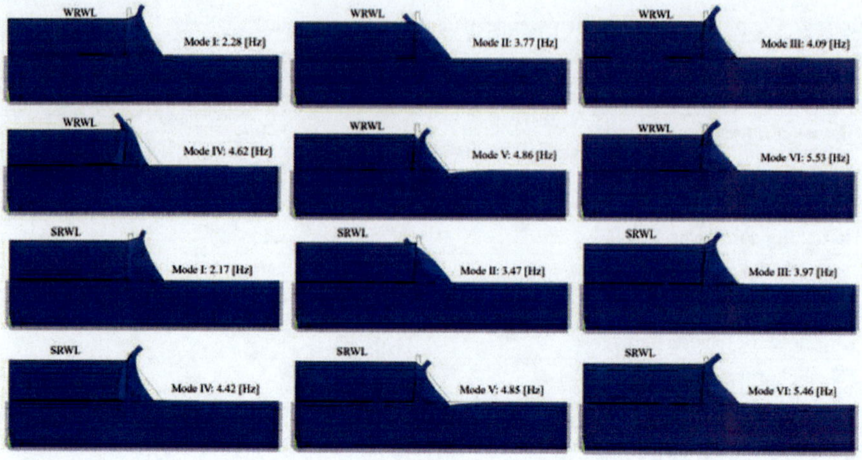

Fig. 4 Computed natural mode shapes for the reservoir-dam-foundation system; Winter Reservoir Water Level (WRWL), Summer Reservoir Water Level (SRWL)

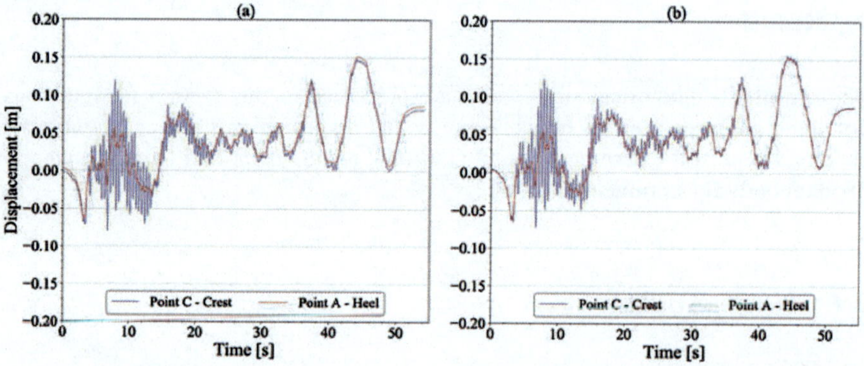

Fig. 5 Computed horizontal displacements at crest and heel of the dam; **a** WRWL, **b** SRWL; Taft record acceleration; Linear analysis

Dynamic analysis. Figiure 5 shows computed horizontal displacements at crest and heel for WRWL and SRWL in the time domain with the Taft Record acceleration. Results show that horizontal displacements follow complete system displacements for both WRWL and SRWL, with a residual displacement of the dam base of approximately 8 cm in the downstream direction at the end of simulation. The amplitude of the maximal oscillations around complete system displacements is in the order of magnitude of 10 cm at crest and 10 μm at heel. Oscillation amplitudes are greater between $t = 4$ and $t = 14$ s, corresponding to greater relative displacement between

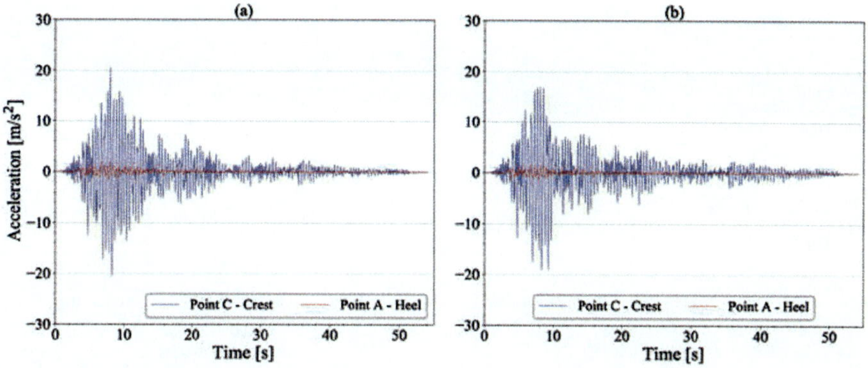

Fig. 6 Computed horizontal acceleration at crest and heel of the dam; **a** WRWL, **b** SRWL; Taft record acceleration; linear analysis

crest and heel. Finally, relative displacements between crest and heel are smaller considering SRWL, particularly at the end of simulation.

Figure 6 shows computed horizontal accelerations at crest and heel for WRWL and SRWL in the time domain with the Taft Record acceleration. Results show that horizontal accelerations at heel approximately match the free-surface Taft Record while horizontal accelerations at crest are substantially greater (amplified), mostly between $t = 4$ and $t = 14$ s. Maximal horizontal accelerations for WRWL vary between 20.70 and -20.56 m/s^2 at crest (approximately twice the gravity) and between 1.86 and -1.66 m/s^2 at heel.

Finally, results show that a higher reservoir water level slightly reduces maximal computed accelerations, with maximal horizontal accelerations varying between 16.94 and -19.3 m/s^2 at crest and between 1.63 and -1.47 m/s^2 at heel.

Massless foundation analysis Fig. 7 shows computed horizontal displacements at crest and heel for WRWL and SRWL in the time domain with the Taft Record acceleration for a massless foundation. Results are similar to those with the non-zero mass foundation model. Horizontal displacements at crest and heel both follow the complete system displacements for both water levels. Oscillation amplitudes are greater between $t = 4$ and $t = 14$ s with maximal amplitude in the order of magnitude of 10 cm at crest and 10 μm at heel. Relative displacements between crest and heel are smaller considering SRWL, particularly at the end of simulation. The main difference with the non-zero mass model lies in the fact that the massless foundation model shows *no residual horizontal displacement* of the dam base at the end of simulation.

Figure 8 shows computed horizontal accelerations at crest and heel for WRWL and SRWL in the time domain with the Taft Record acceleration for a massless foundation. The main difference with regards to the non-zero mass foundation is a smaller attenuation of the oscillation amplitudes in the time domain and greater

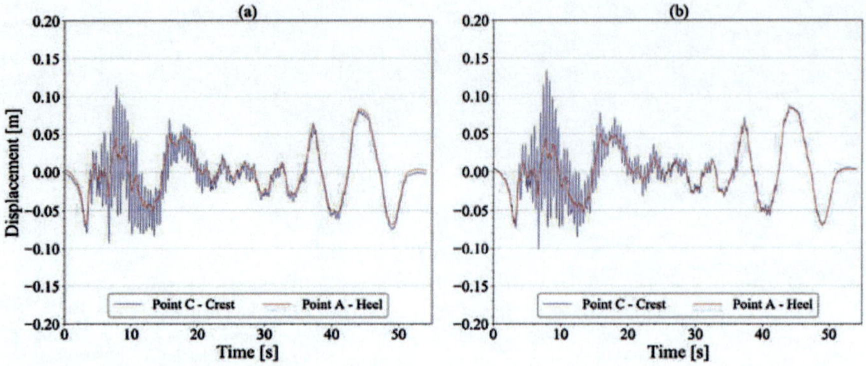

Fig. 7 Computed horizontal displacements at crest and heel of the dam; **a** WRWL, **b** SRWL; Taft Record acceleration; Massless foundation; linear analysis

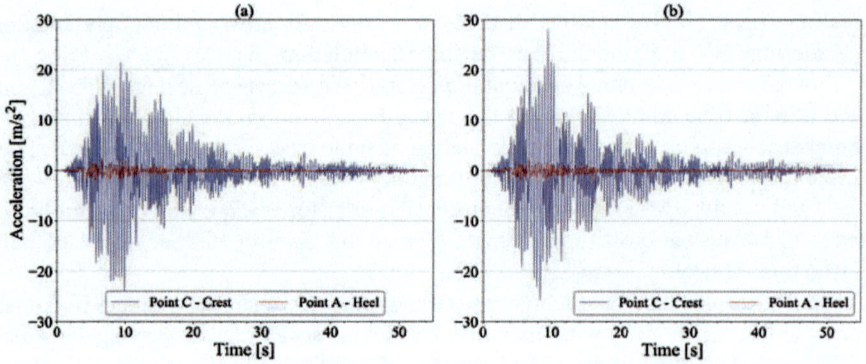

Fig. 8 Computed horizontal acceleration at crest and heel of the dam; **a** WRWL, **b** SRWL; Taft Record acceleration; massless foundation; linear analysis

maximal horizontal accelerations. Results show that maximal horizontal accelerations for WRWL vary between 21.50 and −24.40 m/s^2 at crest (approximately twice the gravity) and between 2.61 and −1.96 m/s^2 at heel. Regarding SRWL, results show that a higher reservoir water level induces greater maximal computed accelerations, with maximal horizontal accelerations varying between 28.24 and −25.71 m/s^2 at crest and between 2.33 and −1.99 m/s^2 at heel.

3.2 Nonlinear Analysis

Figure 9 presents the computed horizontal displacements for WRWL in the time domain with the TAFT Record acceleration and the ETAF acceleration for the

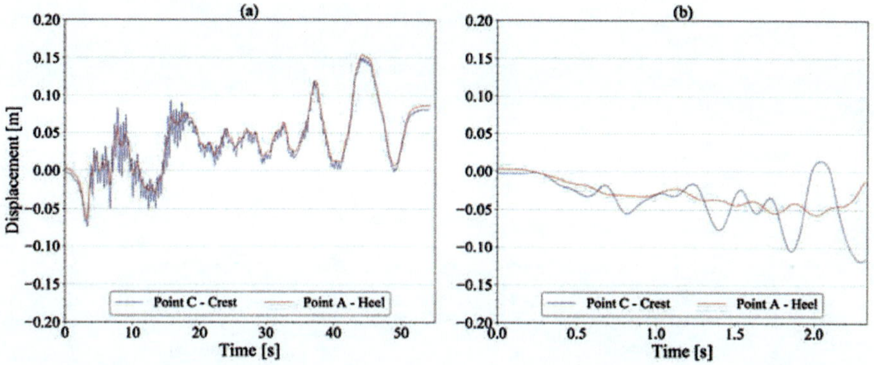

Fig. 9 Computed horizontal displacements at crest and heel of the dam; **a** Taft Record acceleration, **b** ETAF acceleration; nonlinear analysis

nonlinear analysis. No local dam failure ($DMG = 1$) is observed with the Taft Record acceleration, while local dam failure occurs at $t = 2.0$ s and failure of the dam body (subsequent large deformations) occurs at $t = 2.33$ s with the ETAF acceleration.

Computed horizontal displacements with the Taft Record acceleration are similar with regards to linear analysis. Horizontal displacements at crest and heel follow complete system displacements. Oscillation amplitudes are greater between $t = 4$ and $t = 14$ s. A residual horizontal displacement of the dam base of approximately 8 cm is observed in the downstream direction at the end of simulation. The main difference compared to linear analyses lies in the lower relative displacements between crest and heel. The amplitude of the maximal oscillations around complete system displacements is in the order of magnitude of 4 cm at crest and 1 μm at heel.

Regarding the nonlinear analysis with the ETAF acceleration, complete system displacements in the upstream direction are observed. Relative displacements between crest and heel increase with time, with a significant rise at local dam failure ($t = 2.0$ s). At failure of the dam body ($t = 2.33$ s), a residual displacement of −11.5 cm at crest and −1.2 cm at heel (in the upstream direction) is seen.

Figure 10 shows the computed horizontal accelerations for WRWL in the time domain considering both Taft Record and ETAF accelerations for the nonlinear analysis. Computed horizontal accelerations considering the Taft Record acceleration have a similar behavior with regards to linear analyses. Results show that accelerations at heel approximately match the free-surface Taft Record while horizontal accelerations at crest are substantially greater (amplified), mostly between $t = 4$ and $t = 14$ s. The main difference lies in the maximal horizontal accelerations at crest, which are much smaller compared to linear analyses, with values varying between 11.53 and −11.31 m/s^2 (approximately half of the computed accelerations with linear analyses).

Results for the nonlinear analysis with the ETAF acceleration show increasing accelerations oscillation, with higher amplitude at crest and at heel of the dam

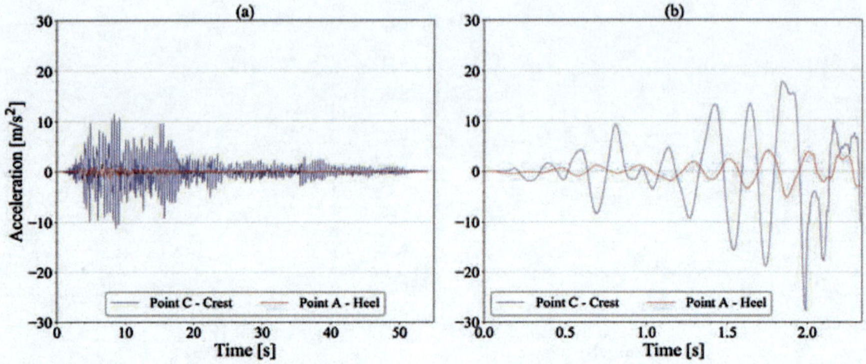

Fig. 10 Computed horizontal acceleration at crest and heel of the dam; **a** Taft Record acceleration, **b** ETAF acceleration; nonlinear analysis

compared to the analysis with the Taft Record acceleration. It is also seen that accelerations at crest and heel have behaviors which may seem to be in phase with peak accelerations towards opposite directions between crest and heel. At local dam failure ($t = 2.0$ s), crest and heel accelerations behavior changes drastically until failure of the dam body.

Figure 11 shows the damage index evolution in the time domain of the dam body ($DMGI_A$) and at the base ($DMGI_B$) for the nonlinear analysis. The damage index is a ratio between the damaged and undamaged area or base length. Results show that damage develops from $t = 4$ s in the dam body and at the base and increases until $DMGI_A = 0.16$ and $DMGI_B = 0.19$. Regarding the ETAF acceleration, results show that damage starts to develop from $t = 1.1$ s and reaches $DMGI_A = 0.47$ and $DMGI_B = 0.41$ at failure of the dam body.

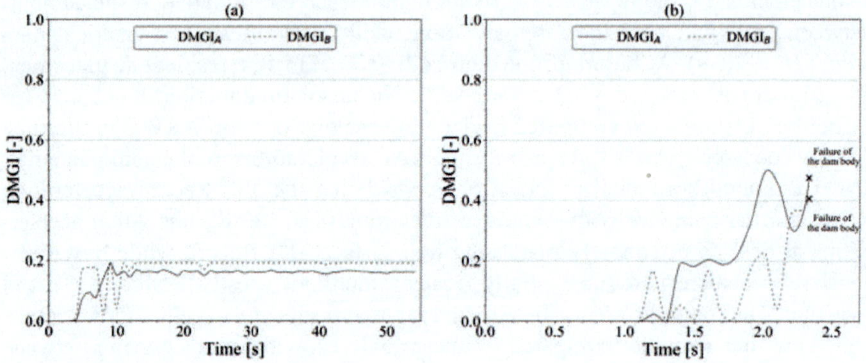

Fig. 11 Damage Index evolution of the dam body and at the base; **a** Taft Record acceleration, **b** ETAF acceleration

For both solicitations, results show oscillations of the damage index $DMGI$, illustrating the ability of the damage-plasticity model to represent cyclic loading conditions with recovery of the stiffness lost during cracking in the transition from tension to compression state. Only damage due to stress peaks in tension state is observed for both TAFT and ETAF accelerations. No damage in compression state occurs.

Figure 12 shows the maximal damage evolution in the time domain of the dam body ($DMG_{\max,A}$) and at the base ($DMG_{\max,B}$) for the nonlinear analysis. As mentioned before, results show that no local dam failure or failure of the dam body occur with the Taft Record acceleration. Regarding the ETAF acceleration, a local dam failure is observed in the dam body at $t = 2.0$ s, explaining the behavior drastic change observed for previous computed accelerations and displacements. Failure of the dam body occurs at $t = 2.33$ s.

Figure 13 shows the computed damage parameter DMG at the end of simulation with the Taft Record acceleration and at failure of the dam body with the ETAF

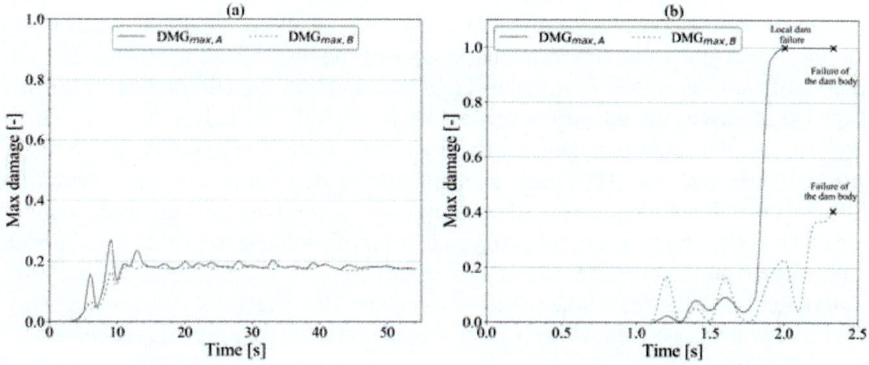

Fig. 12 Maximal damage evolution of the dam body and at the base; **a** Taft Record acceleration, **b** ETAF acceleration

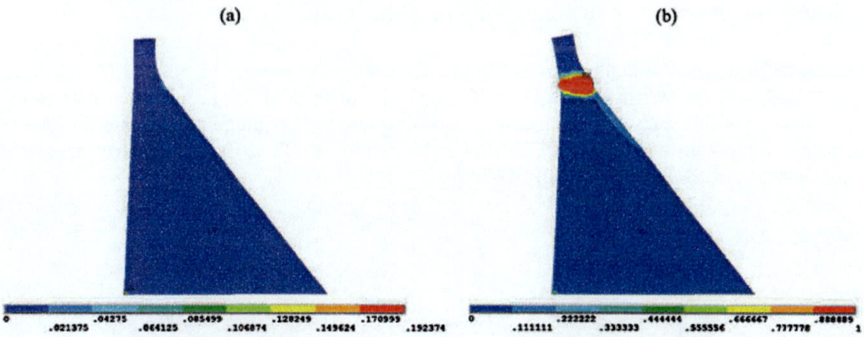

Fig. 13 Computed damage parameter DMG; **a** Taft Record acceleration, **b** ETAF acceleration

acceleration. Regarding the Taft Record acceleration, most damaged elements are located at the heel and toe of the dam ($DMG_{\max,B} = 0.05$) and at the base of the crown on the downstream face of the dam ($DMG_{\max,A} = 0.06$). Regarding the ETAF acceleration, completely damaged elements are located at the base of the crown where local dam failure occurs.

Elements with the highest computed damage at the base are similarly located at the heel and toe of the dam. No local dam failure is observed at the base of the dam, as the maximal damage is $DMG_{\max,B} = 0.58$. Failure of the dam body is observed as subsequent large deformations develop on completely damaged elements, across the width of the dam from the upstream face to the downstream face.

4 Conclusion

The study presents a seismic analysis of Pine Flat Dam considering different dynamic loads and modelling approaches. Several numerical models of the tallest non-overflow monolith consider the effect of fluid-structure interaction for both linear and nonlinear analysis using accelerations records of the historical Taft earthquake and of an artificially designed Endurance Time Acceleration Function (ETAF). Results show the ability of the numerical models to: (i) reproduce adequately the dynamic properties of the reservoir-dam-foundation system and (ii) conduct dynamic linear and nonlinear analysis. The linear analysis allows describing and quantifying the different behavior of the reservoir-dam-foundation system for non-zero mass foundation and massless foundation, for two typical reservoir water levels. Whilst the former is physically more complete, it leads to an effective horizontal displacement after the dynamic loading (the whole model is somewhat deformed towards downstream), whereas the massless foundation model shows no residual horizontal displacement of the dam base. The nonlinear analysis investigates the extent of damage in dam body using a damage-plasticity model based on microplane formulation. Results show the ability of the model to represent cyclic loading conditions, with recovery of the stiffness lost during cracking in the transition from tension to compression state, and subsequent failure of the dam body near the crest.

Acknowledgements The authors would like to acknowledge Dr. Thomas Menouillard (CADFEM, Switzerland) and Dr. Mário Pimentel (FEUP University of Porto, Portugal) for their help in the concretization of this study, and the financial support of the Swiss Competence Center for Energy Research - Supply of Electricity (SCCER-SoE, under contract CTI/2013/0288) and of Innosuisse (under contract FLEXTSOR 17902.3PFIW-IW).

References

1. Calayir Y, Karaton M (2005) A: continuum damage concrete model for earthquake analysis of concrete gravity dam-reservoir systems. Soil Dyn Earthq Eng 25(11):857–869
2. Brühwiler E, Wittmann F (1990) Failure of dam concrete subjected to seismic loading conditions. Eng Fract Mech 35(11/2/3):565–571
3. ANSYS (2009) Theory reference for the mechanical APDL and mechanical applications. Southpointe, Canonsburg, PA, p 15137
4. Chopra AK (2012) Dynamic of structures: theory and applications to earthquake engineering, 4th edn. Pearson Education Inc., Boston
5. Fenves G, Chopra AK (1983) Effects of reservoir bottom absorption on earthquake response of concrete gravity dams. Earthq Eng Struct Dyn 11:809–829
6. Fenves G, Chopra AK (1984) Earthquake analysis of concrete gravity dams including reservoir bottom absorption and dam-water-foundation rock interaction. Earthq Eng Struct Dyn 12(12):663–680
7. Khosravi S, Heydari M (2015) Design and modal analysis of gravity dams by Ansys parametric design language. Walailak J Sci Technol (WJST) 12(12):167–180
8. Liu Y, Chen D (2013) Earthquake response analysis of gravity dam considering the radiation damping of infinite foundation. APCOM & ISCM, Singapore
9. Zreid I, Kaliske M (2018) A gradient enhanced plasticity-damage microplane model for concrete. Comput Mech 62(15)
10. Bazant Z, Bambarova P (1984) Crack shear in concrete: crack band microplane model. J Struct Eng 10(19):201–2035
11. Bazant Z, Oh B (1985) Microplane model for progressive fracture of concrete and rock. J Eng Mech 11(115):559–582

Earthquake Soil Structure Interaction Analysis of a Gravity Dam

H. Yang, H. Wang, J. W. Salamon, and B. Jeremic

Abstract This paper presents numerical modeling and simulation issues seen in static and dynamic analysis of dam-foundation systems. Domain reduction method (DRM) is chosen as the preferred approach for seismic motion input. Proper modeling of inelastic material behavior and seismic energy dissipation is discussed. Verification and validation are essential to ensure reliability of numerical results. Numerical modeling and simulation of a soil/rock—concrete dam system subjected to static and dynamic earthquake loading are presented. It is shown that, in order to make safe and economical design decisions, engineers should have complete understanding and control of the numerical analysis process.

Keywords Earthquake soil structure interaction · Dam modeling · Nonlinear finite element · Seismic energy dissipation · Verification and validation

1 Introduction

Credibility of numerical analysis results relies to a great extent on a number of numerical modeling and numerical simulation issues. Those issues need to be properly addressed and resolved so that Earthquake, Soil/Rock, Structure Interaction (ESSI) analysis for dams produce rational, reliable results that can be used for design and assessment. Some of those issues are briefly presented in this short paper. More detailed account of modeling and simulation of ESSI behavior of a soil-structure systems is presented by Jeremić et al. [1].

H. Yang · H. Wang · B. Jeremic (✉)
University of California, Davis, CA 95616, USA
e-mail: jeremic@ucdavis.edu

J. W. Salamon
United States Bureau of Reclamation, Denver, CO 80225, USA

G. Bolzon et al. (eds.), *Numerical Analysis of Dams*, Lecture Notes in Civil Engineering 91, https://doi.org/10.1007/978-3-030-51085-5_20

2 Modeling and Simulation of Earthquake Soil Structure Interaction for Concrete Dams

A number of modeling and simulation details, issues, control credibility, accuracy of numerical results for seismic analysis of soil/rock—structure/dam system. Few of those details, issues that are relevant to modeling of earthquake, soil/rock, structure interaction for concrete dams are described below. Particular concrete dam, Pine Flat dam on Kings river in California was used for modeling. This dam has been studied earlier [2, 3], and has significant amount of data about dynamic behavior collected during a number of earthquakes.

In this section, some of the more important issues related to credible and accurate modeling and simulation of seismic behavior of soil/rock-dam systems are presented. Next section present select results for Pine Flat dam modeling, emphasizing issues with nonlinear/inelastic modeling.

2.1 *Seismic Input*

Input of earthquake motions to soil/rock, dam, reservoir models is a very important component of ESSI modeling. Earthquake motions almost always consist of three translational and three rotational components and for tall and stiff structures, like dams, it is important to be able to input all components properly into the soil/rock—structure model. Three translational and three rotational motions are measured in the field [4], yet they are rarely used in modeling [5–8]. In addition, radiation damping, propagation of waves away from the soil/rock—dam system needs to be modeled as well. Arriving/incoming seismic waves will reflect from the free surface, from dam/structural foundations, and travel back into the half space. Moreover, dam structure will oscillate due to earthquake excitations, and will radiate waves into the half space as well. All of these dynamic effects need to be modeled as they can influence dynamic response results.

Seismic input and radiation damping are best modeled using the Domain Reduction Method (DRM) developed by Bielak and coworkers [9, 10]. The DRM uses free field seismic wave field, for developing the so-called effective forces, given by Eq. 1:

$$P^{eff} = \begin{Bmatrix} P_i^{eff} \\ P_b^{eff} \\ P_e^{eff} \end{Bmatrix} = \begin{Bmatrix} 0 \\ -M_{be}^{\Omega+} \ddot{u}_e^{0} - K_{be}^{\Omega+} u_e^{0} \\ M_{eb}^{\Omega+} \ddot{u}_b^{0} + K_{eb}^{\Omega+} u_b^{0} \end{Bmatrix}. \quad (1)$$

Few important observations are in order:

- Effective forces are a consistent dynamic replacement for the seismic forces at the earthquake source;

- Single layer of finite elements, encompassing soil/rock—dam system, is loaded with time varying effective forces;
- Soil/rock—dam system inside the single layer of finite elements can be nonlinear/inelastic;
- Only residual waves are outgoing from the soil/rock—dam system, and can be effectively dumped out;
- All types of earthquake wave fields are modeled, including body (P, SV, SH) and surface waves (Rayleigh, Love, etc.).

The Domain Reduction Method requires availability of the free field earthquake wave field, at the surface and at depth. Free field earthquake wave field can be obtained in a number of different ways. For example, for one component (1C) wave field, deconvolution of surface earthquake motions can be performed. Full three component (3C) wave field can be obtained through deconvolution of each component of surface recorded earthquake motions. However, this approach has to be used with care as it produces the so-called $3 \times 1C$ wave field, essentially a 3 translational components of the earthquake wave field, comprised of body waves only, without any surface waves. Alternatively, analytic solutions for inclined plane waves [11, 12] or regional scale models [13–15] can be used to create earthquake wave fields that are then used to develop effective forces for the DRM.

2.2 *Inelastic Behavior of Materials and Seismic Energy Dissipation*

Inelastic/nonlinear behavior of materials is frequently present and consists of inelasticity of rock and soil under the dam, inelasticity of interfaces/contacts between dam and foundations, within rock mass surrounding dam site, within concrete dam joints, and within concrete and steel of dam structure and components. Modeling of inelastic behavior of materials is important as it represents the main energy dissipation mechanism for seismic excitations, and significantly contributes to the damage of structures. A number of inelastic, nonlinear, elastic-plastic material models do exist for soil, rock, interfaces/contacts/joints, concrete and steel behavior [1].

Inelastic response of soil/rock—dam/structure systems frequently results in reduction of some components of accelerations and increase in some components of displacements within the dam structure and components. Above observation of reduction of some components of accelerations and increase in some components of displacements is not universal and each soil/rock—dam/structure system has to be analyzed in order to assess potential benefits and detriments of full nonlinear ESSI response.

Inelastic, nonlinear, elastic-plastic behavior of all the components of soil/rock—structure/dam system results in energy dissipation, conversion from mechanical, seismic waves to other energy forms, heat, sound. etc. Modeling of energy dissipation is important as it provides engineers information about locations and timing of energy dissipation within soil/rock—dam/structure system. Energy dissipation within inelastic/elastic-plastic soil/rock—dam/structure system has to be properly modeled so that results are credible and principles of thermodynamics are observed. For example, it is noted that in general plastic energy dissipation is different from plastic work, as discovered experimentally by Farren and Taylor [16] and explained theoretically by Taylor and Quinney [17]. There exist a number of publications that do not distinguish between plastic work and plastic dissipation, leading to violation of the second law of thermodynamics. Based on earlier work by [17], recently Yang et al. [18–21] developed theory and simulation methodology for analyzing plastic energy dissipation within soil/rock—dam/structure system. Accurate evaluation of energy dissipation during earthquakes can be used by engineers during design and assessment to control energy dissipation locations and timing, thus improving designs and retrofits.

2.3 *Verification and Validation*

Computer programs that are used for static and dynamic analysis of ESSI behavior of dams need to develop appropriate, significant Quality Assurance (QA) components. Quality Assurance is important, in the form of Verification and Validation (V&V), as confidence in, credibility of, and accuracy of numerical results depend significantly on proper QA and V&V activities [22–26]. Verification is a process of determining that a model implementation accurately represents the developer's conceptual description and specification. Verification is a mathematics issue.

Verification provides evidence that the model is solved correctly. Validation is a process of determining the degree to which a model is an accurate representation of the real world, mechanics, from the perspective of the intended uses of the model. Validation is a physics issue. Validation provides evidence that the correct model is solved. Every component of the numerical analysis program that is used in ESSI analysis, finite elements, constitutive integration algorithms, static and dynamic solution advancement algorithms, etc., need to be verified against available very accurate solutions. High quality laboratory and/or field test results can then be used to validate modeling.

3 Numerical Modeling of a Soil/Rock-Dam System

Presented here is numerical modeling of a soil/rock—concrete dam system subjected to static and dynamic, earthquake loading. Modeling is for a particular, real dam, the Pine Flat concrete dam in California. While a number of results were developed for static and dynamic behavior of Pine Flat dam, using the Real-ESSI Simulator program (http://real-essi.info/) [27], presented here are only select results that aim to illustrate some of the issues in soil/rock—dam modeling.

It is noted that modeling and simulation of Pine Flat dam was performed both with and without reservoir present. For modeling with reservoir present, Real-ESSI Simulator is coupled with the OpenFOAM fluid dynamics program [28, 29].

3.1 Soil/Rock—Dam Numerical Model

The mesh for foundation and dam, is shown in Fig. 1. Full 3D mesh was used, using 8 node brick finite elements, however displacements out plane were restricted so that the model was effectively a 2D, plane strain finite element model. Material properties for linear elastic modeling were provided by workshop organizers. In addition to linear elastic modeling of the soil/rock—dam system, limited inelastic/nonlinear modeling was also performed in order to illustrate some interesting behavior. For inelastic/nonlinear modeling of interface/contact between concrete dam and the foundation rock, material properties specified in NAVFAC [30] and Jeremić et al. [1] were used. In addition, since viscous damping used in the analysis was defined by the workshop organizers, a variation of numerical, algorithmic damping was used to illustrate the important influence this damping can have on dynamic response of the dam. Seismic motions defined by the workshop organizers, that were comprised of Taft earthquake, impulse motions were used for modeling. Components of surface motions were deconvoluted to a depth and then propagated upward using DRM seismic input into Real-ESSI Simulator.

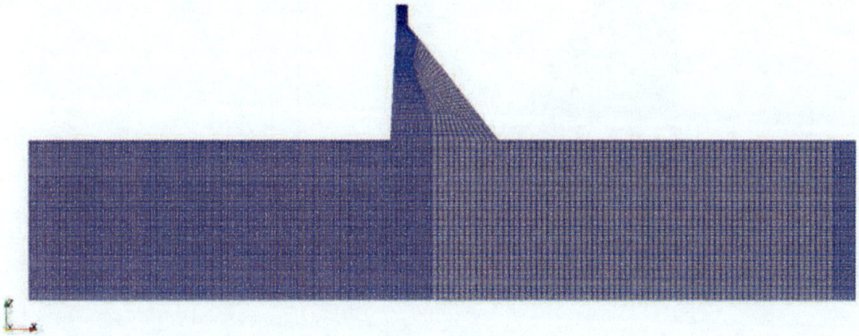

Fig. 1 Pine Flat Dam model mesh without reservoir

3.2 Analysis Results: Static Analysis

Two stages of loading were used. First stage is dam and foundation rock self-weight, the second stage is water pressure on the side of dam and on the bottom of reservoir. While load staging is not important in linear elastic analysis, as principle of superposition holds, for nonlinear/inelastic analysis, load stages are essential.

Mesh refinement effects. It is always beneficial to assess the effects of mesh size on results. A simple mesh refinement study was performed in order to assess these effects. Mesh was refined 4 times in each direction to create 16 times more finite elements. Figure 2 shows refined finite element mesh. Static self-weight analysis was performed and differences in results are presented in Table 1.

Differences are small and it is noted that the largest difference is for vertical displacements at the dam heel, with the differences of 0.001 m. It is also noted that results with refined mesh are softer, displacements are larger, which is a good

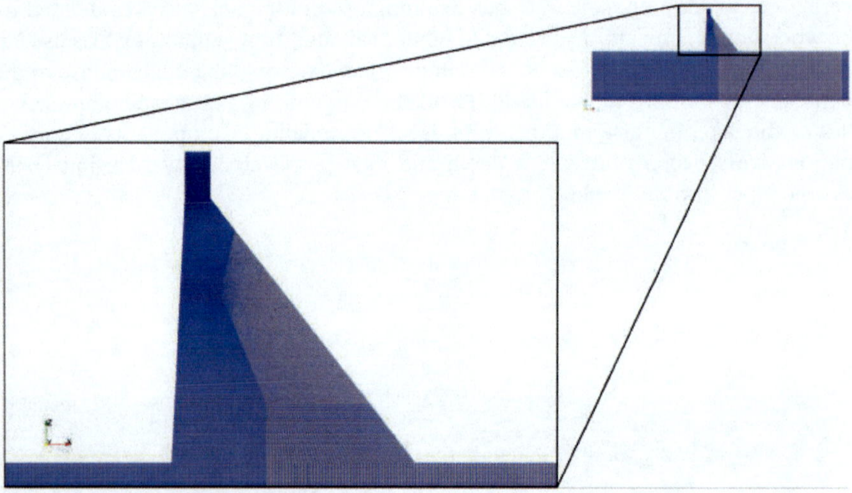

Fig. 2 Model with refined mesh

Table 1 Differences in static displacements due to self-weight, at the dam heel, for the original and refined finite element mesh

		Displacement (m)		
		Original	Refined	Difference (%)
Dam top	Horizontal	0.0121	0.0121	0.0
	Vertical	−0.0095	−0.0098	3.0
Dam heel	Horizontal	0.0031	0.0033	6.0
	Vertical	−0.0060	−0.0070	14.3

Table 2 Case A: displacements for two stiffness cases

		Displacement (m)		
		Dam top	Dam heel	Relative
Material properties I (Softer Foundation)	Horizontal	0.012	0.003	0.009
	Vertical	−0.009	−0.006	−0.003
Material properties II (Stiffer Foundation)	Horizontal	0.010	0.001	0.009
	Vertical	−0.005	−0.002	−0.003

indication of convergence of displacement based finite elements toward accurate solution. Based on these results, it was decided to use the original mesh for all subsequent analysis.

Displacement Results. Relative displacement between dam top and heel are shown in Table 2, for different material properties, as prescribed by workshop organizers.

3.3 Analysis Results: Dynamic Analysis, Taft Earthquake

Analysis results for vertical stress the dam heel, for a case of Tuft earthquake record, without the presence of a reservoir, are shown in Fig. 3. It is noted that there are tensile stresses at dam heel, at some time instances, even without reservoir loading.

Figure 4 shows distribution of vertical stress at the dam-foundation interface for case of maximum tension at the dam heel, at $t = 8.75$ s (Fig. 4a) and case of maximum compression at the dam heel, at $t = 8.98$ s (Fig. 4b). All of the presented maximum and minimum interface stress cases are for the linear elastic, bonded interface model where there is no possibility of gap opening when tensile stress occurs.

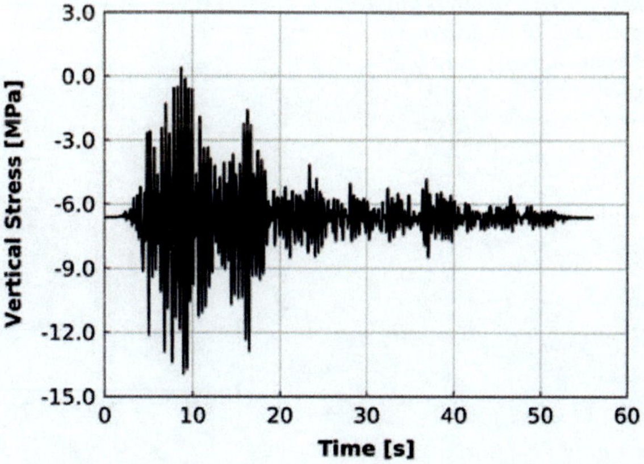

Fig. 3 Vertical stress time history at the dam heel, no reservoir

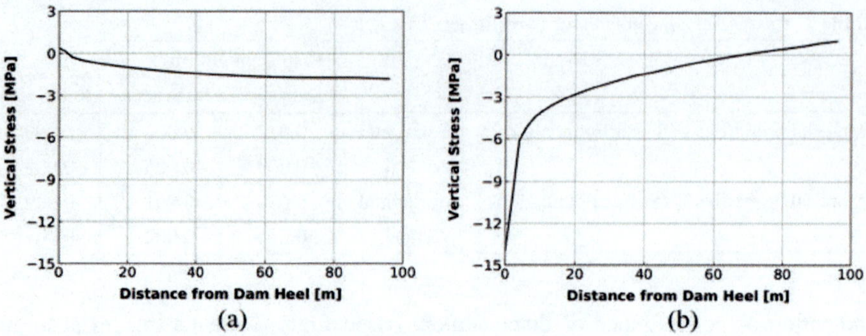

Fig. 4 Distribution of vertical stress at the dam-foundation interface, no reservoir. **a** Maximum tension at the dam heel; **b** Maximum compression at the dam heel

3.4 *Numerical Damping Effects*

Previous dynamic results were obtained using very small numerical/algorithmic damping, using Newmark time integration algorithm [31], with Newmark coefficient $\gamma = 0.505$ [32, 33]. It is common to use numerical/algorithmic damping in dynamic finite element simulations, in order to damp out higher frequencies that are introduced through finite element discretization process, and are not physical [32].

However, numerical/algorithmic damping has to be introduced carefully as it can also affect overall results, particularly lower frequencies that it was not intended for. For example, Fig. 5 shows the same vertical stress response at the dam heel as in Fig. 3, with different values of numerical/algorithmic (non-physical) damping. Stress response is quite different, and for higher values of numerical damping, tensile stress at the heel disappears altogether.

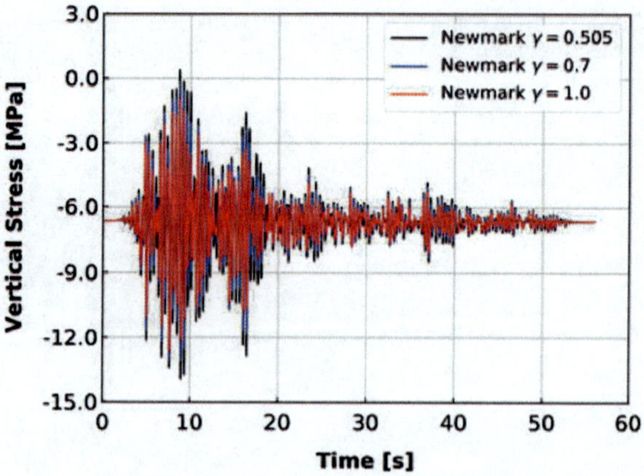

Fig. 5 Vertical stress at the dam heel when different Newmark coefficient γ is used

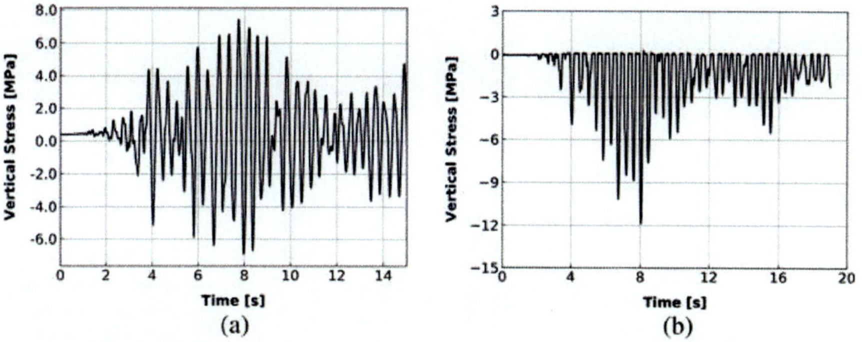

Fig. 6 Vertical stress time history at the dam heel for Taft earthquake, hydrostatic pressure. **a** Fully bonded contact; **b** Interface/contact gap allowed to open and close

3.5 *Analysis Results: Dynamic Analysis, Taft Earthquake, Hydrostatic Pressure, Soft Inelastic Contact/Interface*

Results of fully coupled solid-fluid analysis do not differ too much from dynamic analysis with hydrostatic pressure, while requiring much longer run times, for this particular dam and earthquake loads It was thus decided that for a nonlinear/inelastic interface/contact analysis with Taft earthquake, hydrostatic pressure from reservoir be used. This analysis allows opening and closing of gaps at the dam-foundation interface. The main goal is to explore nonlinear/inelastic behavior effects on dam response, and in particular vertical stress at the dam heel. Figure 6 shows vertical stress results for dam heel, with bonded, linear elastic contact, and with inelastic, gap open/close interface.

One of the main observations is that maximum compressive stress significantly increases at the dam heel, while there are prolonged time instances where gap is opened, potentially allowing water to infiltrate a gap. It is noted that water infiltration/movement was not modeled in this preliminary analysis.

In addition to significant differences in results with and without allowing inelastic, gap open/close interface effects, numerical damping also significantly influence stress results at the damp heel. Figure 7 shows time histories of vertical stress at dam heel as a function of numerical/algorithmic damping as influenced by Newmark time integration coefficient γ. It is obvious that numerical damping can significantly change the response of a dam-foundation system, and as such deserves special attention during numerical modeling.

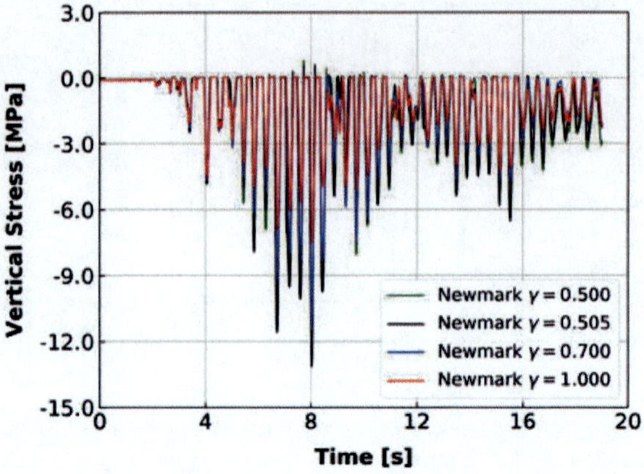

Fig. 7 Vertical stress at the dam heel as a function of Newmark coefficient γ

4 Summary

Presented in this paper were few modeling and simulation issues for static and dynamic analysis of dam-foundation systems. While a full set of results for ICOLD workshop will be presented at the conference, a small set of numerical modeling and simulation issues that can and will affect results need to be appreciated. Practicing engineers, working on designs and assessment of dams, need to be able to have a complete control of the numerical analysis process, so that their decisions are safe and economical.

References

1. Jeremić B, Yang Z, Cheng Z, Jie G, Tafazzoli N, Preisig M, Tasiopoulou P, Pisanò F, Abell J, Watanabe K, Feng Y, Sinha SK, Behbehani F, Yang H, Wang H (1989–Present) Nonlinear finite elements: modeling and simulation of earthquakes, soils, structures and their interaction. Self-Published, University of California, Davis, CA, USA, and Lawrence Berkeley National Laboratory, Berkeley, CA, USA). ISBN 978-0-692-19875-9
2. Chavez JW, Fenves GL (1993) Earthquake analysis and response of concrete gravity dams including base sliding. Tech Rep 93/07, EERC/UCB
3. Fenves G, Chopra AK (1984) EAGD-84 a computer program for earthquake analysis concrete gravity dams. Tech Rep 84/11, EERC/UCB
4. Oliveira CS, Bolt BA (1989) Rotational components of strong surface ground motions. Earthq Eng Struct Dyn 18:517–526
5. Nazarov YP, Poznyak E, Filimonov AV (2015) A brief theory and computing of seismic ground rotations for structural analyses. Soil Dyn Earthq Eng 71:31–41
6. Tomasin M, Domaneschi M, Guerini C, Martinelli L, Perotti F (2018) A comprehensive approach to small and large-scale effects of earthquake motion variability. Comput Struct 207:155–170
7. Bońkowski PA, Zembaty Z, Minch MY (2019) Engineering analysis of strong ground rocking and its effect on tall structures. Soil Dynamics and Earthquake Engineering 116:358–370

8. Vicencio F, Alexander NA (2019) A parametric study on the effect of rotational ground motions on building structural responses. Soil Dyn Earthq Eng 118:191–206
9. Bielak J, Loukakis K, Hisada Y, Yoshimura C (2003) Domain reduction method for three-dimensional earthquake modeling in localized regions. Part I: theory. Bull Seismol Soc Am 93(2):817–824
10. Yoshimura C, Bielak J, Hisada Y (2003) Domain reduction method for three–dimensional earthquake modeling in localized regions. Part II: verification and examples. Bull Seismol Soc Am 93(2):825–840
11. Thomson WT (1950) Transmission of elastic waves through a stratified solid medium. J Appl Phys 21(2):89–93
12. Haskell NA (1953) The dispersion of surface waves on multilayered media. Bull Seismol Soc Am 43(1):17–34
13. Rodgers A, Petersson NA, Nilsson S, Sjögreen B, McCandless K (2008) Broadband waveform modeling of moderate earthquakes in the San Francisco bay area and preliminary assessment of the USGS 3D seismic velocity model. Bull Seismol Soc Am 98(2):969–988
14. Petersson NA, Sjögreen B (2018) High order accurate finite difference modeling of seismo-acoustic wave propagation in a moving atmosphere and a heterogeneous earth model coupled across a realistic topography. J Sci Comput (14):290–323
15. Abell JA, Orbović N, McCallen DB, Jeremić B (2018) Earthquake soil structure interaction of nuclear power plants, differences in response to 3-D, 3 × 1-D, and 1-D excitations. Earthq Eng Struct Dyn 47(6):1478–1495
16. Farren W, Taylor G (1925) The heat developed during plastic extension of metals. Proc R Soc London A Math Phys Eng Sci 107(743):422–451
17. Taylor GI, Quinney H (1934) The latent energy remaining in a metal after cold working. Proc R Soc London A Ser Contain Papers Math Phys Character 143(849):307–326
18. Yang H, Sinha SK, Feng Y, McCallen DB, Jeremić B (2018) Energy dissipation analysis of elastic-plastic materials. Comput Methods Appl Mech Eng 331:309–326
19. Yang H, Feng Y, Wang H, Jeremić B (2019) Energy dissipation analysis for inelastic reinforced concrete and steel beam-columns. Eng Struct 197:109431
20. Yang H, Wang H, Feng Y, Jeremić B (2020) Plastic energy dissipation in pressure-dependent materials. ASCE J Eng Mech 146(3):04019139
21. Yang H, Wang H, Feng Y, Wang F, Jeremić B (2019) Energy dissipation in solids due to material inelasticity, viscous coupling, and algorithmic damping. ASCE J Eng Mech 145(9):04019060
22. Roache PJ (1998) Verification and Validation in Computational Science and Engineering. Hermosa Publishers, Albuquerque, New Mexico
23. Oberkampf WL, Trucano TG, Hirsch C (2002) Verification, validation and predictive capability in computational engineering and physics. In: Proceedings of the foundations for verification and validation on the 21st century workshop, pp 1–74. Johns Hopkins University/Applied Physics Laboratory, Laurel, Maryland
24. Oden JT, Babuška I, Nobile F, Feng Y, Tempone R (2005) Theory and methodology for estimation and control of errors due to modeling, approximation, and uncertainty. Comput Methods Appl Mech Eng 194(2–5):195–204
25. Babuška I, Oden JT (2004) Verification and validation in computational engineering and science: basic concepts. Comput Methods Appl Mech Eng 193(36–38):4057–4066
26. Roy CJ, Oberkampf WL (2011) A comprehensive framework for verification, validation, and uncertainty quantification in scientific computing. Comput Methods Appl Mech Eng 200(25–28):2131–2144
27. Jeremić B, Yang Z, Cheng Z, Jie G, Tafazzoli N, Preisig M, Tasiopoulou P, Pisanò F, Abell J, Watanabe K, Feng Y, Sinha SK, Behbehani F, Yang H, Wang H (1989–2020) The real-ESSI simulator system. University of California, Davis and Lawrence Berkeley National Laboratory. http://real-essi.us/
28. Weller H, Greenshields C, Santos B, Bainbridge W, Janssens M OpenFOAM. The OpenFOAM Foundation. https://openfoam.org/

29. Weller HG, Tabor G, Jasak H, Fureby C (1998) A tensorial approach to computational continuum mechanics using object-oriented techniques. Comput Phys 12(6):620–631
30. NAVFAC. Foundations & Earth Structures, DESIGN MANUAL 7.02 (1986) Naval Facilities Engineering Command, 200 Stovall Street, Alexandria, Virginia, 22332–2300, USA
31. Newmark NM (1959) A method of computation for structural dynamics. ASCE J Eng Mech Div 85:67–94
32. Argyris J, Mlejnek H-Pn (1991) Dynamics of structures. North Holland Elsevier
33. Hughes T (1987) The finite element method; linear static and dynamic finite element analysis. Prentice Hall, Upper Saddle River

Dynamic Foundation-Fluid-Structure Interaction of a Concrete Gravity Dam: Influence of Input Signal Frequency Content and of Water Modelling on Wave Propagations—Example of Pine Flat Dam

V. Mouy and X. Molin

Abstract This paper presents a contribution to the analysis of the Pine flat concrete dam, for the 15th ICOLD benchmark. Models developed and main results are presented, as well as additional developments, regarding the influence of some factors on the response of the model to the high frequency impulse and the low frequency impulse defined by the formulators. In order to investigate this frequential behaviour, frequency response analysis is performed. The effect of mesh size, of compressibility of water, and of reflection factor of water in the reservoir is investigated.

Keywords Concrete dam · Numerical modelling · Perfectly match layer (PML) · Fluid-structure interaction · Impulse response analysis · Modal analysis

1 Introduction

This article aims to present a contribution for the 15th ICOLD International Benchmark Workshop on Numerical Analysis of Dams. The first part presents numerical models that have been performed (mesh, material properties, boundaries, loads). The second part presents impulse load response, in order to assess the behaviour of the model regarding frequency content of the signal. In the third part the influence of water on model behavior is analysed.

The following calculations have been performed:

- Modal analysis (cases A1 and A2),
- EMVG simulations, i.e. simulation of and excited mass on the crest of the dam (cases A3 and A4, not presented in this paper),
- Impulse response, with 2 different horizontal signals applied at the bottom of the model (one high frequency signal, and one low frequency signal): cases B1, B2, C1, C2, C3, and C4 (with or without dam, and with or without water),

V. Mouy (✉) · X. Molin
Tractebel Engineering France, Gennevilliers, France
e-mail: vincent.mouy@tractebel.engie.com

G. Bolzon et al. (eds.), *Numerical Analysis of Dams*, Lecture Notes in Civil Engineering 91, https://doi.org/10.1007/978-3-030-51085-5_21

- Response to a seismic signal, with linear (D1, D2, D3) or nonlinear (E1) concrete behavior (not presented in this paper).

2 Presentation of the Model

2.1 Geometry and Mesh

Geometry is in conformity with requirements of the formulator, with a 700 m long model (foundation), PML boundaries excluded.

The model is 2-dimensional, with plain strain elements. Since the mesh acts as a filter and cuts high frequencies, element size has to be adjusted for each simulation on the objectives and on frequency content. Besides, sensitivity analysis on mesh size have been performed:

- For impulse simulations (simulations B and C), as required by formulators, mesh length is of 1.5 m, but calculations were made with larger meshes: 3, 6, 12, and 24 m.
- For other simulations, mesh length is of 2.5 m for the dam and 5 m for other elements.

2.2 Material Properties

Linear properties for ground and concrete are those given by the formulators (Table 1).

For most calculations presented in this paper, no damping has been considered (for response to impulse loads). When considered, damping has been defined with Rayleigh coefficients. Two sets of damping coefficients have been considered: one is similar to the formulators' instructions (below 2% between 5 → 8 Hz), the other is adjusted for higher frequency content, and below 2% in the range 10 → 60 Hz (Fig. 1).

Table 1 Material properties (linear)

Type of text	Concrete	Foundation	Water
Modulus of elasticity (MPa)	22410	22410	–
Density (km/m^3)	2483	2483	1000
Poisson's ratio	0.2	0.2	–
Compressional wave velocity (m/s)	3167	3167	1439
Shear wave velocity (m/s)	1939	1939	

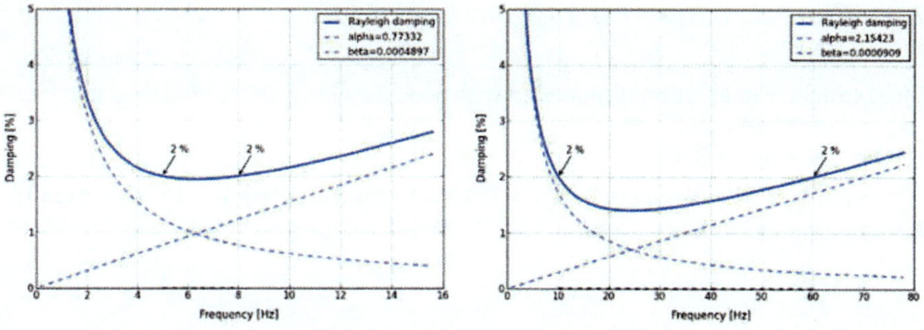

Fig. 1 Rayleigh damping curves

2.3 *Boundaries*

Various kind of boundaries have been considered:

- Massless foundation with fixed boundaries condition. for the modal analysis,
- The use of Perfectly Match Layer elements (PML), that absorbs seismic waves and prevents their reflection, for simulation of excited mass (EMVG), and on lateral borders of other simulations,
- Far-field elements for the bottom of most simulations, since formulator imposed the used of deconvoluted force, which is only valid for far-field elements, and not for PML (Models including PML boundaries require frequency analysis to be made on the basis for force loads, and then a specific deconvolution to be performed; input signal is thus affected by characteristics (thickness, weight…) of PML elements).

These far-field elements have the same characteristics as the foundation, with a 10,000 m thickness (for the calculation of global mass); damping interface have a normal or shear damping coefficient of 7.863 MNs/m. Interface elements don't have any normal or shear stiffness (Fig. 2).

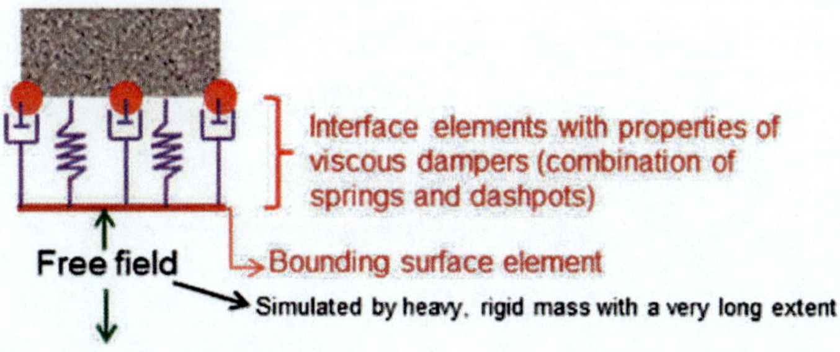

Fig. 2 Principle of far-field boundaries

2.4 Fluid Structure Interaction

Fluid structure interaction is modelled with acoustic elements, having the same mesh size as the foundation.

The fluid is considered incompressible, except in the last paragraph where a comparison has been performed between incompressible and compressible assumptions of fluid behaviour.

2.5 Loads

Only impulse loads are presented in this paragraph. As it has been explained in the "boundaries" paragraph, input signal is entered as a linear force, with the values given by the formulators. Two different signals are given, similar, with a 10 ratio on time (one of them is 2 s long, and the other 20 s long) (Fig. 3).

Table 2 decribes the frequency content of these two signals.

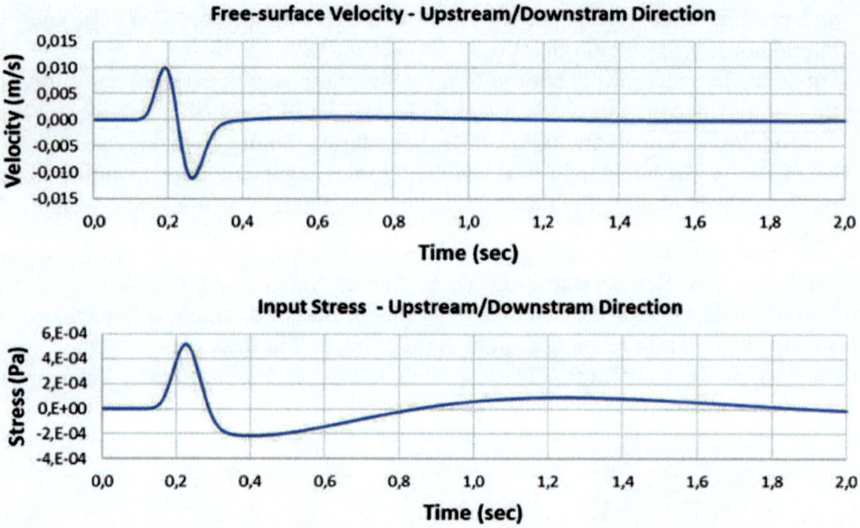

Fig. 3 Velocities and stress for first 2 s of Low freq. impulse

Table 2 Frequency content of low and high frequency impulses

Type of text	Low freq impulse	High freq impulse
Maximum value of the spectrum	4.35 Hz	43.5 Hz
Range over 50% of spectrum	0.93 → 8.54 Hz	9.28 → 85.4 Hz
Range over 20% of spectrum	0.34 → 10.9 Hz	3.42 → 108.9 Hz
Range over 5% of spectrum	0.24 → 13.4 Hz	2.44 → 134.3 Hz

2.6 *Time Steps, Mesh Size, and Foundation Dimensions Regarding Frequencies*

This paper will deal with frequency response related aspects of the model. It is therefore relevant to recall usual recommandations:

- Mesh size: recommendations vary from a maximum dimension of 1/8 of wavelength, down to a maximum dimension of 1/12 of wavelength,
- Shannon theory gives a maximum time step of $\frac{1}{2f}$, where f is the frequency of the signal; but no recommandation dedicated to such calculations is known from the author,
- Foundation size: a usual recommendation for finite element calculations is to have at least 5 times the height of the dam, regardless the signal frequency; besides, the "quarter wavelength approximation" prescribes a depth of at least ¼ of the wavelength of the lowest frequencies, for the computation of site amplification.

Table 3 provides a correspondence between mesh sizes, wavelength, and frequencies.

Table 3 Mesh sizes studied, with corresponding wavelength and frequencies for 1/8 criterion

Mesh size (m)	Corresponding wavelength (m)	Corresponding frequency (Hz)
1.5	12	160
3	24	80
6	48	40
12	96	20
24	192	10

Table 4 Frequencies of 6 first modes (Hz)

Mode	Low level water	Mid-level water	Sup-level water
1	2.30	2.21 (−3.91%)	2.08 (−9.57%)
2	4.11	4.10 (−0.24%)	4.06 (−1.22%)
3	4.86	4.85 (−0.21%)	4.83 (−0.62%)
4	5.53	5.51 (−0.36%)	5.43 (−1.81%)
5 (8th calculated)	6.92	6.92 (0.00%)	6.91 (−0.14%)
6 (10th calculated)	7.45	7.45 (0.00%)	7.44 (−0.13%)

2.7 Modal Analysis

Modal deformations have similar same shapes for all studied cases. The first most relevant modal frequencies, according to an engineering analysis, are reported in Table 4. Indicated percentage of variation is the percentage of variation in comparison with low water level, considered as a reference case. It appears that the 1st mode is strongly influenced by the water level, and that modes 2 and 4 are also slightly affected, others are hardly affected.

3 Impulse Response Analysis (Cases B and C)

3.1 Description of Impulse Response

The points's velocities in Fig. 4 have been post-treated.

In Fig. 5 are presented the velocities at these points for high and low frequency impulses.

In both cases, there is a first impulse, ocurring at the bottom of the model, then a second impulse, approximately 70 ms later, and a third impulse, 70 ms later, at the bottom of the model (return of the wave ofter its reflection on the surface). Theoretically, the amplitude of the surface impulse should be twice the amplitude of the input impulse.

For the high frequency impulse, the duration of the signal is short enough to avoid any interfecrences between the input and reflected signals at the bottom of the model, and these 3 signals can clearly be identified. Also, the amplitude of the signals for points *a* and *b* perfectly fits with the theory.

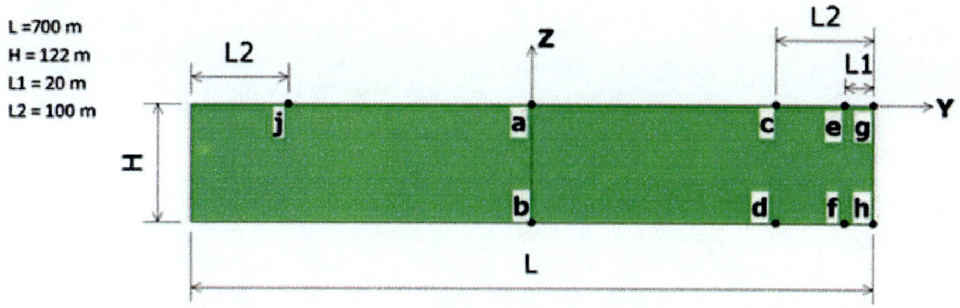

Fig. 4 Position of post-treated points for response analysis simulations

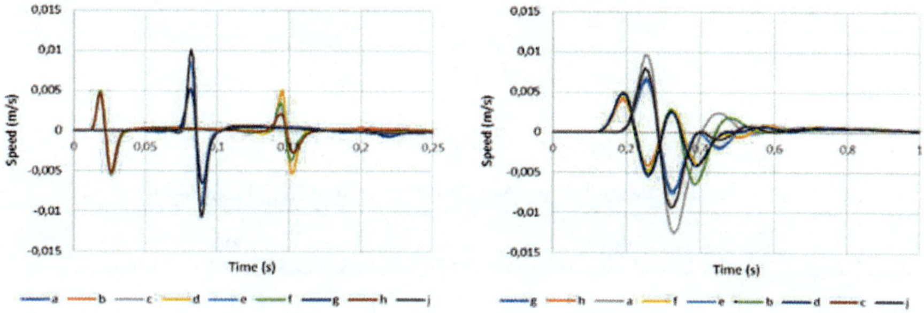

Fig. 5 Velocities at reference points of the benchmark (left: High freq; right: Low freq)

For the low frequency impusle, time between peaks is of approximately 70 ms, so that the max peak at the surface occurs at the same time as the min peak of the input, and so that, also, the arrival of the reflected impulse on the bottom of the model occurs in the continuity of the end if the input impulse. This potential source of interference could explain the inappropriate amplitude of the reflected signal.

The theoretical 2 factor between the amplitude "far from the surface" and the amplitude at the surface is similar to well know results for reflection of stationary waves (such as, for example, the reflection of waves on a vertical wall). The distance on which this amplification can be observed is diretly related with the wavelength, as it is illustrated in Figs. 6 and 7, with:

- Time history velocities of 5 points on the vertical line,
- Curves presenting the maximum and minimum amplitudes of the signals at each level.

It can be confirmed that, for the first input signal, the factor 2 is perfectly obtained for the three simulations; the length of the amplification zone can be estimated of:

- High freq. impulse: 7→10 m (1/4 of wavelength: 11 m),
- Mid freq. impulse: 20→30 m (1/4 of wavelength: 36 m),
- Low freq. impulse: 60→100 m (1/4 of wavelength: 110 m).

Quarter wavelength approximation can thus be justified and illustrated.

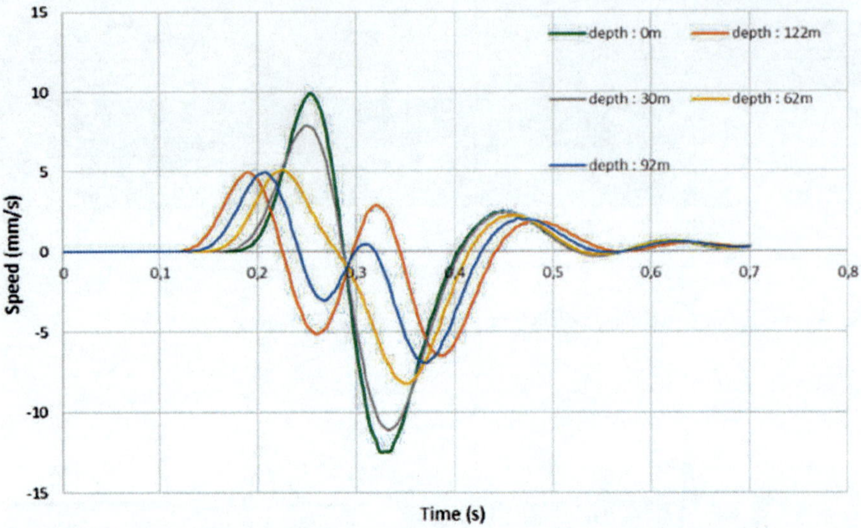

Fig. 6 Velocities at 5 points located along the vertical line (High freq. impulse)

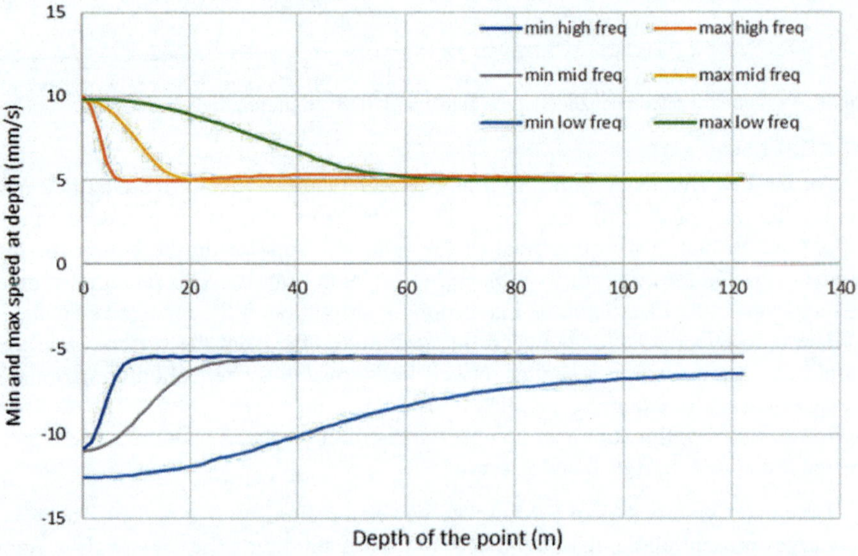

Fig. 7 Amplitude of the signals depending with the depth. *Note* A mid frequency signal has been added above, with a max frequency of 13 Hz. Time step used for calculations is of 2/1/0.5 ms resp. for low, mid, and high frequency impulses

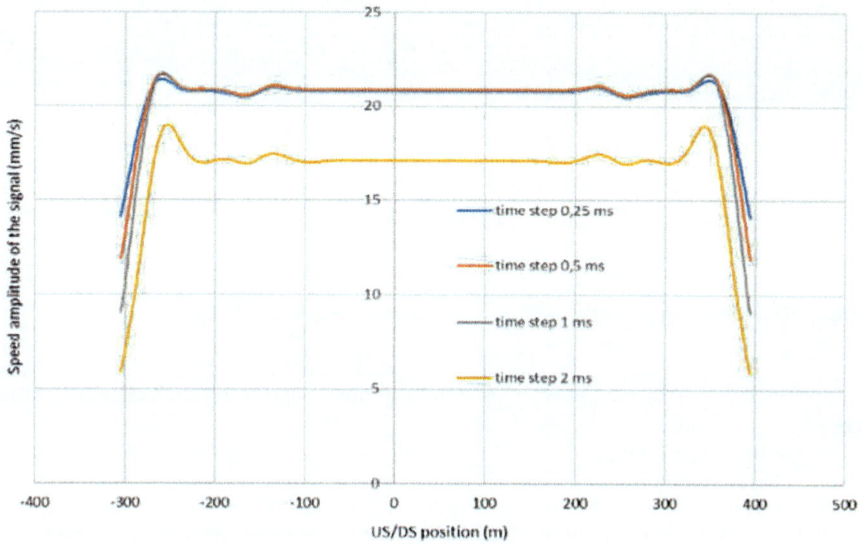

Fig. 8 Amplitude of the surface signal, depending on US/DS position (various time steps)

3.2 *Effect of Lateral Borders*

Some attenuation of the input signal is observed between center points (a and b) and the others. In order to investigate this attenuation, are hereunder presented the curves showing the amplitude of the surface peaks depending on the US/DS position of the point. Sensitivity analysis have been made on PML coefficient (on one hand) and on time step calculations. No significant difference has been observed on PML coefficient (Fig. 8).

It can be observed that, under 1 ms, there is almost no impact of a modification of the time step, whereas the impact becomes significant for a 2 ms time step. Regarding the frequency content of the signal (mostly under 85 Hz), It could be suggested that a factor $\frac{1}{6f}$ to $\frac{1}{10f}$ would be more appropriate, in order to represent properly such a signal.

3.3 *Influence of Mesh Size (with and Without Dam)*

No influence of mesh is expected for low frequency impulse; therefore, only high frequency impulse in discussed in the present paragraph.

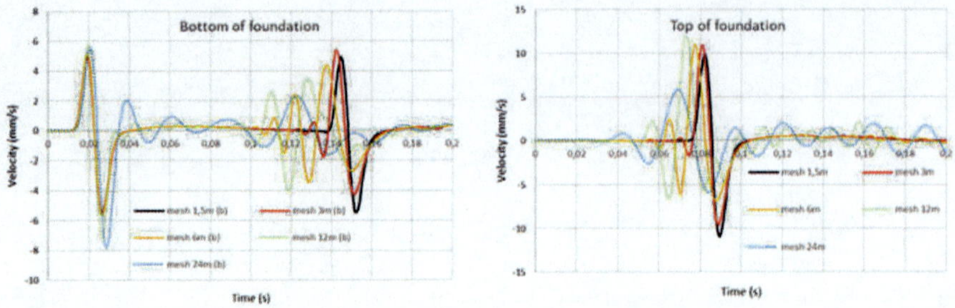

Fig. 9 Time evolution of top and bottom velocities depending on mesh size

In Fig. 9 are presented time velocities for points a and b, for various mesh sizes (from 1.5 to 24 m).

It can be observed that:

- There is only a small difference between 1.5 and 3 m meshes,
- For 3 and 6 m meshes, there is an attenuation of the peak, and the apparition of a small oscillation before the main peak; this secondary oscillation is probably due to the higher frequency content of the signal,
- For 12 m, and moreover for 24 m mesh, the signal is totally misrepresented.

In Fig. 10 are presented similar calculations, with the dam (no water), and speeds are observed at the foot and at the crest of the dam.

It appears that, at the top of the sam, there is no major difference between 1.5, 3, and 6 m meshes, although the impact of the first peak is visible for the first peak.

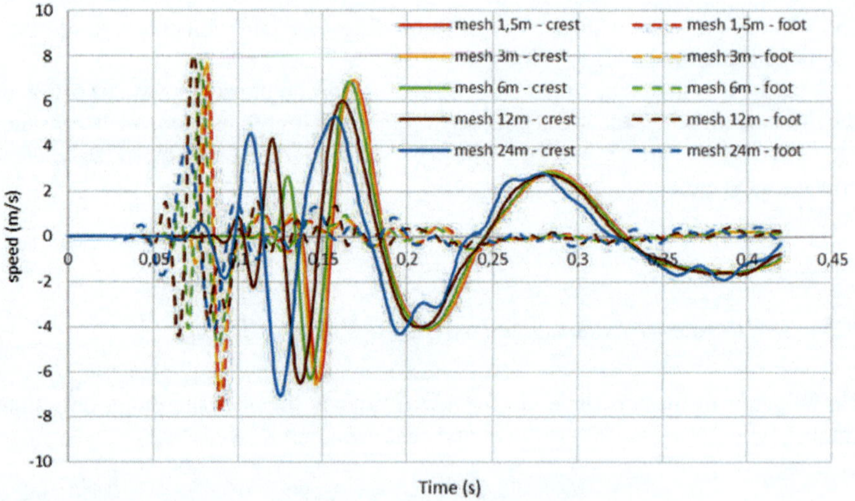

Fig. 10 Velocities at the foot and crest of the dam, depending on mesh size

Once the main peaks have been reached, even model with a 12 m mesh behaves similarily to the others, and only 25 m mesh has an inappropriate behaviour. This last constatation can be explained by the fact that main eigenfrequencies are in the same range as the maximum frequencies regarding the size mesh (eigen frequencies: 2→8 Hz; max frequency for a 24 m mesh: 6.5→10 Hz depending on the criterion).

It appears clearly that the application of the mesh criterion to the main eigenfrequencies allow a good reprensentation of the global displacement of the dam, but allows only a poor reprensentation of the main peaks, for which the high frequency content of the signal plays a significant role. Nevertheless, it appears clearly that mesh criterion can be less strict for a dam analysis than it should be taken for a good representation of HF signals in the foundation.

3.4 Analysis of Transfer Functions

Calculations presented in this paragraph are ratio between frequency responses of the top and of the bottom of the foundation. For this calculation, a 2% damping has been considered, with 2 ranges of parameters (high and low frequency targets).

It can be observed that discrepancies between theory and numerical results appear:

- Over 15 and 75 Hz approximately for low freq. and high freq. damping, which is quite logical (high limits of the damping are respectively 8 and 60 Hz),
- Over about 35/40 Hz for 6 m large mesh (with an 8 ratio, maximum validity is considered to be 40 Hz),
- Over about 70/75 Hz, discrepancies occur between 1.5 and 3 m meshes, independently from theoretical curve (which is affected by the damping problem).

The 8 factors between wavelength and mesh size seems relevant, but possibly a bit low, since small discrepancies start occurring, in the present case, under approximately a 9 factor (Figs. 11 and 12).

3.5 Impact of Integration Scheme

In order to investigate a possible numerical damping of integration scheme, HF simulation has been ran with 5 different integration schemes (Fig. 13).

It appears that Euler backward induces a very important attenuation, and Wilson scheme a little attenuation. The 3 other integration schemes are equivalent, without any damping (in conformity with theory).

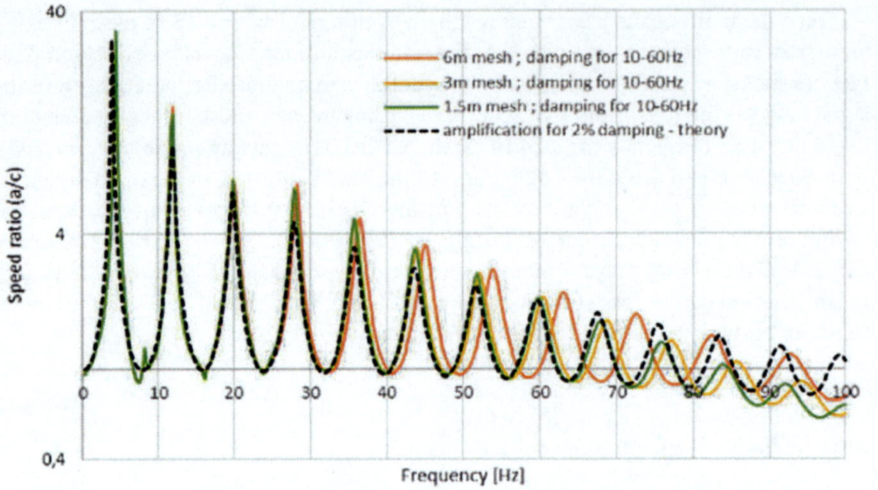

Fig. 11 Transfer function of the foundation (modulus), for various meshes

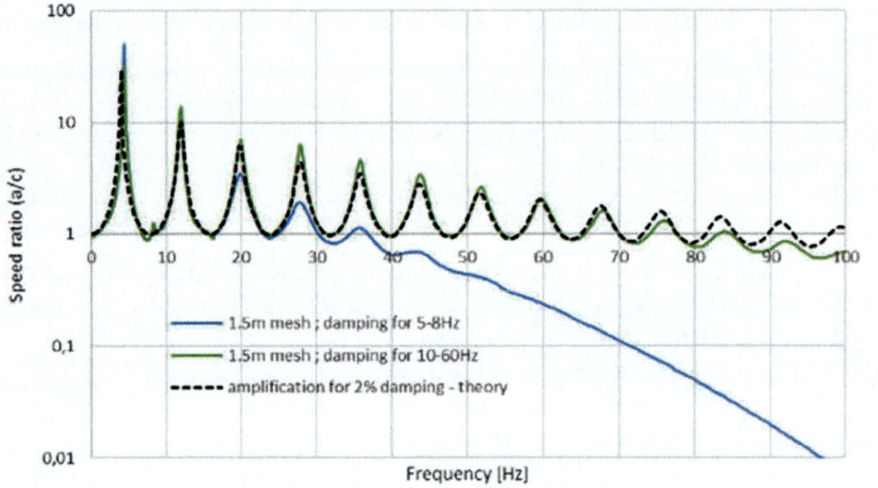

Fig. 12 Transfer function of the foundation (modulus), for two different damping

4 Influence of Water-Structure Interaction

4.1 Influence of Water Level in the Reservoir

Results are presented in Fig. 14 only for low frequency impulse, since no difference can visually be made beween various water levels for high frequency impulse.

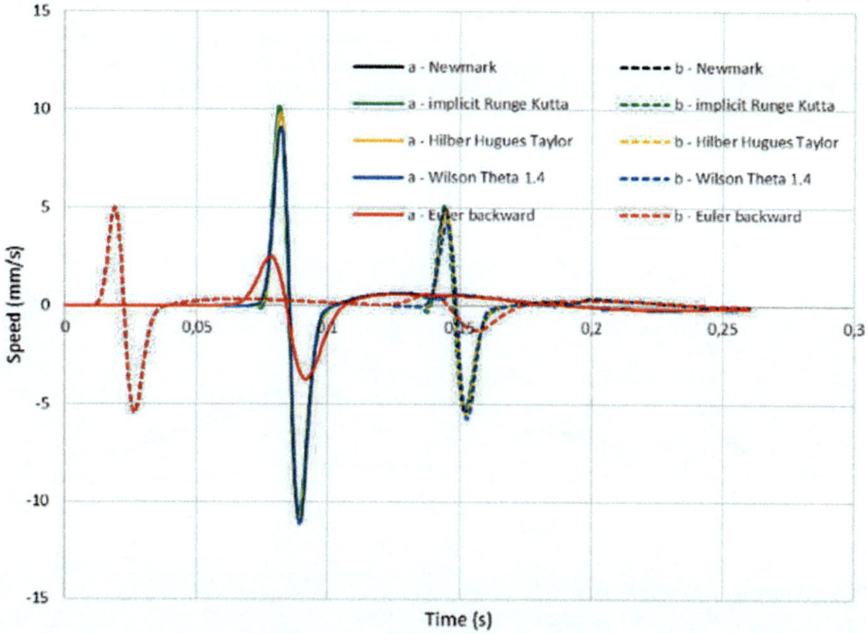

Fig. 13 Velocities at top and bottom of the foundation, depending on integration scheme

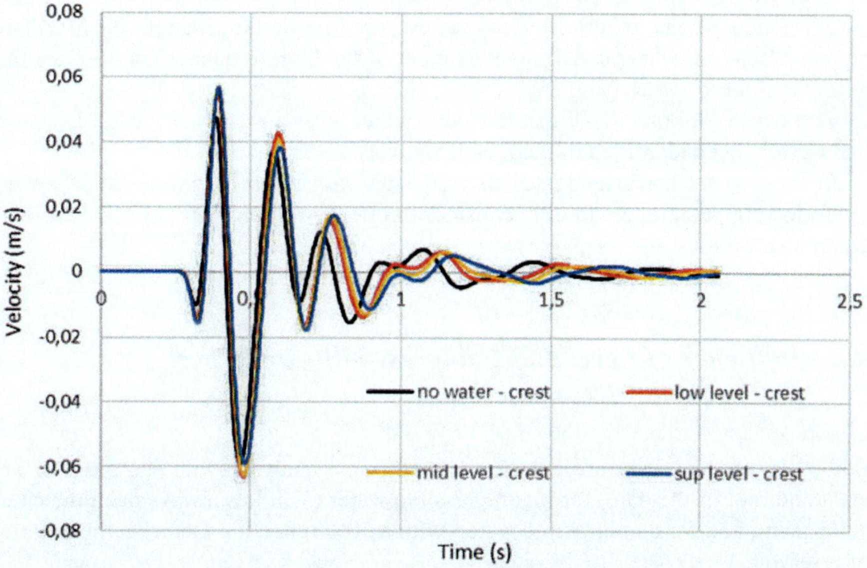

Fig. 14 Velocities the crest of the dam for various water levels and LF impulse

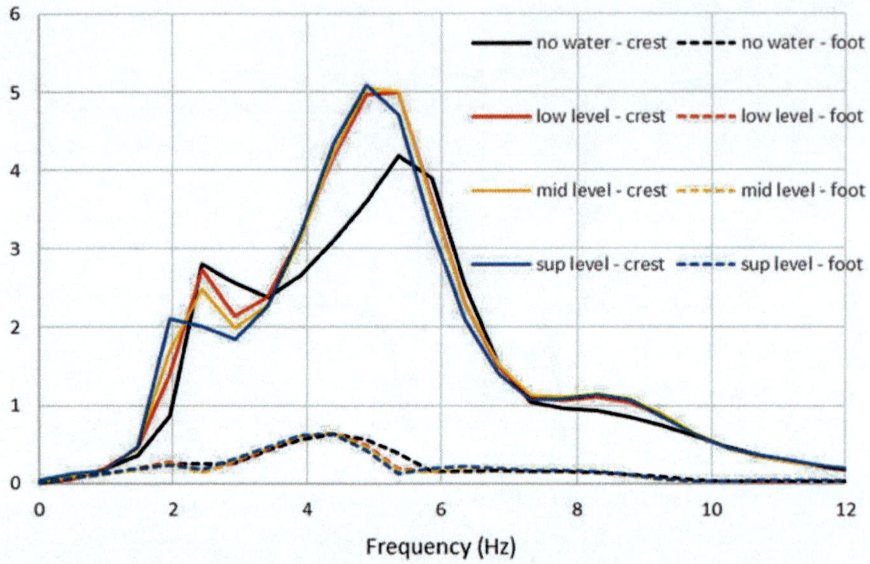

Fig. 15 FFT transform of velocities of the dam for various water levels and LF impulse

Maximum crest speed is of almost 60 mm/s for the low frequency impulse, but of hardly more than 8 mm/s for the high frequency impulse, which is quite logical, since frequency content of the low frequency impulse is in the range for the first dam eigenfrequencies, whereas frequency content of the high frequency impulse are far away from these values (Fig. 15).

This curve confirms that the influence of water level is mainly aroud 2/3 Hz, as it had already been identified in the modal analysis.

In order to further investigate this frequency impact on the influence of water, the following paragraph will only consider frequency response analysis, for various water assumptions.

4.2 Influence of Level and Compressibility and Level on Frequency Response

In the Fig. 16 the curves present the speed response analysis (with damping) at the crest and foot of the dam, for incompressible water (as it has always been the case since then), and for compressible water, with a 50% reflection factor on the bottom of reservoir.

In all the coming figures, level of water will be represented with the following color code: Low-level water: red; Mid-level water: blue; Sup-level water: yellow/orange. It can be observed that:

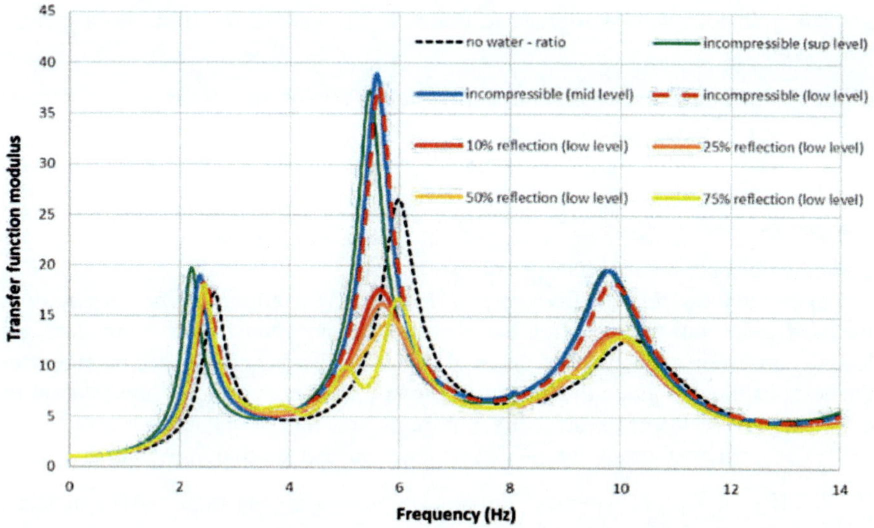

Fig. 16 Transfer function between foot and crest for various water levels (incompressible water) and various reflection factors (compressible water)

- For the 1st peak (about 2.5 Hz), the effect of water mass is to reduce the frequency of the peak. Frequency is reduced with the increase of water level. But there is no effect of water compressibility on this peak,
- For the second main acceleration peak (between approximately 5 and 6.5 Hz), the effect of water is to considerably increase the amplification, but with, surprisingly, a limited effect of variations in water level. For this peak, the impact of compressibility is major. It is such important that the amplification is less than the amplification of the dam without water. The effect of reflection factor remains minor, in comparison with the introduction of compressibility,
- For the third main peak (between approximately 9 and 11 Hz), there is, as well, almost no effect of water level, and amplification is reduced by the introduction of compressibility, so that it cancels the effect of water.

5 Conclusions

Analysis performed confirm the relevance of usual recommendations for mesh size and foundation thickness:

- foundation thickness larger than 1/4 of the wavelength for the lower relevant frequencies (f1), known as "quarter wavelength approximation",
- mesh size lower than 1/10 to 1/8 of wavelength for higher relevant frequencies (f2).

Two well documented sources of potential misleading results have also been pointed out:

- the choice of integration scheme shall be taken care of, some of them being likely to introduce numerical damping,
- the choice of damping parameters shall also be taken special care of, regarding relevant frequencies for the model.

Besides, analysis performed suggest that a criterion on maximal time steps for calculations could be of about $\frac{1}{10.f_2}$ to $\frac{1}{6.f_2}$.

For the choice of relevant frequencies, f1 and f2, there remains an indetermination: the most conservative approach is based on frequency content of the signal. It might lead to non-useful refinement of the mesh. The other approach is based on eigenfrequencies of the structures to analyze. It broadly allows a correct representation of the main movements of the structure, but might attenuate some peaks.

Besides, water structure interaction has been investigated. It appears that:

- The effect of water is to increase the acceleration for already amplified frequencies, but has no significant effect for other frequencies,
- For the lowest frequencies, usually the most critical, the effect of water level increase is to reduce the frequency, but there is no significant impact of water compressibility,
- For the other frequencies, there is no significant impact of water level, but the introduction of water compressibility introduces a very important damping, reducing the amplifications,
- In any case, the introduction of compressibility doesn't affect significantly the frequencies for which amplification can be noted.

Theme B: Seismic Analysis of Menta Embankment Dam

Seismic Analyses of Menta Embankment Dam: Formulation and Synthesis of Results

G. Russo, M. Cecconi, A. Vecchietti, V. Pane, A. Fiorino, and S. De Marco

Abstract Menta Embankment Dam has been selected as Theme B of the International Benchmark Workshop of the International Committee on Large Dams held in Milan in September 2019. The static and seismic behaviour of the dam under seismic input has been evaluated by 13 groups of Contributors coming all around the world. The Menta Dam is a rockfill dam with upstream bituminous facing located in the South of Italy. The major challenge faced by the Contributors was the simulation of the complex dynamic behaviour of rockfill material in order to predict the stress-strain behaviour of the embankment under seismic conditions. Despite the different modeling assumptions proposed by the Contributors, the numerical analyses showed a satisfactorily dynamic performance of the embankment under the proposed seismic input.

Keywords Embankment dams · Rockfill · Seismic analysis · Constitutive models

1 Seismic Behaviour of Rockfill Embankment Dams

Large dams in Europe have been designed systematically against earthquakes since the 1930s. Nevertheless, most existing dams built before the 1990s were designed against earthquakes using either seismic design criteria and/or methods of dynamic

---

G. Russo (✉)
Department of Earth Science, Environment and Resources, University of Napoli Federico II, Naples, Italy
e-mail: giarusso@unina.it

Department of Civil and Mechanical Engineering, University of Cassino and Southern Lazio, Cassino, Italy

M. Cecconi · A. Vecchietti · V. Pane
Department of Engineering, University of Perugia, Perugia, Italy
e-mail: manuela.cecconi@unipg.it

A. Fiorino · S. De Marco
So.Ri.Cal. s.p.a., Catanzaro, Italy

G. Bolzon et al. (eds.), *Numerical Analysis of Dams*, Lecture Notes in Civil Engineering 91, https://doi.org/10.1007/978-3-030-51085-5_22

analysis which may be considered obsolete [20]. The great development of computational methods may thus provide an up-to-date assessment of the level of safety of such dams, according to advanced knowledge and modern criteria.

Rockfill embankment dams are constituted more than 50% of compacted or dumped pervious fill and their watertightness depends on an impervious upstream blanket or an impervious core. Like earth dams they are composed of fragmental materials, with each particle independent of the others. The mass stability is guaranteed by the friction and interaction between particles. Rockfill dams are recognized for their safety, adaptability to widely varying site conditions, practicality of construction, and economy [12, 6]. Several rockfill dams around the world have demonstrated the good performance for both normal and exceptional loading cases, such as during earthquake events (e.g. Zipingpu Concrete Faced Rock-fill Dam). In fact, nil or limited damages have been recorded after strong earthquakes for this kind of dams in the literature (e.g. Foster et al. [8]). The effects of strong shaking have manifested primarily as surface cracking or local slides near the crest of the dam [11].

For the seismic performance and safety of rockfill dams, material degradation due to pore water pressure build-up is not a significant factor affecting safety and liquefaction is not one of the expected modes of failure. Instead, permanent deformations, cracking, and other types of local failures are of concern, as they may endanger serviceability [9].

The focus of Theme B in the Benchmark Workshop is to analyse the seismic behaviour of a bituminous-faced rockfill dam (BFRD), the Menta Dam (De Marco et al. [1, 2, 5], Vecchietti et al. [18]. The aim is to predict stress/strain behaviour and to estimate potential damages.

In the theme, the geometry, material properties and seismic loads are provided by the Formulators. Some aspects of the numerical modelling are intentionally not defined so that the participants could make their own assumptions and choose suitable approaches to solve the problem. In particular, participants may choose their own constitutive model to describe cyclic, non-linear behaviour of rockfill material; consequently, they may select the analysis approach they believe to be appropriate for the case at hand. Thereby, by comparing the different solutions, it will be possible to draw conclusions regarding how different assumptions and approaches influence the results.

2 Formulation of Theme B

2.1 Menta Dam

The Menta dam (Fig. 1) is a bituminous-faced rockfill dam located in Southern Italy (Latitude 38.1234°N, Longitude 15.8998°E), lying in the heart of Aspromonte Massif at an elevation of about 1400 m a.s.l. The embankment is about 90 m high

Fig. 1 Menta Dam

at its deepest point, and the reservoir impounds 1.8×10^7 m^3 of water. The crest height is 1431.75 m a.s.l and the normal reservoir level is at 1424.75 m a.s.l.. It was designed in the late Seventies, and constructed between 1987 and 2000.

The geometry of the embankment base is adapted to natural morphology of the rock basement and river Menta canyon: the dam is thus characterized by a complex shape, with a main embankment closing the Menta valley which is about 325 m long at the crest, and a curved portion (of about 125 m) that rests on a saddle of rock that closes the valley on the right side. Plan view of the embankment is reported in Fig. 2.

In Fig. 3 the typical cross-section of the embankment has been reported. The embankment slopes have 1.8:1 inclination, and the downstream face has two berms at the elevation 1404.5 m and 1377.5 m a.s.l.

The embankment is made of compacted rockfill excavated by blasting along the rivers of the reservoir and was planned to be divided in five zones at the design stage. Each zone was characterized by different grain size distribution. A draining layer with a thickness of about 3.0 m was placed at the embankment-foundation interface, as a safety layer to prevent positive water pressures rise in the embankment.

The rock mass forming the foundation of Menta dam is a fractured metamorphic rock belonging to the "Calabrian Complex". Cement-water injections reaching a depth of about 50 m were realized departing from the inner perimeter tunnel to create a grout curtain on the upstream side of the embankment (Fig. 4).

The impervious bituminous concrete facing is placed on the upstream side. The facing is formed by superposing different layers of bituminous concrete upon a regularly packed basement of compacted sandy gravel. The facing is anchored to the reinforced concrete tunnel running along the inner perimeter of the dam (Fig. 5).

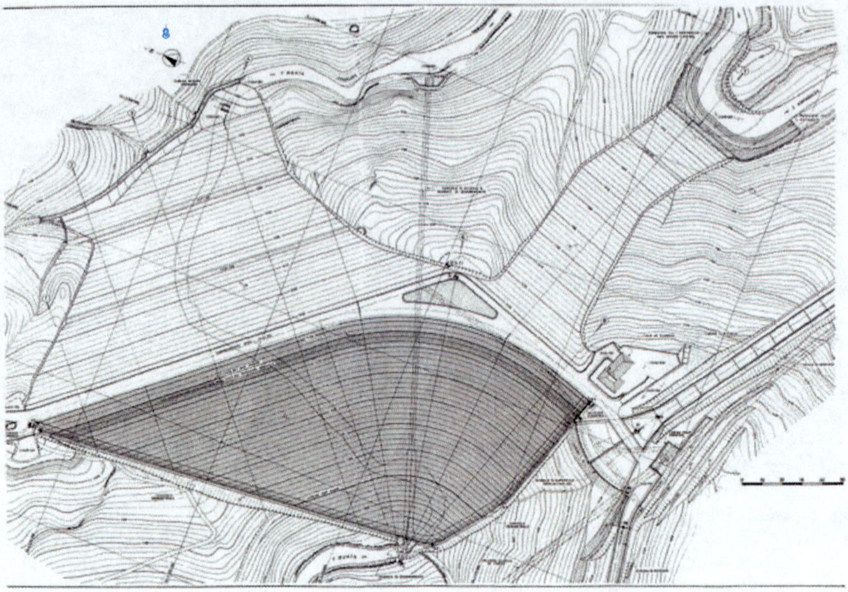

Fig. 2 Menta Dam—plan view

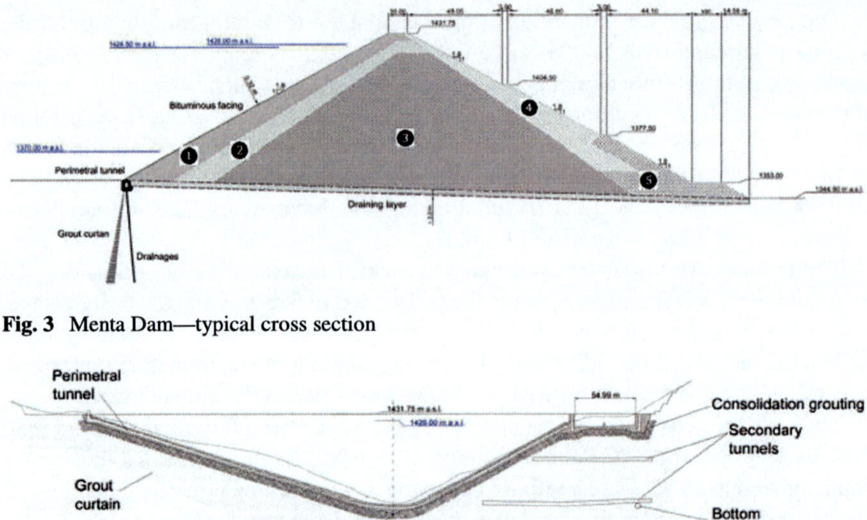

Fig. 3 Menta Dam—typical cross section

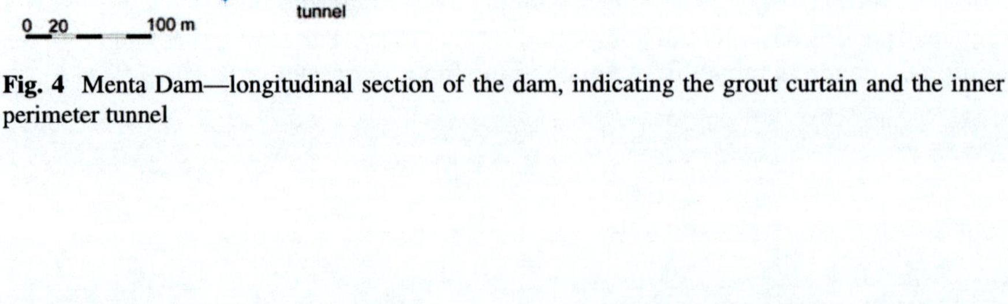

Fig. 4 Menta Dam—longitudinal section of the dam, indicating the grout curtain and the inner perimeter tunnel

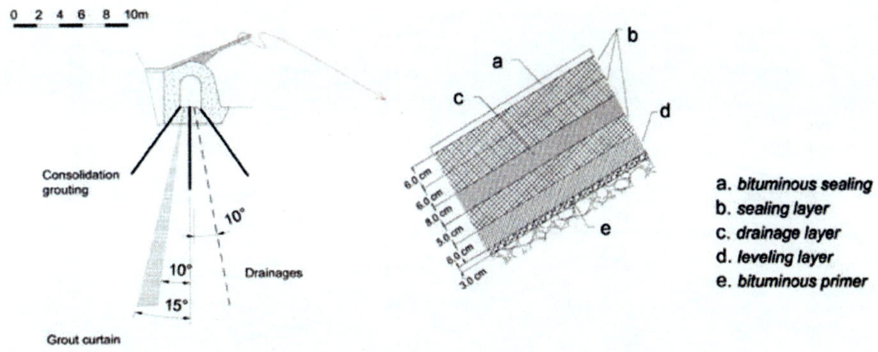

Fig. 5 Menta Dam—cross section of the perimeter tunnel and bituminous facing

2.2 *Materials*

A main concern of Theme B has been the modelling of the complex behaviour of rockfill forming the embankment. Rockfill stress-strain behaviour depends on different parameters, such as mineralogy, grain size distribution, particle size, stress level, and particle breakage, if any. The mechanical behaviour of rockfill is highly non-linear and, in general, both strength and stiffness are found to be dependent on applied stress level [14]. Under seismic loading, a decrease of stiffness with increasing strain level is expected.

Mechanical characterization refers to material of Zone 3 (see Fig. 3), therefore the embankment has been considered as formed only by this material. Available data regarding the main physical and mechanical properties of the rockfill are summarized in Table 1 and Fig. 6: unit bulk weight and hydraulic conductivity, results from triaxial-K_0 consolidated drained tests, initial stiffness from conventional triaxial tests,

Table 1 Physical and mechanical properties of rockfill material

γ (kN/m^3)	k (m/s)
23	1.0×10^{-5}

γ = bulk unit weight; k = hydraulic conductivity.

Summary of TXCD-K_0 laboratory tests.

#	σ'_3 (kPa)	σ'_1/σ'_3 (-)	ϕ' (°)	ε_a (-)	ε_{vol} (-)	B_g (%)
1	100	7.1	48.8	-0.02	-0.0135	5
2	300	5.2	42.6	-0.06	-0.0405	6
3	900	4.2	38	-0.10	-0.114	7

σ'_1, σ'_3 max. and min. effective stress at failure; ϕ' = friction angle at failure; ε_a, ε_{vol} = axial and volumetric strain at failure; B_g = Breakage index, computed as:

$$B_g = \frac{\sum_{k=1}^{n} \Delta w_k}{n}$$

where Δw_k is the difference between initial and final weight of k-th grain size fraction (Marsal 1972).

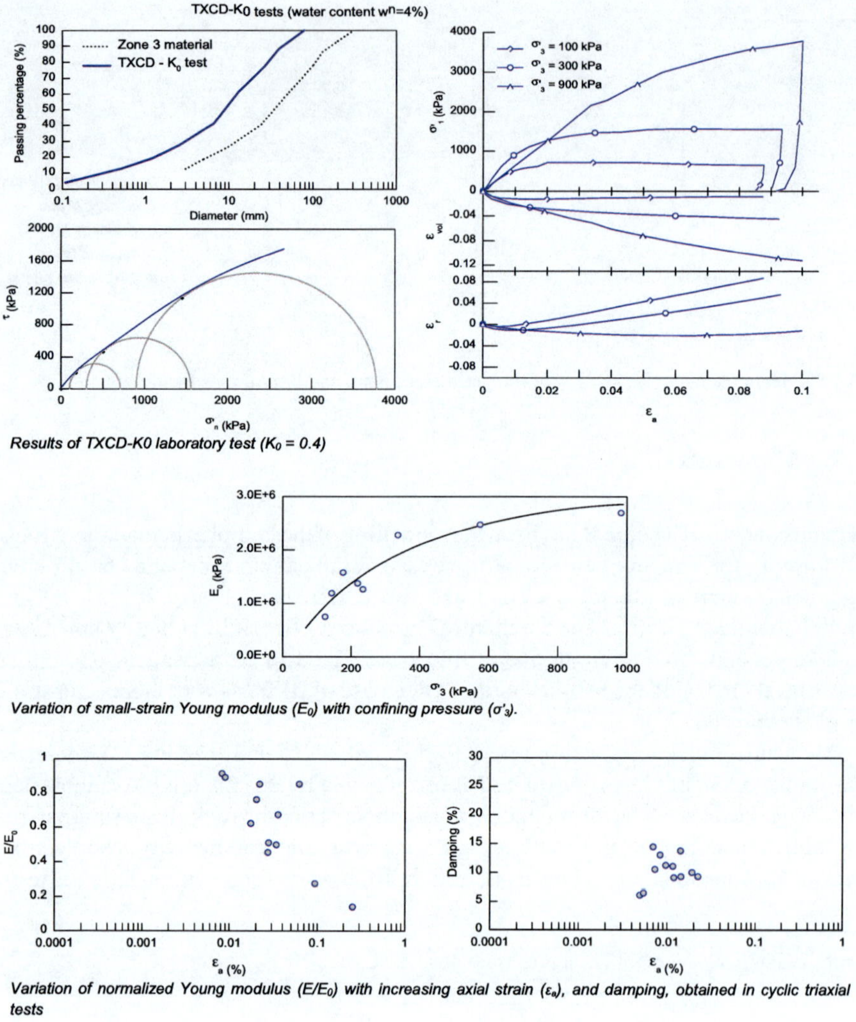

Fig. 6 Available results from laboratory tests: mechanical properties of rockfill material

results from cyclic triaxial tests. The value of the lateral earth pressure coefficient at rest $K_0 = 0.3–0.4$ derives from pressure cells measurements.

The rock mass foundation stress-strain behaviour has been assumed elastic. The main physical and mechanical properties are summarised in Table 2.

The mechanical behaviour of bituminous concrete is visco-elastic and depends on load magnitude and loading rate, as well as on temperature. The facing layers can be modelled as single equivalent layer having the mechanical characteristics of the sealing layer. The elastic stiffness has been computed based on the provisional

Table 2 Physical and mechanical properties of rock mass foundation

γ (kN/m³)	*E* (MPa)	ν (-)	*k* (m/s)
27	7×10^3	0.25	1.0×10^{-7}

γ = bulk unit weight; E = Young modulus; ν = Poisson ratio; k = hydraulic conductivity.

Table 3 Physical and mechanical properties of bituminous impervious facing

Characteristics of the bituminous concrete mixture (as single equivalent layer):

γ (kN/m³)	*Temperature*	*E** (MPa)	ν * (-)	*k* (m/s)
24	1°C	14000	0.16	*Impervious*
	26°C	2800	0.33	

γ = bulk unit weight; E = complex Young modulus; ν* = complex Poisson ratio; k = hydraulic conductivity.*

Strength characteristics of the bituminous facing, deduced from Marshall test results (sealing layer):

Compressive strength f_t (kPa)	Tensile strength f_t (kPa)
1490	950

model of University of Maryland [3]. This methodology provides complex Young modulus, E*, and Poisson ratio, v^*, accounting for bituminous-concrete particle size distribution (PSD) of aggregates, mixture quantities, viscosity of the bitumen. The main physical and mechanical properties are reported in Table 3.

2.3 Construction Stages

According to construction reports, the different portions of the embankment have been put in place following the temporal sequence of Fig. 7. Following the real construction sequence, the construction of the embankment has been suggested to be divided into three subsequent steps.

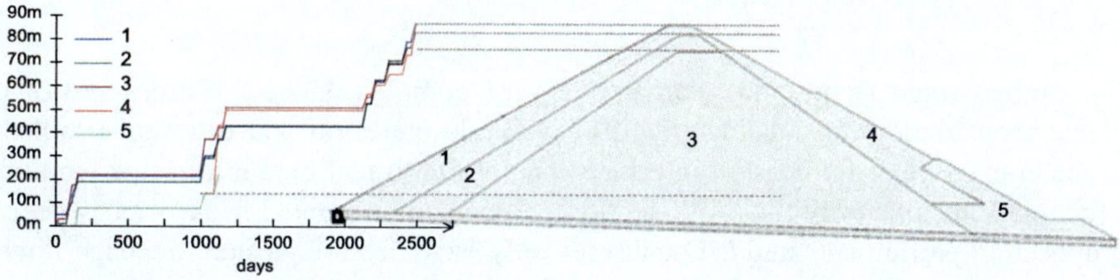

Fig. 7 Embankment construction process

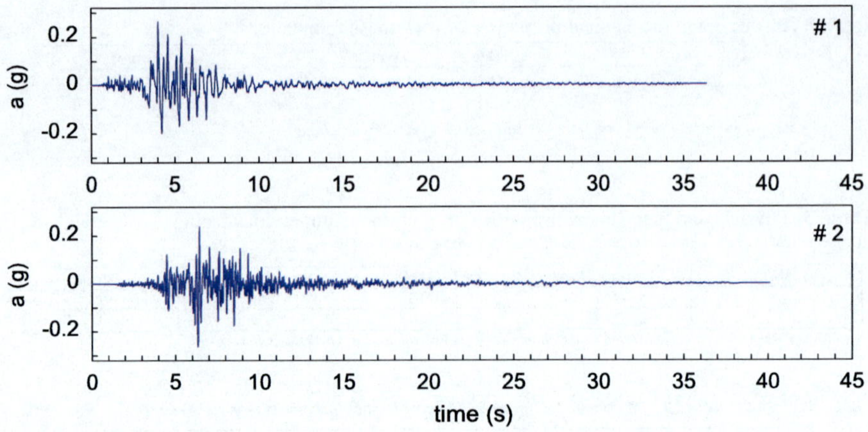

Fig. 8 Acceleration time series representing the dynamic input of the analyses

2.4 Seismic Input

Two target acceleration time series (acceleration vs. time) represent the seismic input scenarios (Fig. 8). These records have been selected in accordance with the criterion of spectro-compatibility with the Italian Seismic Hazard map for the site of Menta Dam and represent two severe events occurred in Italy (Friuli earthquake occurred in 1976 and Central Italy earthquake sequence of 2016). The records are scaled to have a reference Peak Ground Acceleration equal to 0.26 g. The three components (2 horizontal and 1 vertical) of the seismic input have been provided to the participants. The target records represent the acceleration recorded at *rock outcrop in free-field conditions.*

2.5 Expected Results from Contributors

The embankment shape and geometry is very complex, since the cross-sections of the embankment are characterized by variable elevation and different profiles of the embankment-foundation interface. Therefore, to facilitate the comparison of benchmark results, only the cross-section geometry represented in Fig. 9 has been proposed to participants and 2-D analyses only have been requested. Seepage flow occurring in the foundation rock mass does not affect the operating conditions of the embankment in terms of stress states nor its seismic behaviour. Therefore the Contributors were recommended to disregard the seepage phenomenon in the lower rock mass and assume the foundation rock mass as impervious.

The Contributors have evaluated the initial stress-strain behaviour of the embankment during the first impoundment with water level reaching the maximum level

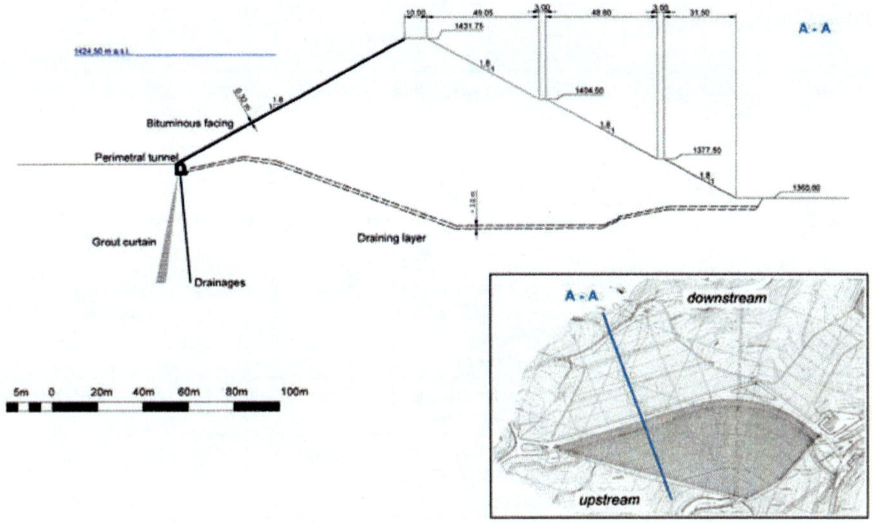

Fig. 9 Reference cross section of Menta Dam for numerical 2D analyses

(1424.5 m a.s.l.) in the reservoir. The Contributors were asked to provide for contour plots of vertical stresses, horizontal stresses, friction angle and Young modulus within the embankment.

The evaluation of the embankment stress-strain behaviour under seismic conditions was requested, taking into account the stiffness dependence on shear strain levels (G-γ) and damping ratio. Analyses were based on the seismic inputs free-field conditions discussed in Sect. 2.4. The rock basement was assumed to behave as compliant base. The Contributors were asked to determine the first natural frequency of the embankment (in both empty and full reservoir condition), acceleration time history at the dam crest, relative displacement time history of the crest with respect to the dam base, deformation patterns of the embankment section. A final assessment of stability conditions (proximity to failure) of the embankment after the earthquakes was requested as well.

The assessment of seismic behaviour of the bituminous facing, based on the analysis of potential damages and cracking of the upstream facing, was requested as an option to the Contributors with reference to deformation pattern of the bituminous facing and cracking occurrence.

3 Contributions on Theme B—Menta Embankment Dam

As summarized in Table 4, thirteen Contributors coming from all over the world proposed their solution. Among them, 3 coming from Italy, 2 from China, 1 from

Table 4 Contributors

Code	Participant	Affiliation	Country	Numerical code
01 Alib	Aliberti, Vecchiotti, Cascone	Università degli Studi di Messina	Italy	FLAC
02 Catal	Catalano, Stucchi, Agosti, Crapp	Lombardi Engineering Ltd.	Switzerland/Italy	FLAC
03 Chugh	Chugh	US Bureau of Reclamation, Denver	Colorado (USA)	FLAC, SHAKE
04 Freius	Freuis, Kainrath, Krstic, Smesnik	Pöyry Austria GmbH, Vienna, AUSTRIA & Tschernutter Consulting Ltd.	Austria	Quake/W Sigma/W
05 Fu	Fu, Mi, Wei	Nanjing Hydraulic Research Institute	China	Own FEM code
06 Glag	Glagovski & Gibyanskaya	Vedeneev Research Institute (VNIIG), St. Petersburg	Russia	Plaxis 2D
07 Liu	Liu J, Zou, Liu H, Wang	Dalian University of Technology, Dalian & Chengdu Engineering Corporation limited, Chengdu	China	Own FEM code (GEODYNA)
08 Lu	Lu & Athanasiu	Multiconsult, Oslo	Norway	Plaxis 2D
09 Mész	Mészáros, Minárik Bakeš	Vodohospodarska Vystavba, Bratislava	Slovakia	Quake/W Sigma/W
10 Mond	Mondoloni, Kolmayer, Alves-Fernandes	Electricité de France-Hydraulic Engineering Center, La Motte-Servolex, Electricité de France R&D, Palaiseau	France	own FEM code (Code_Aster)
11 Petk	Petkovski, Mitovski, Panovska	University "Sts Cyril and Methodius", Civil Engineering Faculty Skopje	Rep. of Macedonia	Quake/W Sigma/W
12 Raggi	Raggi	ELC Electroconsult SpA, Milano	Italy	FLAC
13 Tzen	Tzenkov & Schwager	TUCKY LTD, Renens & Swiss Federal Office of Energy SFOE, Bern	Switzerland	DIANA

Norway, 1 Slovakia, 1 France, 1 from Switzerland/Italy, 1 from the Republic of Macedonia, 1 from Austria, 1 from USA and 1 from Russia. Overall, the Benchmark Workshop has involved the contribution of five research groups, including Universities and Research Institutes, and eight Companies and Consultants. This extended participation demonstrates the extreme interest of both researchers and engineers for such specific and up-to date topic.

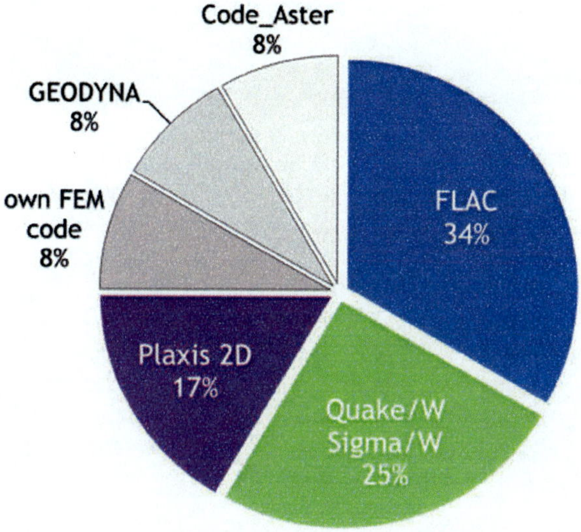

Fig. 10 Distribution of used numerical codes

Figure 10 shows the percentages of users adopting different numerical codes. It is noted that the majority of Contributors used the DFM numerical code FLAC, fol-lowed by GEOSTUDIO software packages and FEM numerical code PLAXIS.

4 Results and Discussion

The main focuses of Theme B were the performance of different constitutive models, all capable to reproduce—to some extent—the rockfill behaviour observed during laboratory tests on soil samples, and the seismic behaviour of the dam. With reference to the first aspect, Contributors adopted non-linear Mohr-Coulomb elasto-plastic constitutive models, as well as advanced constitutive models with multiple parameters to be calibrated. The second aspect was faced by Contributors using different integration techniques, namely FDM and FEM, in assessing Menta dam behaviour under prescribed seismic loadings.

4.1 Static Conditions

Vertical and horizontal in situ stresses. The initial stress state of the embankment has been assumed as the stress field after the staged construction and reservoir impounding. The ratio between vertical and horizontal stresses at the embankment base, on the central axis, has been evaluated. Such ratio value (K_0) varies from 0.25 to 0.43 among the Contributors, which is fairly consistent with load cell measurements

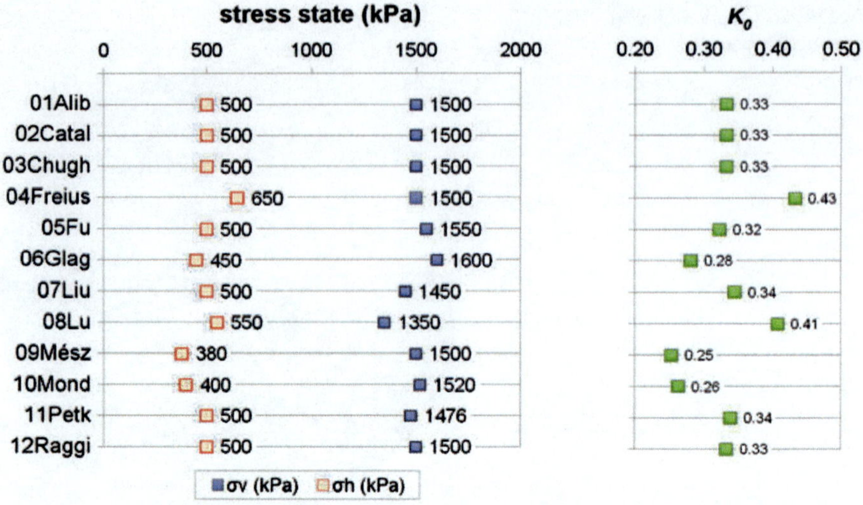

Fig. 11 Stress states and K_0 values determined in each contribution

($K_O = 0.3 \div 0.4$). In Fig. 11 a summary of the stress states for each contribution has been reported.

Rockfill mechanical properties. The Contributors used several different constitutive models to represent rockfill non-linear behavior. Some of them considered strength and/or stiffness to be dependent on applied stress level. Starting from interpretation of triaxial tests and literature data (including national Regulations and best practices), and depending on the chosen constitutive model formulation, the Contributors derived the small strain Young modulus, E_O, and eventually an operational static Young modulus E, as well as friction angle ϕ distribution within the rockfill embankment. The range of values for E_O, E and ϕ retrieved from contour plots reported by Contributors is depicted in Fig. 12.

Under seismic loading, a decrease of stiffness with increasing strain level and hysteretic damping has been implemented by most of the Contributors. Viscous, Rayleigh damping formulation up to 10% has been used by few Contributors.

4.2 Fundamental Frequency

Natural frequencies (eigenfrequency) are the frequencies at which the system tends to oscillate in the absence of any driving or damping force [13]. The lowest natural frequencies, and particularly the fundamental one, f, contribute mostly to the total response of the system: comparing f with the external force frequency content can be an indicator for envisaging larger deformation/amplification phenomena for

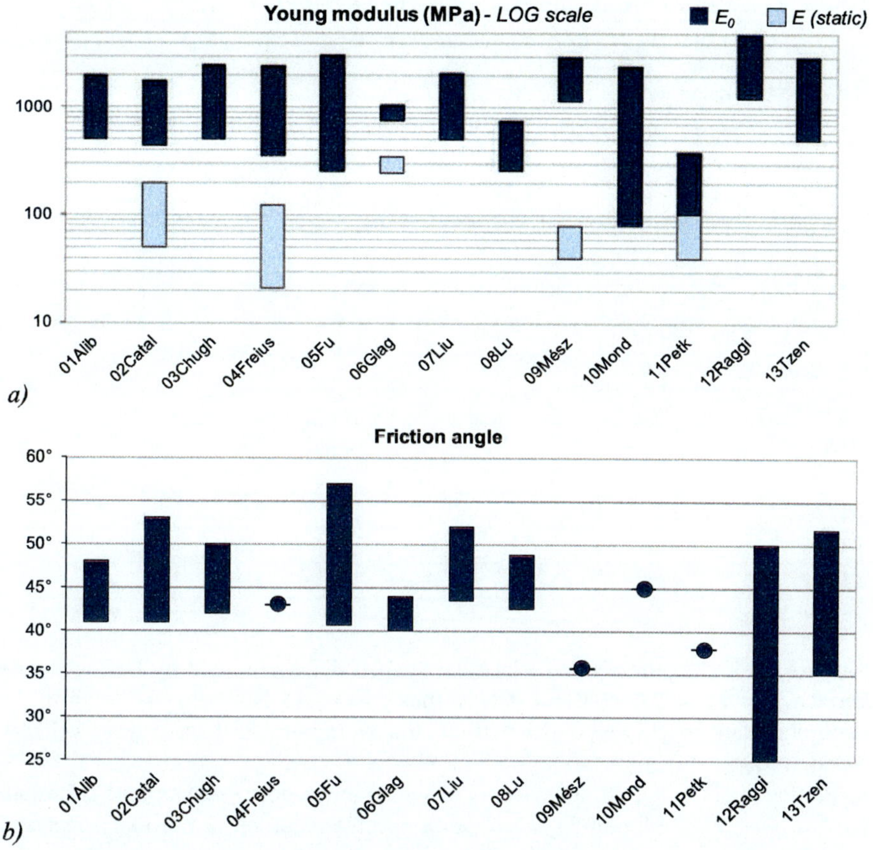

Fig. 12 Ranges of values of Young modulus, **a** and friction angle, **b** used by contributors

any engineering structure. Table 5 shows the frequency values obtained from the Contributors.

Participants 01, 02, 03, 06, 07, 10, 12, 13 provided evaluation of f by modeling the dam as undamped system, conducting specific modal analyses. Such evaluation refers to elastic, undamped behavior and does not depend on the input excitation and strain amplitude. In general, the fundamental frequency under empty reservoir condition is lower than that observed for full reservoir, since the system response is stiffer due to reservoir effect: at one hand, the reservoir pressure modifies the stress state on the upstream face and, in case of constitutive models with stress-dependent stiffness, this leads to an increase in dam initial stiffness; on the other hand, water mass actually increases the inertia of the whole system. Nevertheless, the effect of hydrodynamic pressure in dynamic analyses is still debated [19], Fenves and Chopra [7], Pelecanos et al. [15] and only a few Contributors accounted for it. Values of f ranges between 0.9 and 2.44 Hz. Values approach the result of elastic solution

Table 5 Values of fundamental frequency reported by Contributors

	Modal Analysis (elastic, undamped f_1)		Ex-post evaluation (amplification effect, f)			
	Empty	Full	Empty (Friuli)	Empty (C. Italy)	Full (Friuli)	Full (C. Italy)
01 Alib	2.32	2.44				
02 Catal	1.85	1.9				
03 Chugh	1.3	1.3				
06 Glag	1.089	1.0				
07 Liu	2.183	2.247				
10 Mond	1.82	1.86				
12 Raggi	0.9	0.9				
13 Tzen	0.95	0.95				
04 Freius			–	–	1.6	3.7
05 Fu			2.07	2.35	2.15	2.43
08 Lu			2.0	0.7	2.0	0.44
09 Mész			1.8	6.0	1.8	6.7
11 Petk			3.03	3.03	3.03	2.7

formula for homogeneous and triangular-shaped dams provided by Dakoulas and Gazetas [4]: $T_1 = 2.6\ H/V_S = 0.47$ s, thus $f_1 = 2.13$ Hz (being $H = 84$ m the maximum dam height and $V_S = 470$ m/s the average rockfill shear wave velocity based on E_0).

Participants 04, 05, 08, 09, 11 only provided an ex-post evaluation of f through crest to base ratio of pseudo-acceleration spectra recorded at the dam crest and the base after performing the dynamic simulations. Such evaluation is based on amplification effect in frequency domain and reflects the response of the system in terms of stiffness decay, damping and strain level generated during each earthquake. Contributors observed that different frequencies are excited by the two seismic inputs, depending on applied strain level. In this case, also the effect of reservoir varies from case to case. A summary of the different fundamental frequency has been represented in Fig. 13.

4.3 Dynamic Conditions

Most of the Contributors performed time-history dynamic analyses, adopting non-linear or equivalent-linear, elasto-plastic formulation. They usually applied combined horizontal and vertical seismic input at the base of the model. Several Contributors used viscous or free-field boundary conditions at the bottom and lateral sides of the model, in order to correctly simulate the compliant base assumption, avoiding

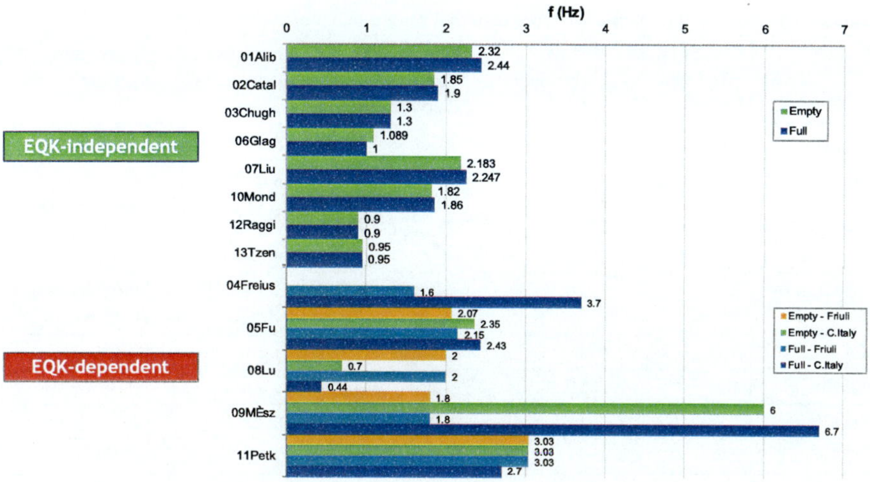

Fig. 13 Values of fundamental frequency *f* determined by each Contributors

reflection of waves from the boundaries; they often performed deconvolution of the acceleration signal to account for location of the model base within the foundation rock mass below the dam. Other Contributors applied the seismic input at the dam/foundation interface, in other words, the foundation rock mass is included in the model as embankment base constraint. In general, the Friuli (1976) earthquake caused more severe results.

Table 6 summarizes the main results of dynamic analyses. The embankment geometry and stiffness caused non-negligible amplification phenomena. Maximum horizontal acceleration at the crest resulted larger than that at the base by a factor ranging from 2 to 3.5, as represented in Fig. 14, where the amplification factor determined by each Contributor has been reported and compared with crest amplification factor computed by Yu et al. [21] for numerous rockfill dams around the world. Interestingly, the value of crest amplification factor predicted using Yu et al. [21] predictive equation is around 2, whereas the mean value of the amplification factors proposed in the benchmark is 2.9.

Crest settlement represents the largest vertical displacement experienced by the model over time. Crest settlement has been used by Swaisgood [16, 17] as an index to evaluate damages caused by earthquakes. For the case at hand, as represented in Fig. 15 crest settlement normalized to dam height resulted in the range 0.01 $\div$ 1%, corresponding to minor to moderate damages.

The maximum residual displacement occurred at the downstream slope and near crest. The distribution looks different for the two different earthquakes. Values reported from Contributors range between few centimeters, i.e. very limited plastic deformations have occurred, up to approximately 40 cm. A damage criterion based on permanent displacements was proposed by Idriss [10] with reference to natural

Table 6 Summary of dynamic analyses results reported by Contributors

	Crest to base amplification	Crest settlement (cm)	Post-seismic horizontal displacements	Post-seismic settlements	Maximum strains
01 Alib	2.6	8	31 cm	13.5 cm	1%
02 Catal	2.7	2	3 cm (up to 10 cm localized on top d/s slope)	2 cm (up to 5.5 cm localized on top d/s slope)	up to 0.45% localized on the d/s slope
03 Chugh	2.5	2	±2 cm	±2 cm	0.05% (local 0.15%)
04 Freius	6.2	1	<10 cm	<1 cm	0.43%
05 Fu	3.3	19	8 cm	19 cm	0.04%
06 Glag	1.7	1	2 cm	1.2 cm	–
07 Liu	2.0	21	50 cm	20 cm	0.40%
08 Lu	3.1	21	37 cm	21 cm	1.30%
09 Mész	1.1	8	>100 cm	–	–
10 Mond	3.8	14	3 cm	14 cm	0.30%
11 Petk	3.0	34	36 cm	34 cm	1.3% (in the facing)
12 Raggi	4.2	2	8 cm	8 cm	1%
13 Tzen	3.4	44	45 cm	40 cm	0.05%

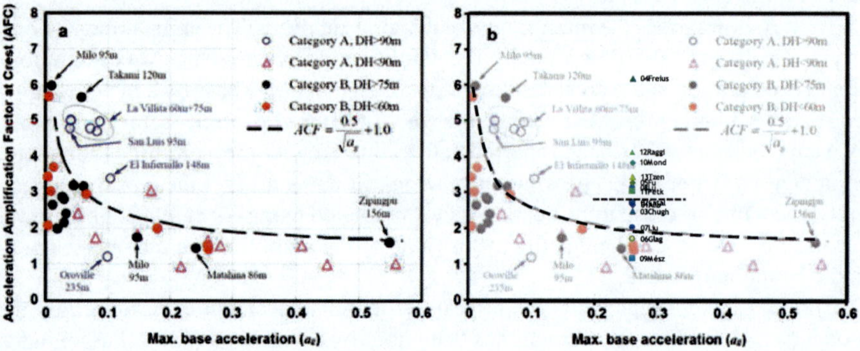

Fig. 14 Benchmarking of amplification factor at crest: **a** Yu et al. [21] relation; **b** results from each contributor compared with Yu et al. [21] relation

slopes: displacements lower than 30 cm denote moderate damages. Dam displacements after seismic shacking develop primarily in the downstream direction. In some cases, the maximum shear strains reached 1%, meaning that significant non-linearity has been experienced by the embankment. No relevant patterns in shear strain nor sliding surfaces have been detected. Despite the different modeling assumptions, the

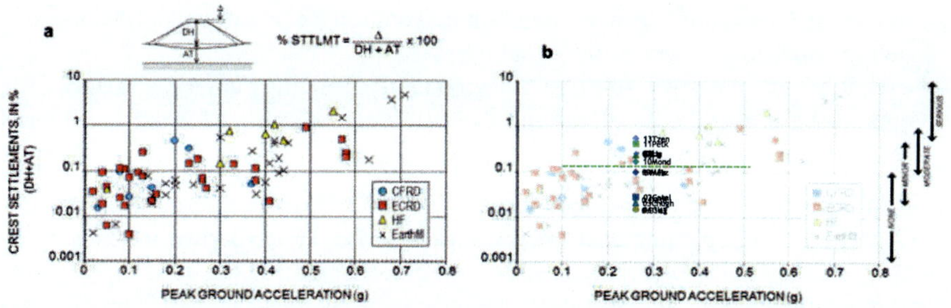

Fig. 15 Benchmarking of crest settlements: **a** Swaisgood [17] relation; **b** results from each Contributor compared with Swaisgood [17] relation

performed analyses show a satisfactorily dynamic performance of the embankment under the proposed seismic input.

4.4 *Performance of the Bituminous Facing*

About half of the Contributors analyzed the seismic behavior of the bituminous concrete facing. Evaluation of the facing performance included calculation of maximum stress, to be checked against provided maximum tensile and compressive strength, or experienced strain level, to be compared with literature values related to cracking development. Few Contributors also considered the effect of temperature on the bitumen stiffness: no clear indication about which of the two temperatures, $T = 1$ °C and $T = 26$ °C favored the facing behavior, as divergent trend in results are observed (cfr. 01Alib and 08Lu). Unsatisfactory performance has been assessed only in two analyses: computed compressive stresses (07Liu) and both tensile and compressive stresses (08Lu) exceeded the material resistance, denoting that even if the dam does not experience stability problems, dam water tightness can be endangered.

5 Conclusions

The comprehensive studies performed by all Contributors have demonstrated the ability of numerical analyses to predict the behavior of Menta dam upon static and seismic conditions. From such studies, some general conclusions can be drawn:

- a good agreement in stress predictions (stress profiles in static conditions) at the end of construction and impoundment was achieved through different numerical methods, codes and constitutive models;

- in seismic conditions, a general trend for first natural frequency and its dependence on the empty/full reservoir was not observed;
- except for one proposed solution, independently from the numerical methods the crest-to base average acceleration ratio for Menta dam and for the considered seismic inputs is about 2.9.

From the inspection of each proposed solution, it is worth to note that geotechnical properties and parameters have been inferred from (i) experimental laboratory avail-able data (from the proposal); (ii) data derived from previous studies; (iii) literature recommendations and Software Manuals; (iv) experience gained from previous investigated case studies and engineering practice, (v) and from their combination.

Apparently, it seems that the obtained results are not strongly affected by the complexity of constitutive soil models and numerical solutions and that the basic dam behaviour trends and mechanisms are fairly captured by such different numerical models. Upon seismic conditions, horizontal displacements were found to vary in the range 2–50 cm, while calculated vertical settlements ranged from 2–20 cm. Some Authors suggest that vertical shaking component plays a primary role in the overall dam behaviour; only two Contributors have considered only horizontal shaking. Hydro-dynamic pressures have been generally neglected. It is believed that this results in a more conservative evaluation.

Finally, as concern the seismic behavior of the bituminous facing (an optional issue), the fifty percent of Contributors have assessed the seismic behavior of the bituminous facing in terms of stress, strain and expected performance. When approaching these analyses, temperature effects were taken into account. However, in this case, the obtained results were found to be not entirely consistent, thus enhancing the need for further and specific research on this subject.

As a final consideration, despite the different modeling assumptions, the performed analyses show a satisfactorily dynamic performance of the Menta Dam under the proposed seismic input.

References

1. Albano M, Modoni G, Croce P, Russo G (2015) Assessment of the seismic performance of a bituminous faced rockfill dam. Soil Dyn Earthq Eng 75:183–198
2. Albano M, Modoni G, Russo G, Croce P, Polcari M, Saroli M, Bignami C, Moro M, Stramondo S (2016) Rockfill dam monitoring through ground-based and remote sensing techniques. In: 1st IMEKO TC4 international workshop on metrology for geotechnics, metrogeotechnics, pp 276–280
3. American Association of State Highway and Transportation Officials (AASHTO) (2008) Mechanistic-empirical pavement design guide: a manual of practice. AASHTO Designation: MEPDG-1. Washington, DC
4. Dakoulas P, Gazetas G (1985) A class of inhomogeneous shear models for seismic response of dams and embankments. Int J Soil Dyn Earthq Eng 9(4):166–182
5. De Marco S, Albano M, Russo G (2013) The menta hydropower plant: water flow analysis during the controlled filling of Menta Dam. In: 9° ICOLD European symposium, 10–12 April 2013, Venice, Italy

6. Fell R, MacGregor P, Bell G, Foster M (2015) Geotechnical Engineering of Dams, 2nd edn. CRC Press, Taylor & Francis Group, London, UK
7. Fenves G, Chopra AK (1987) Simplified earthquake analysis of concrete gravity dams. J Struct Eng 113(8):1688–1708
8. Foster M, Fell R, Spannagle M (2000) The statistics of embankment dam failures and accidents. Can Geotech J 37(5):1000–1024
9. Gazetas G, Dakulas P (1992) Seismic analysis and design of rockfill dams: state-of-the-art. Soil Dynam Earthq Eng 11:27–61
10. Idriss IM (1985) Evaluating seismic risk in engineering practice. In: Proceedings of 11th ICSMFE, San Francisco, pp 255–320
11. Ishihara, K (2010) Performances of rockfill dams during recent large earthquakes. In: 5th International Conference on Recent Advances in Geotechnical Earthquake Engineering and Soil Dynamics, San Diego (California), pp 1–11
12. Jansen RB (ed) (1988) Advanced dam engineering for design, construction, and rehabilitation. In: Van Nostrand Reinhold (ed), New York
13. Kramer SL (1996) Geotechnical earthquake engineering. Prentice Hall, Upper Saddle River
14. Marsal RJ (1972) Mechanical properties of rockfill in embankment dam engineering. Casagrande Volume, Wiley, pp 109–200
15. Pelecanos L, Kontoe S, Zdravković L (2018) The effects of dam–reservoir interaction on the nonlinear seismic response of earth dams. J Earthq Eng (in press), https://doi.org/10.1080/13632469.2018.1453409
16. Swaisgood JR (2014) Behavior of embankment dams during earthquake. J Dam Saf ASDSO 12(2):35–44
17. Swaisgood JR (2003) Embankment dam deformations caused by earthquakes. In: Proceedings 7th Pacific conference on earthquake engineering, 13–15 February, Christchurch, New Zealand
18. Vecchietti A, Cecconi M, Russo G, Pane V (2019) Seismic vulnerability of a rockfill dam through different displacement-based approaches. J Earthq Eng (in press). https://doi.org/10.1080/13632469.2019.1662346
19. Westergaard HM (1933) Water pressures on dams during earthquakes. Transactions 98(1):418–33
20. Wieland M (2006) Earthquake safety of existing dams. In: Proceeding First European Conference on Earthquake Engineering and Seismology, Geneva (Switzerland) pp 3–8
21. Yu L, Kong X, Xu B (2012) Seismic response characteristics of earth and Rockfill Dams. In: Proceedings 15th world conference on earthquake engineering, Lisboa, Portugal

Numerical Analysis of the Seismic Behavior of the Menta BFR Dam

D. Aliberti, M. Vecchiotti, E. Cascone, and G. Biondi

Abstract The paper describes the results of numerical analyses carried out for the assessment of the seismic behavior of the Menta Bituminous Faced Rockfill Dam, located in Italy. The computer code FLAC based on the finite difference method was used to build a numerical model using the standard input data provided in the framework of the 15th ICOLD International Benchmark Workshop on Numerical Analysis of Dams. The behavior of the rockfill material was modelled using an elasto-plastic constitutive model, taking into account the shear modulus reduction and the hysteretic damping caused by the development of cyclic shear strains in the dam body. Constitutive parameters for rockfill were calibrated using the available laboratory data. Dam seismic behavior with respect to the assigned input accelerograms was inspected; in particular, the seismic performance of the dam was assessed evaluating its response in terms of dam crest acceleration and displacements: the latter turned out to be relatively small at the selected shaking scenario. Finally, the dam seismic behavior was analyzed focusing also on the role of the temperature-dependent bituminous facing stiffness and the return period of the seismic input.

Keywords Dam · Numerical analysis · Dynamic · FDM · Rockfill · Bituminous facing

1 Introduction

The paper describes the numerical analyses carried out for the Menta dam in the framework of the 15th ICOLD International Benchmark Workshop on Numerical Analysis of Dams.

D. Aliberti · E. Cascone · G. Biondi
Department of Engineering, University of Messina, Messina, Italy

M. Vecchiotti (✉)
Geotechnical Engineer, Milan, Italy
e-mail: mauro.vecchiotti@gmail.com

G. Bolzon et al. (eds.), *Numerical Analysis of Dams*, Lecture Notes in Civil Engineering 91, https://doi.org/10.1007/978-3-030-51085-5_23

The Menta dam is a Bituminous Faced Rockfill Dam (BFRD) located in southern Italy on the Menta river. The crest elevation is 1431.75 m a.s.l. while the foundation elevation, at the reference section for this study, is 1347.88 m a.s.l.; the normal reservoir level is assumed 1424.50 m a.s.l.

Being a BFRD, the upstream dam face is made impervious by a series of thin layers (total thickness is approximately 30 cm) of bituminous concrete, starting from the reinforced concrete inner tunnel and reaching the crest of the dam.

The dam geometry is actually very complex, since it is curved and composed by two sections having different orientations: one main section 325 m long, and one secondary smaller section 125 m long (Fig. 1a); this latter is meant to close the small valley on the right side and is built alongside the spillway. Hence, the geometrical settings would require a very large and demanding 3D numerical model [1] but, according to the instructions by the Formulators, a simpler 2D plain strain model is analyzed.

The scope of the dynamic numerical analysis of such dam type is twofold: not only it is important to investigate the seismic behavior of the rockfill dam body, but also the integrity and water tightness of the upstream facing. Therefore, in addition to the evaluation of seismic-induced permanent displacements, possibly associated to the development of local or general failure mechanisms, possible damaging (i.e.

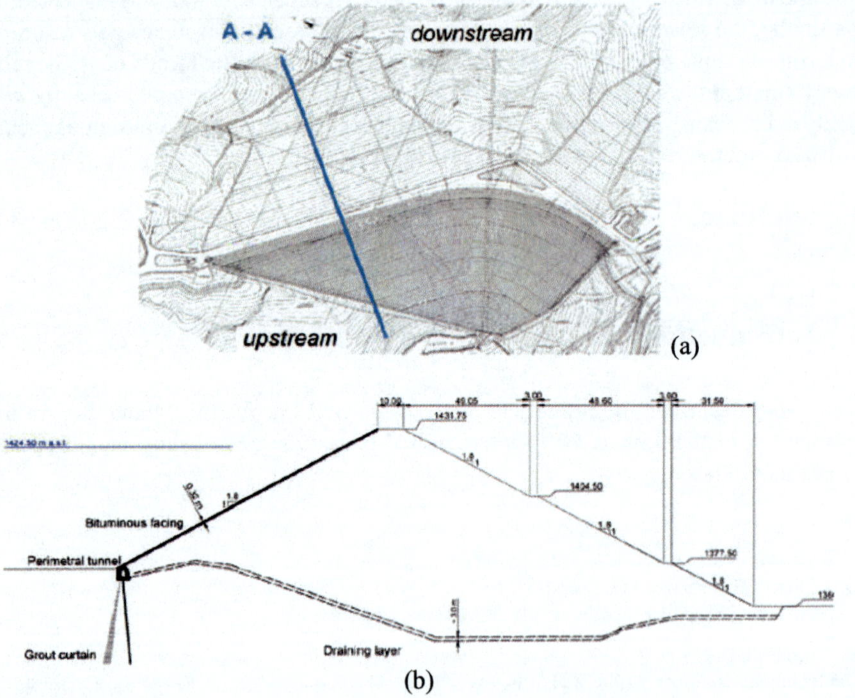

Fig. 1 Plan view of the dam (**a**) and cross section considered in the analyses (**b**)

cracking) of the upstream facing, caused by unacceptable displacements or by high axial stresses arising during the earthquake should also be investigated.

This latter is a key point especially for the Menta dam, since no particular dam zoning is implemented to control seepage through the dam body, in case of massive leakage from the upstream face. In this paper the results of dynamic numerical analyses are presented, addressing both the above-mentioned issues and taking into account different shaking intensities. Comments relevant to the static dam behavior after the first impounding are provided as well.

2 Numerical Model

2.1 Input Data

The Menta dam has a total body volume of about 2.1 millions of cubic meters of compacted rockfill, excavated by blasting along the Menta river. The 2D geometrical section chosen for the analyses is shown in Fig. 1b, and is characterized by a maximum height of about 84 m above the foundation layer. The bedrock consists of fractured metamorphic rock belonging to the "Calabrian Complex": according to the Formulation document, the rock is assumed as homogeneous and linear elastic ($\gamma = 27$ kN/m^3; $E = 7$ GPa; $\nu = 0.25$). The dam embankment is also assumed to be homogeneous (i.e. no dam zoning is taken into account in the numerical model, even if 5 different zones can be distinguished based on grain size distribution) and is characterized by the mechanical and physical properties reported in Table 1 and Fig. 2.

Data made available by the Formulators comprise results of drained triaxial laboratory tests (TXCD-K0), as well as variation of the small strain Young Modulus E_0 with minor effective principal stress σ'_3 and variation of the normalized Young Modulus E/E_0 and damping with axial strain ε_a, as obtained in cyclic triaxial tests. Bituminous facing can be modelled as an equivalent single layer, for which complex Young Modulus E^* and Poisson ratio ν^* have been provided based on the model of the University of Maryland [2], making reference to a temperature range of 1°–26 °C (Table 2). Maximum compressive and tensile stresses were

Table 1 Summary of TXCD-K0 tests as per formulation document

σ'_3 (kPa)	$\sigma'_{1F}/\sigma'_{3F}$	φ' (°)	ε_a (–)	ε_v (–)	$\lvert d\varepsilon_r - d\varepsilon_a \rvert$	B_g (%)
100	7.10	48.80	−0.02	−0.0135	0.33	5
300	5.20	42.60	−0.06	−0.0405	0.39	6
900	4.20	38.00	−0.10	−0.114	0.16	7

σ'_3: minimum principal stress; σ'_{1F} and σ'_{3F}: maximum and minimum principal stresses at failure; φ': soil friction angle; ε_a, ε_r, ε_v: axial, radial and volumetric strain at failure; B_g: breakage index

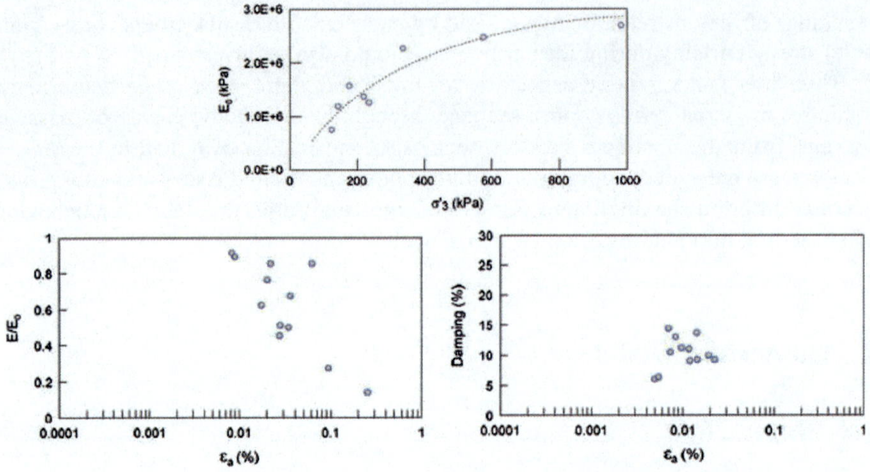

Fig. 2 Results of laboratory tests on rockfill as reported in the formulation document

Table 2 Elastic properties for bituminous facing

Temperature	E^* (MPa)	ν^* (–)
1 °C	14000	0.16
26 °C	2800	0.33

also provided ($f_c = 1490$ kPa, $f_t = 950$ kPa). As recommended by the Formulators seepage in the bedrock and dam body is disregarded, since it is prevented by the bituminous facing and by a grout curtain injected in the rock mass under the upstream toe of the dam.

The prescribed seismic input is represented by two groups of acceleration time histories (each consisting of two horizontal and one vertical components) selected to guarantee the spectro-compatibility with the Italian seismic hazard map for the Menta dam site, according to the Italian seismic code [3, 4]. Records belong to two significant Italian earthquakes, namely Friuli 1976 (HNN-HNE-HNZ) and Central Italy 2016 (HGN-HGE-HGZ) and have been provided with the horizontal component characterized by the highest amplitude scaled to a peak horizontal acceleration *PHA* = 0.26 g, and the other components accordingly scaled using the same scaling factor. For each (scaled) accelerogram Fig. 3 shows the relevant Fourier spectrum and Table 3 lists the main parameters; the values of a_g (the maximum acceleration expected in free field conditions on a stiff and horizontal soil deposit) at the dam site for different values of the return period T_R are reported in Table 4.

Thus, it is apparent that the selected target $PHA = 0.26\ g$ matches a return period $T_R = 475$ years. The Menta dam is an existing strategic dam, therefore as required by the Italian seismic codes [4, 5] the reference return period corresponding to a selected limit state is also shown in Table 4; accordingly it can be inferred that the provided seismic input is in between the Damage (*SLD*) and the Life Safety (*SLV*)

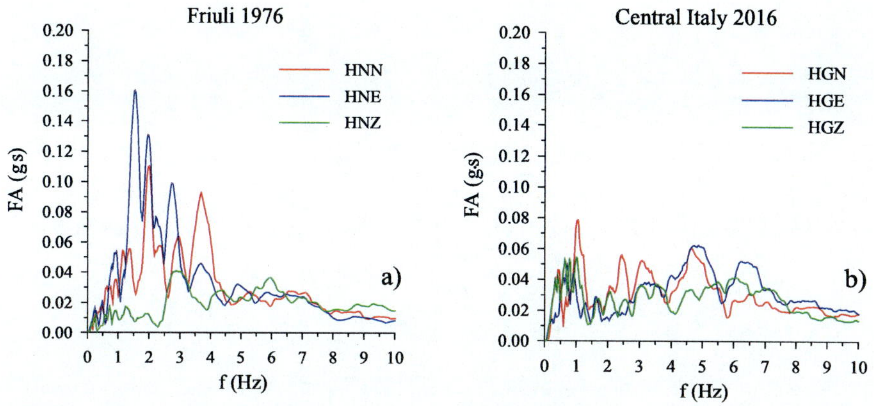

Fig. 3 Fourier Spectra for the scaled accelerograms: (**a**) Friuli 1976; (**b**) Central Italy 2016

limit states. For completeness, additional analyses with reference to the Collapse limit state (*SLC*, T_R = 1950 yrs) are presented and results compared with those requested by the Formulators.

Such check is performed using the Friuli 1976 input motion (Tolmezzo Station, May 06th, 1976 h.20.00.12 s), which has been scaled by the Formulators by a factor of about 0.74. At the *CLS* according to the Italian Seismic code, the expected peak horizontal and vertical accelerations are *PHA* = 0.454 *g* and *PVA* = 0.413 *g*, respectively. Thus, the scale factors to be applied to the original East-West (*HNE*) and vertical (*HNZ*) records, in order to reach the target *PHA* and *PVA*, are 1.29 and 1.51. Such values fall within the limits (0.5 ÷ 2) suggested in the literature for a proper scaling procedure.

2.2 *Discretization and Modelling*

The numerical analyses of the dam are performed by means of the finite difference (*FD*) code FLAC [6]. The reference section is discretized using approximately 13400 elements and includes the dam and the foundation rock mass that extends vertically to a depth of 40 m below the upstream ground level and horizontally for 150 m in the upstream and downstream directions. In order to achieve good quality results, the height *l* of the elements is set very small with respect to the minimum wavelength λ_{min} associated to the maximum frequency of the propagating waves, so that the condition $l \le \lambda_{min}/10$ [7, 8] is satisfied. Viscous and free-field boundary conditions are applied to the bottom and lateral sides of the model respectively, in order to correctly simulate the infinite extension of the bedrock in the vertical and horizontal direction. The prescribed seismic input is applied to the bottom boundary of the

Table 3 Summary of input earthquake data

Record		*PGA* (g)	*PGV* (cm/s)	*PGD* (cm)	I_A (cm/s)	P_d (10^{-4}g·s^3)	D_{5-95} (s)	T_m (s)	T_p (s)	*HI* (m)
#1	Friuli 1976 HNN	0.260	16.69	2.93	44.34	10.42	4.25	0.40	0.50	0.55
#2	Friuli 1976 HNE	0.237	22.60	4.77	67.79	26.89	4.93	0.50	0.64	0.66
#3	C. Italy 2016 HGN	0.186	18.55	5.55	40.44	1.57	7.35	0.41	0.21	0.63
#4	C. Italy 2016 HGE	0.260	16.03	4.47	40.12	1.53	6.90	0.28	0.22	0.51
	Friuli 1976 HNZ	0.198	7.97	1.99	17.51	0.60	4.77	0.22	0.33	0.17
	C. Italy 2016 HGZ	0.207	15.32	7.70	29.28	1.75	7.73	0.43	0.18	0.56

PGA, *PGV*, *PGD*: peak ground acceleration, velocity and displacement; I_A: Arias intensity; P_d: destructiveness potential factor; $D_{5\text{-}95}$: Trifunac & Brady duration; T_m, T_p: mean and predominant period; *HI*: Housner intensity

Table 4 Summary of a_g and limit states for the Menta dam site

T_R (yrs)	a_g (g)	Limit state	T_R (yrs)	a_g (g)
30	0.065	*SLO*	60	0.097
101	0.127	*SLD*	101	0.127
475	0.262	*SLV*	949	0.347
2475	0.496	*SLC*	1950	0.454

model, and the rock basement is assumed as a compliant base. The deconvoluted velocity time history is converted into a shear stress time-history in FLAC.

In order to capture the dam behavior during the earthquake shaking, the rockfill was assimilated to an elasto–plastic material obeying to the Mohr–Coulomb failure criterion while the non-linear and dissipative behavior was represented using the hysteretic model *Sigmoidal 3*, available in the FLAC library. Rockfill friction angle is assumed to vary with the minor principal stress σ'_3 as $\varphi' = \varphi'_0 \cdot (\sigma'_3/p_{REF})^{-n}$, the parameters $\varphi'_0 = 48.8°$ and $n = 0.11$ being calibrated on the results of the TXCD-K0 tests (Fig. 4); the cut-off value $\varphi' = 48.80°$ is introduced for very small confining pressures and dilatancy is neglected. Likewise, small strain Young Modulus is assumed to vary as $E_0 = 2 \cdot (1 + \nu) \cdot S \cdot p_{REF}\ (\sigma'_3/p_{REF})^{\beta}$, where the parameters $\nu = 0.25$, $S = 3840$ and $\beta = 0.50$ are calibrated on available laboratory tests data (Fig. 4). *Sigmoidal 3* model parameters are set to fit the lower bound $G/G_0-\gamma$ curve proposed by Rollins et al. [9] for gravel and sandy gravel (Fig. 5a) that fairly reproduces the normalized stiffness curves $G/G_0-\gamma$ available in the literature [10] obtained through large scale cyclic triaxial tests on rockfill dam materials (Fig. 5a) with grain size distribution similar to that of the specimen tested in the TXCD-K0 tests (Fig. 5b). The grain size interval considered by [9] for the definition of upper and lower bound $G/G_0-\gamma$ curves includes the grain size distribution of the soils of the Menta dam used in the laboratory tests (Fig. 5b); moreover the lower bound $G/G_0-\gamma$ curve allows a conservative evaluation of soil stiffness.

The bituminous facing is modelled by using liner elements having 3 degrees of freedom for each terminal node and an elasto-plastic constitutive model; a specific M-N domain is used to determine the maximum axial load as a function of the bending moment. Possible damage of the facing is detected by comparing the maximum dynamic load occurring during the earthquake and the liner resistance (Fig. 4c). Complex Young modulus E^* and Poisson coefficient ν^* are provided in the formulation document with reference to two values of the temperature: $T = 1$ °C and $T = 26$ °C (Table 2). Analyses in the present study are referred to both these extreme values and also to an average temperature $T_s = 13$ °C. At such temperature the complex Young modulus was evaluated by means of the empirical relationship:

$$E^*(\mathrm{T}) = E^*_R \cdot 10^{\alpha \cdot (T_R^2 - T^2)} \qquad (1)$$

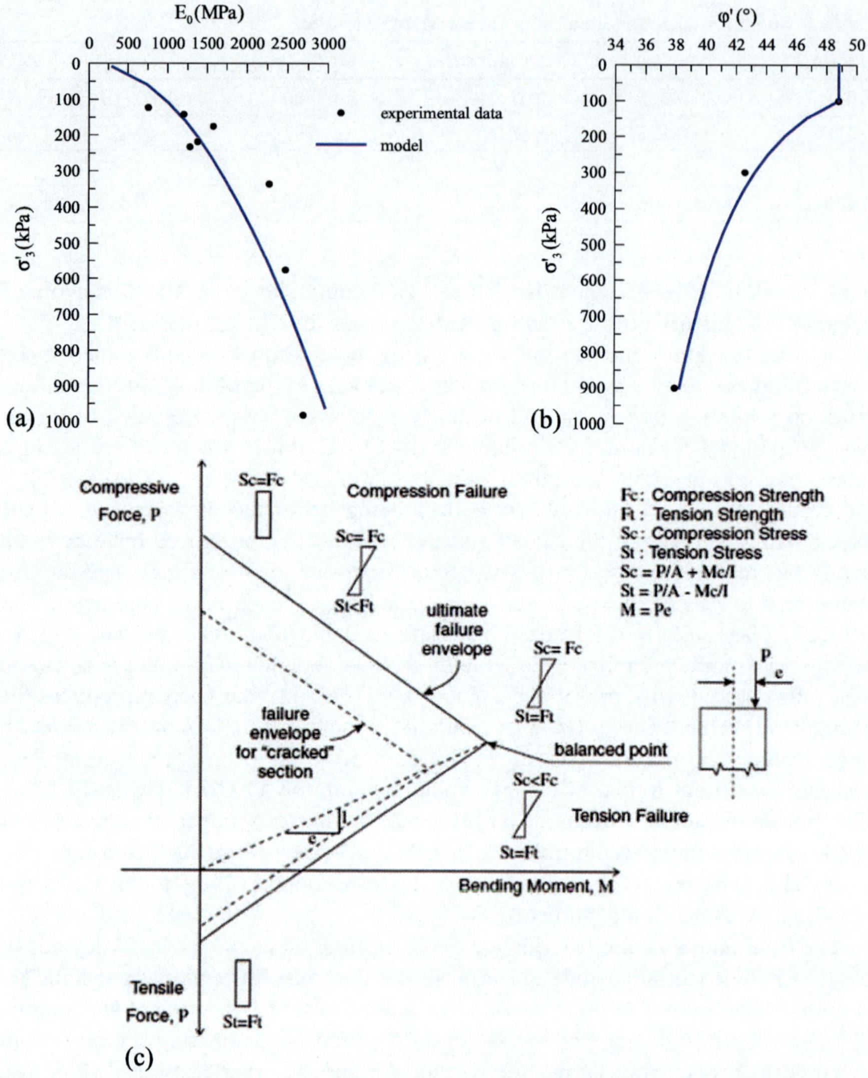

Fig. 4 Rockfill small strain Young modulus (**a**) and friction angle profiles (**b**); M-N domain for bituminous facing (**c**) [6]

where T is the temperature expressed in Fahrenheit degrees, $E*_{R} = 28000$ MPa is the reference Young modulus at $T = 26$ °C and $\alpha = 0.0001$ is the susceptibility parameter. Complex Poisson coefficient ν* is calculated based on the formulation [2]:

$$\upsilon * = 0.15 + \left[0.35/\left(1 + e^{-7500+2291 \cdot log|E*|}\right)\right] \quad (2)$$

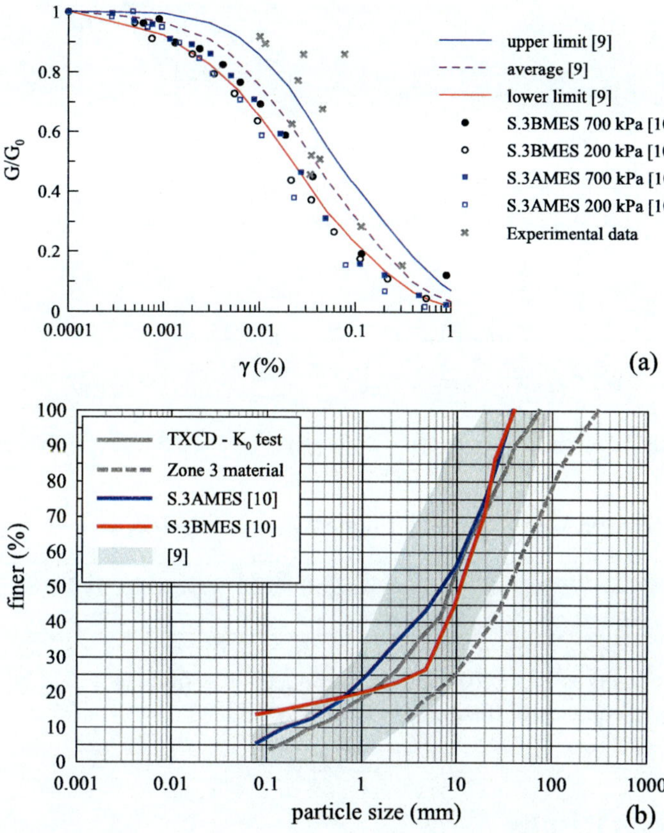

Fig. 5 Stiffness reduction (**a**) and grain size distribution for rockfill (**b**)

The obtained values are $E^*(13\ °C) = 5770$ MPa and $\nu^*(13\ °C) = 0.24$. Since seepage is neglected, the reservoir is taken into account by means of a hydrostatic load applied on the dam upstream face, starting from the normal water level 1424.50 m a.s.l.

This assumption is consistent with the results by Kontoe et al. [11], who showed that the effect of hydrodynamic pressures of the reservoir on the acceleration response of the dam is comparable to that obtained when hydrostatic pressures are applied as a boundary condition.

Preliminarily the first natural frequency of the system is evaluated by running an elastic dynamic undamped analysis, using a white noise as input motion; Fig. 6 shows the amplification function along the dam axis, obtained as the ratio between the Fourier amplitude spectra of the acceleration time histories at the crest and at the foundation elevations.

The condition of empty and full reservoir are both simulated, leading to $f_{1,\text{empty}} = 2.32$ Hz and $f_{1,\text{full}} = 2.44$ Hz respectively. It is worth noting that such values are

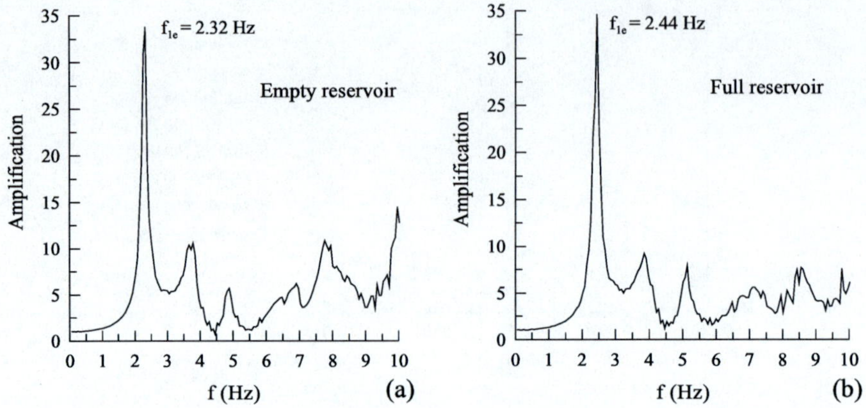

Fig. 6 Amplification functions at the dam axis for empty (**a**) and full reservoir (**b**) conditions

in good agreement with the elastic solution $T_1 = 2.59\ H/V_s = 0.46$ s applicable to homogeneous and triangular-shaped dams (being $H = 84$ m the maximum dam height and $V_s = 475$ m/s the average rockfill shear wave velocity based on the E_0 profile).

Finally, to avoid numerical instability, for all materials a viscous Rayleigh damping as small as $\xi_{min} = 0.5\%$ was introduced at a frequency f_{min} equal to the fundamental frequency of the dam ($f_1 = 2.44$ Hz).

3 Analyses Results

3.1 *Static Behavior of the Dam*

Dam static behavior is investigated by running the model and calculating the stress field and dam displacements after the reservoir impounding.

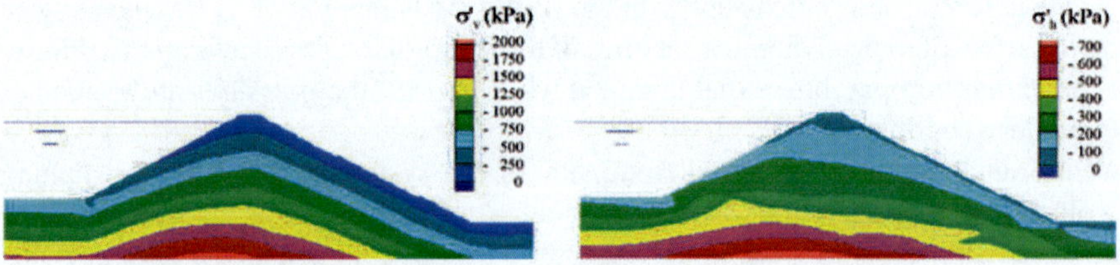

Fig. 7 Results of dam static analysis: contours of vertical (**a**) and horizontal (**b**) effective stresses after reservoir impounding functions

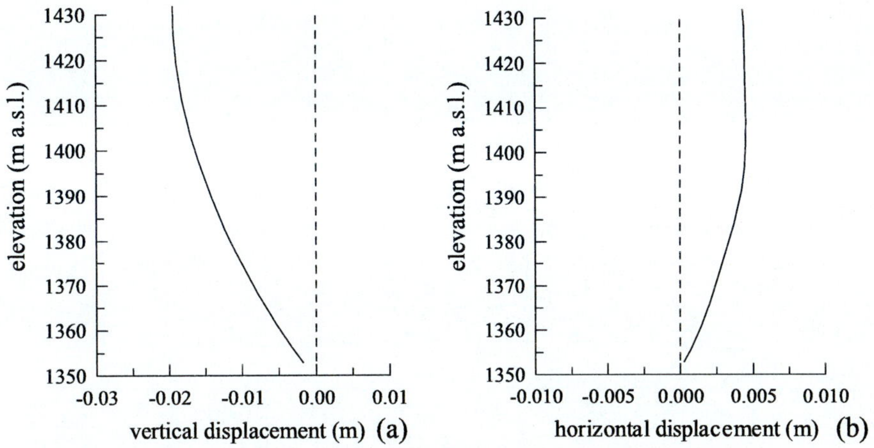

Fig. 8 Vertical (**a**) and horizontal (**b**) displacement along the dam axis after impounding

Figure 7 shows the contours of the vertical and horizontal effective stresses in the dam body that are barely affected by the irregular profile of the bedrock.

Horizontal displacements caused by the first impounding are very small and therefore negligible, as shown in Fig. 8 with reference to the vertical axis of the dam.

3.2 Dynamic Behavior of the Dam

A first set of dynamic analyses was run assuming as input motion the four horizontal components of the acceleration records; successively, in order to assess the effect of the vertical acceleration on the response of the dam, analyses accounting for the vertical components *HNZ* and *HGZ* were also carried out. Facing stiffness was assumed at the reference temperature $T = 13$ °C.

Table 5 summarizes the main results in terms of maximum horizontal and vertical crest acceleration, $a^c_{x,max}$ and $a^c_{y,max}$, ratios of crest to base maximum horizontal ($a^b_{x,max}$) and vertical ($a^b_{y,max}$) accelerations, maximum horizontal and vertical displacements in the dam, $u_{x,\max}$ and $u_{y,\max}$, and at the crest, $u_{x,\text{crest}}$ and $u_{y,\text{crest}}$. In addition, Fig. 9 shows the profile of the computed horizontal ($a_{x,\max}/a^b_{x,max}$; Fig. 9a, c) and vertical ($a_{y,\max}/a^b_{x,max}$; Fig. 9b, d) amplification factors along the dam axis, which is in good agreement with data from literature [12]. The introduction of the vertical acceleration in the analysis causes severe amplification of the shaking in both the horizontal and vertical directions: maximum calculated acceleration along the dam axis increases by a factor of about 1.50 and 2.00 in the horizontal and vertical direction respectively. Calculated displacements are generally negligible when vertical shaking is neglected, except for those caused by the Friuli HNE earthquake (Fig. 10);

Table 5 Summary of results

Record	$a^c_{x,max}$ (g)	$a^c_{y,max}$ (g)	$\frac{a^c_{x,max}}{a^b_{x,max}}$	$\frac{a^c_{y,max}}{a^b_{y,max}}$	$u_{y,max}$ (cm)	$u_{x,max}$ (cm)	$u_{y,crest}$ (cm)	$u_{x,crest}$ (cm)
#1 (*)	0.325	0.285	1.73	1.52	−0.40	1.25	0.26	0.23
#2 (*)	0.289	0.270	1.01	0.94	−2.50	6.00	−0.90	4.64
#3 (*)	0.291	0.352	1.73	2.09	−0.30	0.80	0.02	0.65
#4 (*)	0.333	0.408	1.82	2.23	−0.30	0.20	−0.15	−0.04
#1	0.461	0.527	2.15	2.46	−5.50	10.50	−0.40	0.94
#2	0.532	0.752	2.64	3.732	−13.50	31.00	−8.03	21.30
#3	0.417	0.443	3.03	3.23	−1.80	2.20	−1.34	1.08
#4	0.423	0.547	2.5245	3.2708	−1.50	3.50	−0.37	1.16

*Vertical earthquake component neglected

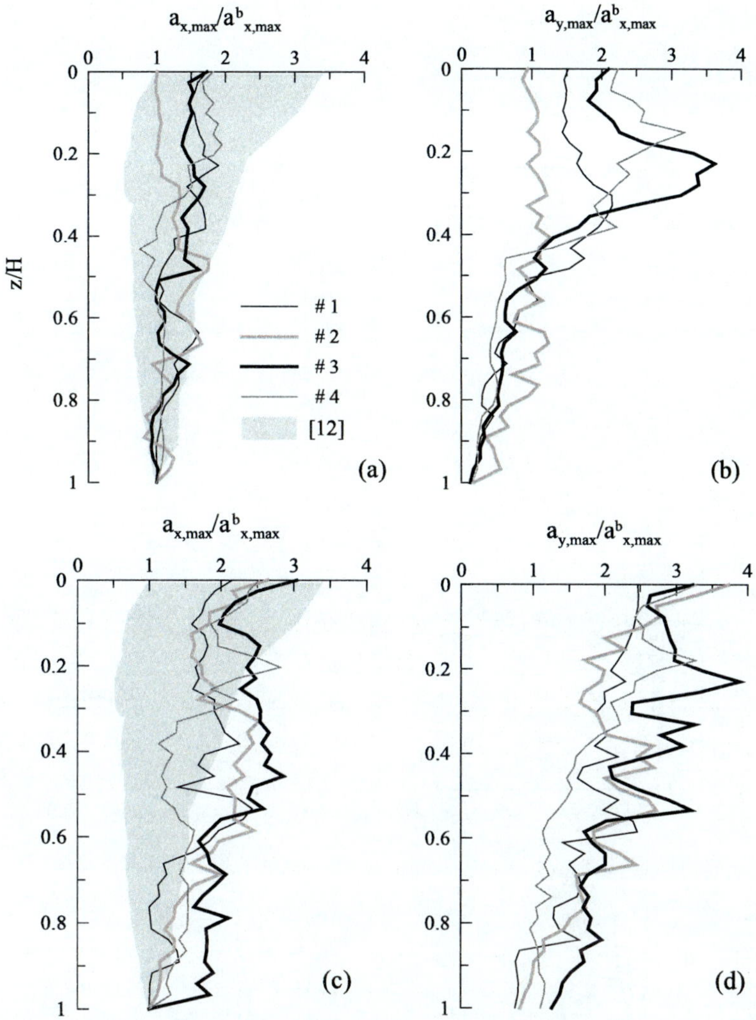

Fig. 9 Maximum horizontal (**a**, **c**) an vertical (**b**, **d**) accelerations computed along the dam axis neglecting (**a**, **b**) or accounting for (**c**, **d**) the vertical earthquake component

this is justified by the Fourier spectrum of such record (Fig. 3), which is particularly severe in a range of frequencies close to the first natural frequency of the dam.

Again, dam crest permanent displacements increase if vertical shaking is accounted for, even if they are acceptable for dam stability and serviceability.

Permanent dam crest displacements caused by Friuli *HNE* earthquake (#2) are the largest, reaching 8 and 21 cm in the vertical and horizontal direction respectively. Shear bands mainly affecting the downstream shell of the dam are observed, characterized by values of the shear strain generally less than 1% (Fig. 11).

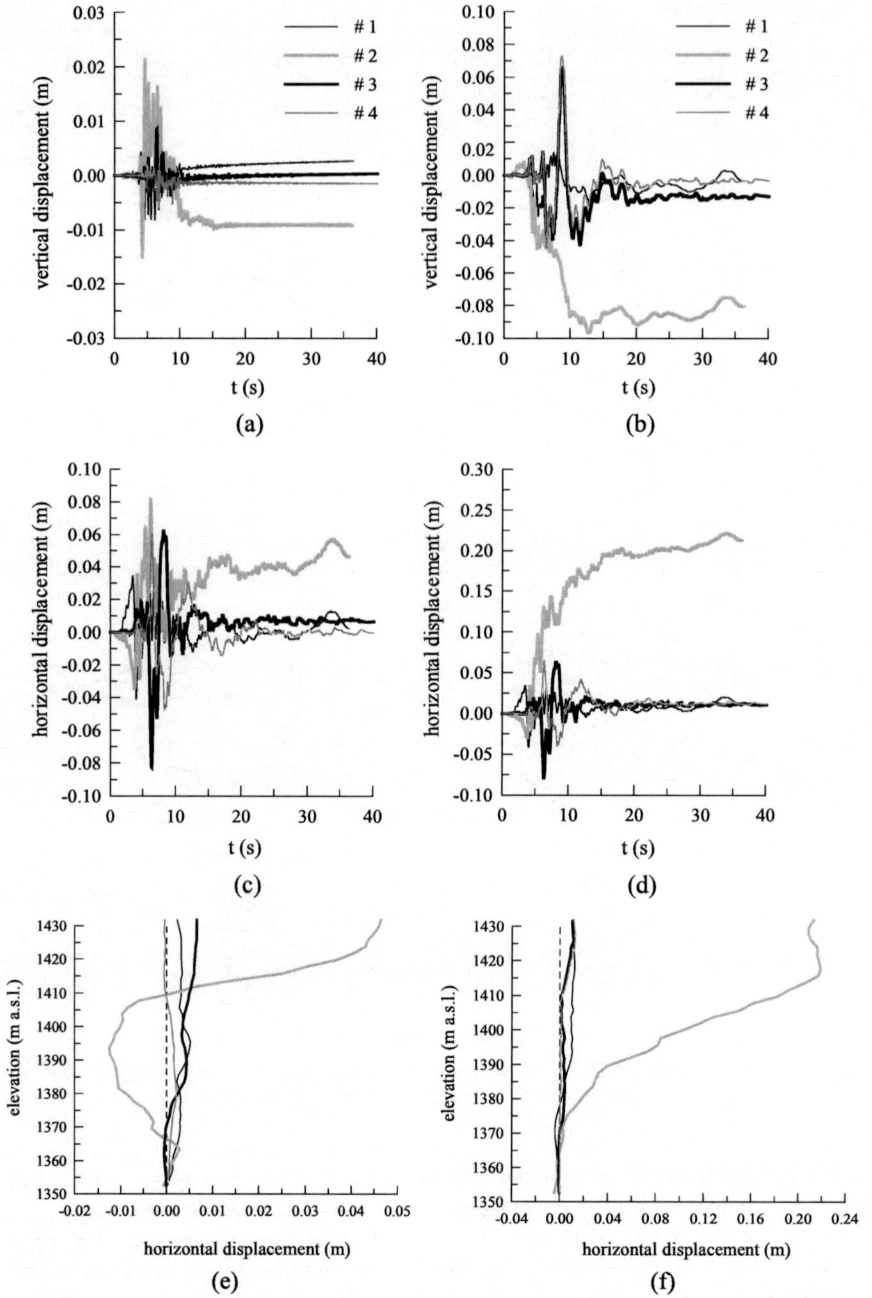

Fig. 10 Displacement time histories at the dam crest computed neglecting (**a**, **c**) or considering (**b**, **d**) the vertical acceleration; horizontal displacement along dam axis computed neglecting (**e**) or considering (**f**) the vertical acceleration

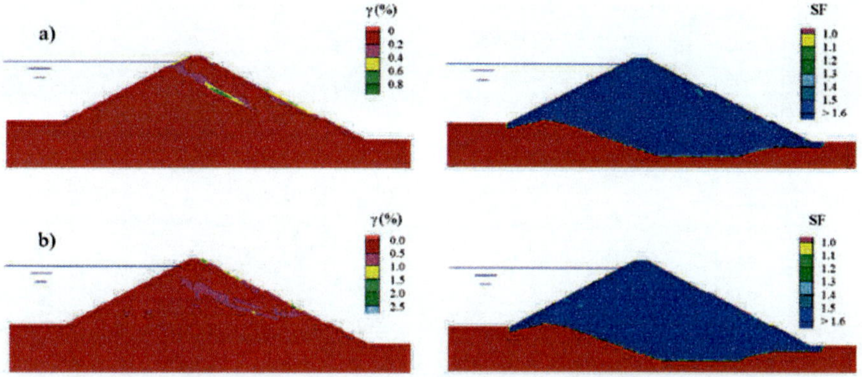

Fig. 11 Maximum shear strain and post seismic safety factor (Friuli *HNE*): (**a**) vertical acceleration neglected, (**b**) vertical acceleration included

Final safety factor (*SF*) calculated as the ratio between the available shear strength as per Mohr-Coulomb criterion and the mobilized shear stress after the earthquake is also quite large, envisaging a satisfactory performance of the dam under the assumed seismic input motions (Fig. 11).

3.3 *Effect of Temperature and Assessment of Bituminous Facing Damage*

Increase in temperature implies reduction of the bituminous facing stiffness (Table 2). The effect of such variation is assessed in this study for the two extreme values of temperature ($T = 1$ °C and $T = 26$ °C) using the most severe earthquake as input motion, i.e. Friuli 1976 *HNE* (#2); vertical shaking is also taken into account.

Figure 12a shows the displacements evaluated at the dam crest: facing stiffness has a minor effect on vertical settlement, but the horizontal displacement is significantly increased (+39%) in the case of $T = 26$ °C, with respect to the case of $T = 1$ °C. Such behavior is also confirmed by displacements (Fig. 13a–d) and maximum shear strains distribution shown in Fig. 13e, f; the same failure mechanism is developing in the dam body, but a stiffer facing contributes in reducing the final horizontal displacement. Permanent displacements seem acceptable in both cases, as also confirmed by the values of the post seismic safety factor (Fig. 13g, h). Since Friuli 1976 *HNE* is the most severe input, it is also used as a reference to check possible damaging of the bituminous facing. If the criterion shown in Fig. 4 is applied at the $T = 13$ °C, damaging caused by axial load occurs in most of the submerged facing (Fig. 14); therefore water flow in the dam body is theoretically possible, but since the adopted criterion does not provide any information concerning possible opening of cracks

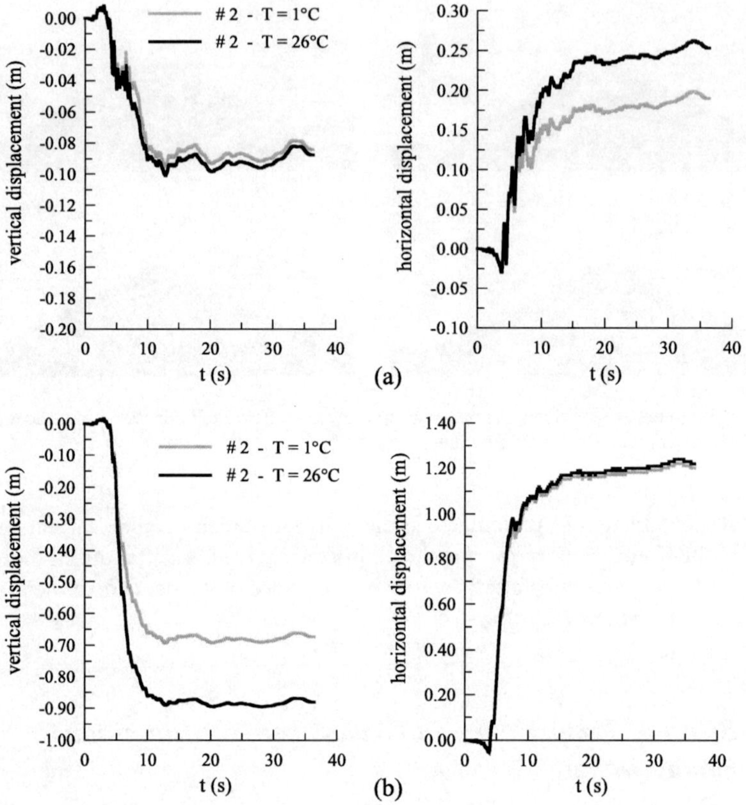

Fig. 12 Dam crest displacements at different temperatures: (**a**) input motion scaled to 0.26 g, (**b**) CLS input motion

in the facing, no clear conclusions can be drawn on this issue and more detailed analyses are required.

3.4 Collapse Limit State Dynamic Analyses

Dynamic analyses referred to the Collapse Limit State were run taking into account the vertical component of the seismic motion, for both the extreme temperatures $T =$ 1 °C and 26 °C. As expected, owing to the much stronger input motion displacements increase dramatically (Fig. 12b).

It is worth noting that contrary to the previous case (Fig. 12a), an increased facing stiffness (lower temperature) has a minor effect on the horizontal displacement while significantly reduces the vertical. Such behavior may possibly be ascribed to the different plastic mechanisms developing in the upper portion in the dam: as shown in

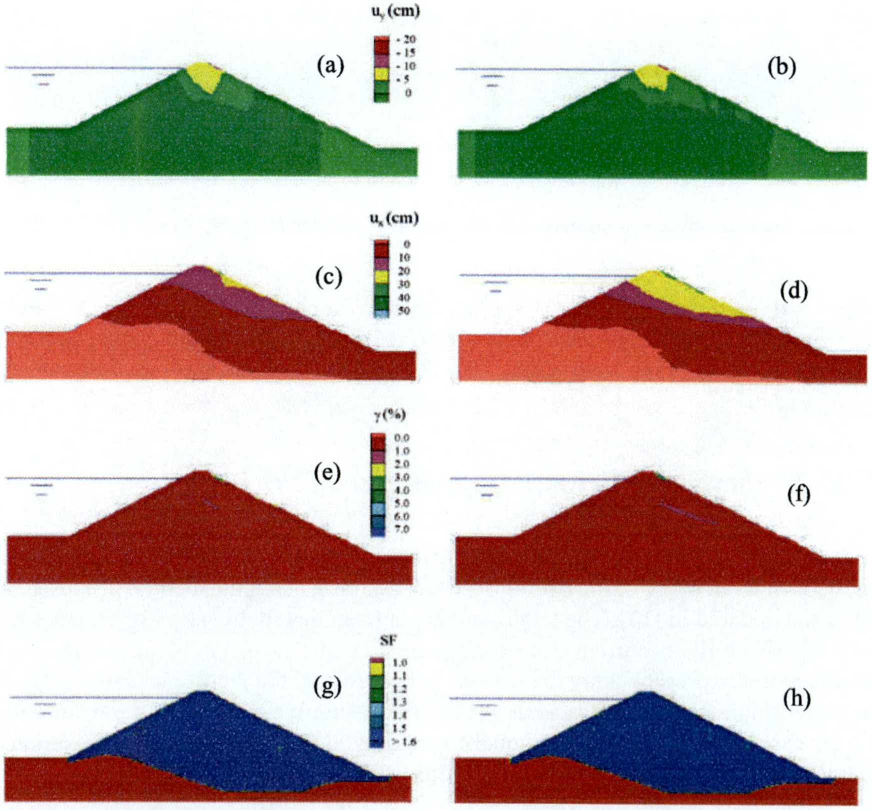

Fig. 13 Contours of vertical and horizontal displacements, maximum shear strain and safety factor at different temperatures: T = 1 °C (**a, c, e, g**); T = 26 °C (**b, d, f, h**)

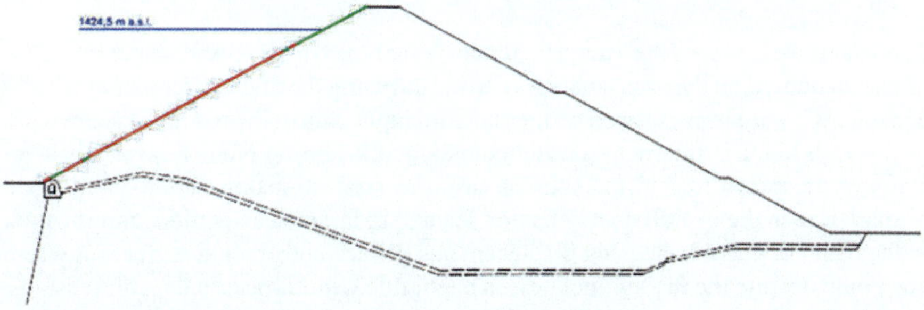

Fig. 14 Damage distribution in the bituminous facing (Friuli *HNE* + *HNZ*, *T* = 13 °C). Red = damaged, green = undamaged

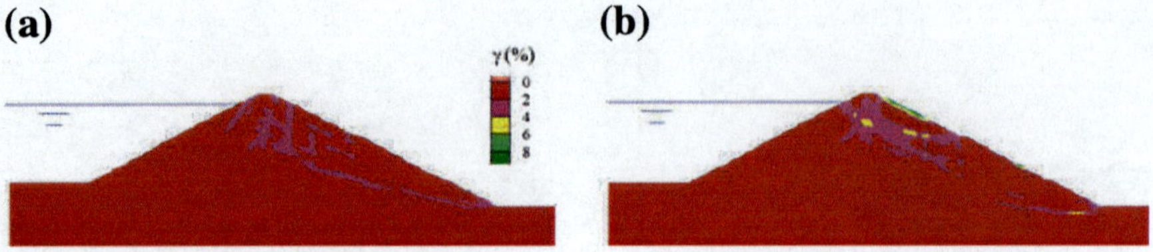

Fig. 15 Contours of maximum shear strain evaluated at *CLS* for $T = 1$ °C (**a**) and $T = 26$ °C (**b**)

Fig. 15, if the facing is stiffer ($T = 1$ °C, Fig. 15a) approximately linear shear bands are formed, whereas in the case of $T = 26$ °C (Fig. 15b) some rotational mechanisms take place.

3.5 Evaluation of Seismic Performance of the Dam

The seismic performance of the dam was finally assessed assuming the dam crest settlement as an index of the overall seismic response, using the method proposed in [13] and updated in [12]. The latter Authors defined threshold values of crest settlement to check the occurrence of possible limit states: accordingly, the settlements obtained in the present study are plotted in Fig. 16. All the records scaled to 0.26 *g* analyzed in the present study lead to minor or negligible damage of the dam; only in the case of a very strong earthquake (red dots in the chart) moderate to serious damage is expected, but no collapse condition is envisaged.

4 Conclusions

The seismic behavior of the Menta BFR dam is analyzed in the present paper by means of a set of numerical dynamic analyses carried out using the finite difference computer code FLAC. The study is based on the standard input data delivered in the framework of the 15th ICOLD International Benchmark Workshop on Numerical Analysis of Dams. It is shown that in the present case the vertical shaking component has a primary role in the overall dam behavior, leading to larger acceleration amplification in the dam body and increasing displacements. Bituminous facing stiffness is also a key point: despite the facing thickness is negligible with respect to dam dimensions, dam displacements can be different depending on temperature, at least in the case of full reservoir here analyzed. On the other hand, it is observed that damaging of the facing is also likely to occur because of axial loads, therefore possible leakage can be expected and the facing should be carefully inspected in case of earthquake. More detailed studies should be developed to assess the response of the bituminous facing, in order to estimate possible seepage through the dam body. Based on the

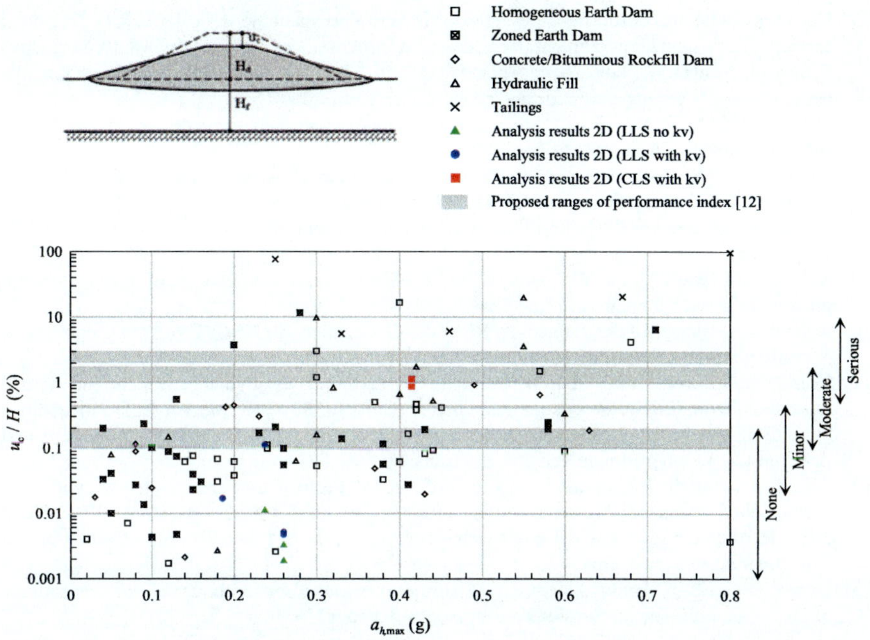

Fig. 16 Evaluation of dam damage as per [12]

computed displacements, it is finally shown that the reference shaking of this study is expected to cause only limited damage to the dam: moderate to serious damage can be expected only if stronger shaking referred to a much higher return period is taken into account.

Acknowledgments This work is part of the research activities carried out in the framework of the research projects of major national interest, PRIN n. 2017YPMBWJ, on "Risk assessment of Earth Dams and River Embankments to Earthquakes and Floods (REDREEF)" funded by the Italian Ministry of Education, University and Research (MIUR).

References

1. Albano M, Modoni G, Croce P, Russo G (2015) Assessment of the seismic performance of a bituminous faced rockfill dam. Soil Dyn Earthq Eng 75:183–198
2. AAVV (2004) Guide for mechanicistic empirical design of new and rehabilitated pavement structures. National Cooperative Highway Research Program, Washington DC
3. Ministero delle Infrastrutture e dei Trasporti. Aggiornamento delle Norme Tecniche per le Costruzioni. D.M. 17.01.2018. G.U. n.42 del 20.02.2018
4. Ministero delle Infrastrutture e dei Trasporti. Norme Tecniche per la Progettazione e la Costruzione degli Sbarramenti di Ritenuta (Dighe e Traverse). Decreto 26.06.2014. G.U. n.156 del 08.07.2014

5. Ministero delle Infrastrutture e dei Trasporti. Verifiche sismiche delle Grandi Dighe, degli scarichi e delle opere complementari e accessorie. Istruzioni per l'applicazione della Normativa Tecnica di cui al D.M. 26.06.2014 (NTD14) e al D.M. 17.01.2008 (NTC18). Direzione Generale per le dighe e le infrastrutture idriche ed elettriche – Luglio 2018
6. ITASCA (2011) FLAC—Fast Lagrangian Analysis of Continua—Version 7.0. User's Guide. Itasca Consulting Group, Minneapolis
7. Kuhlemeyer RL, Lysmer J (1973) Finite element method accuracy for wave propagation problems. J Soil Mech Found. ASCE 99(SM5):421–427
8. Lysmer J, Kuhlemeyer RL (1969) Finite dynamic model for infinite media. J Eng Mech. ASCE 95(EM4):859–877
9. Rollins KM, Evans M, Diehl N, Daily W (1998) Shear modulus and damping relationships for gravels. J Geotech Geoenviron Eng 124(5):396–405
10. Aghaei AA, Razeghi HR, Tabatabaei SH, Ghalandarzadeh A (2010) Dynamic properties of gravelly materials. Civil Eng 17(4):245–261
11. Kontoe S, Zdravković L, Han B, Pelecanos L (2019) Hydrodynamic effects and hydro-mechanical coupling in the seismic response of dams. In: Silvestri, Moraci (eds) Proceedings 7th ICEGE, Rome 17–20 June 2019, Earthquake geotechnical engineering for protection and development of environment and constructions, pp 506–521
12. Aliberti D, Biondi G, Cascone E, Rampello S (2019) Performance indexes for seismic analyses of earth dams. In: Silvestri, Moraci (eds) Proceedings 7th ICEGE, Rome 17–20 June 2019, Earthquake geotechnical engineering for protection and development of environment and constructions, pp 1066–1073
13. Swaisgood J (2003) Embankment dam deformations caused by earthquakes. In: Proceedings of Pacific Conference on Earthquake Engineering, Paper No. 14

Seismic Analysis of Menta Embankment Dam

E. Catalano, R. Stucchi, M. Agosti, and R. Crapp

Abstract The paper presents a study carried out to evaluate the mechanical behaviour, in static and dynamic conditions, of a 90 m-high bituminous-faced rockfill dam, the Menta dam. This calculation exercise was proposed in the frame of the 15th International Benchmark Workshop on Numerical Analysis of Dams.

Keywords Dam numerical analysis · Seismic analysis · ICOLD benchmark

1 Introduction

The Theme B of the 15th International Benchmark Workshop on Numerical Analysis of Dams dealt with the analysis of the seismic behaviour of a 90 m high bituminous-faced rockfill dam, the Menta dam. The scope of the analysis was to predict the stress-strain behaviour of the dam under seismic loading and to estimate potential damages.

The paper presents the main aspects of the numerical model that was prepared and the constitutive model that was chosen to reproduce the behaviour of the modelled materials in static and dynamic conditions. The results in terms of permanent displacements of the dam at the end of the earthquake shaking, together with some comments on the general safety of the dam, are presented and commented.

E. Catalano (✉) · R. Stucchi · R. Crapp
Lombardi Engineering Ltd, Minusio, Switzerland
e-mail: emanuele.catalano@lombardi.group

M. Agosti
Lombardi Engineering Ltd, Milano, Italy

G. Bolzon et al. (eds.), *Numerical Analysis of Dams*, Lecture Notes in Civil Engineering 91, https://doi.org/10.1007/978-3-030-51085-5_24

2 The Dam and the Seismic Input

2.1 General Description

The Menta dam (Fig. 1) is a bituminous-faced rockfill dam located in Southern Italy at an elevation of about 1400 m a.s.l. The embankment is about 90 m high at its deepest point, and the reservoir impounds 1.8×10^7 m^3 of water. The crest elevation is 1431.75 m a.s.l. and the normal reservoir level is at 1424.75 m a.s.l. The upstream and the downstream slope of the dam is 1.8:1.0 (H:V).

The impervious bituminous concrete facing is placed on the upstream side. The facing is formed by superposing different layers of bituminous concrete upon a regularly packed basement of compacted sandy gravel. The facing is anchored to the reinforced concrete tunnel running along the inner perimeter of the dam.

2.2 Material Properties

The stress-strain specific behaviour of this material depends on different parameters, such as mineralogy, grain size distribution, size of particles, stress level, and particle breakage, if any. The mechanical behaviour of rockfill is highly non-linear and, in general, both strength and stiffness are found to be dependent on applied stress level. Under seismic loading, a decrease of stiffness with increasing strain level is expected. Results from TXCD-K0 laboratory tests were made available by the Formulators (see Fig. 2). The following considerations can be formulated:

- Taking into account the principles of similitude [1], the information on the mechanical behaviour of the dam material can be directly obtained from TX tests on a material which has a similar grain size distribution curve and similar sample size ratios. This is the case for the present study.

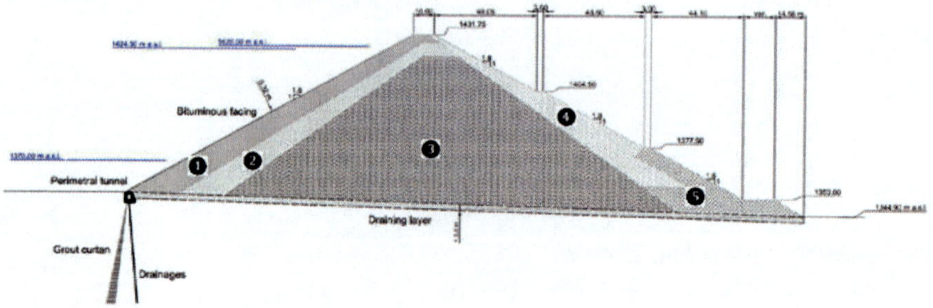

Fig. 1 Menta Dam—Typological section of the embankment

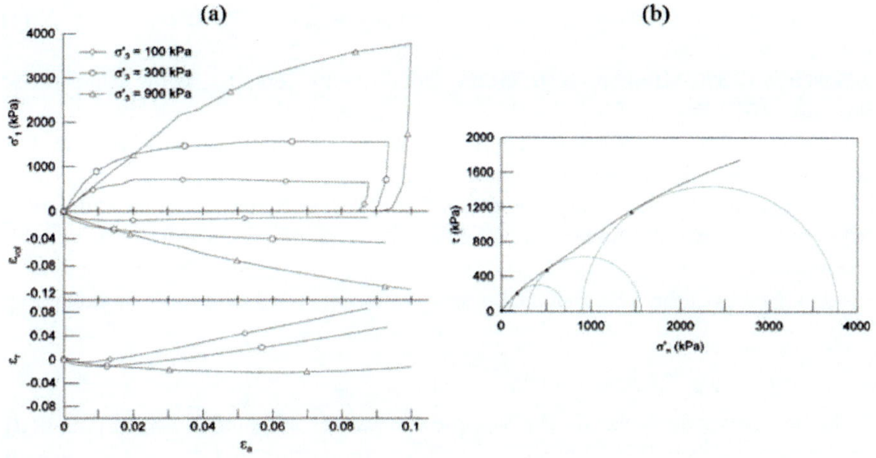

Fig. 2 TXCD-K0 laboratory tests (formulation documentation)

- The evolution of the radial deformation and volumetric deformation indicates that the mechanical behaviour of the material is strongly influenced by particle breakage. This would explain the mainly contractive behaviour (see Fig. 2a).
- The friction angle depends on the level of confinement, as well as on the particle breakage (see Fig. 2b).

On the basis of this information a stress-dependent Young modulus and friction, as described here below, was defined. The definition of a stress-dependent elastic modulus is based on the variation of small-strain Young modulus (E_0) with confining stress (σ_{conf}), issued from cyclic triaxial tests. The evolution of E_0 and the proposed interpolation are given in Fig. 3. The equation that describes the interpolation reads:

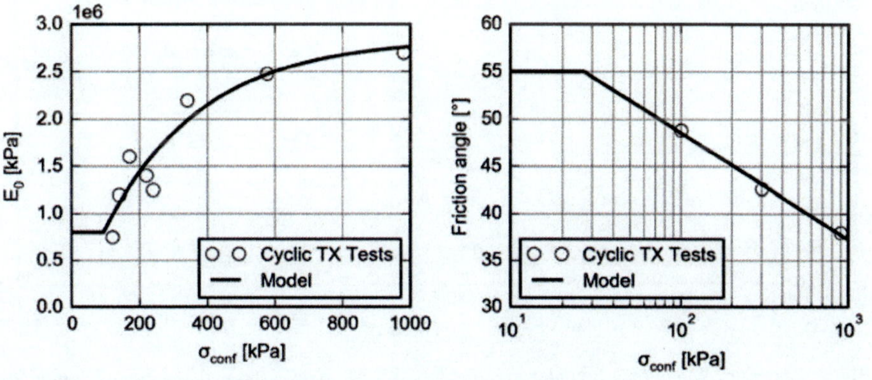

Fig. 3 Small-strain Young modulus (E_0) and friction angle as function of confining stress

$$E_0 = E_{ref} \cdot (1 - exp(-a \cdot \sigma_{conf/}p_{atm}) > 800\text{MPa} \quad (1)$$

where E_{ref} (2850 MPa) and a (0.35) are model parameters; p_{atm} is the atmospheric pressure (101 kPa).

The computed small-strain Young modulus is limited to 800 MPa at low confinement stress, as it can be seen in Fig. 3.

It is assumed that the Young's modulus E, that describes the behaviour of the rockfill in quasi-static conditions, can be taken as a fraction of the small-strain Young's modulus E_0.

On the basis of the TXCD-K0 laboratory tests, a good agreement is found taking:

$$E = 0.1 \cdot E_0 \quad (2)$$

In the next sections we will see how the computed static moduli for the dam body are between 50 and 200 MPa. In Fig. 4 a direct comparison of this range of values is made with literature data of measured moduli during construction and/or first filling, showing the soundness of the numerical assessment.

Following the ICOLD recommendations [3], on the basis of the TXCD-K0 laboratory tests (see Fig. 2b), the dependency of the friction angle on the confinement stress is expressed by the following expression:

$$\varphi = \Delta\varphi \cdot log\,(\sigma_{conf}) + \varphi_0 < 55° \quad (3)$$

where $\Delta\varphi$ (−11.3°) and φ_0 (71.2°) are model parameters. The computed friction angle is limited to 55° at low confinement stress, as it can be seen in Fig. 3.

The foundation is modelled as an elastic (E = 7 GPa, ν = 0.25) and perfectly impervious material. Monitoring data show there are no positive pore water pressures

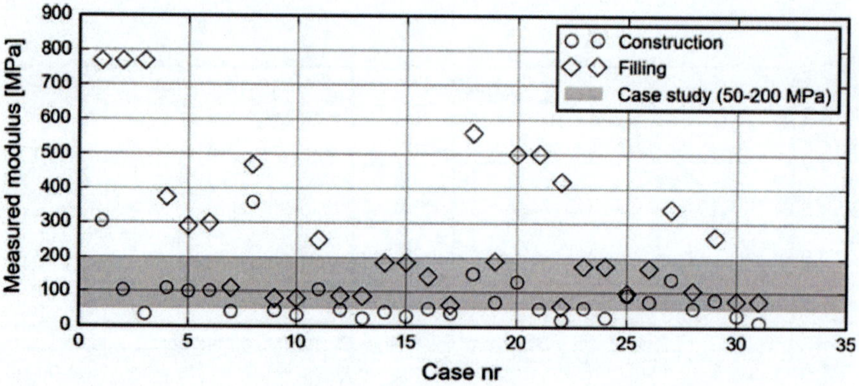

Fig. 4 Literature data of measured deformation moduli and comparison with the moduli computed in our case study [2]

Table 1 Mechanical parameters of the modelled materials

Material	Density [t/m^3]	Friction [°]	Cohesion [kPa]	Dilatancy [°]	k_0 [−]	E [MPa]	ν [−]
Dam	2.3	Var.	0.0	0.0	0.35	Var.	0.20
Foundation	2.7	elastic	elastic	0.0	1.0	7000	0.25

Table 2 Seismic input—Signal properties

Seismic input	Arias intensity [m/s]	PGA [g]	Duration [s]
Central V	0.30	0.21	40.13
Central H1	0.41	0.26	40.13
Central H2	0.41	0.19	40.13
Friuli V	0.18	0.20	36.39
Friuli H1	0.69	0.24	36.39
Friuli H2	0.45	0.26	36.39

within the embankment. The seepage phenomenon is disregarded, as recommended by the Formulators. The mechanical parameters are summarized in Table 1.

2.3 The Seismic Input

Two target acceleration time series (acceleration vs. time) represent the seismic input scenarios. These records were selected by the Formulators according to the criterion of spectro-compatibility with the Italian Seismic Hazard map for the site of the Menta Dam and represent two severe events occurred in Italy (Friuli earthquake occurred in 1976 and Central Italy earthquake sequence of 2016). The records are scaled to have a reference Peak Ground Acceleration equal to 0.26 g. The three components (2 horizontal and 1 vertical) of the seismic input were provided. The target records represent the acceleration recorded at rock outcrop in free-field conditions.

The main characteristics of the provided signals are presented in Table 2.

3 The Numerical Model

3.1 The Model

The software FLAC v. 8.00 [4] is used to perform the numerical simulations.

The model has a lateral extension of 650 m from the central axis of the dam. The bottom boundary is at about 450 m from the dam foundation (Fig. 5).

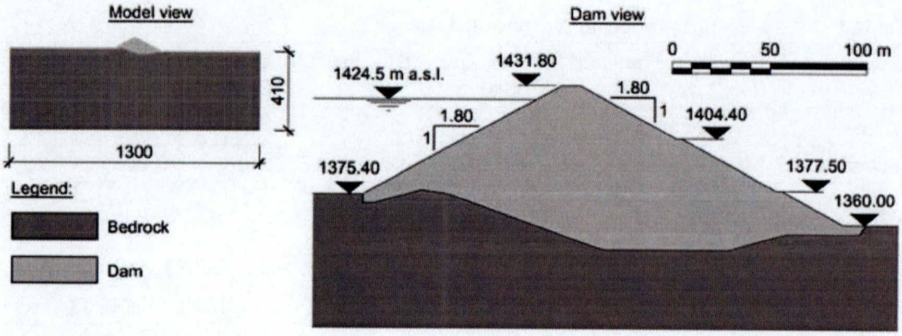

Fig. 5 The numerical model

The mesh was prepared for an accurate representation of the wave transmission with frequency up to 15 Hz, with a minimum size of 2.5 m for dam zones. The spatial element size must be smaller than approximately one-tenth to one-eighth of the wavelength associated with the highest frequency component of the input wave or accurate representation of wave transmission through a model [4]. Therefore, the choice of limiting the frequency transmission to 15 Hz is based on computing needs and on the fact that a rockfill dam is usually less affected by the signal content at high frequencies.

3.2 Constitutive Model and Boundary Conditions

The numerical model adopts a nonlinear Mohr-Coulomb constitutive model. The dependency of the material stiffness in the local value of the confinement stress is introduced, as described in Sect. 2.2.

The energy dissipation during dynamic events is performed in FLAC by means of the hysteretic damping (HD) algorithm, which reduces the shear modulus according to the strain state of the material. The use of this algorithm requires the definition of the strain dependent modulus function. This is done in agreement with the information provided by the Formulators, issued from cyclic TX tests.

In Fig. 6, laboratory measurements and model interpolation are presented. Both the modulus decay and the damping ratio are given as a function of the developed axial strain. It is observed that E/E_0 measurements are rather dispersed in the range $0.01\% < \varepsilon_a < 0.1\%$. Following a conservative approach, it was chosen to fit laboratory data with a "lower bound" curve.

During the quasi-static simulation (dam construction and reservoir filling), all displacements normal to bottom and lateral boundaries are imposed to be none.

During the dynamic simulation, quiet boundaries have been applied to avoid seismic wave reflections at the model bottom. Free-field boundary conditions are

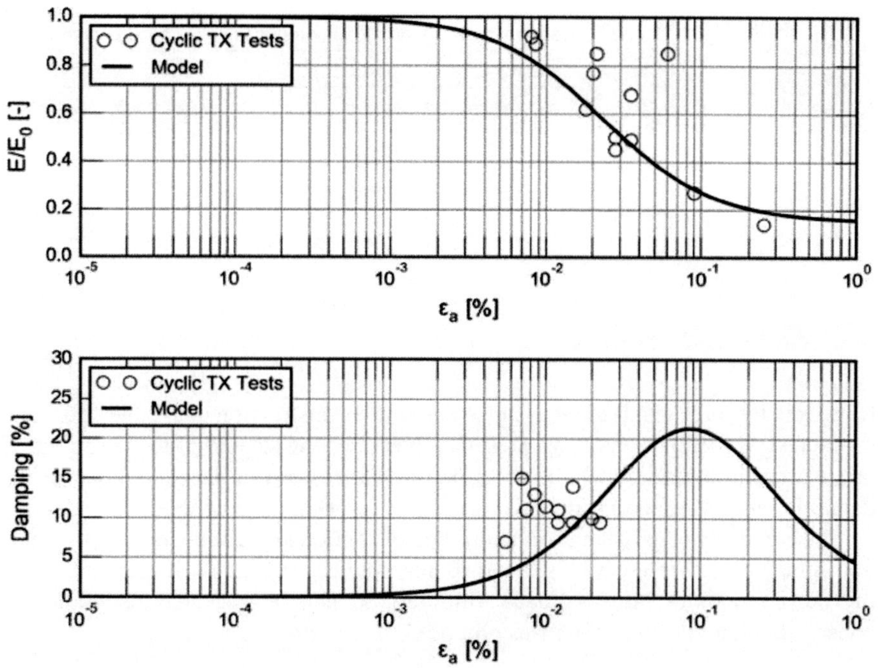

Fig. 6 Modulus decay and damping with increasing axial strain ε_a

defined at lateral boundaries to "enforce" the free-field motion, with the boundaries retaining their non-reflecting properties.

3.3 The Simulated Phasing

The following phasing has been simulated:

- Layered dam construction: ten layers (about 10 m thick) are considered; an equilibrium configuration is computed for each layer; the dam moduli and the friction angle are updated for each element at the end of each step according to the computed confinement stress.
- Reservoir filling: the maximum water level (1424.5 m a.s.l.) is reached in six steps; an equilibrium configuration is computed for each step; the dam moduli and the friction angle are updated at the end of each step.

The configuration computed at the end of the reservoir filling is the initial state for the dynamic simulation. The dynamic input combinations as listed in Table 3 are simulated.

Table 3 Dynamic input combinations

Combination	Horizontal component	Vertical component
Central VH1	Central H1	Central V
Central VH2	Central H2	Central V
Friuli VH1	Friuli H1	Friuli V
Friuli VH2	Friuli H2	Friuli V

4 Results and Discussion

4.1 Static Behaviour

Results in terms of displacements at the end of the construction phase and at the end of the reservoir filling phase are presented hereafter. At the end of the construction phase a maximum settlement of about 30 cm is measured in the central part of the dam (Fig. 7). The reservoir filling phase induces displacements of about 8 cm (Fig. 8).

The contour plot of Young's modulus and of friction angle at the end of the reservoir filling are presented in Figs. 9 and 10, respectively. In Fig. 4 a direct comparison of Young's modulus values is made with literature data of measured moduli during construction and/or first filling was proposed.

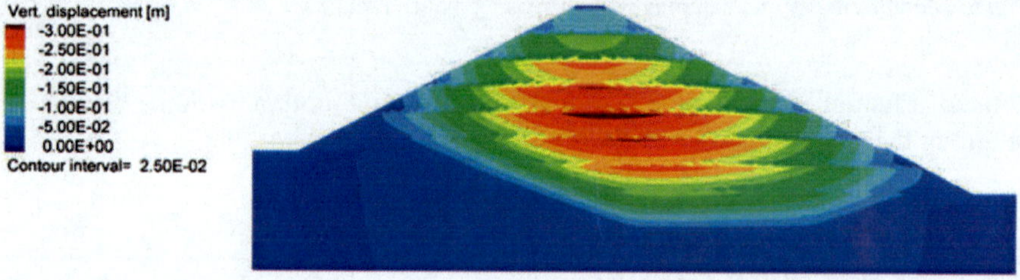

Fig. 7 End of construction—Vertical displacement [m]

Fig. 8 First impoundment behavior—Total displacement [m]

Fig. 9 First impoundment behavior—Young's modulus plot contour [MPa]

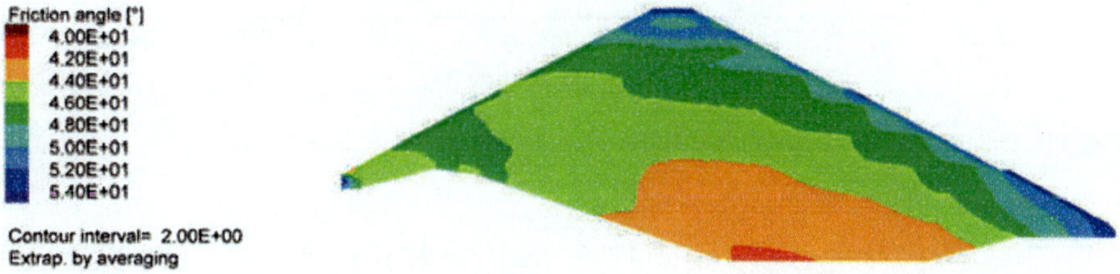

Fig. 10 First impoundment behaviour—Friction angle [°]

4.2 Seismic Behaviour

The natural frequency of oscillation of the system is estimated following the procedure described hereafter:

- The dynamic simulation is performed assuming elastic behaviour of materials.
- During the simulation the acceleration is recorded at the dam crest and in the foundation far from the dam footprint (free-field).
- The ratio between the Fast Fourier Transform (FFT) of the horizontal acceleration at the dam crest and the FFT of the horizontal acceleration recorded in the foundation is computed and plotted.

The obtained result is presented in Fig. 11, for both, empty and full reservoir conditions, respectively. The curves are found to perfectly coincide for the four seismic inputs. The first natural frequency of the dam is 1.85 Hz and 1.90 Hz (0.54 s and 0.53 s) for the empty and the full reservoir case, respectively. The shift on the right (higher stiffness due to higher confinement stresses) of frequency peaks for the full reservoir case is more evident for higher frequencies.

Analytical expression for the natural frequency (f_1) of a homogeneous 1-D shear beam [5] would give a result in good agreement with the obtained value:

$$f_1 = V_s/2.61\,H = 1.92\,\text{Hz} \tag{4}$$

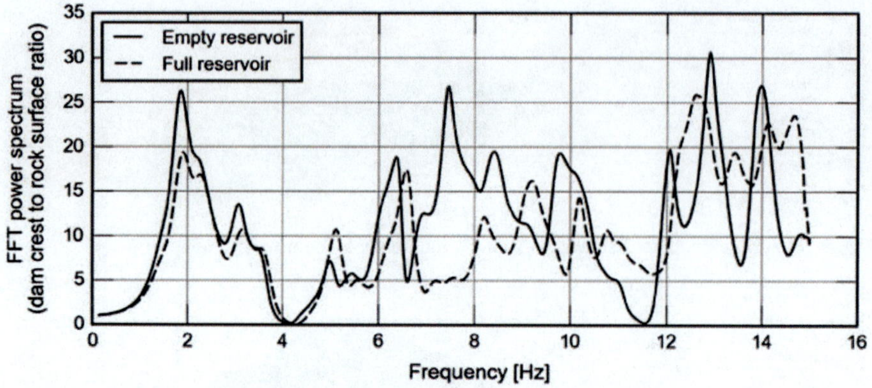

Fig. 11 Transfer function of Menta dam for empty and full reservoir conditions

where V_s (450 m/s) is the average shear wave velocity (the dynamic Young modulus is assumed to be 1100 MPa, which corresponds to an average value of the dynamic Young modulus of the dam); H is the "average" dam height (80 m).

The computed values for the first natural frequency are representative for a small strain behaviour of the dam. In case of higher strains induced by a stronger earthquake, the assumption of an elastic behaviour of the materials is no longer valid and the plasticization of the materials would cause a reduction of the first natural frequency. The maximum crest acceleration and the crest-to-base amplification that were computed for the four seismic inputs are listed in Table 4.

The crest vertical settlement computed during the dynamic simulations is plotted in Fig. 12. The maximum settlement during the simulation is computed for the Central VH1 and Central VH2 inputs and is of about 2 cm. The maximum final settlement is computed for the Friuli VH1 and Friuli VH2 inputs and is of about 1 cm.

The relative horizontal displacement at the crest is plotted in Fig. 12. The higher horizontal displacement is computed for the Friuli VH1 case.

As expected, the most significant deformation of the dam is computed for the Friuli VH1 input, the one with the highest Arias intensity, which is found to be a key indicator of the earthquake severity. However, the seismic loading does not induce significant plasticity events within the dam, resulting in very low permanent displacements which does not affect the dam safety in terms of freeboard. This result

Table 4 Crest acceleration and crest-to-base amplification

Combination	Max crest horizontal acceleration (g)	Crest-to-base amplification
Central VH1	0.45	1.8
Central VH2	0.38	2.6
Friuli VH1	0.56	2.7
Friuli VH2	0.62	2.4

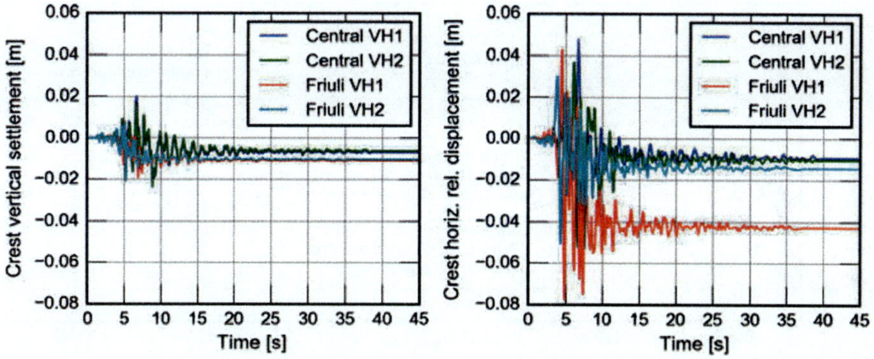

Fig. 12 Crest vertical settlement and horizontal relative displacement during the dynamic simulation

is due to the relatively low inclination of the dam faces and to the low value of the PGA. In fact, the assumed value for the PGA corresponds to a return period of only about 600 years according to the location of the dam. An MCE earthquake with a return period of 5000–10000 years would have a PGA of 0.63–0.84 g, thus significantly higher than the assumed PGA.

5 Asphalt Facing

Given the limited knowledge on the effective composition of the bituminous facing and the behaviour of the interface between the facing and the rockfill a conservative approach is followed by modelling the asphalt facing as a membrane glued to the upstream slope. A very low elastic modulus (E = 10 MPa) and moment of inertia ($I = 8*10^{-8}$ cm^4, which corresponds to a thickness of 0.01 cm) are considered for the structural element representing the asphalt, which is activated at the end of the construction phase. The intention is to conduct a conservative analysis and let the asphalt deform as the dam body does, acting like a glued membrane not resisting to bending moments.

The obtained result is represented in Fig. 13. The minimum and the maximum axial strain (please note that compression is positive here) that was experienced by the facing during the dynamic simulations is represented. The maximum computed tensile axial strain in dynamic conditions is below 0.2%. The result obtained for the first impoundment configuration is also plotted. In static conditions the asphalt facing is under tension in the bottom part (approx. 0.1%), and in the central part of the dam (approx. 0.05%).

A reference on the maximum allowed tensile strain in quasi-static conditions is given in Fig. 14 [6]. At low temperature (0 °C, E ≈ 14 GPa), the maximum allowable tensile strain is of about 1%. At high temperatures (25 °C, E ≈ 3 GPa),

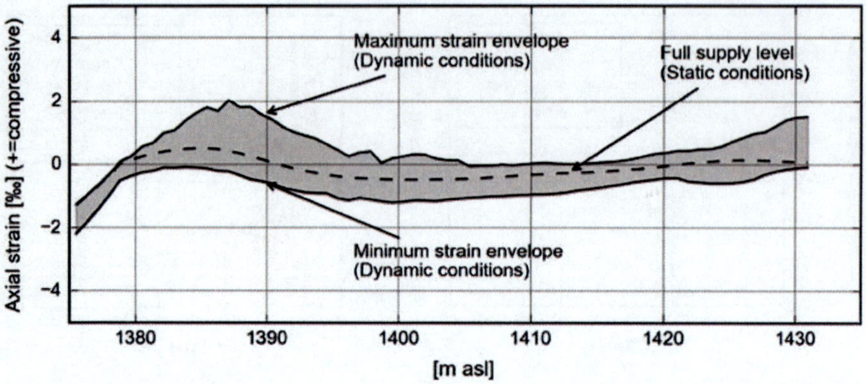

Fig. 13 Asphalt facing—Minimum and maximum axial strain experienced by the facing during dynamic simulations and comparison with static conditions (asphalt glued to the dam)

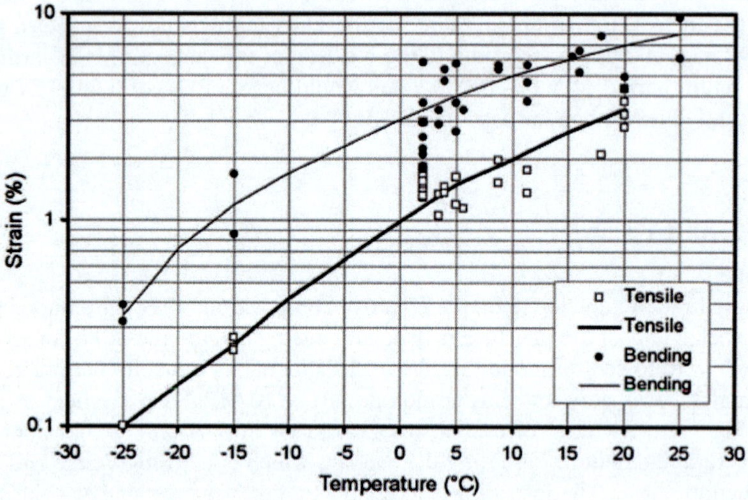

Fig. 14 Asphalt facing—Direct tensile strain at cracking versus temperature for nine different asphalt mixes [6]

the maximum allowable tensile strain is of about 7–10%. It is worth noticing that in dynamic conditions, i.e. at higher strain rates, the tensile capacity is generally lower. According to the representation given in Fig. 15 [7], the tensile capacity is of about 0.3% at −5 °C for a strain rate of 0.02 1/s.

On the basis of the collected information, and of the computed strain of the asphalt for the given seismic input, the safety of the asphalt facing can be guaranteed.

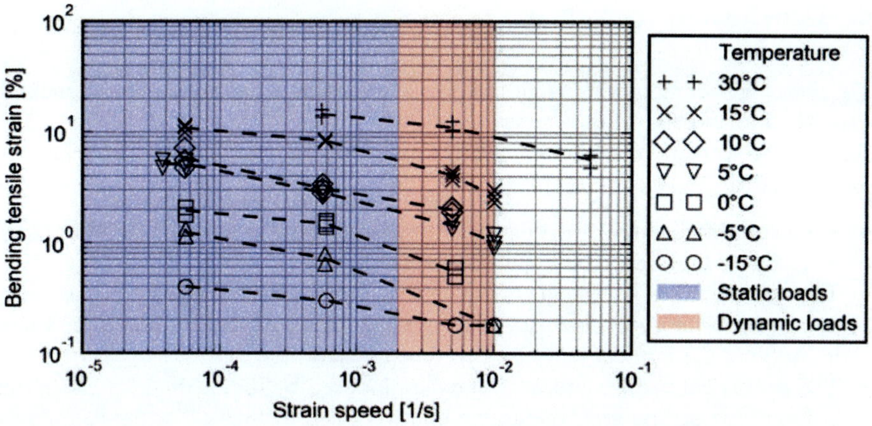

Fig. 15 Asphalt facing—Max. tensile strain versus strain rate for various temperatures [7]

For comparison, an analysis is conducted considering the actual properties of the asphalt face, in terms of stiffness (E = 14 GPa), thickness (30 cm) and possibility of relative displacement between the face and the dam (Mohr-Coulomb interface with a friction angle of 32°). The strains of the asphalt face in this condition are considerably lower, by approximately one order of magnitude, both in static and dynamic conditions, confirming the conservativeness of the proposed simplified approach (Fig. 16).

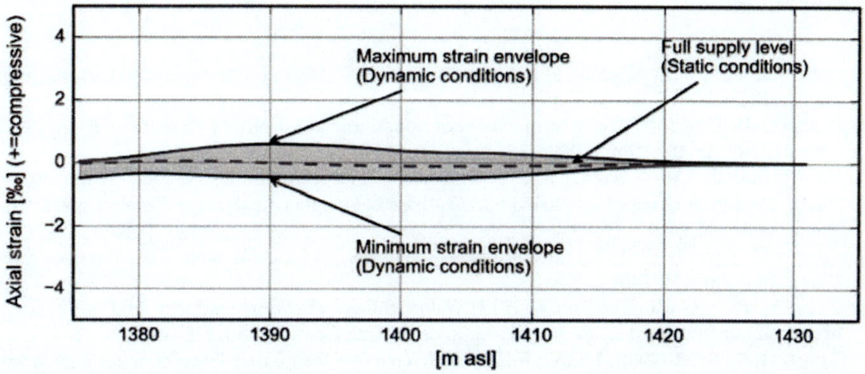

Fig. 16 Asphalt facing—Minimum and maximum axial strain experienced by the facing during the dynamic simulations and comparison with static conditions (asphalt face with actual material properties)

6 Conclusions

The present paper reports a seismic analysis of a 90 m high bituminous-faced rockfill dam, the Menta dam, in the framework of the Theme B of the 15th International Benchmark Workshop on Numerical Analysis of Dams. The following conclusions may be stated:

- The natural frequency of the dam is estimated to be 1.85 Hz/1.90 Hz, for empty reservoir and full reservoir conditions, respectively.
- The permanent displacements of the dam induced by the earthquake is positively correlated to the Arias intensity of the seismic input, which is found to be a key indicator of the input severity for this type of analysis.
- The permanent displacements of the dam induced by the earthquake are in the order of few centimetres, due to the relatively low inclination of the dam faces and to the low value of the PGA, which corresponds to a return period of about 600 years according to the location of the dam.
- The verification of the water tightness of the asphalt face is a crucial aspect for this type of dam. The conservative approach adopted in this paper assumes that the asphalt face is glued to the dam, i.e. no relative displacement can occur between the face and the dam, in this way maximizing the face strains. It could be demonstrated that the obtained maximum tensile strain is below the asphalt yield strain even for high strain rates under low temperatures.

References

1. Indraratna B, Wijewardena LSS, Balasubramaniam AS (1993) Large scale triaxial testing of greywacke rockfill. Géotechnique 43(1):37–51
2. Hunter G, Fell R (2002) The deformation behaviour of rockfill, UNICIV Report No. R-405, The University of the New South Wales, Sydney, Australia
3. ICOLD Bulletin 141—Concrete face rock-fill dams: concepts for design and construction (2010)
4. Itasca Consulting Group, Inc. (2016) FLAC—Fast Lagrangian Analysis of Continua, Ver. 8.0. Minneapolis: Itasca
5. Gazetas G, Dakoulas P (1992) Seismic analysis and design of rockfill dams: state of the art. Soil Dyn Earthq Eng 11:27–61
6. Zhang Y, Höeg K, Wang W, Zhu Y (2013) Watertightness, cracking resistance, and self-healing of asphalt concrete used as a water barrier in dams. Can Geotech J 50:275–287
7. Ishii K, Kamijo M (1988) Design for asphaltic concrete facing of Sabigawa upper dam, 16th Cong. Large Dams San Francisco 2:327–357

Numerical Simulation of Seismic Behavior of Menta Dam, Italy

A. K. Chugh

Abstract Results of static and dynamic analysis of Menta Dam, Italy, for two pre-selected seismic excitations are presented. Static analysis results include effects of: (a) mechanical and fluid flow through the weathered metamorphic rock foundation, (b) staged construction of the dam, and (c) one-step filling of the reservoir. Dynamic analysis results include effects of: (i) natural vibration characteristics of the dam and foundation, (ii) deconvolutions of the seismic accelerations, and (iii) two seismic excitations. Computer programs used include: FLAC (Fast Lagrangian Analysis of Continua) for the static and dynamic analyses; and SHAKE for the deconvolution analyses. Comments on the results presented are included.

Keywords Numerical analysis · Continuum mechanics · Menta dam · Seismic behavior · Vibration characteristics

1 Introduction

Our participation in this 15th International Benchmark Workshop on Numerical Analysis of Dams was motivated by two objectives: (a) check on our procedures and practices in analysis of earth dams located in high seismic regions of western U.S., and (b) learn from the collective knowledge and experiences of others engaged in similar activities. Ideally, we like to compare computed response of a dam with its measured response for known static, and seismic loadings; however, such is not the case with Menta Dam. Nonetheless, this workshop provides an opportunity to see how our computed results compare with those of others for a specific facility and specified seismic loading conditions.

Figure 1a shows the location map of the dam, and Fig. 1b shows an aerial view of the dam and appurtenant structures. Figure 2 shows the 2-D cross section of the dam selected for the benchmark. The dam height at the selected location is approximately (~) 85.5 m. Figure 3a shows the numerical model of the dam and foundation, and

A. K. Chugh (✉)
U.S. Bureau of Reclamation, Denver, CO 80225, USA
e-mail: achugh@usbr.gov

G. Bolzon et al. (eds.), *Numerical Analysis of Dams*, Lecture Notes in Civil Engineering 91, https://doi.org/10.1007/978-3-030-51085-5_25

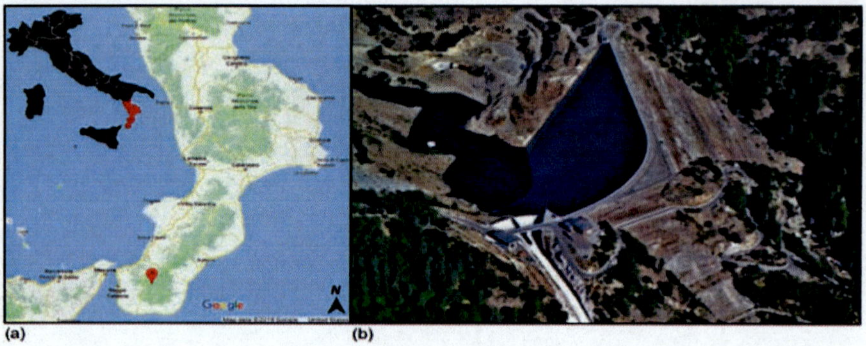

Fig. 1 Menta Dam, Italy—**a** location map (38.123336°N, 15.899301°E); **b** aerial view

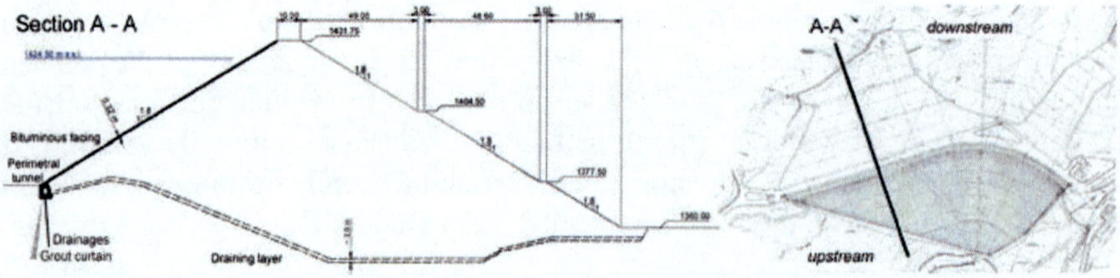

Fig. 2 Menta Dam cross section selected for the 2019 benchmark workshop

boundary conditions (b.c.) used in analyses. The planar geometry is described in x–y plane with the x-axis normal to the dam centerline and the y-axis pointing upward, see Fig. 3a. The model is 531 m wide × 186 m high and is discretized into 10,974 zones—177 along the x-axis and 62 along the y-axis; each zone is ~ 3 m × 3 m. Construction of the dam is modeled using twenty-eight ~3-m thick lifts as shown in Fig. 3b. Based on the Menta River elevations near the upstream (u/s) and downstream (d/s) toes of the dam, the regional ground water table (prior to the construction of the dam) is taken to be at elevation (el.) ~1350 m near the u/s end, and at el. ~1345 m near the downstream (d/s) end of the model.

Computer programs used for analyses include: (a) FLAC (an acronym for Fast Lagrangian Analysis of Continua) for static and seismic analyses, and (b) SHAKE for deconvolution of the rock outcrop seismic acceleration data to the bottom of the model. FLAC uses standard continuum-mechanics sign conventions for right-hand Cartesian coordinate system; SHAKE uses sign conventions commonly used in civil engineering, and Imperial units; see Refs. [8, 9] for details. Results included herein are in SI units; the polarity (±) of stress and deformation results are according to the sign convention used in the applicable computer program.

Special purpose functions included in FLAC were used when needed; additional functions were written to facilitate execution of data files and processing of results. Copies of data files used can be obtained from the author on request.

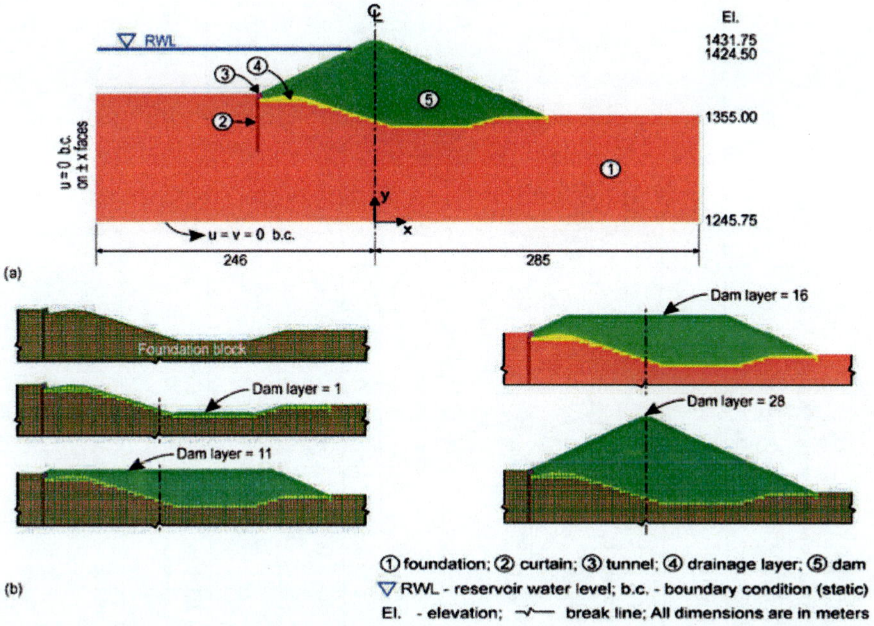

Fig. 3 Numerical model of Menta Dam cross section under study: **a** model details; **b** illustration of staged construction modeled

2 Material Properties

Table 1 shows the material properties used in the analyses; sources for the selected values are shown by reference #—these references include additional citations for sources of data (they are not repeated herein to conserve space). Shear and bulk modulus values used in FLAC and shear wave velocities were derived using the elasticity relations, Eqs. (5)–(7). In numerical analyses, the reservoir water was considered as mechanical/hydraulic pressure.

2.1 Deconvolution of Ground Motions

Figure 4a shows the SHAKE column for a 100 m deep foundation near the centerline of Menta dam; it is modeled using ten 10 m-thick horizontal layers (# 1 to 10) overlying half-space (# 11). The input data included: (i) unit weight, (ii) shear wave velocity, (iii) relation between shear modulus and damping with respect to cyclic shear strain, and (iv) rock outcrop motions. The material property data are shown in Fig. 4a. The outcrop motions data for Friuli and Central Italy events are shown in

Table 1 (a) Material properties used in static and seismic analyses of Menta Dam. (b) Modulus and damping factors used in seismic analyses

Item name, group name	Unit weight, γ (kN/m^3)	Young's modulus, E (kPa)	Poisson ratio, ν	Permeability, k (m/sec)	Angle of internal friction, φ (°)	Porosity	References #
Foundation, fdn	27	$E_{rock} = 7 \times 10^6$ in Eqs. (1, 2)	0.25	1×10^{-7}	–	0.10	[1, 10]
Grout curtain, ctn	21	1.4×10^4	0.21	0	–	0.01	[3]
Concrete tunnel, tun	24	3.2×10^7	0.16	1×10^{-9}	–	0.01	
Drainage layer, drn	20	$E_{rock} = 7 \times 10^6$ in Eq. (1)	0.25	1×10^{-3}	–	0.30	
Rockfill, dam	23	$25164 \times (p')^{0.6686}$	0.20	2×10^{-5}	$\phi'_{crit} = 36.45$; m = 3; Q = 10; $D_r = 0.9$ in Eq. (3)	0.30	[2, 7, 10]
Bituminous facing, fac	24	1.5×10^5	0.44	0	–	0.03	[5, 10]

Cyclic shear strain, γ_c (%)	1×10^{-4}	3×10^{-4}	1×10^{-3}	3×10^{-3}	1×10^{-2}	3×10^{-2}	1×10^{-1}	3×10^{-1}	1	References #
Shear modulus reduction factor, G/G_{max}	1.0	0.95	0.88	0.76	0.56	0.36	0.20	0.11	0.04	[1]
Damping ratio, D (%)	0.7	1.2	3.0	5.5	11.0	15.0	19.0	21.0	22.5	
Rayleigh Damping, (%)	0.05									
Implemen-tation in FLAC	3-parameter sigmoidal function, M_s, with parameters a = 1.03, b = −0.65, $x_0 = -1.9$ in Eq. (4). For Rayleigh damping, the specified damping is percent of critical damping, and central frequency is 1.2 Hz									

– not used; p' = mean principal effective stress in kPa; z in Eq. (2) is depth measured from top of the Foundation where z = 0 and increases in the negative y-coordinate direction, see Fig. 4a

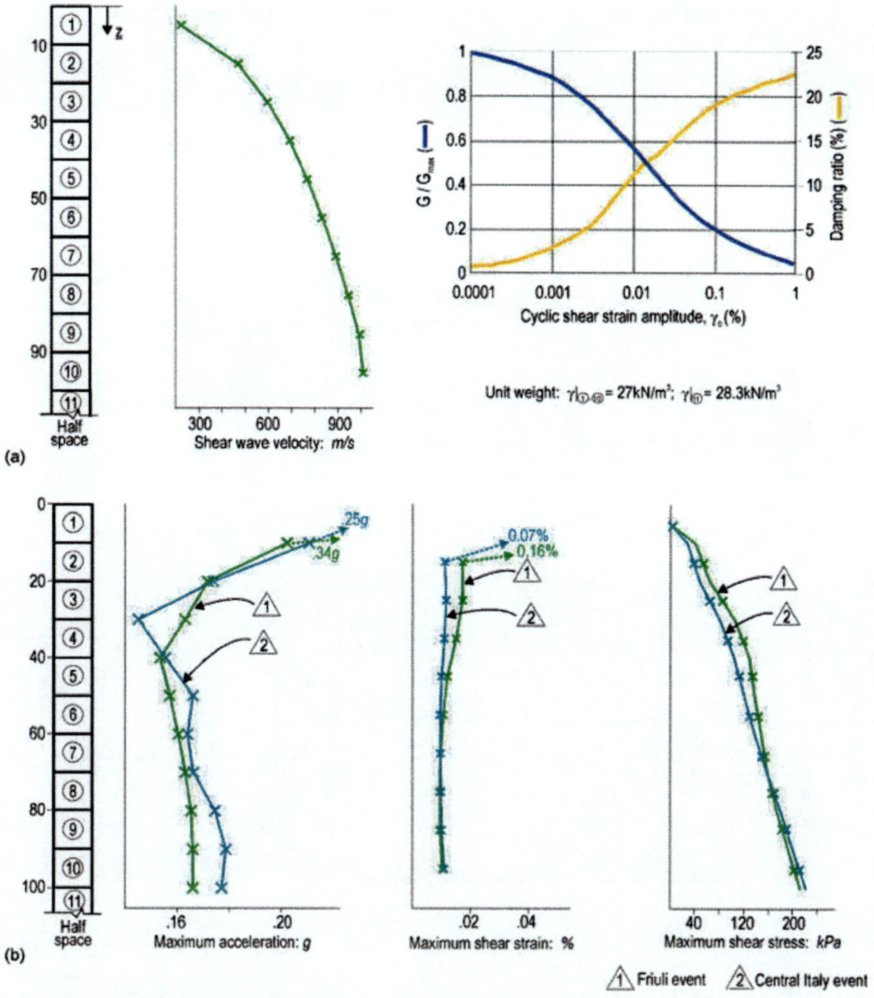

Fig. 4 Numerical details for one-dimensional wave propagation analysis: **a** input data used; **b** results of interest

Fig. 5a; sampling rate per sec for each of the two events is 200. All frequencies in the outcrop motions were retained in the deconvolution analyses.

SHAKE analysis results are shown in Fig. 4b and include: (i) maximum (max.) horizontal acceleration (a_{max}), (ii) shear strain ($\gamma_{c\,max}$), and (iii) shear stress (τ_{max}) profiles. Values of a_{max} and $\gamma_{c\,max}$ for layers # 1 and 2 are not plotted in Fig. 4b because of their relatively larger magnitude—instead their values for the layer # 1 are shown. The SHAKE analyses results correspond to the location within each layer of the SHAKE column. The computed motions at the top of layer # 11 for the Friuli and Central Italy events are shown in Fig. 5b and are identified as deconvolved motions.

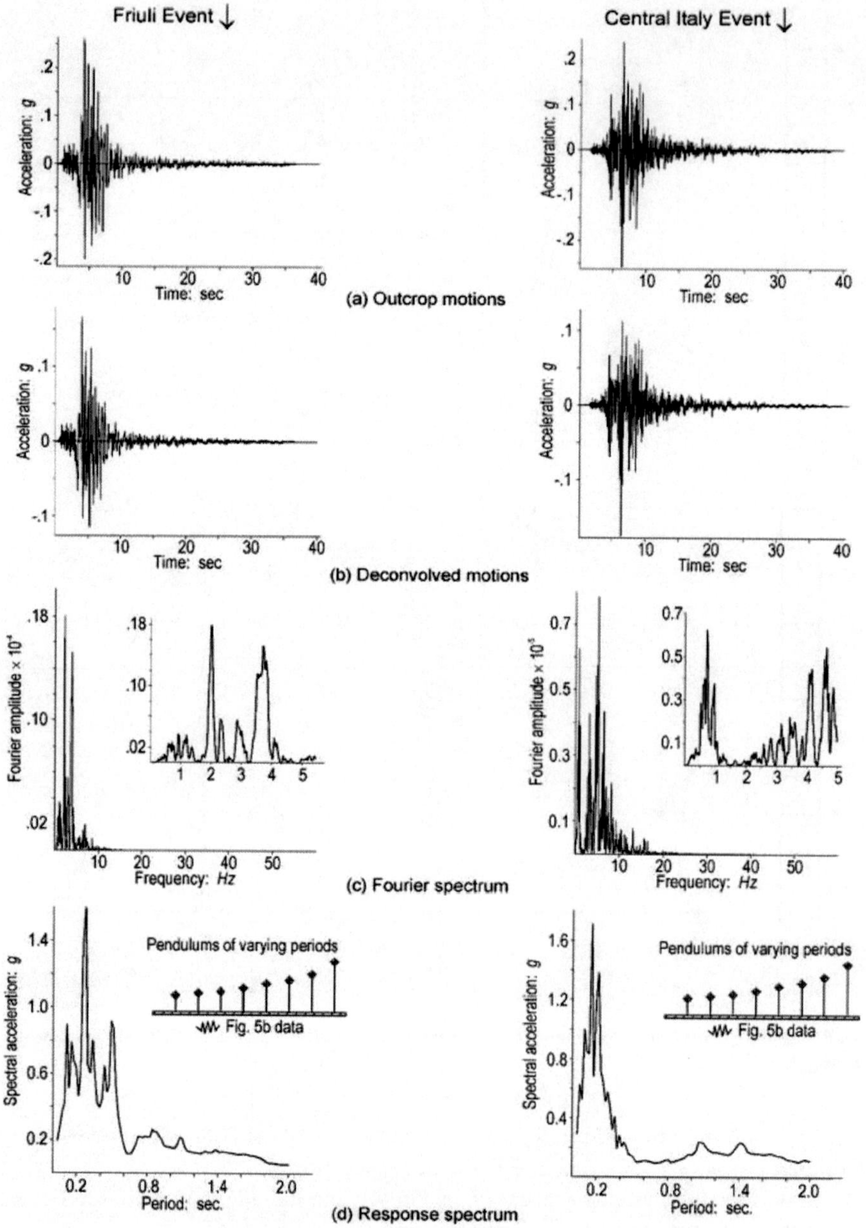

Fig. 5 Ground motions details Friuli and Central Italy events: **a** outcrop motions; **b** deconvolved motions; **c** Fourier power spectrum of deconvolved motions in (**b**); **d** acceleration response spectrum of deconvolved motions in (**b**)—damping ratio = 0

The outcrop motions and their deconvolved counterpart motions were checked to be baseline correct.

$$E_{rock-mass} = E_{rock} \times 10^{0.0186\times RQD - 1.91} \tag{1}$$

$$RQD = 17.536 \times \ln(z) + 24.847 \tag{2}$$

$$\phi' = \phi'_{crit} + m \times \left[D_r \times \left(Q - \ln(p')\right) - 1\right] (3) \tag{3}$$

$$M_s = a / (1 + e^{expo});\ expo = -((L - x_0) / b);\ L = \log(\gamma_c)\ (4) \tag{4}$$

$$\text{Shear modulus, } G = E / 2(1 + \nu) \tag{5}$$

$$\text{Bulk modulus, } K = E / 3(1 - 2\nu)\ (6) \tag{6}$$

$$\text{Shear wave velocity, } V_s = (G / \rho)^{1/2};\ \rho = \gamma / g\ (7) \tag{7}$$

2.2 *Fourier Analysis of the Deconvolved Motions*

Frequency contents of deconvolved Friuli and Central Italy events were assessed using a Fast Fourier Transform (FFT) of each. These results are shown in Fig. 5c—the insets shown are enlarged view of the FFT results up to ~5 Hz frequency. Frequencies in these events which are close to the natural vibration frequencies of Menta Dam (presented in Sect. 4.1 of this paper) are summarized in Table 2. These are termed resonant frequencies.

Table 2 First ten resonant frequencies in the deconvolved motions

Seismic data ID	Frequency, Hz									
	f_1	f_2	f_3	f_4	f_5	f_6	f_7	f_8	f_9	f_{10}
Friuli event	1.127	1.210	1.347	1.567	1.732	2.007	2.337	2.557	2.612	2.887
Central Italy event	1.022	1.146	1.246	1.296	1.346	1.445	1.620	1.794	1.944	2.044

2.3 Response Spectrum Analysis of the Deconvolved Motions

Response spectrum analyses of the deconvolved motions were performed using null damping. The acceleration response spectrum of each of the two motions are shown in Fig. 5d. The predominant period (T) for Friuli and Central Italy events is: 0.2752 s (freq. f = 3.63 Hz) and 0.1576 s (f = 6.35 Hz), respectively; f = 1/T.

3 Static Analyses

Static stresses in the dam and foundation were calculated in a progressive manner using the material properties shown in Table 1, ground water conditions mentioned in Sect. 1, and boundary conditions shown in Fig. 3a, b shows the dam construction sequence used; incremental impoundment was not used due to lack of information. The following is a summary of steps used in the static analyses: (a) mechanical stresses in the foundation block → (b) ground water flow through the foundation block → (c) staged construction of the dam in twenty-eight ~ 3-m thick lifts → (d) seepage and mechanical analyses for the reservoir water. Steps (a and b) were for the pre-existing conditions prior to the construction of the dam; presence of the grout curtain, drainage gallery, and drainage layer were included in steps (c and d). The fluid flow and mechanical analyses were intertwined to account for the effects of one on the other. The unbalance force ratio of 1×10^{-3} and fluid unbalance ratio of 1×10^{-3} or 20,000 computational steps were used in concluding a pseudo-static analysis in FLAC.

The results of static analysis at the end of impoundment are shown in Fig. 6. Results shown in Fig. 6a–d are self-explanatory and do not need elaboration.

However, (i) tension zones in the weathered foundation rock u/s of the dam shown in Fig. 6(e), and (ii) arching ratio ($\sigma yy/\gamma z$) profile along the centerline of the rockfill dam shown in Fig. 6f are unexpected. The irregular dam-foundation contact, and low shear modulus of the weathered rock near the dam foundation contact are the likely contributing factors to the unexpected results. Total stresses in the embankment are increased due to impoundment resulting in a decrease in the arching action as shown in Fig. 6f.

4 Seismic Analyses

4.1 Natural Vibration Characteristics of Menta Dam

Natural vibration characteristics included herein are: (i) natural frequencies, and (ii) associated mode shapes. Natural frequencies were calculated using the following steps: (a) at the end of a static solution, apply a horizontal force, $F_x = 1000$ kN to

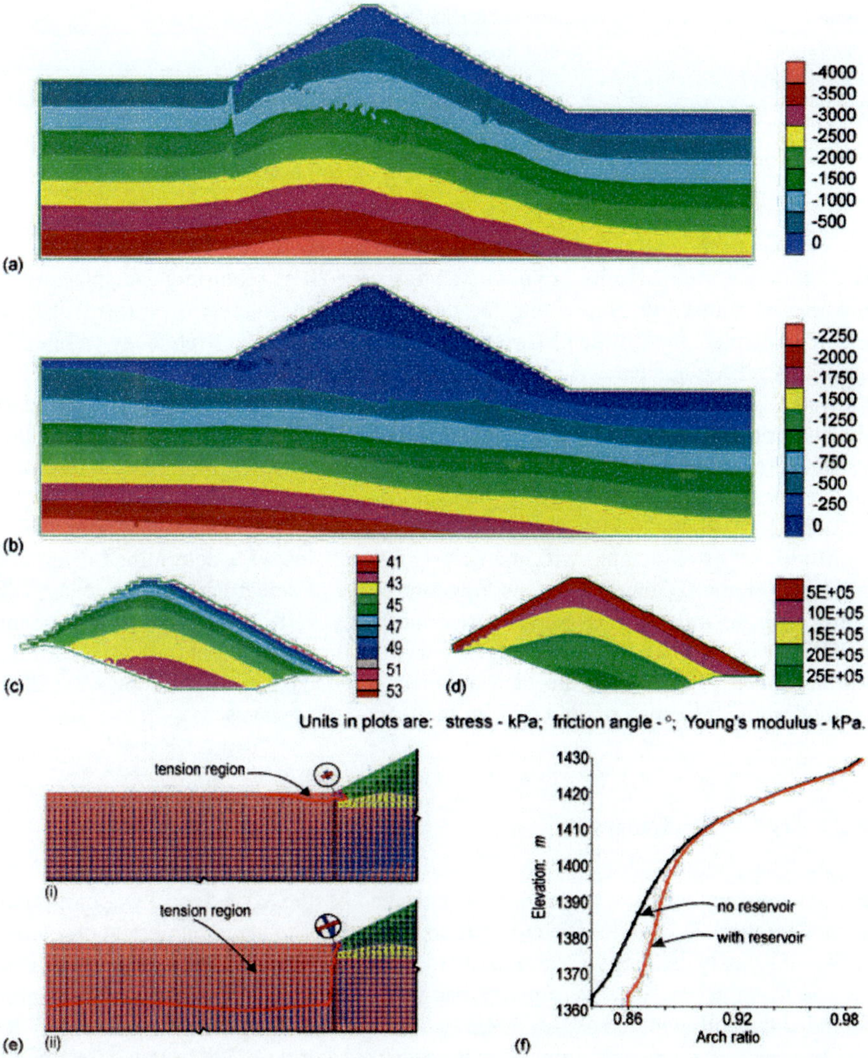

Fig. 6 Static analysis results at the end of impoundment: **a**, **b** contours of vertical and horizontal total stresses in the model; **c**, **d** contours of angle of internal friction and Young's modulus in the embankment; **e** zero contour of minor principal effective stress separating regions of tension and compression at the end of (i) construction and (ii) impoundment; **f** arching ratio along the dam centerline at the end of construction and impoundment

Table 3 Computed natural frequencies of Menta Dam

Dam stage	Natural frequency, Hz				
	f_1	f_2	f_3	f_4	f_5
End of construction	1.30	2.05	2.80	3.15	3.50
End of impoundment	1.30	2.10	2.80	3.20	3.80

each node at the dam crest and cycle the analysis for 10,000 computational steps; (b) set $F_x = 0$ and perform a dynamic analysis for 20 s., recording x-displacement histories at select locations along the centerline of the dam; (c) perform FFT of recorded displacement-time histories; (d) plot the FFT results (frequency vs Fourier amplitude). Frequencies corresponding to the spikes in the plots of results in (d) are taken to represent the natural frequencies of the dam. This procedure was applied to the numerical model at the end of dam construction; and at the end of impoundment. Frequencies corresponding to the first five spikes are shown in Table 3. Resonant frequencies in the deconvolved motions of Friuli and Central Italy events were shown in Table 2.

Mode shapes associated with the natural frequencies were determined using the following steps: (i) generate a 10 s long sine wave of select natural frequency; (ii) apply the sine wave as a dynamic horizontal force, $F_x(t)$, to each node at the dam crest; (iii) perform a dynamic analysis for 10,000 computational steps, recording the deformed geometry of the dam at select intervals. Figure 7c shows the dam configuration at the end of step (iii) for the f_1, f_2, f_3 Hz waves.

4.2 *Dynamic Analysis*

For dynamic analysis, the static solution at the end of step (d) in Sect. 3 was carried to a smaller tolerance of unbalance force ratio (1×10^{-14}) to avoid numerical problems in dynamic analysis. This was advised by FLAC support staff [9] on an earlier project.

The dynamic analyses were performed in continuation of the static solution and included the following steps: (e) integrated the deconvolved acceleration history to create a velocity history; (f) initialized the dynamic damping (Table 1b); (g) changed static b.c. to dynamic b.c. shown in Fig. 8; (h) applied shear stress history derived from the velocity history in step (e) to the base of the numerical model; (i) reset the static displacement and velocities to null; and (j) performed dynamic analysis for a dynamic time (dyt) of 50 s. This procedure was used for Friuli and Central Italy events.

The dynamic analysis results for the Friuli and Central Italy events are shown in Fig. 9. The results shown in Fig. 9 demonstrate that the model is responding smoothly to the seismic excitations with no spurious response(s). The results needing an explanation are: (a) the post-seismic horizontal displacement of the central axis of the dam in Fig. 9c; and (b) the maximum shear strain plots shown in Fig. 9f. For

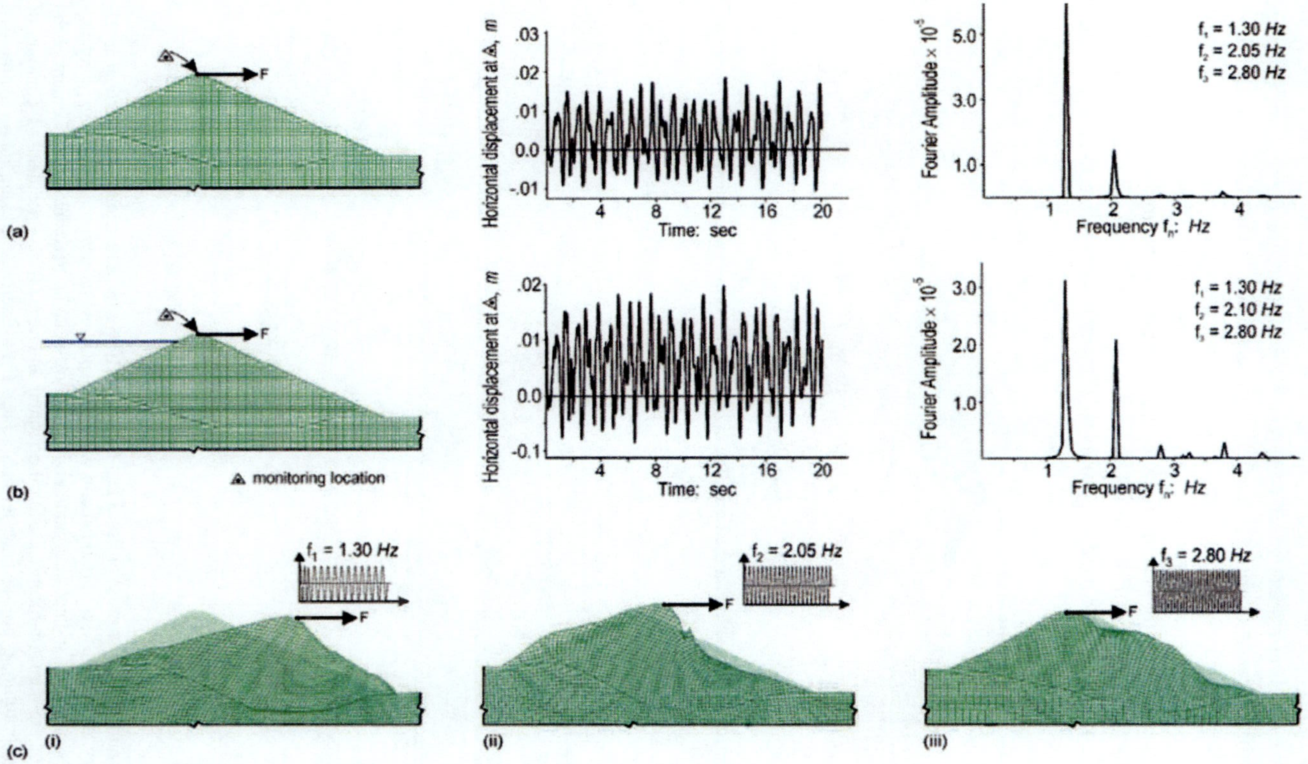

Fig. 7 Natural vibration characteristics of Menta Dam in transverse (u/s−d/s) direction: **a** problem setup and results for the end-of-construction condition; **b** problem setup and results for RWL = 1424.5 condition; **c** mode shapes for the end-of-construction condition

(a), the kink in the plots occurs near ~ El. 1350; it is because of the change from low Young's modulus (stiffness) of the weathered foundation rock to a higher value in the dam. For (b), maximum shear strain contours in the dam are localized in the drainage layer for similar reason as for (a). Tension regions shown in Fig. 9g are an amplified version of the ones shown in Fig. 6e–ii for the static analysis results. The degree of arching results at the end of seismic analysis are shown in Fig. 9h; these results show that the degree of arching is more than its value at the end of impoundment.

It should be pointed out that in the dynamic analyses of Menta Dam, static b.c. (u = 0 on ± x faces) were used. Attempts to enforce free-field b.c. on the ± x faces

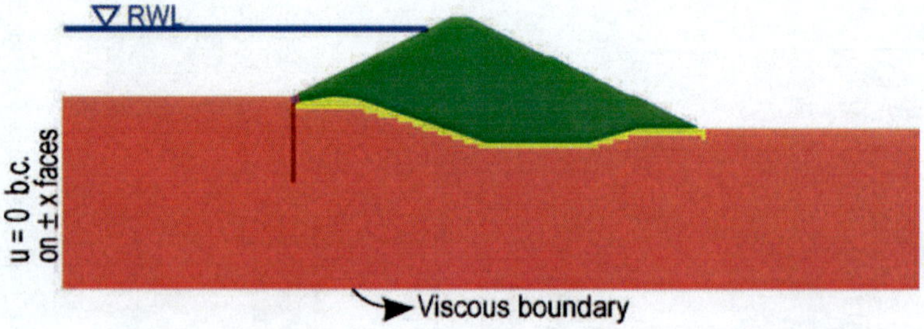

Fig. 8 Numerical model showing the boundary conditions used in dynamic analysis

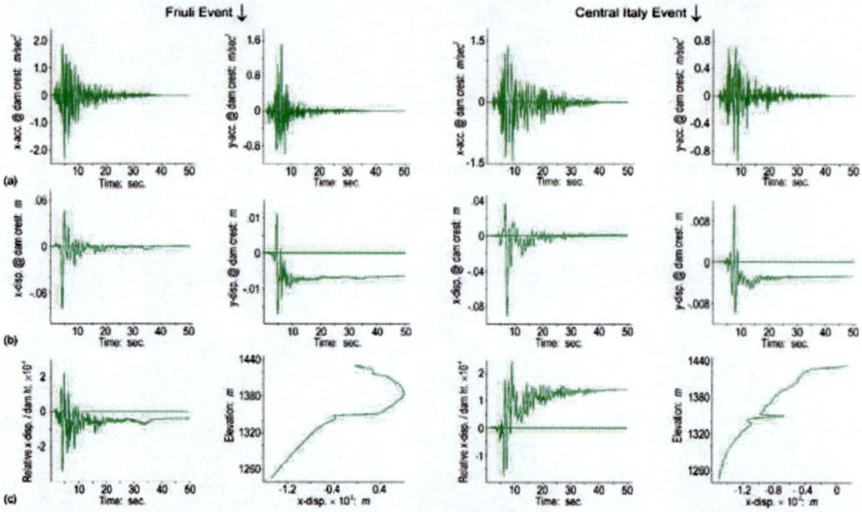

Fig. 9 Dynamic analysis results for Friuli and Central Italy events: **a** horizontal and vertical acceleration history at dam crest; **b** horizontal and vertical displacement history at dam crest; **c** relative horizontal displacement at dam crest/dam height; and post-seismic horizontal displacement of the central axis of the dam; (d–e) post-seismic horizontal and vertical displacements; **f** contours of maximum shear strain; **g** post-seismic tension zones; **h** post-seismic arching ratios for the two seismic events; and comparison with their counterparts in static analysis results (Fig. 6f)

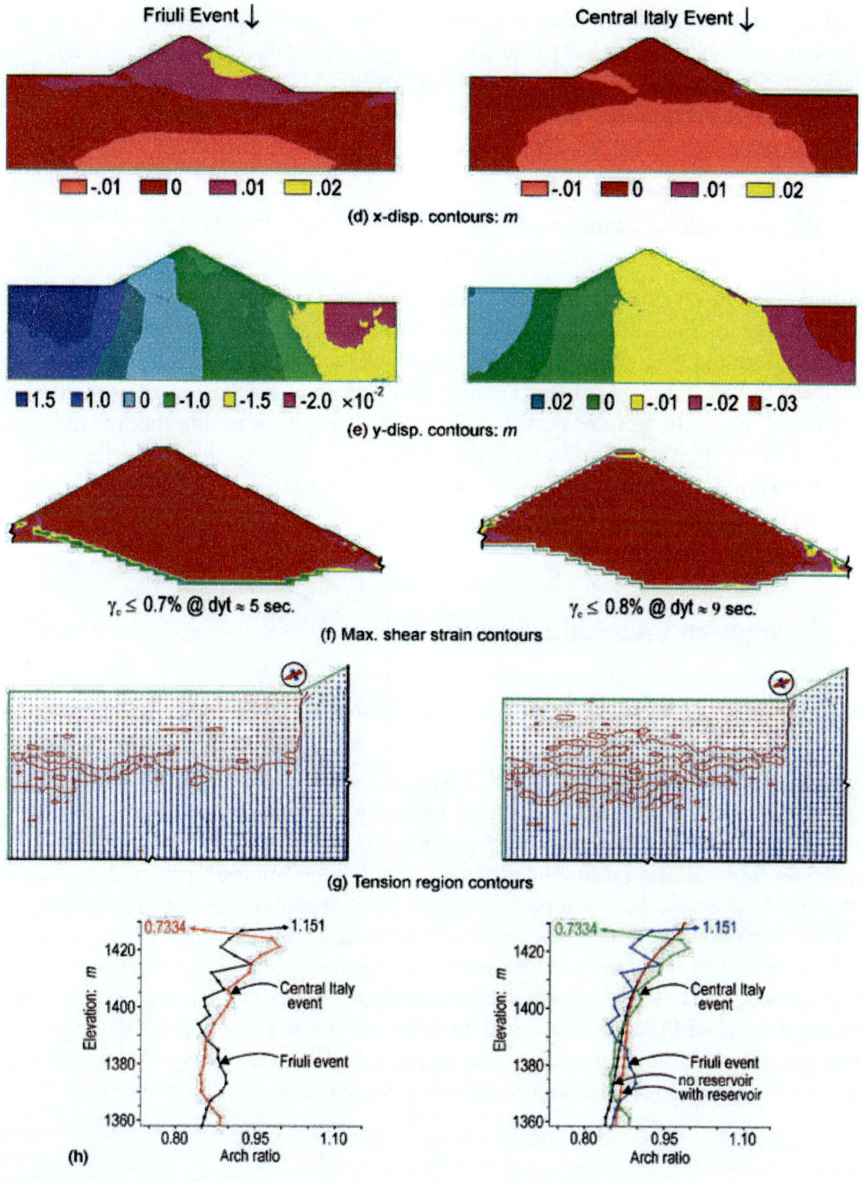

Fig. 9 (continued)

lead to numerical problems due to excessive (flow-like) deformations of the model. The large tension region in the foundation u/s of the grout curtain; Fig. 6(e-ii) is the likely cause of numerical difficulties when free field b.c. is applied to the ± x face boundaries of the model in dynamic analysis.

5 Bituminous Facing

Figure 10 shows (i) the bituminous facing zones, and (ii) the computed post-seismic axial stresses and resultant displacements at the end of each of the two seismic events. For both seismic events, the maximum axial stresses in the deck occur in the zones adjacent to the drainage tunnel. History plots of axial stresses in the three zones that attach to the tunnel are shown in Fig. 10d. It is worth noting that (a) the axial stresses in the deck are compressive and their magnitudes are less than the strength of 1490 kPa, and (b) the deck zones attached to the tunnel do not experience axial tension during the seismic loadings.

6 Tension and Arching Results

6.1 *Tension in the Foundation Upstream of the Grout Curtain*

In the static analyses of Sect. 3, initial stresses in the foundation (prior to the construction of the dam) were allowed to develop naturally under the action of gravity loading using the assigned properties for the foundation shown in Table 1a; the Poisson ratio, ν, is 0.25. The resulting ratio of computed in- and out-of-plane horizontal stresses to the vertical stresses, (k_o), is approximately 0.33. A repeat of the analyses using $k_o = 0.5$ to establish initial horizontal stresses in the foundation resulted in tension zones which are identical in horizontal extent but shallower in depth to the ones shown in Figs. 6e and 9g. However, $k_o = 0.5$ corresponds to $\nu = 0.33$ which is greater than the assigned value of 0.25. The repeated analyses results suggest that the development of tension in the foundation is not due to the initial stresses therein being too low. The results from repeated analyses are not included herein to conserve space.

6.2 *Arching in the Rockfill Dam*

Based in elastic solutions of homogeneous rockfill dams founded on firm and level foundation media, no arch action would be expected in the Menta Dam. However, in this study, the Young's modulus, E, of the fractured foundation rock increases from its lowest value at the dam-foundation contact to full value of 7×10^6 kPa at depth

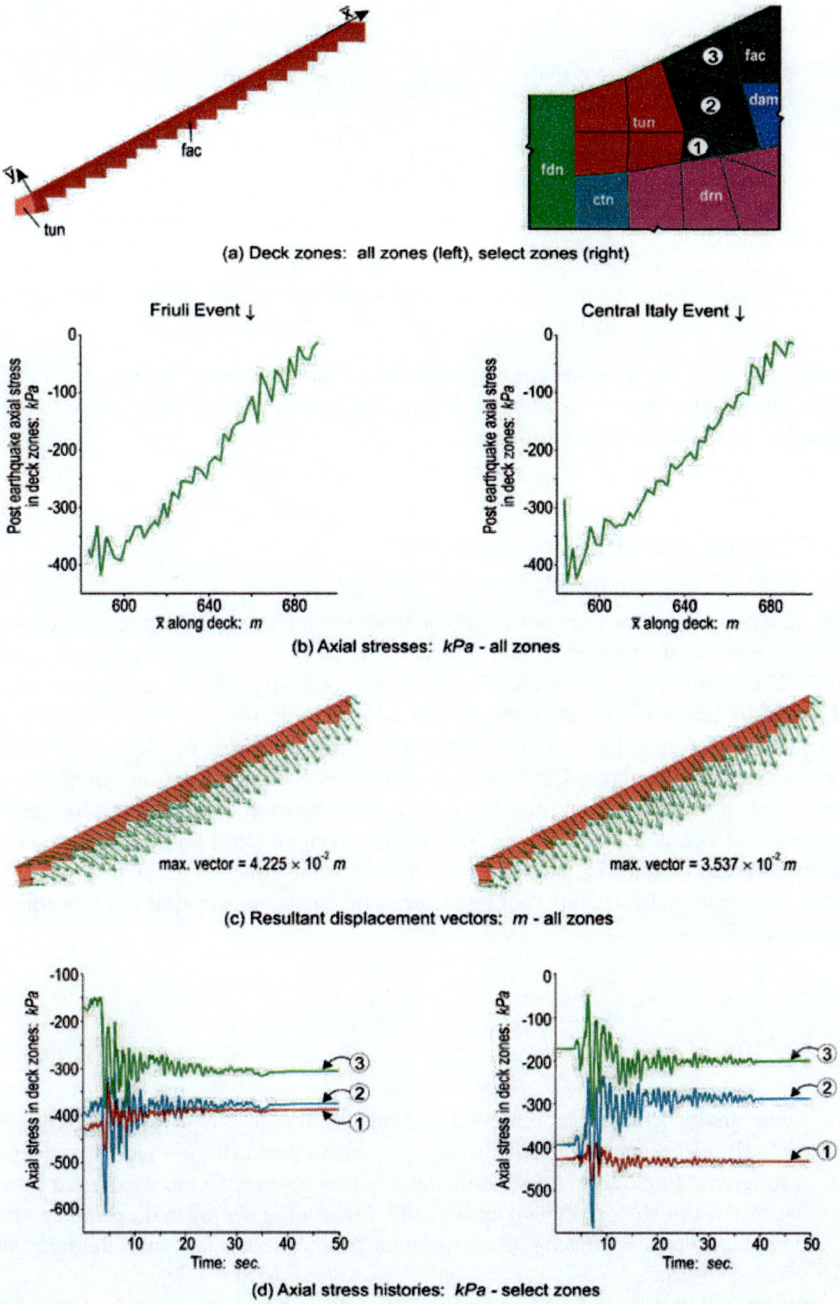

Fig. 10 Post-seismic axial stress and displacement results in bituminous facing

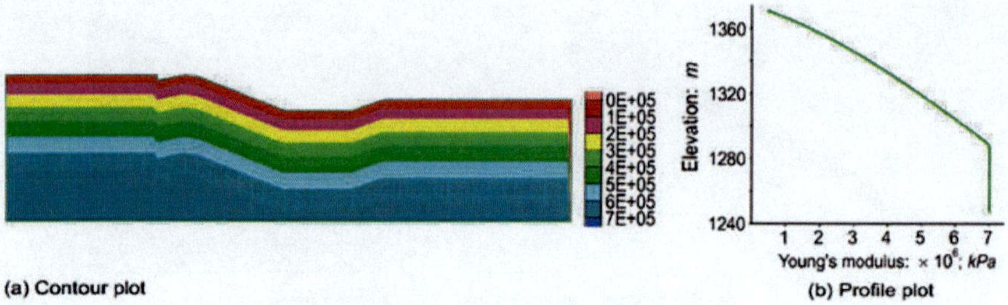

Fig. 11 Variation in Young's modulus with depth in foundation block: kPa; see Fig. 3a for elevations

as shown in Fig. 11. Also, the dam-foundation contact is not level as shown in Fig. 2. Thus, the arching ratio results shown in Figs. 6f and 9h are not comparable to their counterparts of ideal elastic solutions.

7 Post-seismic Stability Analysis

The material properties data assigned for the benchmark workshop are for elasto-plastic behavior of the rockfill and linear elastic behavior for other units shown in Fig. 2. The computer program, FLAC, allows factor of safety (FoS) calculations when all materials in the model are defined with Mohr-Coulomb strength parameters. A meaningful assessment of stability of the dam should include the dam foundation. Thus, formal FoS analyses for the pre- and post-seismic stress conditions were not performed. It suffices to say that the dam and foundation show essentially elastic response for pre and post-seismic loading for each of the two seismic events—during the seismic loading, computed stresses in some zones in the dam exceed the Mohr-Coulomb yield strength, but revert back to elastic state during the subsequent seismic loading.

8 Summary

(a) Static analysis results indicate: (i) no seepage flow occurs through the rockfill dam; (ii) no seepage flow exits from the drainage layer; (iii) seepage flow occurs through the foundation rock; (iv) tension zones develop in the weathered foundation rock upstream of the dam toe; and (iv) arching occurs in the dam, i.e., the computed vertical stresses along the dam centerline are less than the nominal vertical stresses.

(b) Computed first natural frequency of the dam-foundation system is 1.3 Hz. Friuli and Central Italy events have frequencies which are close to the computed natural frequencies of the dam-foundation-reservoir system.

(c) Use of free field boundary conditions along the ±x face boundaries of the model leads to large deformations which result in bad geometry and termination of the dynamic analysis. The cause of this numerical instability was not well understood.
(d) Use of static boundary conditions on the ±x face boundaries allows dynamic analysis to run full duration of the event. Dynamic analyses were extended to run for 100 s. with no sign of numerical difficulty or change in post-seismic analysis results.
(e) The dynamic analysis results included herein show that tension zones develop in the foundation upstream of the grout curtain and the tunnel—the region encapsulating tension zones is identified by a single contour which separates regions of tension and compression. The principal stress tensors are shown as crosses—the red-cross lines indicate tensile stress, the blue-cross lines indicate compression. The orientations of the red-cross lines in Fig. 9g imply that the cracks are essentially in the vertical (up–down) direction.

9 Items of Interest

It will be helpful to seek guidance and clarity on the following items of interest:

(a) Reliance on computed results: Characteristics of seismic events, and likely performance of constructed facilities affected by seismic events are not known as a priori. Reliance on computed results from numerical analyses requires expert knowledge, engineering judgement, and field experience—a combination which is hard to find and is declining.
(b) Simple versus complex numerical models: Based on observed cracks and observed performance of cracked dams, it was concluded that high rockfill dams, built with best available construction practices, develop tension cracks [4]. Comparison of theoretical results with observed cracks and observed performance of cracked dams gave validity to the use of elastic material models to investigate a potentially serious problem in rock- and earth-fill dams [6].

Acknowledgments The author would like to express his sincere thanks to Prof. George Dounias for his helpful comments on the initial draft of the paper. Also, thanks are due to Cindy Gray for her assistance in preparing the graphics and for final formatting of this paper.

References

1. Albano M (2013) Numerical modeling of the seismic performance of bituminous faced rockfill dams. University of Cassino and Southern Lazio, Italy, Thesis
2. Albano M, Modoni G, Croce P, Russo G (2015) Assessment of the seismic performance of a bituminous faced rockfill dam. Soil Dyn Earthq Eng 75:183–198

3. Allan ML, Philippacopoulos AJ (1999) Properties and performance of cement-based grouts for geothermal heat pump applications. U.S. Department of energy, Washington, District of Columbia
4. Casagrande A (1950) Notes on the design of earth dams. J Boston Soc Civil Eng 37(4):231–255
5. CEA General Development, Dams Division. Asphalt Dams. Internet website: https://issuu.com/agostiniassa/docs/cea_sarti_catalogo_eng. Accessed March 11, 2019
6. Covarrubias SW (1969) Cracking of earth and rockfill dams—a theoretical investigation by means of the finite element method. Harvard University, Cambridge, Massachusetts
7. De Marco S, Albano M, Russo G (2013) The Menta Hydropower plant: Water flow analysis during the controlled filling of Menta Dam. 9th ICOLD European Club Symposium, Venice, Italy, April 10–12
8. Idriss IM, Sun JI (1992) SHAKE91: a computer program for conducting equivalent linear seismic response analyses of horizontally layered soil deposits. University of California, Davis, California
9. Itasca consulting Group (2006) FLAC: fast lagrangian analysis of continua. Itasca Consulting Group Inc, Minneapolis, Minnesota
10. Russo G, Vecchietti A, Cecconi M, Pane V, De Marco S, Fiorino A (2019) 15th International Benchmark Workshop on Numerical Analysis of Dams. Theme B: Seismic analyses of Menta embankment dam. International Commission on Large Dams

Static and Dynamic Analysis of a Bituminous Faced Rockfill Dam

A. Freuis, A. Kainrath, S. Krstić, and M. Smesnik

Abstract Rockfill embankment dams are widely used in areas with high risk of seismic activity because of their good performance for both normal and exceptional loading cases. The paper presents results of 2D numerical analysis with focus on the seismic behaviour of the Menta rockfill dam. The cyclic and highly non-linear mechanical behaviour of rockfill material is one of the main objectives in the simulation. Therefore, the hyperbolic material model has been chosen to model the mechanical behaviour of rockfill material in order to analyse the stress/strain behaviour and potential damages of Menta dam.

Keywords Embankment dam · Bituminous faced rockfill dam · 2D numerical analysis · Non-linear Stress-Strain behaviour · Hyperbolic material model · Seismic behaviour

1 Introduction

The Menta dam is located in the middle of Aspromonte Massif in the south of Italy. The construction of this bituminous faced rockfill dam (BFRD) was completed in 2000. The embankment was constructed with rockfill material which was derived by blasting excavation along the rivers. The sealing element is a bituminous concrete facing which was placed in multiple layers on the upstream side. The reservoir has a capacity of $1.8 \times 107\ m^3$ which leads to a normal reservoir water level of 1424.75 m a.s.l. The crest height is 1431.75 m a.s.l. At the deepest point of the foundation, the dam has a maximum height of 90 m. The slope inclination on upstream as well as on downstream side is 1.8:1.0 (H:V). Fractured metamorphic rock belonging to the "Calabrian Complex" forms the foundation for the rockfill dam. The comprehensive

A. Freuis · S. Krstić (✉) · M. Smesnik
Pöyry Austria GmbH, Vienna, Austria
e-mail: sanja.krstic@poyry.com

A. Kainrath
Tschernutter Consulting Ltd, Villach, Austria

G. Bolzon et al. (eds.), *Numerical Analysis of Dams*, Lecture Notes in Civil Engineering 91, https://doi.org/10.1007/978-3-030-51085-5_26

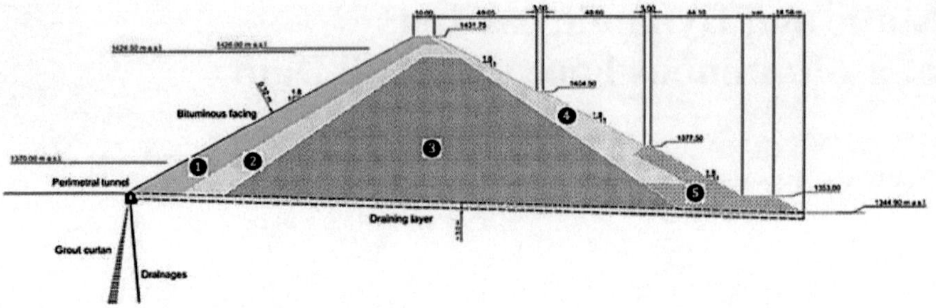

Fig. 1 Section of Menta Dam [1]

monitoring system, which is installed in the Menta dam, is used to observe the dams behaviour in natural and exceptional loading situations. A cross section of Menta dam is shown in Fig. 1.

The main aim of this workshop is to analyse the seismic behaviour of Menta BFED. This includes prediction of the stress/strain behaviour of Menta dam and estimation of the damage potential.

To get the initial stresses for further analysis, the stress/strain patterns have been calculated for different loading situations (e.g. with and without taking the reservoir into account, etc.). In the second step, the stress/strain behaviour under seismic loading has been evaluated based on initial static stress/strain results.

1.1 Numerical Model and Parameters

The finite element software GeoStudio 2019 has been used for the various analysis presented in this paper. In the first step, the static load-deformation calculation has been performed. In the next step, by using the load-deformation calculation as a parent, the dynamic calculations with the equivalent linear dynamic method have been performed. To gain further knowledge about the post seismic deformations of the embankment dam, a dynamic deformation analysis has been carried out.

1. Load/Deformation Analysis with SIGMA/W
2. Dynamic Analysis (Equivalent Linear Dynamic Analysis) QUAKE/W
3. Dynamic Deformation Analysis with SIGMA/W.

The 2D geometrical model used for numerical analysis, presented in Fig. 2, has been derived from the main dam cross section shown in Fig. 1.

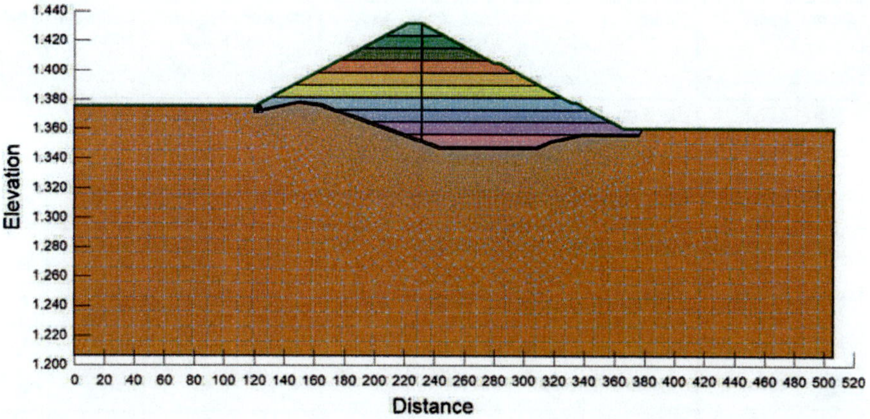

Fig. 2 Finite element model

1.2 Soil Parameters

In order to describe the cyclic, non-linear behaviour of the rockfill material, the hyperbolic material model has been chosen. The hyperbolic constitutive model implemented in GeoStudio 2019 software (described by Duncan et al. [2]) is referred to as the hyperbolic E-B constitutive model because, the bulk modulus (B) is considered to be constant while loading and the elastic modulus (E) varies according to the Duncan and Chang hyperbolic relationship [3]. For the purpose of taking into account the stress-dependency of rockfill material, an initial Young's modulus has been determined according to the average static stress level for each lift. The resulting constant Young's Modulus were applied to the corresponding lift (see Fig. 2). As it can be seen in Fig. 3, the embankment body was divided into 10 lifts. For other remaining materials presented in Fig. 3, the linear elastic material model has been used (Fig. 4).

The material parameters used for both dynamic analyses (QUAKE/W and SIGMA/W) are presented in Figs. 5, 6 and 11.

As it can be seen in Figs. 5 and 6, the equivalent linear material model has been used for the embankment material during the dynamic analysis (QUAKE/W). The equivalent material model implemented in GeoStudio 2019 is similar to the linear-elastic model with the difference that the soil stiffness G is modified in response to computed strains [4] (Figs. 7 and 8).

The basis for the dynamic G modulus curve presented in Fig. 9 was the Young's modulus function presented in Fig. 10. For the average stress level along each lift, the Young's modulus value was selected out of the Young's modulus function curve. The corresponding dynamic G modulus was calculated according to the following formula (Poisson's ratio ν was set to 0.3):

$$G = \frac{E}{2(1 + \gamma)} \tag{1}$$

Color	Name	Model	Effective Young's Modulus (E') (kPa)	Unit Weight (kN/m³)	Rf	Kur	Poisson's Ratio	Cohesion' (kPa)	Phi' (°)
	Asphalt	Linear Elastic (Effective)	14.000	24			0,16		
	Drainage	Linear Elastic (Effective)	60.000	21			0,2		
	Gallery	Linear Elastic (Effective)	30.000.000	24			0,2		
	Rock mass foundation	Linear Elastic (Effective)	7.000.000	27			0,25		
	Rockfill 1	Hyperbolic (Effective)	21.100	23	0,7	2,5	0,3	0,5	43
	Rockfill 10	Hyperbolic (Effective)	133.000	23	0,7	2,5	0,3	0,5	43
	Rockfill 2	Hyperbolic (Effective)	34.000	23	0,7	2,5	0,3	0,5	43
	Rockfill 3	Hyperbolic (Effective)	47.000	23	0,7	2,5	0,3	0,5	43
	Rockfill 4	Hyperbolic (Effective)	60.000	23	0,7	2,5	0,3	0,5	43
	Rockfill 5	Hyperbolic (Effective)	77.000	23	0,7	2,5	0,3	0,5	43
	Rockfill 6	Hyperbolic (Effective)	90.000	23	0,7	2,5	0,3	0,5	43
	Rockfill 7	Hyperbolic (Effective)	103.000	23	0,7	2,5	0,3	0,5	43
	Rockfill 8	Hyperbolic (Effective)	116.000	23	0,7	2,5	0,3	0,5	43
	Rockfill 9	Hyperbolic (Effective)	124.000	23	0,7	2,5	0,3	0,5	43

Fig. 3 Stiffness material parameters

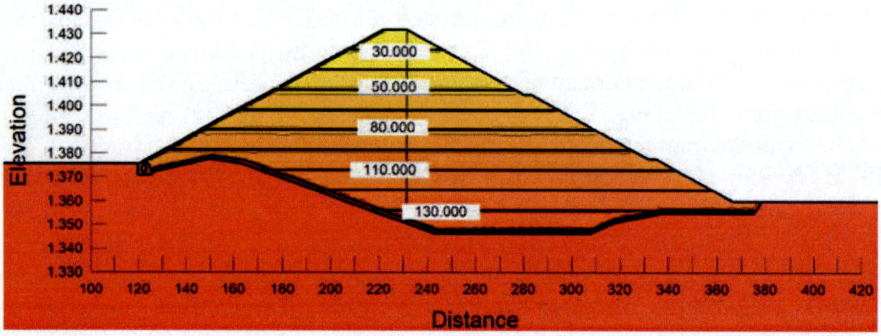

Fig. 4 Stress dependent Young's Modulus [kPa]

For other remaining materials presented in Figs. 5 and 6, the linear elastic material model has been used. The imported acceleration time history series have been applied in the dam foundation. To model this accurately in GeoStudio 2019, the rock mass foundation has been modelled as infinitely rigid and massless.

Color	Name	Model	Unit Weight (kN/m³)	Poisson's Ratio	Cohesion' (kPa)	Phi' (°)
■	Asphalt	Linear Elastic	24	0,16		
■	Drainage	Equivalent Linear	21	0,2	1	37
■	Gallery	Linear Elastic	24	0,2		
■	Rock mass foundation dyn	Linear Elastic	1e-20	0,25		
■	Rockfill 1 dyn	Equivalent Linear	23	0,3	0,5	43
■	Rockfill 10 dyn	Equivalent Linear	23	0,3	0,5	43
■	Rockfill 2 dyn	Equivalent Linear	23	0,3	0,5	43
■	Rockfill 3 dyn	Equivalent Linear	23	0,3	0,5	43
■	Rockfill 4 dyn	Equivalent Linear	23	0,3	0,5	43
■	Rockfill 5 dyn	Equivalent Linear	23	0,3	0,5	43
■	Rockfill 6 dyn	Equivalent Linear	23	0,3	0,5	43
■	Rockfill 7 dyn	Equivalent Linear	23	0,3	0,5	43
■	Rockfill 8 dyn	Equivalent Linear	23	0,3	0,5	43
■	Rockfill 9 dyn	Equivalent Linear	23	0,3	0,5	43

Fig. 5 Dynamic material parameters QUAKE/W, Part 1

For the dynamic deformation analysis, the hyperbolic material model has been used for the rockfill material. The value of the dynamic stiffness is, as commonly known, higher than the static stiffness. As previously mentioned for the static analysis, in order to take into account the stress-dependency of rockfill material, an initial Young's modulus has been determined according to the average static stress level for each lift. The resulting constant Young's Modulus were applied to the corresponding lift (see Fig. 2). The presented approach is based on the Young's modulus curve presented in Fig. 10. For other remaining materials shown in Fig. 11, the linear elastic material model has been used.

Color	Name	Dynamic G-Reduction Function	Dynamic Damping Ratio Function	GMax Function	Damping Ratio	Gmax Modulus (kPa)
	Asphalt				0,1	276.000
	Drainage	Gravel - Seed et al.	Seed & Idriss (Sand - average)	Bodenaustausch		
	Gallery				0,05	17.000.000
	Rock mass foundation dyn				0,0001	1e+100
	Rockfill 1 dyn	Gravel - Seed et al.	Seed & Idriss (Sand - average)	Rockfill-Icold		
	Rockfill 10 dyn	Gravel - Seed et al.	Seed & Idriss (Sand - average)	Rockfill-Icold		
	Rockfill 2 dyn	Gravel - Seed et al.	Seed & Idriss (Sand - average)	Rockfill-Icold		
	Rockfill 3 dyn	Gravel - Seed et al.	Seed & Idriss (Sand - average)	Rockfill-Icold		
	Rockfill 4 dyn	Gravel - Seed et al.	Seed & Idriss (Sand - average)	Rockfill-Icold		
	Rockfill 5 dyn	Gravel - Seed et al.	Seed & Idriss (Sand - average)	Rockfill-Icold		
	Rockfill 6 dyn	Gravel - Seed et al.	Seed & Idriss (Sand - average)	Rockfill-Icold		
	Rockfill 7 dyn	Gravel - Seed et al.	Seed & Idriss (Sand - average)	Rockfill-Icold		
	Rockfill 8 dyn	Gravel - Seed et al.	Seed & Idriss (Sand - average)	Rockfill-Icold		
	Rockfill 9 dyn	Gravel - Seed et al.	Seed & Idriss (Sand - average)	Rockfill-Icold		

Fig. 6 Dynamic material parameters QUAKE/W, Part 2

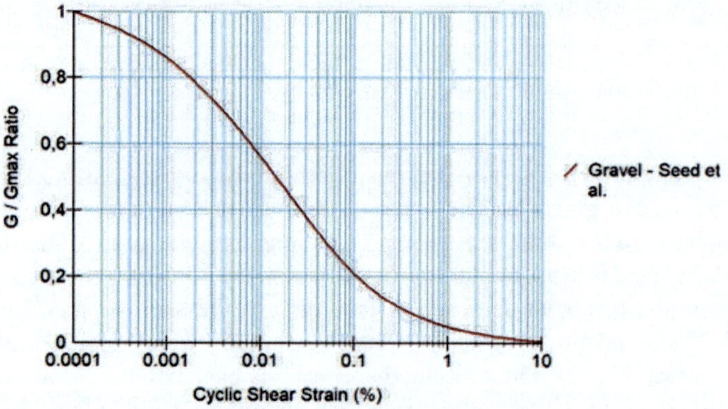

Fig. 7 Dynamic G-reduction function [5]—QUAKE/W

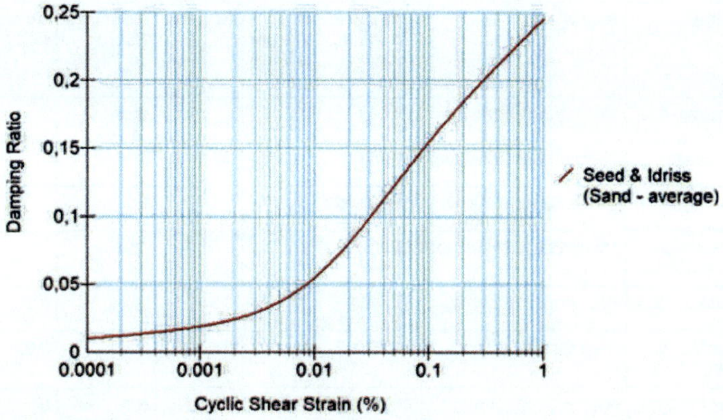

Fig. 8 Dynamic damping ratio function [6]—QUAKE/W

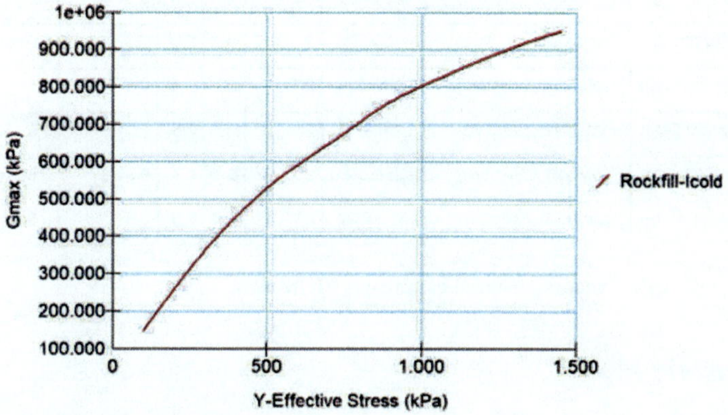

Fig. 9 Gmax Modulus—QUAKE/W

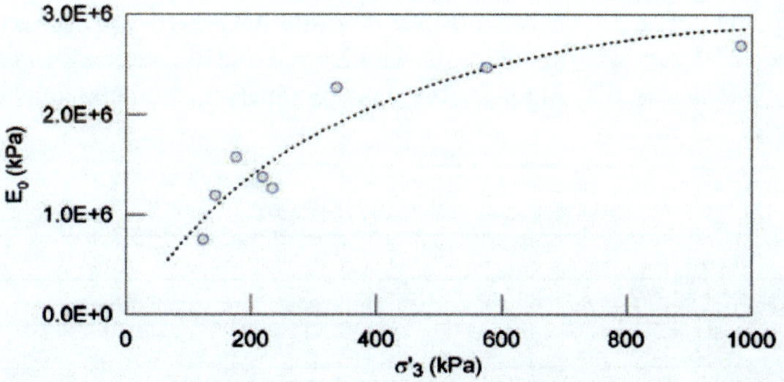

Fig. 10 Variation of small-strain Young's modulus with confining pressure (SIGMA/W) for the rockfill material [5]

Color	Name	Model	Effective Young's Modulus (E') (kPa)	Unit Weight (kN/m³)	Rf	Kur	Poisson's Ratio	Cohesion' (kPa)	Phi' (°)
	Asphalt	Linear Elastic (Effective)	14.000	24			0,16		
	Drainage	Linear Elastic (Effective)	60.000	21			0,2		
	Gallery	Linear Elastic (Effective)	30.000.000	24			0,2		
	Rock mass foundation	Linear Elastic (Effective)	7.000.000	27			0,25		
	Rockfill 1 dyn	Hyperbolic (Effective)	350.000	23	0,7	2,5	0,3	0,5	43
	Rockfill 10 dyn	Hyperbolic (Effective)	2.475.000	23	0,7	2,5	0,3	0,5	43
	Rockfill 2 dyn	Hyperbolic (Effective)	775.000	23	0,7	2,5	0,3	0,5	43
	Rockfill 3 dyn	Hyperbolic (Effective)	1.175.000	23	0,7	2,5	0,3	0,5	43
	Rockfill 4 dyn	Hyperbolic (Effective)	1.450.000	23	0,7	2,5	0,3	0,5	43
	Rockfill 5 dyn	Hyperbolic (Effective)	1.750.000	23	0,7	2,5	0,3	0,5	43
	Rockfill 6 dyn	Hyperbolic (Effective)	2.000.000	23	0,7	2,5	0,3	0,5	43
	Rockfill 7 dyn	Hyperbolic (Effective)	2.125.000	23	0,7	2,5	0,3	0,5	43
	Rockfill 8 dyn	Hyperbolic (Effective)	2.250.000	23	0,7	2,5	0,3	0,5	43
	Rockfill 9 dyn	Hyperbolic (Effective)	2.425.000	23	0,7	2,5	0,3	0,5	43

Fig. 11 Dynamic deformation material parameters SIGMA/W

1.3 Seismic Input

Two different target acceleration time series have been given as seismic input scenarios. In Fig. 12, the acceleration time series for the seismic event Friuli 1976 is depicted and in Fig. 13, the seismic event Central Italy 2016. For each earthquake event three acceleration components, two horizontal and one vertical, were given.

The earthquake used as input for the dynamic analysis are defined as shown in Table 1 (Figs. 14 and 15).

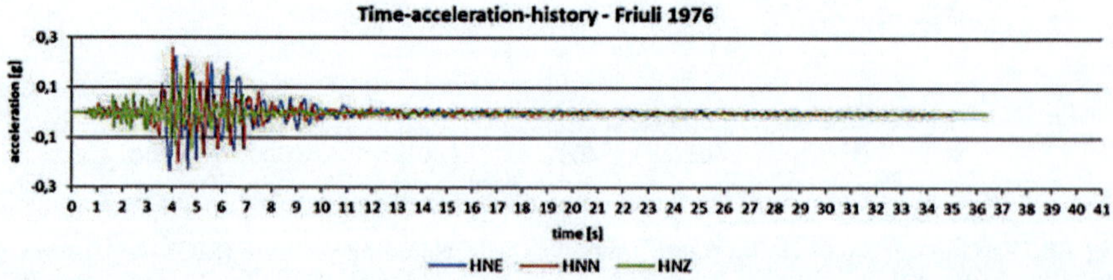

Fig. 12 Acceleration versus time—Friuli 1976

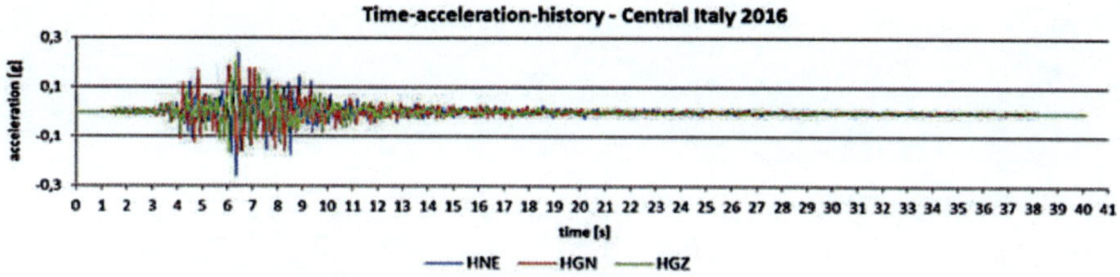

Fig. 13 Acceleration versus time—Central Italy 2016

Table 1 Definition Earthquakes

Name	Vertical	Horizontal
EQ1	HNZ Friuli 1976	HNE Friuli 1976
EQ1–0.26 g	HNZ Friuli 1976	HNN Friuli 1976
EQ2	HGZ Central Italy 2016	HGN Central Italy 2016
EQ2–0.26 g	HGZ Central Italy 2016	HGE Central Italy 2016
EQ1	HNZ Friuli 1976	HNE Friuli 1976

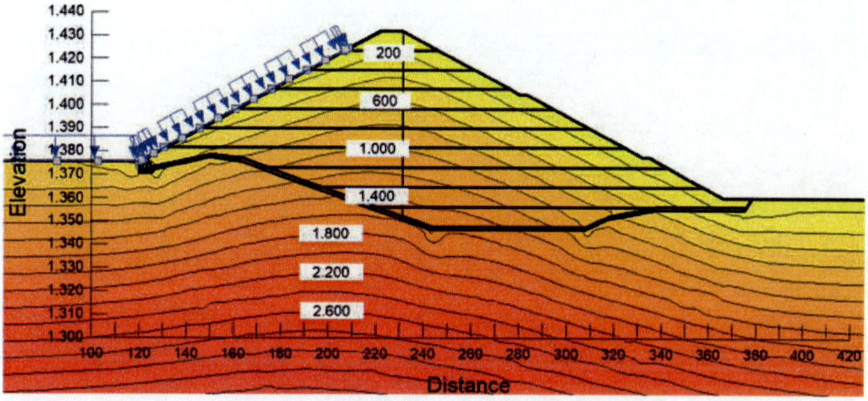

Fig. 14 Total vertical stress [kPa]—Maximum reservoir level

2 Results

2.1 *Total Stresses and Displacements*

As the comparison of the vertical displacements in Figs. 16 and 17 and the horizontal displacements in Figs. 18 and 19 shows only a minor influence of the reservoir level on displacements of the embankment.

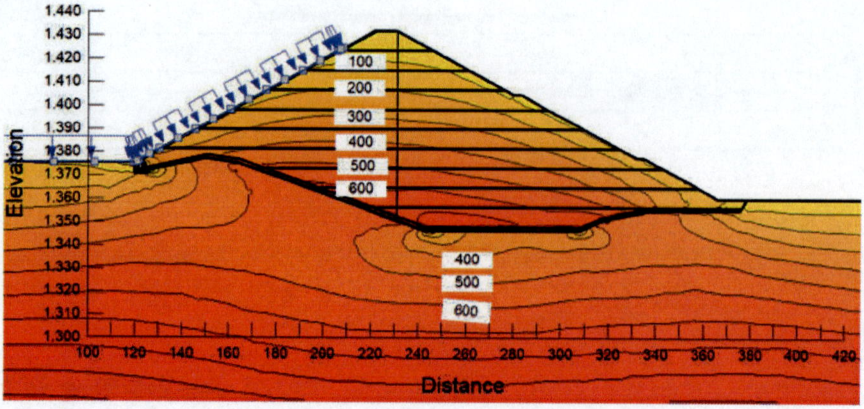

Fig. 15 Total horizontal stress [kPa]—Maximum reservoir level

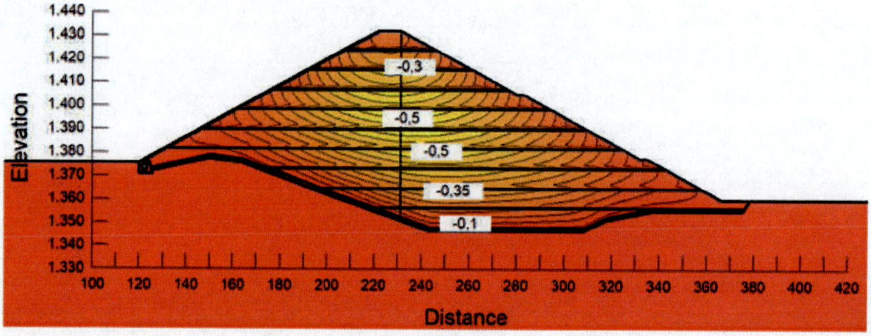

Fig. 16 Vertical displacements [m]—End of construction

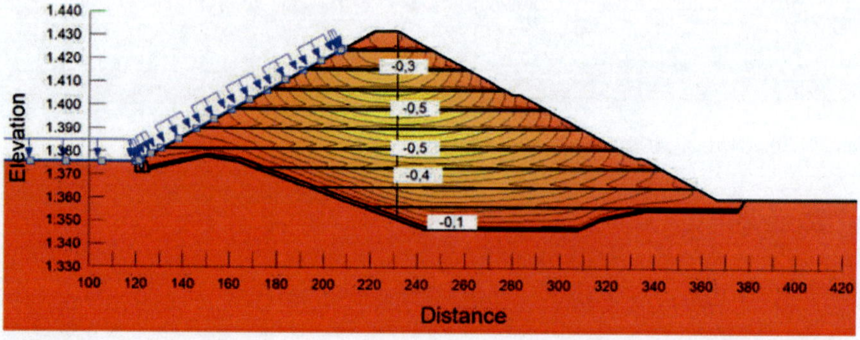

Fig. 17 Vertical displacements [m]—Maximum reservoir level

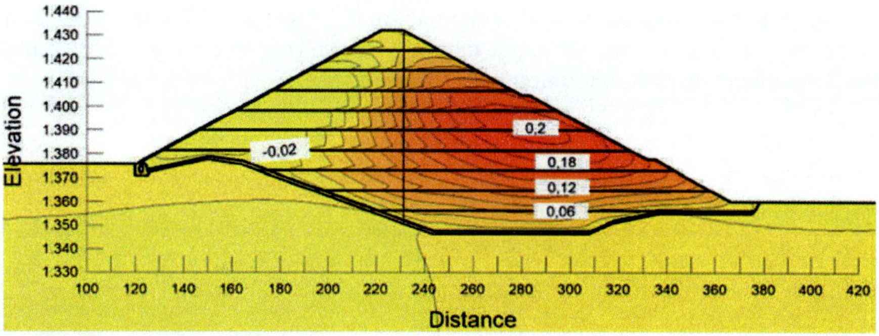

Fig. 18 Horizontal displacements [m]—End of construction

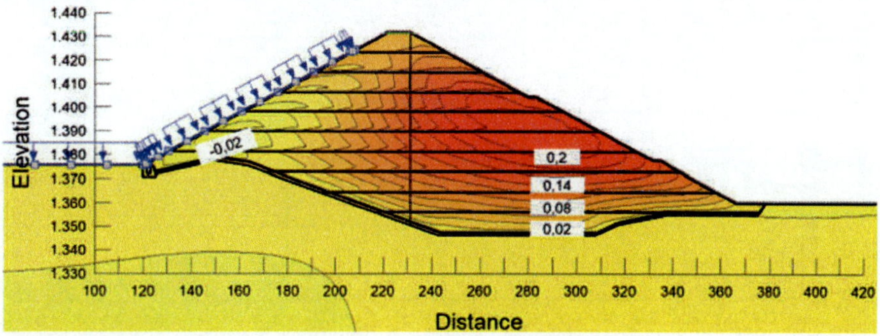

Fig. 19 Horizontal displacements [m]—Maximum reservoir level

Table 2 Frequency calculated out of the predominant period

	Full reservoir [Hz]
EQ1	1.82
EQ1–0.26 g	2.04
EQ2	3.70
EQ2–0.26 g	3.70

2.2 Dynamic Analysis and Post Seismic Deformations

The predominant period is the period during which, maximum spectral magnitudes are shown on response spectra. Usually, acceleration response spectra are used to determine the predominant period(s) of the earthquake ground motion. The maximum response values are expressed as a function of natural period for a given damping [7].

The response spectrum describes the maximum response of a single degree-of-freedom system to a particular input motion [4]. The frequencies calculated out of the predominant periods are presented in Table 2.

EQ1-Friuli

The horizontal spectral response graph for the earthquake Friuli 1967 (EQ1) is shown in Fig. 20 (Figs. 21, 22, 23 and 24).

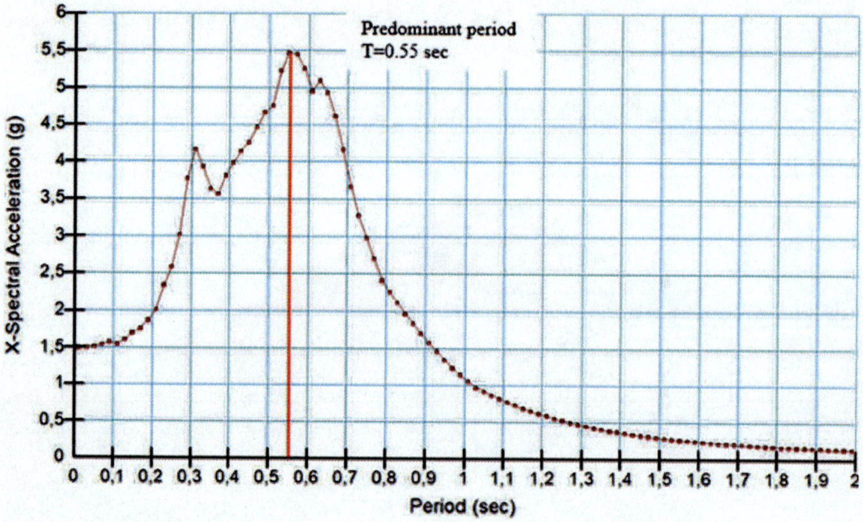

Fig. 20 Spectral acceleration versus Period at dam crest—EQ1-full reservoir

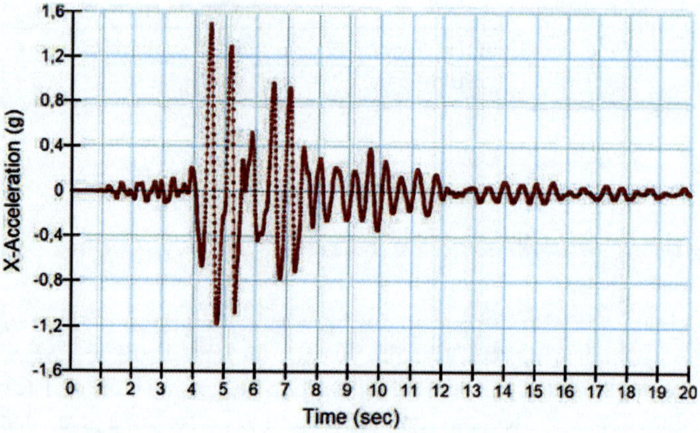

Fig. 21 Horizontal acceleration at dam crest—EQ1-full reservoir

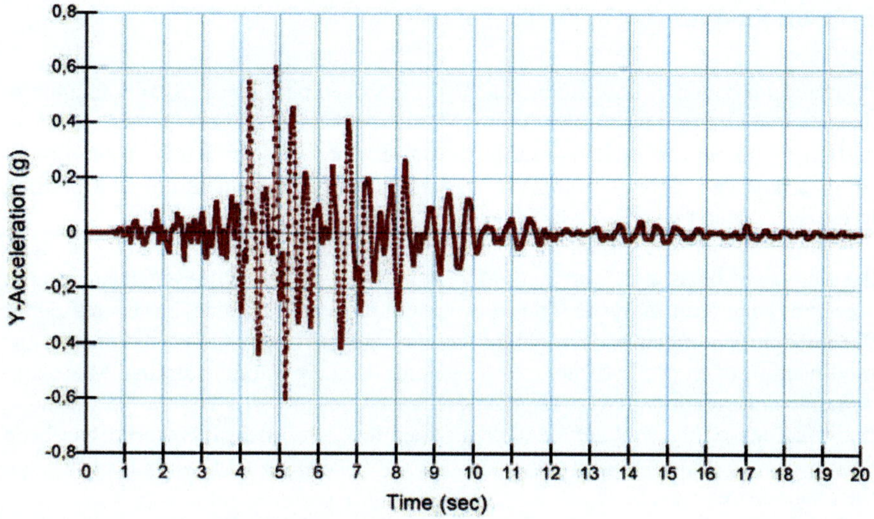

Fig. 22 Vertical acceleration at dam crest—EQ1-full reservoir

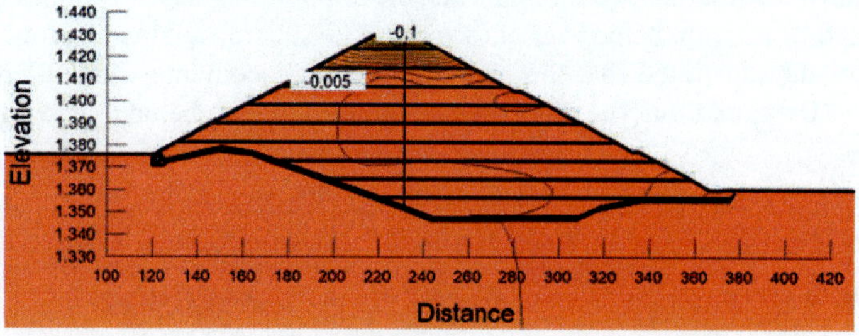

Fig. 23 Post-seismic horizontal displacements [m]—EQ1-Maximum reservoir level

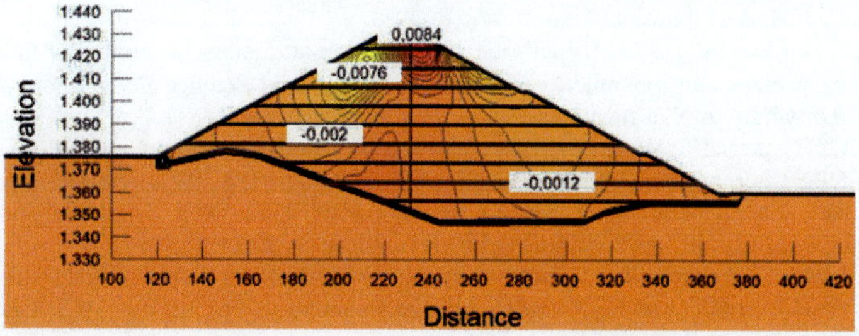

Fig. 24 Post-seismic vertical displacements [m]—EQ1-Maximum reservoir level

3 Conclusions

A static and dynamic finite element analysis has been performed in order to analyse the stress and deformation behaviour of Menta rockfill dam. The hyperbolic Duncan and Chang model was used to describe the non-linear stress-strain behaviour of rockfill material. Therefore, an initial Young's modulus has been determined according to the average static stress level for each lift.

Using the load-deformation calculation as a parent, dynamic calculations with the equivalent linear dynamic method were performed. Four different seismic input scenarios have been analysed. The equivalent linear material model has been used for the embankment material during the dynamic analysis (QUAKE/W) to consider the modification of the soil stiffness G in response to the computed strains. Moreover, dynamic deformation analyses have been carried out to determine the post-seismic deformations of the embankment dam. Therefore, the rockfill material has been described with the hyperbolic material model and the stiffness was varied considering the applied stress level.

The results of static and dynamic analysis show typical deformation behaviour for dams with bituminous membranes, which is outlined in literature as well [8]. The construction deformations predominate and the impounding has only minor effects on the general deformation behaviour of the dam. The seismic analysis of four different earthquakes demonstrated that the post seismic displacements and settlements are lower than 10 cm and that the dam shows a good dynamic behaviour with low risk of damage.

References

1. Russo G, Vecchietti A, Cecconi M, Pane V, De Marco S, Fiorino A (2019) Formulation of theme B: seismic analyses of menta embankment dam. In: 15th international benchmark workshop on numerical analysis of dams, Milano, Italy
2. Wong KS, Duncan JM (1974) Hyperbolic stress-strain parameters for nonlinear finite element analyses of stresses and movements in soil masses. National Science Foundation, Report No. TE-74-3. University of California, Berkeley
3. GEO-SLOPE International: Stress Strain (2004–2017) Modelling with Geostudio. Ltd
4. GEO-SLOPE International: Dynamic Earthquake (2004–2017) Modelling with Geostudio. Ltd
5. Rollins KM, Evans MD, Diehl NB, Daily WD III (1998) Shear Modulus and damping relationships for gravels. ASCE J Geotechn Geoenviron Eng 124(5):396–405
6. Seed HB, Wong RT, Idriss IM, Tokimatsu K (1984) Moduli and damping factors for dynamic analyses of cohesionless soils. National Science Foundation, Report No. UCB/EERC-84/14. University of California, Berkeley
7. FEMA: Prepared by the Interagency Committee on Dam Safety (2004) Federal Guidelines for Dam Safety. Glossary of Terms
8. Kutzner C (1997) Earth and Rockfill Dams Principles of Design and Construction. Balkema. Rotterdam, Netherlands

Numerical Investigation on the Seismic Responses of the Menta Embankment Dam

Z. Z. Fu, Z. K. Mi, and K. M. Wei

Abstract Plane strain (2D) finite element method (FEM) simulations are performed in studying the seismic responses of the Menta embankment dam in Italy. An elasto-plastic constitutive model is used to account for the static stress-strain behavior of the rockfill materials used, and a Hardin-type dynamic constitute model is employed to consider the dependence of the stiffness and damping on the mean effective stress and the cyclic shear strain amplitude. An empirical permanent strain model is incorporated into the traditional equivalent linear method to capture the earthquake-induced permanent deformation. The involved parameters are calibrated by fitting the available testing results or referring to those of similar materials. The stress and deformation within the dam after reservoir impounding, the acceleration responses during seismic excitation as well as the earthquake-induced permanent deformation are analyzed based on numerical simulations.

Keywords Reservoir impounding · Seismic excitation · Factor of safety

1 Introduction

Rockfill dams are increasingly constructed all over the world due to their good adaptability to varying site conditions, practicality of construction and great economy compared with other types of dams [1]. Rockfill dams constructed with modern vibratory compaction method are considered possessing adequate resistance to earthquakes, because of the high permeability of rockfill and consequently the low probability of pore pressure buildup and liquefaction [2]. The recent Zipingpu case in China, however, demonstrated that the impermeable system (e.g. concrete face slabs) of a rockfill dam could be severely damaged by a strong earthquake, and the dam itself might experience a considerable settlement as a result of earthquake shaking [3].

Z. Z. Fu (✉) · Z. K. Mi · K. M. Wei
Geotechnical Engineering Department, Nanjing Hydraulic Research Institute, Nanjing 210024, People's Republic of China
e-mail: fu_zhongzhi@yahoo.com

G. Bolzon et al. (eds.), *Numerical Analysis of Dams*, Lecture Notes in Civil Engineering 91, https://doi.org/10.1007/978-3-030-51085-5_27

Due to the very limited number of cases that have ever experienced an earthquake and the lack of reliable observations made during and after an earthquake, an in-depth understanding of the seismic behavior of a rockfill dam is, nowadays, increasingly facilitated by numerical simulations [4, 5]. In this paper, the results obtained from a series of 2D FEM simulations performed for the Menta embankment dam are reported, with the attention focused primarily on the stress and deformation within the dam after construction and reservoir impounding, the acceleration responses during seismic excitation as well as the earthquake-induced permanent deformation. The safety of the downstream slope is also investigated using the histories of the factor of safety against sliding.

The studied Menta dam is a bituminous-faced rockfill dam located in southern Italy, lying in the heart of Aspromonte Massif. The crest height of the dam is 1431.75 m, and the normal reservoir level is at 1424.75 m. The embankment is about 90 m high at its deepest point, and the reservoir impounds 1.8×10^7 m^3 of water. This dam was designed in the late 1970s and was constructed between 1987 and 2000. The dam was filled with compacted rockfill excavated by blasting along the rivers of the reservoir, and has five main different rockfill zones (Fig. 1). Both the upstream and downstream slopes have a ratio of 1.8:1. Two berms (3.0 m) were designed along the downstream slope at the elevation 1404.5 m and 1377.5 m, respectively. A draining layer with a thickness of about 3.0 m was placed at the embankment-foundation interface, as a safety layer to prevent positive water pressures rise in the embankment.

The rock mass forming the dam foundation is a fractured metamorphic rock belonging to the "Calabrian Complex". Cement-water injections reaching a depth of ~50 m were realized departing from the inner perimeter tunnel to create a grout curtain on the upstream side of the embankment. The impervious bituminous concrete facing is placed on the upstream side. The facing is formed by superposing different layers of bituminous concrete upon a regularly packed basement of compacted sandy gravel. The bituminous concrete facing is anchored to the reinforced concrete tunnel running along the inner perimeter of the dam (Fig. 1).

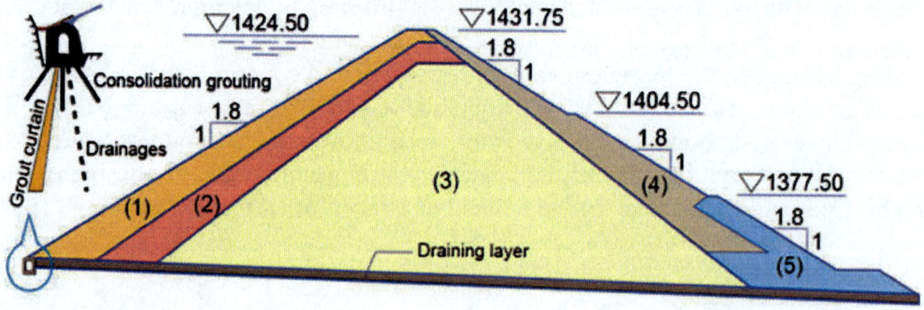

Fig. 1 Topographical section of the dam

2 Material Models and Analytical Methods

The layer-by-layer construction of the dam and the impounding processes are simulated with an elastoplastic constitutive model for the rockfill [6, 7]. The seismic responses are studied with a Hardin-type dynamic constitutive model along with a permanent displacement prediction module [8]. In this part, the models and methods used are briefly described.

2.1 Static Constitutive Model

The elastoplastic constitutive equation that relates the incremental stress tensor ($\mathrm{d}\boldsymbol{\sigma}$) and the incremental strain tensor ($\mathrm{d}\boldsymbol{\varepsilon}$) reads:

$$\mathrm{d}\sigma = \left[\mathbf{D}^{\mathrm{e}} - \frac{(\mathbf{D}^{\mathrm{e}} : \mathbf{n}_{\mathrm{g}}) \otimes (\mathbf{n}_{\mathrm{f}} : \mathbf{D}^{\mathrm{e}})}{H + \mathbf{n}_{\mathrm{f}} : \mathbf{D}^{\mathrm{e}} : \mathbf{n}_{\mathrm{g}}}\right] : \mathrm{d}\boldsymbol{\varepsilon} \tag{1}$$

in which $\boldsymbol{D}^{\mathrm{e}}$ denotes the fourth-order elasticity tensor. $\boldsymbol{n}_{\mathrm{f}}$ and $\boldsymbol{n}_{\mathrm{g}}$ are the second-order tensors for loading direction and plastic strain direction, respectively. H denotes the plastic modulus.

The plastic strain direction tensor, $\boldsymbol{n}_{\mathrm{g}}$, can be constructed using the stress-dilatancy equation and the coaxiality assumption [6, 7], i.e.,

$$\mathbf{n}_{\mathrm{g}} = \left(d_{\mathrm{g}}\frac{\mathbf{I}}{3} + \frac{3\mathbf{s}}{2q}\right) \Big/ \sqrt{\frac{1}{3}d_{\mathrm{g}}^2 + \frac{3}{2}} \tag{2}$$

in which d_{g} is the dilatancy ratio that represents the ratio between the incremental volumetric and deviatoric plastic strains, i.e., $d_{\mathrm{g}} = \mathrm{d}\varepsilon_{\mathrm{v}}^{\mathrm{p}}/\mathrm{d}\varepsilon_{\mathrm{s}}^{\mathrm{p}}$. $\boldsymbol{I}$ and $\boldsymbol{s}$ denote the second-order unit tensor and the stress deviator, respectively. The following stress-dilatancy equation is adopted:

$$d_{\mathrm{g}} = \frac{M_{\mathrm{d}}^2 - \eta^2}{\beta\eta} \tag{3}$$

in which $\eta = q/p$ is the ratio between the deviatoric stress (q) and the mean effective stress (p). β is a positive constant controlling the shear-induced dilatancy ratio. The so-called constant volume stress ratio, M_{d}, is a function of the constant volume friction angle [6], ψ, i.e.,

$$M_{\mathrm{d}} = \frac{6\sin\psi}{3 - \sin\psi} \tag{4}$$

in which ψ depends on the mean effective stress, p, as follows:

$$\psi = \psi_0 - \Delta\psi \lg \frac{p}{p_a} \tag{5}$$

where p_a denotes the atmospheric pressure, and ψ_0 and $\Delta\psi$ are two parameters.

In the elastoplasticity theory, the loading direction is defined using a yield function, and the flow rule is associative if the plastic potential function is the same as the yield function or if the plastic direction tensor is identical to the loading direction tensor. Many experiments have shown that a non-associative flow rule might be more suitable for granular materials such as rockfill materials than an associative flow rule [9]. In this study, however, the plastic flow direction tensor is used as the loading direction tensor to ensure a symmetric elastoplastic tensor and thus a symmetric global stiffness matrix.

The simplest isotropic elastic model is used herein to describe the elastic behavior, i.e.,

$$D^{e}_{ijkl} = \frac{v_e \cdot E_e}{(1+v_e)(1-2v_e)}\delta_{ij}\delta_{kl} + \frac{E_e}{2(1+v_e)}\left(\delta_{ik}\delta_{jl} + \delta_{il}\delta_{jk}\right) \tag{6}$$

in which E_e and v_e denote the Young's modulus and the Poisson's ratio, respectively. δ_{ij} is the Kronecker sign. The Young's modulus is assumed only dependent of the mean effective stress:

$$E_e = k_e \cdot p_a \cdot \left(\frac{p}{p_a}\right)^n \tag{7}$$

in which k_e and n are two parameters. The Poisson's ratio of coarse rockfill materials is relatively difficult to calibrate. For practical problems, however, it can be assumed to be a constant for simplicity, e.g., $v_e = 0.2$–0.3.

The plastic modulus, H, can be expressed indirectly by the tangential modulus (E_t) and the elastic modulus (E_e) as a function of the stress states [6, 7], i.e.,

$$H = \frac{(1+d_g/3)^2}{\left(3/2+d_g^2/3\right)}\left(\frac{1}{E_t} - \frac{1}{E_e}\right)^{-1} \tag{8}$$

in which the tangential modulus depends on the mean effective stress and the stress ratio, i.e.,

$$E_t = \left[1 - \left(\frac{\eta}{M_f}\right)^{\alpha}\right]^{\frac{1}{\alpha}} \cdot k \cdot p_a \cdot \left(\frac{p}{p_a}\right)^n \tag{9}$$

where k and α are another two parameters. M_f is the peak stress ratio at which the tested material fails. Under a triaxial compression stress state, the peak stress ratio reads:

$$M_{\mathrm{f}} = \frac{6 \sin \varphi}{3 - \sin \varphi} \tag{10}$$

in which the peak friction angle depends on the mean effective stress, p, as follows:

$$\varphi = \varphi_0 - \Delta\varphi \lg \frac{p}{p_{\mathrm{a}}} \tag{11}$$

There are 10 parameters in the model used, all of which can be determined using conventional triaxial compression experiments.

2.2 Dynamic Responses Analysis Method

Traditional equivalent linear analysis method is used for the seismic responses. The maximum shear modulus (G_0) of rockfill materials depends on the mean effective stress (p), i.e.,

$$G_0 = k_0 \cdot p_{\mathrm{a}} \cdot \left(\frac{p}{p_{\mathrm{a}}}\right)^m \tag{12}$$

in which k_0 and m are two parameters. The equivalent shear modulus (G) is a function of the cyclic shear strain amplitude (γ^{c}), and the following degradation equation is used herein [8]:

$$\frac{G}{G_0} = \frac{1}{1 + k_1 \bar{\gamma}^{\mathrm{c}}} \tag{13}$$

in which k_1 is a material constant, and the normalized cyclic shear strain amplitude reads:

$$\bar{\gamma}^{\mathrm{c}} = \frac{\gamma^{\mathrm{c}}}{(p/p_{\mathrm{a}})^{1-m}} = \frac{\left(\sqrt{2}/3\right)\sqrt{\left(\varepsilon_1^{\mathrm{c}} - \varepsilon_2^{\mathrm{c}}\right)^2 + \left(\varepsilon_2^{\mathrm{c}} - \varepsilon_3^{\mathrm{c}}\right)^2 + \left(\varepsilon_1^{\mathrm{c}} - \varepsilon_3^{\mathrm{c}}\right)^2}}{(p/p_{\mathrm{a}})^{1-m}} \tag{14}$$

where $\varepsilon_{\mathrm{i}}^{\mathrm{c}}$ (i = 1–3) denotes the principal cyclic strains.

The equivalent damping ratio (λ) also depends on the strain amplitude as follows:

$$\lambda = \lambda_{\mathrm{min}} + (\lambda_{\mathrm{max}} - \lambda_{\mathrm{min}})\frac{k_1 \bar{\gamma}^{\mathrm{c}}}{1 + k_1 \bar{\gamma}^{\mathrm{c}}} = \lambda_{\mathrm{min}} + (\lambda_{\mathrm{max}} - \lambda_{\mathrm{min}})\left(1 - \frac{G}{G_0}\right) \tag{15}$$

in which λ_{min} and λ_{max} are two parameters, corresponding to the minimum and maximum damping ratio when $\gamma^{\mathrm{c}} \to 0$ and $\gamma^{\mathrm{c}} \to \infty$, respectively.

Considerable residual strains can be accumulated for rockfill materials subjected to cyclic loading. By conducting a series of large-scale dynamical triaxial tests on a typical rockfill, the authors found that the amounts of residual strains depend on the stress state prior to cyclic loading, the cyclic shear strain amplitude (γ^{c}), and the number of loading cycles (N) [10]. The following empirical equation is used for the residual shear strain (γ^{p}):

$$\gamma^{p} = \gamma_{1}^{p} \cdot N^{n_{\gamma}} \tag{16}$$

in which the superscript 'p' is used to signify the plastic and irreversible nature of residual strains, and

$$\left.\begin{aligned} \gamma_{1}^{p} &= c_{\gamma} \cdot (\gamma^{c})^{\alpha_{\gamma}} \cdot \frac{\eta}{\sqrt{p/p_{a}}} \\ n^{\gamma} &= d_{\gamma} \cdot (\gamma^{c})^{-\beta_{\gamma}} \cdot \sqrt{p/p_{a}} \end{aligned}\right\} \tag{17}$$

Herein, c_{γ}, α_{γ}, d_{γ}, β_{γ} are four parameters for permanent shear strain. Similarly, an exponential function is used to model the accumulation of the plastic volumetric strain, ε_{v}^{p}, i.e.,

$$\varepsilon_{v}^{p} = \varepsilon_{v}^{f}\left[1 - \exp\left(-\frac{N}{N_{v}}\right)\right] \tag{18}$$

in which

$$\left.\begin{aligned} \varepsilon_{v}^{f} &= c_{v}(\gamma^{c})^{\alpha_{v}} \\ N_{v} &= d_{v} \cdot (\gamma^{c})^{-\beta_{v}} \cdot \sqrt{p_{0}/p_{a}} \end{aligned}\right\} \tag{19}$$

where c_{v}, α_{v}, d_{v}, β_{v} are another group of four parameters for plastic volumetric strain.

In Eqs. (16) and (18), the number of loading cycles is introduced as an independent variable. For practical problems, the loading history is generally irregular and it is necessary to translate the irregular process into an equivalent regular one. In this study, the equivalent number of regular loading cycles, N_{eq}, for an irregular process is determined using the equal energy dissipation assumption [8], i.e.,

$$N_{eq} = \frac{\int (\dot{\gamma}^{c})dt}{\pi\omega\left(\gamma_{eq}\right)^{2}} \tag{20}$$

in which ω denotes the fundamental frequency of the concerned structure, and the equivalent amplitude of cyclic shear strain, γ_{eq}^{c}, can be approximated by a fraction of the maximum shear strain, γ_{max}^{c}, as used by Idriss and Sun (1992) in SHAKE91 program [11], i.e.,

$$\gamma_{\mathrm{eq}}^{\mathrm{c}} = R_{\gamma} \cdot \gamma_{\max}^{\mathrm{c}} \tag{21}$$

Generally, R_{γ} ($0 < R_{\gamma} < 1$) is a constant depending on the magnitude of earthquakes. In this study, R_{γ} is set to 0.65 in all the simulations.

3 Material Parameters and Ground Excitations

3.1 Material Parameters

The parameters of the static elastoplastic constitutive model are calibrated using the triaxial-K_0 consolidated drained test results as shown in Fig. 2. This type of experimental results cannot be used directly as the model used herein is based on conventional triaxial experiments. Nevertheless, constitutive parameters can also be calibrated by best fitting the experimental data taking into account their normal ranges. Table 1 lists the calibrated parameters and Fig. 2 compares the test data and the relevant model predictions. Most of the experimental trends are reproduced satisfactorily.

The parameters for the maximum shear modulus are calibrated using the initial stiffness data obtained from conventional triaxial tests as shown in Fig. 3. Since the initial deformation behavior is essentially elastic, the maximum shear modulus (G_0) can be deduced from the initial Young's modulus with the aid of the following relation [12]:

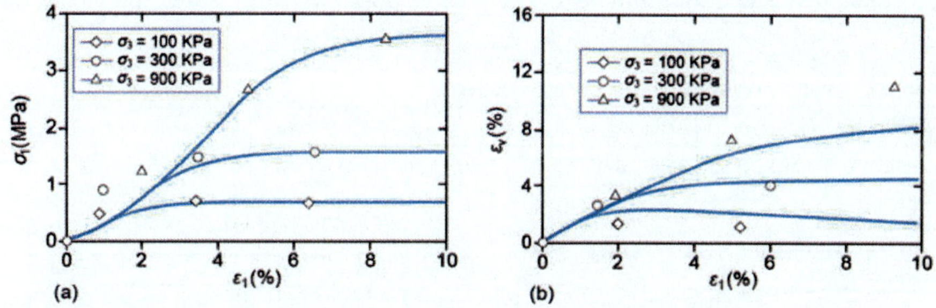

Fig. 2 Triaxial-K_0 consolidated drained test results and model predictions

Table 1 Parameters for static constitutive model

Parameter	k	k_e	n	α	φ_0 (°)	$\Delta\varphi$ (°)	ψ_0 (°)	$\Delta\psi$ (°)	β	ν_e
Value	500	1000	0.40	1.25	54.1	13.8	50.0	8.4	1.0	0.25

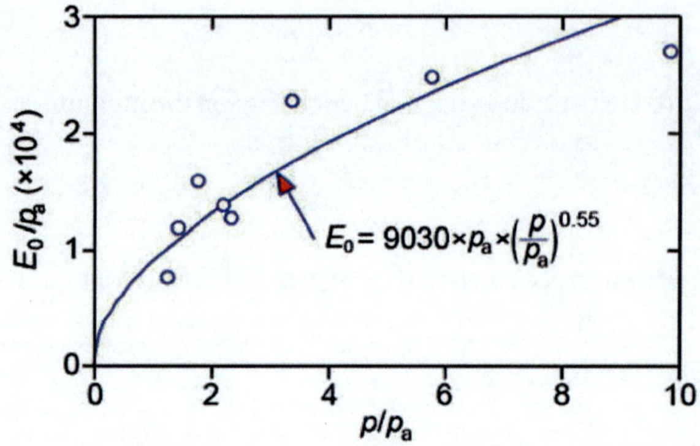

Fig. 3 Relationship between the Young's modulus and mean effective stress

$$G_0 = \frac{E_0}{2(1 + v_e)} \tag{22}$$

It is clear that the parameter k_0 depends on the Poisson's ratio. Cyclic triaxial tests are usually conducted under undrained condition. Therefore, $v_e = 0.5$ is assumed for parameter calibration temporarily. The values of k_0 and m are given in Table 2.

The shear modulus and damping ratio obtained from cyclic triaxial experiments are plotted against the shear strain amplitude in Fig. 4. Note that no information about the initial stress states before the cyclic triaxial tests are reported. Herein, we assume that the mean effective stress varies from 0.1–1.0 MPa, which results in a band in each panel as shown in Fig. 4. Suitable parameters are those ones that the predicted bands enclose as many data points as possible. Table 2 lists the values of

Table 2 Parameters for dynamic responses analysis

Parameter	k_0	m	k_1	λ_{min}	λ_{max}	c_γ	α_γ	d_γ	β_γ	c_v	α_v	d_v	β_v
Value	3010	0.55	35.0	0.05	0.30	1.80	1.47	0.015	0.43	0.15	0.87	0.68	0.56

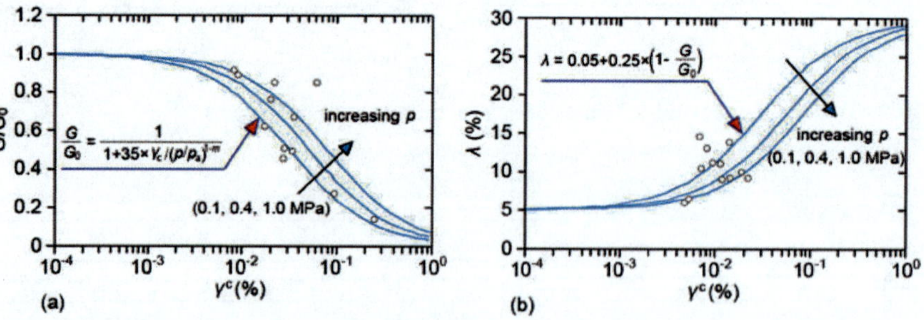

Fig. 4 Cyclic triaxial test results and model predictions

the relevant parameters. Also listed in Table 2 are residual strain model parameters. Since no experiments are reported for the permanent deformation of rockfill materials in Menta dam, these parameters are adapted from a similar rockfill [8].

Throughout the static and dynamic analyses, the rock mass foundation is not considered due to its considerably higher stiffness compared with the rockfill materials used. An average of 23 kN/m^3 is used for the unit weight of rockfill materials. In addition, the rockfill materials are assume to have a Poisson's ratio of 0.25 hereafter.

3.2 Ground Excitations

Two acceleration time series represent the seismic input scenarios. The acceleration histories and the corresponding response spectra (obtained with $\lambda = 0.05$) are shown in Fig. 5. These records are selected in accordance with the criterion of spectro-compatibility with the Italian Seismic Hazard map for the site of Menta Dam and represent two severe events occurred in Italy (Friuli earthquake occurred in 1976 and Central Italy earthquake sequence of 2016). The records are scaled to have a reference Peak Ground Acceleration (PGA) equal to 0.26 g (g = 9.81 m/s^2). In seismic responses analyses, the accelerations are specified to the boundary of the FEM model along the horizontal direction, i.e., a possible vertical excitation is neglected.

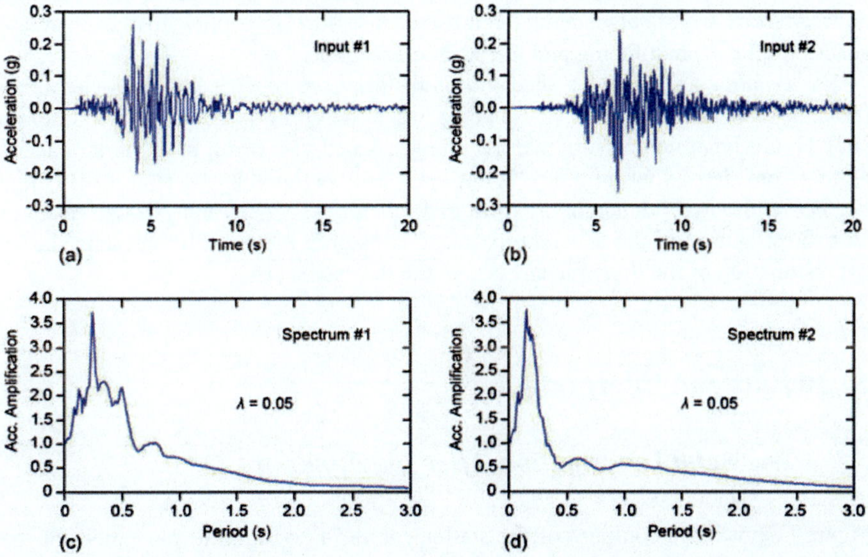

Fig. 5 Input acceleration histories and corresponding response spectra

4 Simulation Steps and Boundary Conditions

4.1 Simulation Steps

The dam is discretized with 884 plane strain solid elements. Rockfill materials are assumed to be filled horizontally layer by layer, and is modeled in 30 steps. In each step several sub-steps are incorporated so as to consider the nonlinear behavior of dam materials. An additional step is used to model the placing of the bituminous facing, and another additional step is used for the modeling of reservoir impounding to the elevation of 1424.75 m.

The considered duration of earthquake excitations (20 s) are divided into 5 stages, and an iterative process is employed for each stage to ensure the compatibility of the shear moduli of rockfill elements and their strain amplitudes, without considering their permanent strains. Once the shear moduli iteration converges, an extra iteration is performed, taking into account the permanent strains of rockfill materials. Details of the algorithm can be found in [8].

4.2 Boundary Conditions

The nodes along the interface between the dam and rock foundation are restricted entirely from any displacement. Hydrostatic pressure is applied to the upstream surface of the asphalt facing and the perimeter tunnel.

For dynamic analyses, the acceleration histories are specified to the nodes along the dam-rock interface. The so-called radiation damping is not considered herein [12]. Hydrodynamic pressure that may be generated as a result of an earthquake is neither considered for simplicity. It has been demonstrated by Fu et al. (2019) that neglecting the hydrodynamic pressure in a seismic responses analysis can result in an overestimation of the acceleration response within the rockfill materials and an overestimation of the dynamic stresses in the face slabs [13].

5 Results and Interpretations

5.1 Stress and Deformation After Impounding

Figure 6 shows the contours of the settlement and horizontal displacement of the embankment after reservoir filling. The settlement follows an elliptical distribution, with the maximum (66.7 cm) presents near the center of the dam. Almost the entire

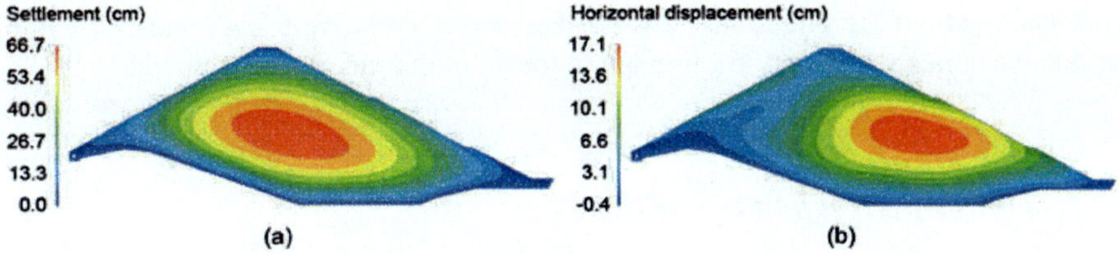

Fig. 6 Contours of displacement after first impounding

dam moves downstream as a result of the hydrostatic pressure and the downstream-inclined dam foundation. The largest horizontal displacement predicted here is about 17.1 cm.

Both the vertical and horizontal stresses increase linearly with the thickness of overlying rockfill materials as shown in Fig. 7. The maximum vertical and horizontal stresses at the deepest place of the dam are approximately 1.6 and 0.6 MPa, respectively, which indicates a stress ratio of about 0.37.

The contours of the internal friction angle and Young's modulus are plotted in Fig. 8. The internal friction angle is predicted using Eq. (11), and the Young's modulus is calculated using the trend curve given in Fig. 3. Since both the vertical and horizontal stresses increase with the depth of rockfill, the mean effective pressure is also expected to follow a layering distribution. Therefore, both the internal frictional angle and Young's modulus also follow layering distributions (Fig. 8); the former decreases with an increasing depth while that latter increases when the buried depth increases. A good feature of rockfill dams can be seen in Fig. 8a, i.e., shallow rockfill

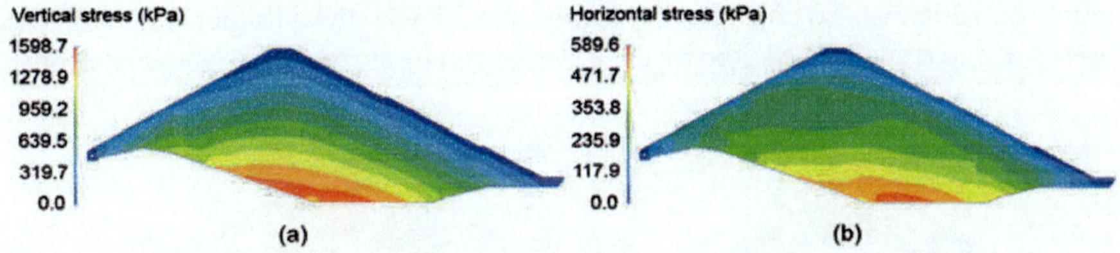

Fig. 7 Contours of stresses after first impounding

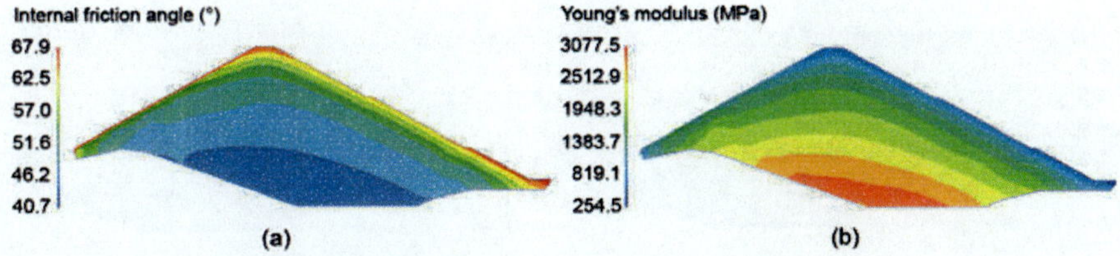

Fig. 8 Contours of friction angle and Young's modulus after first impounding

materials, which are potential sliding zones, have a relatively high shear strength, and those zones with a low friction angle locate at the inner part of the dam.

5.2 Dynamic Responses of the Dam

The seismic responses obtained for the Friuli earthquake (1976, Input #1) are plotted in Fig. 9. There is a decrease in the peak acceleration within the lower part of the dam, as indicated by the zones with an acceleration amplification factor (AAF) below 1.0 (Fig. 9a). However, the AAF increases abruptly within the upper one third of the dam to a maximum of 3.25 at the crest. This is the so-called whipping effect in structural dynamics. The shear strains of rockfill elements, defined by the numerator in Eq. (14), are recorded and the maximum ones during the earthquake are plotted in Fig. 9b. Generally, the maximum shear strain increases with an increasing elevation. The seismic excitation also results in a downstream horizontal displacement up to 8.2 cm at the downstream slope near the crest (Fig. 9c) and a settlement up to 19.4 cm at the crest. Both the horizontal displacement and vertical settlement increase with an increasing elevation, and this trend is in good agreement with the field observations made in the Zipingpu concrete face rockfill dam during the Wenchuan earthquake [3].

The seismic responses excited by the Central Italy earthquake (2016, Input #2) are shown in Fig. 10. Although the PGA of this input is the same as that of the Friuli earthquake input, the responses are considerably lower than those shown in Fig. 9. For instance, the permanent displacements cause by the #2 earthquake input are 4.5 cm along the horizontal direction and 10.6 cm along the vertical direction, which are only 55% of the previous ones. The maximum AAF for the #2 input is only 2.12, which is also considerably lower than 3.25 for the #1 input (Fig. 9a). The lower seismic responses excited by the #2 input can be attributed to its slim response

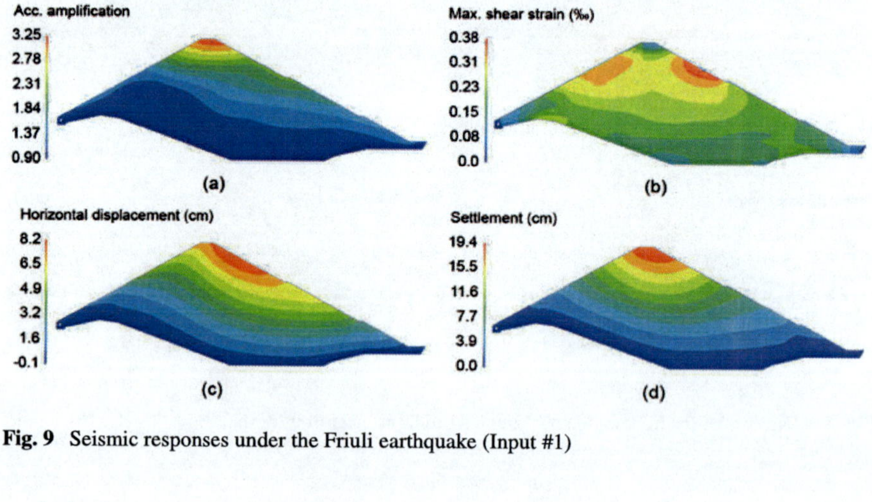

Fig. 9 Seismic responses under the Friuli earthquake (Input #1)

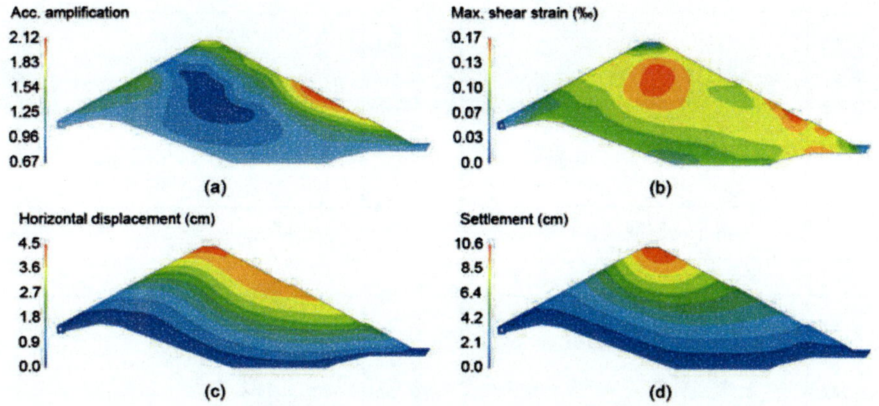

Fig. 10 Seismic responses under the Central Italy earthquake (Input #2)

spectrum as shown in Fig. 5d compared with the fat one for the 1# input (Fig. 5c). Figure 10a also demonstrates that the maximum AAF may be achieved along the downstream slope rather than the crest. This strong acceleration response along the dam slope may be responsible for the loosening, rolling and falling of protective surface rock particles observed in some cases (e.g. the Zipingpu case [3]).

The time histories of the acceleration and displacement at the crest are plotted in Fig. 11. The central node at the crest level is selected to represent the responses of the dam crest. The period of strong acceleration response coincides well with the strong ground excitation period (Fig. 11a, d). A slow accumulation of permanent displacements accompanying a low amplitude oscillation can be observed before the strong shaking period. The horizontal and vertical displacements, however, accumulate much faster during the strong earthquake period, and the amplitude of horizontal oscillation increases considerably as a result of the strong shaking. After the strong shaking period, the amplitude of oscillation decreases to a low level and the accumulation of permanent displacements slows down. It deserves to point out that no vertical excitation is input to the model in the numerical simulations. Therefore, only small vertical oscillation can be observed in Fig. 11c, f. Nevertheless, the magnitude of the earthquake-induced settlement is considerably higher than that of the horizontal displacement, as was observed in the Zipingpu dam [3]. The results shown in Fig. 11 also demonstrate the capability of the enriched equivalent linear method in capturing the deformation history of rockfill dams during earthquakes, which may also be obtained by using advanced elastoplastic constitutive models. Enriching the traditional equivalent linear method enables us to obtain the accumulation of permanent deformation without adding an extra computational cost [8].

The safety of the downstream slope is studied by considering two possible failure surfaces, and Fig. 12 shows the histories of the obtained factor of safety. Herein, the factor of safety, F_s, is evaluated by the following equation:

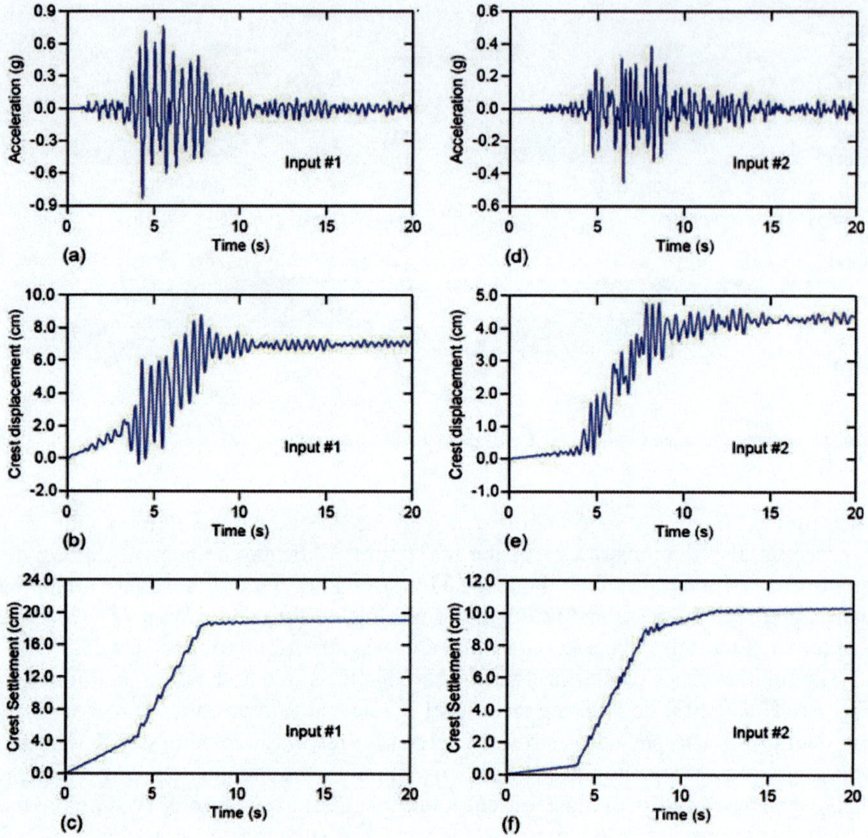

Fig. 11 Response histories of the dam crest

$$F_s = \frac{\sum (c_i + \sigma_{ni} \tan \varphi_i) l_i}{\sum \tau_i l_i} \tag{23}$$

in which c_i and φ_i are the cohesion and friction angle of the ith element that is intersected by the failure surface. l_i denotes the length of the intersecting segment, and σ_{ni} and τ_i denote the normal and shear stresses applied upon the segment. The factor of safety is always above 1.0, indicating that the downstream slope is of adequate resistance to sliding instability during the two earthquakes. Newmark's sliding displacement analyses are also performed, but no sliding displacement is accumulated for both excitations, which agree well with the results in Fig. 12.

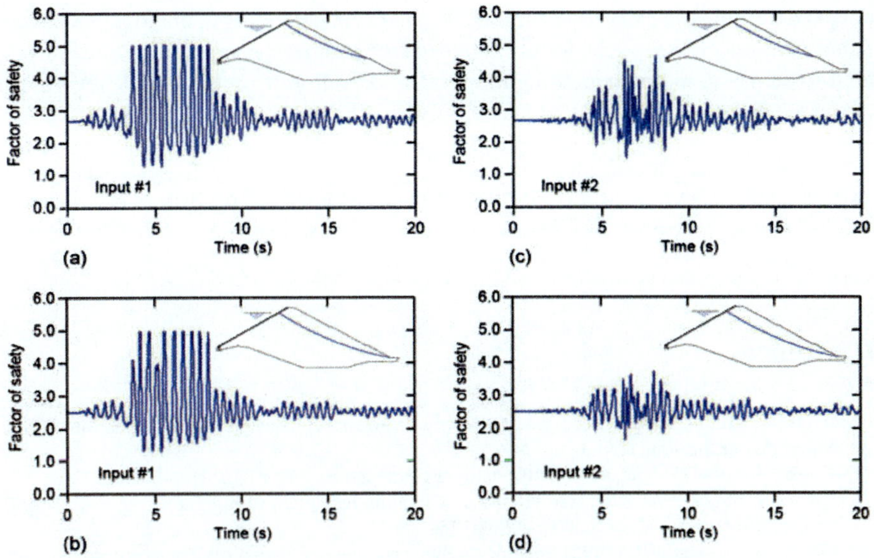

Fig. 12 Histories of the factor of safety against sliding for two failure surfaces

6 Conclusions

A series of plane strain FEM simulations are performed in studying the seismic responses of the Menta embankment dam in Italy. The following conclusions can be summarized based on the obtained numerical results:

(1) Both the vertical and horizontal stresses follow similar layering distributions, where the stresses are roughly proportional to the thickness of the overlying rockfill. As a result of their pressure-dependence, the internal friction angle and the elastic Young's modulus within the dam also follow layering distributions.
(2) The input acceleration is not amplified within the lower part of the dam. The acceleration amplification factor, however, increases abruptly within the upper part of the dam, particular within zones near the crest. This whipping effect is similar as in other structures. Numerical results show that strong acceleration response may also present along the downstream slope.
(3) Permanent deformation accumulates fast during the strong earthquake shaking period. Under a full reservoir condition, almost the entire dam moves towards the downstream, the amount of which increases with an increasing elevation. The earthquake-induced settlement is considerably larger than the horizontal component even if no vertical excitation is input.

It deserves to point out finally that the model parameters used in the static and dynamic responses analyses are calibrated based on very limited amounts of test data and a single material zone assumed in the FEM modelling is obviously an over-simplification. However, the trends obtained from numerical simulations are considered to be useful in understanding the seismic behavior of rockfill dams, regardless of the magnitudes predicted.

Acknowledgements This work is partially supported by the National Key Research and Development Program of China (No. 2017YFC0404800) and the National Natural Science Foundation of China (Nos. 51779152 & U1765203).

References

1. ICOLD (2010) Concrete Face Rockfill Dams, concepts for design and construction, China Water Power, Beijing, China
2. Cooke JB (1984) Progress in rockfill dams. J Geotechn Eng 110(10):1383–1414
3. Chen SS, Fu ZZ, Wei KM, Han HQ (2016) Seismic responses of high concrete face rockfill dams: a case study. Water Scie Eng 9(3):195–204
4. Gazetas G, Dakoulas P (1992) Seismic analysis and design of rockfill dams: state-of-the-art. Soil Dyn Earthq Eng 11(1):27–61
5. Finn WDL (2000) State-of-the-art of geotechnical earthquake engineering practice. Soil Dyn Earthq Eng 20(1–4):1–15
6. Fu ZZ, Chen SS, Peng C (2014) Modeling cyclic behavior of rockfill materials in a framework of generalized plasticity. Int J Geomech ASCE 14(2):191–204
7. Fu ZZ, Chen SS, Wei KM (2019) A generalized plasticity model for the stress-strain and creep behavior of rockfill materials. Sci China Technol Sci, https://doi.org/10.1007/s11431-018-9362-3(2019)
8. Fu ZZ, Chen SS, Wang TB (2017) Predicting the earthquake-induced permanent deformation of concrete face rockfill dams using the strain potential concept in the finite element method. Int J Geomech ASCE, https://doi.org/10.1061/(asce)GM.1943-5622.0001010(2017)
9. Kim MK, Lade PV (1988) Single hardening constitutive model for frictional materials I. Plastic Potent Funct Comput Geotech 5(4):307–324
10. Fu ZZ, Chen SS, Han HQ (2016) Experimental investigations on the residual strain behavior of a rockfill material subjected to dynamic loading. J Mater Civil Eng ASCE, https://doi.org/10.1061/(asce)mt.1943-5533.0001816(2016)
11. Idriss IM, Sun JI (1992) User's manual for SHAKE91, a computer program for conducting equivalent linear seismic response analyses of horizontally layered soil deposits, University of California, Davis, California. November
12. Zienkiewicz OC, Taylor RL (2000) the finite element method. Fifth edition. Oxford, Vol. 1: The Basis
13. Fu ZZ, Chen SS, Li GY (2019) Hydrodynamic pressure problems in concrete face rockfill dams subjected to earthquakes. J Hydrodyn 31(1):152–168

Seismic Analyses of Menta Embankment Dam

V. B. Glagovsky and E. D. Gibyanskaya

Abstract The article presents some results of calculations of the stress-strain state of the Menta embankment dam after filling the reservoir, as well as under possible seismic impacts, given by accelerograms of two earthquakes observed in Italy. The calculations wcrc carried out by the finite element method using the software package Plaxis 2D. The dam is divided into three zones according to the stages of construction. Static deformation modules of these zones were considered to be subject to stepwise change during the transition to the next stage of erection. The angle of internal friction was also assumed to be constant for each zone and subject to stepwise change between zones. The dynamic characteristics of the rockfill material for the HS small Plaxis model also are taken different in these three zones. As a result of the calculations, in particular, the residual displacements of the dam after the earthquake and the acceleration on the crest of the dam during seismic impact were determined. The results of the calculations show that the Menta embankment dam is capable of withstanding the considered earthquakes without significant damage.

Keywords Menta Embankment dam · Stress-Strain state · Construction sequence · Seismic analyses · Finite element method · Plaxis 2D

1 Introduction

The Menta dam is a bituminous-faced rockfill dam located in Southern Italy, lying in the heart of Aspromonte Massif at an elevation of about 1400 m a.s.l. The embankment is about 90 m high at its deepest point, and the reservoir impounds $1.8 \times 107\ m^3$ of water [1].

A homogeneous bituminous-faced rockfill dam, located on a rock mass foundation, was considered. The calculations were carried out by the finite element method using the software package Plaxis 2D [2].

V. B. Glagovsky (✉) · E. D. Gibyanskaya
Vedeneev Research Institute (VNIIG), St. Petersburg, Russia
e-mail: vglag@mail.ru

G. Bolzon et al. (eds.), *Numerical Analysis of Dams*, Lecture Notes in Civil Engineering 91, https://doi.org/10.1007/978-3-030-51085-5_28

2 Design Model

The design model of the dam and the finite element mesh are shown in Fig. 1. The number of model elements is 1283, number of nodes—10,599. Bituminous screen was not considered in the design model, but the upper face was taken waterproof, and the dam material has natural moisture, respectively.

The size of the base area was chosen in such a way as to exclude the influence of the lateral boundaries of the area on the static stress-strain state (SSS) of the dam. Coordinates on the design scheme: Y = absolute coordinate minus 1200 m; X = 0 in the center of the crest (crest from X = −4.9 to X = 5.1 m). In dynamic calculations, the lateral boundaries of the base area were assumed to be viscous in order to avoid the reflection of waves from these boundaries.

At the first stage, the SSS of the dam after construction and first filling was determined. The calculation takes into account the stepwise construction of the dam: 1st stage—from the foot of the dam (146.5 m) to elevation 180 m; 2nd—up to an altitude of 200 m; 3rd—up to crest—231.75 m. The water level in the reservoir after the filling—224.5 m (49.5 m from the foot of the dam).

3 Characteristics of the Dam Material

The choice of physical and mechanical characteristics of rockfill material for static and dynamic calculations of rockfill dams is a complex and not fully investigated issue. Calculations of the stress-strain state of earth dams according to design characteristics are preliminary in this sense and provide only a basic estimate of SSS of the dam during operation. Mathematical models developed at the design stage, as a rule, require refinement and calibration of the model while taking into account the data of field observations.

In [3–7] the deformation properties of the rockfill material for several dams in different ranges of compressive stresses were studied. However, the soils of lower density than used in the Menta Dam were tested. In a relatively recent work [8], coarse-grained soils were tested at a density of 2.2 t/m^3 of rock mass at compressive stresses up to 10 MPa. The values of deformation modules up to 500 MPa (an average of about 250 MPa) were obtained. Characteristics of soils under seismic impacts are

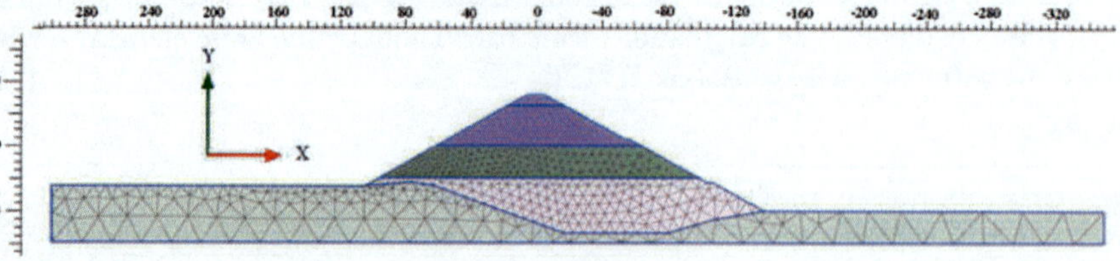

Fig. 1 Design model of the dam

Table 1 Characteristics of rockfill layers

Stages of construction	1			2			3		
Layers	1	2	3	1	2	3	1	2	3
Friction angle, degrees	44	–	–	42	44	–	40	42	44
Young's modulus, MPa	250	–	–	300	250	–	350	300	250

discussed in detail in the monograph [9]. Our Institute also conducts laboratory studies of static and dynamic characteristics of soils on triaxial compression devices for several embankment dams and has accumulated some experience in assessing the static characteristics coarse-grained soils [10].

From the data given in the formulation of theme B, in our opinion, there are no unambiguous answers about the assignment of the characteristics of the dam material, and data for the calibration of the model are not available. With this in mind, we have made the following assumptions. The dam is divided into three zones according to the stages of construction (Fig. 1). Static deformation modules of these zones were considered to be subject to stepwise change during the transition to the next stage of erection (Table 1). The angle of internal friction was also assumed to be constant for each zone and subject to stepwise change between zones (Table 1). The dynamic characteristics of the rockfill material for the HS small Plaxis model also are taken different in these three zones and will be given below.

4 Results of Calculations

4.1 *Static Calculations*

The static calculation uses the Coulomb-More model. The material of the dam body is rockfill with the characteristics shown in Table 1. For an approximate account of the dependence of the characteristics of E and φ on the stress state, their values in the layers are assumed to be different when calculating at different stages of construction. Layer number in the Table 1 corresponds to the construction stages.

Upon completion of construction, the deformation and strength characteristics of the dam layers have the values given in column 3 of the Table 1. Bulk unit weight of rockfill is 23 kN/m^3 and Poisson's ratio is 0.2.

The rock mass foundation is assumed to be linearly elastic; the characteristics are given in Table 2.

Table 2 Characteristics of rock mass foundation

Foundation	γ, kN/m^3	ν	E, MPa
Rock mass	27, 0	0, 25	7000

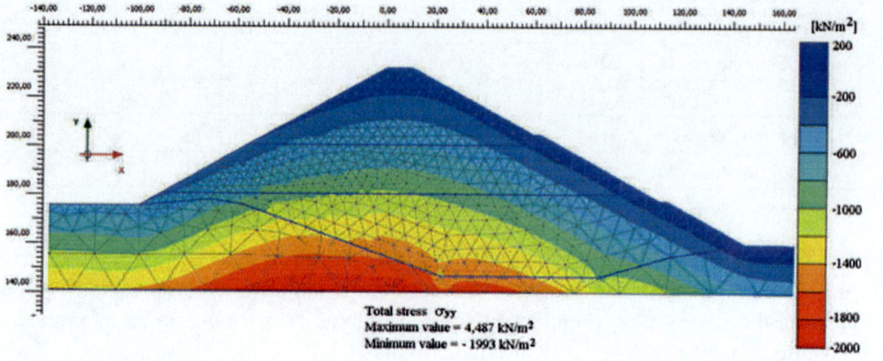

Fig. 2 Vertical stresses σ_{yy}

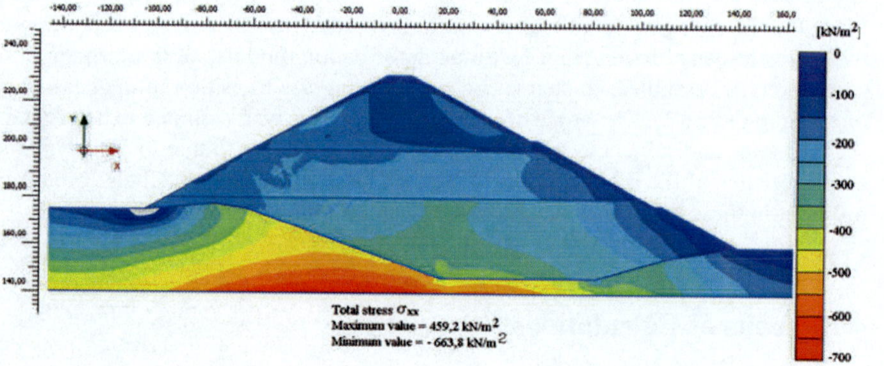

Fig. 3 Horizontal stresses σ_{xx}

After determining the SSS of the dam at the end of construction, SSS was calculated when the reservoir was filled to the design level. Hydrostatic water pressure was represented by a corresponding distributed load on the upstream slope and the foundation of the dam. The calculation results are presented in Figs. 2 and 3.

4.2 Calculation of the Dam Under Seismic Impacts

Calculation of the dam under seismic impacts, represented by records of two strong earthquakes that occurred in Italy (the Friuli earthquake occurred in 1976 and the earthquake in Central Italy in 2016) was carried out. Records are scaled to have a maximum acceleration of 0.26 g. The records represent the acceleration recorded at rock outcrop in free-field conditions.

Table 3 The dynamic characteristics of the rockfill material

Layer number	E_{50}^{ref}, MPa	E_{oed}^{ref}, MPa	E_{ur}^{ref}, MPa	G_0^{ref}, MPa	$\gamma_{0,7}$
1	350	350	1050	933	2×10^{-3}
2	300	300	900	800	2×10^{-3}
3	250	250	750	670	2×10^{-3}

In these calculations, the following assumptions are made. The rock base is assumed to be weightless, linearly elastic, the Young's modulus is E = 1400 MPa, the damping is 5%.

For the dam material, an elastoplastic model with hardening at small strains, implemented in the Plaxis program, is adopted. The following characteristics of the dam material were taken into calculations (Table 3).

Here E_{50}^{ref} is the secant modulus of stiffness in standard drained triaxial test of soil; E_{oed}^{ref}—the tangent modulus of stiffness for primary loading in the odometer; E_{ur}^{ref}—stiffness during unloading/re-loading; G_0^{ref}—the value of shear modulus at very low strains; $\gamma_{0,7}$—shear strain at which the shear modulus $G = 0,722\ G_0^{ref}$.

For all layers damping was taken to be 10%. Other characteristics are the same as in static calculation.

The seismic impact was modeled by accelerograms specified at the lower boundary of the computational domain. For ease of calculation, it was assumed that the records of accelerograms of earthquakes correspond to the lower boundary of the computational domain. Records of the accelerogram were presented for the three coordinates, but only two were used—vertical and one of horizontal, having a larger amplitude.

Accounting for the effect of the reservoir on the SSS of the dam during seismic impact was carried out by introducing the added mass in accordance with Russian standards [11]. The results of the calculations showed that the dynamic interaction of the earth dam with the reservoir has little effect on the SSS of the dam and, specifically, on the magnitude of the crest acceleration. This was noted earlier by many researchers, for instance, in the well-known work [12]. Therefore, in further calculations, the presence of the reservoir was taken into account only by the applied load from the hydrostatic pressure on the upstream slope and the foundation.

The results of seismic calculations are given below. As a result of the calculations, in particular, the residual displacements of the dam after the earthquake and the acceleration on the crest of the dam during seismic impact were determined.

4.3 The Earthquake Friuli, 1976

The records of earthquake Friuli are shown in Fig. 4.

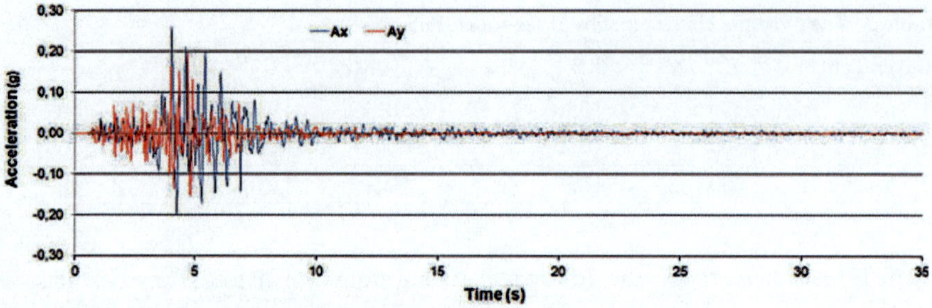

Fig. 4 Accelerograms of the Friuli earthquake, 1976

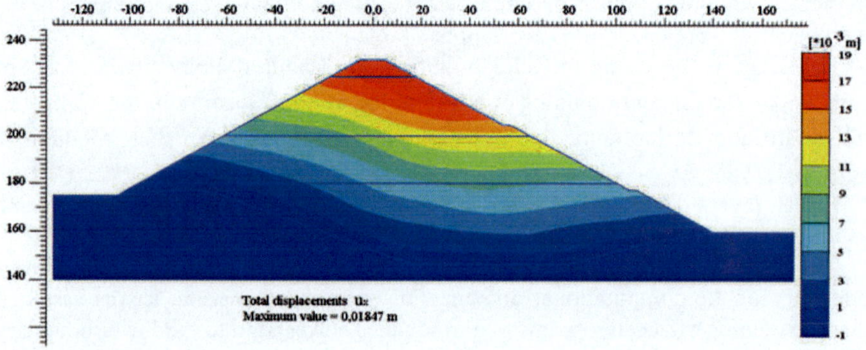

Fig. 5 Displacements u_x

A horizontal displacement of central axis on the crest is 1.78 cm and vertical displacement of central axis on the crest is −1.23 cm.

The residual displacements in the body of the dam after earthquake are shown in Fig. 5, 6, and 7.

Figures 8 and 9 represent accelerations and displacements of the point on the crest during the earthquake.

4.4 The Central Italy Earthquake, 2016

The records of the Central Italy earthquake are shown in Fig. 10.

The residual displacements in the body of the dam after earthquake are shown in Figs. 11, 12, and 13.

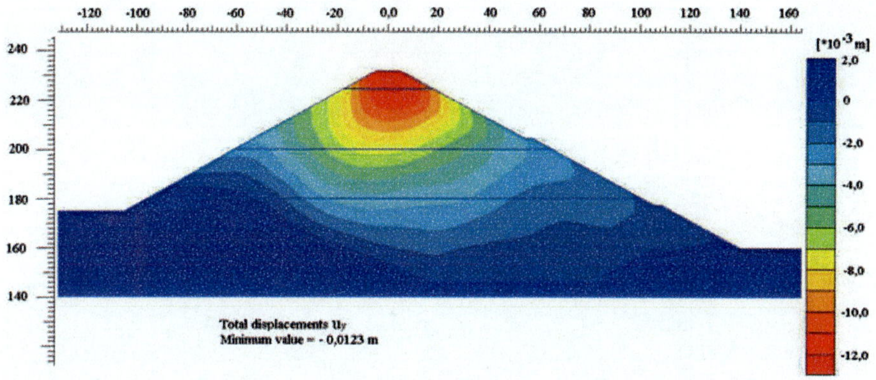

Fig. 6 Displacements u_y

Fig. 7 Horizontal and vertical displacements of central axis

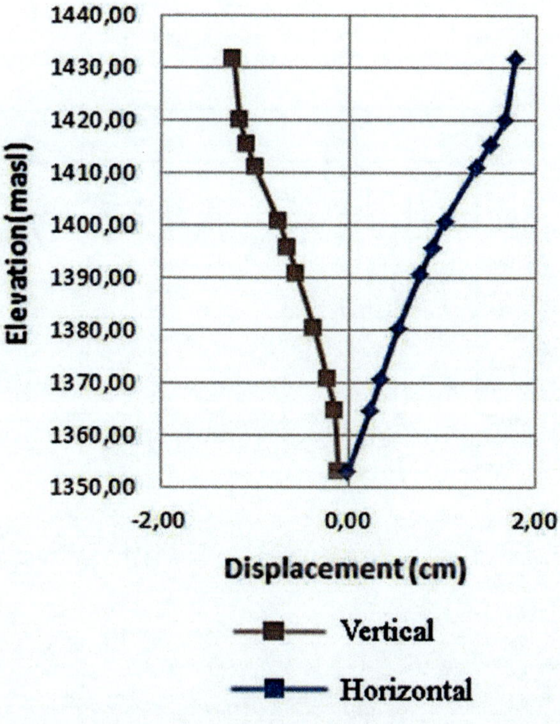

A horizontal displacement of central axis on the crest is 0.43 cm and vertical displacement of central axis on the crest is −0.58 cm.

Figure 14 and 15 represent accelerations and displacements of the point on the crest during the earthquake.

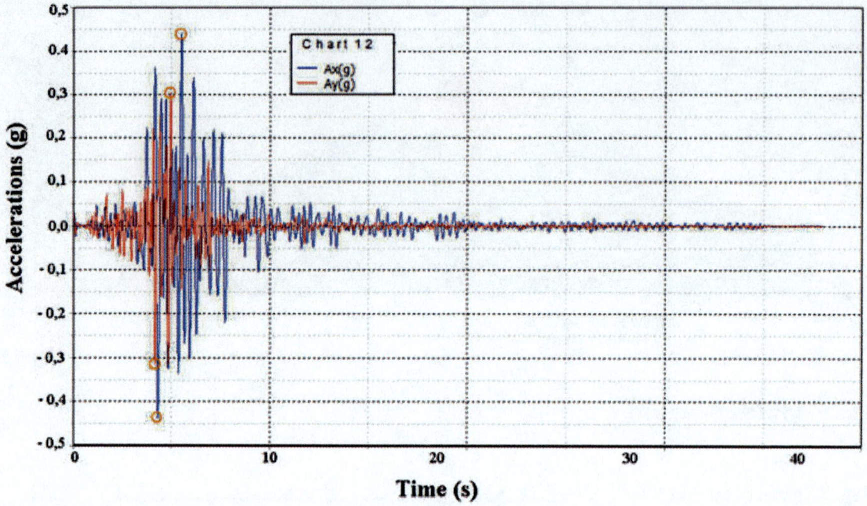

Fig. 8 Accelerations of a point on the crest during an earthquake

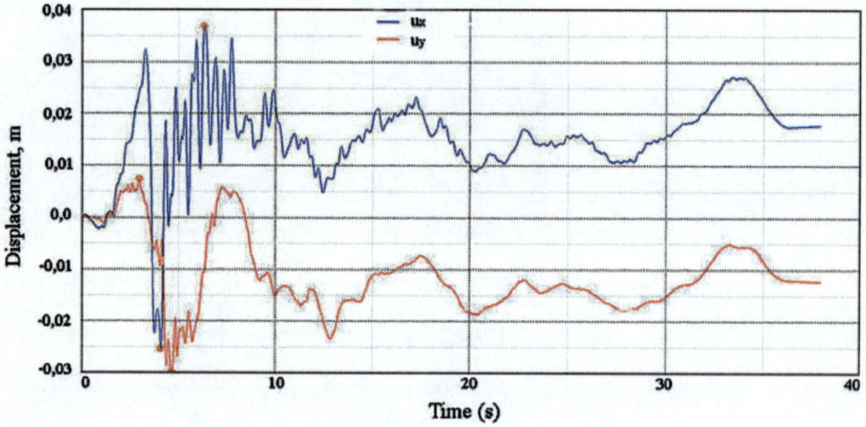

Fig. 9 Displacements of a point on the crest during an earthquake

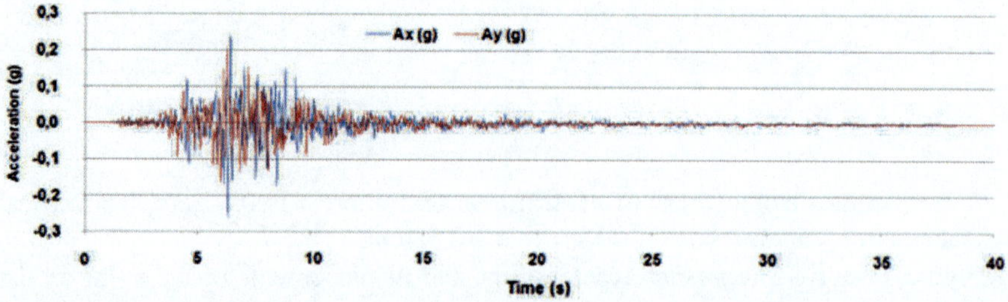

Fig. 10 Accelerograms of the Central Italy earthquake, 2016

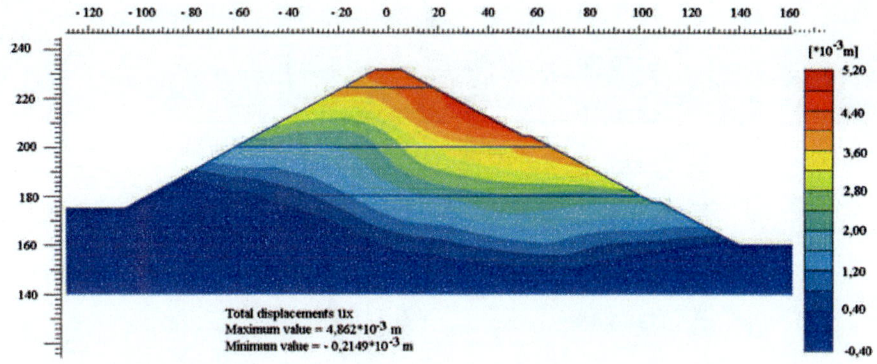

Fig. 11 Displacements u_x

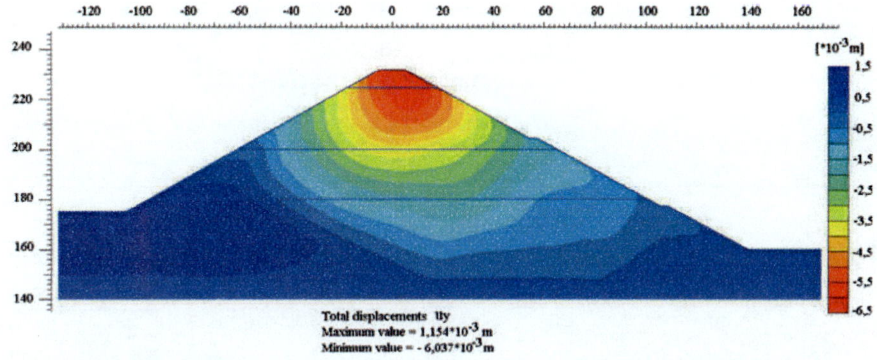

Fig. 12 Displacements u_y

5 Conclusions

The results of the calculations show that the Menta embankment dam is capable of withstanding the considered earthquakes without significant damage.

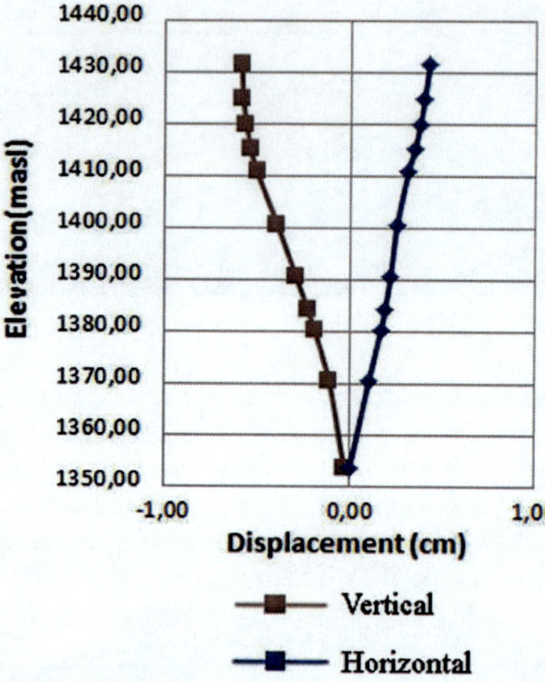

Fig. 13 Horizontal and vertical displacements of central axis

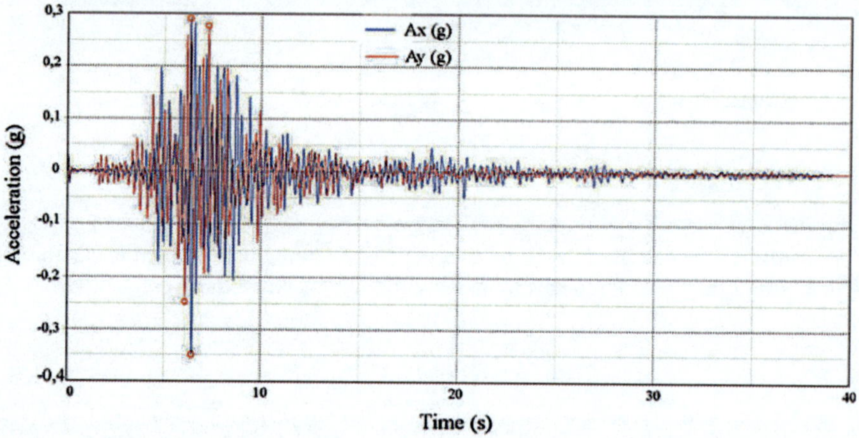

Fig. 14 Accelerations of a point on the crest during an earthquake

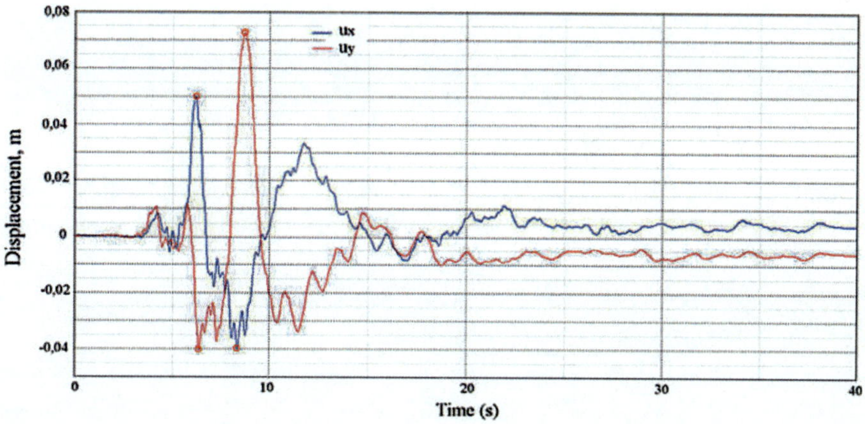

Fig. 15 Displacements of a point on the crest during an earthquake

References

1. Vecchietti A, Russo G, De Marco S, Fiorino A, Cecconi M, Pane V, Mazzà G (2019) Theme B formulation. Seismic analysis of Menta embankment dam. In: 15th ICOLD benchmark workshop on the numerical analysis of dams, Milano, Italy
2. PLAXIS 2D Reference Manual, CONNECT Edition, V20, 523 p
3. Marsal RJ (1967) Large scale testing of rockfill materials. ASCE J Soil Mech Found Div 93(2):27–43
4. Marachi ND, Chan CK, Seed HB (1972) Evaluation of properties of rockfill materials. ASCE J Soil Mech Found Eng 98(1):95–114
5. Gupta AK (2009) Triaxial behaviour of rockfill materials. Electr J Geotech Eng Bund J 14:1–18
6. Linero S, Palma C, Apablaza R (2007) Geotechnical characterization of waste material in very high dumps with large-scale triaxial testing. In: Proceedings, international symposium on rock slope stability in open pit mining and civil engineering, Perth, Australia, pp 59–76
7. Chavez C, Alonso EA (2003) Constitutive model for crushed granular aggregates which includes suction effects. ASCE J Soil Mech Found Div 43(4):215–227
8. Jia Y, Xu B, Chi S, Xiang B, Zhou Y (2017) Research on the particle breakage of rockfill materials during triaxial tests. Int J Geomech 17(10):04017085, 1–11
9. Ishihara K (2003) Soil Behaviour in earthquake geotechnics. Oxford University Press, New York
10. Glagovsky VB, Goldin AL, Radchenko VG (2011) Research in the construction of embankment dams (in Russian). Gidrotekhnicheskoe stroitel'stvo (Hydraulic Engineering) 9:43–46
11. Seismic building design code. SP 14.13330.2018 (in Russian)
12. Hall JF, Chopra AK (1982) Hydrodynamic effects in earthquake response of embankment dams. J Geotech Eng Div 108(4):591–597

Elasto-Plastic Finite Element Analysis of Menta Dam Under Two Earthquake Excitations

J. Liu, D. Zou, H. Liu, and F. Wang

Abstract As a part of seismic analyses of Menta embankment dam (Theme B) in 15th International Benchmark Workshop on Numerical Analysis of Dams, a dynamic elasto-plastic analysis was conducted to study the seismic behavior of Menta dam. An advanced elasto-plastic model to better capture complex loading history, implemented in the finite element procedure Geotechnical Nonlinear Dynamic Analysis (GEODYNA), was adopted to reproduce the dynamic behavior of the rockfill materials. The model parameters of rockfill materials were obtained based on the test results of dynamic modulus and damping ratio. An elasto-plastic soil-structure interface model that can trace the separation and re-contact was used to simulate the interface behaviors between the face slabs and cushion layer, and the model parameters were determined by empirical data published in the literature. The Friuli earthquake occurred in 1976 (1#) and Central Italy earthquake sequence of 2016 (2#) were applied at the bottom boundary of the mesh, respectively. The horizontal records were scaled to a peak acceleration of 0.26 g. The numerical results-initial stress state, acceleration, displacement and slab stress were plotted and analyzed.

Keywords Elasto-Plastic model · Rockfill materials · Dynamic behavior · Soil-Structure interface model · Acceleration · Slab stress

J. Liu (✉) · D. Zou · H. Liu
State Key Laboratory of Coastal and Offshore Engineering, Dalian University of Technology, Dalian 116024, China
e-mail: liujm@dlut.edu.cn

School of Hydraulic Engineering, Institute of Earthquake Engineering, Dalian University of Technology, Dalian 116024, China

F. Wang
Chengdu Engineering Corporation Limited, Chengdu 611130, China

G. Bolzon et al. (eds.), *Numerical Analysis of Dams*, Lecture Notes in Civil Engineering 91, https://doi.org/10.1007/978-3-030-51085-5_29

1 Introduction

Earthquake-induced damage of geotechnical engineering (such as rockfill dam) continues to be an active topic, driven by strong earthquake happened during the past decade. Due to the big size of dam, the dynamic behavior of prototype dam is still hard to capture by current experimental technology (such as large centrifuge test). Numerical analysis is still the most powerful method to evaluate the safety of the dam under earthquake. Currently, two different types of numerical analysis method have been proposed to assess the dynamic responses of dam. The first type is based on the widely used empirical equivalent linear theory [1]. The second type is based on the increment theory of plasticity [2, 3]. Due to the natural drawback of equivalent linear theory, the cyclic residual deformation cannot be reproduced directly. The increment theory of plasticity is more reasonable in theory than equivalent linear theory. Many scholars had made important contributions in the dynamic numerical analysis of practical engineering with increment theory of plasticity [4–7], but it still has far to go to verify the reliability of the advanced elasto-plastic models under complex conditions. In this study, an advanced elasto-plastic model to better capture complex loading history was adopted to reproduce the dynamic responses of the rockfill materials [8]. An elasto-plastic interface model that can trace the separation and re-contact was used to simulate the interaction between the face slab and the cushion layer [9].

The two models have been implemented into the finite element procedure Geotechnical Nonlinear Dynamic Analysis (GEODYNA) developed by the second author, which is based on GPU + CPU parallel computing [6, 7, 10–13]. Finally, a dynamic elasto-plastic analysis on Menta dam was conducted with GEODYNA to study the dynamic behavior under earthquake, and the effects of earthquake type and intensity on the distribution of acceleration, displacement and slab stress were analyzed and discussed.

2 Menta Dam

The Menta dam is a bituminous-faced rockfill dam located in Southern Italy. The crest height is 1431.75 m and the normal reservoir level is at 1424.75 m. The height of the dam is about 92 m at its deepest point, with a crest of 325 m in length and 10 m in width. The thickness of face slab is 0.32 m. The facing is formed by superposing different layers of bituminous concrete upon a regularly packed basement of compacted sandy gravel. The embankment slopes have 1.8:1 inclination. The embankment is made of compacted rockfill and was planned to be divided in five zones at the design stage (it was considered uniform in this study). The foundation of Menta dam is a fractured metamorphic rock. The cross-section of the embankment analyzed will be introduced later in the article.

3 Constitutive Models and Parameters

3.1 Rockfill Material

In the study, an advanced elasto-plastic model to better capture complex loading history was used for the rockfill material [8]. The distinctive feature of the model is that the concept of scaled memory is extended and incorporated into the plastic modulus and dilatancy. The influence of any loading and unloading history on stress-strain behaviors can be captured for complex loading paths, particularly for irregular cyclic loading. Brief introduction of the model was illustrated in Appendix A. The parameters of the model were obtained based on the test results of dynamic modulus and damping ratio. Figure 1 shows the comparison between the test and the simulated results.

The difference between the two could be induced by the discreteness of the tests. But the model is able to satisfactorily reproduce the variation of normalized Young modulus and damping ratio with increasing cyclic axial strain [14, 15]. The rockfill model parameters were calibrated listed in Table 1.

3.2 Interface Between Face Slab and Rockfill Material

An interface model was employed to simulate the interaction between the face slab and rockfill material [9]. The model could properly reproduce the monotonic and cyclic loading behaviors and could trace the separation and re-contact between the soil and structure. The model parameters were determined by empirical data published in the literature [7].

3.3 Slab

A linear elastic model was adopted to reproduce the bituminous-face slab with elastic modulus of 2800 MPa and Poisson's ratio of 0.33 (at the temperature of 26 °C) [16].

3.4 Peripheral Joint

A joint was set between face slab and foundation. The model and its parameters for peripheral joints were consistent with Zou et al. [6]. The compressive stiffness of the joints was assumed to be a constant value of 25,000 MPa/m. The shear stiffness was assumed to be a constant value of 1 MPa/m.

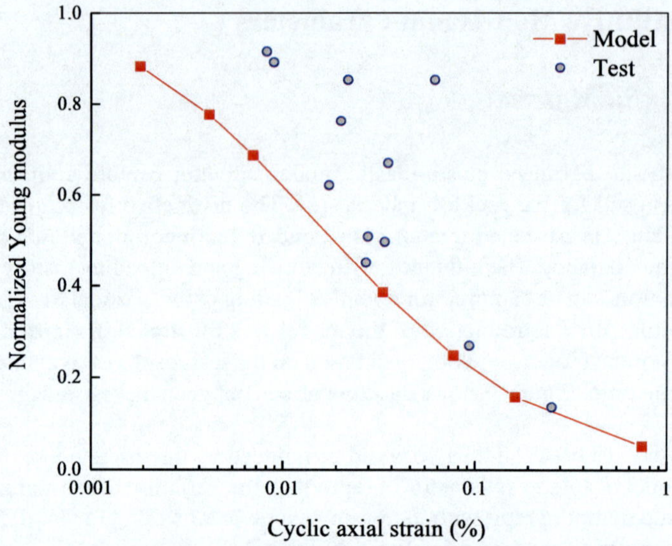

(a) Normalized Young modulus

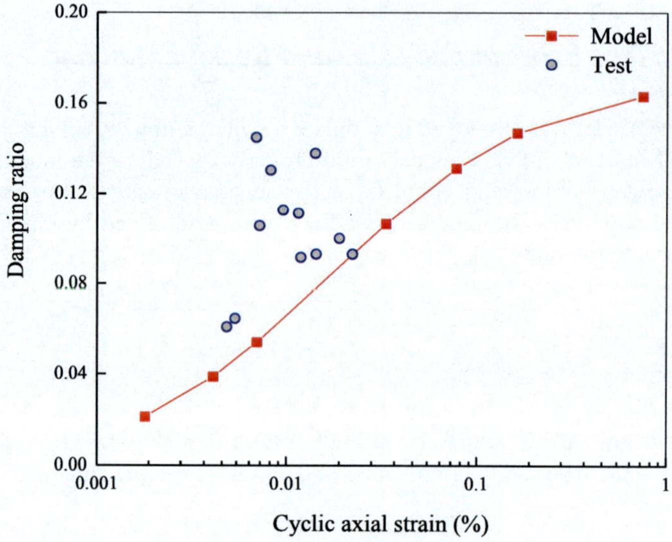

(b) Damping ratio

Fig. 1 Comparison of test and simulated results. **a** Normalized young modulus. **b** Damping ratio

Table 1 Model parameters of rockfill material

Elastic moduli		Critical state			Dilatancy				Plastic modulus						
G_0	ν	$e_{\tau 0}$	λ	M_c	α_1	α_2	k_m	r_d	h_1	h_2	m	κ	k_p	r_v	r_n
590	0.2	0.52	0.080	1.70	0.85	1.0	0.0	0	30	10	0.2	0.1	5.0	1.2	0.0

4 Elasto-Plastic Finite Element Analysis

4.1 Finite Element Program

A finite element procedure, Geotechnical Nonlinear Dynamic Analysis (GEODYNA), was used to conduct static and dynamic elasto-plastic finite element analysis of Menta dam. The above material models has been incorporated into GEODYNA.

4.2 Finite Element Mesh

The two-dimensional (2D) finite element mesh of Menta dam used in this study, as shown in Fig. 2. There are 425 total elements. The rockfill material and slab were simulated using spatial four-node isoparametric solid elements. The interface and peripheral joint were simulated using slip element. The bottom boundary of the dam was fixed in the x and y directions.

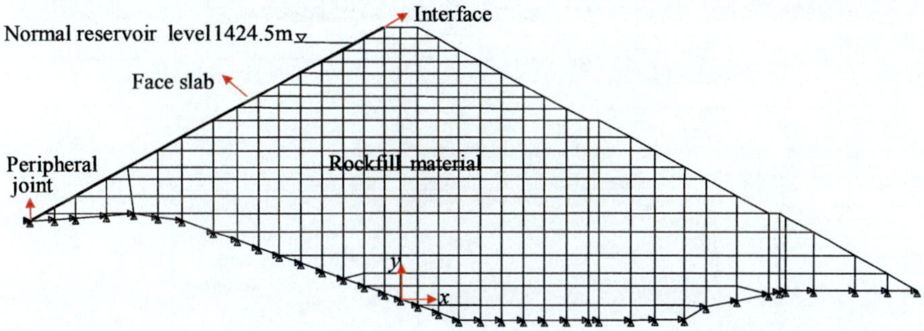

Fig. 2 Typical section of Menta dam and mesh grid

4.3 Construction and Impounding Processes

The construction was simulated with FE (finite element) steps ranging from 1 to 21 for construction. The impounding process was simulated with FE steps ranging from 22 to 45. The water pressure was simulated as a surface force on the slab.

4.4 Earthquake Process

The Friuli earthquake occurred in 1976 (1#) and Central Italy earthquake sequence of 2016 (2#) recorded at rock outcrop were selected as the seismic inputs for the dynamic analysis of Menta Dam. The input ground motion and acceleration response spectrums are plotted in Fig. 3. The horizontal records are scaled to a peak acceleration axbase of 0.26 g. The vertical peak acceleration aybase are 0.763 and 0.797 times of axbase for 1# and 2#, respectively.

In the dynamic analysis, the hydrodynamic pressure acting on the face slab was simulated using the adding mass method. Viscous damping of the Rayleigh type was included in the dynamic simulations, and a damping ratio of 3% was assumed for

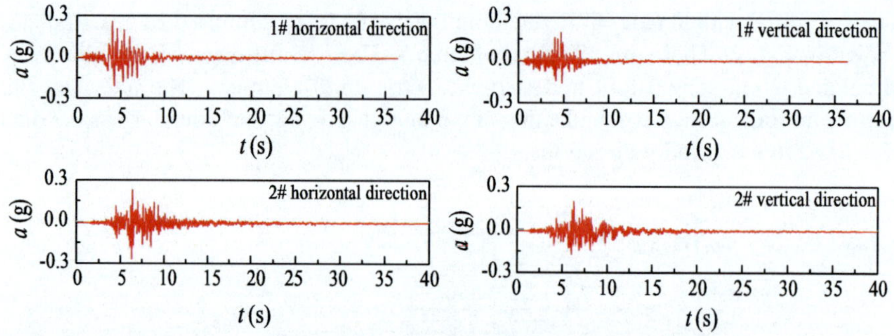

(a) Input earthquake motion

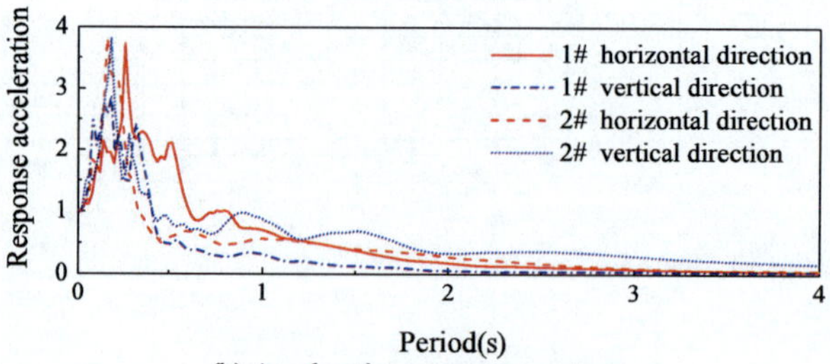

(b) Acceleration response spectrum

Fig. 3 Input earthquake motion and acceleration response spectrum. **a** Input earthquake motion. **b** Acceleration response spectrum

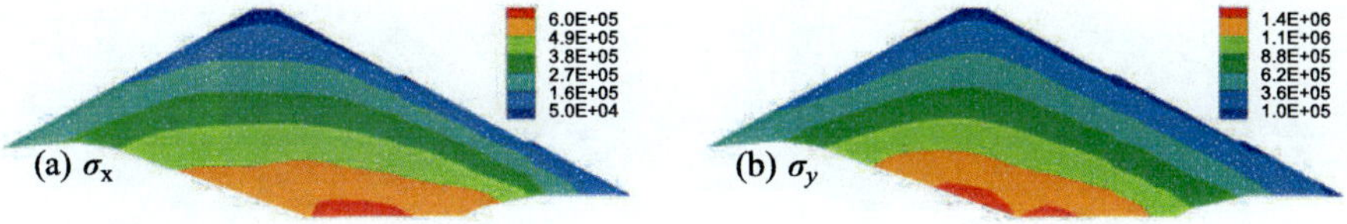

Fig. 4 Distribution of stress after impounding (unit: Pa)

the rockfill material. The dynamic response of the dam was analyzed using a time step of 0.005 s equal to the time interval of the earthquake records.

5 Results and Discussions

5.1 Initial Stress State

The simulated contour plot of horizontal stress σx and vertical stress σy within the embankment after impounding are plotted in Fig. 4. The simulated contour plot of principal stress ratio σ3/σ1 in most regions is ranging from 0.30–0.46, close to the pressure cells measurements (σ3/σ1 = 0.3–0.4).

5.2 Dam Acceleration

The typical horizontal and vertical acceleration time history of 1# and 2# at mid-crest are presented in Fig. 5. Figures 6 and 7 show the distribution of the horizontal acceleration amplification (axmax/axbase) and vertical acceleration amplification (aymax/aybase), and the horizontal and vertical maximum acceleration amplifications occurred at the crest were about 1.9 and 2.5 for 1# and 2.0 and 2.1 for 2#.

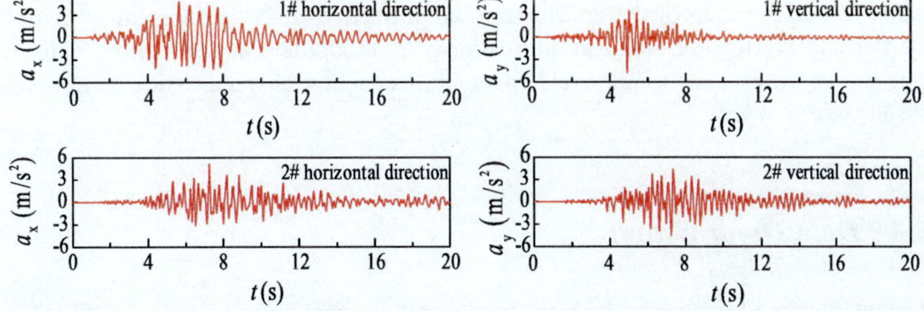

Fig. 5 Acceleration time history at mid-crest

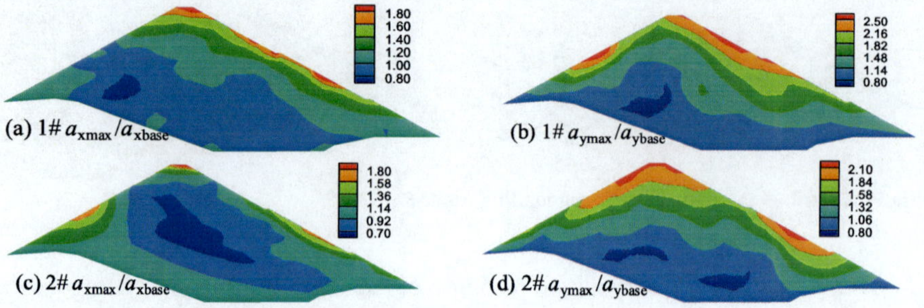

Fig. 6 Distribution of acceleration amplification

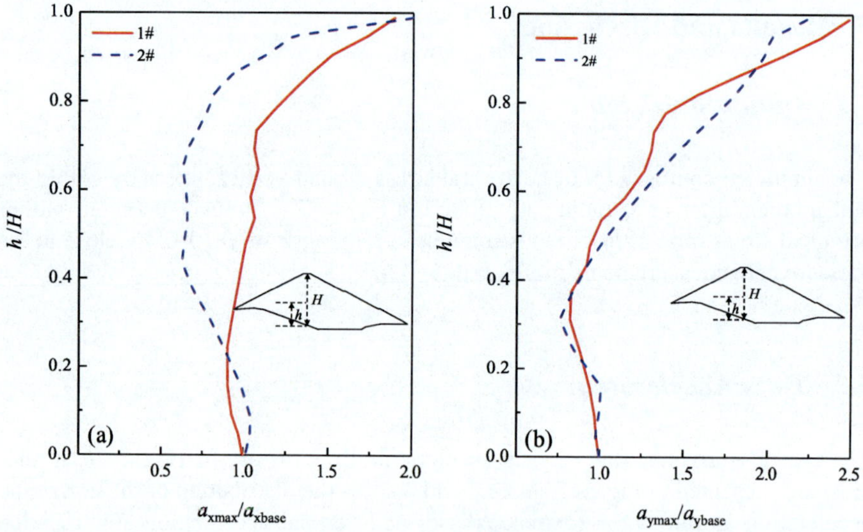

Fig. 7 Relation between acceleration amplification and normalized dam height

The high horizontal peak accelerations mainly developed in the upper part of the dam. The distribution of the acceleration amplification of 2# is different to that of 1#. Particularly, the horizontal amplification of 2# is smaller than 1# in most regions. The above differences mainly related to the frequency characteristics of the two earthquake.

5.3 *Dam Deformation*

The typical time histories of the horizontal and the vertical displacements, u and v, at mid-crest are plotted in Fig. 8. The horizontal displacement developed primarily in

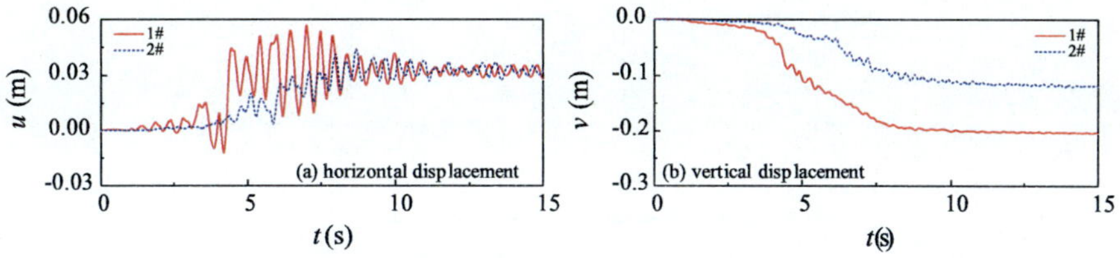

Fig. 8 Displacement time history at mid-crest

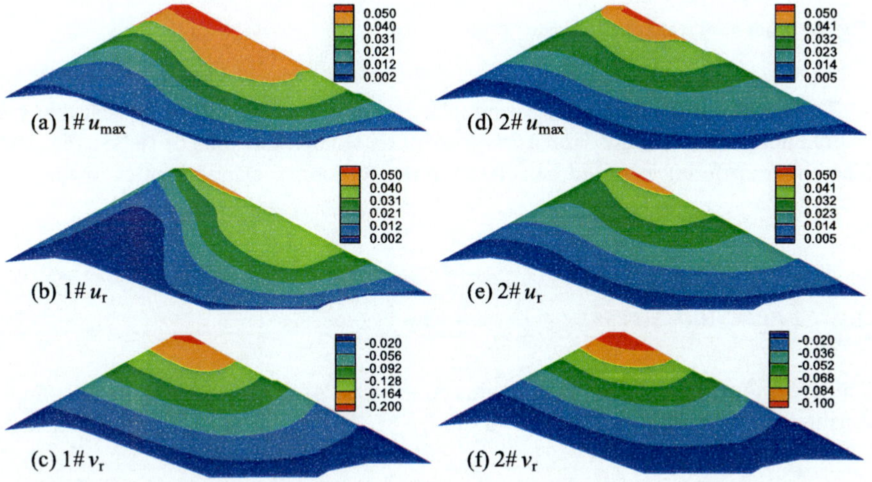

Fig. 9 Distribution of displacements (Unit: m)

the downstream direction. The nonlinear and irreversible response can be realistically reproduced. The irreversible displacements stabilized at the end of the shaking.

Figure 9 shows the contour plot of predicted maximum and residual (post-seismic) horizontal (umax, ur) and vertical (vmax, vr) displacements. The maximum and residual vertical displacements are nearly the same (only residual vertical displacement is plotted), but there exists some differences on maximum and residual horizontal displacements.

The maximum residual vertical displacement occurred at the crest. The residual vertical displacement of 2# with maximum value of 0.121 m is obviously smaller than that of 1# with maximum value of 0.206 m, but the distribution rule of vertical displacements of 2# is similar to that of 1# within the embankment (see Fig. 9c and f). The maximum residual horizontal displacement occurred at the downstream slope near crest, and it is nearly the same with value of about 0.05 m for 1# and 2#, but there existed a clear difference on the distribution rule of horizontal displacements within the embankment between them (see Figs. 9b and e).

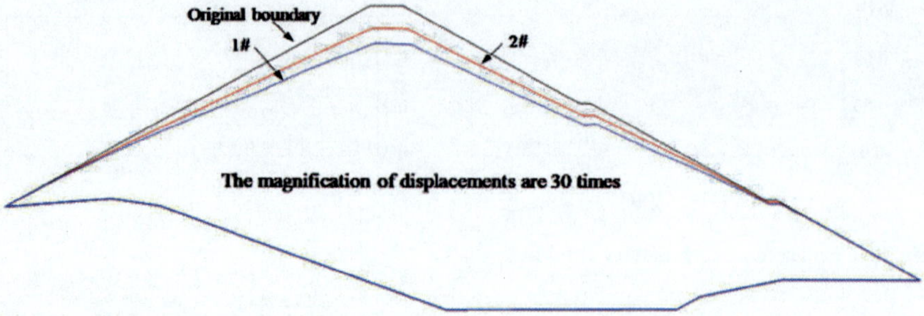

Fig. 10 Dam profile after earthquake

The profile of the dam after earthquake are plotted in Fig. 10. It is clear that the overall deformation of the dam of 2# is obviously smaller than that of 1# (see Fig. 10). The above differences could be related to the frequency characteristics of the two earthquake.

5.4 Face Slab Stress

The minimum and maximum slope-direction stress of face slab, σsmin and σsmax, during the two earthquake are shown in Fig. 11. The tensile stress zone of face slab

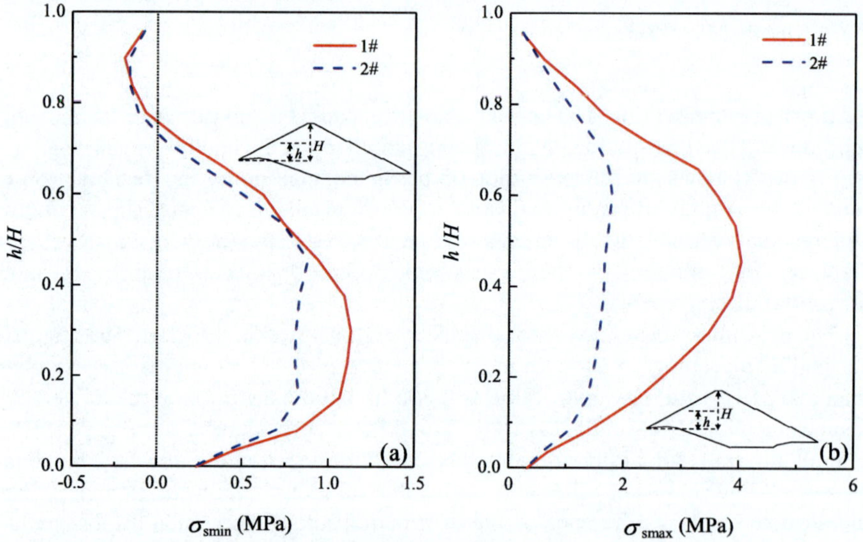

Fig. 11 Minimum and maximum slope-direction stress of face slab

occurred ranging from 0.73 to 1.0 H (H is dam height), with peak value of −0.19 MPa at 0.90 H for 1# and −0.16 MPa at 0.90 H for 2#. The tensile strength of sealing layer is 0.95 MPa, and larger than σsmin mentioned above.

The maximum slope-direction stress of face slab for 1# is obviously larger that of 2#, similar to the above deformation behavior. The maximum slope-direction stress of face slab occurred at 0.46H with peak value of 4.03 MPa for 1# and 0.60H with peak value of 1.78 MPa for 2#. The maximum slope-direction stress of face slab in most regions (the maximum slope-direction stress is 1.30 MPa after impounding) is larger than the compressive strength 1.49 MPa, and protective measures should be applied in compressive stress zone to improve the aseismic capability of the slab.

It should be mentioned that face slab of Menta dam is formed by superposing different layers of bituminous concrete, and it was modelled as single equivalent layer. Meanwhile, the mechanical behavior of bituminous concrete depends on temperature, which was not considered in this study.

6 Conclusions

An advanced elasto-plastic model to better capture complex loading history, implemented in the finite element procedure Geotechnical Nonlinear Dynamic Analysis (GEODYNA), was adopted to predict the dynamic responses of Menta dam under two different earthquake (1# and 2#). The numerical results-initial stress state, acceleration, displacement and slab stress were analyzed. The following conclusions were drawn:

(1) The distribution of the acceleration amplification of 1# is different to that of 2#. The horizontal maximum acceleration amplifications occurred at the crest were about 1.9 for 1# and 2.0 for 2#, but the horizontal amplification of 2# is smaller than 1# in most regions within the embankment.
(2) The nonlinear and irreversible response of the dam can be realistically reproduced. The irreversible displacements stabilized at the end of the shaking. The maximum residual vertical displacement occurred at the crest. The distribution rule of vertical settlement of 2# is similar to that of 1#, but the vertical deformation of 2# with maximum value of 0.121 m is obviously smaller than that of 1# with maximum value of 0.206 m. The maximum residual horizontal displacement occurred at the downstream slope near crest. Although the maximum horizontal displacement is nearly the same with value of about 0.05 m for 1# and 2#, a clear difference on the distribution rule of horizontal displacements exists between them.
(3) The tensile stress zone mainly occurred in the upper of the face slab ranging from 0.73 to 1.0H, with maximum value at 0.90H smaller than tensile strength of sealing layer. The maximum slope-direction stress of face slab occurred at 0.46H with peak value of 4.03 MPa for 1# and 0.60H with peak value of 1.78 MPa for 2#, which are larger than the compressive strength of sealing layer.

Protective measures should be applied in compressive stress zone to improve the compressive strength of the slab. The effect of temperature on the mechanical behavior of bituminous concrete was not considered in the static and dynamic analysis, which would have a strong effect on the face slab stress.

Acknowledgements This work was supported by the National Natural Science Foundation of China (Grant Nos.U1965206 & 51779034), the Fundamental Research Funds for Central Universities (Grant No. DUT19ZD216), and Technology Project of China Huaneng Group (Grant No. HNKJ18-H25).

Appendix A

The elastic shear modulus G is specified as:

$$G = G_0 p_a \frac{(2.97 - e)^2}{1 + e} \sqrt{\frac{p'}{p_a}} \tag{1}$$

where G_0 are material constants, e is the current void ratio, p' is the mean effective pressure, and p_a is the atmospheric pressure. The elastic bulk modulus can be calculated with Poisson's ratio v.

The plastic modulus H_1 corresponding to a stress ratio increment is specified as

$$H_1 = h_1 p_a \beta(e) \left(\frac{p'}{p_a}\right)^m \left\{ \left[\frac{\delta_1 |\delta_1|^{r_x/S_h}}{(1 - \delta_1)^{r_y S_h}} \right] \sqrt{\left(\frac{\rho_1^{max}}{\rho_1}\right)} \right\} \tag{2}$$

with $(e) = \exp\left(\frac{1}{e + \Delta e - \kappa}\right)$, $\Delta e = W_p / \left(a + bW_p\right)$, $S_h = exp\left(r_v \Delta\varepsilon_{vc} \frac{\beta(e)}{\left(R_c^{N-1}\right)^{r_n}}\right)$.

where h_1, m, , a, b, r_v, r_n are model parameters. r_x and r_y can set to default value, equal to 1.0 and 1.5 for rockfill material, respectively. W_p is plastic work. ρ_1 and ρ_1^{max} are the distances between the two stress points in deviator stress space defined by bounding surface theory. δ_1 is scaled memory parameter, and can be expressed as $\delta_1 = 1 - \eta/(M_p g(\theta))$ under virgin loading. $g(\theta)$ is the interpolation function. θ is Lode's angle. η is stress ratio.

$$M_p = M_c - k_p \psi \tag{3}$$

with $\psi = e - \left[e_{\tau 0} - \Delta e - \lambda \ln\left(p'/p_a\right)\right]$.

The dilatancy D_1 associated with a stress ratio increment is defined as

$$D_1 = \frac{\alpha_1}{S_d}\left[M_d - M_p\left(1 - \delta_1^{k=2}\right)\frac{\rho_1}{\rho_1^{max}}\right]g\left(\tilde{\theta}\right) \tag{4}$$

with $M_d = M_c + k_d\psi$, $S_d = exp\left(r_d\Delta\varepsilon_{vc}\frac{\beta(e)}{(R_c^{N-1})^{r_n}}\right)$,

where $M_c, k_p, k_d, r_d, e_{\tau 0}, \lambda$ are model constants. $\delta_1^{k=2}$ is the scaled memory parameter at the maximum stress state. $\tilde{\theta}$ is the Lode's angle determined by scaled memory theory.

The plastic modulus H_2 under a constant ratio is defined as

$$H_2 = h_2 p_a \beta(e)\left(\frac{p'_{ult}\delta_2}{p_a}\right)^m\sqrt{\frac{\rho_2^{max}}{\rho_2}}exp\left(\frac{c_0}{\eta}\right)\text{sign}\left(dp'\right) \tag{5}$$

where h_2, m, c_0 are model constants. c_0 can be set to a default value of 0.5. p'_{ult} is a reference mean effective stress, which is set to a large constant value of 1e20. $\text{sign}\left(dp'\right)$ is the sign of the increment of mean effective stress. ρ_2 and ρ_2^{max} are the distances between the two stress points in hydrostatic stress space defined by bounding surface theory.

The dilatancy D_2 under a constant ratio path is defined as

$$D_2 = \frac{\alpha_2}{S_d}\left\langle M_c g\left(\tilde{\theta}\right) - \eta\right\rangle exp\left(\frac{c_0}{\eta}\right)\text{sign}\left(dp'\right) \tag{6}$$

where α_2 is a model constant. The Macaulay brackets <> define the operations <x> = *x* if $x \geq 0$ and <x> = 0 if $x < 0$.

The model for rockfills requires 16 material constants for monotonic and cyclic loading, including two elastic constants G_0 and K_0; three critical state constants e_τ, λ, and M_c; four dilatancy constants α_1, k_m, α_2, and r_d; seven plastic modulus constants h_1, h_2, , m, k_p, r_v and r_n; and two additional particle breakage constants *a* and *b*, if necessary. The three constants r_v, r_n, and r_d were used to capture the cyclic loading behavior. Most of the parameters have a definite physical meaning and can be directly determined using conventional experimental results. The detailed calibration procedure can be found in Liu et al. [8].

References

1. Idriss IM, Seed HB (1968) Seismic response of horizontal soil layers. J Soil Mech Found Div 94(SM4):1003–1031
2. Lacy SJ, Prevost JH (1987) Nonlinear seismic response analysis of earth dams. Soil Dyn Earthq Eng 6(1):48–63
3. Zienkiewicz OC, Chang CT, Hinton E (1978) Non-linear seismic response and liquefaction. Int J Numer Anal Meth Geomech 2(4):381–404

4. Zienkiewicz OC, Chan A, Pastor M, Schrefler BA, Shiomi T (1999) Computational geomechanics. Wiley Chichester, New York
5. Alyami M, Rouainia M, Wilkinson SM (2009) Numerical analysis of deformation behaviour of quay walls under earthquake loading. Soil Dyn Earthq Eng 29(3):525–536
6. Zou D, Xu B, Kong X, Liu H, Zhou Y (2013) Numerical simulation of the seismic response of the Zipingpu concrete face rockfill dam during the Wenchuan earthquake based on a generalized plasticity model. Comput Geotech 49:111–122
7. Kong X, Liu J, Zou D (2016) Numerical simulation of the separation between concrete face slabs and cushion layer of Zipingpu dam during the Wenchuan earthquake. Sci China Technol Sci 59(4):531–539
8. Liu J, Zou D, Kong X (2018) Three-dimensional scaled memory model for gravelly soils subject to cyclic loading. J Eng Mech 144(3):4018001
9. Liu J, Zou D, Kong X (2014) A three-dimensional state-dependent model of soil–structure interface for monotonic and cyclic loadings. Comput Geotech 61:166–177
10. Xu B, Zou D, Liu H (2012) Three-dimensional simulation of the construction process of the Zipingpu concrete face rockfill dam based on a generalized plasticity model. Comput Geotech 43:143–154
11. Chen K, Zou D, Kong X, Chan A, Hu Z (2017) A novel nonlinear solution for the polygon scaled boundary finite element method and its application to geotechnical structures. Comput Geotech 82:201–210
12. Zou D, Han H, Liu J, Yang D, Kong X (2017) Seismic failure analysis for a high concrete face rockfill dam subjected to near-fault pulse-like ground motions. Soil Dyn Earthq Eng 98:235–243
13. Kong X, Zhou Y, Zou D, Xu B, Yu L (2011) Numerical analysis of dislocations of the face slabs of the Zipingpu concrete faced Rockfill dam during the Wenchuan earthquake. Earthq Eng Eng Vibr 10(4):581–589
14. Rollins KM, Evans MD, Diehl NB, Daily WD III (1998) Shear modulus and damping relationships for gravels. J Geotech Geoenviron Eng 124(5):396–405
15. Zou DG, Gong T, Liu JM, Kong XJ (2012) Shear modulus and damping ratio of gravel material. Appl Mech Mater 105–107:1426–1432
16. https://www.eko.polimi.it/index.php/icold-bw2019/2019/about/editorialPolicies#custom-3

3 FEM Model and Inputs

3.1 FEM Model

The layout of the Plaxis2D FEM model is shown in Fig. 1.

For the mesh generation, the element size is recommended to fulfil the requirement suggested by Kuhlmeyer and Lysmer [2] in Formula (1). With estimated lowest $V_{s,min} = 400$ m/s, $f_{max} = 5$ Hz, average element size should be <10 m. The element size can be referred to Fig. 1. The average mesh size is 4.84 m.

$$Average\ Element\ Size \leq \frac{\lambda}{8} = \frac{V_{s,min}}{8f_{max}} \tag{1}$$

3.2 Earthquake, Material and Structure Inputs

Time Acceleration Histories. Both Friuli (1976) and Central Italy earthquake (2016) are input into the analysis. They are shown in Fig. 2. The horizontal time acceleration histories are scaled to the design PGA (Peak Ground Acceleration) = 0.26 g. The peak accelerations for the vertical time acceleration histories are around 0.2 g. The duration of the Friuli and Central Italy earthquakes are selected as 20 s and 18 s, respectively. Both horizontal and vertical time acceleration histories are applied at the model base.

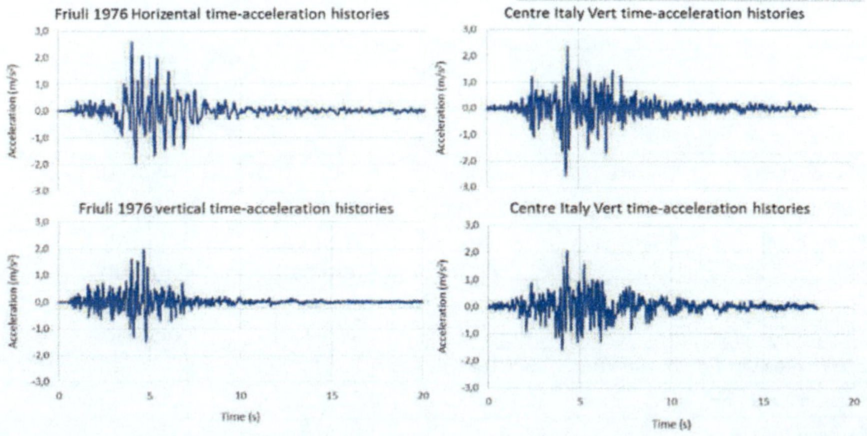

Fig. 2 Input time acceleration histories: upper-left horizontal Friuli EQ (PGA = 0.26 g); upper-right horizontal Central Italy EQ (PGA = 0.26 g); lower-left vertical Friuli EQ; lower-right vertical Central Italy EQ

Material Properties for Rockfill, Concrete and Rock Mass. Since many input parameters for the rockfill are depending on the consolidation stress, σ'_{3c}, a simple static model is made (Fig. 3). It is observed that the σ'_{3c} for the rockfill is generally ranging between 0 and 600 kPa. It is decided, therefore, to separate the rockfill into two materials. For the material with σ'_{3c} ranging from 0 to 200 kPa, the input parameters are selected based on the triaxial test at $\sigma'_{3c} = 100$ kPa; for the material with σ'_{3c} ranging between 200 and 600 kPa, the input parameters are selected based on the triaxial test at $\sigma'_{3c} = 300$ kPa.

The common strength, stiffness, hydraulic and damping material properties are summarised in Table 2. All the parameters are either specified by the proposer or well-established in the engineering practice.

According to ICOLD B148 [3], the fundamental frequency for the embankment dam is typically in range between 0.67–2 Hz. The material Rayleigh damping coefficient α and β are assigned to make sure that the frequency in the range out of 0.1–10 Hz has over 10% of extra damping ratio. So, the analysis result will be focused on the dominated frequency (Fig. 4). HS-small model is selected to simulate the behaviour of rockfill material. The reason for this is that while HS model

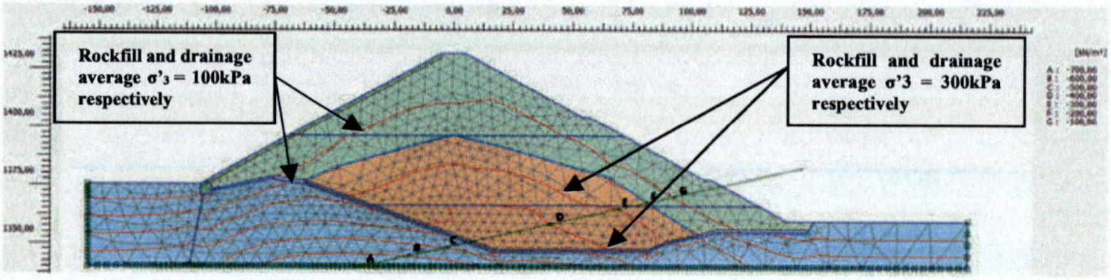

Fig. 3 σ'_3 distribution and material adopt to it

Table 2 Common strength, stiffness and hydraulic material inputs

Parameters	Rock mass	Concrete gallery	Rockfill/Drainage average $\sigma'_{3c} = 100$ kPa	Rockfill/Drainage average $\sigma'_{3c} = 300$ kPa
Material type	Linear-elastic	Linear-elastic	HS-small, drained	HS-small, drained
Unit weight γ	27 kN/m^3	24 kN/m^3	23 kN/m^3	23 kN/m^3
Permeability k	1E−07 m/s	1E−10 m/s	1E−05 m/s (rockfill)	1E−05 m/s (rockfill)
			1E−04 m/s (drainage)	1E−04 m/s (drainage)
Elastic modulus E	1E+06 kPa	30E+06 kPa	–	–
Poisson's ratio ν	0.25	0.2	–	–
Cohesion c'	–	–	1 kPa	1 kPa
Friction φ'	–	–	48.8°	42.6°

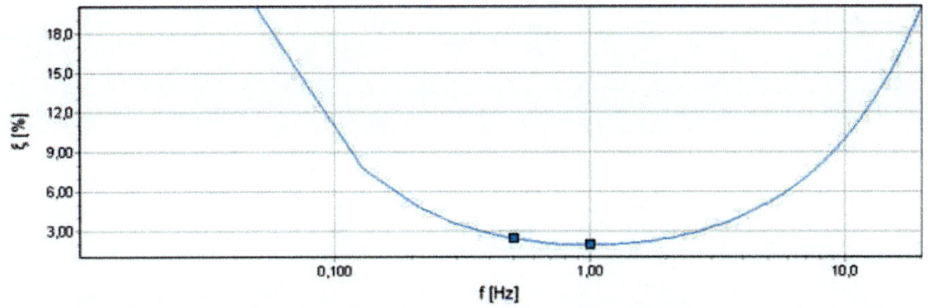

Fig. 4 Rayleigh damping of the Rockfill material to damp out the vibration outside of the 0.1–10 Hz dominated frequency

assumes elastic behaviour during unloading/reloading in small strain range, the HS-small model assumes the elasto-plastic behaviour of soils during loading, unloading and reloading in small strain range. This ensures a realistic stress-strain hysteresis curves and damping ratios during earthquake analysis. Input parameters are shown in Table 3.

The HS-small model parameter estimation is based on combination of the given test results [1], Plaxis soil model manual [4] and our best engineering practice. The E_{50} is calculated from $\sigma_1' - \varepsilon_a$ curve from the triaxial tests. E_{50}^{ref} is calculated by E_{50}, c, φ, σ_{3c}' and m according Formula (2). The parameter m is estimated as 0.5 as the best practice of the granular soil. Since the loading and reloading oedometer test and void ratio are not available, E_{oed}^{ref} is estimated as $0.8 \times E_{50}^{ref}$ suggested by Plaxis as the Formula (3). It is also suggested by Plaxis to estimate E_{ur}^{ref} as three times of E_{50}^{ref}. $\gamma_{0.7}$ is obtained from the G_0 and corresponding t_{ult} and G_{ult} obtained from σ_1' and ε_a curve suggested by Formula (5).

$$E_{50} = E_{50}^{ref}\left(\frac{c\cos\varphi - \sigma_3'\sin\varphi}{c\cos\varphi + p^{ref}\sin\varphi}\right)^m \tag{2}$$

Table 3 Detail HS-small model input parameters for the rock fill and drainage layer

Detail HS-small model input parameters	Rockfill/drainage average: $\sigma_{3c} =$ 100 kPa	Rockfill/drainage average: $\sigma_{3c} =$ 300 kPa
E_{50}^{ref}	100E + 03 kPa	86.92E + 03 kPa
E_{ode}^{ref}	80E + 03 kPa	69.53E + 03 kPa
E_{ur}^{ref}	300E + 03 kPa	260.74E + 03 kPa
m	0.5	0.5
ν_{ur}	0.2	0.2
G_0^{ref}	375E + 03 kPa	446.5E + 03 kPa
$\gamma_{0.7}$	0.000223	0.0002783

$$E_{50}^{ref} = 1.25 E_{oed}^{ref} \tag{3}$$

$$G_0 = G_0^{ref} \left(\frac{c \cos \varphi - \sigma_3' \sin \varphi}{c \cos \varphi + p^{ref} \sin \varphi} \right)^m \tag{4}$$

$$\tau = G_s \gamma = \frac{G_0 \gamma}{1 + 0.385 \frac{\gamma}{\gamma_{0.7}}} \tag{5}$$

G_0^{ref} is calculated from the G_0, c, φ, σ_3' and m as Formula (4) shown. G_0 is obtained from the $E_0 - \sigma_3'$ curve (Fig. 5), using $\nu_{ur} = 0.2$. The $m = 0.5$ is verified and well fit with the given test results. The G/G_0 function and damping ratio function is controlled by G_0^{ref} and $\gamma_{0.7}$ in Plaxis HS-small model. The G/G_0 functions are converted to E_0/E function and axial strain ε_a. They are presented in Fig. 6. The function curves are

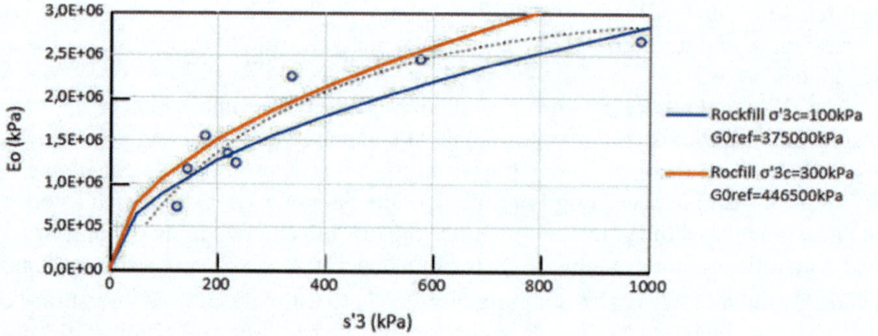

Fig. 5 Selection of G_0^{ref} for materials 1 ($G_0^{ref} = 375{,}000$ kPa) and 2 ($G_0^{ref} = 446{,}500$ kPa) based on cyclic triaxial test results; $m = 0.5$

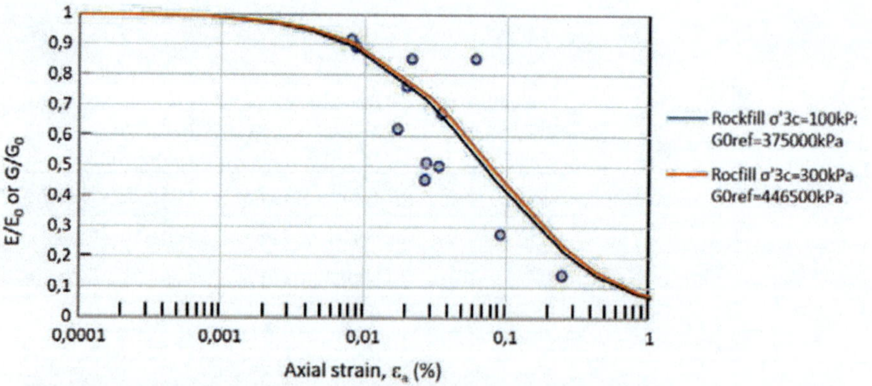

Fig. 6 Normalized E/E0 functions obtained from Plaxis HS-small model from the corresponding input parameters as Table 2 comparing with the given cyclic triaxial test results

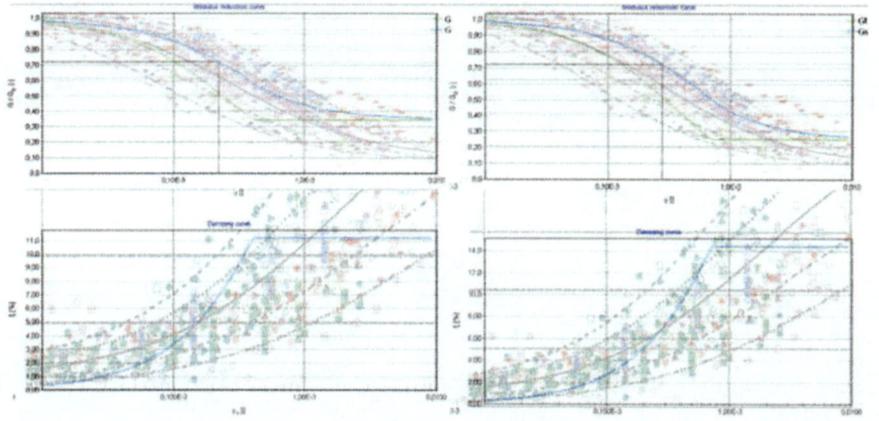

Fig. 7 G/G$_0$ functions and damping ratio function obtained from the HS-small model (Rockfill $\sigma_1' = 100$ kPa on the left, Rockfill $\sigma_1' = 300$ kPa on the right). They are compared with cyclic triaxial test results of rockfill from 35 large rockfill dams from China [10]

overlapped and comparable with the given cyclic triaxial test results [1]. Moreover, G/G_0 functions and damping ratio functions are overlapped and compared with the cyclic triaxial test data from 35 large rockfill dams in China [5] shown in Fig. 7. It indicates that the inputs are generally in the correct range.

Moreover, the relative magnitude of E_0^{ref}, E_{50}^{ref} and E_{ur}^{ref} is also compared to experience and recommendations from literature for stiff clays and cohesion less, high relative density soils shown in Fig. 8. According to Vermeer [6] and Cox and Mayne [7] the ratio E_{50}/E_0 and E_{ur}/E_0 are 0.1 and 0.4 respectively. The corresponding inputs for rockfill material are comparable to them.

Bituminous Facing. Bituminous facing is assumed isotropic elastic perfectly-plastic material. EA and EI is calculated from complex Young's modulus E* with assuming the bituminous facing is 0.35 m thick. Since the isotropic model is assumed, the lower tensile strength 950 kPa is used and the compressive strength is assumed the same as the tensile strength. Maximum allowed bending moment is calculated 5/6 * EI suggested by Plaxis [4]. The Rayleigh damping coefficient is assumed the same as the Rockfill. The detail inputs for the bituminous facing is presented in Table 4.

Boundary Condition. For the seepage analysis at the fully supplied operation condition, hydraulic head boundary EL 1424.5 masl is applied to the dam upstream boundaries. Bituminous facing is simulated as the impermeable interface layer. Seepage review boundary condition is given to the dam downstream surfaces. No-flow boundary condition is given to the model left, right and bottom boundaries.

For the dynamic analysis, free-field boundary conditions are applied on the both sides of the model. The compliant base boundary condition is applied at the bottom of the model.

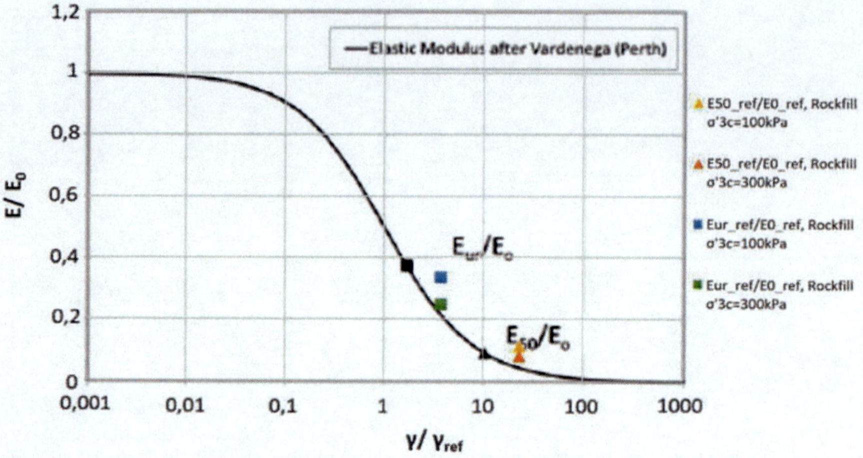

Fig. 8 Stiffness degradation curve, Recommended ratios for stiff soils: $E_{50}^{ref} \big/ E_{0}^{ref} = 0.1$, $E_{ur}^{ref} \big/ E_{0}^{ref} = 0.4$. Comparison with input data

Table 4 Material inputs for the bituminous facing with difference temperatures

Parameters	Bituminous facing 1 °C	Bituminous facing 28 °C
Material type	Isotropic elastic perfectly-plastic	
Unit weight γ	24 kN/m^2	
Complex young modulus E*	14E + 06 kPa	2.8E + 06 kPa
EA	4.9E + 06 kN/m	980E + 03 kN/m
EI	50.02E + 03 kN * m^2/m	10E + 03 kN * m^2/m
Complex poisson's ratio ν*	0.16	0.33
Tensile strength $N_{p,1}$	950 kN/m	950 kN/m
Maximum allowed bending movement M_P	41.68E + 03 kN * m/m	8.34E + 03 kN * m/m
Rayleigh damping coeff. α	0.124	
Rayleigh damping coeff. β	0.003152	

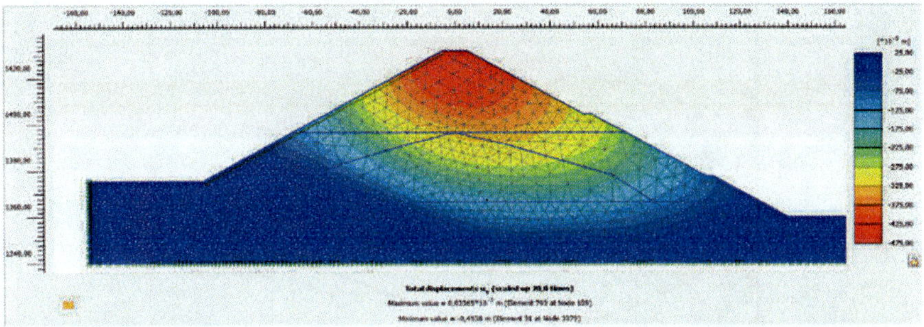

Fig. 9 Total settlement contour plot at the end of the construction. The maximum settlement is about 0.46 m

4 Analysis Results

4.1 End of Construction

Three construction stages are considered. At the end of the construction, the total settlement is about 0.46 m (Fig. 9), which is in line with the total settlement (0.5–1% of the dam height) of a well-constructed rockfill dam according to the engineering practice.

4.2 Pore Pressure Distribution in the Operation Condition

The steady-state seepage analysis is undertaken to simulate the initial stress condition at the full supplied water level (EL 1424.5 masl) during the operation.

The pore pressure distribution is presented in Fig. 10. In general, the pore pressure is not built up in the rockfill. The functionality of the bituminous facing and the drainage layers are well simulated in the model.

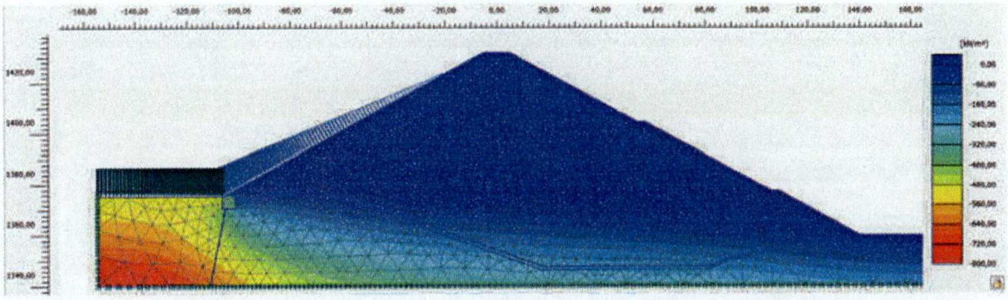

Fig. 10 Pore pressure after the fully supplied water level (EL 1424.5 masl) after the steady-state seepage analysis

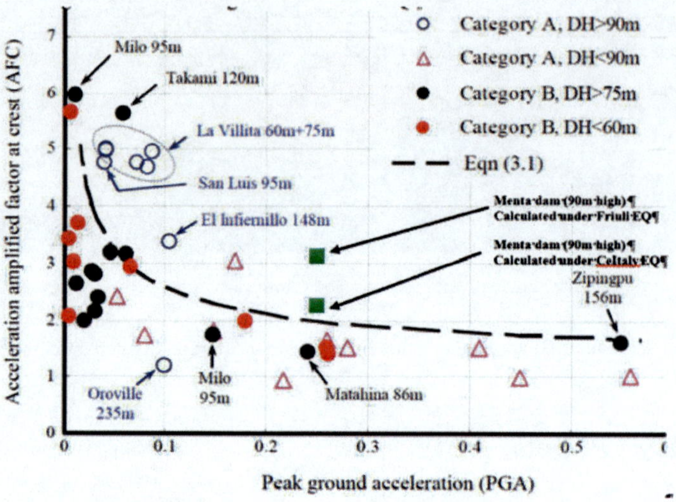

Fig. 11 Variation of acceleration amplified factor at crest against PGA from case histories (Category A are the old dams, the Category B are the new well-compacted rockfill dams) [8]

4.3 *Dynamic Behaviour*

Crest Acceleration. The response spectrum at the model base and dam crest centre are plotted in Fig. 12. The peak crest acceleration is 0.808 g and 0.606 g respectively for these two earthquakes. The crest to base acceleration amplification factor is 3.11 and 2.33 respectively. According to the existing actual measurements summarised by Yu et al. [8] in Fig. 11, the crest acceleration amplification factor for the Central Italy EQ is in line with the measurements, but it is higher above the average under the Friuli EQ.

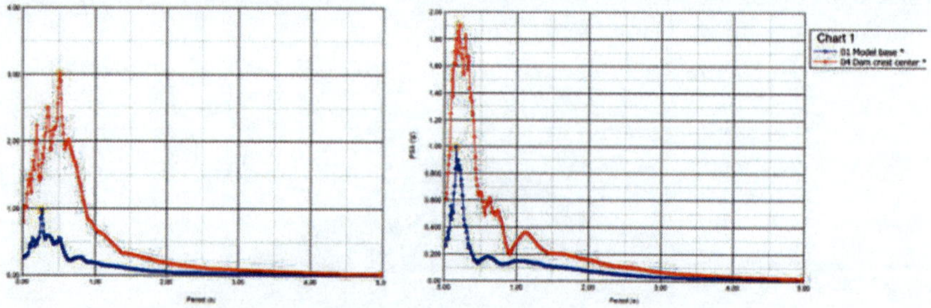

Fig. 12 Response spectrum at the model base and dam crest for EQ1 Friuli (left) and EQ2 Central Italy (right)

Dam Fundamental Frequency. Dam fundamental frequency is obtained from the Fourier power spectrum of the x-direction velocity at the dam crest centre (Fig. 13). The fundament frequency for the dam under Friuli and Central Italy earthquake is 2.00 and 0.72 Hz respectively when the reservoir is at EL 1424.5 masl. The fundament frequency for the empty reservoir is 2.00 and 0.44 Hz respectively. The typical fundamental frequency for the embankment dam is in the range (0.67–2 Hz) discussed by ICOLD [3]. Different fundament frequency of the same dam at the different earthquake is likely due to the input shear strain depending stiffness for the rockfill.

Maximum Shear Strain. The maximum shear strain is 1.3% and 0.8% respectively under these two earthquake loads (Fig. 14). According to the common engineering practice and given triaxial tests, the failure shear strain is generally well above 2% for the rockfill materials. Moreover, a failure pattern has not been formed as indicated in the maximum shear strain plots. It indicates that the overall stability of the dam is acceptable.

Dam Centre Line Deflection and Post-earthquake Deformation. The dam maximum centre line deflection and post-earthquake centre line deflection are presented in the Fig. 15. The highest horizontal deflection is as expected at the dam crest, and they are 0.37 m and 0.19 m respectively. The horizontal displacements

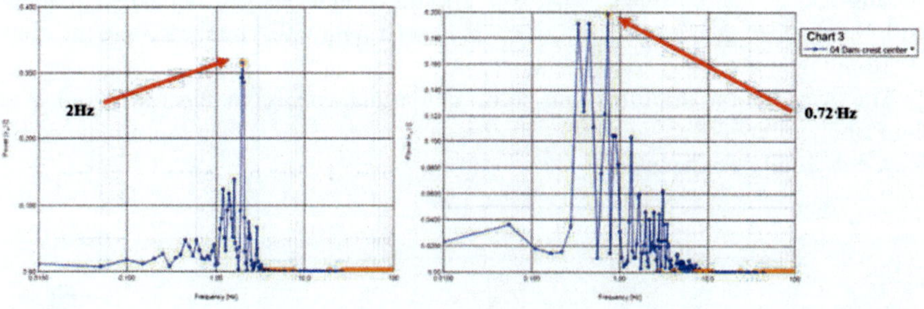

Fig. 13 Fourier spectrum of the x-direction velocity, the fundamental frequency for Menta dam under Friuli EQ and Central Italy EQ is about 2.0 Hz (left) and 0.72 Hz (right) respectively

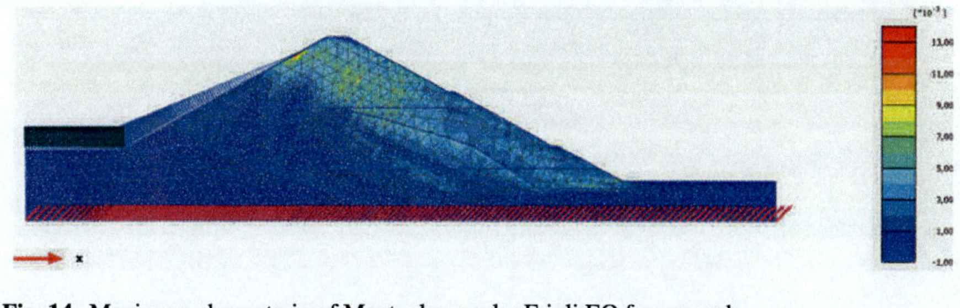

Fig. 14 Maximum shear strain of Menta dam under Friuli EQ for example

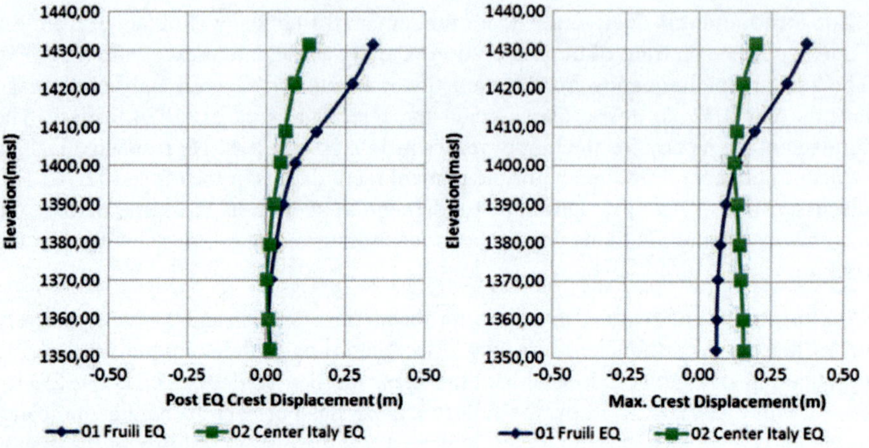

Fig. 15 Dam maximum centre line deflection (left) and the dam post-earthquake centre line deflection

due to the Central Italy earthquake are significantly lower than those due to Friuli earthquake.

The post-earthquake deformation is generally vertical and towards the downstream direction as expected (Fig. 16). The maximum post-earthquake settlement is 0.21 m under the Friuli earthquake. It is well below the freeboard height, thus overtopping is not likely to happen.

The acceleration amplification, deformation and maximum shear strain, due to the Friuli earthquake are generally significantly more critical than those induced by the Central Italy earthquake.

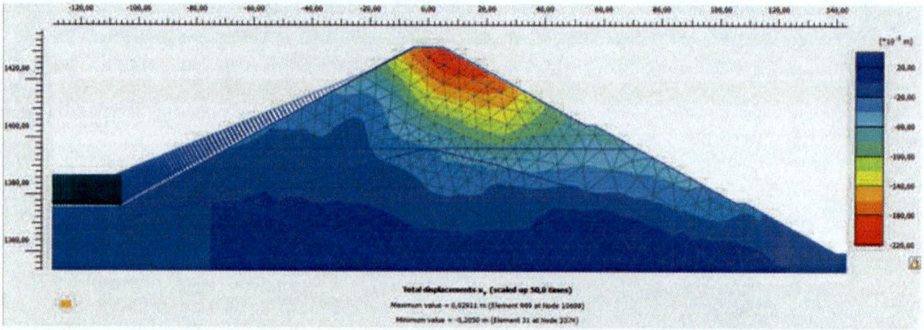

Fig. 16 Typical post-earthquake settlement under Friuli EQ for example

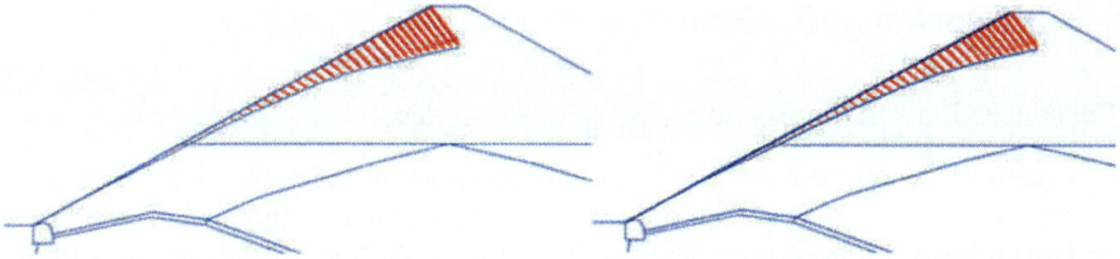

Fig. 17 Bituminous facing deformation for different temperatures (1 °C at the left and 28 °C at the right)

4.4 Bituminous Facing

Bituminous Facing Deformation. Shown in Fig. 17, the deformation of the bituminous facing is about 0.284 m and 0.281 m respectively for the same Friuli earthquake but different temperature. The bituminous facing near the dam crest has the largest deformation since the acceleration and deformation of the dam crest is the highest. So the deformation pattern is as expected.

Bituminous Facing Normal Force. The maximum experienced and the last step normal force in the bituminous facing is presented in Fig. 18. It can be seen that when temperature is 1 °C, both the tensile and compressive stress has reached to the capacity of 950 kPa. It is over 1/2 of length in the middle of the bituminous facing that load is equal to the capacity. When temperature is 28 °C, the length the load is equal to capacity is reduced to 1/3 of bituminous facing length: Moreover, the compressive stress is never of the strength for the entire bituminous facing length.

It is quite difficult to assess if the bituminous facing will crack even through the force is over the capacity. The reason is that the bituminous is viscous-elastic-plastic material with self-healing ability. However, it is almost certain that the situation bituminous facing shall be favourable at 28 °C than 1 °C.

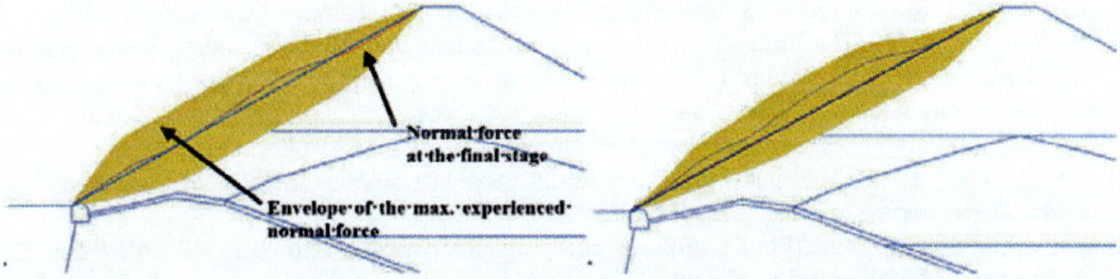

Fig. 18 Maximum experienced and the last step normal force of the bituminous facing at the different temperatures (1 °C at the left and 28 °C at the right)

5 Conclusion and Discussion

The conclusion and discussion are presented as follows:

- The fundamental frequency of the dam is believed around 0.72–2 Hz. It is comparable with the ICOLD description. Different fundamental frequency for the same dam under different earthquake is likely due to the shear strain depending stiffness.
- The dam crest to base amplification factors are 3.11 and 2.33 respectively for two earthquakes. Amplification factor is well in line with the actual measurements.
- Deformation and the dam centre line deflection pattern is as expected. The post-earthquake deformation and maximum settlement are 0.37 m and 0.21 m respectively. They are generally low and acceptable compared with the freeboard.
- Regarding to acceleration amplification, deformation and maximum shear strain, the Friuli earthquake is believed more critical than the Central Italy earthquake.
- It is quite difficult to comment if the bituminous facing will crack. It is almost certain that the bituminous facing shall be favourable at 28 °C than 1 °C.

Acknowledgments The authors would like to thank section leaders Robin Wood and Steffen Giese in section dam and underground and marine geotechnics respectively in Multiconsult for giving us the opportunity to undertake this study.

References

1. Russo G, Vecchietti A, Cecconi M, Pane V, Marco SD, Fiorino A (2019) 15th international benchmark workshop on numerical analysis of dams, Theme B seismic analyses of Menta Embankment damhttps://www.eko.polimi.it/index.php/icold-bw2019/2019
2. Kuhlmeyer RL, Lysmer J (1973) Finite element method accuracy for wave propagation problems. J Soil Mech Found Div 99:421–427
3. ICOLD (2016) Bulletin 148, selecting seismic parameters for large dams guidelineshttps://www.icold-cigb.org/GB/publications/bulletins.asp
4. Plaxis (2019) Material models manual https://www.plaxis.com/support/?category=1100
5. Jia YF, Chi SC (2012) Application of Rockfill dynamical characteristic statistic curve in mid-small scale concrete face dam dynamic analysis. In: Proceedings of 15th world conference on earthquake engineering, Lisbon
6. Vermeer P.A (2001) On single anchored retaining walls. PLAXIS Bulletin No 10
7. Cox C, Mayne P (2015) Soil stiffness constitutive model parameters for geotechnical problems: a dilatometer testing approach. In: Proceedings DMT-2015
8. Yu L, Kong X, Xu B (2012) Seismic response characteristics of earth and Rockfill dams. In: Proceedings of 15th world conference on earthquake engineering, Lisbon

Numerical Analysis of the Menta Dam

T. Mészáros, M. Bakeš, and M. Minárik

Abstract The main objective of the reported manuscript is to analyze the seismic behavior of a bituminous-faced rockfill dam (BFRD), specifically the Menta dam. The Menta dam is a rockfill embankment, located in Southern Italy, lying in the heart of Aspromonte Massif. For the purposes of our work we used the GEOSTUDIO 2019 software, were we conducted a series of numerical nonlinear seismic analyses. As a seismic input, recordings from two seismic events (Friuli 1976 and Central Italy 2016) have been considered. Calculated results of the seismic stress/strain behavior, and their brief discussion, are included in the presented paper.

Keywords Seismic behavior · Numerical analysis · Rockfill embankment · Bituminous-faced · Menta dam

1 Introduction

The work reported in this paper was motivated by formulators long-standing interest in numerical evaluation of rockfill dams for assessing disservice of their bituminous facing due to seismic activity. In an attempt to do so, we performed a numerical study of the Menta dam (Fig. 1).

The Menta dam is a bituminous-faced rockfill embankment, built between 1987 and 2000. The embankment, with a height about 90 m, lies in the heart of Aspromonte Massif in Southern Italy at an elevation about 1400 m a.s.l. The dam foundation, formed from a fractured metamorphic rock, belongs to the "Calabrian Complex". The geometry of the dam base is adapted to natural morphology, resulting in a 325 m long complex shape with a curved portion of about 125 m. The embankment is divided in five zones, characterized by different grain size of compacted rockfill excavated at the river site. The crest height is 1431.75 m a.s.l. and the normal reservoir level is at 1424.75 m a.s.l. The upstream slope, with the bituminous concrete facing, has a

T. Mészáros · M. Bakeš · M. Minárik (✉)
Department of Dam Safety, Vodohospodárska Výstavba, SOE, Bratislava, Slovakia
e-mail: marian.minarik@vvb.sk

G. Bolzon et al. (eds.), *Numerical Analysis of Dams*, Lecture Notes in Civil Engineering 91, https://doi.org/10.1007/978-3-030-51085-5_31

Fig. 1 Menta embankment dam

1.8:1 inclination, while the downstream face has two berms at the elevation 1404.5 m and 1377.5 m a.s.l.

2 Numerical Model

In order to predict and describe the seismic stress/strain behavior of the Menta embankment, and to estimate potential damages, a series of numerical analyses have been conducted [1], [2]. For the numerical analyses, the Geostudio 2019 software have been utilized. The task was divided into 2 stages (empty/full reservoir), represented by a number of interlinked analyses within the implemented numerical solution (Fig. 2):

1. Stage, empty reservoir—Initial conditions (0 s)

- Analysis—nonlinear dynamic (0.005—12.5 s)
- Analysis—dynamic deformation (permanent deformation) [3]
- Analysis—Newmann deformation (slope stability) [4]

2. Stage, full reservoir after first impoundment—Initial conditions (1 s)

- Analysis—nonlinear dynamic (1.005–13.5 s)
- Analysis—dynamic deformation (permanent deformation)
- Analysis—Newmann deformation (slope stability).

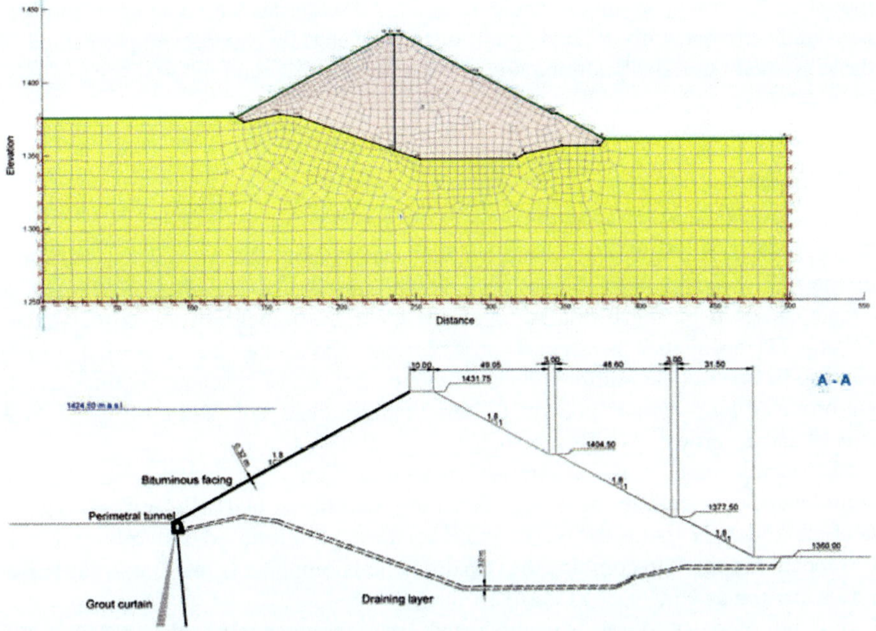

Fig. 2 Cross-section A-A (bottom); FEM Model (top)

2.1 Geometry and Meshing

The boundary conditions and geometry of the modelled cross-section corresponds with the provided cross-section A-A. Figure 2 shows the 2D model of the numerical solution and the cross-section A-A of the Menta dam. The model is 500 m wide, 181.75 m high, and it was discretized into 1270 quad and triangle shaped finite elements with 1343 nodes. The size of the finite element's ranges from 3.0 m to 10.0 m, with an average size of 5.0 m. The 2D model is considered.

2.2 Material Properties and Constitutive Models

Mechanical characterization of the dam refers to material of zone 3, therefore the embankment is considered as formed only by this material. The mechanical behavior of rockfill materials is highly non-linear and complex (dependence on mineralogy, grain size distribution, particle size, stress level, etc.). In an attempt to model this complex behavior, we applied the hyperbolic E-B model [5] for static conditions and a nonlinear constitutive relation [6] for the dynamic analyses. As for the foundation rock mass, in order to avoid further numerical complexity, elastic behavior was modelled, assuming an impermeable material. All materials were assumed homogenous and isotropic with drained conditions [7]. Material properties implemented in the presented numerical solution are given in Tables 1, 2 and Fig. 3 (Fig. 4).

2.3 Seismic Input

The seismic inputs are represented by two severe events occurred in Italy (Friuli, earthquake occurred in 1976 and Central Italy, earthquake sequence of 2016). The events, selected by the formulators, are in spectral compatibility with the Italian Seismic Hazard map. The target records represent the acceleration recorded at rock outcrop in free field conditions. Seismic inputs, in form of acceleration time series, are provided by three components (2 horizontal, marked as N and E—Figs. 7, 8, 9 and 10 and 1 vertical, Z—Figs. 5 and 6).

As seismic input in the 2D numerical analyses, only two components (1 horizontal and 1 vertical) are possible. Assuming, that horizontal acceleration component marked as N stands for North and E for East, we proposed a merging scheme (presented in Fig. 4) to combine the two horizontal components in a single one (time series marked as NE, Figs. 11 and 12).

For reference purposes, we conducted six separate numerical solutions with different seismic inputs: (1) Friuli Z + N; (2) Friuli Z + E; (3) Friuli Z + NE; (4) Central Italy Z + N; (5) Central Italy Z + E, (6) Central Italy Z + NE.

3 Computed Results

Initial condition of the performed numerical analyses are presented in form of different contour plots from Figs. 13, 14, 15, 16, 17 and 18. The results of the conducted dynamic analyses are reported in Table 3 and in Figs. 19, 20, 21, 22, 23 and 24. Figures 25 and 26 shows the results of the dynamic deformation analyses (stress redistribution analyses to evaluate permanent post-seismic deformations). Table 4 summarizes minimal FoS evaluated during the modeled seismic events.

4 Conclusions

In the framework of the reported numerical solutions, considering two seismic events, we made a prediction of the stress/strain seismic behavior of the Italian Menta dam. The main conclusions of the conducted analyses are as follows:

Table 1 Constitutive models and material properties applied in the numerical model

Col.	Name	Model	γ	c´	φ´	E´	K_0	K_{ur}	R_f	υ
[-]	[-]	[-]	[kN/m^3]	[kPa]	[^{0}]	[MPa]	[-]	[-]	[-]	[-]
	Rockf.	Hyperb.	23	93	35.7	$E = f(\sigma_{vert})$	0.408	5	0.8	0.42
	Found.	Elastic	27			7000	0.333			0.25

Table 2 Dynamic material properties

Col.	Name	Model	γ	c´	φ´	G_{max}	Damp.	Max DR	υ
[-]	[-]	[-]	[kN/m^3]	[kPa]	[°]	[MPa]	Ratio [-]	[-]	[-]
	Rockf.	Non-li.	23	93	35.7	$G_{max} = f(\sigma_{vert})$	0.06	0.15	0.42
	Found.	El.-lin.	27			2800	0.05		0.25

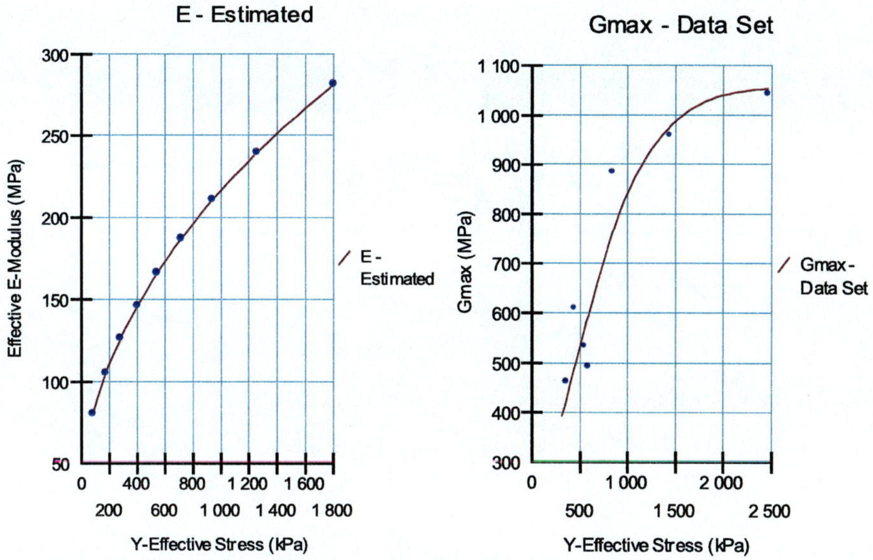

Fig. 3 Data set functions applied as material properties

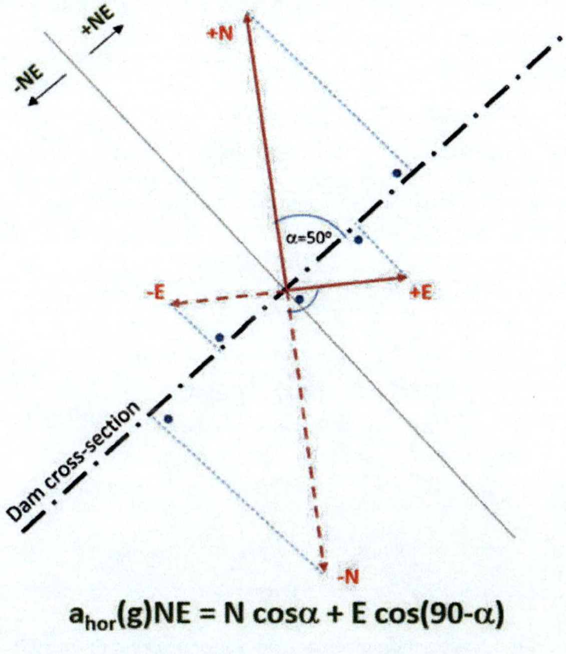

Fig. 4 Merging scheme of E and N horizontal acceleration components

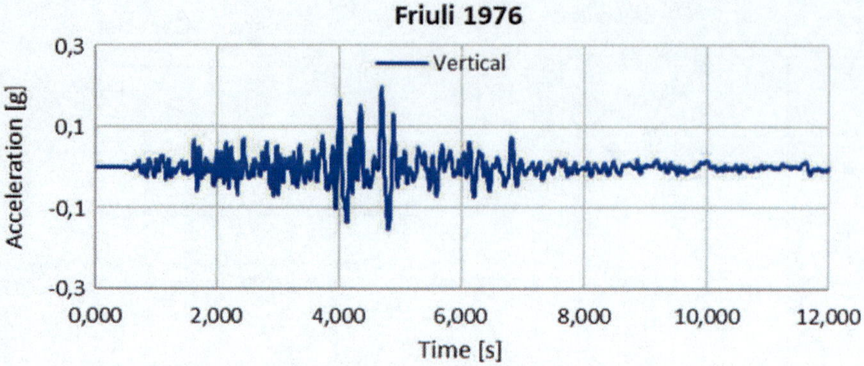

Fig. 5 Vertical acceleration time series—seismic event Friuli 1976

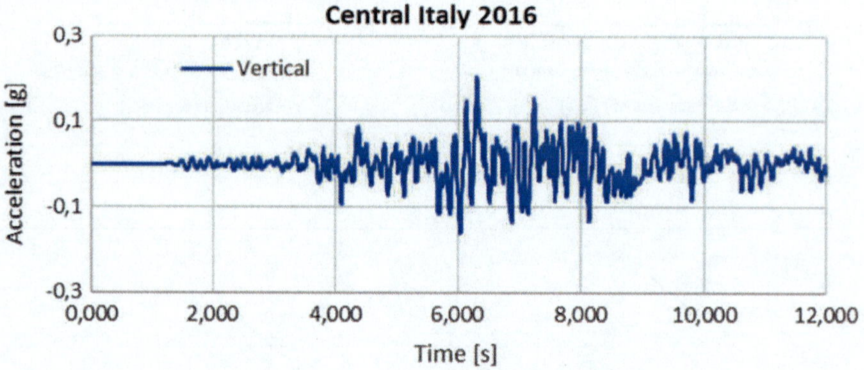

Fig. 6 Vertical acceleration time series—seismic event Central Italy 2016

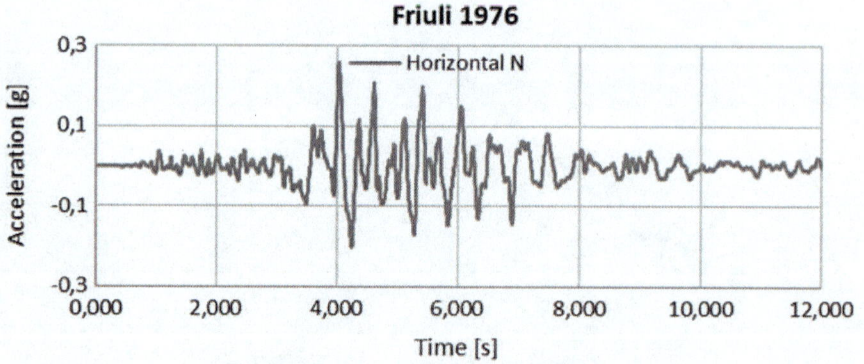

Fig. 7 Time series of the N acceleration component—seismic event Friuli 1976

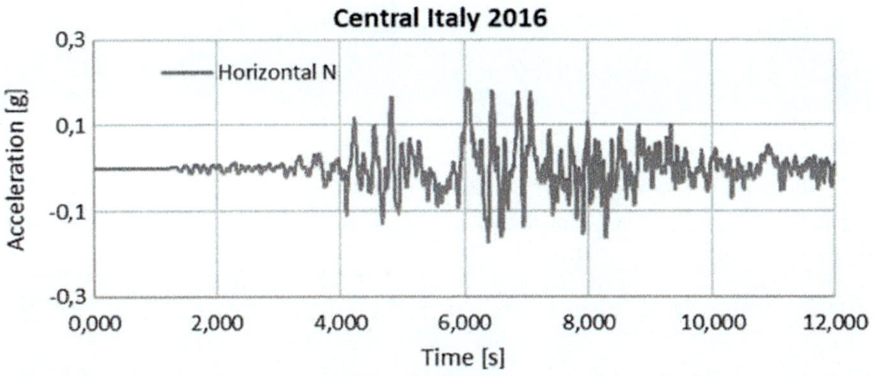

Fig. 8 Time series of the N acceleration component—seismic event C. Italy 2016

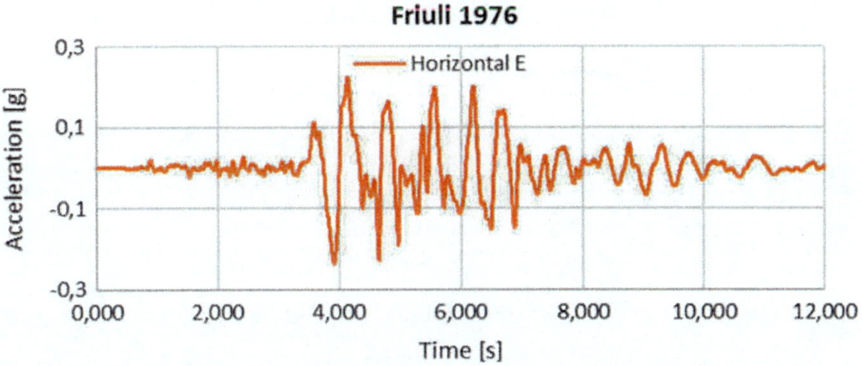

Fig. 9 Time series of the E acceleration component—seismic event Friuli 1976

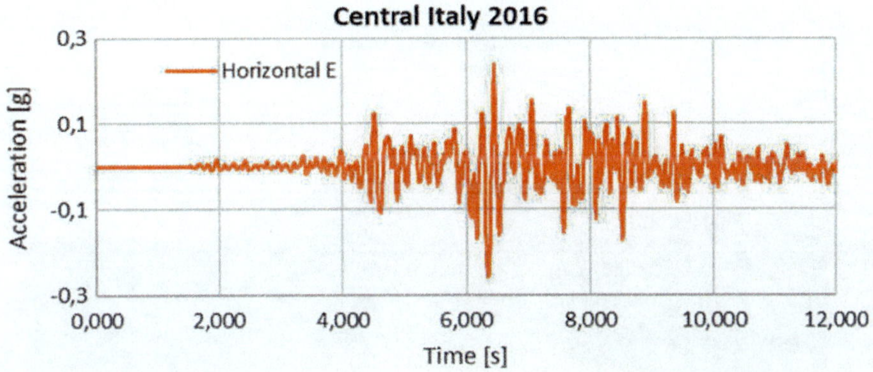

Fig. 10 Time series of the E acceleration component—seismic event C. Italy 2016

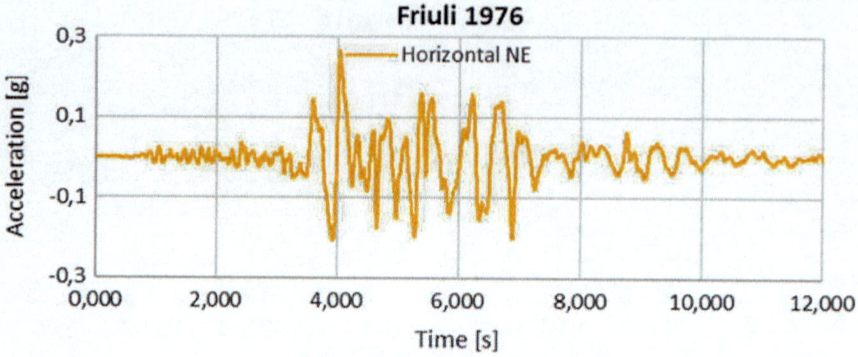

Fig. 11 Time series of the merged E and N horizontal acceleration components—seismic event Friuli 1976

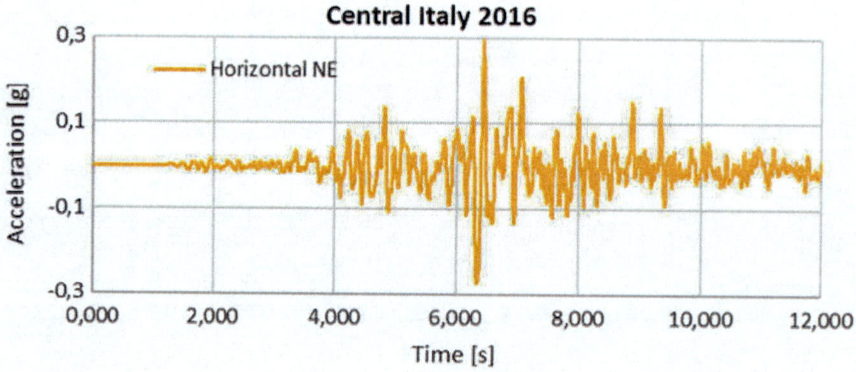

Fig. 12 Time series of the merged E and N horizontal acceleration componentsvseismic event Central Italy 2016

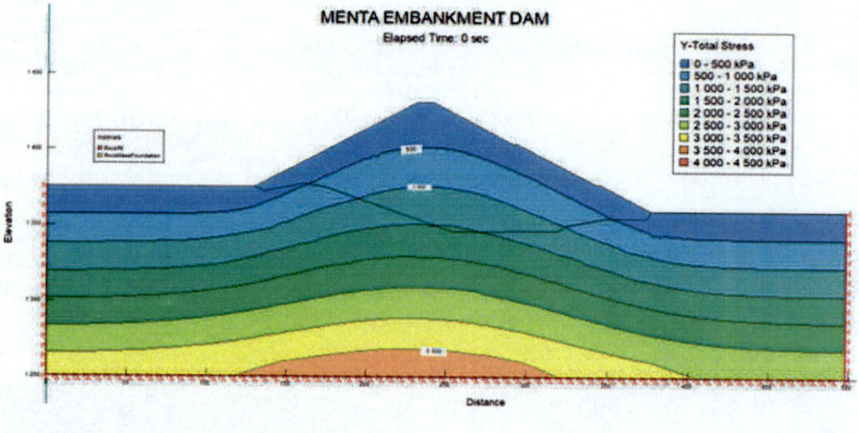

Fig. 13 Initial state—vertical total stresses

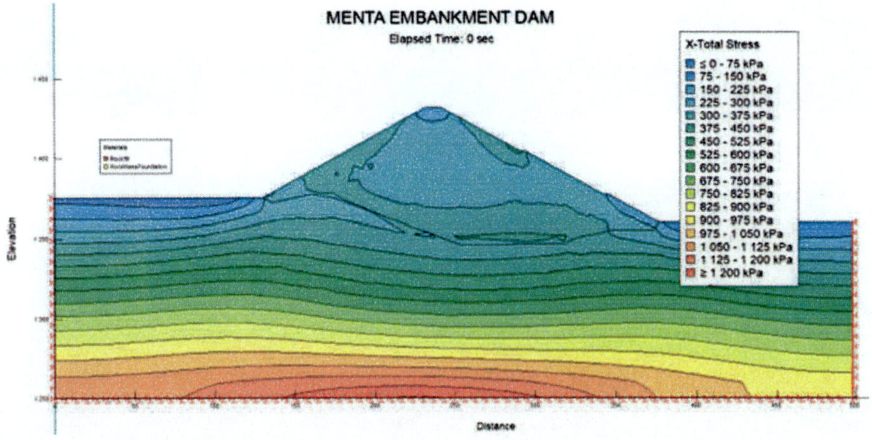

Fig. 14 Initial state—horizontal total stresses

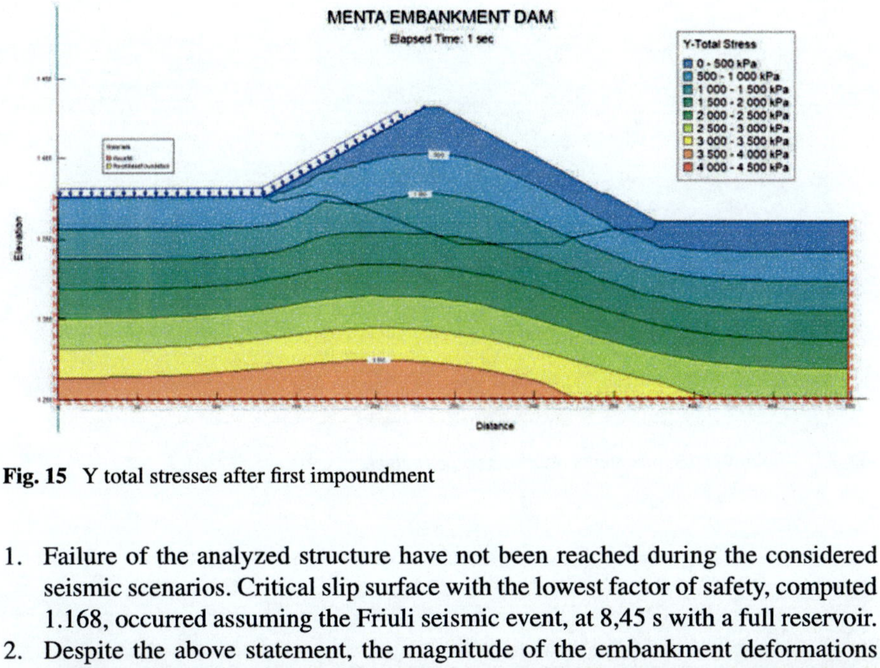

Fig. 15 Y total stresses after first impoundment

1. Failure of the analyzed structure have not been reached during the considered seismic scenarios. Critical slip surface with the lowest factor of safety, computed 1.168, occurred assuming the Friuli seismic event, at 8,45 s with a full reservoir.
2. Despite the above statement, the magnitude of the embankment deformations indicate, that disservice of the bituminous-facing could have been assessed. However, this is only the authors' assumption. No analysis regarding the bituminous-facing have been made.
3. In order to reduce model complexity, the dynamic effect of the full reservoir (waves induced in the reservoir) and the curved shape of the embankment have been neglected, despite their impact on the dams seismic behavior.

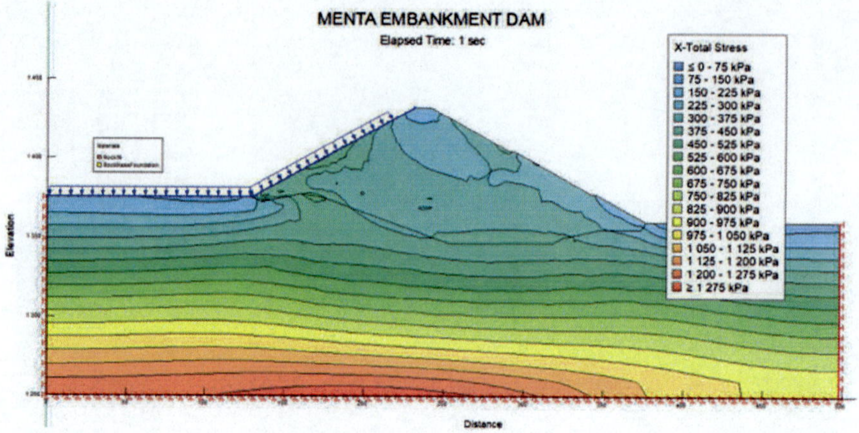

Fig. 16 Horizontal total stresses after first impoundment

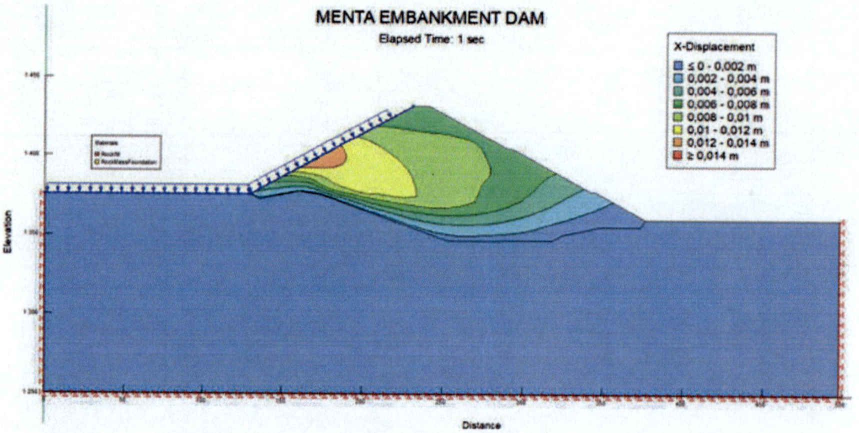

Fig. 17 Horizontal displacements after first impoundment

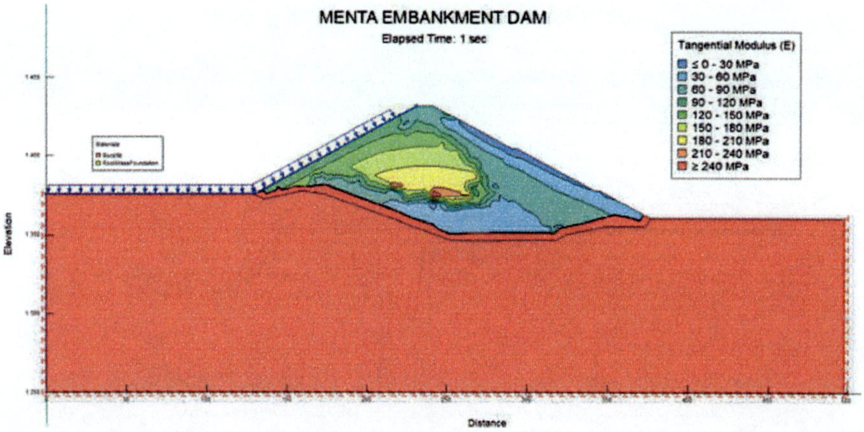

Fig. 18 Young's modulus after first impoundment

Table 3 Natural frequencies corresponding to different seismic inputs

Reservoir	Phys. Prop.	Horizontal seismic input					
		Fr. N	Fr. E	Fr. NE	CI. N	CI. E	CI. NE
Empty	Nat. freq. [Hz]	2.0	1.6	1.6	2.1	1.7	2.4
Full	Nat. freq. [Hz]	2.0	1.6	1.6	2.1	1.7	2.4

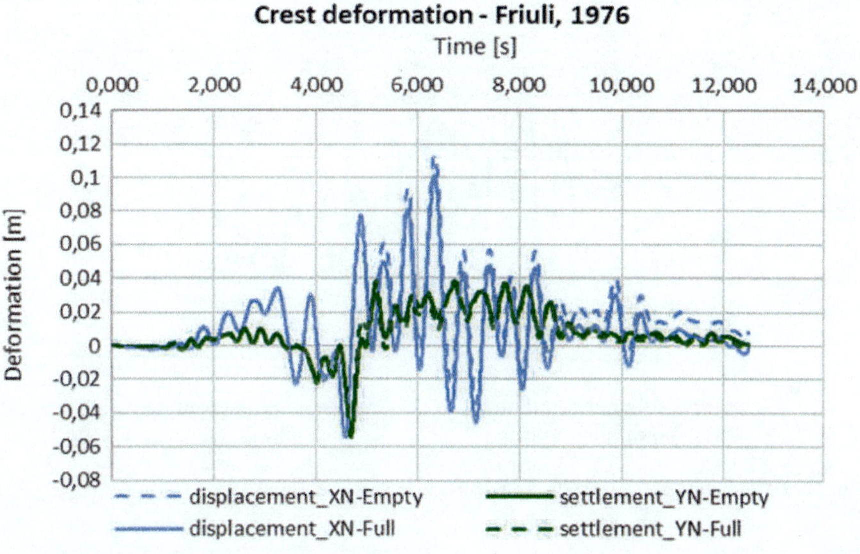

Fig. 19 Deformation at the crest due to seismic load comp. N, Friuli 1976

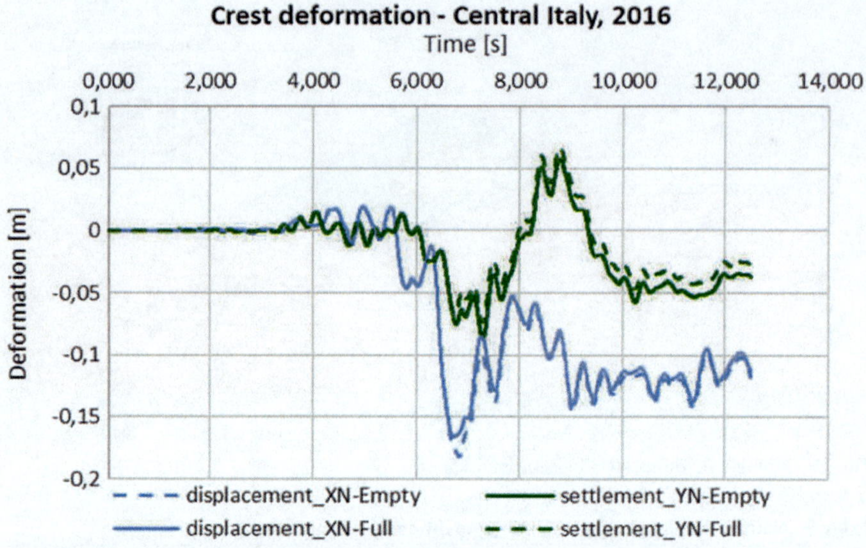

Fig. 20 Deformation at the crest due to seismic load comp. N, C. Italy 2016

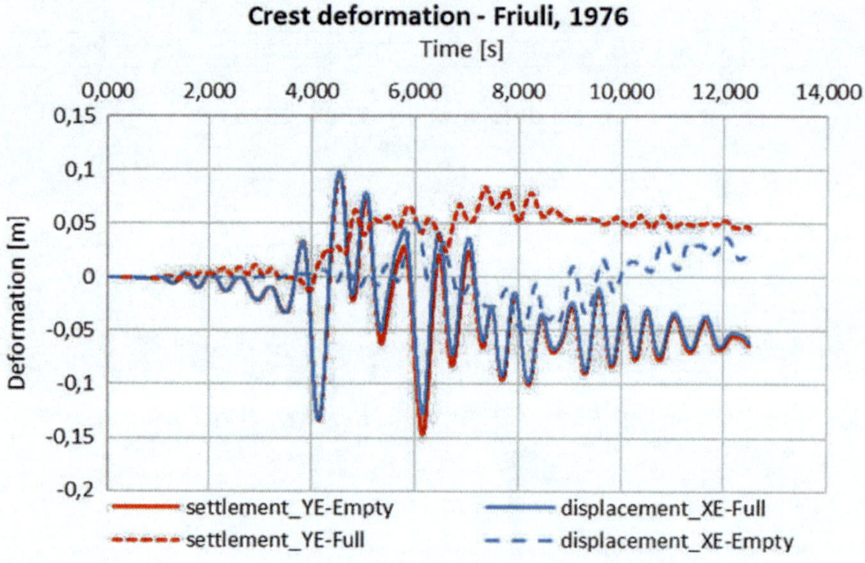

Fig. 21 Deformation at the crest due to seismic load comp. E, Friuli 1976

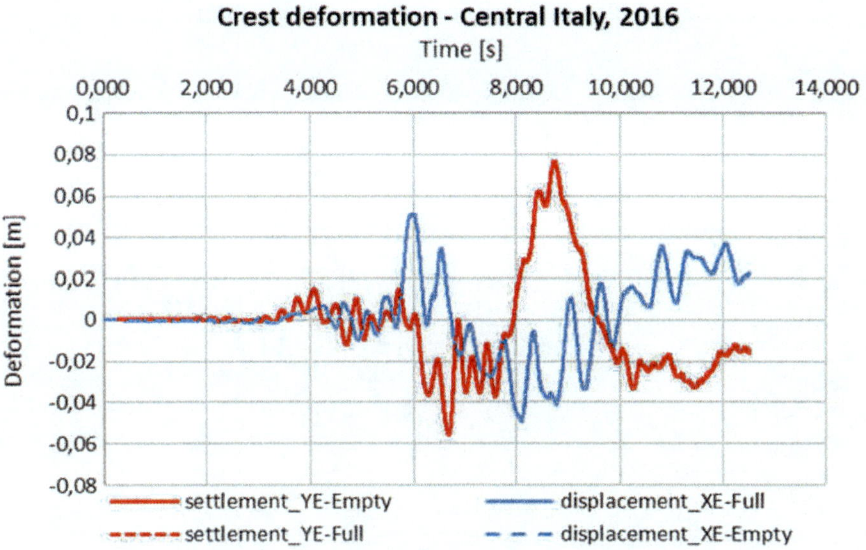

Fig. 22 Deformation at the crest due to seismic load comp. E, C. Italy 2016

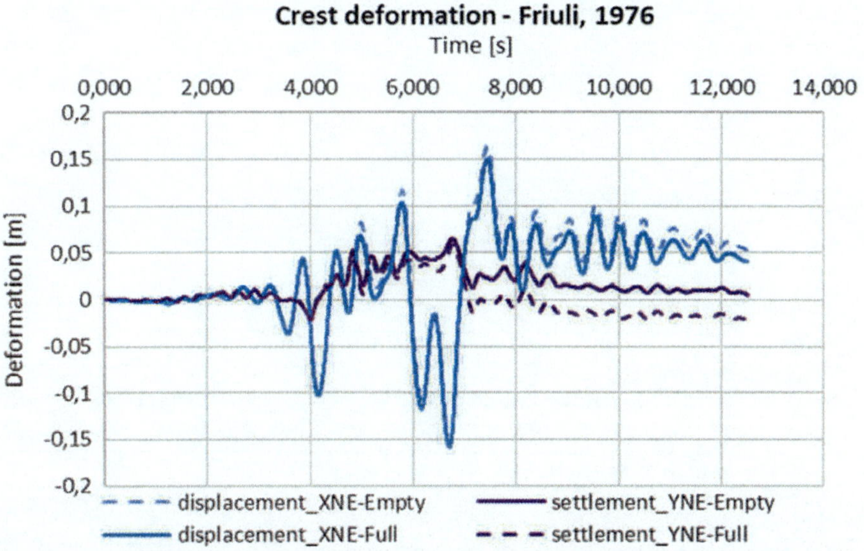

Fig. 23 Deformation at the crest due to seismic load comp. NE, Friuli 1976

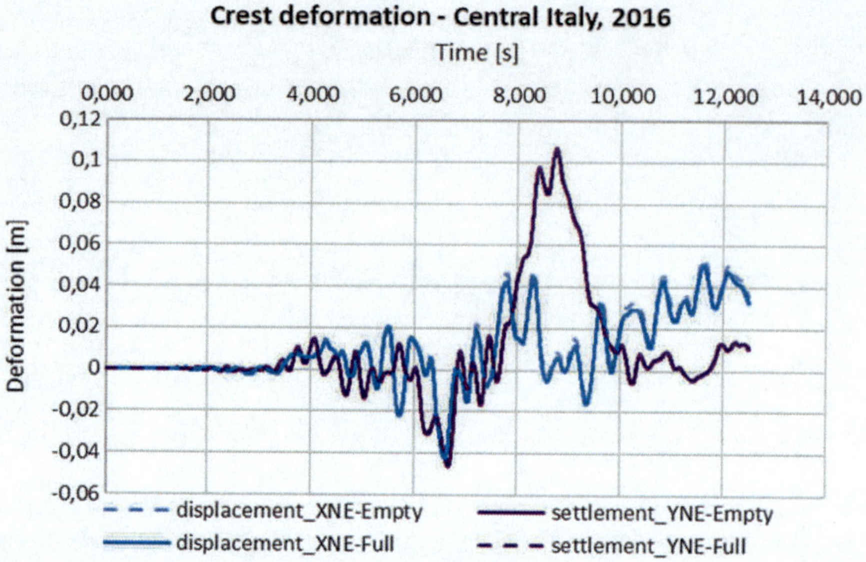

Fig. 24 Deformation at the crest due to seismic load comp. NE, C. Italy 2016

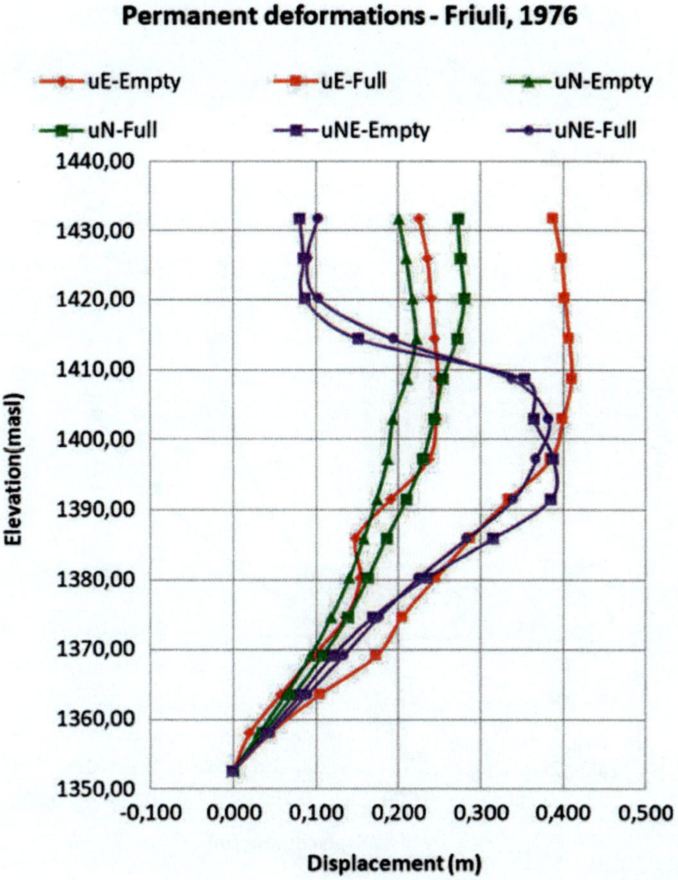

Fig. 25 Permanent horizontal deformation along the embankment's vertical axis due to seismic event: Friuli, 1976

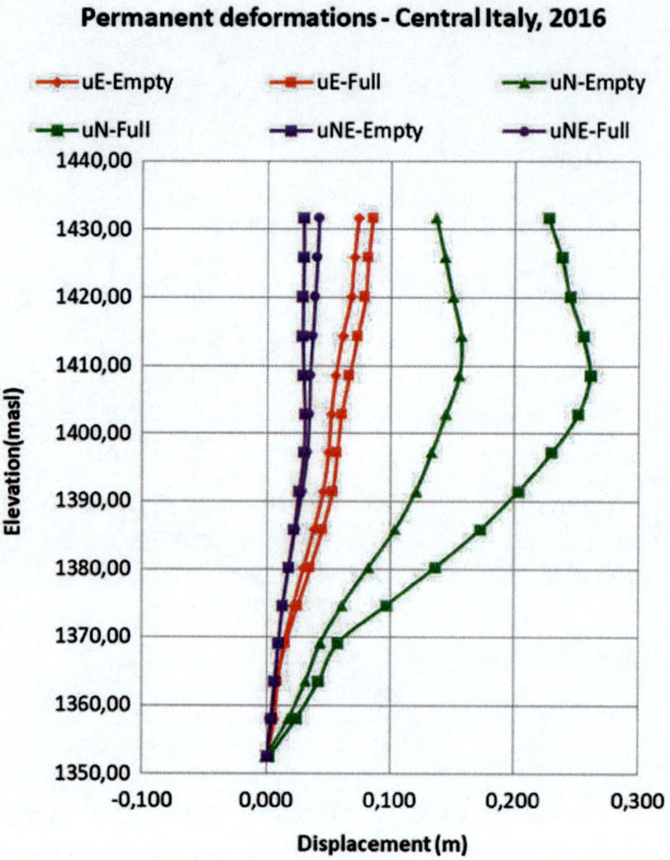

Fig. 26 Permanent horizontal deformation along the embankment's vertical axis due to seismic event: Central Italy, 2016

Table 4 Minimal factor of safety of the critical slip surface from the entire seismic event

Reservoir	Crit. slip surface	Horizontal seismic input					
		Fr. N	Fr. E	Fr. NE	CI. N	CI. E	CI. NE
Empty	Min. FoS [-]	1.305	1.263	1.256	1.348	1.299	1.497
Full	Min. FoS [-]	1.221	1.238	1.168	1.307	1.307	1.511

Acknowledgments We gratefully acknowledge the Management of VODOHOSPODÁRSKA VÝSTAVBA, SOE for the given support.

References

1. Zienkiewicz OC, Taylor RL, Zhu JZ (2013) The finite element method: its basis and fundamentals, 7th edn. Butterworth-Heinemann, Oxford
2. Zienkiewicz OC, Chan AHC, Pastor M, Schrefler BA, Shiomi T (1998) Computational geomechanics with special reference to earthquake engineering. Wiley, Oxford
3. Dynamic Earthquake Modeling with GeoStudio, An Engineering Methodology, July 2017 Edition, GEO-SLOPE International Ltd., pp. 173, (2017)
4. Chowdhury R, Flentje P, Bhattacharya G (2010) Geotechnical Slope analysis, 2nd edn. CRC Press, Wollongong
5. Stress-Deformation Modeling with SIGMA/W, An Engineering Methodology, July 2013 Edition, GEO-SLOPE International Ltd., pp. 85, (2013)
6. Kramer SL (1996) Geotechnical earthquake engineering. Prentice-Hall, New Jersey
7. Head KH (1998) Manual of soil laboratory testing, Volume 3, Effective stress test, Second Edition, Wiley, London

Modeling the Seismic Behavior of Bituminous Faced Rockfill Dam with Hujeux Model Implemented in Code_Aster

A. Mondoloni, P. Kolmayer, and V. Alves Fernandes

Abstract This article presents a simulation of the static and dynamic behaviour of a bituminous faced rockfill dam with the Hujeux constitutive model and the methodology originally developed on code_aster for dams made of saturated granular materials. The calibration procedure of the parameters is carried out on the basis of available tests and static and dynamic results are presented for two seismic signals with different frequency contents. Results are compared with correlations available in the literature. The discrepancies are analysed in terms of the functioning of the behaviour model.

Keywords Bitouminous faced rockfill dam · Hujeux constitutive model · Code_aster

1 Introduction

For a long time, the seismic analysis of existing dams in France has concerned only works protecting classified installations. These were mainly embankment dams presenting risks of pore pressure increase induced by the shaking for which EDF developed a methodology based on the Hujeux elastoplastic constitutive model that is well adapted for granular soils. From now on, French regulations require a seismic analysis for of all major dams in the most seismic areas of the French territory and EDF will therefore have to produce these studies for 8 CFRD in the coming years. The proposed benchmark on Menta is an opportunity to test the ability of the existing methodology for analyzing the earthquake behavior of upstream faced rockfill dams.

A. Mondoloni · P. Kolmayer
Electricité de France-Hydraulic Engineering Center, La Motte-Servolex, France

V. Alves Fernandes (✉)
Electricité de France R&D, Palaiseau, France
e-mail: vinicius.alves-fernandes@edf.fr

IMSIA, UMR EDF-CNRS-CEA-ENSTA 9219, Palaiseau, France

G. Bolzon et al. (eds.), *Numerical Analysis of Dams*, Lecture Notes in Civil Engineering 91, https://doi.org/10.1007/978-3-030-51085-5_32

The constitutive model and its calibration are developed in the first part of this paper. The static and seismic results are presented and discussed in the following parts.

2 Numerical Hypothesis

2.1 Constitutive Model

The ECP elastoplastic multi-mechanism model [1] is developed to represent soil behavior under cyclic loading in a large range of loading paths. It is use for very coarse rockfill material must be done carefully.

The representation of all irreversible phenomena is idealized by four coupled elementary plastic mechanisms for the monotonic behavior and four for the cyclic behaviour: three plane-strain deviatoric plastic deformation mechanisms in three orthogonal planes (Eq. 1) and an isotropic one (Eq. 2). The model is written in terms of Terzaghi effective stresses and it uses a Coulomb-type failure criterion and the critical state concept.

$$f_i\left(q_i, p_i', \varepsilon_v^p, r_i\right) = q_i - p_i' \sin\varphi\left(1 - b\ln\left(\frac{p_i'}{p_c}\right)\right)r_i, i \in [1,3] \quad (1)$$

$$f_{iso}\left(p', \varepsilon_v^p, r_{iso}\right) = p' - p_c\, d\, r_{iso} \quad (2)$$

$$p_c = p_{co}\exp\left(\beta\varepsilon_v^p\right) \quad (3)$$

$$\frac{\partial\varepsilon_{vi}^p}{\partial\varepsilon_i^p} = \alpha\left(\sin\varphi - \frac{q_i}{p_i'}\right) \quad (4)$$

The evolution of hardening is based on plastic strain, deviatoric and volumetric strain for the deviatoric mechanisms and volumetric strain for the isotropic one (Eq. 5).

$$r_i = r_i^{ela} + \frac{\int \varepsilon_{di}^p dt}{a_{cyc} + \left(a_{mon} - a_{cyc}\right)\alpha(r_i)\int \varepsilon_{di}^p dt}, i\varepsilon[1,3] \quad (5)$$

The model takes into account non-linear elasticity and to account for cyclic behavior both isotropic and kinematical hardenings are used. Soil behavior is divided in four sub-domains:

- Elastic domain: no energy dissipation and reversible deformations (Eq. 6).

$$E = E_o \left(\frac{p'}{p_{ref}} \right)^n \tag{6}$$

- Hysteretic or pseudo-elastic domain: energy dissipation when plastic deformations appear, but the volumetric plastic strain variation is neglected (Eq. 7).

$$\alpha(r_i) = 0 \text{ if } r_i < r^{hyst} \tag{7}$$

- Intermediate domain: the volumetric plastic strain variation is no longer neglected (Eq. 8).

$$\alpha(r_i) = \left(\frac{r_i - r^{hyst}}{r^{mob} - r^{hyst}} \right) \text{ if } r_i < r^{hyst} \tag{8}$$

- Mobilized domain: the shear stress mobilizes completely the shear resistance of the solid phase (Eq. 9)

$$\alpha(r_i) = 1 \text{ if } r_i > r^{mob} \tag{9}$$

The list of the parameters and their description are given in Table 1. The model is integrated into code_aster and an implicit integration scheme is used [2]. Further analytical details concerning the model can be found in the specific reference [1].

2.2 *Model Calibration*

Hujeux model is based on the critical state concept and does not allow introducing an explicit dependency of the maximal friction angle on the confining stress. But according to Eq. (1), the friction angle can evaluate with the ratio p_c/p' where p_c is the critical mean pressure and p' the mean effective pressure. Friction angle is fixed to 45°, the shape parameter b is chosen equal to 0.3 which is acceptable for sand (b ≈ 1 for clay) and an optimization process leads to the identification of parameters p_{c0} and β in order to reproduce the observed evolution of the friction angle with respect of Eq. (1) giving the yield surface and Eq. (3) for the evolution of the critical pressure and the measured volumetric strain.

The comparison between the experimental triaxial tests and their simulation with Hujeux's model are given on Figs. 1 and 2. One can notice the difficulties of the model to represent correctly the stress evolution and the volumetric strain evolution on a wide range of confining pressures (defined in Table 2). In particular, the model

Table 1 Hujeux's model parameters

Parameter		Value	Unit
E_0	Young modulus at reference pressure	2.45 10^9	Pa
ν_0	Poisson ratio	0.25	–
n	Exponent regulating the stress dependency of the Young modulus	0.55	–
p_{ref}	Mean reference pressure	106	Pa
p_{c0}	Initial critical pressure	1.25 10^6	Pa
β	Plastic compressibility in the isotropic plane	5	–
φ_{pp}	Friction angle at critical state	45	(°)
ψ	Angle of dilatancy	45	(°)
b	Numerical parameter which control the shape of the loading surface in the deviatoric stress space	0.3	–
d	Distance between the isotropic consolidation line and the critical state line in the (p,ln(e)) plane	2.7	–
r_{dela}	Extend of elastic domain of the deviatoric yield surface	0.12	–
r_{iela}	Extend of elastic domain of the isotropic yield surface	0.001	–
r_{hyst}	Extend of hysteretic domain	0.3	–
r_{mob}	Extend of mobilized domain	0.9	–
a_{mon}; a_{cycl}	Numerical parameter regulating the deviatoric hardening	0.008; 0.0008	–
c_{mon}; c_{cycl}	Numerical parameter regulating the volumetric deformation according the normalized deviatoric stress	0.14; 0.07	–
χ_m	Numerical parameter regulating the dilatancy according the deviatoric stress	1.0	–
α	Numerical parameter regulating the dilatancy	0.4	–

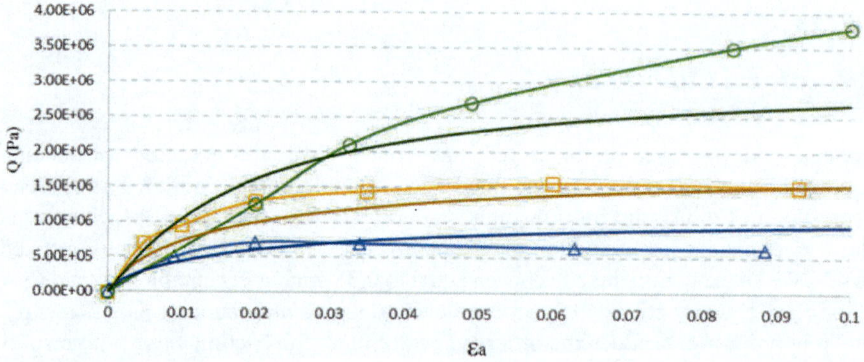

Fig. 1 Comparison of the simulation with laboratory tests on triaxial tests for deviatoric stresses

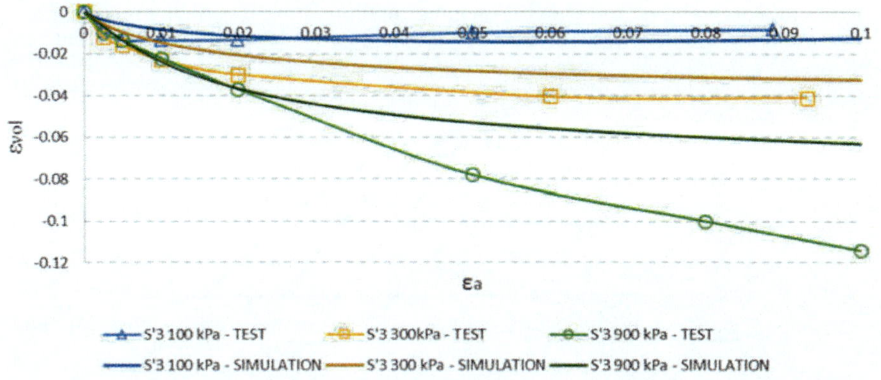

Fig. 2 Comparison of the simulation with laboratory tests on triaxial tests for volumetric strains

Table 2 Comparison of experimental and modelled maximum friction angle on triaxial tests for β = 5 et $p_{c0} = 1250$ kPa

σ_3 [kPa]	ε_{vol} (experimental)	P_c [kPa]	σ_1 [kPa]	Φ [°] (model)	Φ [°] (experimental)
100	0.01	1284	707	48.8	48.8
300	0.04	1466	1495	41.7	42.6
900	0.11	2080	3492	36.2	38.8

underestimates the volumetric strain at small deformation (less than 1%) for 100 and 300 kPa confining pressure. This may lead to underestimate the final settlement of the dam, in particular close to the perimeter of the dam where stresses are small.

In the Hujeux model, the stiffness depends on the mean effective stress according to a power law (Eq. 6). The mean pressure is calculated with K_0 equal to 0.4 and the variation of the stiffness is given in Fig. 3. Final values, $n = 0.55$ and $E_0 = 2.45$ GPa considering $p_{ref} = 1$ MPa, are close to that given by the automatic regression

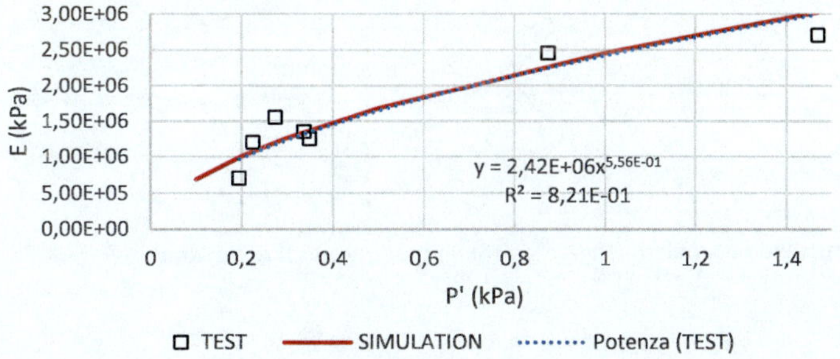

Fig. 3 Stiffness dependency on mean effective pressure

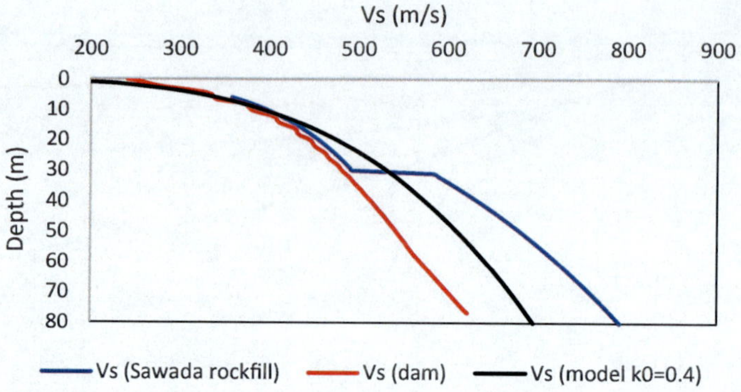

Fig. 4 Comparison of the calculated Vs-profile with Sawada's formulation for dry rockfill

process. The corresponding Vs-profile is close to that given by Sawada [3] for dry rockfill even if the velocity is over-estimated in the upper part of the dam (Fig. 4).

Degradation of E/E_0 and damping evolution with the shear strain are not input data of the model, but must be simulated on cyclic shear strain tests. The decrease of the stiffness modulus depends not only on the different radii r_{ela} and r_{hyst} but also on the hardening parameters a_{mon} and a_{cyc} and the shape parameter b. The curves produced with the final set of parameters are represented on Figs. 5 and 6. The evolution of the stiffness might be a little high compared to the experimental data, but it corresponds with the lower bound given by Seed for gravels. Damping is null for very small distorsion and a little strong for distorsion higher than 10^{-3} compared to laboratory results.

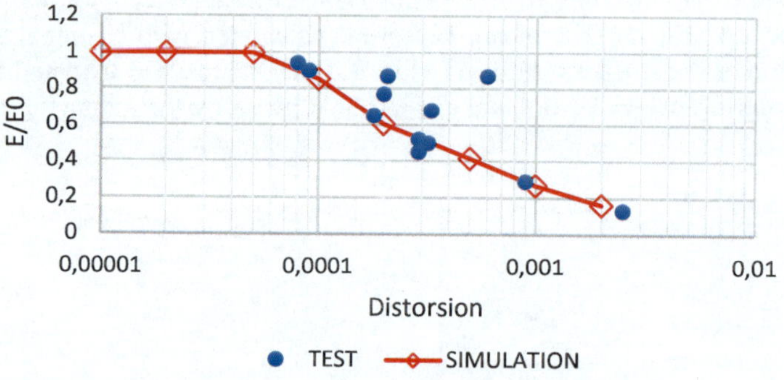

Fig. 5 Normalized degradation curve of Young modulus with shear strain

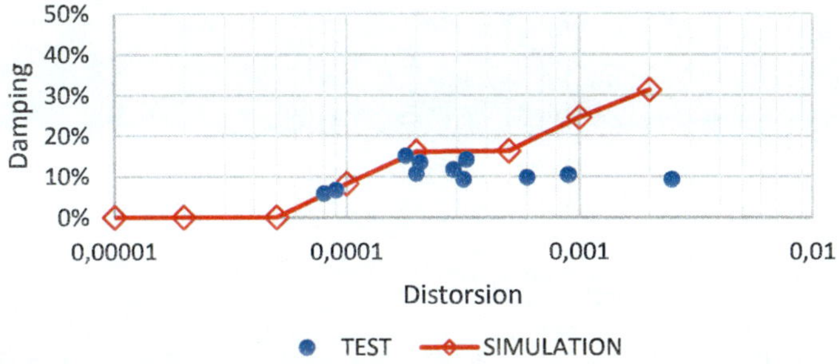

Fig. 6 Damping evolution with shear strain

2.3 *Numerical Modeling*

The dam mesh is bi-dimensional (plane strain condition). It is constituted by the dam body (green) and the rock foundation (blue). There is 8720 quadratic elements and 15097 nodes. The detailed mesh and the principal dimensions are given on Fig. 7.

The dam is represented by only one material and the bituminous facing is not considered. The Hujeux constitutive model is used for the dam body (see Sect. 2.1 and 2.2 for calibration) and the bedrock foundation is considered as elastic linear with the parameters of Table 3.

As there are no positive water pressure within the embankment and the foundation is assumed to be impervious, the numerical modelling is mechanical only (no hydro-mechanical coupling). The simulations are performed with code_aster software developed by EDF Lab [4].

During the static phases, the base of the foundation is embedded and the two laterals frontiers are horizontally blocked. Static phases aims to determine initial

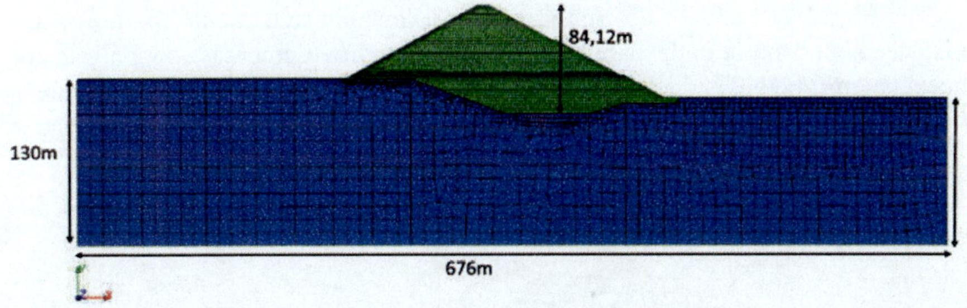

Fig. 7 Menta dam mesh

Table 3 Properties of the rock mass foundation

γ [kN/m^3]	E [MPa]	ν
27	7.10^3	0.25

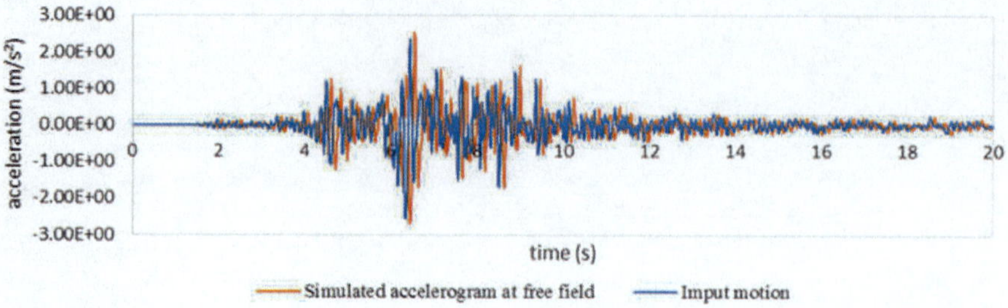

Fig. 8 Comparison between input motion and computed accelerogram at free field for Central Italy earthquake

state for the dynamic phase. For this one, the cinematic limit conditions of the foundation are transformed in strength conditions by using spring system. Furthermore, boundaries elements are used to absorb reflected waves.

The seismic signal is introduced to the base of the foundation as a shear wave with vertical direction. On the lateral boundaries, free field conditions are applied to guarantee the wave propagation accuracy in all the foundation. A verification of the signal is carried out at surface in both cases (Fig. 8).

3 Static Results

3.1 End of Construction

The numerical modelling of construction phase was performed using staged construction in 15 layers. The weight of the materials is the only loading of the dam.

The Figs. 9, 10 and 11 illustrate the final calculated settlements and horizontal displacement contour in the dam body. The maximum settlement is about 66 cm and occurs at mid-height and slightly downstream the dam axis. The maximum value of horizontal displacement reaches 20 cm on the both upstream and downstream facing.

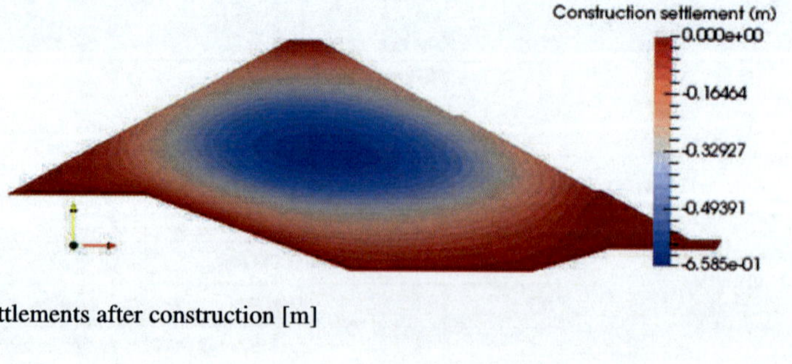

Fig. 9 Settlements after construction [m]

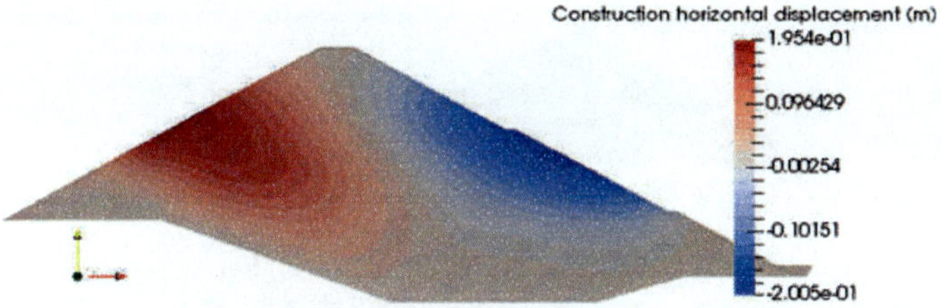

Fig. 10 Horizontal displacement after construction [m]

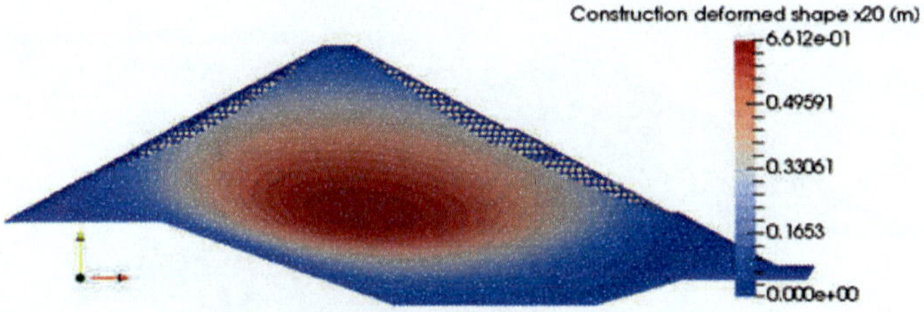

Fig. 11 Deformed shape × 20 after construction [m]

3.2 *After the First Impoundment*

The increase of the water level in the reservoir is simulated by applying a function that depends on height and time until the water level reachs 1425.75 m a.s.l (crest at 1431.75 m a.s.l). The settlements and horizontal displacement for the impounding only are plotted on Figs. 12 and 13.

The settlements are maximum (about 11 cm) on the upstream facing at the mid-height of the water level. The horizontal displacement generated by the impounding is negligible. Indeed, the calculated value obtained by the end of the impounding only is lower than 6 cm. The deformed shape (Fig. 14) shows that the lower part of the upstream facing is moving to downstream.

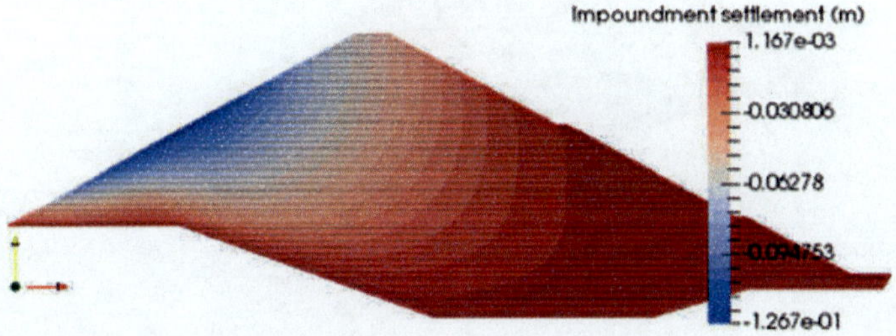

Fig. 12 Settlements after impounding only [m]

Fig. 13 Horizontal displacement after impounding only [m]

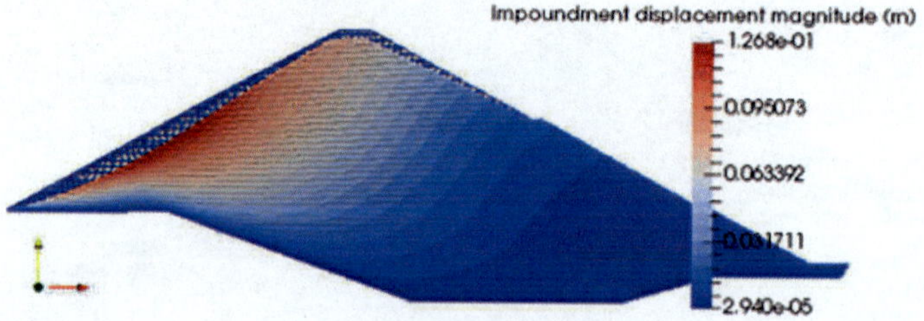

Fig. 14 Deformed shape - × 50) after impounding only [m]

The aim of the static phases is to obtain the good stress profile before the seismic motion. The stress fields computed at the end of the first impoundment is given on Figs. 15 and 16. This stresses are varying from the top to the bottom. The maximum value of vertical stress is about 1.5 MPa at the bottom.

As described by Eq. (6), the Young modulus profile depends on the mean effective stress p' with exponent "*n*". So, the stress field (Fig. 15 and Fig. 16) is used to determine the Young modulus profile shown in Fig. 17. This real profile is the initial state for performing the modal analysis (Sect. 4.1).

Fig. 15 Vertical stress profile [Pa] after first impoundment

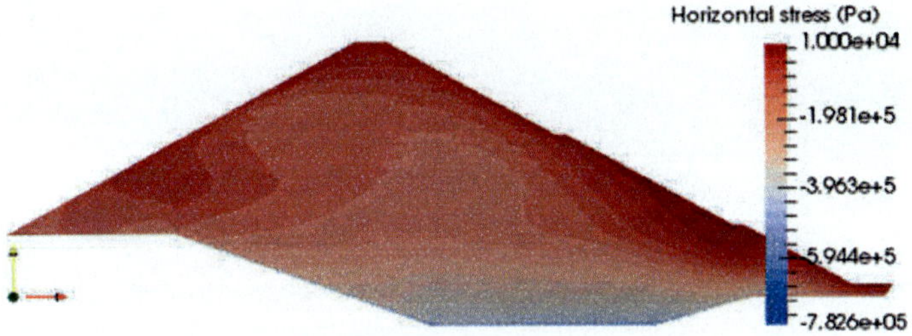

Fig. 16 Horizontal stress profile [Pa] after first impoundment

Fig. 17 Young modulus [Pa] within the embankment

Table 4 Assessment of f_0

Condition	f_0 [Hz]
Empty reservoir	1.82
Full reservoir	1.86

4 Dynamic and Seismic Analysis

4.1 Modal Analysis

The assessment of f_0 gives lightly different values between empty and full reservoir condition (Table 4). The impoundment increases the stress state close to the upstream facing, hence, the Young modulus profile is slightly different between full and empty reservoir condition. With full reservoir condition, the dam is more rigid so, its first natural frequency is higher.

4.2 Input Motions

The Fig. 18 shows the frequency content of the two input seismic signals. The dominant frequencies are respectively close to 4 and 7 Hz for Friuli and Central Italy

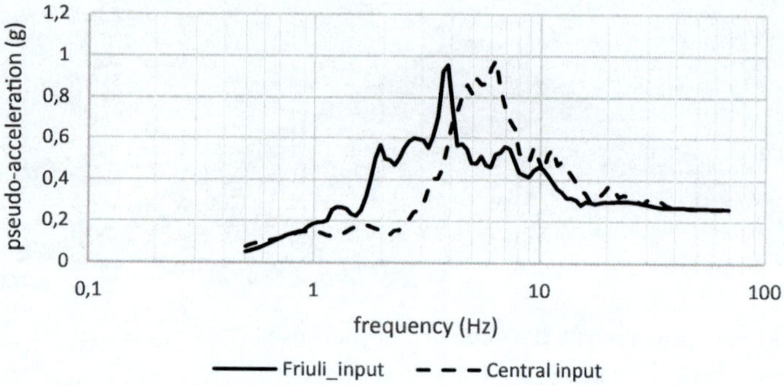

Fig. 18 Frequency content of Friuli and Central Italy input motions

ground motions and are well above the fundamental frequency of the dam close to 1.86 Hz. However, the Friuli earthquake has a higher frequency content between 2 and 4 Hz than the Central Italy earthquake whose frequency content is between 4 and 8 Hz.

Given these observations, it is likely that the Friuli earthquake is more damaging to the dam than the Central Italy earthquake. It should be noted, however, that the frequency content of the Friuli earthquake decreases rapidly between 2 Hz and 1.5 Hz.

4.3 Simulation Results

Time histories of the horizontal acceleration, the spectral amplifications and crest settlement produced by both earthquakes are given on Figs. 19, 20 and 21.

A higher amplification is observed for the Central Italy ground motion whereas no amplification is observed close to the central frequency of both signals (from 3 to 4 Hz). For Central Italy ground motion, the amplification is observed at 2.2 Hz

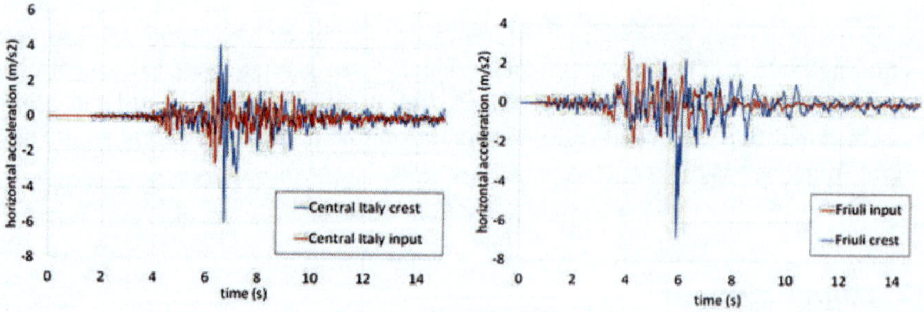

Fig. 19 Acceleration time histories at the crest (blue line) and at the basis (red line) (left: Central Italy; right: Friuli)

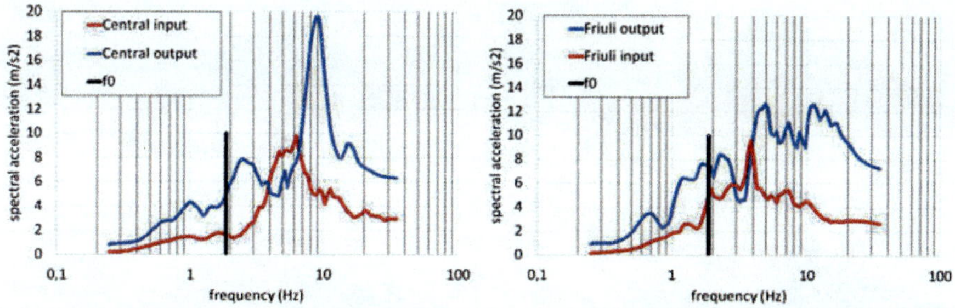

Fig. 20 Spectral acceleration at the crest (blue line) and at the basis (red line) (left: Central Italy; right: Friuli)

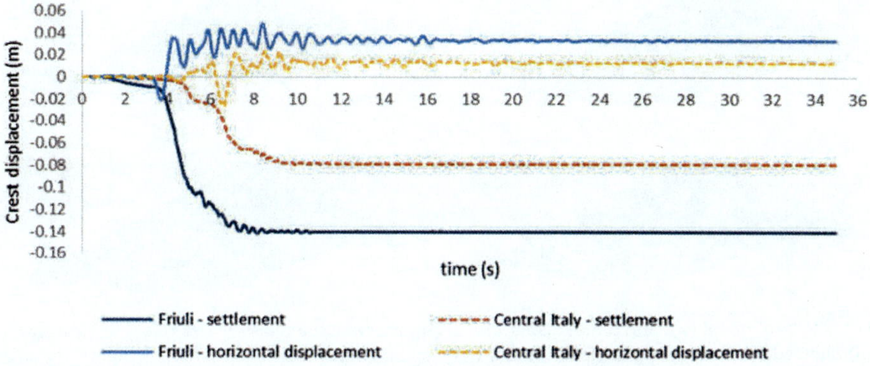

Fig. 21 Time histories of crest settlement and horizontal displacement

but it drops rapidly for lower frequencies. For Friuli earthquake, the amplification is observed close to the fundamental frequency and for lower value [1.5–2 Hz]. This shifting of the fundamental frequency of the dam can be related with the decay of stiffness with increasing shear strains which are more important for the Friuli ground motion compared to the Central Italy ground motion (see Fig. 22c).

The maximum deviatoric deformations (Fig. 22c) during the earthquake are of the order of 10^{-3} inside the embankment. They correspond to a degradation of the Young modulus up to 60% of the maximum modulus and to a damping ranging from 15 to 20%. These deformations are more important for the Friuli input motion than for the Central Italy ground motion. This explains a lower amplification of the acceleration at the crest of the dam for the former signal and the switch of the fundamental frequency to lower value.

The Fig. 22d shows the vertical deformations induced by the earthquake. They evolve linearly from the base of the dam to a few meters below the crest where they increase significantly.

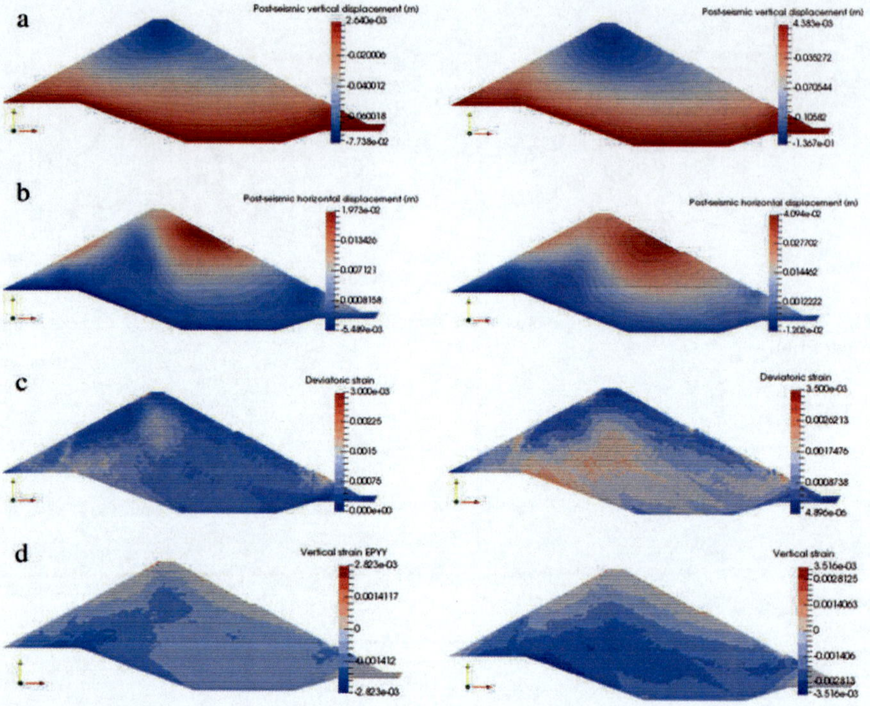

Fig. 22 Settlement (**a**), horizontal displacement (**b**), maximum deviatoric strain (**c**), vertical strain (**d**) induced by the earthquake (left: Central Italy; right: Friuli)

The settlement profiles (Fig. 23) in the axis of the dam present a linear evolution from the basis of the dam up to the crest where it reaches 14 cm for the Friuli earthquake and 8 cm for Central Italy earthquake. The profiles of the permanent displacement evolve in the same way as the settlements. The final horizontal displacement is centimetric for both signals and from 6 to 7 times lower than the settlements.

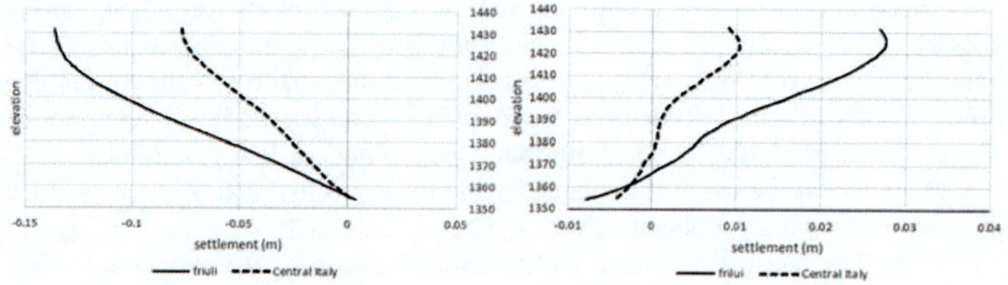

Fig. 23 Vertical profile of settlement (left) and horizontal displacement (right)

5 Discussions

Despite the maximum settlement is observed at the crest of the dam, the vertical deformation are more important in the core of the embankment than at the crest. This result does not agree with the observations made on the centrifuge tests by Okamoto [5]. This reduction in the simulations of the vertical deformation close to the crest is a consequence of the model which reduces the compressibility according the increase of the p_c/p' ratio where p_c is the critical pressure related with the density of the material and p' the mean pressure affecting the material. Figure 24 shows the vertical profile of K_0 in the axis of the dam. The low values observed at the core of the fill explain why the maximum settlement induced by the earthquake is observed there. On the other hand, the high values of K_0 observed on the crest lead to small crest settlements. The rather loose stress state in the embankment body is also an explanation for the low frequencies of the empty and full dam.

The model was originally developed for granular soils and is perfectly suited to account for the degradation of the G/G_{max} curve using elastic and deviatoric plastic mechanisms. But the calibration of the parameters was carried out by assimilating the shape of the young modulus degradation curve E/E_{max} to that of the shear modulus G/G_{max}. This simplification seems to be not correct because the isotropic and deviatory mechanisms of the Hujeux model are activated and evolve separately according to the hardening parameters of the mechanisms and the critical pressure. In addition, the cyclic shearing leads to positive hardening of the isotropic mechanism corresponding to the densification of the rockfill (increase of the compressibility modulus) and not to a reduction of the compressibility modulus.

Curves giving the volumetric deformations as a function of shear cycles would be of great help to validate the calibration of the model for large rockfill as pore pressure

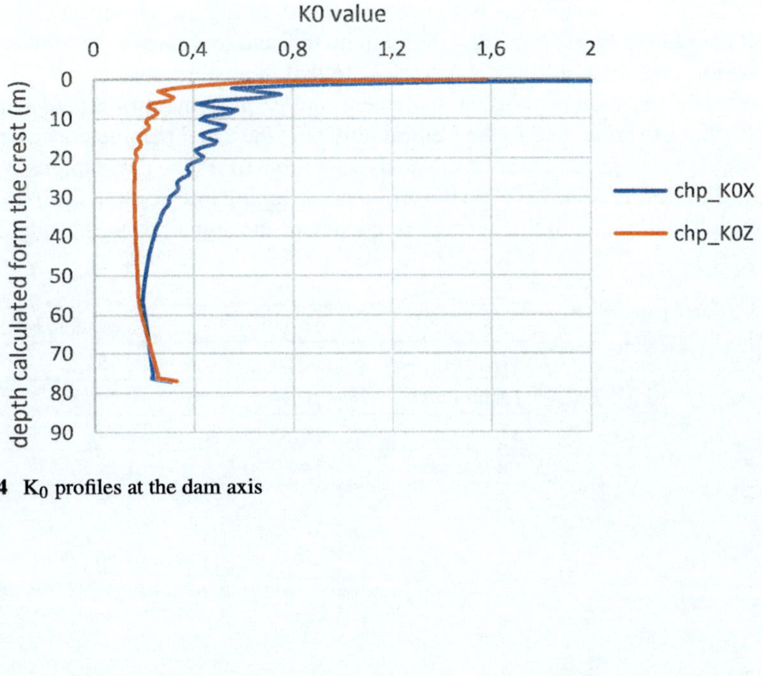

Fig. 24 K_0 profiles at the dam axis

Table 5 Settlement of the crest estimated by different published formulas or studies

	dH/H		Settlement (m)	
	Mean value (%)	Lower value (%)	Mean value	Lower value
Friuli	0.18		0.14	
Central Italy	0.1		0.08	
Swaisgood	0.06		0.05	
Bureau	0.43	0.18	0.34	0.15
Okamoto	0.68	0.55	0.54	0.44
Okamoto (slope correction)	0.32	0.28	0.26	0.22

increase or CSR curves as a function of the number (and amplitude) of cycles are useful to validate the calibration of the model for sandy materials.

In order to check the efficiency of the simulations, the calculated settlements are compared to the settlements given by different formulas of the literature [5–7] in Table 5.

The calculated settlements are in good agreements with Swaisgood formula [7] but under-estimate by several decimeters the settlement predicted by the earthquake severity Index given by Bureau [4] and also the settlements observed by Okamoto [5] on centrifuge tests. These tests may appear conservative because they were operated for 10 sinus waves on rockfill dams with 1 V:1.4H slopes whereas the slope of Menta dam are 1 V:1.8H and the number of equivalent cycles ranges between 7 and 8 corresponding to a 6.4 magnitude earthquake.

Despite corrections related to the slope and the magnitude, the calculated values are within the lower range of the literature formulas. This is probably related to the insufficiency of volumetric deformations under monotonous loading and for small shear strains at low confining pressure (cf. triaxial calibration curves on Figs. 1 and 2). A better knowledge of the volumetric deformations during a cyclic shear test would make it possible to recalibrate the set of parameters and to improve the efficiency of the constitutive model on a stress path close to that of an earthquake.

To explain the more important settlement and displacement observed with the Friuli earthquake compared to the Central Italy one, the usual parameters describing the noxiousness of ground motion are analyzed (Table 6). PGA, Arias intensity I_{Arias} or phase duration T_f do not explain the differences. Taking into account the frequency content, for example with the average frequency of the signal, as in the calculation

Table 6 Global parameters describing the ground motions

	Friuli	Central Italy
PGA (g)	0.26	0.26
IArias (m/s)	0.44	0.40
Tf (s)	4.25	6.90
Pd (Saragoni)	4.89 10-3	1.78 10–3

of the Saragoni's index Pd, allows to find a better correlation between the predicted settlements and the indicator of the earthquake.

6 Conclusions

The study shows the possibility of carrying out the static and dynamic study of a dam type BFRD with the Hujeux model.

The model predicts settlements of several tens of centimeters in the core of the fill and confirms that the filling of the hold has little influence with horizontal displacements of about fifteen centimeters in the middle of the upstream facing. The level of the reservoir leads to a slight increase in the fundamental frequency of the dam.

Seismic induced settlements of the dam are very progressive with the signal and follow the evolution of the Arias intensity curve, but this indicator alone cannot explain the difference in settlement and displacement observed between the two input signals. A better correlation with the permanent displacements is obtained by taking into account the frequency content of the input signal like the Saragoni index. The amplification of the acceleration at the crest depends on the frequency content of the input signal and the dynamic characteristics of the dam and its foundation.

The crest settlements are generally modest for the two earthquakes, less than twenty centimeters. This corresponds to about 0.2% of the height of the fill, a value within the lower range of that given by the literature. According the simulations, the maximum deformations induced by the earthquake are located in the core of the dam which does not seem to agree with the observations on small-scale models [5].

These last two remarks led to a more detailed analysis of the behaviour of the Hujeux model and the stress distribution in the embankment. Results are consistent with the K_0 profile and the evolution of the critical pressure leading to a low stress level inside the dam and to "overconsolidated" condition at the crest. Finally, the isotropic and deviatoric mechanisms of the Hujeux model, which was initially developed for saturated granular soils, seem less suitable for reproducing the Young modulus evolution curve than that of the shear modulus. The calibration of the model could be improved with curves giving the volumetric deformations as a function of shear cycles.

References

1. Aubry D, Hujeux JC, Lassoudière F, Meimon Y (1982) A double memory model with multiple mechanism for cyclic soil behavior. Inte Symp Numerical Models Geomech Zurich 1:3–13
2. Code_Aster Reference document R7.01.23.Hujeux model for the cyclic behavior of soils
3. Sawada Y, Takahashi T (1975) Study on the material properties and the earthquake behavior of rockfill dams. In: Proceedings of 4th Japan earthquake engineering symposium, pp 695–702
4. code_aster, general public licensed structural mechanics finite element software, internet site: http://www.code-aster.org

5. Okamoto T (2006) Seismic stability evaluation and criteria of rockfill dam basing on residual deformation. In: Proceedings 16th International conference soil mechanics and geotechnical engineering, Osaka
6. Bureau G (1997) Evaluation methods and acceptability of seismic deformation in embankment dams. ICOLD 19th congress of large dams, Florence, Italy
7. Swaisgood JR (2003) Embankment dam deformations caused by earthquake. Pacific Conference on Earthquake Engineering, Christchurch, New Zealand

Seismic Behavior of a Bituminous-Faced Rock-Fill Dam: The Menta Dam

L. Petkovski, S. Mitovski, and F. Panovska

Abstract In the analysis of the dynamic response of the bituminous-faced rock-fill dam (BFRD) Menta, is applied non-linear model, where the rock material is approximated by variable sliding modulus, and the asphaltic facing with thickness of 32 cm is applied by joint elements with linear elastic constitutive law. Permanent displacements during the seismic excitation are determined by Dynamic deformation analysis, where the incremental forces are calculated by the difference of the effective stresses in two successive time steps, resulting in adequate deformations. For the state of reservoir rapid filling, as pre-earthquake state by normal water elevation in the reservoir, is used elastoplastic model by variable modulus of elasticity for the rock material. By the dynamic analysis is verified the seismic resistance of the fill dam at action of design earthquakes by PGA of 0.26 g, without disruption of the water impermeability of the asphaltic facing and without danger for rapid and uncontrolled emptying of the reservoir, because the settlements in the dam crest caused by dynamic inertial forces for the earthquake duration amounts 40 cm, i.e. they are much lower than the height above normal elevation in the reservoir till the dam crest, with value of 7.25 m.

Keywords Menta dam · Dynamic analysis · Asphaltic facing

1 Dam Model and Material Parameters for the Structural Analysis

The representative cross section and parameters of the local materials for the fill dam Menta are adopted according to the available data bases [1]. The model for structural (static and dynamic) analysis of the dam is composed of two materials: rock material in the dam's body and asphaltic facing (Fig. 1). For the static analysis,

L. Petkovski (✉) · S. Mitovski · F. Panovska
Civil Engineering Faculty, University Ss. Cyril and Methodius, Skopje, North Macedonia
e-mail: petkovski@gf.ukim.edu.mk

Macedonian Committee on Large Dams, Skopje, North Macedonia

G. Bolzon et al. (eds.), *Numerical Analysis of Dams*, Lecture Notes in Civil Engineering 91, https://doi.org/10.1007/978-3-030-51085-5_33

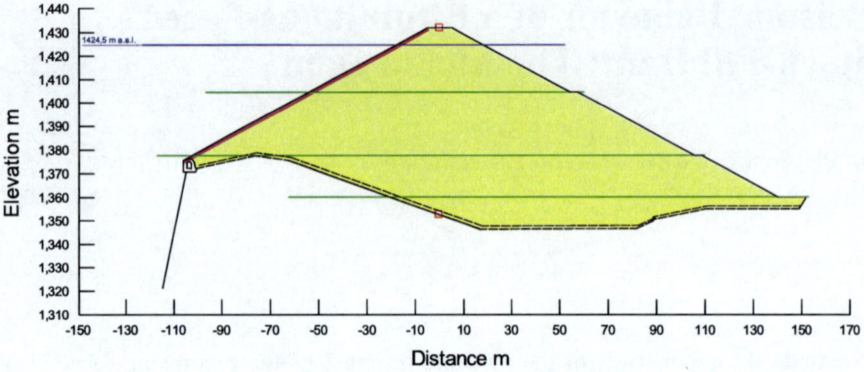

Fig. 1 Model for structural analysis, by specified elevations: dam crest at 1,431.75 m asl, normal water elevation at 1,424.5 m asl, first berm on 1,404.5 m asl, second berm on 1,377.5 m asl, upstream toe at 1,375.53 m asl (above grouting gallery), rock at cross section axis 1,352.57 m asl, downstream toe at 1,360.0 m asl. Structural height of 1,431.75–1,352.57 = 79.18 m

for the asphaltic facing, with thickness of 0.32 m is applied linear elastic law, by following material parameters: E = 150,000 kPa, $\nu = 0.44$, $\gamma = 24$ kN/m^3, while for the rock material is applied elastoplastic constitutive law with variable modulus of elasticity (Fig. 2), with following geo-mechanical parameters: $E = E(\sigma_y')$, $\nu = 0.27$, $\gamma = 23$ kN/m^3, $\varphi = 38^\circ$ and c = 0.0 kN/m^2.

For the dynamic analysis, for the asphalt facing with thickness of 0.32 m is applied linear elastic law, by following material parameters: $G_{max} = 2.0$ GPa, $\nu = 0.44$, $\gamma =$ 24 kN/m^3, DR = 0.1, while for the rock material is applied non-linear constitutive

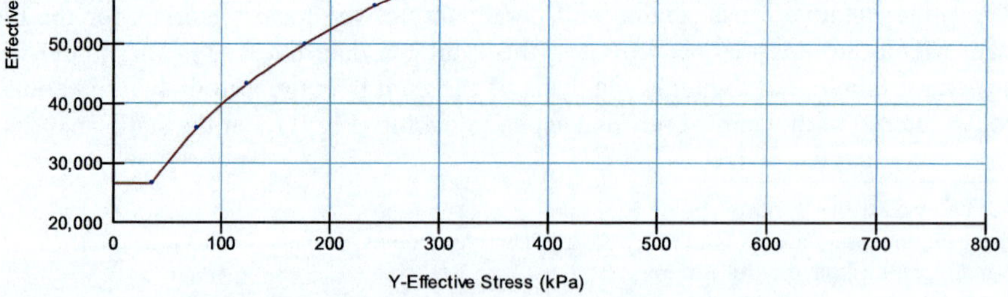

Fig. 2 Variable static modulus of elasticity for rock material $E = E(\sigma_y')$

law with variable maximal sliding modulus (Fig. 3), by following geo-mechanical parameters: $G_{max} = G(\sigma y')$, $\nu = 0.27$, $\gamma = 23$ kN/m3, DR = 0.05–0.25, $\varphi = 38^{\circ}$ and c = 0.0 kN/m^2. The applied dependence $G_{max} = G(\sigma_y')$ is adopted by using of variation of small-strain Young modulus (Eo) with confining pressure (σ'3), obtained in cyclic triaxial tests, (from Table 1, page 10). Available results from laboratory tests: physical and mechanical properties of rock-fill material (Fig. 4) and using of expressions: $k_o = 1 - \sin\varphi$, where $\varphi = 38^{\circ}$, $\nu = k_o/(1+ k_o)$, $G_{max} = Edyn/[2(1+\nu)]$, $\sigma'_y = \sigma'_x/k_o$.

The plane model for dam Menta is discretized with finite element mesh, Fig. 5.

For determination of the internal static values (bending moments M, axial forces N and transversal forces Q) in the asphaltic facing, it has been replaced with structural

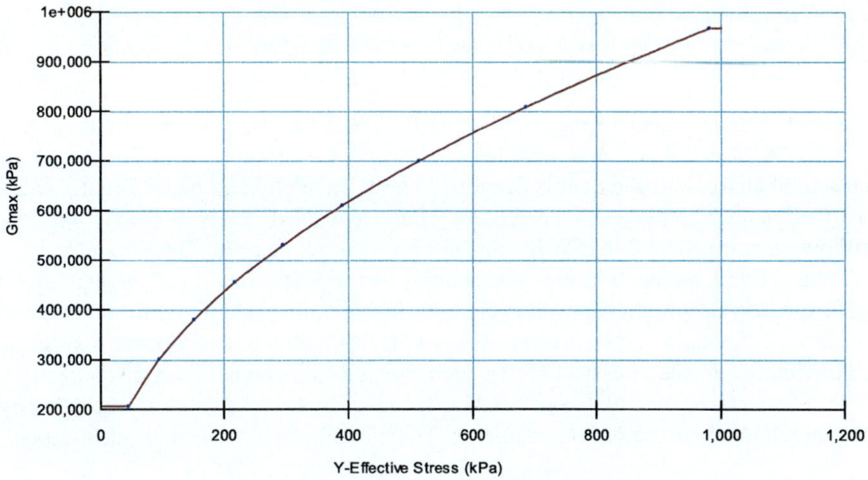

Fig. 3 Variable maximal sliding modulus for rock material $G_{max} = G(\sigma_y')$

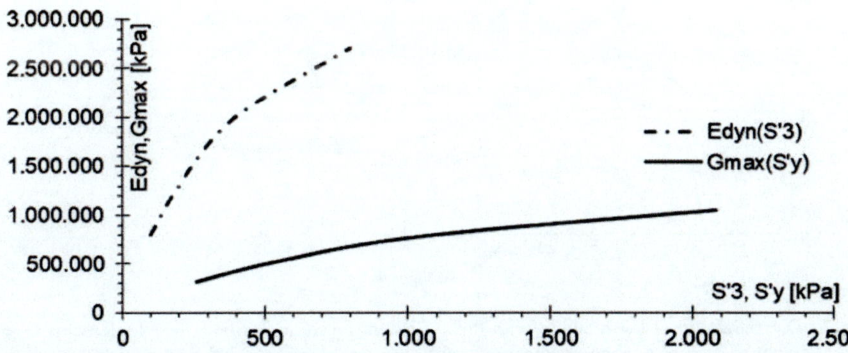

Fig. 4 Dynamic material parameters for the rock material, $E_{dyn}(\sigma'_3)$—given (from Table 1, page 10) and $G_{max}(\sigma'_y)$—estimated

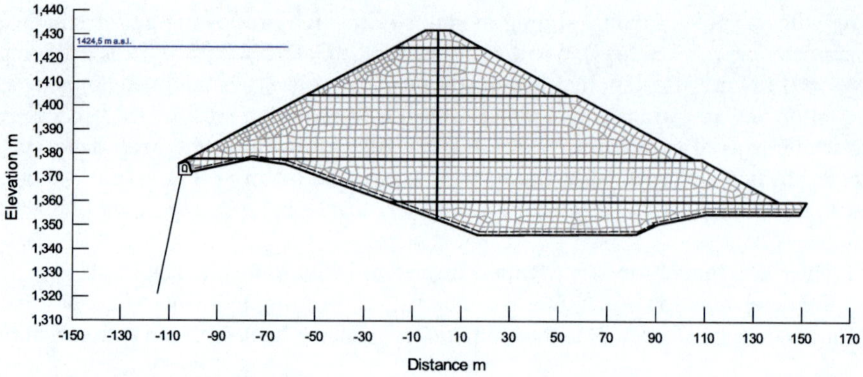

Fig. 5 Discretization of the medium with finite element mesh, nodes N = 1,444, elements E = 1,373, by application of non-linear model and joint elements for the asphaltic facing with d = 0.32 m

element with: moment of inertia I = 1.0*0.323/12 = 0.002731 m^4, cross section F = 1.0*0.32 = 0.32 m^2 and modulus of elasticity E = 150,000 kPa. The model for structural analysis of the dam is discretized with finite element mesh, Fig. 6, where as the dynamic analysis is conveyed by equivalent linear analysis (ELA) with the following non-elastic dynamic parameters for the rock material, Figs. 7 and 8.

The applied approach in the present analysis for determination of the permanent deformations during the seismic excitation, for any node within the fill dam, is the method of "Dynamic Deformation Analysis" (DDA), which is successive non-linear redistribution of the stresses [2]. By such method, for geo-medium discretized by finite elements, are calculated deformations caused by forces in nodes, calculated by the incremental stresses in the elements. Thus, by application of non-linear model,

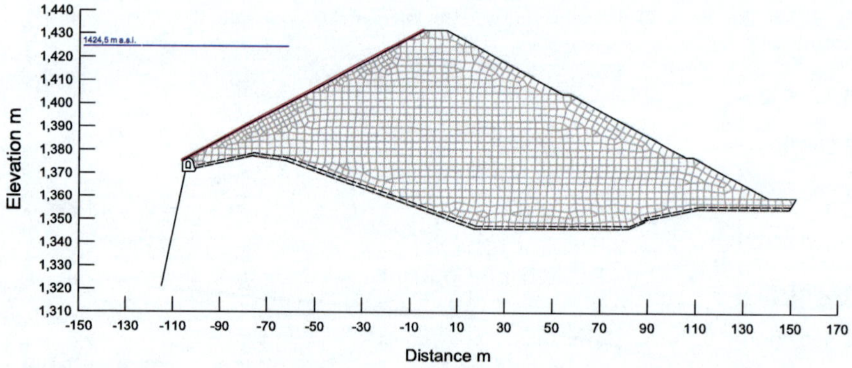

Fig. 6 Discretization of the medium with finite element mesh, nodes N = 1,384, elements E = 1,313, by application of equivalent linear law with structural element for the asphaltic facing with h = 0.32 m

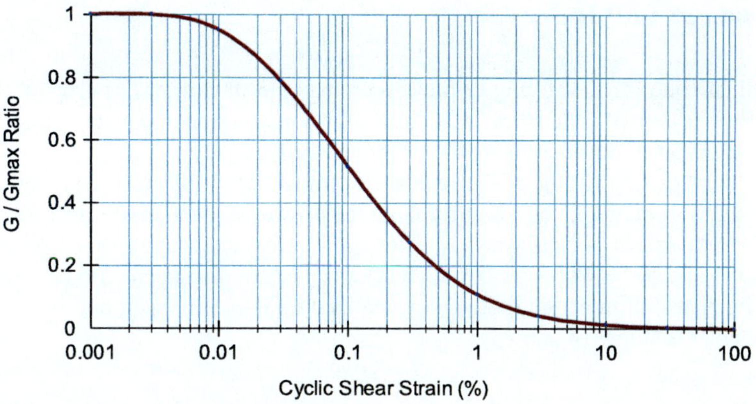

Fig. 7 Rcduction of the sliding modulus by rise of the tangential strains for the rock material by application of equivalent linear analysis

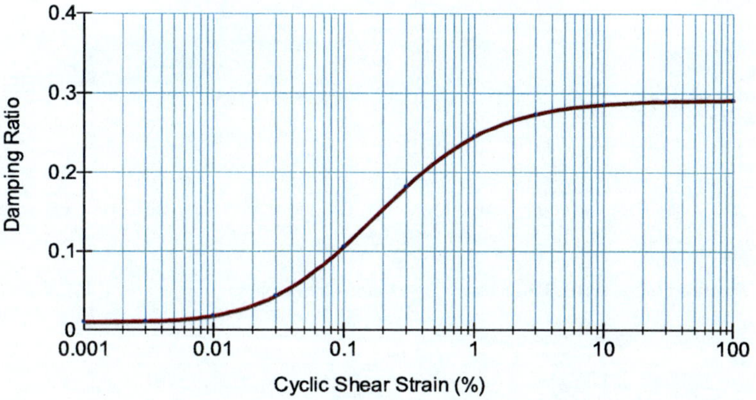

Fig. 8 Rise of the damping coefficient by rise of the tangential strains for the rock material by application of equivalent linear law

for each time step of the dynamic response of the structure [3] is obtained new state of the total stresses and pore pressure.

By the differences of the effective stresses in two successive time steps are obtained incremental forces, resulting in deformations, in accordance with the chosen constitutive law for dependence stress—strain. So, for each loading case during the dam's dynamic response are produced elastic and eventual plastic strains. If dynamic inertial forces cause plastic strains, then in the geo-medium will occur permanent deformations. The permanent displacements, at any point in the dam and at end of the seismic excitation, are cumulative sum of the plastic deformations.

2 State at Full Reservoir

The initial state for first filling of the reservoir is the state after construction of the dam, Figs. 9 and 10. The stress state after dam construction and state at first filling of the reservoir is obtained using the program Sigma/W [2].

The state at first filling of the reservoir is modeled in 10 increments, by linear rise of the water level in the reservoir from the upstream toe at elevation 1,375.53 m asl, till elevation of normal water level at 1,424.5 m asl, apropos for water head of 1,424.5–1,375.53 = 48.97 m. For such stress state are displayed stresses (Figs. 11 and 12), distribution of the modulus of elasticity, Fig. 13, as well and incremental displacements (Figs. 14 and 15).

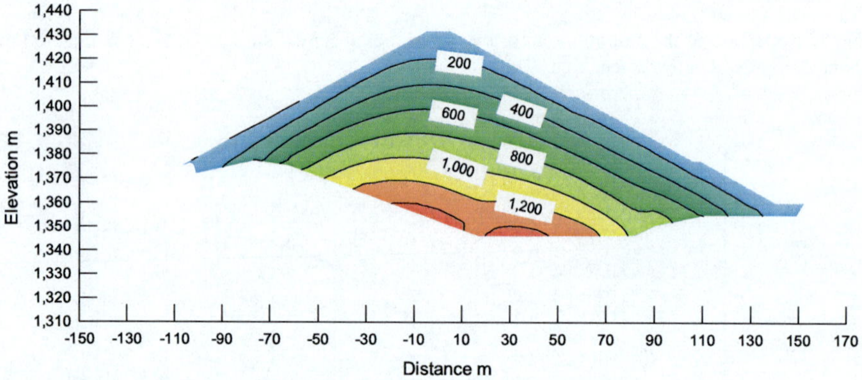

Fig. 9 Distribution of vertical effective stresses after dam construction, σ'$_y$ = 1,461.3 kPa

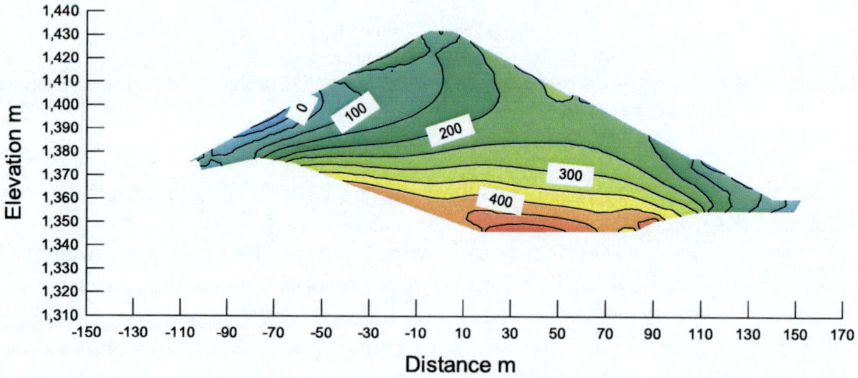

Fig. 10 Distribution of horizontal effective stresses after dam construction, σ'$_{x, max}$ = 540.78 kPa

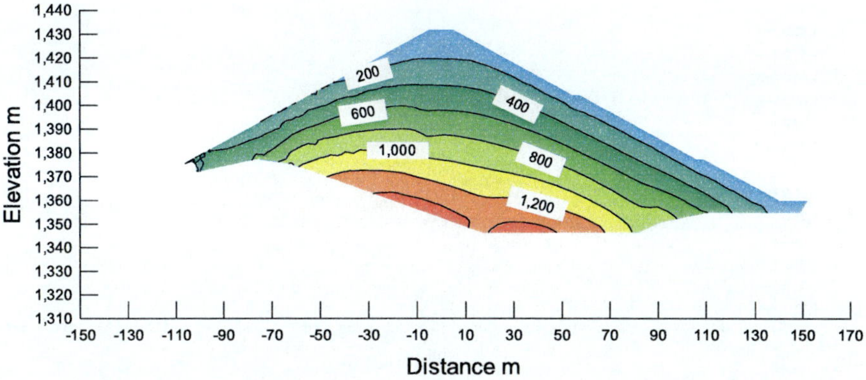

Fig. 11 Distribution of vertical effective stresses after reservoir filling, $\sigma'_{y,\,max} = 1{,}492$ kPa

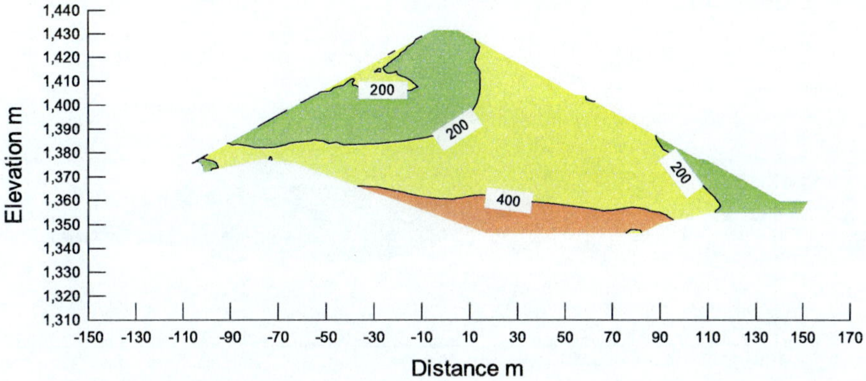

Fig. 12 Distribution of horizontal effective stresses after reservoir filling, $\sigma'_{x,\,max} = 700.0$ kPa

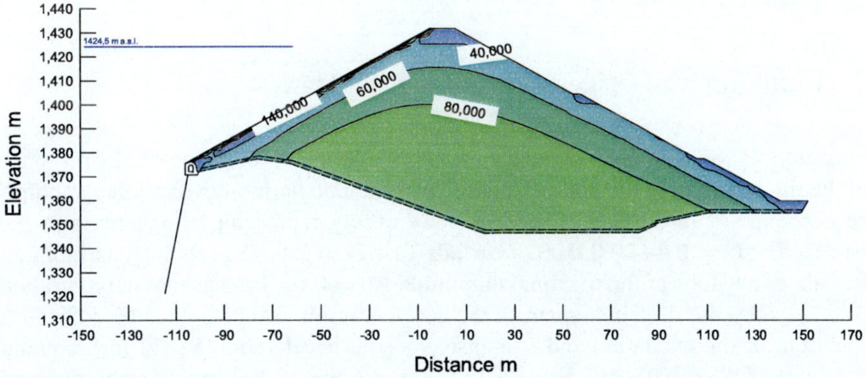

Fig. 13 Distribution of modulus of elasticity after reservoir filling, $E_{max} = 150{,}000.0$ kPa

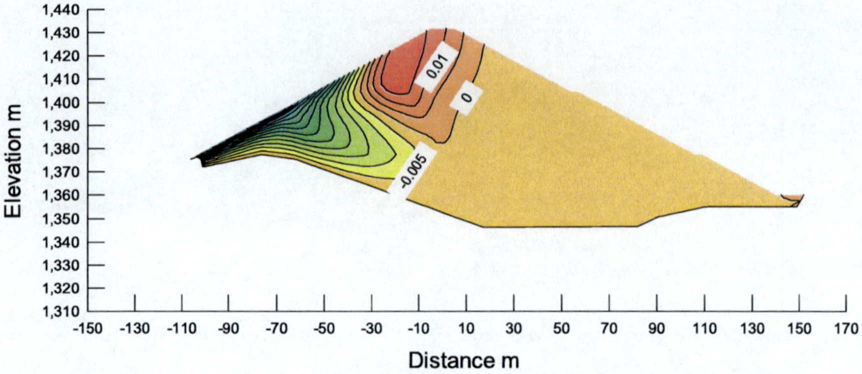

Fig. 14 Distribution of incremental vertical displacements after reservoir filling, dY = 0.074 + 0.019 m

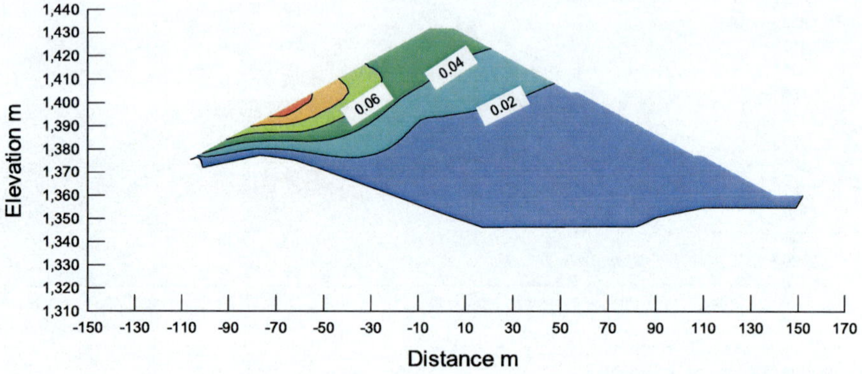

Fig. 15 Distribution of incremental horizontal displacements after reservoir filling, dX = 0.000, +0.110 m

3 Fundamental (Eigen) Periods of the Dam

For determination of the fundamental periods for certain level of non-elastic response of the fill dam is used dynamic excitation of synthetic harmony vibration by continuous change of the frequencies apropos by evenly represented frequencies in the interval f1÷f2 = 0.4÷10.0 [Hz] or periods T1÷T2 = 2.5÷0.1 [s]. Such harmonium is with: two values of the maximal amplitude Ao = 0.05, 0.30 g, summary duration St = 12 s, by the time increment in the accelegram dt = 0.01 s, Fig. 16. Response spectrum of the excitation and response, spectral acceleration Sa [g] for damping coefficient DR = 0.05, are displayed in Figs. 17 and 18 (for empty reservoir) and Figs. 19 and 20 (for full reservoir). The dynamic response of the dam is determined by use of the program Quake/W [3].

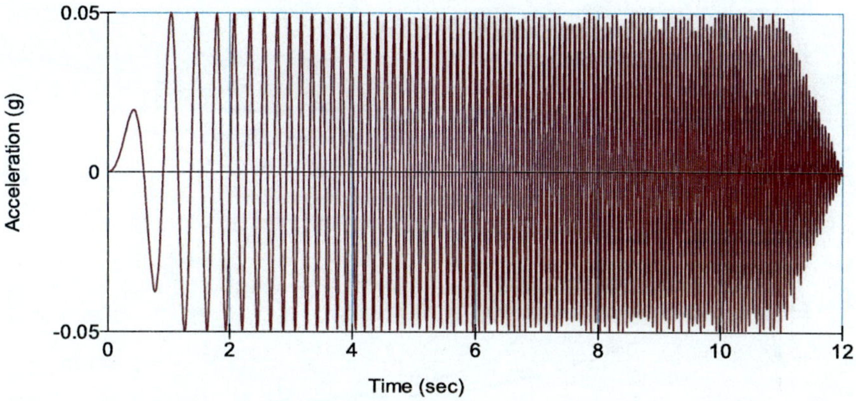

Fig. 16 Time history of horizontal displacements of harmonic vibration by evenly represented frequencies f1÷f2 = 0.4÷10.0 [Hz], scaled by Ao = 0.05 g

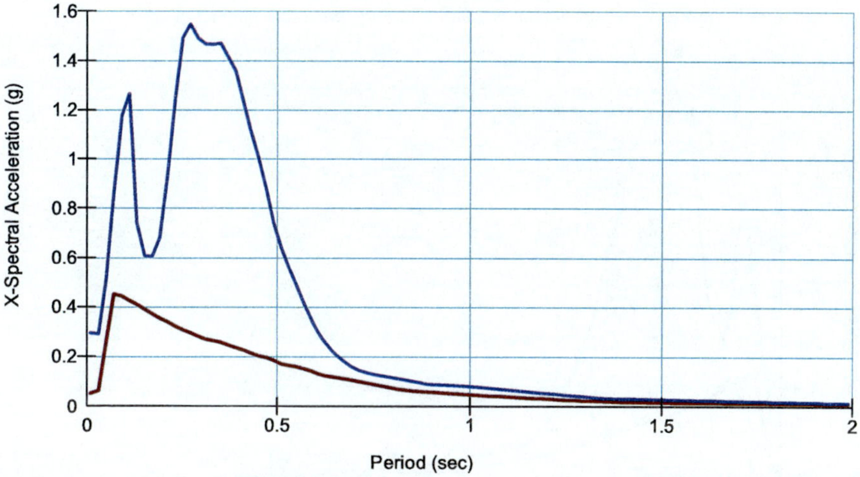

Fig. 17 Response spectrum of absolute accelerations in the dam crest at empty reservoir, caused by harmonic oscillation with low intensity PGA = 0.05 g, with first Eigen period T1 = 0.32 s

4 Response at Action of Friuli Earthquake 1976

The accelerogram of the horizontal and vertical component of the Friuli earthquake 1976 are displayed on Figs. 21 and 22.

Horizontal displacements in the dam crest at action of such earthquake are displayed on Fig. 23, and the response spectra of the accelerations are displayed on Fig. 24.

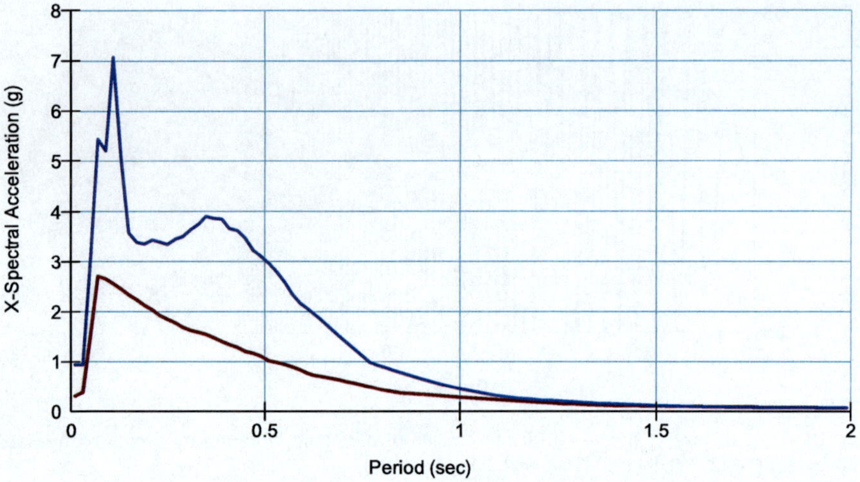

Fig. 18 Response spectrum of absolute accelerations in the dam crest at empty reservoir, caused by harmonic oscillation with high intensity PGA = 0.30 g, with first Eigen period T1 = 0.39 s

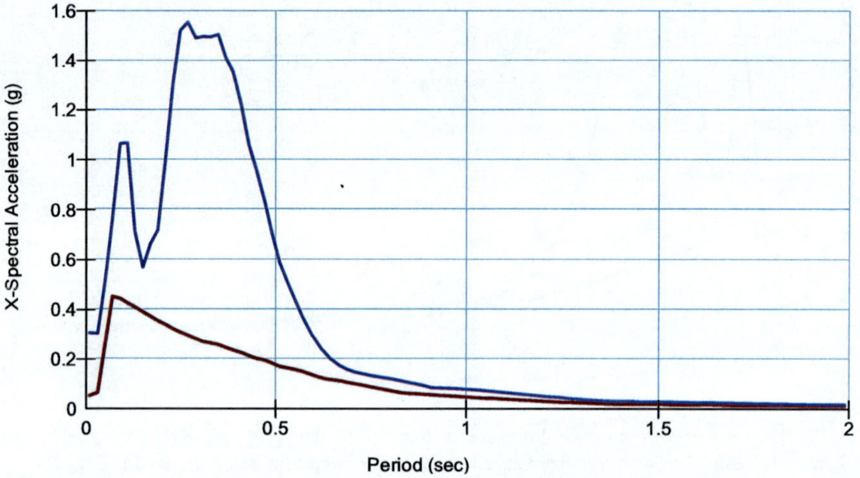

Fig. 19 Response spectrum of absolute accelerations in the dam crest at full reservoir, caused by harmonic oscillation with low intensity PGA = 0.05 g, with first Eigen period T1 = 0.31 s

Relative displacements at the dam crest at action of such earthquake are displayed on Fig. 25.

Permanent horizontal and vertical displacements at the dam crest during action of such earthquake are displayed on Figs. 26 and 27.

Permanent horizontal and vertical displacements at the dam axis after action of such earthquake are displayed on Figs. 28 and 29. Maximal shear and volumetric

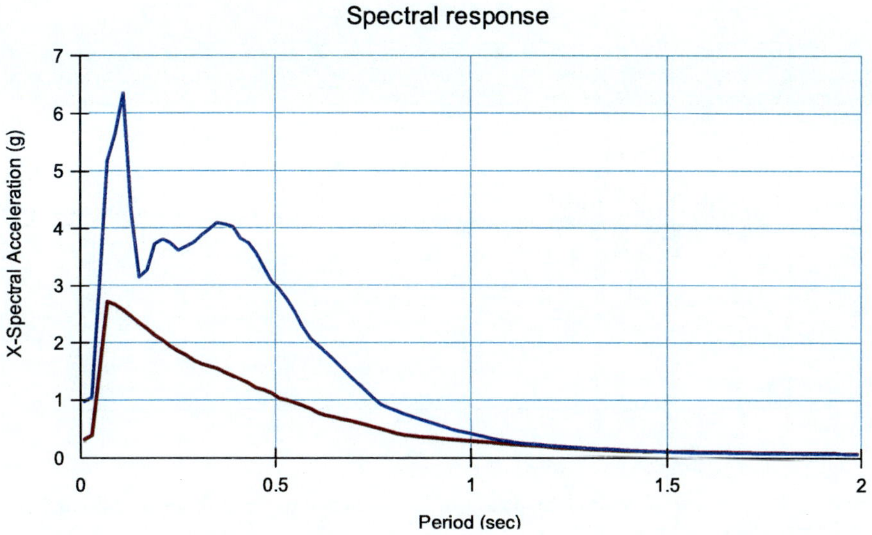

Fig. 20 Response spectrum of absolute accelerations in the dam crest at full reservoir, caused by harmonic oscillation with low intensity PGA = 0.30 g, with first Eigen period T1 = 0.37 s

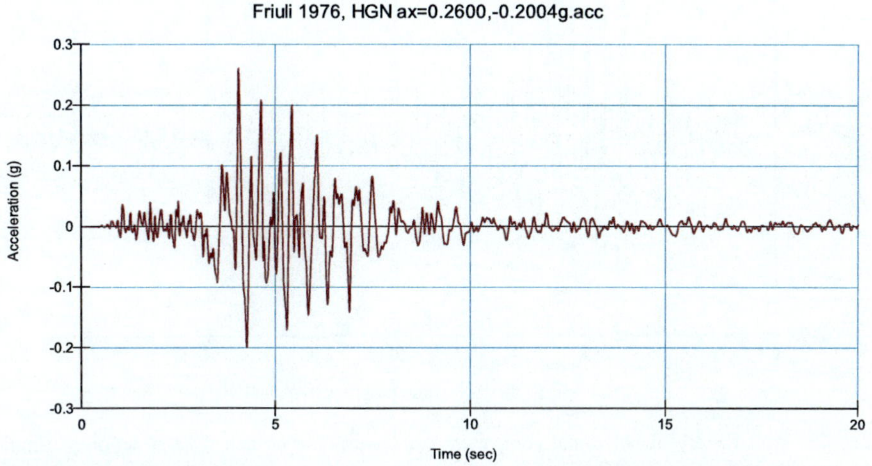

Fig. 21 Time history of horizontal component of the excitation in the rock foundation for Friuli earthquake 1976, PGA = 0.26 g

strains in the upstream slope of the dam, at asphaltic facing during the earthquake are displayed on Figs. 30 and 31.

Permanent displacements in XY direction in the dam after action of such earthquake are displayed in Fig. 32.

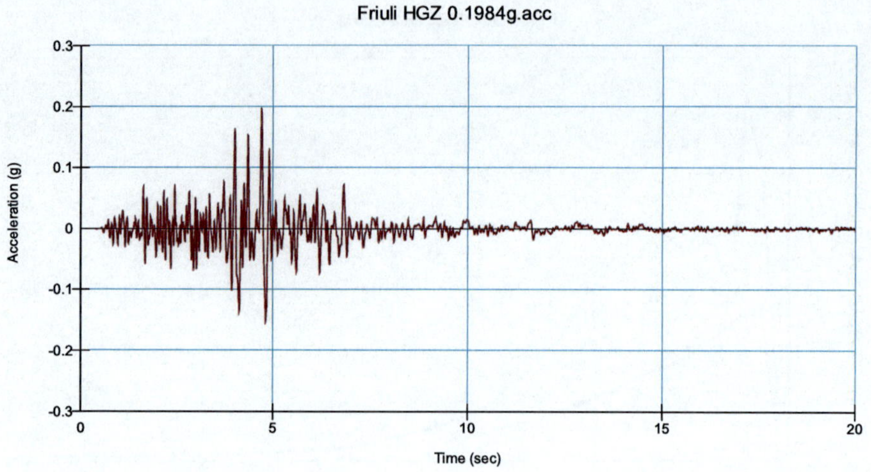

Fig. 22 Time history of vertical component of the excitation in the rock foundation for Friuli earthquake 1976

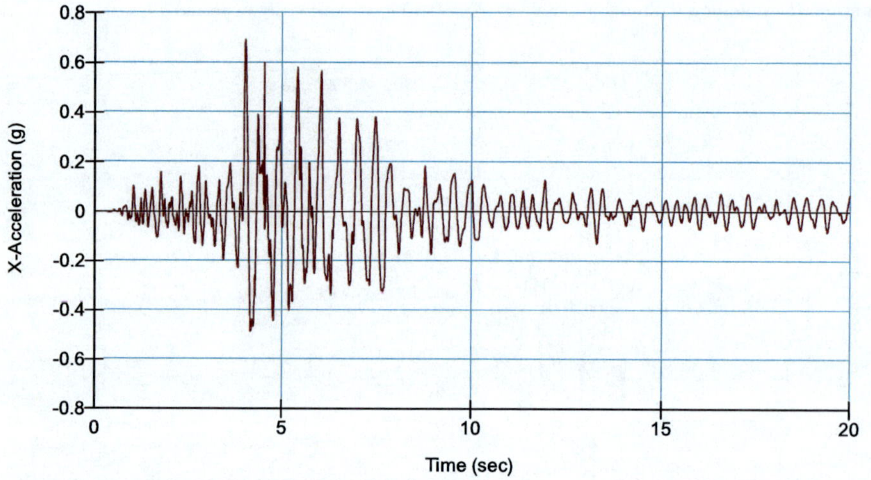

Fig. 23 Time history of horizontal component of the response at dam crest at action of Friuli earthquake 1976, PCE = 0.693 g

By application of equivalent linear model are obtained similar relative displacements at dam crest (compared to nonlinear model), Fig. 33, and the value of the axial forces in the asphaltic facing is displayed on Fig. 34.

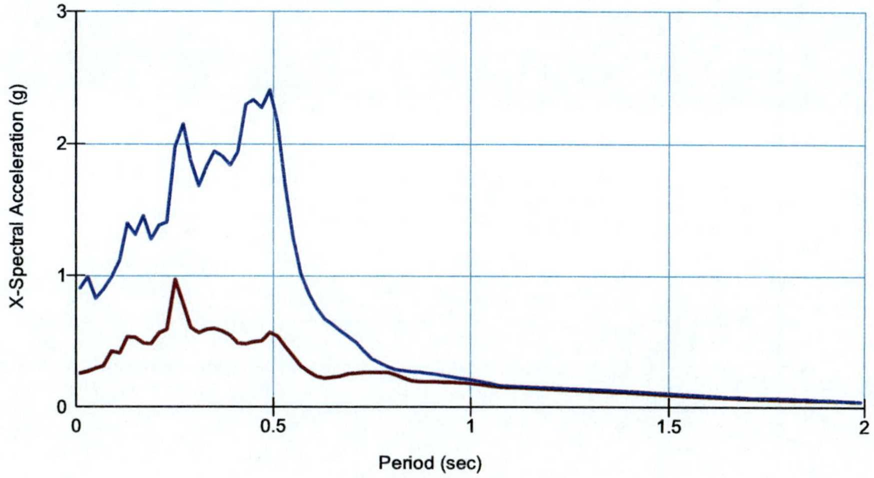

Fig. 24 Response spectra for accelerations in the foundation and at the dam crest at action of Friuli earthquake 1976

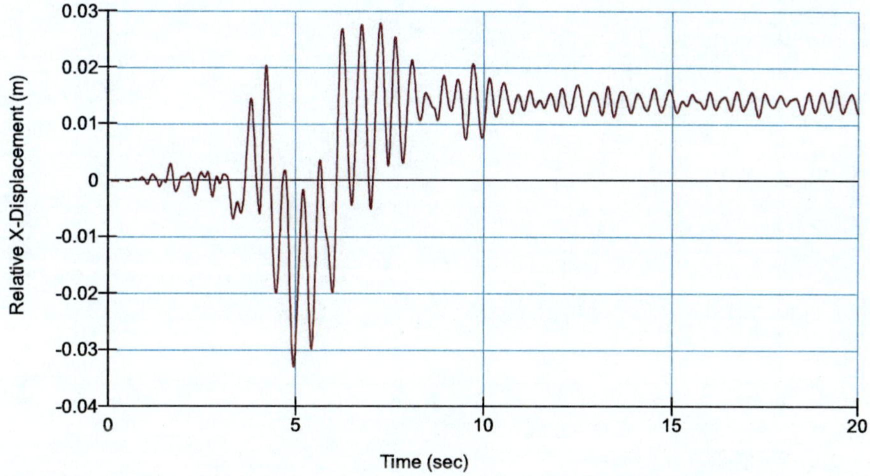

Fig. 25 Relative horizontal displacements at dam crest at action of Friuli earthquake 1976

5 Response at Action of Central Italy Earthquake 2016

Accelerograms of the horizontal and vertical component of Central Italy earthquake 2016 are displayed on Figs. 35 and 36.

Horizontal accelerations in the dam crest at action of such earthquake are displayed on Fig. 37, and the response spectra of the accelerations are displayed on Fig. 38.

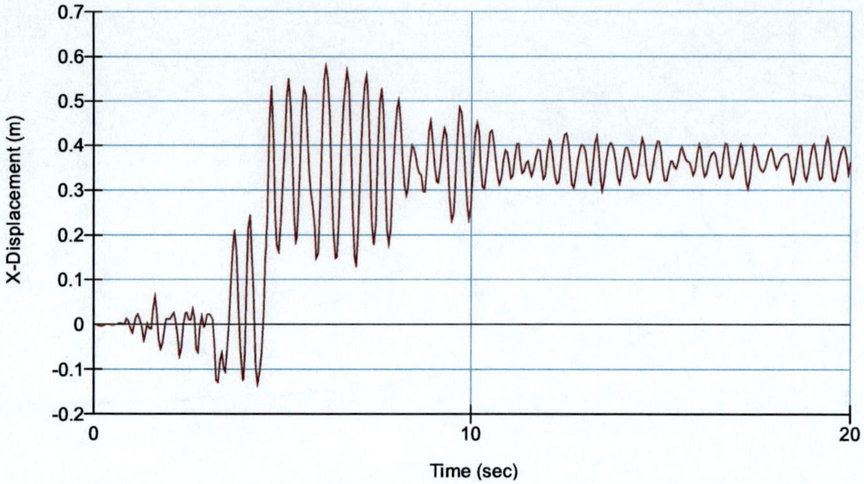

Fig. 26 Permanent horizontal displacements at dam crest at action of Friuli earthquake 1976

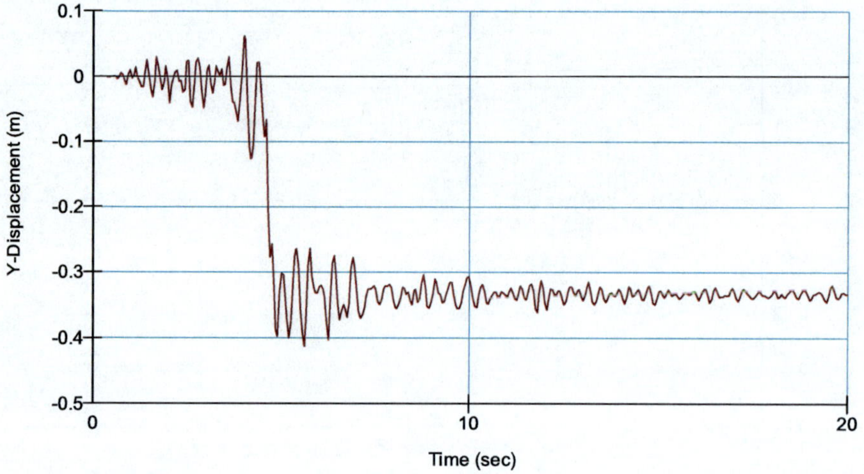

Fig. 27 Permanent vertical displacements at dam crest at action of Friuli earthquake 1976

Relative displacements at dam crest at action of such earthquake are displayed on Fig. 39.

Permanent horizontal and vertical displacements at dam crest at action of such earthquake are displayed on Figs. 40 and 41.

Permanent horizontal and vertical displacements at dam crest after action of such earthquake are displayed on Figs. 42 and 43. Maximal shear and volumetric strains in

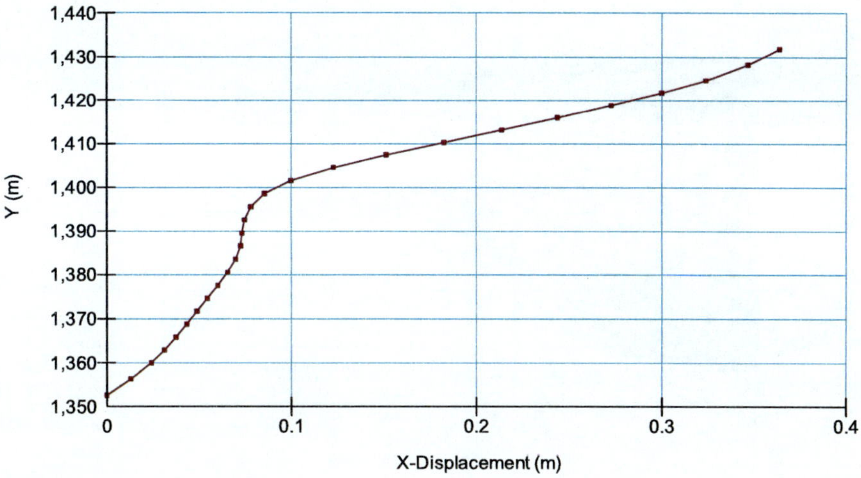

Fig. 28 Permanent horizontal displacements at dam crest axis after action of Friuli earthquake 1976, X = 0.363 m (downstream)

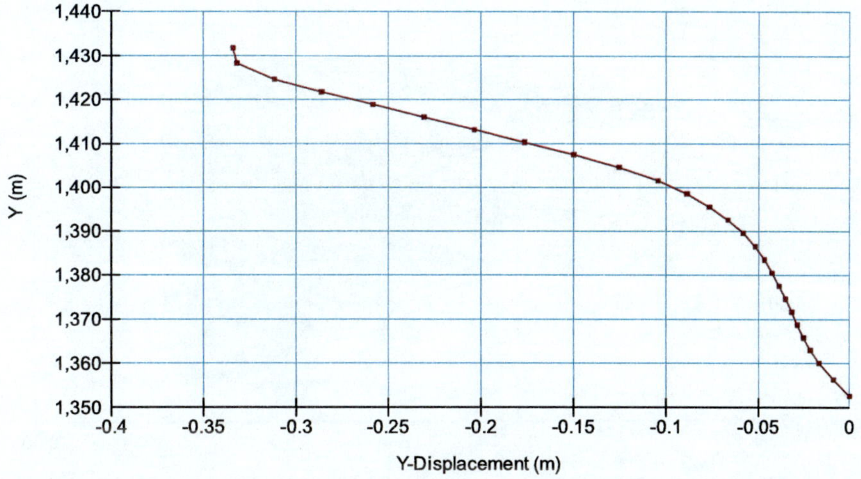

Fig. 29 Permanent vertical displacements at dam crest axis after action of Friuli earthquake 1976, Y = −0.334 m (settlement)

upstream slope of the dam, at asphaltic facing during such earthquake are displayed on Figs. 44 and 45.

Permanent displacements in XY direction after action of such earthquake are displayed on Fig. 46.

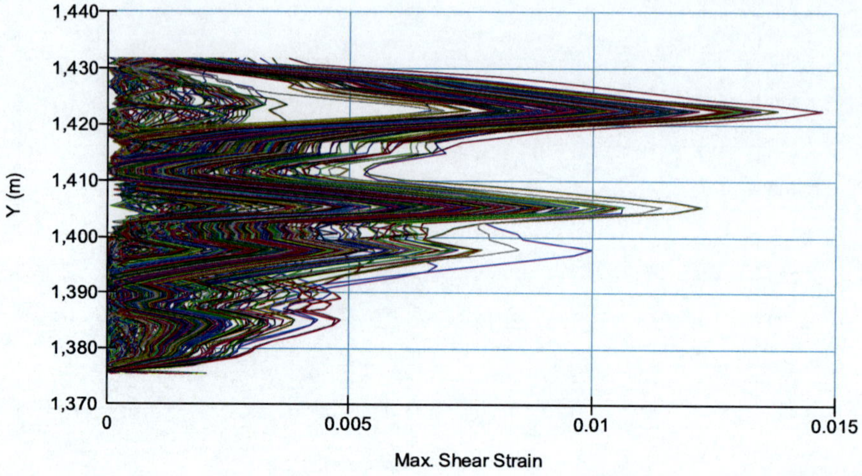

Fig. 30 Maximal shear strains in upstream slope of the dam, at asphaltic facing during Friuli earthquake 1976

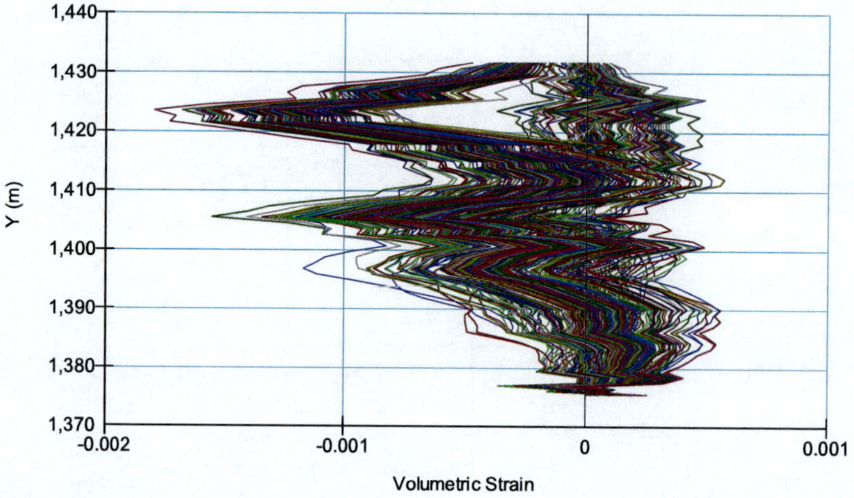

Fig. 31 Volumetric strains in upstream slope of the dam during Friuli earthquake 1976

By application of equivalent linear model (ELM) are obtained similar relative displacements at dam crest (compared with nonlinear model NLM), Fig. 47, and the value of the axial forces in the asphaltic facing is displayed on Fig. 48.

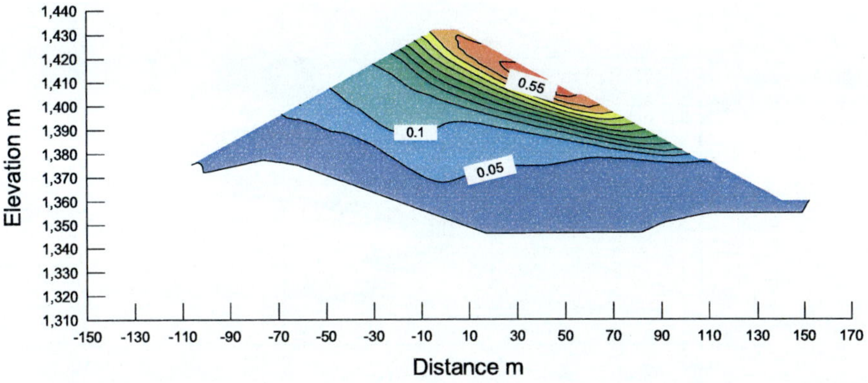

Fig. 32 Permanent displacements in XY direction in the dam after action of Friuli earthquake 1976, $XY_{max} = 0.580$ m

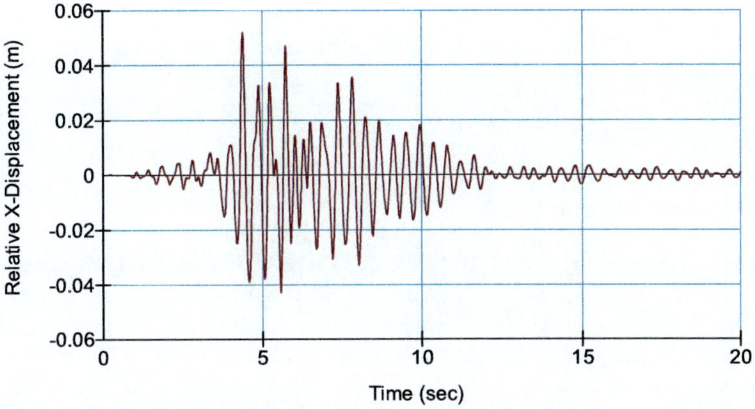

Fig. 33 Relative horizontal displacements at the dam crest at action of Friuli earthquake 1976, by application of equivalent linear model (ELM)

6 Conclusions

Eigen periods of the dam are determined by the response spectrum excited by harmonic vibration, by evenly represented frequencies of 0.4 to 10.0 Hz, scaled by PGA of 0.05 g (low excitement—L) to 0.30 g (high excitement—H). For low excitement (L), for initial state of stress at empty reservoir (E), first Eigen period is T1LE = 0.32 s, and for high excitement (H), is obtained T1HE = 0.39 s. In case of higher excitement, the response of the fill dam is more nonlinear, the stiffness of the rock material is decreased by increased non-elastic deformations, thus conditioning

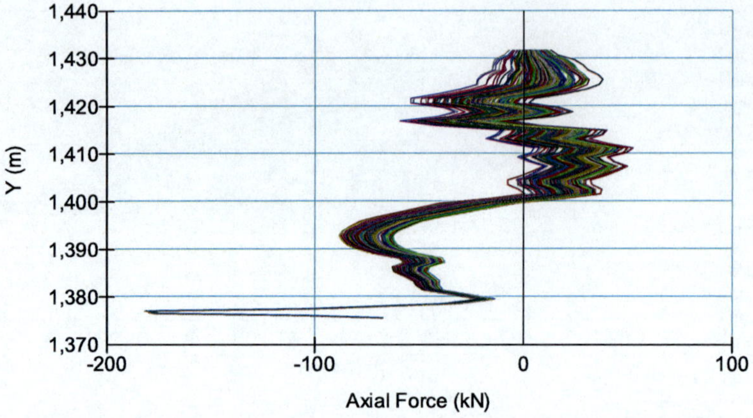

Fig. 34 Axial forces in the asphaltic facing at action of Friuli earthquake 1976, by application of equivalent linear model (ELM), N = −181.3, 52.6 kN

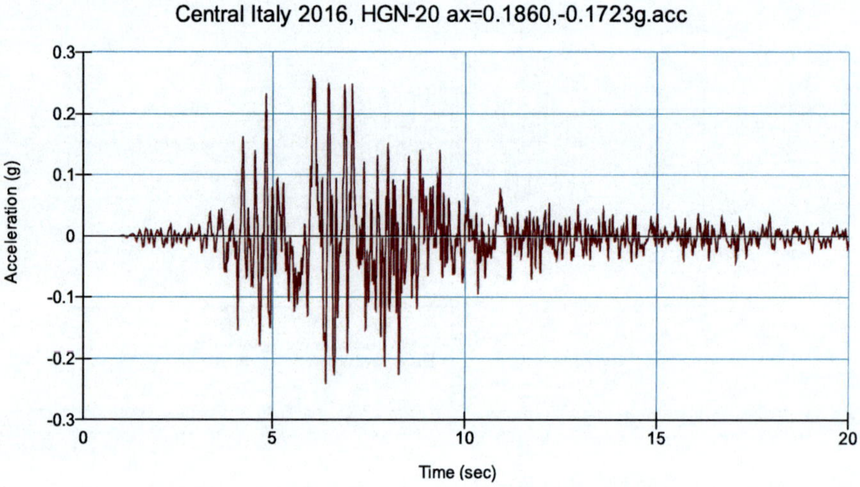

Fig. 35 Time history of horizontal component of the excitation in the rock foundation for Central Italy earthquake 2016, scaled on 0.26 g

rise of the period of the basic tone [4]. For initial stress state at full reservoir (F), below the hydrostatic pressure of the asphaltic facing the normal stresses are increased in the upstream part of the dam. As result of the increased stiffness of the rock material, for such state are obtained lower values for the basic tone (T1), apropos T1LF = 0.31 s and T1HF = 0.37 s. The values for the basic tone (T1), determined in the analysis, matches with the measured value for dams exposed on strong earthquakes

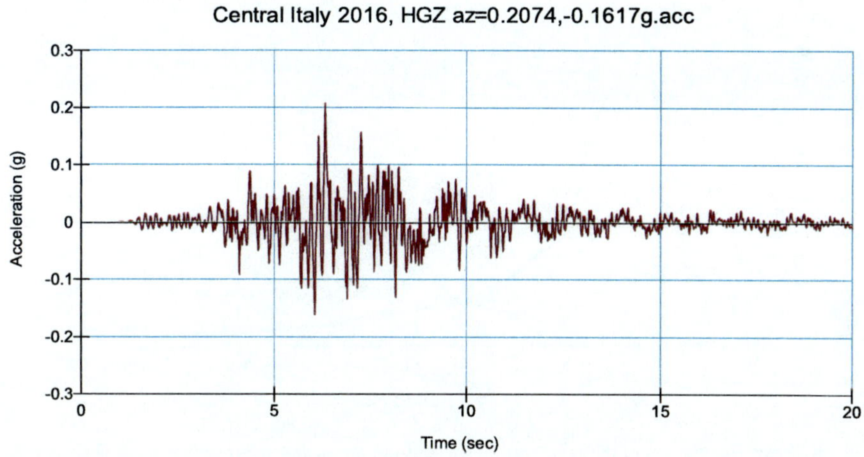

Fig. 36 Time history of vertical component of the excitation in the rock foundation for Central Italy earthquake 2016, scaled on 0.26 g

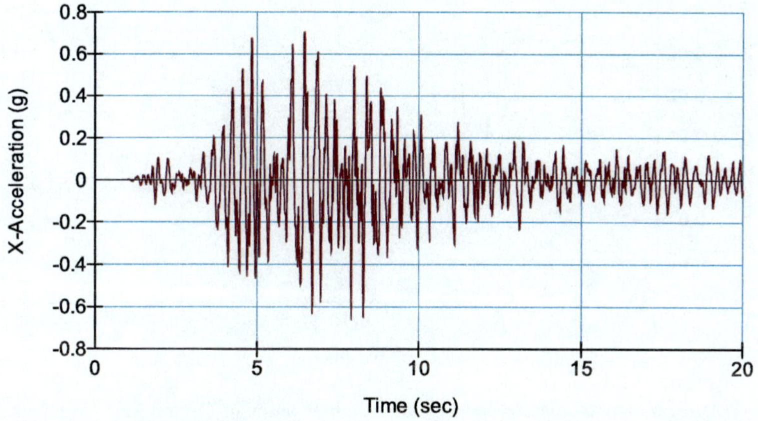

Fig. 37 Time history of the horizontal component of response at dam crest at action of Central Italy earthquake 2016, PCE = 0.698 g

in Japan [5, 6], that is best verification for the properly adopted dynamic material parameters for nonlinear dynamic analysis.

The values for Dynamic Amplification Factor DAF = PCA/PGA, where PGA is Peak Ground Acceleration (in horizontal direction), and PCA is Peak Crest Acceleration (in horizontal direction) are: 0.693/0.26 = 2.67 in case of Friuli earthquake 1976 and 0.698/0.26 = 2.68 in case of Central Italy earthquake 2016. The response at dam crest matches the registered data for the degree of dynamic amplification for

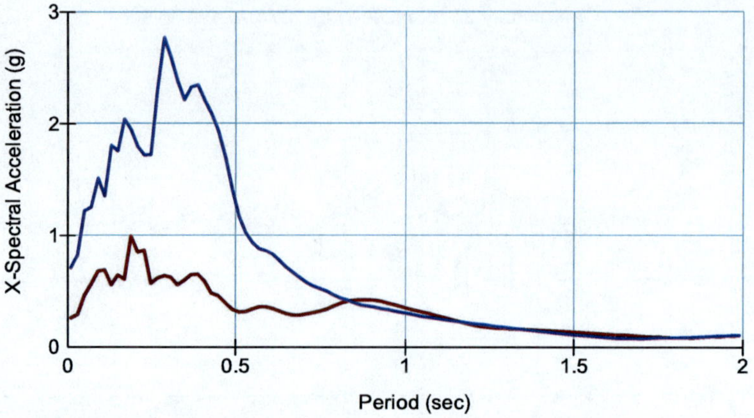

Fig. 38 Response spectra of the accelerations in the foundation and at dam crest at action of Central Italy earthquake 2016

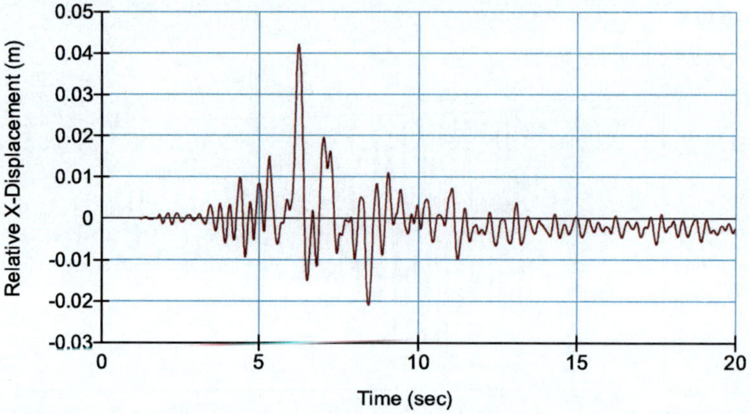

Fig. 39 Relative horizontal displacements at dam crest at action of Central Italy earthquake 2016

such structures under action of strong earthquakes, [7], apropos is the key indicator for the correctness of the dynamic analysis.

The permanent settlements at dam crest, caused by dynamic inertial forces for the duration of the earthquake, determined by the method of Dynamic Deformation Analysis, are $Y = -0.334$ m in case of Friuli earthquake 1976, and $Y = -0.399$ m in case of Central Italy earthquake 2016. Independently that in the analysis are not taken into account the settlements from additional compaction and decreased stiffness at materials exposed on cyclic action, the total settlement can not overcome the height

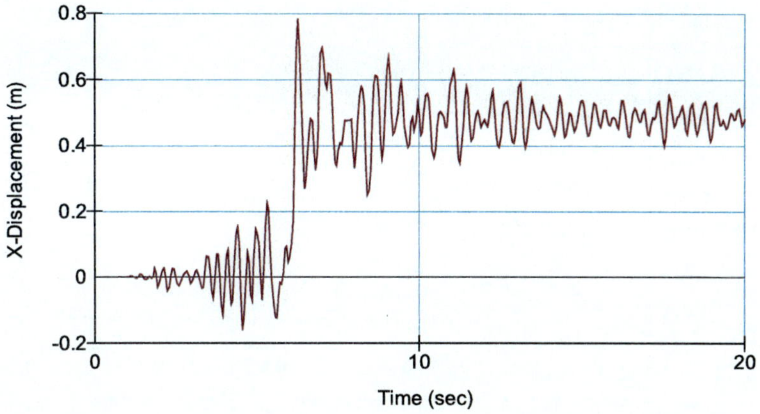

Fig. 40 Permanent horizontal displacements at dam crest at action of Central Italy earthquake 2016, X = 0.480 m

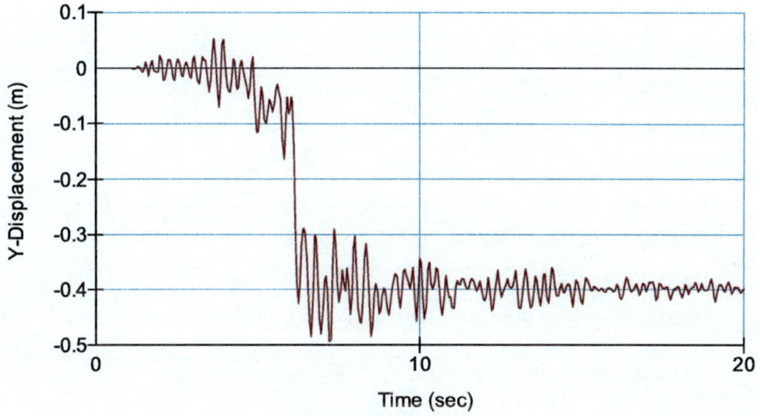

Fig. 41 Permanent vertical displacements at dam crest at action of Central Italy earthquake 2016, Y = −0.399 m

from dam crest (1431.75 m asl) till normal water elevation in the reservoir (1424.5 m asl), apropos height of 7.25 m.

The increase in permeability only occurred as fissures opened for shear deformations close to the failure level (strength) for the asphaltic facing. No significant increase was detectable until about 80% of the strength was mobilized [8]. Under moderate earthquake loading with PGA = 0.26 g, using the Friuli earthquake 1976 and Central Italy earthquake 2016, the induced deformations in dam body are small (XY_{max} ranges from 58.0 cm to 67.4 cm in the downstream slope of the dam), and

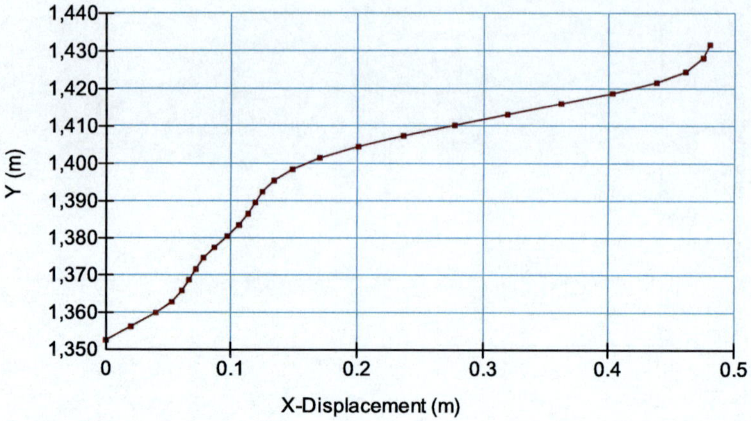

Fig. 42 Permanent horizontal displacements at dam crest axis after action of Central Italy earthquake 2016, at crest X = 0.480 m

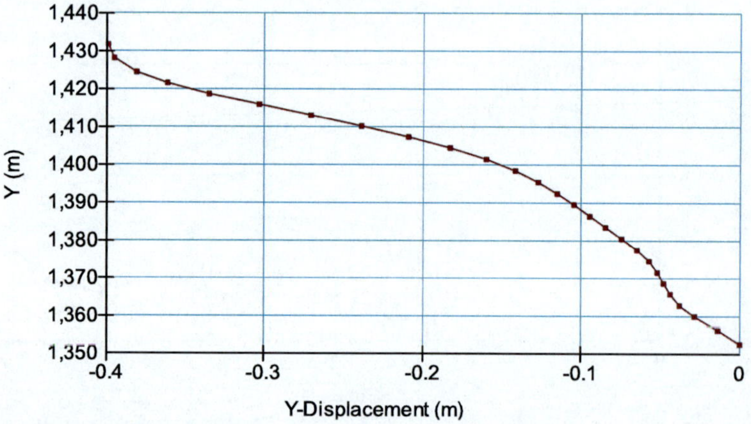

Fig. 43 Permanent vertical displacements at dam crest axis after action of Central Italy earthquake 2016, at crest Y = −0.399 m

the maximum shear strains in the asphalt facing are less than 1.5% and 1.8%, respectively. In these ranges of deformations, according to the results of permeability tests, the asphalt facing remains watertight. Therefore, it is expected that no damage will occur at Menta dam during the seismic excitation. At the same time, by application of equivalent linear analysis the asphaltic facing with thickness of 0.32 m is approximated by structural element and axial forces during the seismic excitation are calculated. By the dynamic response for both earthquakes is obtained maximal tension force of 181.3 kN (182.6 kN) and compression force of 52.6 kN (67.3 kN),

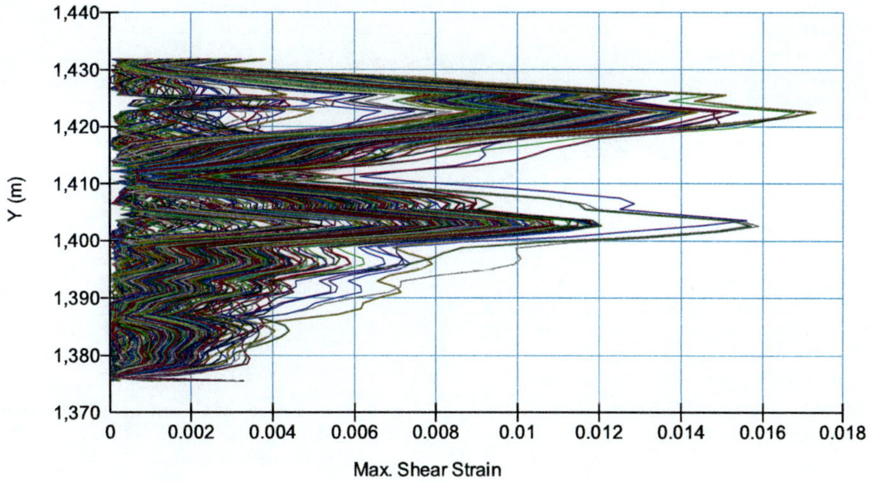

Fig. 44 Maximal shear strains in upstream slope of the dam at asphaltic facing during Central Italy earthquake 2016

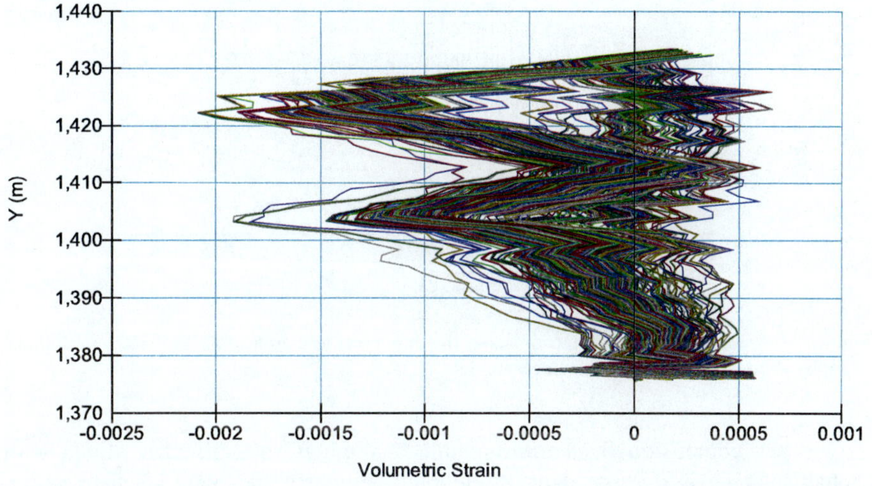

Fig. 45 Volumetric strains in upstream slope of the dam at asphaltic facing during Central Italy earthquake 2016

thus obtaining stresses much lower than the specific strength of the asphalt, such as compression strength of fc = 1.46 GPa and tension strength of ft = 0.95 GPa.

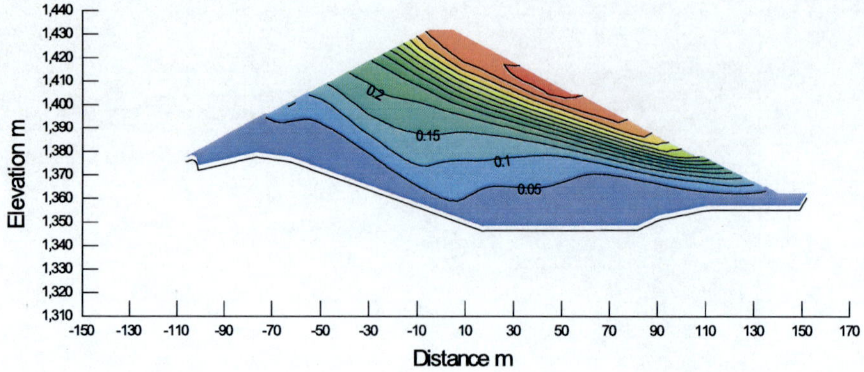

Fig. 46 Permanent displacements in XY direction in the dam after action of Central Italy earthquake 2016, $XY_{max} = 0.674$ m

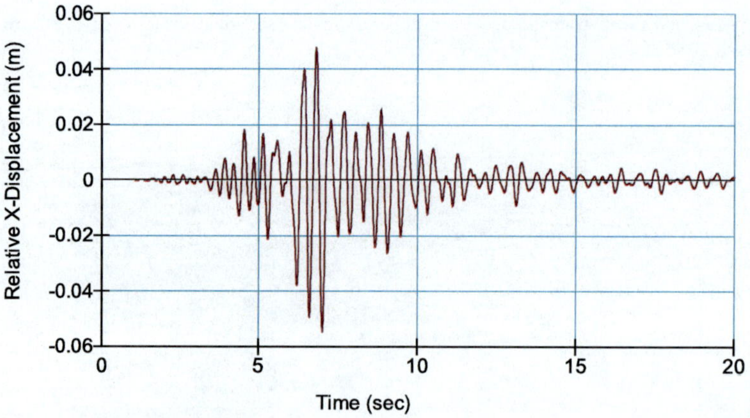

Fig. 47 Relative horizontal displacements at the dam crest at action of Central Italy earthquake 2016, by application of equivalent linear model (ELM)

The key conclusion from the dynamic analysis if rock-fill dam Menta with asphaltic facing is that the dam, by adopted geometry and material distribution, possesses satisfactory seismic resistance. There is no case of disruption of the water impermeability of the watertight element (asphaltic facing with thickness of 0.32 m), nor there is danger of rapid and uncontrolled emptying of the artificial lake, because by the calculated settlements for design earthquakes with PGA of 0.26 g the protection height of 7.25 m is not exceeded.

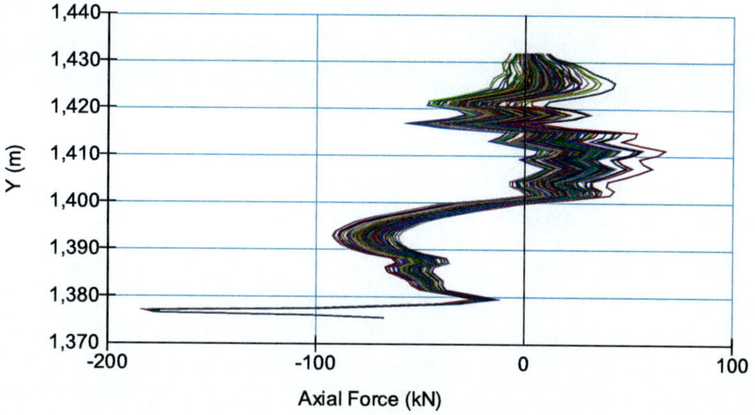

Fig. 48 Axial forces in the asphaltic facing at action of Central Italy earthquake 2016, by application of equivalent linear model (ELM), N = −182.6, 67.3 kN

References

1. Russo G, Vecchietti A, Cecconi M, Pane V, De Marco S, Florino A (2019) 15th international benchmark workshop on numerical analysis of dams, Theme B
2. Geo-Slope: SIGMA/W v8 (2017) User's Guide for finite element stress/deformation analysis. GEO-SLOPE International Ltd., Calgary, Alberta, Canada
3. Geo-Slope: QUAKE/W v8 (2017) Dynamic Modeling. GEO-SLOPE International Ltd., Calgary, Alberta, Canada
4. Park DS (2018) Fundamental period of embankment dams based on strong motion records, topic earthquakes—embankments. In: USSD 38th annual meeting and conference, a balancing act: dams, levees and ecosystems, April 3- May 4, 2018, Miami, Florida, USA, CD Proceedings
5. Matsumoto N, Ohmachi T, Yasuda N, Yamaguchi Y, Sasaki T, Kurahashi H (2005) Analysis of strong motions recorded at dams during earthquakes. In: 73rd annual meeting of ICOLD, Tehran, IRAN, Paper No.: 094-W
6. Fry JJ, Matsumoto N (2018) Validation of dynamic analyses of dams and their equipment, CRC Press/Balkema. Tailor & Francis Group, London, New York
7. ICOLD, Bulletin 113. Seismic observation of dams—Guidelines and case studies (1999)
8. Hoeg K (1993) Asphaltic Concrete cores for embankment dams—experience and practice. Norwegian Geotechnical Institute of Technology, Oslo, Norway

Seismic Analyses of Menta Embankment Dam: An Elasto-Plastic Model with a Stress-Strain Dependent Stiffness and a Composite Yielding Surface

F. Raggi

Abstract In this paper, a seismic analysis of the Menta dam is briefly described. The Menta dam is a bituminous-faced rockfill dam located in Southern Italy, lying in the heart of Aspromonte Massif at an elevation of about 1400 m a.s.l. The embankment is about 90 m high at its deepest point, and the reservoir impounds $1.8 \times 107\ \text{m}^3$ of water. This dam was designed in the late seventies and built between 1987 and 2000. In the seismic analysis, the dam material has been modelled with an elasto-plastic constitutive law which is characterized by a Young's modulus dependent on the stress level and on the distortional strain and by a curvilinear yielding surface. Both the calibration of the model and the dynamic analysis has been carried out by means of the FDM code FLAC (Fast Lagrangian Analysis of Continua [1]) version 8.1 which adopts an explicit time-integration scheme.

Keywords Rockfill · Dam · Hardening · Hysteretic · Damping

1 Rockfill Constitutive Law

The adopted constitutive model is the FLAC built-in Double-Yield model characterized by two yielding surfaces: one Mohr-Coulomb yield surface and one volumetric yield surface (cap surface).

While in the built-in constitutive law the cohesion and the friction angle are fixed, in the present study both the Mohr-Coulomb strength parameters (friction angle and cohesion) are dependent on the mean stress and on a hardening parameter. In fact, the shear yielding surface is divided in three different parts, depending on the stress level: an initial parabolic stretch characterized by an initial peak friction angle $\varphi_{0,\,pk} = 58.5°$ and zero initial peak cohesion, a central parabolic stretch with an increasing tangent peak cohesion and a decreasing tangent peak friction angle, a final straight stretch, tangent to the second parabola, in which the peak friction angle and the peak cohesion are constant. Here below the equations for each stretch are

F. Raggi (✉)
ELC-Electroconsult, Milan, Italy
e-mail: francesco.raggi@elc-electroconsult.com

G. Bolzon et al. (eds.), *Numerical Analysis of Dams*, Lecture Notes in Civil Engineering 91, https://doi.org/10.1007/978-3-030-51085-5_34

reported (compressive stress is negative):

$$\tau_n = \begin{cases} A_1\sigma_n^2 + B_1\sigma_n, & \mid \sigma_m \geq \sigma_a \\ A_2\sigma_n^2 + B_2\sigma_n + C_2, & \mid \sigma_a > \sigma_m \geq \sigma_b \\ \tan\varphi_{f,pk}\sigma_n + c_{f,pk}, & \mid \sigma_b > \sigma_m \end{cases} \tag{1}$$

where A_i, B_i and C_i ($i = 1, 2$, $C_1 = 0$) are the parabola parameters and $\varphi_{f,pk}$ and $c_{f,pk}$ are the final peak friction angle and peak cohesion defining the tangent line to the second parabola at $\sigma_b = \sigma_m$. The stresses σ_a and σ_b define the limit of the validity of each equation and σ_m is the mean stress defined as:

$$\sigma_m = \frac{\sigma_1 + \sigma_3}{2} \tag{2}$$

where σ_1 and σ_3 are the maximum and the minimum principal stresses.

In the parabolic parts, the peak friction angle and the peak cohesion are obtained, in the Mohr plane, taking the derivative of τ_n with respect to σ_n and writing the equation of the tangent line at the point (τ_{nt}, σ_{nt}):

$$\tan\varphi_{t,pk} = (2A_i\sigma_{nt} + B_i) \tag{3}$$

$$c_{t,pk} = C_i - A_i\sigma_{nt}^2 \tag{4}$$

where $\varphi_{t,pk}$ and $c_{t,pk}$ are respectively the tangent peak friction angle and peak cohesion. Noting that, at the point (τ_{nt}, σ_{nt}), the tangent line to a parabola and the tangent line to the Mohr circle are the same line, σ_{nt} can be obtained solving the following cubic equation:

$$2A_i^2\sigma_{nt}^3 + 3A_iB_i\sigma_{nt}^2 + \left(1 + 2A_iC_i + B_i^2\right)\sigma_{nt} + B_iC_i - \sigma_m = 0 \tag{5}$$

where σ_m is defined by Eq. (2).

In the parabolic stretches, in terms of (q, p), the yielding surface can be written as:

$$q_f = \frac{-2\tau_{nt}}{\cos\varphi_{t,pk}} \tag{6}$$

$$p_f = \sigma_{nt} - \frac{q}{6} + \frac{q}{2}\sin\varphi_{t,pk} \tag{7}$$

where $q = \sigma_1 - \sigma_3$ and $p = (\sigma_1 + 2\sigma_3)/3$.

The mobilized friction angle φ_m and cohesion c_m are calculated with the following equations:

$$\tan\varphi_m = \tan\varphi_{t,pk}\left(1 - k_2e^{-k_1\varepsilon_{ps}}\right) \tag{8}$$

$$c_m = c_{t,pk}\left(1 - k_2 e^{-k_1 \varepsilon_{ps}}\right) \tag{9}$$

where k_1, k_2 are model parameters and ε_{ps} is the deviatoric hardening parameters which, in triaxial conditions, is the plastic part of the following deviatoric strain:

$$\varepsilon_s = \frac{|\varepsilon_1 - \varepsilon_3|}{\sqrt{3}} \tag{10}$$

where ε_1 and ε_3 are respectively the axial and the radial principal strains.

On the volumetric yielding surface, the behavior under a spherical stress p is controlled by the following equations:

$$p = p_c \tag{11}$$

$$p_c = p_{c0} + k_3 \varepsilon_{pv} \tag{12}$$

where p_c is the cap pressure, p_{c0} is the initial cap pressure, k_3 is a model parameter and ε_{pv} is the volumetric hardening parameter which, in triaxial conditions, is the plastic part of the following volumetric strain:

$$\varepsilon_v = |\varepsilon_1 + 2\varepsilon_3| \tag{13}$$

where ε_1 and ε_3 are respectively the principal axial and radial strain.

Other features of the adopted model are the Young's modulus E dependent on a measure of distortional strain and the small-strain Young's modulus E_0 dependent on the stress level. The equation characterizing this non-linear behavior are (elongation is positive):

$$E = \begin{cases} E_0 \frac{1}{1+\frac{\varepsilon_s}{\varepsilon_{ref}}}, \ primaryloading \\ E_0 \frac{1}{1+\frac{|\varepsilon_s - \varepsilon_{srev}|}{2\varepsilon_{ref}}}, \ unloading - reloading \end{cases} \tag{14}$$

$$E_0 = \begin{cases} E_{0i} + \beta \left|\frac{\sigma_2}{\sigma_{ref}}\right|, \sigma_2 \geq \sigma_c \\ E_{0ref} + \alpha \ln\left(\left|\frac{\sigma_2}{\sigma_{ref}}\right|\right), \sigma_2 < \sigma_c \end{cases} \tag{15}$$

where ε_{ref}, E_{0i}, E_{0ref}, α, β, σ_{ref}, σ_c are model parameters and σ_2 is the intermediate principal stress. The strain ε_{srev} is the deviatoric strain at the reversal point. A cut-off equal to 5% on the ratio E/E_0 for a deviatoric strain larger than 0.01 has been adopted

2 Calibration of the Rockfill Model

The values of the model parameters have been assessed by means of a strain-driven 1-zone FDM model, trying to reproduce the rockfill behavior as obtained by the laboratory triaxial tests. In Table 1, all these values are documented (the Poisson's ratio is $\upsilon = 0.25$).

In the following figures (Figs. 1, 2 and 3), the results of the calibration process are reported. The best-fitting of the laboratory triaxial data, for all considered confining pressures, has been obtained considering the values $k_1 = 380$, $k_2 = 0.95$ and $k_3 = 2.5 \times 10^4$ kPa. The stress path simulated with the FDM model is reported in Fig. 3b (right): the deviatoric stress has been applied after a $K_0 = 0.4$ has been reached in the consolidation phase.

The model is capable to catch the hysteretic behavior in an unloading-reloading stress path (Fig. 1a) and the relevant damping curve is reported in Fig. 2a. This curve is strictly valid only in elastic conditions so, it doesn't take into account the energy dissipation due to plastic flow: as the plastic strain increases, the dissipated energy in an elastic hysteresis becomes neglectable. Since $k_2 = 0.95$, the initial value for

Table 1 Values of model parameters

Symbol	Value	Unit
σ_a	−300	kPa
σ_b	−3300	kPa
A_1	-2.637×10^{-3}	1/kPa
B_1	−1.630	–
C_1	0	kPa
A_2	-1.011×10^{-4}	1/kPa
B_2	-9.070×10^{-1}	–
C_2	$5.157 \times 10^{+1}$	kPa
$\varphi_{f,pk}$	20.3	°
$c_{f,pk}$	766	kPa
k_1	380	–
k_2	0.95	–
σ_c	−100	kPa
E_{0i}	$4.875 \times 10^{+4}$	kPa
E_{0ref}	$7.845 \times 10^{+5}$	kPa
β	$7.340 \times 10^{+5}$	kPa
α	$9.365 \times 10^{+5}$	kPa
σ_{ref}	−100	kPa
ε_{ref}	2.474×10^{-4}	–
p_{c0}	0	kPa
k_3	$2.500 \times 10^{+4}$	kPa

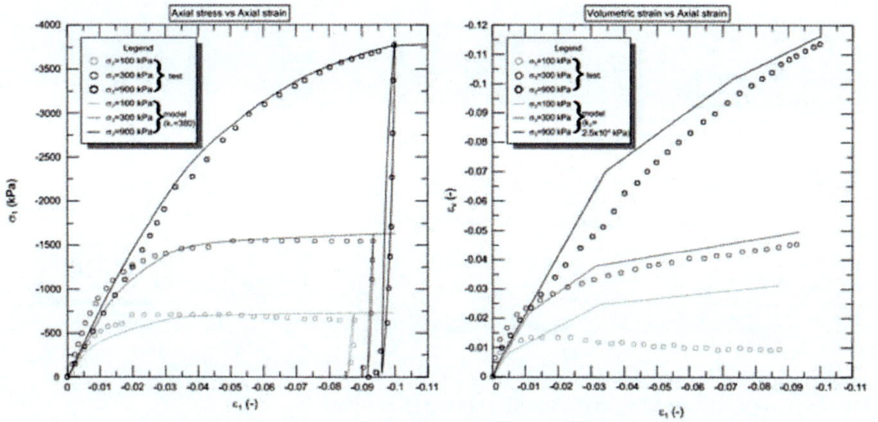

Fig. 1 σ_1 versus ε_1 (1.a, left) and ε_v versus ε_1 (1.b, right)

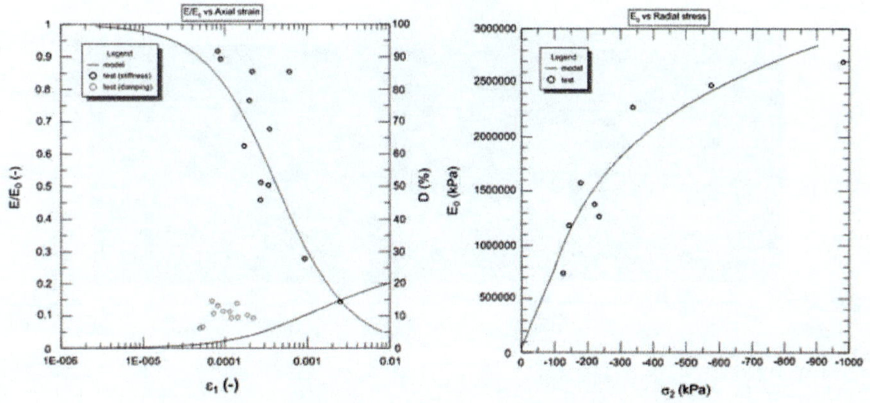

Fig. 2 E/E_0 and damping ratio versus ε_1 (2.a, left) and E_0 versus σ_2 (2.b, right)

c_m and φ_m are very low and the plastic strain develops almost immediately as the model undergoes to stress therefore, the damping curve reported in Fig. 2a is mostly theoretical.

3 Dynamic Analyses

The seismic analyses have been carried out by the FDM model depicted in Fig. 4. The grid is characterized by 242 × 156 zones, the largest one is about $\Delta l_{max} = 6$ m. Along the bottom boundary and along the side boundaries the vertical and the horizontal degree of freedom has been respectively fixed.

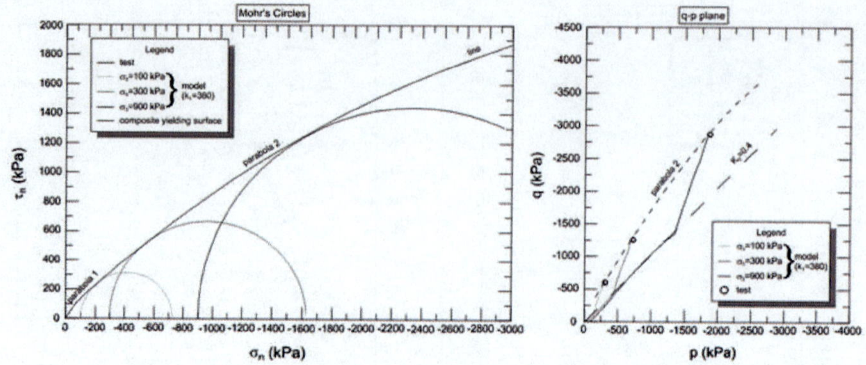

Fig. 3 (τ_{nt}, σ_{nt}) plane (3.a, left) and (q, p) plane (3.b, right)

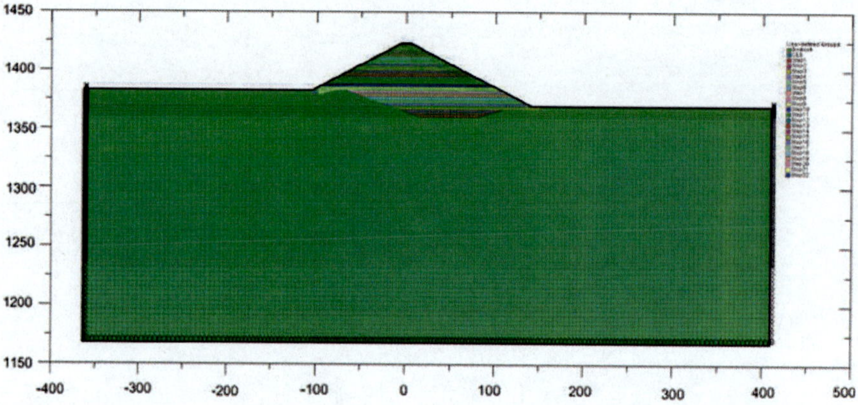

Fig. 4 FDM model of the Menta dam

Before the seismic analysis, the construction of the dam has been modelled by 22 construction steps. In Fig. 5 and in Fig. 6 the vertical displacement at the end of construction and the horizontal displacement after the impounding are respectively depicted: the maximum settlement is about 50 cm and the maximum horizontal displacement is about 6 cm. In Fig. 7 and in Fig. 8, the mobilized friction angle and cohesion at the impounding completion are shown.

Since the seismic inputs (Central Italy 2016 and Friuli 1976) have been recorded at the outcrop, a deconvolution process has been carried out with the code EERA (Equivalent-linear Earthquake site Response Analyses [2]). However, as the bedrock has been modelled with a linear elastic constitutive law, with a damping ratio $D = 0$, the effect of the deconvolution is just a translation in time of the ground motion recorded at the outcrop. So, the shear stress σ_s applied at the compliant base, at the bottom of the model, has been calculated as:

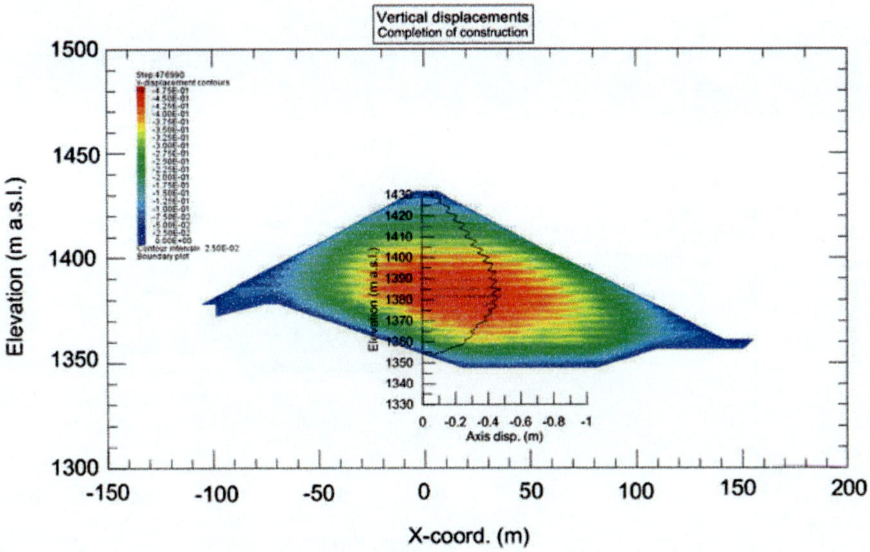

Fig. 5 Vertical displacements at the end of construction

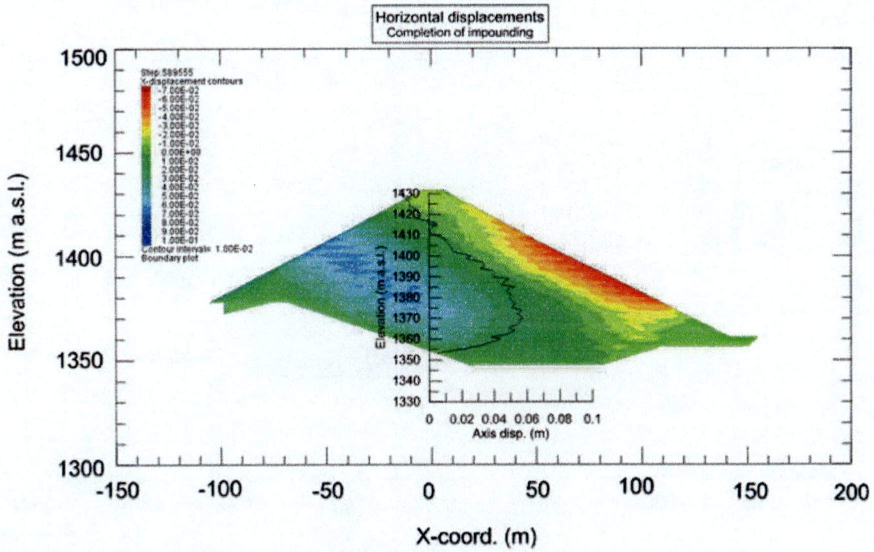

Fig. 6 Horizontal displacements at the impounding completion

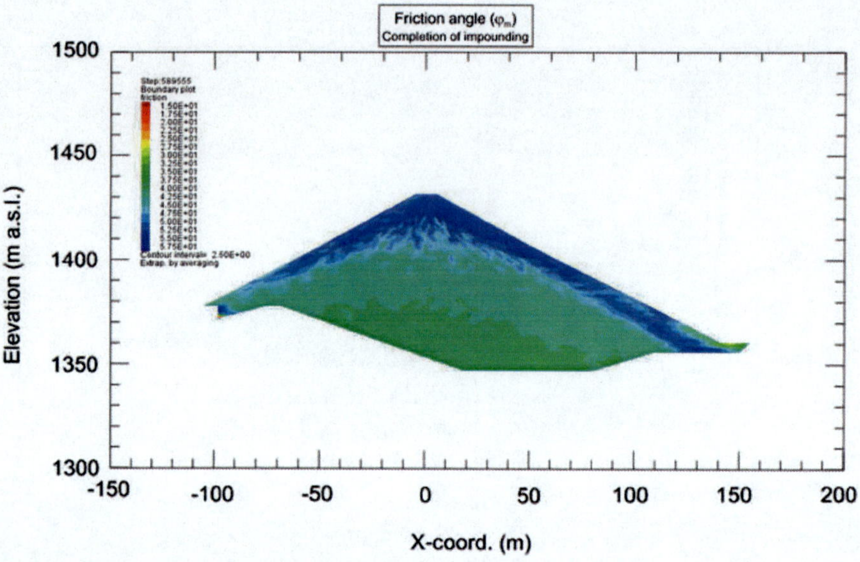

Fig. 7 Friction at the impounding completion

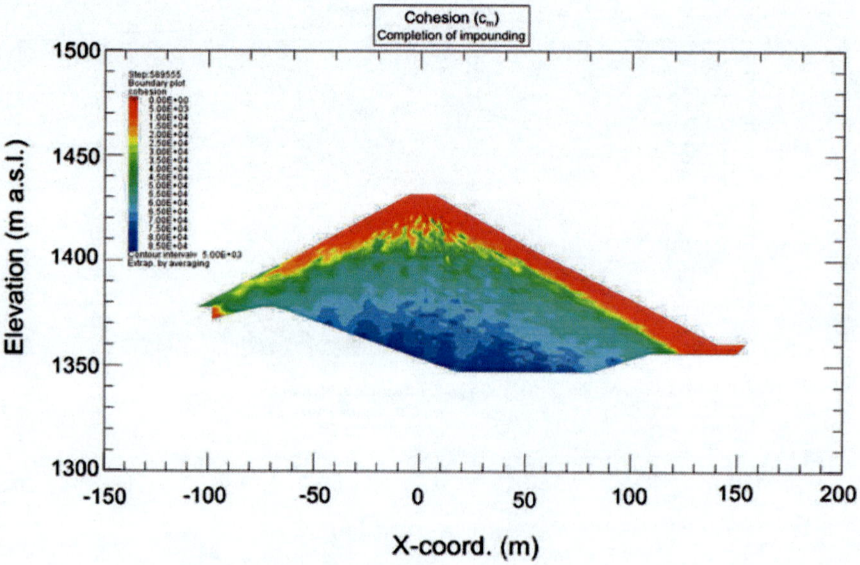

Fig. 8 Cohesion at the impounding completion

$$\sigma_s = \sqrt{\rho G} V_{s,out} \tag{16}$$

where $V_{s,out}$ is the velocity at the outcrop (see Fig. 9), ρ and G are respectively the density and the shear modulus of the bedrock. In this case, $\sqrt{\rho G} = 2.7 \times 10^{+6} \frac{kg}{sm^2}$.

The FFT (Fast Fourier Transform) analysis of the two outcrop acceleration inputs is reported in Fig. 10: considering that the maximum frequency which can be modelled accurately depend on the following ratio:

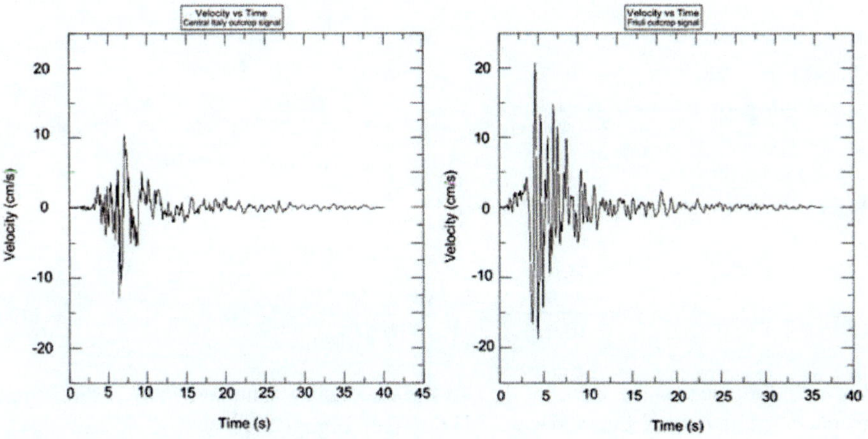

Fig. 9 Velocity at the outcrop ($V_{s,out}$)

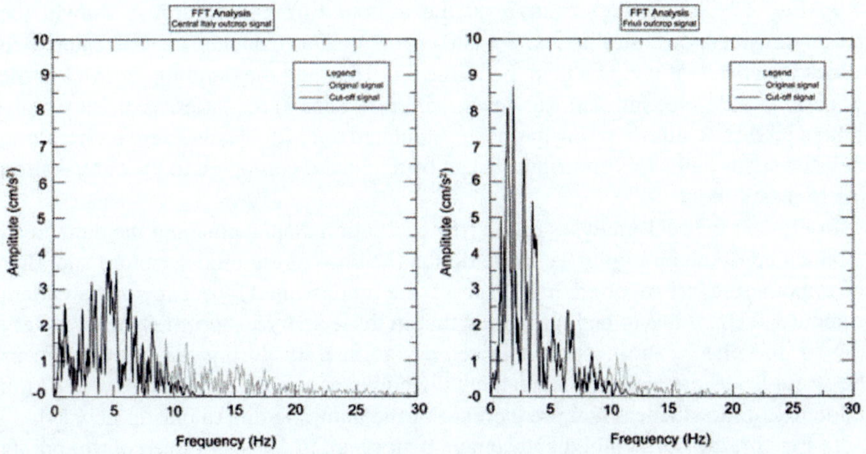

Fig. 10 FFT (Fast Fourier Transform) of the outcrop signal

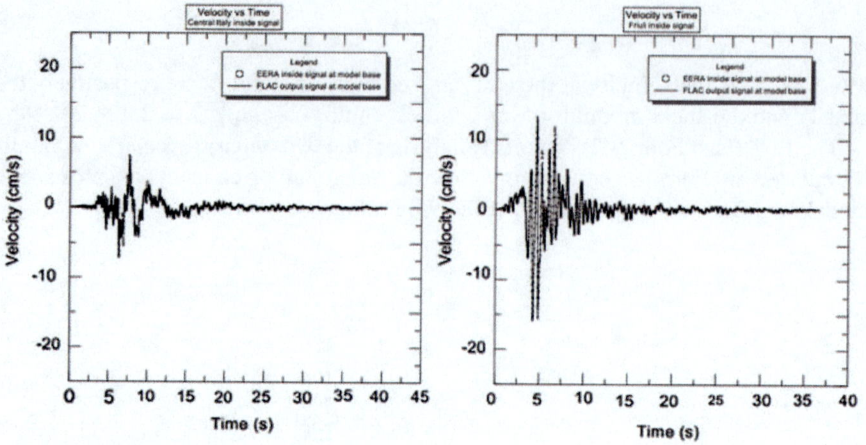

Fig. 11 Velocity output at the base of the model

$$f = \frac{C_s}{10\Delta l_{max}} = \sqrt{\frac{G}{\rho}} \frac{1}{10\Delta l_{max}} \approx 15\,\text{Hz} \tag{17}$$

between the shear wave velocity C_s and the maximum zone size Δl_{max}, the signal frequencies have been cut-off over 15 Hz by (see Fig. 10).

In Fig. 11, the velocity recorded at the base of the FLAC model and the velocity obtained by EERA for the inner signal at the bedrock base are compared: these signals take into account both the upward and the downward propagating waves.

In Fig. 12, the deviatoric strain at the impounding completion is shown: the maximum deviatoric strain is $\varepsilon_s = 3.0 \times 10^{-2}$ kPa. Considering the calculated values for ε_s the ratio E/E_0 is in the range $5 \div 15\%$ and the damping ratio D in the range $15 \div 25\%$ (see Fig. 2a). However, with reference to the latter, as already highlighted in Sect. 2, this damping has been calculated only for elastic stress paths: along inelastic stress paths this contribution can be neglected compared to the contribution due to plastic flow.

In Fig. 13, the FFT analyses of the free oscillation displacement at the dam crest, obtained applying an impulsive excitation at the base of the model, before and after the impounding, are reported: in both cases, the fundamental frequency of the system is about 1.0 Hz. It has to be highlighted that in these analyses the stress level is very low so, the plastic strain doesn't increase significantly during the calculation: as the stress level increases the fundamental frequency could decrease because of the reduction of the stiffness and the increase of the damping due to plastic flow [3].

In Fig. 14, the horizontal displacement history, up to 16 s, at three control points is depicted: the point A is at the foundation level (about 1353 m a.s.l.), the point B is at about 1390 m a.s.l. and the point C is at the dam crest (about 1430m a.s.l.). The maximum horizontal displacement is about 20 cm recorded during Friuli earthquake between 8 and 10 s.

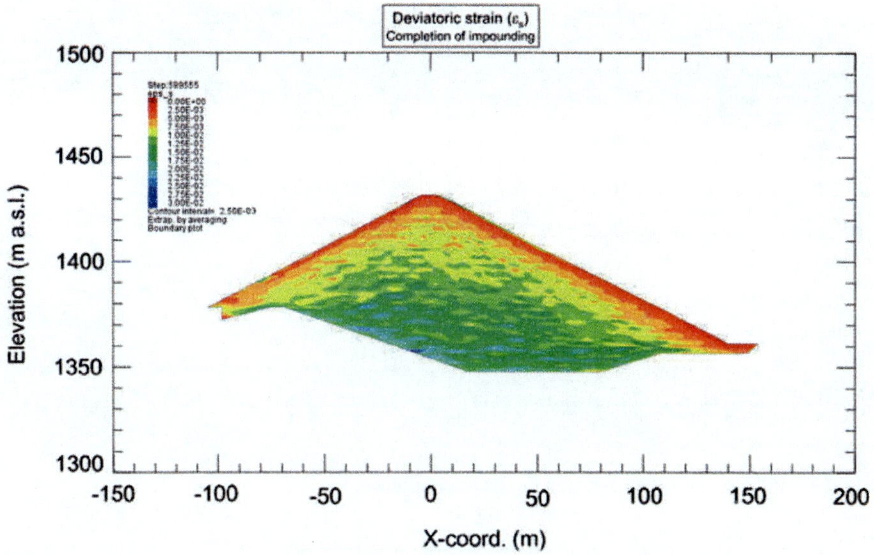

Fig. 12 Deviatoric strain at the impounding completion

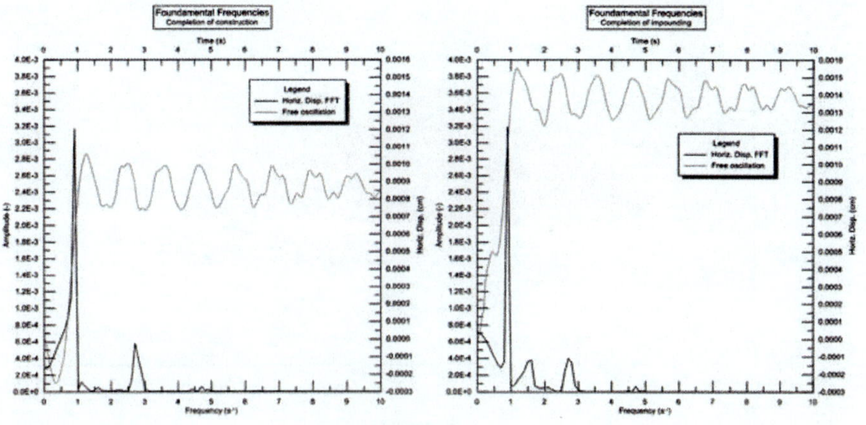

Fig. 13 Fundamental frequencies of the system

In Fig. 15 and in Fig. 16, both for the Central Italy seism and for the Friuli seism, the Safety Factor and the maximum shear strain increment (with respect to the post-seismic state), calculated with a $c - \varphi$ reduction procedure, are shown. These figures highlight the development of a global instable zone with a minimum Safety factor $SF = 2.21$.

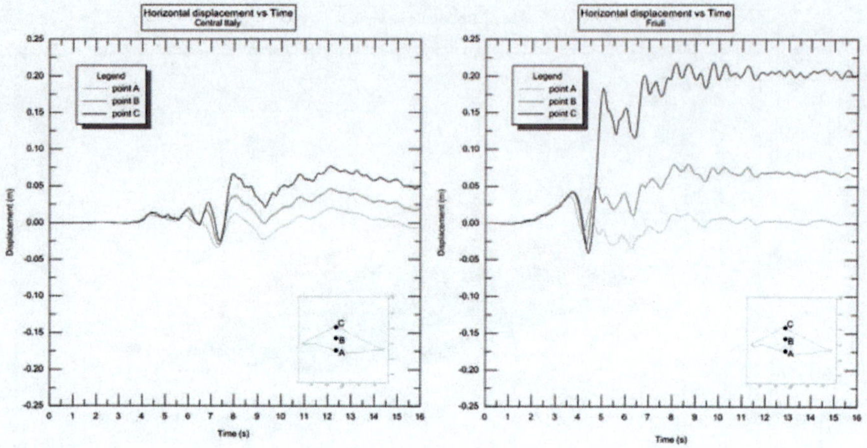

Fig. 14 Horizontal displacements in three control points

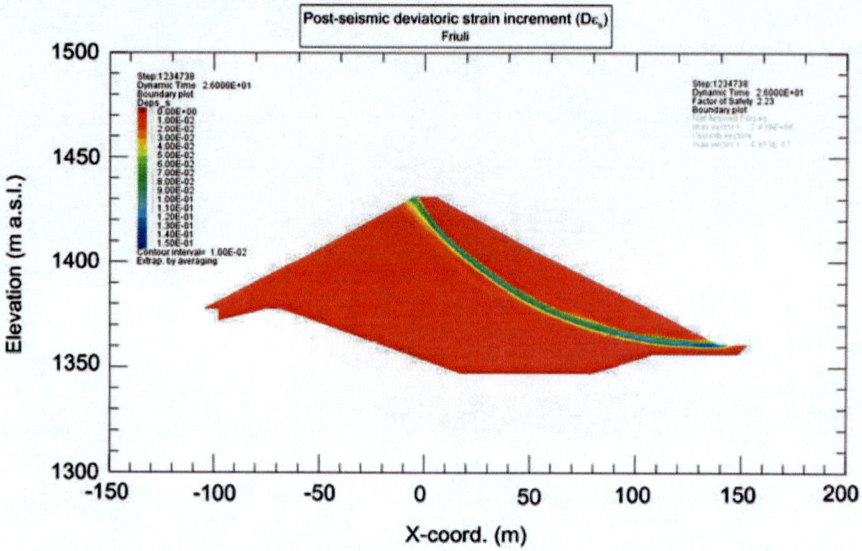

Fig. 15 Safety Factor assessment (Central Italy)

4 Conclusions

Starting from the laboratory tests, an elasto-plastic model, characterized by a composite yielding surface and by a stress-strain dependent Young's modulus, has been first calibrated and then adopted for the seismic analysis of the Menta dam. After the construction and the impounding, the maximum settlement is about 50 cm

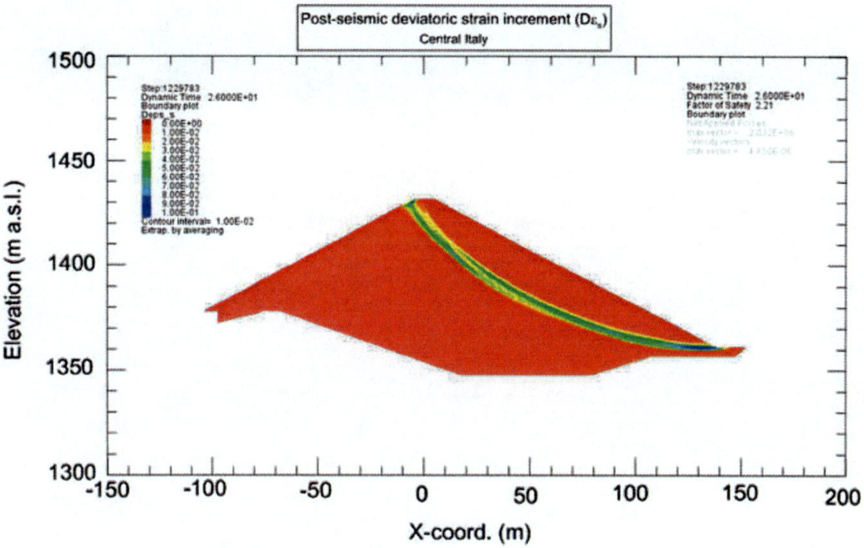

Fig. 16 Safety Factor assessment (Friuli)

(about 0.6% of the height of the dam) and the decay of the Young's modulus, ranging in the interval $5 \div 15\%$, determines a fundamental frequency around 1.0 Hz.

The seismic behavior of the dam is characterized by a maximum horizontal displacement equal to about 20cm, without the activation of instable zones.

With the stability analysis in post-seismic conditions, the calculated minimum Safety Factor is $SF = 2.21$ relevant to a global instable zone so, the dam could be considered safe under the hypotheses taken into account.

References

1. Itasca (2019) User manual for FLAC, version 8.1. Itasca Consulting Group Inc., Minnesota, EEUU
2. Bardet JP, Ichii K, Lin CH (2000) EERA. A computer program for Equivalent-linear Earthquake site Response Analyses of layered soil deposits, University of Southern California, Department of Civil Engineering, User's manual
3. Boulanger R, Bray J, Merry S, Mejia L (2011) Three-dimensional dynamic response analyses of Cogswell Dam. Can Geotech J 32:452–464

Nonlinear FEM Analysis of the Seismic Behavior of the Menta Bituminous-Face Rockfill Dam

A. D. Tzenkov and M. V. Schwager

Abstract Large embankment dams are characterized by highly nonlinear material behavior. A main challenge to the numerical investigations of such type of dams is the proper selection of constitutive material models and of the required laboratory and field investigations and tests. It is equally important to model realistically the compaction of the fill layers to account for the pre-consolidation effects that fill material undergoes. This paper presents a FEM nonlinear static and seismic analysis of Menta Dam which is located in Southern Italy and is under operation since 2000. The analysis is aimed at considering in a realistic and precise manner the above-described effects. The FEM model is based on the input data provided in the formulation of Theme B of the 15th ICOLD Benchmark Workshop. The Hardening Soil model combined with the Small Strain Stiffness model is used to simulate the behavior of the rockfill for the considered load combinations. The hydrodynamic pressures caused by the reservoir during earthquake is approximated by means of fluid-structure interaction with incompressible fluid. The foundation is assumed massless. The maximum vertical displacement at the end of the construction is approximately eighty centimeters and is in the central zone of the dam. The first impounding causes re-orientation of the principal stresses in the dam body zone underlying the bituminous face, but it does not entail significant additional displacements. The permanent displacements in case of earthquake are moderate, as the permanent settlements are well within the freeboard of the dam. Based on the results of the analyses performed, recommendations are formulated regarding the constitutive models and the type of analysis to be considered for this dam type.

Keywords Bituminous face · Rockfill dam · Nonlinear seismic analysis

A. D. Tzenkov (✉)
Stucky Ltd, Renens, Switzerland
e-mail: anton.tzenkov@stucky.ch

M. V. Schwager
Swiss Federal Office of Energy SFOE, Bern, Switzerland

G. Bolzon et al. (eds.), *Numerical Analysis of Dams*, Lecture Notes in Civil Engineering 91, https://doi.org/10.1007/978-3-030-51085-5_35

1 Introduction

The current work investigates the static and seismic behavior of Menta Dam, a 90 m high bituminous-face rockfill dam built in Southern Italy from 1987 to 2000. The investigations are carried out by means of the FEM software code DIANA [1] on a two-dimensional FEM model which was built in accordance to the formulation of Theme B of the 15th ICOLD Benchmark Workshop, Milan 2019 [2].

The mechanical behavior of the rockfill material is defined by means of the Hardening Soil/Modified Mohr-Coulomb model combined with the Small Strain Stiffness model as implemented in DIANA. The associated material model parameters are calibrated on the results of the laboratory tests (see Sect. 3) in order to be used for a stress-strain static and earthquake analysis. The effect of pre-consolidation of the rockfill material due to compaction during the rockfill layers placement is taken into account by assuming a reasonable compaction effort.

According to the problem formulation, no pore-water pressure in the embankment is considered.

Linear-elastic material models are associated to the rock foundation, as well as to the bituminous face. The rock foundation is assumed massless.

The stages of construction of the dam embankment are simulated by modelling seven horizonal layers that are activated stepwise. Each layer is approximately 10 m high.

Similarly, the first impounding is modelled in two phases.

The solution of the dynamic equilibrium equations is carried out in the time-domain. The investigated earthquake loading is introduced at the end of the impounding as base excitation loads to which are associated a horizontal and a vertical acceleration time-histories.

The fluid-structure interaction during earthquake is taken into account by modelling part of the reservoir as incompressible fluid.

This paper presents the results of the dam behaviour in terms of displacements, strains and stresses for the investigated static and dynamic load combinations. Recommendations regarding the constitutive models and the type of analysis to be considered for bituminous-face dams are given in the conclusions.

2 Finite Element Model

2.1 Geometry

The analyses are performed on a two-dimensional FEM model of the dam-foundation-reservoir system, which is defined on the basis of the geometry given by the Formulator of the current Theme B [2]. A small modification of the rock foundation line is introduced in the middle of the dam-foundation interface in order to provide for a more regular FEM mesh. The downstream face of the embankment is approximated by a straight line, as the berms are not explicitly modelled.

The considered part of the rock foundation extends approximately two times the dam height towards upstream, downstream and in depth as measured, respectively, from the upstream toe, the downstream toe and the lowest point of the dam body.

The length of the reservoir considered is approximately five times its depth at its normal level.

2.2 *Boundary Conditions*

Structural system boundary conditions. Kinematic constraints in direction normal to the rock foundation vertical and bottom boundaries are considered.

Fluid system boundary conditions. The fluid is specified as incompressible by setting the sonic wave velocity c tending to infinity, assigning hydrodynamic pressure $p = 0$ at the far-field and at the free surface of the reservoir, and setting the hydrodynamic pressure gradient equal to zero in the normal direction of the reservoir bottom. Therefore, the hydrodynamic effect is represented by a consistent mass matrix that is added to the mass matrix of the structural system.

2.3 *Finite Element Mesh*

The structural system (the dam-foundation system) model is meshed with second order plane-strain elements of quadrilateral shape (8 nodes). The reservoir is modelled by means of second order quadrilateral potential flow elements and the interface between two system is represented by fluid-structure interface elements.

The finite element mesh and the mechanical boundary conditions are given in Fig. 1. A zoom on the dam's mesh is presented in Fig. 2.

3 Material Characterization

In the finite element model, three different materials are considered: the rockfill material, the bedrock and the asphalt concrete of the lining. This section gives the constitutive parameters used for the three different materials.

An emphasis is put on the constitutive behavior of the rockfill material, since it is governing the overall seismic response of the dam. Therefore, a full elasto-plastic constitutive relationship accounting as well for small strain nonlinearity is used for the rockfill material. The so-called Hardening Soil Model [3] is used (see

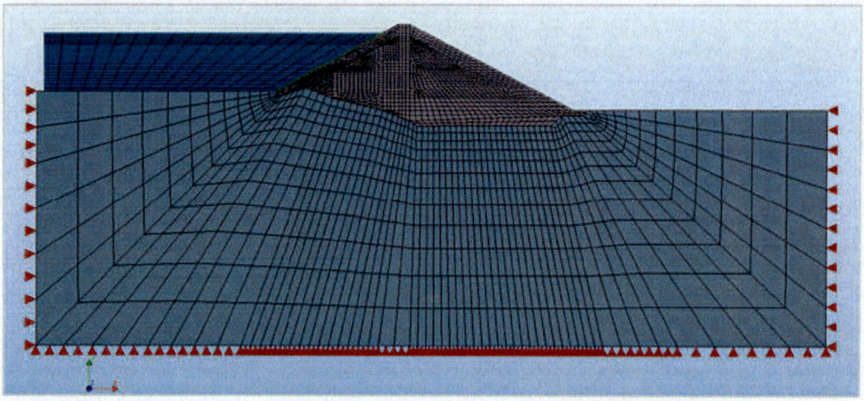

Fig. 1 Finite element mesh of the dam-foundation-reservoir system and mechanical boundary conditions

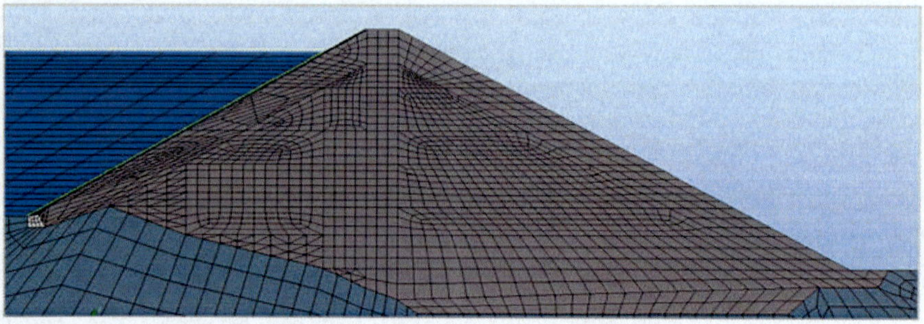

Fig. 2 Finite element mesh of the dam

Sect. 3.1) in combination with the Hardin-Drnevich relationship [4] for the small strain nonlinearity (see Sect. 3.2).

Furthermore, potential scale effects due to the different particle size of the rockfill material compared to the one of the materials tested in the laboratory are investigated (see Sect. 3.4).

3.1 Elasto-Plastic Behaviour of the Rockfill Material

The parameters of the elasto-plastic Hardening Soil Model are derived from triaxial shear tests performed on rockfill material having a modified grain size distribution, see Fig. 7. For the purpose of derivation of parameters, the test data given is separated into its consolidation ($K_0 = 0.4$) and shearing phase (i.e. triaxial compression). The constitutive parameters are obtained in several consecutive steps:

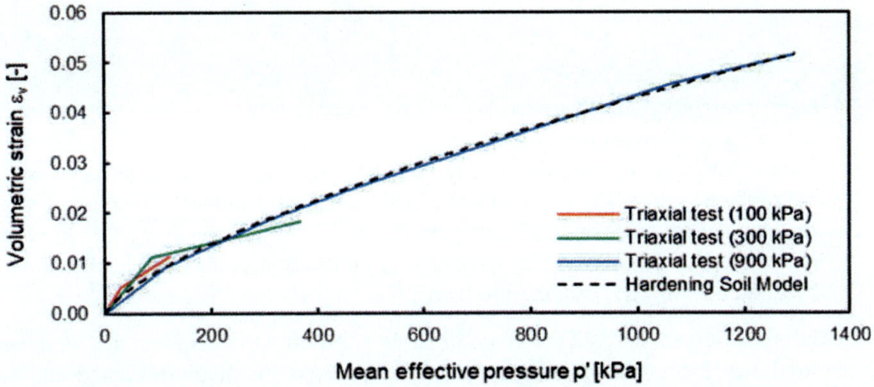

Fig. 3 Comparison of the results of triaxial tests with the Hardening Soil Model: volumetric behavior in K_0-consolidation

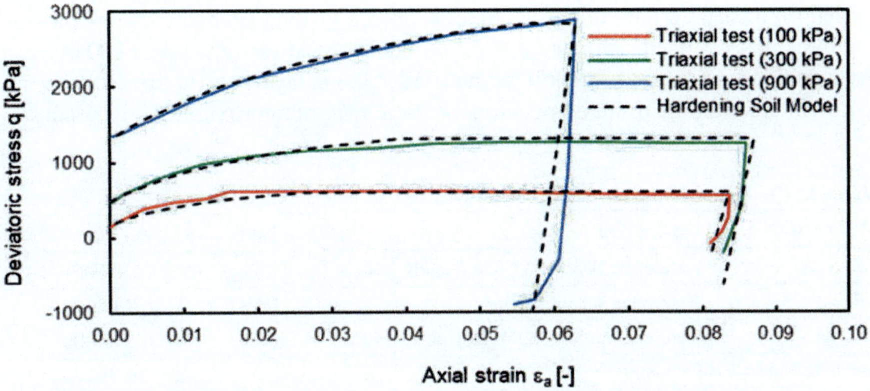

Fig. 4 Comparison of the results of triaxial tests with the Hardening Soil Model: deviatoric behavior in triaxial compression

(1) The parameters governing the elastic behaviour (i.e. $E_{\mathrm{ur,\,ref}}$, m) are calibrated on the unloading path performed at the end of the shearing phase (see Fig. 4).
(2) The parameters governing the volumetric yielding (i.e. $E_{\mathrm{oed,\,ref}}$, α) are calibrated on the change in volume measured during the K_0-consolidation phase (see Fig. 3).
(3) The parameters governing the shear hardening (i.e. $E_{\mathrm{sec,\,ref}}$, R_f) are calibrated on the deviatoric behaviour measured in triaxial compression (see Fig. 4).
(4) The parameters describing the failure envelope are derived on the stress state when reaching peak strength. Assuming the Mohr-Coulomb failure envelope, the angle of internal friction is found to be stress dependent for the large range of pressures concerned. Therefore, the secant friction angle φ is derived as a function of the mean effective pressure at failure p_f' in accordance with [5].

$$\varphi(p'_{\mathrm{f}}) = A \log\left(\frac{p'_{\mathrm{f}}}{p'_{\mathrm{ref}}}\right) + B \qquad (1)$$

The parameters A and B are calibrated on the test data (see Fig. 4). It is assumed that there is no cohesion.

(5) The parameter governing the plastic flow, i.e. the dilatancy angle ψ, is derived from the change in volume measured in shearing after failure is reached. Even for the smallest level of confining pressure of 100 kPa, only very little dilation is observed. At higher confining pressure, no dilation is measured. Therefore, the dilatancy angle is assumed to be nil.

All the parameters derived are given in Table 1. A comparison of test data and the integrated incremental response of the constitutive model is given in Figs. 3 and 4. The volumetric behaviour is shown for the consolidation phase, while the deviatoric behaviour is shown for the shearing phase.

An additional parameter that is introduced is the pre-consolidation pressure prescribing the initial position of the cap yield surface in terms of hydrostatic pressure. Based on measured data obtained directly from a leading producer of compaction equipment, it has been estimated that a reasonably conservative value of 200 kPa can be specified for the case of rockfill compacted at 0.8 m layers and at least four passes with vibratory single drum compactors of static load of approximately 10 tonnes.

Table 1 Parameters of the Hardening Soil Model as implemented in DIANA

Parameters	Description	Unit	Value
$E_{\mathrm{oed,ref}}$	Reference oedometer tangent stiffness	[kPa]	28 000
$E_{\mathrm{sec,ref}}$	Reference secant stiffness	[kPa]	34 000
$E_{\mathrm{ur,ref}}$	Reference unloading reloading stiffness	[kPa]	340 000
m	Exponent m	[-]	0.70
p'_{ref}	Reference pressure	[kPa]	100
α	Cap shape parameter	[-]	0.3
ν_{ur}	Poisson's ratio	[-]	0.25
R_{f}	Failure ratio $q_{\mathrm{f}}/q_{\mathrm{a}}$	[-]	0.70
ψ	Dilatancy angle	[°]	0
c	Cohesion	[kPa]	0
A	see Eq. (1)	[-]	−13.67
B	see Eq. (1)	[°]	55.0
γ_{d}	Dry density	[kN/m^3]	23.0

3.2 Small-Strain Nonlinearity of the Rockfill Material

The deviatoric behaviour at small strains is assumed according to the Hardin-Drnevich relationship [4]. It is considered within in the framework of the elasto-plastic constitutive model as long as its stiffness is higher than the elastic stiffness given by the pure Hardening Soil Model. In the latter case, the shear modulus G depending on the shear strain γ is assumed as follows:

$$G = \frac{G_0}{1 + a\frac{\gamma}{\gamma_{0.7}}} \tag{2}$$

where G_0 is the maximum shear modulus and $\gamma_{0.7}$ is shear strain corresponding to the shear modulus of $G = 0.72\ G_0$. Owing to the definition of $\gamma_{0.7}$, the parameter α is equal to 0.385.

In addition, the maximum shear modulus is considered to be dependent on the effective confining stress level p':

$$G_0 = G_{0,\mathrm{ref}}\left(\frac{p'}{p'_{\mathrm{ref}}}\right)^n \tag{3}$$

where $G_{0,\,ref}$ is the reference shear modulus at the reference $p'_{ref} = 100$ kPa and n is a fitting parameter.

The parameters $\gamma_{0.7}$, $G_{0,\,ref}$ and n are calibrated by adjusting Eqs. (2) and (3) to the test data provided. The comparison of the measured Young's moduli with the Hardin-Drnevich relationship is shown in Fig. 5. The obtained parameters are given in Table 2.

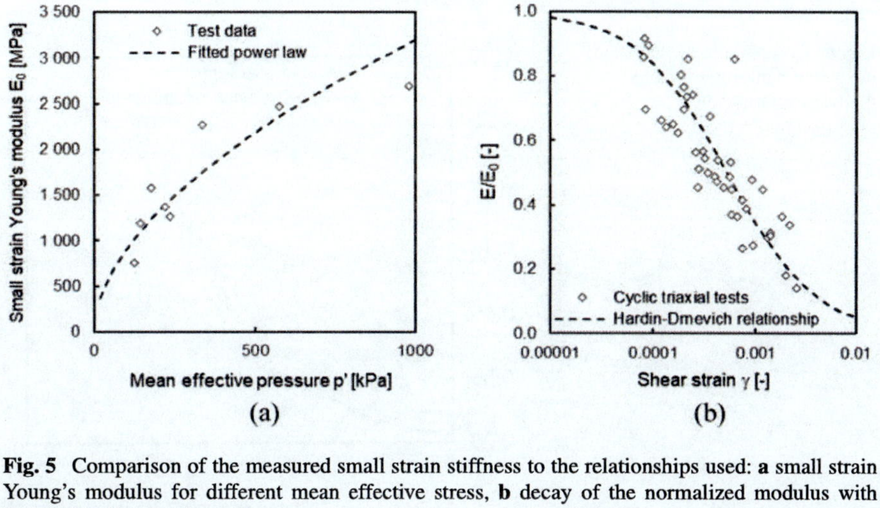

Fig. 5 Comparison of the measured small strain stiffness to the relationships used: **a** small strain Young's modulus for different mean effective stress, **b** decay of the normalized modulus with increasing shear strain

Table 2 Parameters describing the shear modulus and the damping ratio

Parameters	Description	Unit	Value
$G_{0,\,ref}$	Reference small strain shear modulus	[kPa]	360 000
n	Exponent n	[-]	0.55
p'_{ref}	Reference pressure	[kPa]	100
D_{max}	Maximum damping ratio	[-]	0.3
$\gamma_{0.7}$	See Eqs. (2) and (4)	[-]	0.0002
a	See Eqs. (2) and (4)	[-]	0.385

3.3 *Material Damping of the Rockfill*

The dissipation of the rockfill material in cyclic loading is measured in triaxial shear tests. Such dissipation is often described by viscous material damping (see Fig. 5) according to the Hardin-Drnevich relationship [4]. Thus, when considering the dissipation in terms of viscous damping, the damping ratio D is assumed to be depending on the shear strain γ:

$$D = \frac{D_{\max} a \frac{\gamma}{\gamma_{0.7}}}{1 + a \frac{\gamma}{\gamma_{0.7}}} \tag{4}$$

where D_{max} is the maximum damping ratio at large strains, which can be derived by adjusting Eq. (4) to the cyclic triaxial test data. The parameter $\gamma_{0.7}$ has already been determined from the decay of the shear modulus (see Sect. 3.2). The comparison of the measured damping ratios with the Hardin-Drnevich relationship is shown in Fig. 6. Its parameters are given in Table 2.

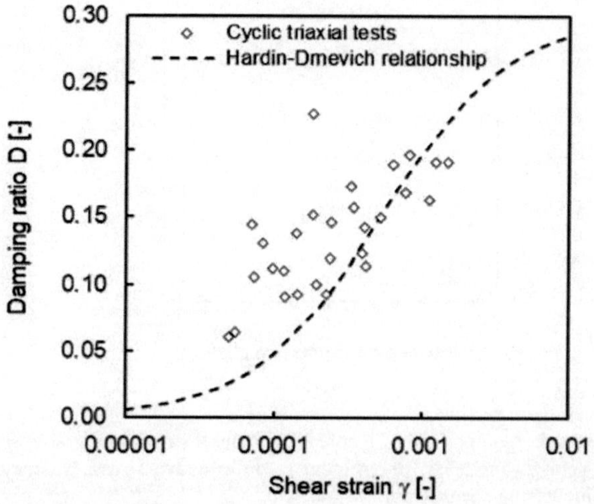

Fig. 6 Comparison of the measured damping ratios to the Hardin-Drnevich relationship

Seismic Analyses for Menta Embankment Dam: Nonlinear Dynamic Analyses with HS-Small (Hardening Small Strain) Model

H. Lu and C. Athanasiu

Abstract Dynamic behaviour of a rockfill dam with bituminous facing is assessed through non-linear time domain FEM (Finite Element Method) analyses. Plaxis HS-small model is used for the rockfill materials. Sensitivities analysis of two different time acceleration histories, bituminous facing stiffness with different temperature and with or without upstream reservoir is undertaken. The dynamic response is generally in line with the expectation. The post-earthquake settlement is estimated 0.21 m, so the dam is unlikely to be overtopped. There could be cracks in the middle part of the bituminous facing. The cracks are more likely to occur when the temperature is 1 °C than 28 °C. Even though a non-linear analysis of this type is more time costly, today's computer hardware makes it possible for the engineer to perform a full nonlinear dynamic analysis within reasonable time.

Keywords Dynamic · Rockfill dam · HS-small model

1 Introduction

Large dams have been designed systematically against earthquakes since the 1930s [1]. With severe failure consequences, the dam safety under the seismic load conditions is always a topic for the dam design, particularly in the earthquake active regions. With the advanced computation technology and modern criteria, it is achievable to detail assess the dam safety under the seismic load in the engineering design practice. This paper addresses the response of the seismic analyses for embankment dam in a workshop case prepared for the 15th ICOLD International Benchmark Workshop on Numerical Analysis of Dams—Theme B Seismic analysis of Menta Embankment dam [1].

Menta Dam is a bituminous-faced rockfill dam located in southern Italy [1]. The dam is 90 m high 325 m long. The dam crest is at 1431.75 masl. The inclination for both upstream and downstream slope inclination is 1:1.8. The downstream face

H. Lu (✉) · C. Athanasiu
Multiconsult, Oslo, Norway
e-mail: hui.lu@multiconsult.no

G. Bolzon et al. (eds.), *Numerical Analysis of Dams*, Lecture Notes in Civil Engineering 91, https://doi.org/10.1007/978-3-030-51085-5_30

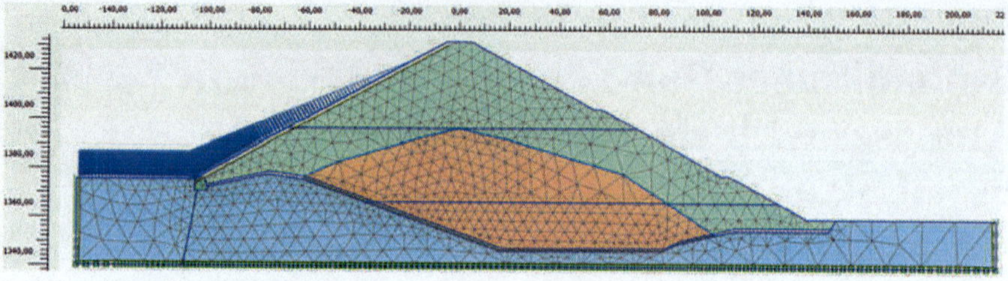

Fig. 1 Overview of the model layout mesh and analysis stage structure

has two berms. Fully supplied water level is at 1424.5 masl. The bituminous face is 0.35 m thick. There is a drainage layer under the bituminous face and embankment foundation interface. There is a concrete gallery at the upstream face of the dam. The Plaxis 2D FEM model layout is shown in Fig. 1.

2 Analysis Methodology

The analysis is undertaken in commercial FEM program Plaxis2D version 2019 with the HS-small soil model for the rockfill. Nonlinear dynamic method is applied to this analysis.

It is assumed that the dam is constructed to EL 1366 masl, EL 1396 masl and to the dam crest EL 1431.75 masl in three stages. It is also assumed that the seismic loads are applied afterwards (Table 1).

Moreover, sensitivity analysis is undertaken for two different time acceleration earthquakes, under fully supplied water level condition versus empty reservoirs and different assumed stiffness of the bituminous facing under the temperature of 1 and 28 °C.

Table 1 Modelling strategy

Construction stage	Description
01 Initial stage	Before any construction
02 Construction stage-1	Assuming the dam is constructed to EL 1366 masl
03 Construction stage-2	Assuming it is constructed to EL 1396 masl
04 Construction stage-3	Assuming it is constructed to the dam crest EL 1431.75 masl
05 Under the operation load condition	Assuming the upstream water level is filled up to EL 1424.5 masl and the steady state condition is achieved

Since the constitutive model used allows for plastic work, most of the dissipation measured in the laboratory tests is expected to be already captured by the plasticity part of the model. Therefore, the viscous damping is considered to be much smaller as if it would be in order to account for the entire dissipation. In order to address dissipation at small strains and potential rate effects, a damping ratio of 5% is considered for the rockfill material.

3.4 Scale Effects Due to the Particle Size of the Rockfill Material

It is widely recognized that the shear strength properties of rockfill are depending on the size of its particles (e.g. [6, 7]). Since the triaxial shear tests on which the strength properties have been calibrated are performed on a modified grain size distribution (see Fig. 7), a correction is applied to the obtained strength properties. Therefore, the failure envelope is modified according to the respective method introduced by [8]. In order to apply the correction, the failure envelope assumed is reformulated in terms of normal stress and shear strength:

$$\tau_{\mathrm{f}} = \tan\left(A\log\left(\frac{\sigma'_{\mathrm{n}}}{p'_{\mathrm{ref}}}\left(\frac{3-\sin(\varphi)}{3\sin^2(\varphi)}\right)\right)+B\right)\sigma'_{\mathrm{n}} \tag{5}$$

The correction factor f to be applied both on the shear strength and on the normal stress is considered according to [8]:

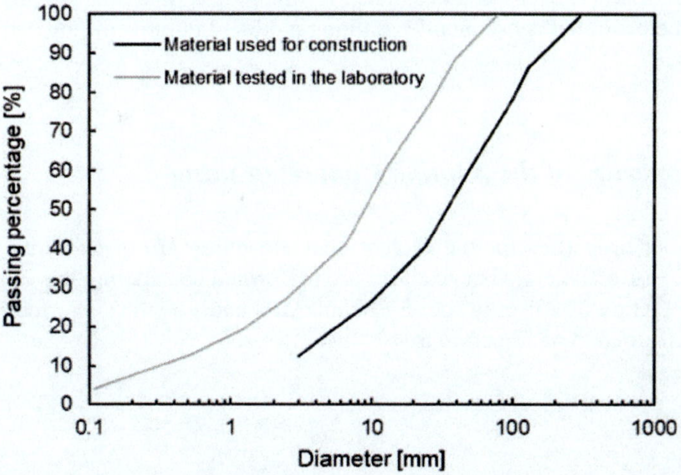

Fig. 7 Grain size distributions of the rockfill materials used for construction and tested in the laboratory

Table 3 Parameters of the modified failure envelope accounting for scale effects

Parameters	Description	Unit	Value
A	See Eq. 1	[-]	−13.67
B	See Eq. 2	[-]	51.70

$$f = \left(\frac{D_{50\text{GSD Field}}}{D_{50\text{GSD Lab}}} \right)^{\frac{3}{m}} \tag{6}$$

where $D_{50\text{ GSD Field}}$ and $D_{50\text{ GSD Lab}}$ are assumed to be the characteristic grain sizes of the grain size distributions used for the dam and tested in the laboratory respectively. The parameter m is a material parameter.

After applying the factor f in Eq. (5) both on the shear strenh and on the normal stress, the pressure dependency of the friction angle is rewritten:

$$(p'_{\text{f}}) = A \log\left(\frac{p'_{\text{f}}}{p'_{\text{ref}}} \right) + B^* \tag{7}$$

introducing the parameter B^* which takes into account the reduction of shear strength due to scale effects of the particle size:

$$B^* = \alpha + A \log\left(\left(\frac{D_{50\text{GSD Field}}}{D_{50\text{GSD Lab}}} \right)^{\frac{3}{m}} \right) \tag{8}$$

The material parameter m is assumed based on the back-analysis for similar dam fill materials tested in triaxial shearing [6, 8]. It is assumed to be equal to 7 which is the value that has been back calculated for the rockfill material of Oroville dam. The parameters of the consequently resulting modified failure envelope are given in Table 3.

3.5 *Properties of the Asphalt Concrete Lining*

The material properties for the asphalt concrete lining are given in the problem formulation (see Table 4). The analyses are performed considering the values given for a temperature of 26° only (i.e. $E = 2800$ MPa and $\nu = 0.33$). Furthermore, the lining is assumed to be impervious.

Table 4 Properties of the asphalt concrete lining as given in the problem formulation

Parameters	Description	Unit	Value
$E_{T=1}$	Complex modulus at 1 °C	[MPa]	14 000
$\nu_{T=1}$	Complex Poisson's ratio at 1 °C	[-]	0.16
$E_{T=26}$	Complex modulus at 26 °C	[MPa]	2 800
$\nu_{T=26}$	Complex Poisson's ratio at 26 °C	[-]	0.33
γ	Density	[kN/m^3]	24.0
fc	Compressive strength	[MPa]	1.48
ft	Tensile strength	[MPa]	0.95

Table 5 Investigated seismic excitation cases

Investigated case	Description	PGA, g	X-direction	Y-direction
Case 1	Friuli 1976 EZ	0.26	HNE	HNZ
Case 2	Friuli 1976 NZ	0.26	HNN	HNZ
Case 3	C. Italy 2016 EZ	0.26	HGE	HGZ
Case 4	C. Italy 2016 NZ	0.26	HGN	HGZ

3.6 Properties of the Bedrock

The bedrock is assumed homogeneous, isotropic and with linear elastic constitutive behaviour. The Young's modulus and the Poisson's ratio considered are 7 000 MPa and 0.25 respectively.

The foundation is assumed massless and its viscous damping is taken 5% of the critical.

4 Seismic Excitation

The strong ground motion input is applied on the outer vertical and on the bottom boundaries of the modelled part of the rock foundation. All the acceleration time-histories provided in [2] are considered in the analysis by defining four seismic excitation cases. Table 5 indicates the acceleration time-histories applied in the horizontal (X) and in the vertical (Y) directions.

5 Static Analysis Results

As already discussed in Sect. 3, the friction angle of the rockfill is introduced depending on the stress state according to Eq. 1 and Table 3. In order to avoid potentially unrealistic high friction angles on the surface, the friction angle is limited to the value of 51.7°, which corresponds to a level of confinement of 100 kPa. Since the stress state in the dam body evolves during the construction and the impounding of the dam, the friction angles in the different zones of the embankment fill are updated for each stage of construction, as well as for the two impounding phases considered.

The sign convention in the software program used (DIANA 10.3) is as follows:

- Strains: contractions have negative values, while extensions have positive ones.
- Stresses: compressive stresses have negative values, while tensile stresses have positive ones.

5.1 End of Construction

Figures 8 and 9 show the contours of the vertical displacements and stresses. Figure 10 gives the trajectories of the principal stresses. Figure 11 presents the contours of the maximum strains of contraction. As evidenced by these figures, the maximum settlements due to the construction process occur in the central part of the dam body and amount to 0.75 m. The vertical stress contours are parallel to the dam outer contour, i.e. to the polygon formed by the upstream face, the crest and the downstream face of the dam. The maximum compressive principal stress trajectories are vertical in the embankment column below the dam crest. In the zone between this column and the dam slopes, they gradually rotate to become approximately parallel to the upstream, respectively the downstream face, in the vicinity of the faces. The total strains, which coincide almost entirely with the plastic ones, amount to a maximum of 3%.

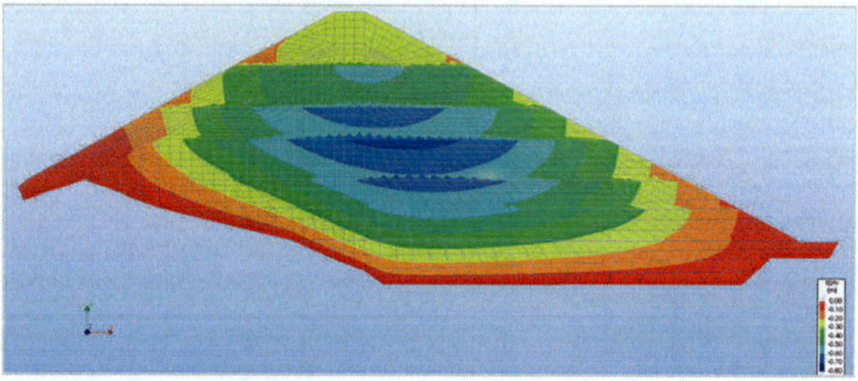

Fig. 8 End of construction vertical displacements, m

Fig. 9 End of construction vertical stresses, Pa

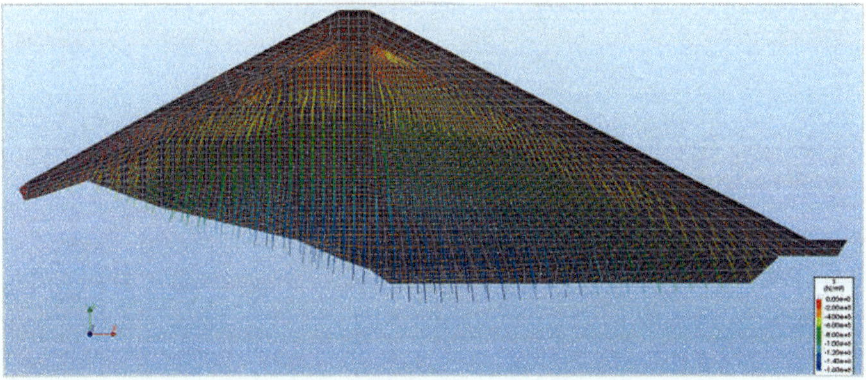

Fig. 10 End of construction principal stress trajectories

Fig. 11 End of construction total strains (contraction)

5.2 End of First Impoundment

The contours of the vertical displacements and the horizontal and vertical stresses are given in Fig. 12, Fig. 13 and Fig. 14, respectively. Figure 15 shows the trajectories of the principal stresses and Fig. 16 gives the contours of the maximum strains of contraction.

The first impoundment does not entail significant increase of the maximum settlements with respect to the end of construction. There is a certain increase of the vertical stresses below the bituminous face, which is accompanied by a pronounced rotation of the stress trajectories in the whole upstream part of the embankment. This evolution of the stress state is due to the hydrostatic pressure exerted on the bituminous face. There is also an increase of the total strains below the upstream face, especially at the lower part of the embankment where the hydrostatic pressures are the highest.

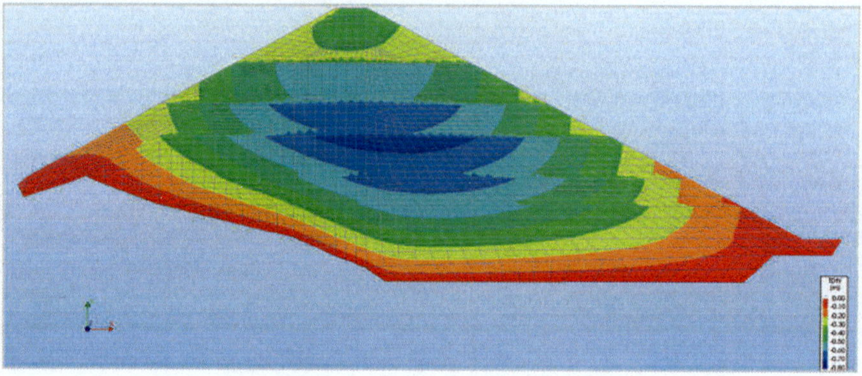

Fig. 12 End of first impoundment vertical displacements, m

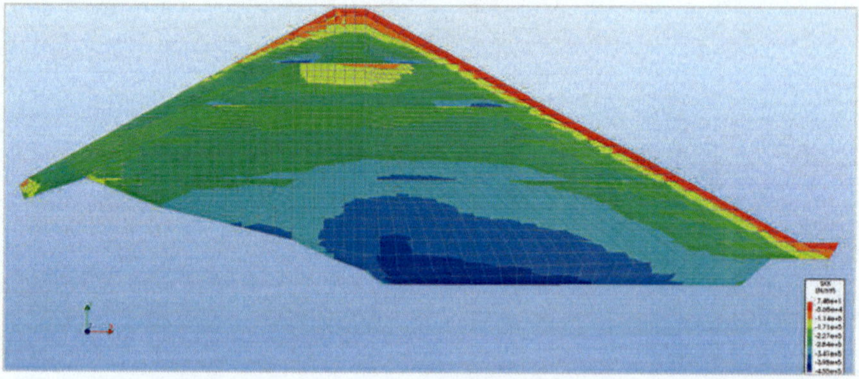

Fig. 13 End of first impoundment horizontal stresses, Pa

Fig. 14 End of first impoundment vertical stresses, Pa

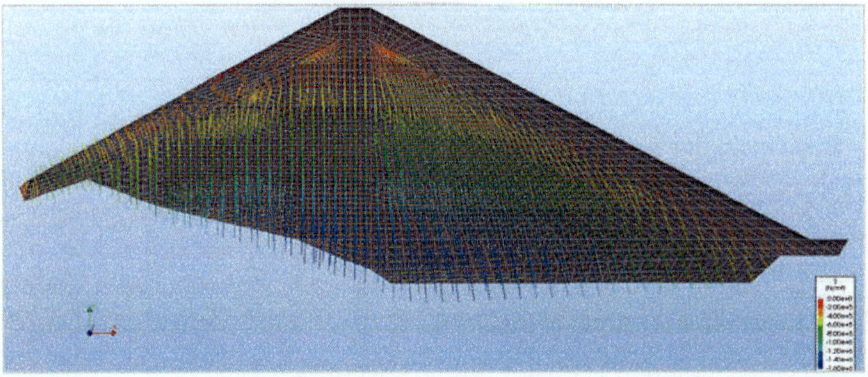

Fig. 15 End of first impoundment principal stress trajectories

Fig. 16 End of first impoundment total strains (contraction)

Table 6 Eigenrequencies and effective mass cumulative percentage

Mode	Empty reservoir			Full reservoir		
	f, Hz	X, %	Y, %	*f*, Hz	X, %	Y, %
1	0.952E + 00	0.696E + 02	0.251E + 01	0.946E + 00	0.705E + 02	0.359E + 01
2	0.135E + 01	0.724E + 02	0.600E + 02	0.13E + 01	0.733E + 02	0.622E + 02
3	0.160E + 01	0.767E + 02	0.601E + 02	0.159E + 01	0.772E + 02	0.622E + 02
4	0.183E + 01	0.774E + 02	0.622E + 02	0.175E + 01	0.779E + 02	0.669E + 02
5	0.209E + 01	0.774E + 02	0.765E + 02	0.207E + 01	0.786E + 02	0.816E + 02
6	0.223E + 01	0.859E + 02	0.769E + 02	0.216E + 01	0.861E + 02	0.816E + 02
7	0.260E + 01	0.865E + 02	0.786E + 02	0.255E + 01	0.863E + 02	0.842E + 02
8	0.279E + 01	0.866E + 02	0.786E + 02	0.270E + 01	0.889E + 02	0.847E + 02
9	0.286E + 01	0.886E + 02	0.809E + 02	0.278E + 01	0.889E + 02	0.849E + 02
10	0.302E + 01	0.892E + 02	0.818E + 02	0.291E + 01	0.906E + 02	0.857E + 02
11	0.327E + 01	0.900E + 02	0.828E + 02	0.325E + 01	0.909E + 02	0.857E + 02
12	0.341E + 01	0.907E + 02	0.845E + 02	0.330E + 01	0.917E + 02	0.908E + 02

In summary, the first impoundment does not lead to a significant increase of the displacements, the strains and the stresses in the dam body. The strains and stresses withstood by the embankment during the construction process remain predominant.

6 Eigenvalue Analysis Results

The computed eigenfrequencies f for the cases of empty and full reservoir with added masses defined for incompressible fluid-structure interaction are given in Table 6.

The eigenfrequencies of the cases with empty and full reservoir are almost all equal to its corresponding one. Hence, the reservoir does not influence significantly the dynamic properties of the investigated system. The results also show that the first two eigenfrequencies activate approximately 70% and 60% of the system's mass in the horizontal and the vertical direction, respectively. The first eigenmode is given in Fig. 17.

7 Seismic Analysis Results

Four nonlinear seismic analysis are performed in the time-domain by means of direct integration of the equations of dynamic equilibrium for the seismic excitation cases defined in Sect. 4. The effect of the hydrodynamic pressures caused by the reservoir during earthquake is approximated as a fluid-structure interaction with incompressible fluid. The analyses are carried out using the α-method of Hughes, Hilbert and

Fig. 17 Dam-foundation-reservoir system first eigenmode

Taylor with $\alpha = -0.3$. The viscous damping of the system is represented by the Rayleigh approach, with proportionality constants computed so as to give a modal damping ratio of 5% in the first and the twelfth vibration modes.

The field of the friction angle is considered as a function of the initial stress state (see Sect. 3). During the seismic analysis, the friction angles remain constant.

Based on the results of the crest acceleration and relative displacements time histories, it is found that Seismic Excitation Case 1 entails the maximum seismic response of the system of all the four investigated cases. Therefore, the results are presented for Case 1 only.

The horizontal and vertical acceleration time-histories of the middle point of the dam crest are given, respectively, in Figs. 18 and 19.

Figures 20 and 21 present the time-histories of the relative displacements of the same point with respect to the base of the dam-foundation system.

Figure 22, Fig. 23 and Fig. 24 present, respectively, contour plots of the permanent horizontal, vertical and total displacements. Figure 25 shows the profile of the permanent post-seismic horizontal displacements along the dam axis.

Figure 26 shows the total shear strains in the embankment after the earthquake. Finally, Fig. 27 presents the maximum tensile strains in the bituminous face due to the strong ground motions only.

The crest—to—base maximum acceleration ratio is evaluated to 3.41 for the horizontal component and 2.88 for the vertical component of the accelerations. The results from the seismic analyses show that the dam will withstand a maximum horizontal seismic displacement of approximately 30 cm. The permanent settlement of the crest due to strong ground motions will be of the same order, i.e. 45 cm. Finally, based on Figs. 22, 23, 24 and 25 one can conclude that the entire embankment is experiencing permanent settlements and permanent displacements in the downstream direction. This finding can be explained by the distribution of the shear strains showing a pronounced downstream mechanism that is considerably more important than the mechanisms forming in the upstream part of the embankment (see Fig. 26).

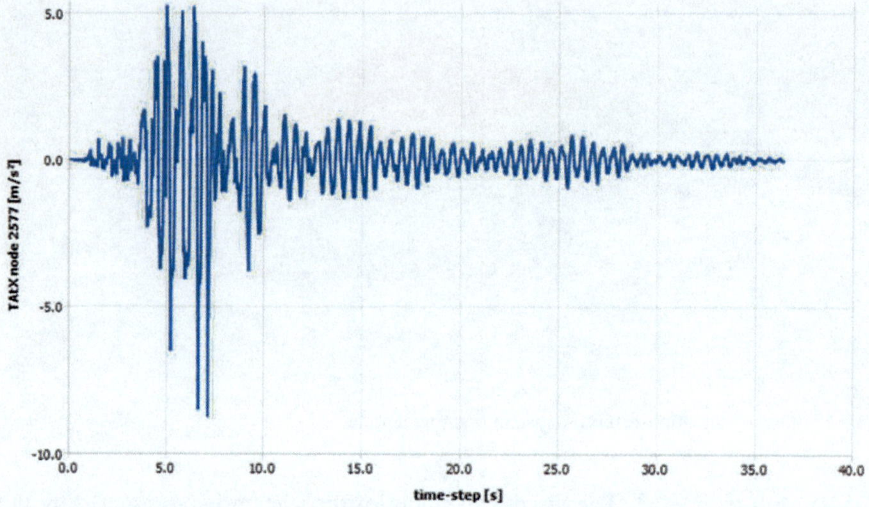

Fig. 18 Horizontal acceleration time histories of the middle point of the dam crest

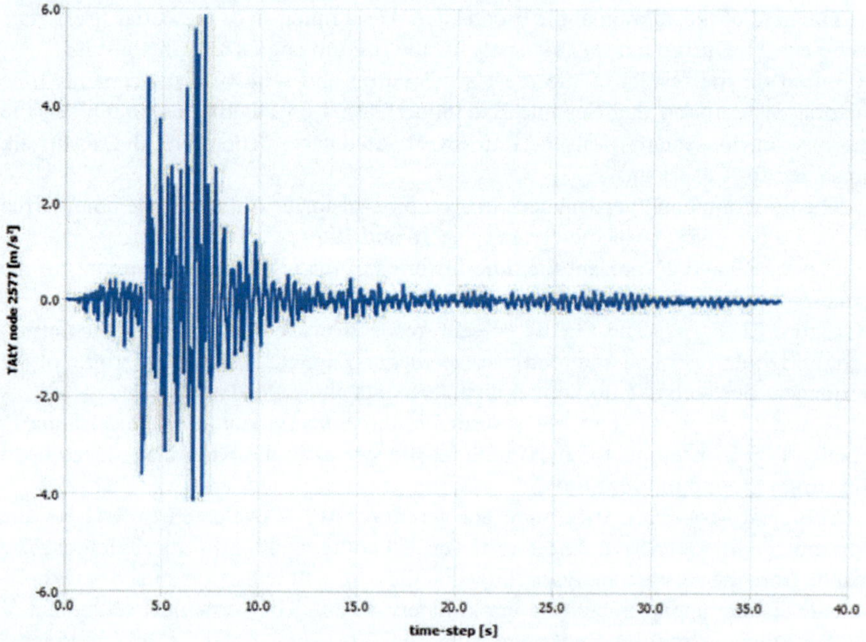

Fig. 19 Vertical acceleration time histories of the middle point of the dam crest

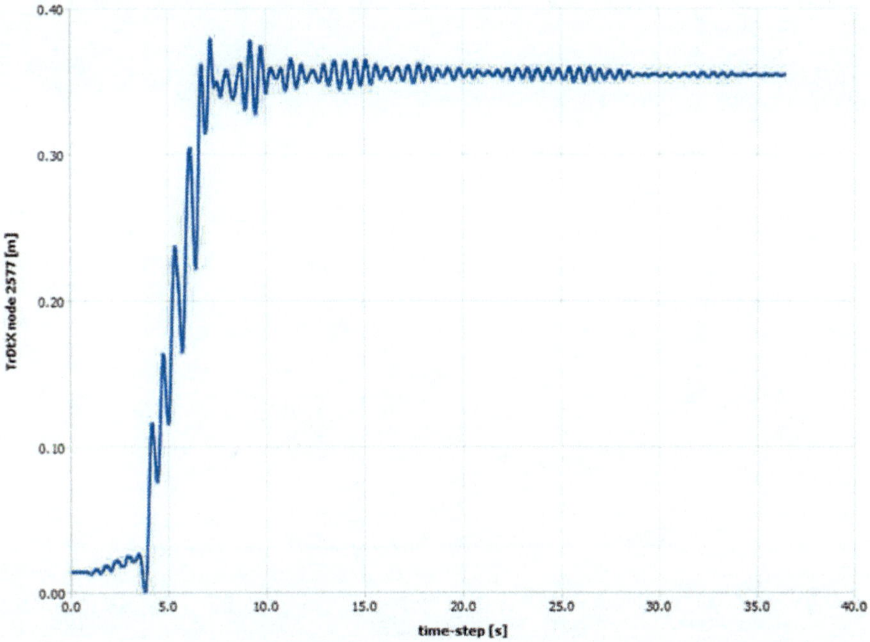

Fig. 20 Horizontal relative displacement time histories of the middle point of the dam crest

As to the bituminous face, the maximum strains of extension due to the strong ground motion are evaluated to amount to 0.0015 at a depth of around 22 m measured from the dam crest. These strains are imposed by the embankment underneath displacing in downstream direction. However, in the top part of the bituminous face, the water pressure acting on it is relatively small creating not enough shear resistance within the embankment material in order to make the bituminous face deforming together with the embankment. As a consequence, shear strains can form within the embankment underneath allowing for slip of the bituminous face (see Fig. 26) which results in low tensile strains in its top part (see Fig. 27).

In general, the tensile strains in the face are found to be within the strain capacity of bituminous concrete. Therefore, no cracking of the bituminous face is expected to occur during strong ground motions due to the seismic impacts investigated in this analysis. However, based on the assumption of linear elasticity for the bituminous concrete material, the corresponding maximum tensile stress (of around 4.2 MPa) is found to be well beyond the tensile limit given in the problem formulation. In order to assess consistently the expected performance of the bituminous face, further investigations are needed in order to take into account the true visco-elasto-plastic constitutive behaviour of the bituminous concrete.

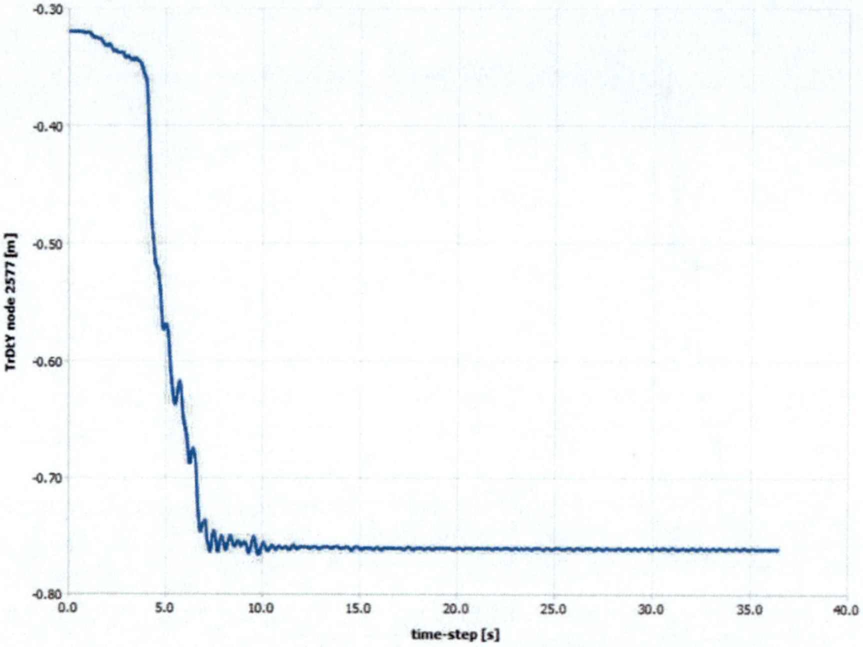

Fig. 21 Vertical relative displacement time histories of the middle point of the dam crest

Fig. 22 Permanent horizontal displacements due to the strong ground motions, m

8 Conclusions

This study presents a comprehensive FEM nonlinear static and seismic analysis of a 90-m high rockfill dam with bituminous face. The Hardening Soil/Modified Mohr-Coulomb model combined with the Small Strain Stiffness model is used in the software program DIANA to model the behaviour of the dam for the investigated static and dynamic load combinations.

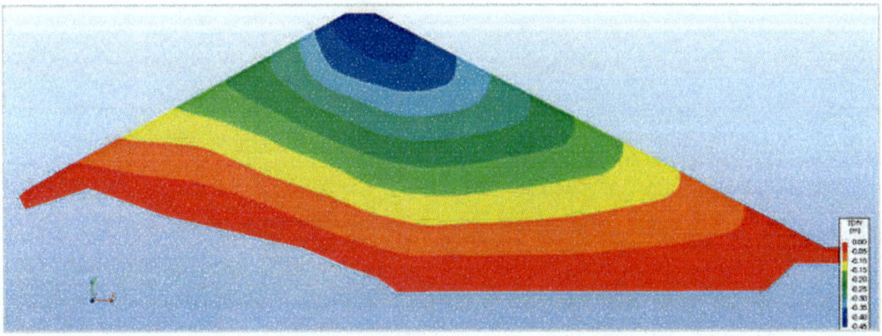

Fig. 23 Permanent vertical displacements due to the strong ground motions, m

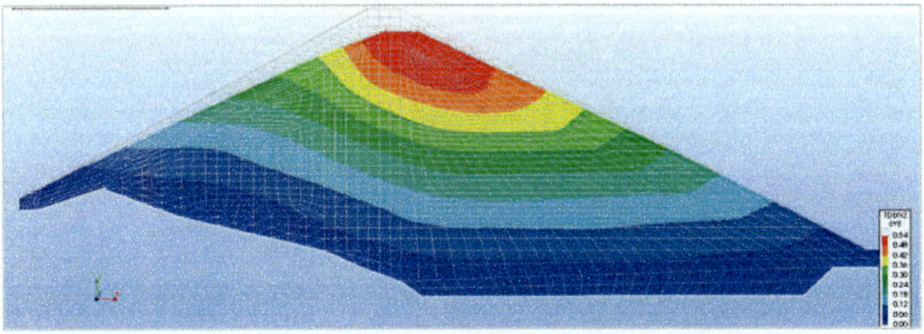

Fig. 24 Permanent total displacements due to the strong ground motions, m

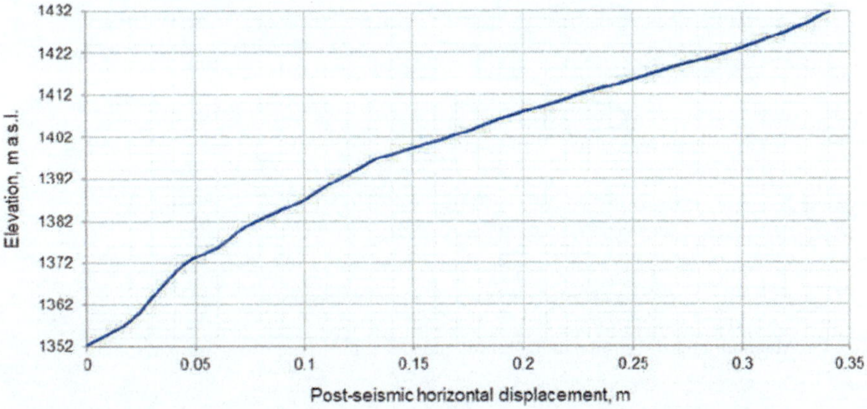

Fig. 25 Permanent horizontal displacements at dam axis due to the strong ground motions, m

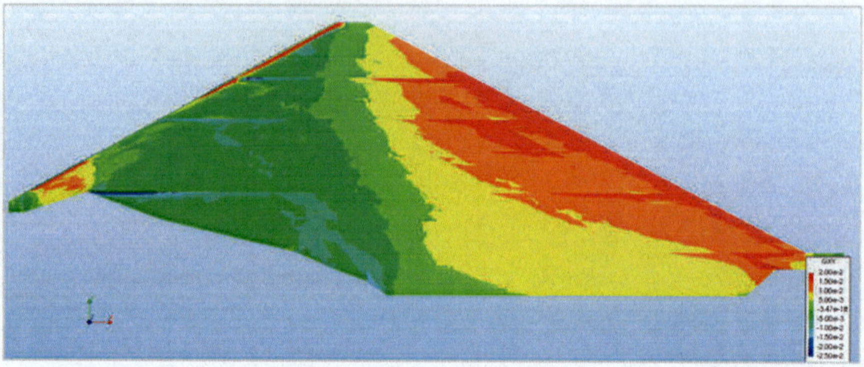

Fig. 26 Total shear strains after the strong ground motions

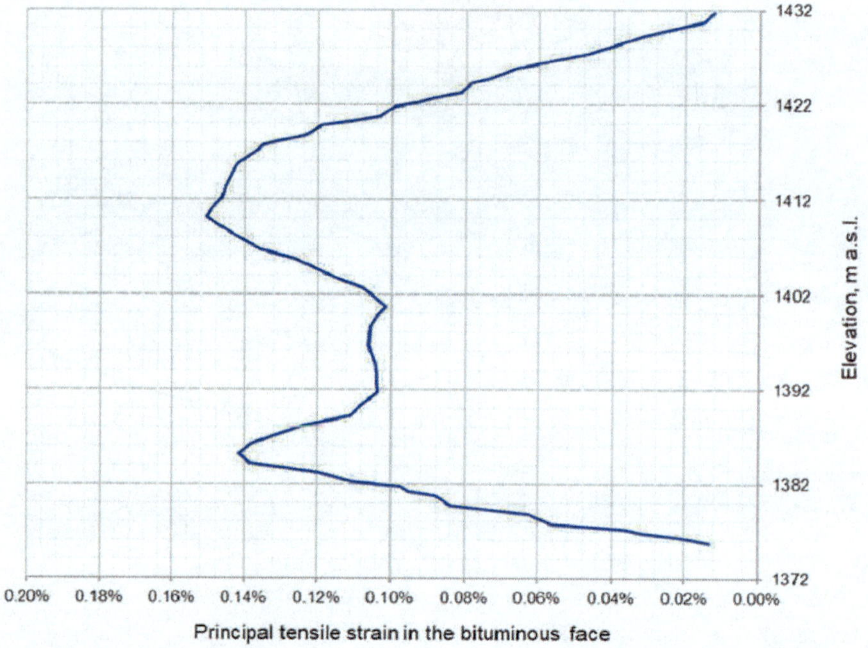

Fig. 27 Maximum tensile strains in the bituminous face due to the strong ground motions

The static analysis shows that the first impoundment entails a limited increase of the displacements, the strains and the stresses in the dam body. The strains and stresses withstood by the embankment during the construction process remain predominant. Nevertheless, the hydrostatic pressure on the dam's watertight barrier leads to an increase of the horizontal and the vertical stresses in the rockfill below the bituminous face, as well as to a rotation of the stress trajectories in the whole upstream part of the embankment. There is also an increase of the total strains, especially at the lower upstream part of the embankment.

As to the investigated seismic impacts, one can conclude that both the permanent displacement and the additional strains caused by the strong ground motions would be moderate and well within the freeboard of the dam. The bituminous face is strained since the embankment underneath is deforming in downstream direction. The maximum additional strains in the bituminous face, i.e. the ones due to the strong ground motions, are moderate and it is expected that they can be withstood without occurring of cracks. However, further efforts are needed in order to assess the expected performance of the bituminous face taking into account its visco-elasto-plastic constitutive behaviour.

Regarding the choice of the constitutive material model for the rockfill, it is recommended to use a full elasto-plastic model material. Incorporating plasticity allows for taking into account realistic stiffness and dissipation within the framework of a nonlinear seismic analysis performed in the time-domain.

In addition, the effects due to the failure envelope being curved in direction of the hydrostatic axis should be considered for rockfill material. Ideally, a constitutive model having such curved failure envelope implemented is to be used. As an approximation when using Mohr-Coulomb failure envelope, the friction angle can be introduced as a function of the pressure level prevailing in the dam body. Furthermore, it is recommended to address potential effects on the shear strength owing to the fact that the material tested is different from the material used for construction in terms of its grain size distribution.

References

1. https://dianafea.com/
2. Russo G, Vecchietti A, Cecconi M, Pane V, De Marco S, Fiorino A (2019) Theme B: seismic analyses of Menta Embankment dam. In: Problem formulation document for the 15th international benchmark workshop on numerical analyses of Dams, Milano, Italy
3. Schanz T, Vermeer PA, Bonnier PG (1999) The hardening soil model: formulation and verification. In: Beyond 2000 in computational geotechnics. Balkema, Rotterdam, pp 281–296
4. Hardin BO, Drnevich VP (1972) Shear modulus and damping in soils: design equations and curves. ASCE J Soil Mech Found Div 98(7):667–692
5. Leps TM (1970) Review of shearing strength of rockfill. ASCE J Soil Mech Found Div 96(4):1159–1170
6. Marachi ND, Chan CK, Seed HB, Duncan JM (1969) Strength and deformation characteristics of rockfills materials, report no TE-69-5. Department of Civil Engineering/Geotechnical Engineering, University of California, Berkeley
7. Marsal RJ (1967) Large-scale testing of rockfills materials. ASCE J Soil Mech Found Div 93(2):27–44
8. Frossard É, Hu W, Dano C, Hicher PY (2012) Rockfill shear strength evaluation: a rational method based on size effects. Géotechnique 62(5):415–427

Theme C: Coupled Hydro-Mechanical Analysis of the Pre-failure and the Failure Behaviour of a Dyke on Soft Subsoil

Coupled Hydro-Mechanical Analysis of the Pre-failure and the Failure Behaviour of a Dyke on Soft Subsoil: Formulation and Synthesis of Results

C. Jommi, D. Sterpi, T. de Gast, S. Muraro, E. Ponzoni, and H. van Hemert

Abstract In an attempt to evaluate current models for the safety assessment of dykes on soft soils, STOWA, the foundation for research on regional dykes in the Netherlands, launched and supported a full scale test on a regional historical dyke, which included observation of the pre-failure response and the design of its failure. The stress test on the dyke included progressive excavation at the toe and rapid drawdown in the ditch next to the toe of the embankment, until failure eventually occurred. The data and the observations collected on site during the test are a unique body of information on the coupled hydro-mechanical pre-failure behaviour and on the resistance of the earth construction. A selection of these data was included in the formulation of the Theme C of the 15th International Benchmark Workshop on Numerical Analysis of Dams, held in September 2019 in Milano, Italy. This contribution presents the main outcomes of the numerical benchmark, coming from the results of the different groups, which analysed the case with current geotechnical constitutive and numerical models.

Keywords Earth construction · Full scale test · Hydro-Mechanical behaviour · Safety assessment

C. Jommi (✉) · D. Sterpi
Department of Civil and Environmental Engineering, Politecnico di Milano, Milan, Italy
e-mail: c.jommi@tudelft.nl

C. Jommi · T. de Gast · S. Muraro · E. Ponzoni
Department of Geoscience and Engineering, TU Delft, Delft, The Netherlands

H. van Hemert
STOWA, Amersfoort, The Netherlands

G. Bolzon et al. (eds.), *Numerical Analysis of Dams*, Lecture Notes in Civil Engineering 91, https://doi.org/10.1007/978-3-030-51085-5_36

1 Introduction

Transport and flood defence infrastructures are crucial assets for the safety and the mobility of people. A relevant part of these geotechnical structures in Europe and overseas, especially in northern and tropical countries, are built on soft organic soils, including a variety of soft organic clays, estuarine sediments and peats.

In an attempt to evaluate current models for the safety assessment of dykes on soft soils, STOWA, the foundation for research on regional dykes in the Netherlands, launched and supported a full scale test on a regional dyke in the Kagerplassen, north of Leiden, which included observation of the pre-failure response and the design of its failure. To the aim of the research the Waterboard HH Rijnland offered a historical dyke, protecting a polder which had to be flooded to become a natural reserve area [1]. The responsibility of the design of the stress and failure test was assigned to TU Delft in the framework of the STW research project Reliable Dykes, and the management of the test was a cooperative effort between HH Rijnland, STOWA, Deltares and TU Delft [2].

To design the test, the site was thoroughly investigated by performing a number of cone penetration tests with pore pressure measurement, CPTu, and collecting undisturbed samples from different boreholes starting at both the crest and the toe of the existing dyke. Standard tests, including triaxial compression and shearing, oedometer compression and direct simple shear, were performed to characterise the soil behaviour of the different layers. In addition, the state of the soil at the beginning of the test was described by means of information on the unit weight of the different formations, on the void ratio and the degree of overconsolidation profiles, and on the daily pore pressure distribution.

The stress test started with an excavation on the polder side of the dyke, and included a sequence of larger excavations and dewatering, until failure eventually occurred during rapid drawdown at the toe of the dyke. During the test, displacements and pore water pressure were measured in three sections of the dyke, to provide information on the dyke response and to enable the design of the final failure stage.

The case is proposed to benchmark existing models for earth water retaining structures, calibrated on laboratory tests currently used in the practice, against the response of the geotechnical system as monitored in the field during the stress test. In the following sections, a synthetic description of the case is provided, the main questions to be answered are formulated, and the field and laboratory data are summarised.

2 Brief Description of the Site and the Stress Test

The site where the stress test was performed is located in The Netherlands, north of Leiden. At the time of the test it was a small polder, dedicated to agricultural activities, facing the small lake Hanepoel and connected to the main land by a small

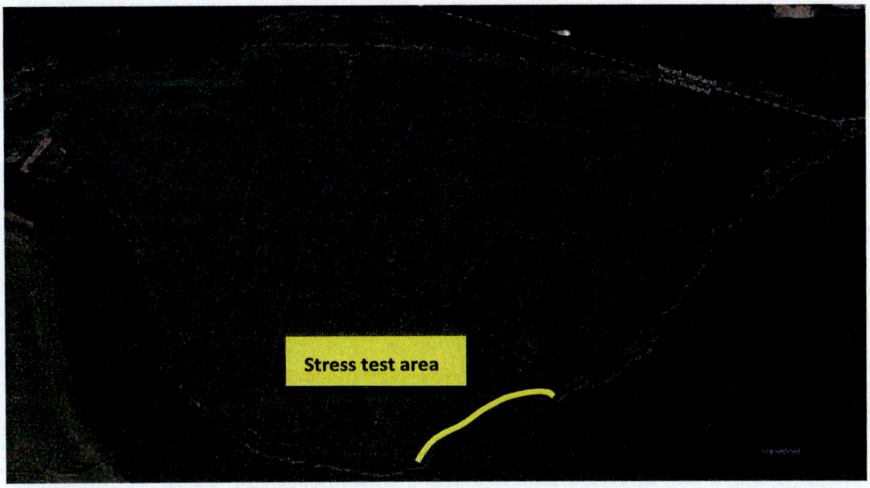

Fig. 1 Overview of the Leendert de Boerspolder test area

path, visible on the right of Fig. 1. The dykes protecting the polder are historical dykes, dating back at least to the XVII century.

The district is managed by the Waterboard HH Rijnland, which offered the polder for research on regional dykes to the STOWA foundation. The latter took the opportunity to perform a stress test including failure, under the responsibility of the Geo-Engineering section of Delft University of Technology, in the framework of the Reliable Dykes STW research project. The final design of the stress test was managed by the researchers at TU Delft together with STOWA, HH Rijnland and Deltares. Among various possible strategies, it was decided to stress and bring to failure the dyke by a series of excavations and dewatering operation at the toe of the dyke (see Fig. 2).

Preliminary information on the dyke and the subsoil was coming from a previous investigation performed by HH Rijnland on a different sector of the dyke.

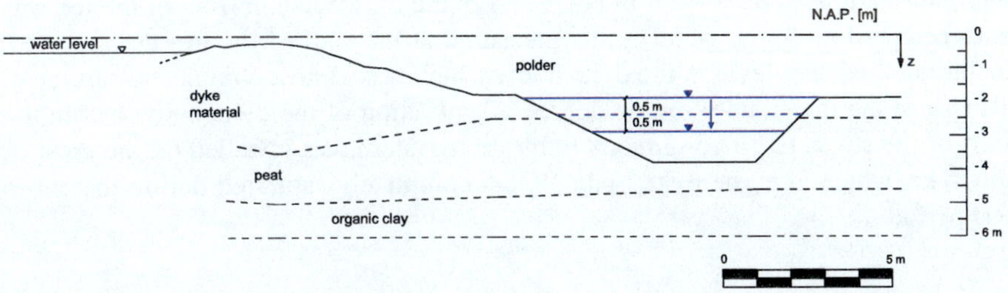

Fig. 2 Schematic drawing of one of the stress test steps, including excavation at the toe and dewatering

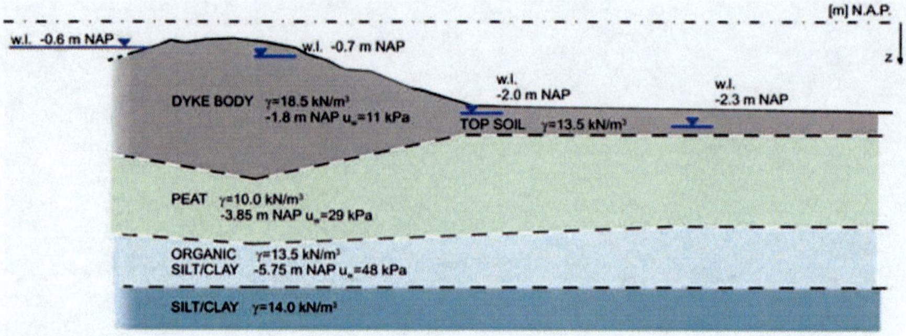

Fig. 3 Schematic representation of the typical soil profile

The specific site investigation designed prior to the test was performed at the end of summer 2014 and included 18 CPTu and 6 boreholes. Samples were collected for laboratory testing with 100 mm diameter samplers, to reduce as much as possible sampling disturbance. The samples were used for a number of laboratory tests, to characterise the soil layer as accurately as possible. It was chosen to use standard laboratory tests to check their representativeness for daily design and assessment.

The CPTu data, together with information from laboratory tests, allowed drawing the typical soil profile schematically indicated in Fig. 3. The dyke body was mostly made of clayey silt and silt and founded over a peat layer, an organic clay layer and a silty clay layer, in turn, resting on the first sandy aquifer at about 15 m depth. The height is measured with reference to the Dutch N.A.P. (Normaal Amsterdams Peil), which will be assumed everywhere as the reference zero for the vertical coordinate.

Only the two first foundation layers, namely the peat and the organic clay layer, were of concern for the deformation and the failure of the dyke, hence, only for these two layers data were given for the benchmark, together with characterisation data of the dyke body material.

On the outer part of the dyke (outer boundary), the regulation system of the canals kept the water level at constant height $h_o = -0.6$ m NAP during the entire duration of the test. On the polder side (inner boundary), the water level was regulated by the pumping stations at $h_i = -2.3$ m NAP, before the drawdown in front of the toe was enforced. In Fig. 3, the water levels measured at the start of the test are reported, including the water level in the dyke body, which was almost completely saturated. To reduce the uncertainty on the degree of saturation of the dyke body, before the start of the stress test a continuous infiltration system was installed on the crest of the dyke, which kept the dyke body almost completely saturated during the entire test period.

3 Questions for the Numerical Benchmark

The numerical benchmark was proposed to evaluate the potential of current models to track the coupled hydro-mechanical response of earth constructions for water defences, both during service life, in view of Serviceability Limit States analyses, and in the prediction of failure, to assess Ultimate Limit States. Different questions of increasing level of complexity were proposed, on purpose leaving the opportunity to benchmark available strategies for the analysis of a time dependent highly non-linear structural response, characterised by dominant hydro-mechanical coupling.

3.1 Prediction of Failure

The analysis of ULS does not require, in general, a fully coupled hydro-mechanical analysis, and can be tackled as a first approximation with a Limit Equilibrium method or with Limit Analysis, provided the ultimate reference drainage conditions (drained or undrained) are properly chosen. However, a Finite Element analysis would allow improving the model and the predictive capabilities by including a back analysis of the monitoring data. For the first list of questions the method of analysis was free.

The following questions were proposed:

(1.a) Given the stress test plan (detailed in Sect. 5), and the geotechnical characterization of the site (given in Sect. 4), when did failure occur?
(1.b) Based on the monitoring data provided (Sect. 5), will the failure be best reproduced by drained or undrained conditions?
(1.c) What was the shape of the failure surface?
(1.d) Which elements of the model are most affected by uncertainty?

3.2 Pre-failure Response

The pre-failure response can be predicted based on the geotechnical characterisation of the site performed before the start of the test, including Cone Penetration Tests with pore pressure measurement (CPTu) and laboratory investigation. The data provided for the benchmark were taken from the data available for the design of the test. Not all the available data were given for the exercise, however, the ones chosen were the most relevant and representative of the conditions of the soil and of its response. The subset of data was chosen in order to provide a clear picture of the embankment and of its foundation, at the same time avoiding overwhelming with huge amount of information. The preferred option to analyse pre-failure response was a fully coupled hydro-mechanical Finite Element analysis, although simpler choices were equally accepted. The benchmark questions were as follows:

(2.a) Which material model can be used to fit the best the subsoil and the material behaviour observed in the laboratory?

(2.b) Given the stress test plan (detailed in Sect. 5), and the geotechnical characterization of the site (given in Sect. 4), how did the soil respond from the start of the test to the beginning of the final drawdown stage? To this aim, provide the time history of:

- 2.b.1. horizontal displacement in correspondence of the two vertical lines monitored by the inclinometers,
- 2.b.2. pore pressure in correspondence of the three locations monitored by piezometers.

(2.c) Given the measurements taken, can the pre-failure displacements and pore pressures be predicted accurately with current models?
(2.d) Are laboratory tests representative of the behaviour of the material in the field?
(2.e) Can the prediction be refined by back-analysing the monitoring data?

To answer the previous questions the authors were requested to provide a first set of results calculated with a model calibrated only on the preliminary information coming from the field investigation and the laboratory tests. The true prediction in the geotechnical practice would follow this procedure. For questions (2.a)–(2.d) the monitoring data were to be used only for the analysis of the accuracy of the numerically predicted response. To answer question (2.e) the authors were invited to refine their model and parameters by back-analysis of the measured data, in order to evaluate the advantages of direct measurement of relevant quantities in the field. Still, some deviation occurs from the measured response. The authors were invited to provide comments on the limitations of current models in accurately predicting the coupled hydro-mechanical response, even in the presence of representative monitoring data.

3.3 Geometry and Geotechnical Model Uncertainties

Although earth embankments in geotechnical engineering are typically analysed with a 2D plane strain scheme, this choice may limit to some extent the accuracy of the prediction. The last set of questions was proposed to evaluate the inherent limitations of a 2D model, together with the typical lack of field information to fully calibrate a 3D spatially variable system. The authors were invited to build up a full 3D model of the dyke, following the geometry provided. CPT data were provided for different sections, to evaluate the spatial variability of soil layers. The laboratory test results could be considered representative of the average response of the soil units. The related questions were:

(3.a) Can we get better prediction of pre-failure and failure including a 3D geometry of the dyke? Therefore, does the true 3D geometry of the dyke influence significantly the response of the dyke during the stress test and at failure?

(3.b) What would be the advantage of including spatial variability in the analysis? To this aim, the variability of the CPTu data could be used as a first measure of heterogeneity.

4 Test Area, Site Investigation and Laboratory Tests

The site was investigated by means of a number of CPTu tests. The plan of the tests is given in Fig. 4, where a view of the geometry is also given. The geometrical data were fully reported in a CAD file.

Typical results of the CPTu tests are given in Fig. 5, giving the schematic profile provided in Fig. 3. Undisturbed samples were retrieved from various locations and depths for laboratory investigation. The initial state of the dyke, including unit weights, initial void ratio, overconsolidation ratio, OCR, and measured pore pressure was derived from a number of measurements on site and in the laboratory, and relevant information was provided to the participants.

The relevant layers were investigated by means of laboratory tests. The locations of the boreholes are reported in Fig. 6, and the chosen test data come from representative stress-strain tests performed on samples retrieved from the depths illustrated in Fig. 7. The laboratory tests included oedometer compression, direct simple shear tests for peat and triaxial tests for the organic clayey silt and the dyke body. The

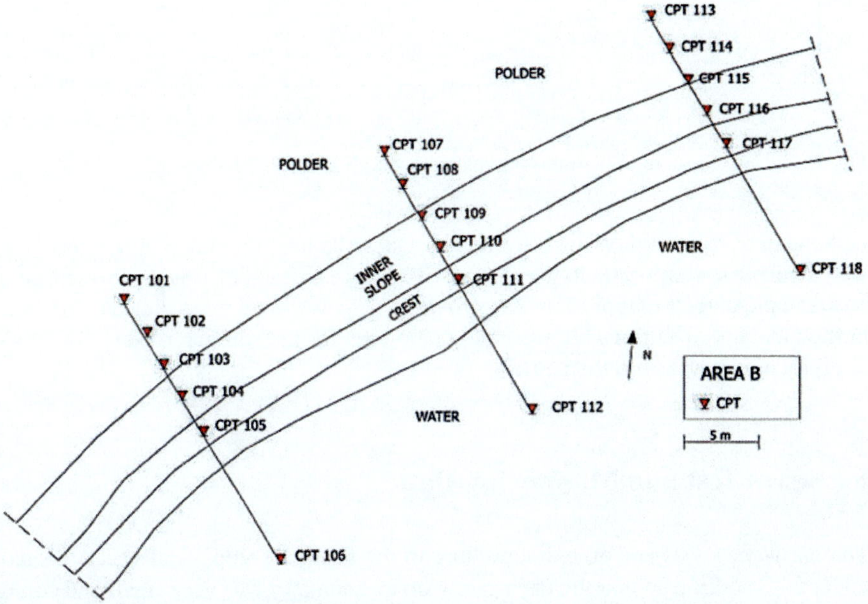

Fig. 4 Overview of the CPTu distribution

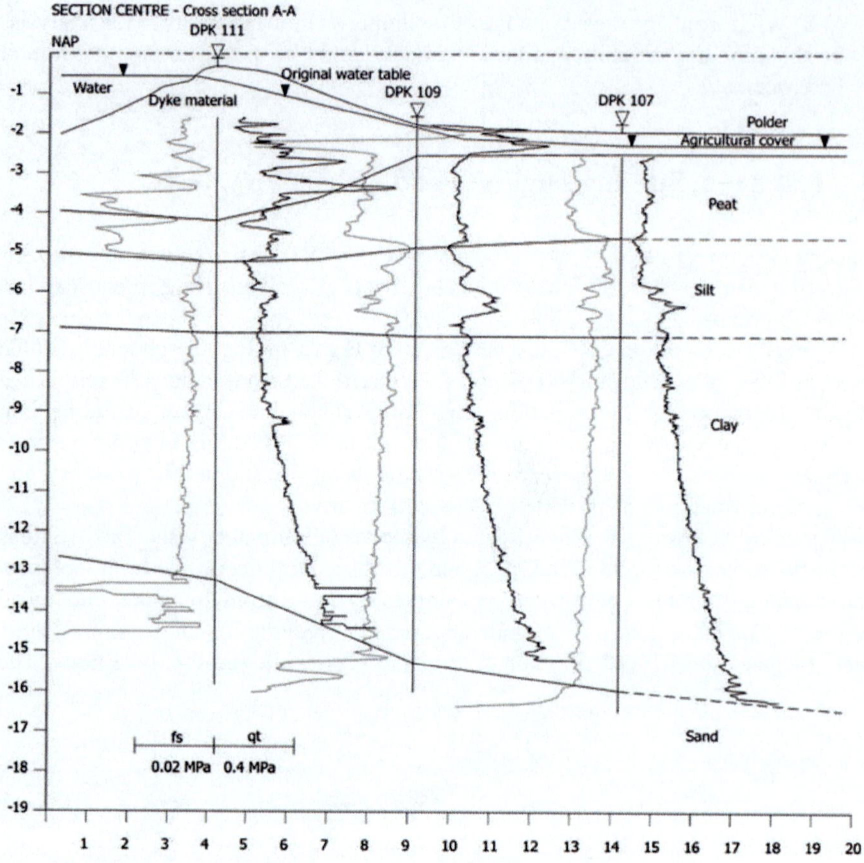

Fig. 5 Site investigation: CPT results for cone resistance and sleeve friction

oedometer tests were mostly performed by using the incremental loading, however, also a constant strain rate compression with K_0 measurement was performed on a peat sample and its elaboration was provided. The test data were cleaned for calibration and initial settings but were not elaborated for the soil parameters. The data interpretation was left to the authors.

5 Stress Test and Monitoring Data

The excavation was performed according to the designed steps, and was executed under GPS control, so that the uncertainty on its geometry was very small and could be disregarded. In Fig. 8 a picture is shown of the site during one of the dewatering stages of the stress test. The entire history of the stress test is reported in Table 1

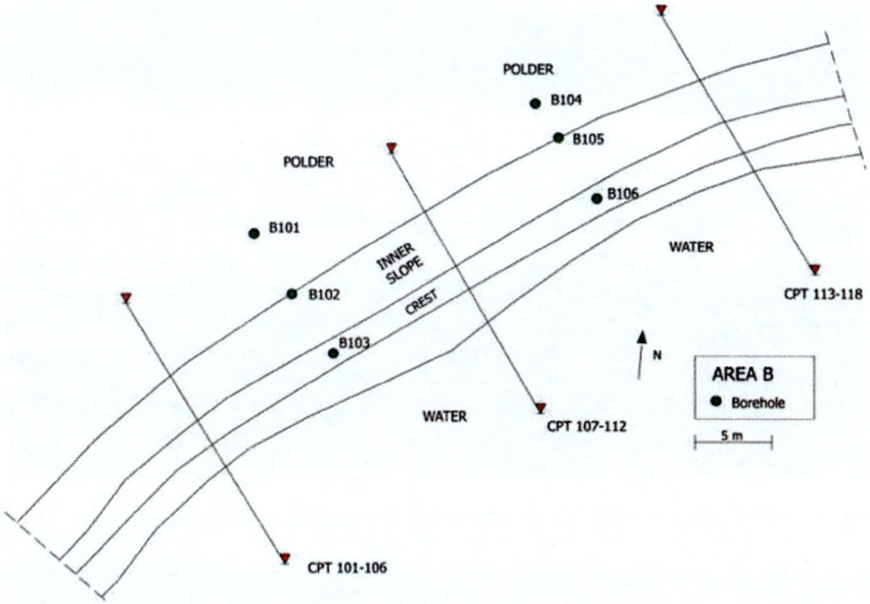

Fig. 6 Location of the boreholes for sampling

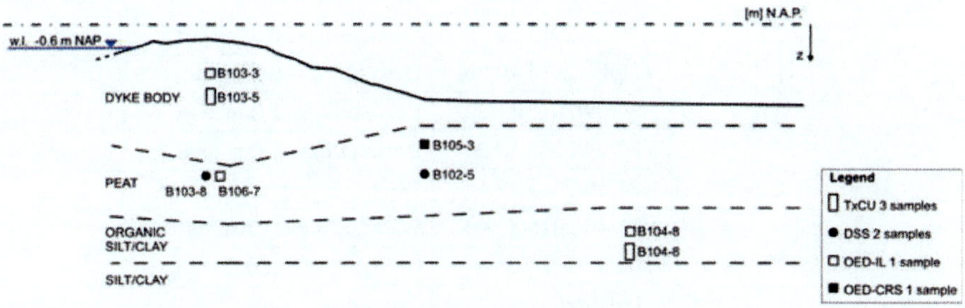

Fig. 7 Position and depths of the samples used for the laboratory investigation

and summarised in detail in the files provided to the authors. The geometry of the excavation is given in Fig. 9, where all the stages are schematically depicted.

It can be disclosed that failure was designed to occur on October 14, during one of the planned drawdown stages. The final outcome was one of the points for discussion of the benchmark exercise.

Before the start of the stress test, monitoring sensors had been installed at various locations in the dyke and the subsoil, including piezometers and inclinometers. The most relevant measurements were given for the scope of this benchmark.

All the monitoring data were elaborated to give directly relevant measurements for comparison with the numerical predictions for the questions in paragraphs 3.1

Fig. 8 View of the test site during one of the dewatering stages included in the stress test

Table 1 History of the stress test

Date	Operation
21/09/2015 15:25	Start wetting from the crest of the dyke
28/09/2015 09:20	Excavation 1 start
28/09/2015 16:15	Excavation 1 end
30/09/2015 09:20	Dewatering—0.5 m start
30/09/2015 10:20	Dewatering—0.5 m end
30/09/2015 11:20	Dewatering—1.0 m start
30/09/2015 12:00	Dewatering—1.0 m end
01/10/2015 12:45	Water filling excavation 1 start
01/10/2015 14:00	Water filling excavation 1 end
05/10/2015 08:00	Excavation 2 start
05/10/2015 17:30	Excavation 2 end
07/10/2015 09:05	Dewatering—0.5 m start
07/10/2015 11:05	Dewatering—0.5 m end
07/10/2015 12:05	Dewatering—1.0 m start
07/10/2015 13:20	Dewatering—1.0 m end
08/10/2015 08:00	Water filling excavation 2 start
08/10/2015 10:05	Water filling excavation 2 end
12/10/2015 07:20	Excavation 3 start
12/10/2015 16:45	Excavation 3 end
14/10/2015 0:38	Pump on—Dewatering start

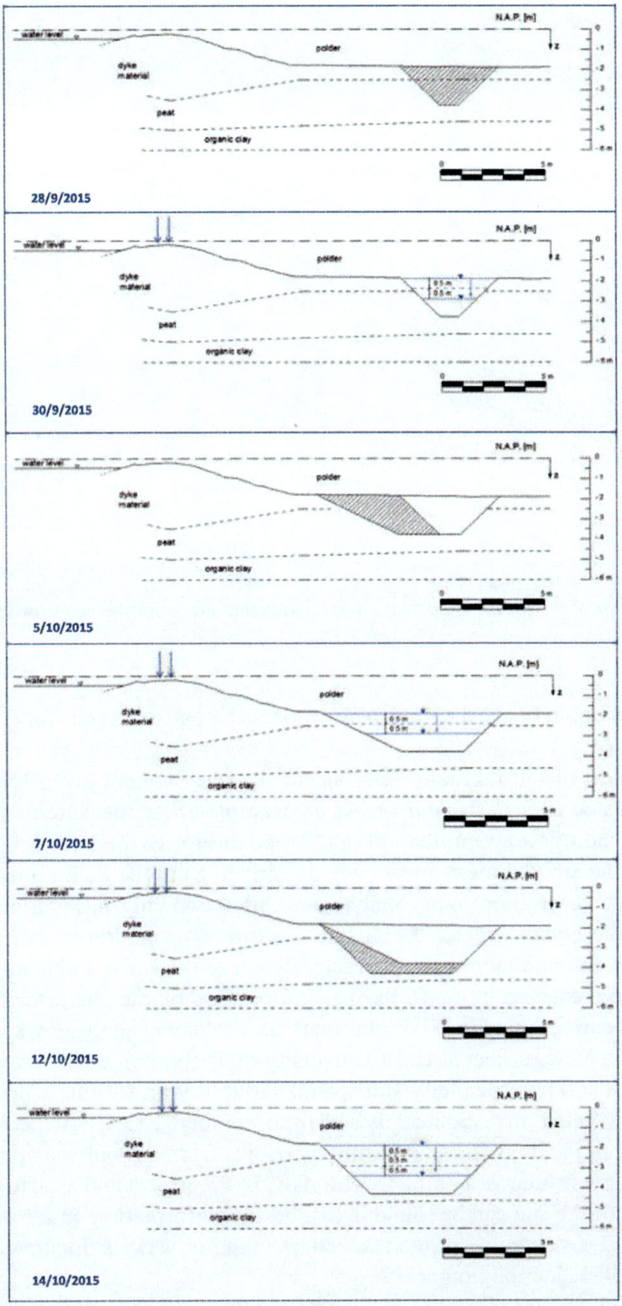

Fig. 9 Schematic drawing of the stress test stages, first excavation, first dewatering, second excavation, second dewatering, third excavation and final dewatering. The excavation is scaled with the geometry of the dyke

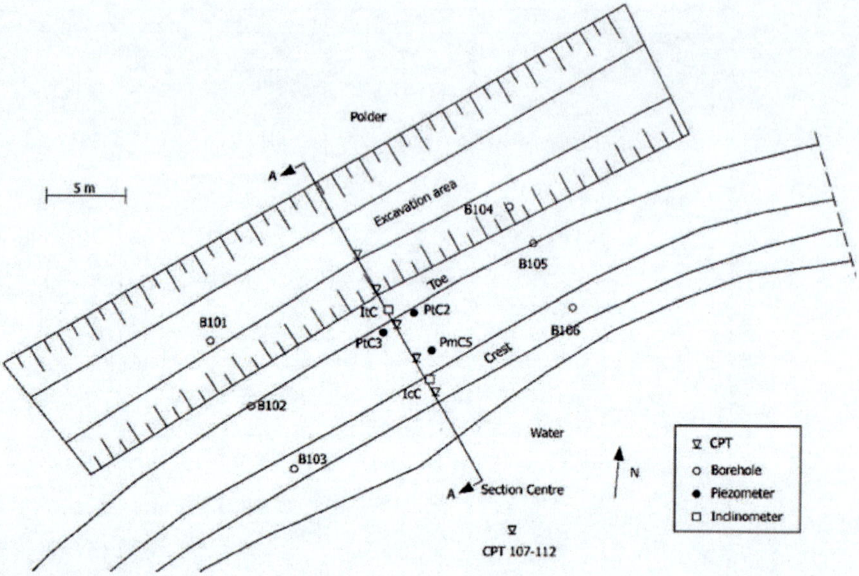

Fig. 10 Position of the piezometers and the inclinometers given for the scope of the benchmark, plan view

and 3.2. They could be used to update the models by back-analysis for question (2.e) and further.

The position of the inclinometers and of the piezometers given for the benchmark is indicated in Figs. 10 and 11. As an example of the measurements provided, piezometer readings are reported in Fig. 12, and in Fig. 13 the lateral displacements at the toe of the embankment are shown, as derived from the inclinometer data.

The stress test was thoroughly analysed as part of two PhD dissertations supported by the Reliable Dykes project. Details on the site investigation and the monitoring design, the in situ and laboratory characterisation of the soils, and detailed analysis of the test are presented in [3, 4]. Further information on the characterisation of the relevant soils can be found in [5]. Preliminary finite element analyses were performed as a result of a MSc project at Delft University of Technology.

The role of soil heterogeneity and spatial variability on various aspects of safety assessment is being investigated, based onto another 100 CPTu performed in a nearby sector of the dyke [6, 7]. Preliminary results from a geophysical investigation performed in the framework of the EU-MAGIC IAPP project and used to broaden the general view of the site can be found in [8]. Further information on the organisation, the results and the outcome of the test can be found in various documents available on the web, [9–12] among others.

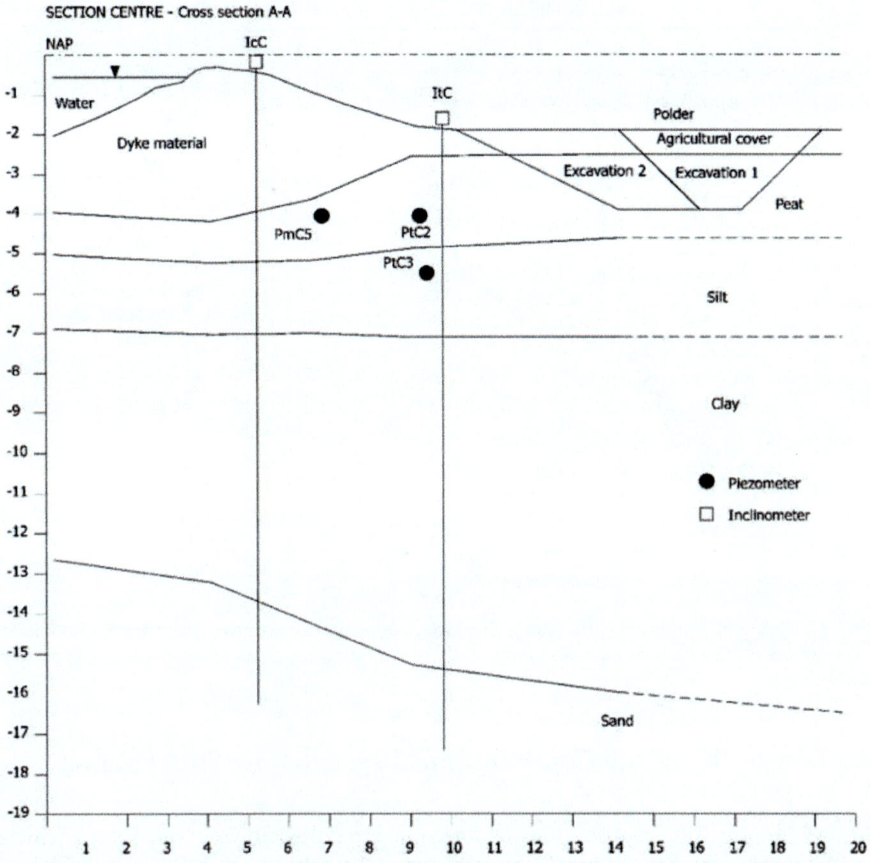

Fig. 11 Position of the piezometers and the inclinometers given for the scope of the benchmark, section view

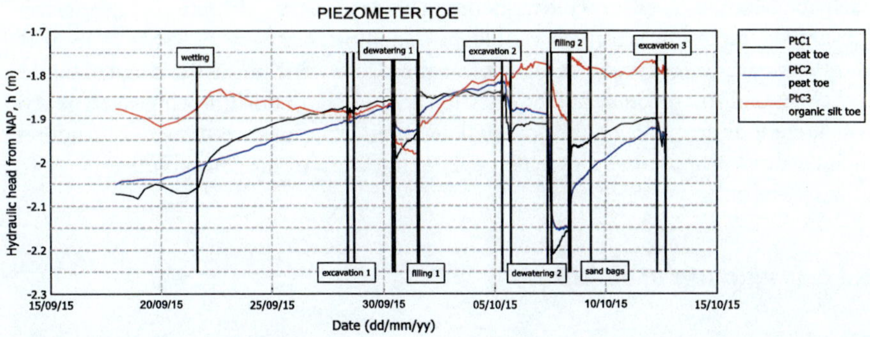

Fig. 12 Readings taken by the piezometer at the toe, reported as hydraulic heads during the stress test until the end of the third excavation stage

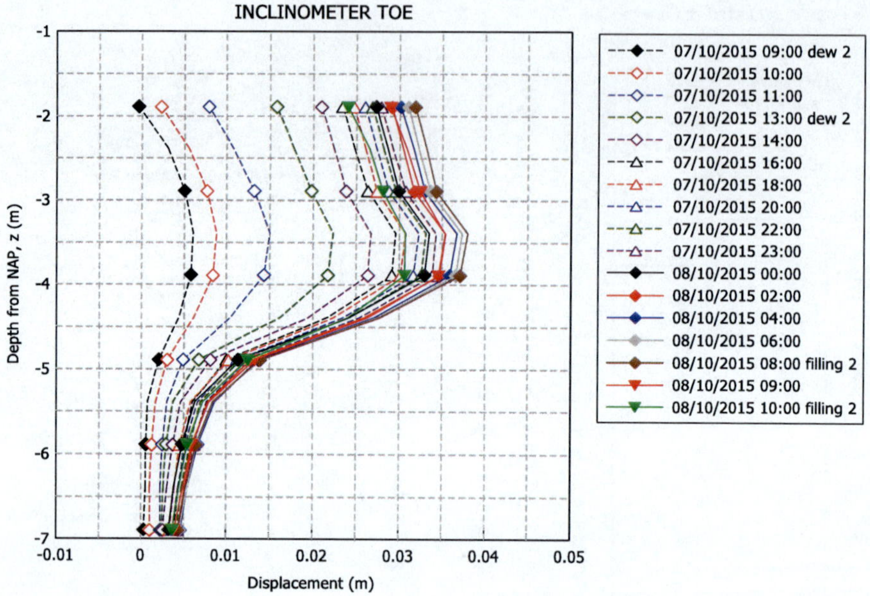

Fig. 13 Selected profiles for the lateral displacement at the toe, derived from the inclinometer measurements. The inclination was automatically recorded over time on sensors spaced 0.5 m over depth

6 Discussion of the Contributions Presented in This Volume

In this volume, four contributions on the topic are collected from the groups which participated to the benchmark. Three of them analyse the pre-failure and the onset of failure of the dyke by means of different commercial FE codes [13–15]. One of them looks into the post-failure response by means of an open-source SPH model [16]. In each contribution, details on the modelling choices, on the calibration of the parameters and on the results are given in a clear and comprehensive way. Table 2 presents a short comparative overview of the chosen approaches, including the adopted numerical codes and soil models, and the question addressed. In the following some general considerations are made on the potential and limitations of current models to address the questions proposed as they emerge from the work of the four groups.

6.1 *Prediction of Failure*

In the current practice, prediction of failure is tackled typically with simplified 2D-plane strain one-phase approaches, assuming either fully undrained or fully drained conditions.

Table 2 Numerical approaches presented

Reference	Numerical approach	Numerical code	Geometry	Type of analysis	Constitutive models	Question addressed
[13]	FEM	GEOSTUDIO	2D	Coupled H-M	LE-PP-MC[a] MCC[b]	/1/2/
[14]	FEM	Midas GTS NX, Abaqus	2D	Coupled H-M	LE-PP-MC[a] MCC[b]	/1/2/
[15]	FEM	DIANA FEA	2D/3D	Undrained/Drained/Coupled H-M	LE-PP-MC[a] MCC[b]	/1/2/3/
[16]	SPH	SPHERA (Open Source)	3D		Equivalent viscous fluid	/1/& post-failure

[a] Linear elastic – perfectly plastic – Mohr-Coulomb failure envelope
[b] Modified Cam-Clay

In contribution [15] this approach is used to provide a preliminary evaluation of the Factor of Safety (FoS) at different test stages and the bounds to the possible response in the field. The case presented was tricky in this respect because, besides excavating, failure was triggered by dewatering at the toe of the embankment. In a one-phase approach this implies assuming a given pore water pressure distribution, which corresponds to the one at the start of the stage for undrained analysis or to the final steady state one for the drained analysis, respectively. A complicating feature of a one-phase undrained analysis is the dependence of the undrained shear strength on the effective stress state and the previous stress history, which typically requires more geotechnical expertise than that needed for choosing effective Mohr-Coulomb shear strength parameters.

The results in [15] show that the critical hydraulic gradient predicted by the undrained analysis is lower than that triggering the failure in drained conditions, suggesting that the undrained approximation would be the normative one. However, the result, together with the shape of the predicted failure surface, dramatically depends on the values of undrained shear strength chosen for the different layers. The case analysed is extremely sensitive to small changes in undrained shear strength due to the very low stresses experienced by the subsoil in the field.

The drained assumption is expected to underestimate the pore pressures during the last test stage, because of two concomitant reasons. On the one hand, pore pressures should decrease because of dewatering, at a rate depending on the allowed drainage, and they are expected to remain higher than those predicted by a fully drained analysis. On the other hand, it is well known that fine soft soils tend to develop shear induced positive pore pressures upon shearing because of their tendency to contract. Therefore, consistently with the results found by [15], in this case a drained analysis is expected to give higher FoS than the undrained one.

Much better results can be obtained if the effective stress history is accounted for by means of a fully coupled hydro-mechanical analysis (Coupled H-M, in Table 2). Not only is this true for the pre-failure behaviour of the dyke, but also for better prediction of failure. Actually, in water defences, the role of pore pressure is of paramount importance in the available resistance of the dyke body and the foundation soils. Disregarding the time history of pore pressure at changing hydraulic heads, or not properly taking them into account, limits the predictive capabilities of numerical models to a large extent. Further limitation in the analyses performed with oversimplified models for the volumetric behaviour of soft soils come from the limitation of these models in properly reproducing the non-linear compressibility of the soil skeleton at varying effective stress and the volumetric-deviatoric coupling. Adopting various versions of Modified Cam Clay, albeit small differences in the chosen combination of effective strength parameters and in spite of poor information on the hydraulic conductivity of the relevant layers, all the groups predicted similar values of the critical water height, which was estimated around −3.0 m NAP.

In reality, failure occurred when the water table was brought to −3.5 m NAP, corresponding to a drawdown of about −1.5 m from the initial water height in the excavation. The higher FoS of the dyke compared to the estimated one is mostly due to 3D geometrical effects, as confirmed by some back-analyses performed at TU

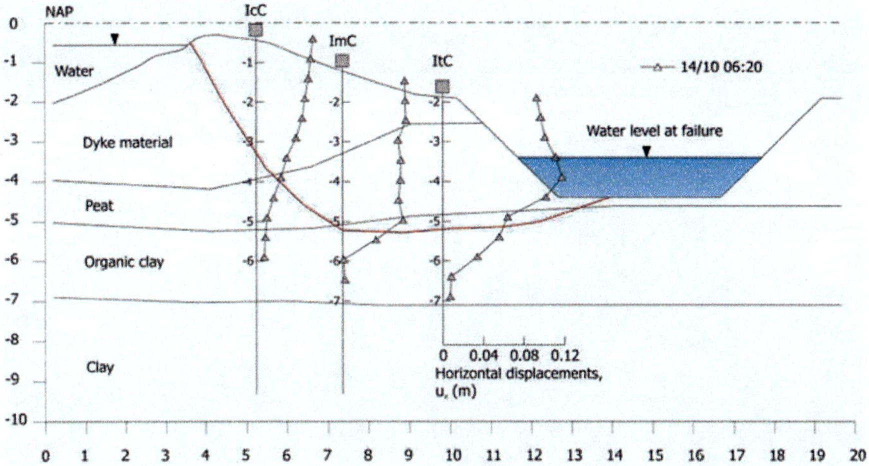

Fig. 14 Schematic profile of the failure surface cross-section, back-analysed from all the sensors data

Delft. The back-analysed failure surface is schematically shown in Fig. 14. The final failure mechanism started at the toe of the embankment and developed backwards inside the dyke body (Fig. 15). The exact depth of the failure surface could not be estimated precisely, as the inclinometer sensors were spaced at a distance of 0.5 m from each other. A failure surface occurring at the interface between the peat and the organic clay layer is a possibility. Nonetheless, the failure surface might have crossed the upper part of the clay layer as well, as indicated in the figure. As a matter of fact, the CPTu data had indicated the upper part of the clay layer as the weakest over depth.

6.2 *Pre-failure Response*

The pre-failure response is analysed by all groups with fully coupled hydro-mechanical analyses, and comparisons are presented between the calculated horizontal displacements and pore pressures and the measured ones.

The field data highlighted small displacements, in the order of mm, before the last excavation stage started. Even after the third excavation, at the start of the final dewatering stage, the maximum horizontal displacements did not overpass few centimetres. Only once failure was triggered, the displacement rate increased significantly, firstly by bulging of the peat layer and then by shearing of the dyke body and of the clay layer behind the peat.

It might be inferred that simple linear models would equally work for such small pre-failure displacements. However, this is not the case, due to the different layers suffering from quite different degree of non-linearity in a compatible strain range.

Fig. 15 Picture taken by Marcel van den Bergh for de Volkskrant, during the failure [9]

With a convenient choice of the average stiffness of the different layers, the order of magnitude of the displacement could be matched. However, the deformed shape was poorly represented. This is the main limitation of simplistic models, as the qualitative deformed shape is important to understand the mechanical response of the construction in view of assessing critical conditions. It is worth noting, in this respect, that the displacements accumulated before the final failure stage were in the same order of magnitude of the displacements recorded over one year service life of the dyke due to seasonal cyclic shrinking and swelling. Using comparable versions of Cam Clay type models, all the groups were able to obtain reasonable results, although not excellent. This is not surprising for embankments on soft soils, for which anisotropic response is expected to play a relevant role in the pre-failure response (see, e.g., [17]).

Actually, the time dependent deformational response of the dyke is the result of both the stiffness and the hydraulic conductivity of the soil layers, due to the different stress stages having a shorter duration than the consolidation period of the impervious layers. This complicates the back-analysis of the data, and hinders the possibility to easily derive uncoupled pieces of information from the field measurements. Moreover, the hydraulic conductivity of soft soils is significantly affected by the void ratio (or, equally, the stress level), to an extent that the compressed layers below the embankment body react differently from the same layers at the toe of the embankment, where they are less compressed. All the groups tackling the benchmark problem indicated hydraulic conductivity as the one most affected by uncertainty among the data which are needed to perform the analysis (e.g., [13]), and one of the groups specifically concentrated the attention of the back-analysis on this uncertainty [14]. A numerical study performed at TU Delft after the test discusses the role of different elastic-plastic constitutive models available in commercial codes on the prediction of the pore pressure evolution [18]. Although some model combinations allowed for reasonable simulations, none results to be completely satisfactory for the entire strain range experienced by the soil in the field.

The influence of spatial variation of shear strength is tackled briefly in [15], by constructing a random field derived from the CPTu data over a cross section of the dyke given as input for the benchmark case, assuming an exponentially decaying correlation function, with a horizontal correlation length 2.5 times the vertical one. The random field was applied to a simplified 3D model, consisting of a slicc of the dyke, and the results show minor effects on the numerical prediction for the model investigated. A more comprehensive study of the implication of heterogeneity on the safety assessment of this case can be found in [3].

6.3 Post-failure Response

In [16] a very interesting preliminary analysis of the post-failure response of the embankment is presented, including run-out of the dyke body, tackled with a Smoothed Particle Hydrodynamic (SPH) approach implemented in a free open source code. The contribution shows the potential of the approach in following the consequences of a failure in terms of displacement of the dyke material after failure and flow rates, adopting a full 3D geometrical model for the dyke and the subsoil. Although no data were given in the benchmark exercise to be compared with the numerical predictions, the results of the analysis match reasonably well with visual observation of the dyke post-failure [19].

It is interesting to note that this type of analysis, which was originally developed mainly for hydraulic applications, has been used recently to analyse the run-out of natural landslides. However, the contribution to this volume is one of the first applications to dyke failure. Modelling the consequences of failure would open a path towards proper risk assessment of water defences, and would help in better managing flood risk.

7 Final Remarks

The proposed benchmark was the first initiative among the ICOLD community on numerical modelling of dykes. The contributions presented in this volume offer a very good overview on the current numerical modelling capabilities for construction and management purpose.

Compared to other types of water retaining structures, difficulties arise in modelling dykes from the inherent hydro-mechanical coupling, from uncertainty and spatial variations of soil properties, also considering that different locally available cheap heterogeneous materials are frequently used in the construction.

Current models seem to be good enough to assess dyke safety, provided information is available on both the mechanical and the hydraulic properties of the soils. Whereas the first ones are typically reasonably characterised with a combination of field investigation and laboratory testing, poor information is usually available on the hydraulic response notwithstanding the role played by pore pressures in the response of the system. Simplifying assumptions on the coupled response, such as fully undrained or fully drained, can be made based on experience. However, the fully drained assumption is frequently on the unsafe side and fully undrained one strongly suffers from the difficulty in choosing representative values for undrained shear strength. A coupled analysis is definitely recommended, although it requires better characterisation of the soil behaviour. The common elastic-plastic models currently implemented in commercial codes are good enough for a preliminary evaluation of the pre-failure response. However, poor description of anisotropy and volumetric–deviatoric coupling typically hinder accurate predictions of the service conditions deformational response.

In this benchmark case, unsaturated conditions were not of relevance, because the test was designed assuring saturated conditions of the dyke body during the entire duration of the field test. However, models for dykes should also be benchmarked against more general situations, typically including different time histories of saturation during time dependent loading.

Besides assessing the failure hazard, the post-failure response is of relevance for better management of flood risk. This benchmark problem also gave the opportunity to broaden the modelling perspective towards methods able to switch to full risk assessment.

Acknowledgments The authors wish to acknowledge STOWA, the foundation for research on regional dykes in the Netherlands, for offering and financing the opportunity to perform the full scale stress test, in the framework of a cooperative effort with different Water Authorities and Provinces in the Netherlands. Thanks are especially due to Ludolph Wentholt from STOWA for his continuous support. André Koelewijn, Dennis Peters and Bernard van der Kolk from Deltares are acknowledged for their expert cooperation in the management of the stress test in the field.

References

1. https://www.rijnland.net/werk-in-uitvoering/overige-werkzaamheden/leendert-de-boersp older
2. https://www.youtube.com/watch?v=uU_HEJkcouo

3. de Gast T (2020) Dykes and embankments: a geostatistical analysis of soft terrain. PhD dissertation, Delft University of Technology, the Netherlands
4. Muraro S (2019) The deviatoric behaviour of peat: a route between past empiricism and future perspectives. PhD dissertation, Delft University of Technology, the Netherlands
5. Ponzoni E (2017) Historical constructions on natural silty soils accounting for the interaction with the atmosphere. PhD dissertation, Università degli Studi di Brescia, Italy
6. de Gast T, Vardon PJ, Hicks MA (2017) Estimating spatial correlations under man-made structures on soft soils. In: Proceedings of the 6th international symposium on geotechnical safety and Risk, 4–6 June, Denver, Colorado, pp 382–389
7. de Gast T, Vardon PJ, Hicks MA (2019) Observations and considerations regarding estimating horizontal scales of fluctuation around linear infrastructure. In: Proceedings of the 7th international symposium on geotechnical Safety and risk, 11–13 Dec, Taipei, Taiwan, pp 340–345
8. Tarantino A, Gallipoli D, Jommi C, Mendes J, Capotosto A, Amabile A, Pedrotti M, Pozzato A, Beneš V, Bottaro F, Denzer H, Boeck F (2016) Advances in the monitoring of geo-structure subjected to climate loading. In: Proceedings E-UNSAT 2016. E3S web of conferences 9, 04001, pp 1–11. https://doi.org/10.1051/e3sconf/20160904001
9. Waardige strijd tussen wetenschappers en dijk—de Volkskrant—14 Oct 2015
10. Geslaagde proef bij Schiphol met dijkdoorbraak—NOS.nl—14 Oct 2015
11. Veendijk bezweken, proef geslaagd—Het Waterschap—N 12, Dec 2015
12. Het succes van een dijkdoorbraak—Contact—Feb 2016
13. Bakeš M, Mészáros T, Minárik M (2020) Analysis of the pre-failure and failure behaviour of a levee on soft subsoil, this Volume
14. Marulanda C, Tello J, León-Vanegas D (2020) Coupled hydro-mechanical analysis of the Leendert de Boerspolder dyke stress-test, this Volume
15. Schwager MV, Tzenkov AD, Schreppers GMA (2020) Analysis of pre-failure and failure of a levee on soft soil with Modified Cam-Clay model in DIANA, this Volume
16. Amicarelli A, Abbate E (2020) SPH modelling of the Kagerplassen dyke failure, this Volume
17. Karstunen M, Krenn H, Wheeler SJ, Koskinen M, Zentar R (2005) Effect of anisotropy and destructuration on the behavior of Murro test embankment. Int J Geomech 5(2):87–97
18. Theodoridis M (2017) Numerical simulation of the coupled hydro-mechanical response of the Leendert de Boerspolder dyke. MSc thesis, Delft University of Technology, the Netherlands
19. https://www.youtube.com/watch?v=5jFiwiw58cM

SPH Modelling of the Kagerplassen Dyke Failure

A. Amicarelli and E. Abbate

Abstract Smoothed Particle Hydrodynamics (SPH) is a mesh-less Computational Fluid Dynamics method suitable for several application fields such as floods, fast landslides and sediment removal from water bodies. As a preliminary demonstration, this method is herein applied to simulate the on-site experiment of the 3D full-scale Kagerplassen dyke failure. The geometries of the granular media and the water reservoir are elaborated from the available measures by means of analytical procedure. Results are provided in terms of: 3D fluid dynamics fields (medium interfaces and velocity); hydrographs (time series) for the medium/fluid level (maximum height), flow rate, cumulated volumes and velocity. The 3D SPH model simulates the triggering and propagation of the sliding surfaces within the dyke and simulates the following run-out of both the granular material and the water flood.

Keywords Smoothed particle hydrodynamics · Computational Fluid Dynamics · Floods · Landslides · Dyke failure · SPHERA software

1 Introduction

The numerical simulation of the principal failure mechanisms of an earth-filled dyke and the following run-out (which might involve both fast landslides and a flood) might provide a precious contribution to secure this type of flood-control works and preserve human health and the surrounding environment from the associated damage. In the next future, Computational Fluid Dynamics (CFD) codes might represent key numerical tools to simulate dyke failure dynamics even though, by now, 3D CFD studies on dyke failure seem rare or absent in the reference literature. The present study reports the preliminary results of a demonstrative study on the application of a CFD-SPH (Smoothed Particle Hydrodynamics) code to the Kagerplassen dyke failure.

A. Amicarelli (✉) · E. Abbate
Department SFE, Ricerca sul Sistema Energetico - RSE SpA, Milan, Italy
e-mail: andrea.amicarelli@rse-web.it

G. Bolzon et al. (eds.), *Numerical Analysis of Dams*, Lecture Notes in Civil Engineering 91, https://doi.org/10.1007/978-3-030-51085-5_37

2 SPH Modelling for Dense Granular Flows

The numerical tool used for the present study is SPHERA v.9.0.0 [1, 2], a research FOSS ("Free/Libre and Open-Source Software") code based on the CFD-SPH method. SPHERA has been applied to floods (with transport of solid bodies, bed-load transport and a domain spatial coverage up to some hundreds of squared kilometers), fast landslides and wave motion, sediment removal from water reservoirs, sloshing tanks, hydrodynamic lubrication.

The features of SPHERA used for this study are synthesized in the following.

SPHERA owns a scheme [3] to represent mixtures of fluid phase and non-cohesive solid granular material, under the "packing limit" of the Kinetic Theory of Granular Flow (KTGF [4]) for dense granular flows. This limit refers to the maximum values of the solid phase volume fraction and is peculiar of bed-load transport (e.g., erosional dam breaks) and fast landslides.

Adopting a Weakly Compressible approach, the continuity equation for the mixture reads:

$$\frac{d\rho}{dt} = -\rho\frac{\partial u_j}{\partial x_j} \tag{1}$$

where ρ (kg/m^3) is density, $\underline{u}$ (m/s) is the velocity vector, t (s) is time and $\underline{x}$ (m) is the vector of the spatial coordinates. The mixture density and velocity (the subscript "$_m$" for the mixture quantities are omitted for simplicity of notation) are defined as follows:

$$\rho \equiv \rho_f\varepsilon_f + \rho_s\varepsilon_s, \quad u_i \equiv \frac{\rho_f\varepsilon_f u_{f,i} + \rho_s\varepsilon_s u_{s,i}}{\rho}, \quad i = 1, 3 \tag{2}$$

The volume fractions (ε) of the fluid ("$_f$") and the solid ("$_s$") phases are constrained to the volume balance equation:

$$\varepsilon_s + \varepsilon_f = 1 \tag{3}$$

The mixture particles do not exchange net mass fluxes with the surrounding environment: this is a reasonable hypothesis for high solid volume fractions in saturated soils, according to the "packing limit" of the Kinetic Theory of Granular Flow [4].

Following the multi-phase approach of [5], the SPH approximation of the continuity equation can be expressed as follows:

$$\frac{d\rho_0}{dt} = \rho_0\sum_b \left(u_{b,j} - u_{0,j}\right)\left.\frac{\partial W}{\partial x_j}\right|_b \omega_b + 2\rho_0 \int_{V_h'} \left[\left(\underline{u}_w - \underline{u}_0\right)\cdot\underline{n}\right] n_j \frac{\partial W}{\partial x_j} dx^3 \tag{4}$$

where W (m^{-3}) is the kernel function [6], ω (m^3) is the SPH particle volume and V_h' (m^3) is the portion of the SPH influence region (i.e. the sphere of radius 2 h-m- around a generic computational particle "$_0$" which influences its balance equations at a fixed time step) truncated by the frontiers of the fluid domain. The subscript "$_b$" refers to the "neighboring particles" (i.e. fluid particles within the influence region of the computational particle). The integral boundary term in (4) is computed according to [7] and represents the effects of fixed wall frontiers (subscript "$_{SA}$" in the following).

The form of the momentum equations for the mixture is identical to Navier-Stokes equations:

$$\frac{du_i}{dt} = -\delta_{i3}g - \frac{1}{\rho}\frac{\partial p}{\partial x_i} + \nu\frac{\partial^2 u_i}{\partial x_j^2} \quad (5)$$

where p (Pa) is the mixture total pressure/stress, g (m/s^2) is gravity acceleration and ν (m^2 × s^{-1}) is the mixture kinematic viscosity. The total stress and density are related by means of a linearized barotropic equation of state:

$$p \cong c_{ref}^2 (\rho - \rho_{ref}) \quad (6)$$

A unique speed of sound (c_{ref}, m/s) can be chosen (i.e. the highest among the SPH particle values, no matter about their phase volume fractions).

The mixture dynamic viscosity $\mu \equiv \upsilon\rho$ (Pa × s) is defined as:

$$\mu \equiv \varepsilon_f \mu_f + H(\varepsilon_s - \varepsilon_{s,p})\mu_{fr} \quad (7)$$

where H is the Heaviside step function.

In the "packing limit" of the KTGF (i.e. for ε_s close enough to the value of $\varepsilon_{s,p}$ = ca.0.59, which is the maximum attainable solid volume fraction for a sheared inelastic hard sphere fluid, [8]), the shear stress gradient term is represented by means of a visco-plastic model for dry granular material based on internal friction [9], by means of a physical quantity named frictional viscosity μ_{fr} (Pa × s):

$$\mu_{fr} \equiv \left(\frac{\sigma_m'(\sin\phi)}{2\sqrt{I_2(e_{ij})}}\right) \quad (8)$$

Herein φ (rad) is the internal friction angle, e_{ij} (s^{-1}) the strain-rate tensor and $I_2(e_{ij})$ (s^{-2}) is its second invariant (formulation for incompressible fluids):

$$e_{ij} \equiv \frac{1}{2}\left(\frac{\partial u_i}{\partial x_j} + \frac{\partial u_j}{\partial x_i}\right),\ |e_{ij}| \equiv \left(\sum_{i,j} e_{ij}^2\right)^{\frac{1}{2}} = \sqrt{2I_2(e_{ij})} \quad (9)$$

The mean effective stress σ_m' (Pa) is computed as the difference between the total stress and the fluid pressure, according to the principle of [10]:

$$p = p_f + \sigma_m^{'} \tag{10}$$

The fluid pressure in the granular material is related to two different soil conditions, as follows:

$$p_f = \begin{cases} p_{f,blt-top} + \rho_f g(z_{blt-top}\big|_{x_0,y_0} - z_0)\cos^2(\alpha_{TBT}), saturated \\ 0, \ dry \end{cases} \tag{11}$$

where α_{TBT} (rad) is the topographic angle at the top of the bed-load transport layer (it lies between the local interface normal and the vertical) and the subscript "$_{blt\text{-}top}$" refers to the top of the bed-load transport layer (or the layer of saturated material). This equation assumes a 1D filtration flow parallel to the slope of the granular material. This simplifying hypothesis is still consistent with SPH conservative particles.

Following the multi-phase approach of [5], with the boundary treatment method proposed by (7), the SPH approximation of the momentum equations becomes:

$$\left\langle \frac{du_i}{dt} \right\rangle_0 = -\delta_{i3} g + \frac{1}{\rho_0} \sum_b (p_b + p_0) \frac{\partial W}{\partial x_i}\bigg|_b \omega_b + 2\frac{p_0}{\rho_0} \int_{V_h'} \frac{\partial W}{\partial x_i}\bigg|_b dx^3 +$$
$$+ 2\upsilon \sum_b \frac{m_b}{\rho_0 r_{0b}} (\underline{u}_b - \underline{u}_0) \frac{\partial W}{\partial r}\bigg|_b +$$
$$- \upsilon_M \sum_b \frac{m_b}{\rho_0 r_{0b}^2} (\underline{u}_b - \underline{u}_0) \cdot (\underline{x}_b - \underline{x}_0) \frac{\partial W}{\partial x_i}\bigg|_b + -\upsilon_M (\underline{u}_{SA} - \underline{u}_0) \cdot \left(\int_{V_h'} \frac{1}{r_{0w}^2} (\underline{x} - \underline{x}_0) \frac{\partial W}{\partial x_i} dx^3 \right) \tag{12}$$

where m (kg) is the particle mass and r (m) the relative inter-element distance.

In order to detect the elasto-plastic regime and avoid the unbounded growth of the frictional viscosity, the threshold υ_{max} ($m^2 \times s^{-1}$) for the mixture viscosity is introduced. Mixture particles with a higher viscosity value are considered in the elasto-plastic regime of soil deformation. As their velocities are negligible (with respect to the velocity scale of the KTGF packing limit), these particles are kept fixed as long as they belong to this regime and their pressure is derived from the mixture particles flowing above them. The viscosity threshold is assumed to be high enough not to influence the simulation results.

Further details on the scheme for dense granular flows are available in [3].

Time integration is ruled by a second-order Leapfrog scheme, as described by [7, 11]:

$$x_i\big|_{t+dt} = x_i\big|_t + u_i\big|_{t+dt/2} dt, \quad i = 1, 2, 3$$
$$u_i\big|_{t+dt/2} = u_i\big|_{t-dt/2} + \left\langle \frac{du_i}{dt} \right\rangle\bigg|_t dt, \quad i = 1, 2, 3$$

$$\rho|_{t+dt} = \rho|_t + \left\langle \frac{d\rho}{dt} \right\rangle \bigg|_{t+dt/2} dt \tag{13}$$

where *dt* (s) is the time step duration. Time integration is constrained by the following stability criteria:

$$dt = \min_0 \left\{ C_\nu \frac{2h^2}{\nu}; CFL \frac{2h}{c + |\underline{u}|} \right\} \tag{14}$$

where *CFL* is the Courant-Friedrichs-Lewy number. Following [12], the viscous term stability parameter is set to $C_\nu = 0.05$.

3 Kagerplassen Dyke Failure: Results

The Kagerplassen benchmark data are provided by [13]. Some official images of the in situ dyke failure are available at [14].

The geometry of the granular media and the water reservoir are obtained by a procedure featured by the following stages: raw data digitization; roto-translation; elaboration of parametric curves; extrusion; surface mesh generation; format conversion.

The digitization stage is carried out by means of Engage Digitizer [15]. Both the top view (Fig. 1, center panel) and the reference section "AA" are digitized. The latter is processed by treating the two images associated with both the second excavation (where the reservoir bottom and other details are accurate) and the third excavation (to provide the complete drill profile). The top view map is georeferenced thanks to the benchmark data on the borehole positions. The following borehole points are used: *B101* (x = 6.900 m; y = 37.150 m); *B103* (x = 12.160; y = 29.020); *B105* (x = 27.010; y = 43.420). The above coordinates refer to the numerical reference system, which is translated with respect to the benchmark reference system (for the top views) by the following offsets: $x_{off,1}$ = 101,200 m; $y_{off,2}$ = 469700 m.

The horizontal coordinates of the vertical section "AA" seem incoherent with the top view data. Thus, a 2D roto-translation is mandatory and does not affect the heights. The roto-translation parameters are assessed considering two selected points, whose coordinates are available on both the maps. These points are: the intersection between "section AA" and the coastline; the intersection between "section AA" and the drill downstream edge. This couple of points (a vector) is selected (as the best combination among the eight points available on both views) because it provides the least deformation errors during the roto-translation stage (both raw data and digitization seem affected by errors). The rotation from the reference system of the vertical and the reference system of the top view is carried out as follows. The rotation angle θ_R is defined by means of the atan$_2$ function:

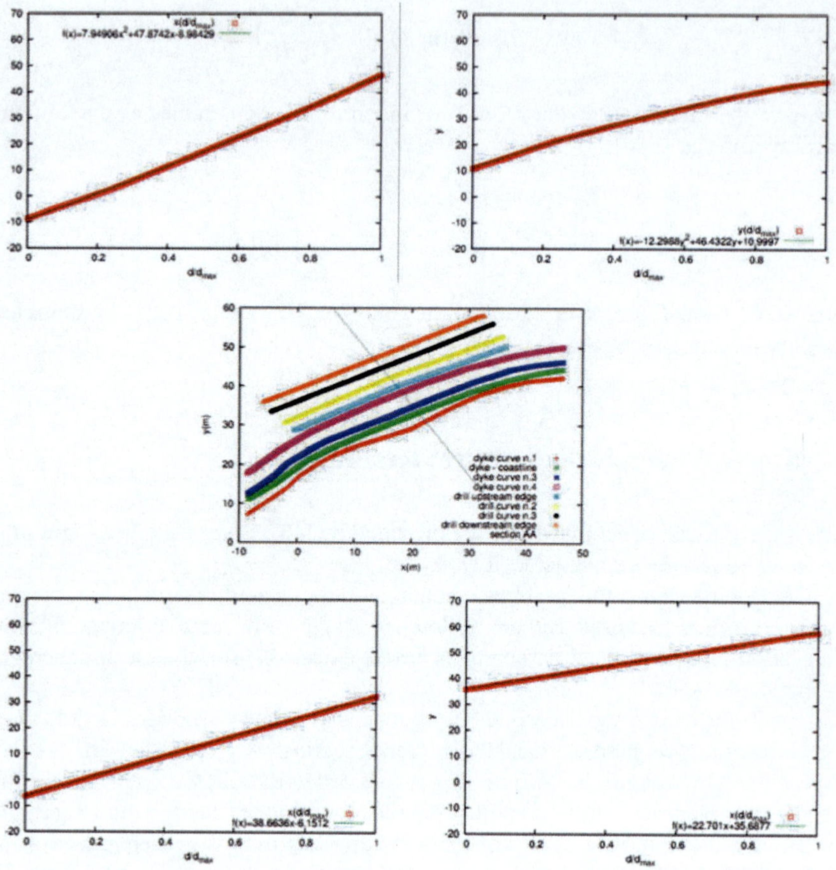

Fig. 1 Digitized data (plane view) and regression parametric curves for the geometries of the involved media. Top row: dyke-coastline intersection. Centre row: digitization of the available curves. Bottom row: excavation downstream edge

$$\theta_R = ATAN_2(\sin\theta_R, \cos\theta_R) = \begin{cases} ATAN\left(\dfrac{\sin\theta_R}{\cos\theta_R}\right), & \cos\theta_R > 0 \\ \pi + ATAN\left(\dfrac{\sin\theta_R}{\cos\theta_R}\right), & \cos\theta_R < 0 \\ \dfrac{\pi}{2}, \quad \cos\theta_R = 0, \quad \sin\theta_R > 0 \\ -\dfrac{\pi}{2}, \quad \cos\theta_R = 0, \quad \sin\theta_R < 0 \end{cases} \tag{15}$$

where the cosine and sine functions are derived from the cross product and the dot product of two vectors ($\underline{v}_A$, $\underline{v}_B$), which represent the relative distance of the couple of selected points on both the reference systems:

$$\sin\theta_R = \frac{|\underline{v_A} \times \underline{v_B}|}{|\underline{v_A}||\underline{v_B}|}, \quad \cos\theta_R = \frac{\underline{v_A} \cdot \underline{v_B}}{|\underline{v_A}||\underline{v_B}|} \tag{16}$$

The resulting roto-translation parameters are: the rotation angle $\theta_R = 119.36776°$ and the offsets $x_{off,2} = 101{,}217.6$ m and $y_{off,2} = 469{,}733.3$ m.

The elaboration of the parametric curves concerns two reference curves (i.e. the coastline and the drill downstream edge) out of the eight curves within the top view map (the digitization is reported in Fig. 1, center panel). This couple of curves is associated with the choice of the two points used for the roto-translation stage. The curves are parameterized with respect to their relative length $d/d_{\max} \in [0, 1]$ and then approximated by means of quadratic regression curves, as reported in Fig. 1 (top and bottom rows).

During the following stage, each point of the vertical section "AA" is used to describe a generic extruded curve in 3D. The x-values of the points which discretize this curve are linear interpolations of the two regression curves for x (from Fig. 1, top left panel and bottom left panel) as function of the horizontal position within the map of the vertical "section AA". Analogously, the y-values of the points which discretize a generic extruded curve are linear interpolations of the two regression curves for y (from Fig. 1, top right panel and bottom right panel).

From the above extruded curves, a 3D Delaunay triangulation is carried out by Paraview [16] within each medium volume. From every 3D triangulation, a surface grid is extracted: this is not a computational mesh, but a positioning grid to define the edges of each medium and set the initial positions of the SPH particles. The surface meshes are exported as ".ply" files. A top view of the resulting numerical domain is shown in Fig. 2.

The ".ply" file is converted into the specific formats suitable for SPHERA.

All the granular media are completely saturated (the dyke granular material is defined as "almost completely saturated" by [13].

The specific weight γ (kN/m^3) and the void ratio e_v are benchmark data, whereas the porosity ε_f and the solid phase density ρ_s (kg/m^3) are obtained by means of the following expressions:

$$\gamma = g\left[\rho_s\left(1-\varepsilon_f\right) + \rho_f\varepsilon_f\right] \Rightarrow \rho_s = \frac{\gamma}{\left(1-\varepsilon_f\right)g} - \rho_f\frac{\varepsilon_f}{\left(1-\varepsilon_f\right)}$$
$$e_v = \frac{\varepsilon_f}{\varepsilon_s} = \frac{\varepsilon_f}{\left(1-\varepsilon_f\right)} \Rightarrow \varepsilon_f = \frac{e_v}{\left(1+e_v\right)} \tag{17}$$

The input quantities for the granular media are synthesized in Table 1.

Initial conditions are defined at the end of the "excavation 3", considering null velocity vectors, whereas hydrostatic conditions are dynamically imposed. The values on the internal friction angles are chosen according to the experimental tests carried out by Politecnico di Milano and Delft University. An outlet section, located at the downstream section of the domain, just above the top soil layer. The maximum viscosity (for every granular media) is set to $\mu_{max} = 1\times10^8$ Pa $\times$ s. This value is

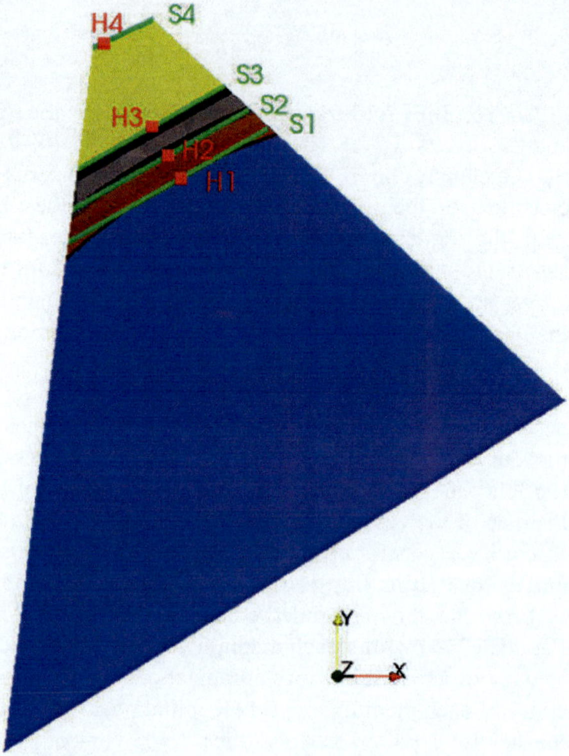

Fig. 2 Geometry of the involved media (blue: water; brown: dyke; grey: peat; yellow: top soil), numerical domain, monitoring lines (green) and sections (red). Top view

Table 1 Input quantities for the granular media

Granular medium	$\varphi(°)$	$\gamma(\text{kN/m}^3)$	$e_{v,mean}$	ε_f	$\rho_s(\text{kg/m}^3)$
Dyke	30ca.	18.5	0.66	0.40	2,478
Peat	30ca.	10.0	9.40	0.90	1,198
Top soil	30ca.	13.5	1.90	0.66	2,108
Organic silt/clay	20ca.	13.5	2.40	0.71	1,068

preliminary as no convergence analysis is carried out. The final simulated time is $t_f = 10$ s.

The SPH spatial resolution is defined by $dx = 0.5$ m and $h/dx = 1.3$. The simulation assumes $CFL = 0.05$ and $C_\nu = 0.05$. Four vertical monitoring lines are used as numerical probes with a vertical spatial resolution of 0.01 m: H_1(21.110 m; 33.288 m); H_2(17.600 m; 39.510 m), H_3(13.310 m; 47.149 m); H_4(0.600 m; y = 69.600 m). Four monitoring sections are defined to monitor the flow rates: S_1 (coastline), S_2 (excavation upstream edge), S_3 (excavation downstream edge) and S_4 (outlet section).

For the preliminary results of this on-going study, several approximations are assumed: the limiting viscosity for the dyke is too low ($\mu_{lim} = 1 \times 10^{-1}$Pa $\times$ s); no convergence analysis is carried for the maximum viscosity; three media are featured by an imposed null kinematics (top soil, peat, organic silt/clay); spatial resolution is relatively coarse; initial conditions for pressure are approximated; the final time is reduced; the surface meshes of the involved media and the water reservoir have many elements of low quality and are not optimized; the actual internal friction angles are under revision; the probe H_4 is deactivated; the geometry of the symmetry planes are approximated; the role of vegetation is not considered; the configuration of the monitoring lines is not optimized.

Nonetheless, the following preliminary results provide a demonstration for the present SPH code on simulating a 3D full scale dyke failure.

Figure 3 reports a time sequence of the 3D field of the involved media. The excavation induces the dyke failure and the subsequent water reservoir overtopping. The mobilized granular material of the dyke accumulates over the excavation bottom. At $t_f = 10$ s, the water flood front reaches around half of the excavation bottom. However, the approximations herein assumed (especially an imposed null kinematics for the peat) strongly slow down the flood dynamics. The 3D model dynamically estimates the sliding surfaces within the dyke and simulates the following run-out of both the dyke material and the water flood.

Figure 4 reports a time sequence of the 3D field of the absolute value of velocity. The maximum fluid velocity is $|\underline{u}|$ = ca.5 m/s. The maxima are located at the upstream edge of the excavation bottom and involve both water and the dyke granular material. During the simulated period the free surface velocity progressively grows within the water reservoir and a regressive velocity wave is detected. At $t_f = 10$ s, a temporary stagnation zone is recorded at the lee side of the excavation bottom, due to the impact of the dyke front with the excavation downstream slope.

Figure 5 reports the flow rate hydrographs for water (left panel) and the dyke saturated granular material (right panel) at the monitoring sections. The maximum water flow rate is recorded along the initial coastline (section S_1), whereas the maximum sediment flow rate is monitored at the excavation upstream edge (section S_2) with a peak arrival time smaller than 2 s. Even if the output time frequency is herein too little at the monitoring sections, the maximum flow rate values lie below $Q = 300$ m^3/s and no relevant flow rate is recorded at the excavation downstream edge (S_3) and the outlet section (S_4).

Figure 6 shows the time series of the cumulative volumes between the monitoring sections and downstream the numerical domain for water (left panel) and the dyke saturated granular material (right panel). The cumulated water volume crossing the reservoir outlet section during the simulated period is $V_{cum, 1\text{–}2, wat}$ = ca.474 m^3 The cumulated volumes within the excavation area at $t_f = 10$ s are $V_{cum, 2\text{–}3, wat} = 206$ m^3 for water and $V_{cum, 2\text{–}3, dyke} = 377$ m^3 for the saturated granular material of the dyke. While the above water time series are interested by positive time derivatives at the end of the simulated period, the dyke volume within the excavation area seem to achieve an almost stationary value.

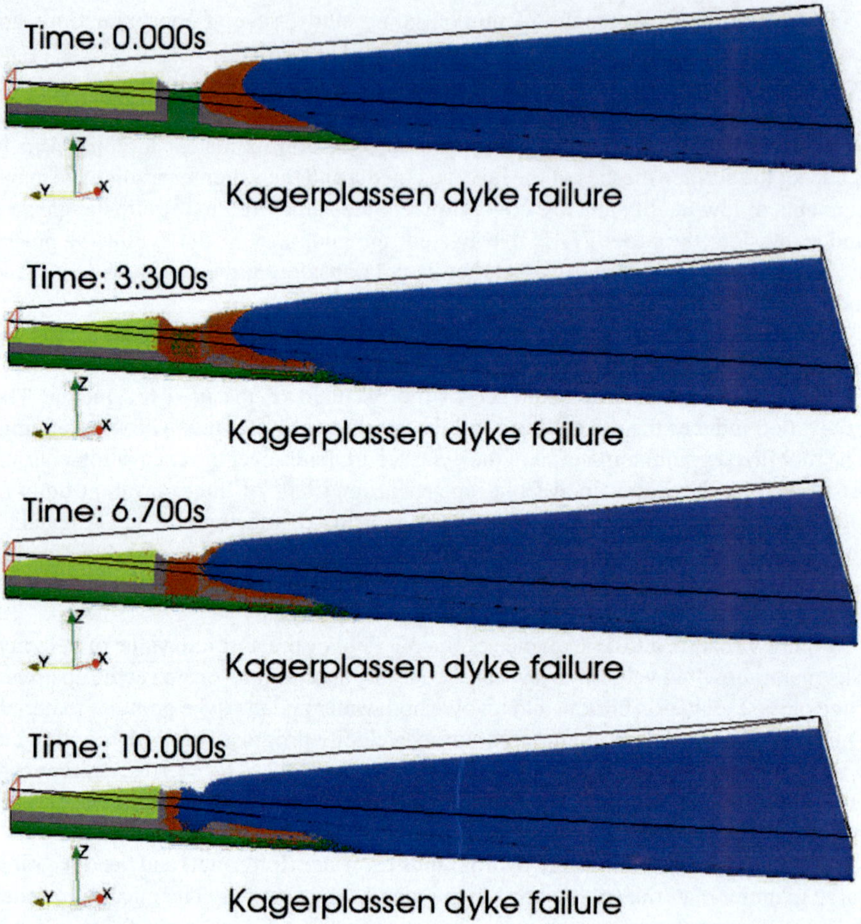

Fig. 3 Time sequence of the 3D field of the media (blue: water; brown: dyke; grey: peat; yellow: top soil; green: organic silt/clay). Preliminary results

Figure 7 reports the hydrographs for the absolute value of velocity (left panel) and the medium maximum height/level (right panel) at the vertical monitoring lines. Within the four monitoring lines, the maximum fluid velocity (ca.4.0 m/s) is recorded for t = ca.1.5s at the excavation upstream edge (probe H2). The free surface at the coastline (probe H1) reduces of ca.2 m during the first 1.5 s. One notices that the present hydrographs are also affected by errors due to non-optimized configurations of the probes.

In the future, SPH simulations will also be performed on a 2D domain built on the vertical "section AA". After its digitization, vertices and edges have been converted to the specific SPHERA format. The resulting domain and the edges of the different media are shown in Fig. 8, where the outer boundaries are considered as solid walls

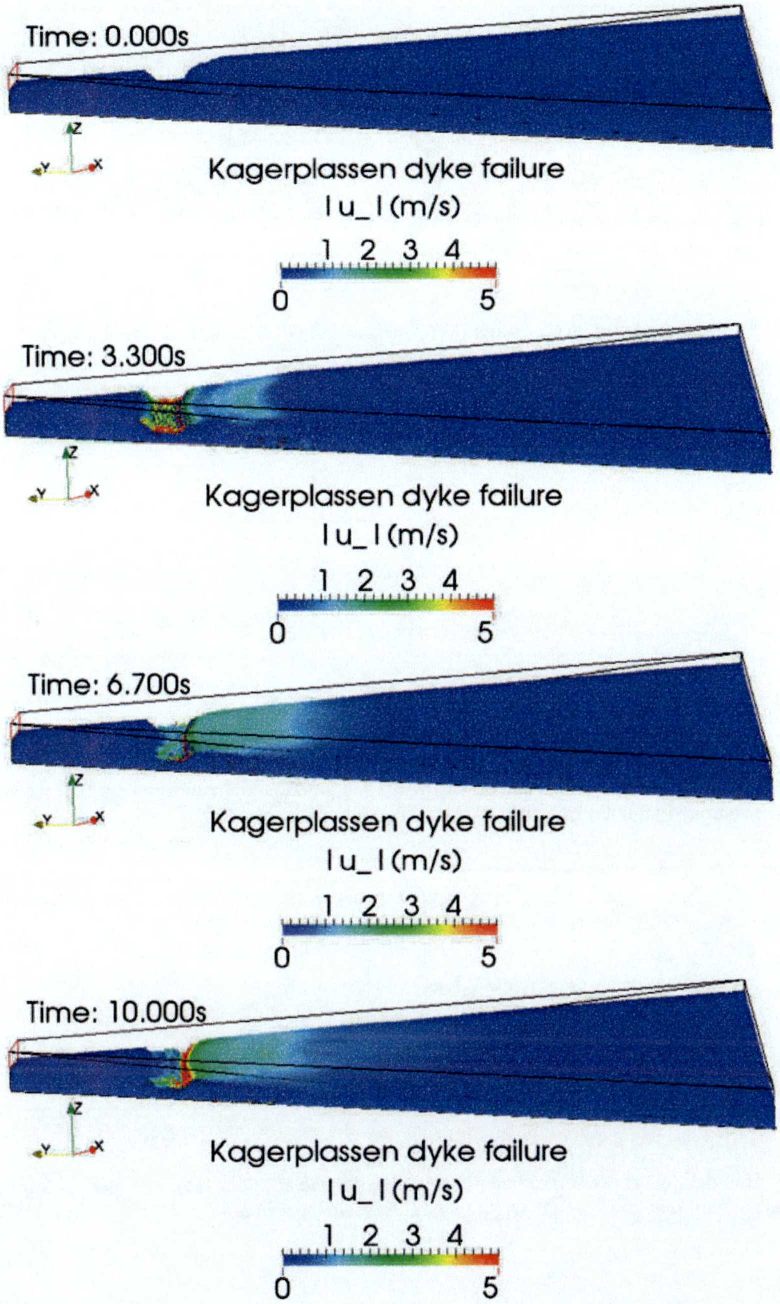

Fig. 4 Time sequence of the 3D field of the absolute value of velocity. Preliminary results

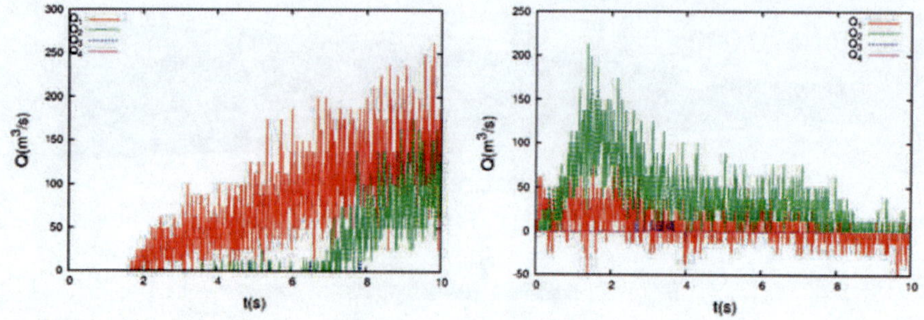

Fig. 5 Flow rate hydrographs for water (left panel) and the dyke saturated granular material (right panel) at the monitoring sections. Preliminary results

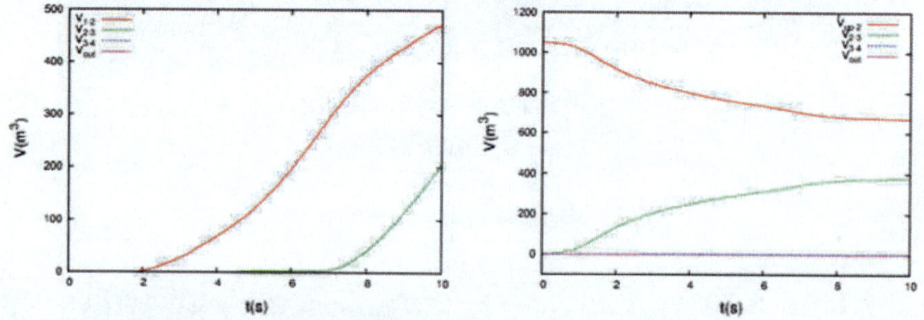

Fig. 6 Flow rate hydrographs for water (left panel) and the dyke saturated granular material (right panel) at the monitoring sections. Preliminary results. Time series of the cumulative volumes between the monitoring sections and downstream the numerical domain for water (left panel) and the dyke saturated granular material (right panel). Preliminary results

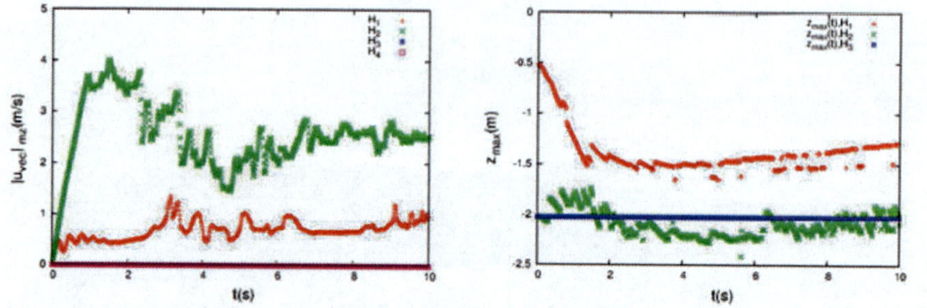

Fig. 7 Hydrographs at the vertical monitoring lines for the absolute value of velocity (left panel) and the medium maximum height (right panel). Preliminary results

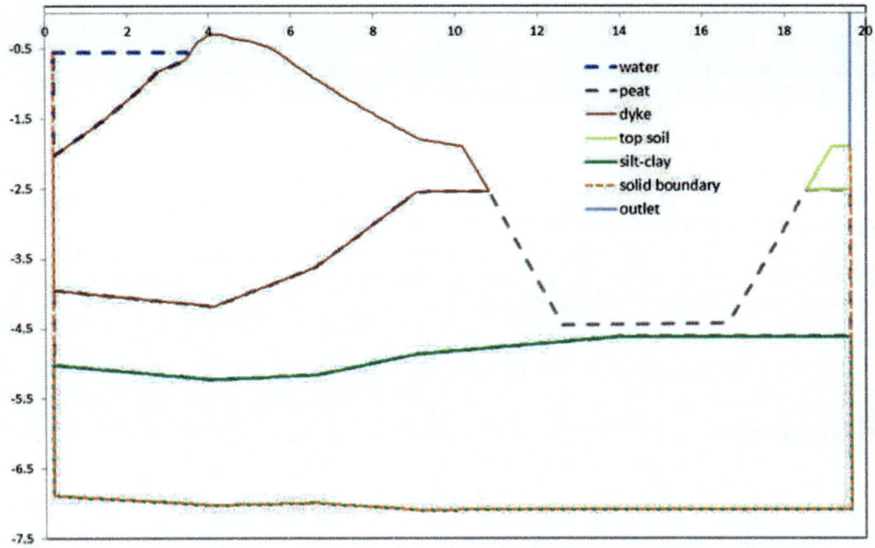

Fig. 8 Numerical domain for 2D simulations. Vertical "section AA": involved media (blue: water; brown: dyke; grey: peat; yellow: top soil; green: organic silt-clay)

and an outlet section is located downstream, as in the 3D case. The aim of this ongoing work is to consistently reduce the computational effort in terms of memory allocation and CPU time. In this perspective, the reduction of the degrees of freedom due to two-dimensionality can also allow to introduce a more refined spatial resolution, in order to produce results with a higher precision. As a complementary study, a quantitative comparison of the 2D and 3D results will be conducted, to assess the impact of the three-dimensional effects on this specific dyke failure simulation.

4 Conclusions

A CFD-SPH code is applied to simulate the 3D full-scale Kagerplassen dyke failure on-site experiment. The 3D geometries of the granular media and the water reservoir are reconstructed from the available measures by means of an analytical procedure. SPH numerical results are provided in terms of: 3D fluid dynamics fields (medium interfaces and velocity); hydrographs (time series) for the medium level (maximum height), flow rate, cumulated volumes and velocity. The 3D model dynamically simulates the triggering and propagation of the sliding surfaces within the dyke, and the following dyke failure run-out and the water flood. A list of possible improvements is reported to provide a more detailed description of the preliminary results of this on-going study. Moreover, a two-dimensional variant of the dyke failure will be thoroughly conducted.

Acknowledgements This study have been financed by the Research Fund for the Italian Electrical System (for "Ricerca di Sistema -RdS-") in compliance with the Decree of Minister of Economic Development April 16, 2018. Reference Project Manager: Antonella Frigerio.

We acknowledge the CINECA award under the ISCRA initiative, for the availability of High-Performance Computing resources and support. Some of the SPH simulations of this study have been financed by the following instrumental funding HPC projects: HSPHER9b, HSPHERA9.

The release of the FOSS versions of SPHERA has been supported in RSE and promoted by the Department Director Michele de Nigris, the Deputy Director and Project Manager Antonella Frigerio, and the Research Team Managers Guido Pirovano (since 2016) and Massimo Meghella (during the period 2015–2016). The first author would like to thank Prof. Cristina Jommi (University of Delft, Politecnico di Milano) for the experimental data provided on the internal friction angles of the granular media for the current test case.

References

1. SPHERA v.9.0.0 (RSE SpA), https://github.com/AndreaAmicarelliRSE/SPHERA. Accessed 15 Jul 2019
2. Amicarelli A, Manenti S, Albano R, Agate G, Paggi M, Longoni L, Mirauda D, Ziane L, Viccione G, Todeschini S, Sole A, Baldini LM, Brambilla D, Papini M, Khellaf MC, Tagliafierro B, Sarno L, Pirovano G (2020) SPHERA v.9.0.0: a Computational fluid dynamics research code, based on the Smoothed Particle Hydrodynamics mesh-less method. Comput Phys Commun, 250:107–157; https://doi.org/10.1016/j.cpc.2020.107157
3. Amicarelli A, Kocak B, Sibilla S, Grabe J (2017) A 3D smoothed particle hydrodynamics model for erosional dam-break floods. Int J Comput Fluid Dyn 31(10):413–434
4. Armstrong LM, Gu S, Luo KH (2010) Study of wall-to-bed heat transfer in a bubbling fluidised bed using the kinetic theory of granular flow. Int J Heat Mass Transf 53(21–22):4949–4959
5. Colagrossi A, Landrini M (2003) Numerical simulation of interfacial flows by smoothed particle hydrodynamics. J Comput Phys 191–2:448–475
6. Monaghan JJ (2005) Smoothed particle hydrodynamics. Rep Prog Phys 68:1703–1759
7. Di Monaco A, Manenti S, Gallati M, Sibilla S, Agate G, Guandalini R (2011) SPH modeling of solid boundaries through a semi-analytic approach. Eng Appl Comput Fluid Mech 5(1):1–15
8. Kumaran V (2015) Kinetic theory for sheared granular flows. C R Phys 16:51–61
9. Schaeffer DG (1987) Instability in the evolution equations describing incompressible granular flow. J Differ Equ 66:19–50
10. Terzaghi K (1943) Theoretical soil mechanics. Wiley, New York
11. Amicarelli A, Albano R, Mirauda D, Agate G, Sole A, Guandalini R (2015) A Smoothed Particle Hydrodynamics model for 3D solid body transport in free surface flows. Comput Fluids 116:205–228
12. Adami S, Hu XY, Adams NA (2012) A generalized wall boundary condition for smoothed particle hydrodynamics. J Comput Phys 231:7057–7075
13. ICOLD Committee on Computational Aspects of Analysis and Design of Dams (2019) Theme C; 15th Benchmark workshop on numerical analysis of dams, Milan, Italy, 9–11 Sept. https://www.eko.polimi.it/index.php/icold-bw2019/2019/about/editorialPolicies#custom-3
14. STOWA: 151014 Kennisfilm, Bezwijkproef-Leendert de Boerspolder. https://www.youtube.com/watch?v=uU_HEJkcouo. Accessed 51 Jul 2019
15. Mitchell et al (2019) Engage digitizer. https://github.com/markummitchell/engauge-digitizer. Accessed 15 Jul 2019
16. Paraview (Kitware). https://github.com/Kitware/ParaView. Accessed 15 Jul 2019

Analysis of the Pre-failure and Failure Behavior of a Levee on Soft Subsoil

M. Bakeš, T. Mészáros, and M. Minárik

Abstract The main aim of Theme C of the 15th International Benchmark Workshop on Numerical Analysis of Dams is the evaluation of currently available computational techniques and numerical models for the assessment of dykes on soft subsoil. For this purpose, the results of a full-scale test on a regional dyke on soft subsoil located in the Kagerplassen, in South Holland has been applied. In this study, a coupled hydro-mechanical analysis was performed in order to predict and describe the pre-failure and failure behavior of this structure. The 2D finite element code GEOSTUDIO 2016 was utilized for the analysis, where within the modelled period the state of failure of the structure was reached. Subsequently the modelled pre-failure behavior was compared to the measured data gained from the performed full-scale test and a back-analysis was conducted with emphasis on the applied material models and properties.

Keywords Dyke · Soft subsoil · Peat · Numerical analysis · Pre-failure behavior

1 Introduction

Motivated by the interest in evaluation of current numerical solutions for safety assessment of dykes on soft soils a full-scale stress test on a historical dyke, offered by the Waterboard HH Rijnland, was launched. The offered historical dykes, located in The Netherlands, north of Leiden in the Kagerplassen, are closing a polder, which had to be flooded to become a natural reserve area, Fig. 1. The test included thorough site investigation, standard laboratory tests, design of failure and subsequently observation of the pre-failure response of the hydraulic structure.

The work reported in this paper implemented the full-scale stress test results to conduct a coupled hydro-mechanical analysis in order to describe the pre-failure behavior of the dyke, and to predict and simulate its failure. In that, this case study

M. Bakeš · T. Mészáros · M. Minárik (✉)
Department of Dam Safety, Vodohospodárska Výstavba, SOE, Bratislava, Slovakia
e-mail: marian.minarik@vvb.sk

G. Bolzon et al. (eds.), *Numerical Analysis of Dams*, Lecture Notes in Civil Engineering 91, https://doi.org/10.1007/978-3-030-51085-5_38

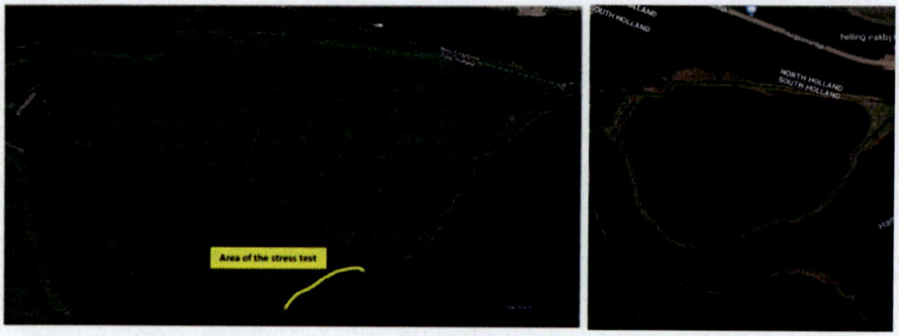

Fig. 1 The historical dyke closing a polder before and after flooding

has all the field data to check the merits of the proposed numerical solution, a comparison of the numerical results to its measured counterparts and a back-analysis was performed.

2 Full-Scale Stress Test

Investigation of the site included 18 CPTu tests and 6 boreholes, from which a number of undisturbed samples were collected for standard laboratory testing. In the investigated area, the body of the dyke is mostly made of clayey silt and silts, and founded on soft subsoil consisting of a peat layer, which is based on a layer of organic clays, Fig. 2 right.

To stress and eventually bring the dyke to failure a series of excavations and dewatering operations has been performed. The test started with an excavation at

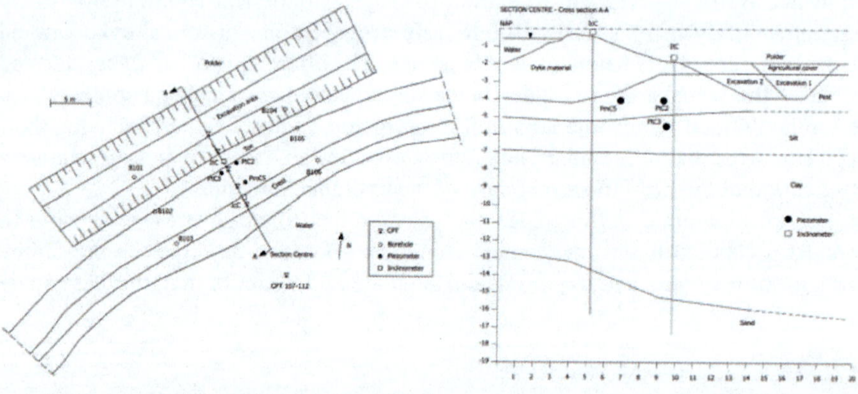

Fig. 2 The location of the excavations (left); The cross section of the dyke, including the position of the installed instrumentation (right)

the toe on the polder side of the dyke, continued with a sequence of two larger excavations and dewaterings (see Fig. 2).

Before the stress test started a continuous infiltration system was installed on the crest of the dyke, and the water level on the outer part of the dyke was kept virtually at the height of h0 = −0.6 m NAP, whilst on the polder side at constant height of hi = −2.3 m NAP. These arrangements kept the initial internal and external boundary conditions constant and the dyke body almost completely saturated during the entire test period in order to reduce the uncertainty on the degree of saturation of the dyke body. Failure of the dyke eventually occurred during rapid drawdown caused by dewatering the excavations at the dyke toe.

3 Numerical Model

For the purpose of this study, the commercial software Geostudio 2016 has been applied, using the finite element method (FEM) [6]. The task was divided, accordingly to the conducted stress test, into 4 stages, represented by 7 interlinked analyses within the implemented numerical solution (Fig. 3):

1. Stage—Initial conditions
   - 1. Analysis—steady-state seepage
   - 2. Analysis—in situ stress
2. Stage—1. Excavation and dewatering—(0–600 000 [s])
   - 3. Analysis—fully coupled hydro-mechanical
3. Stage—2. Excavation and dewatering—(600 000–1 202 400 [s])

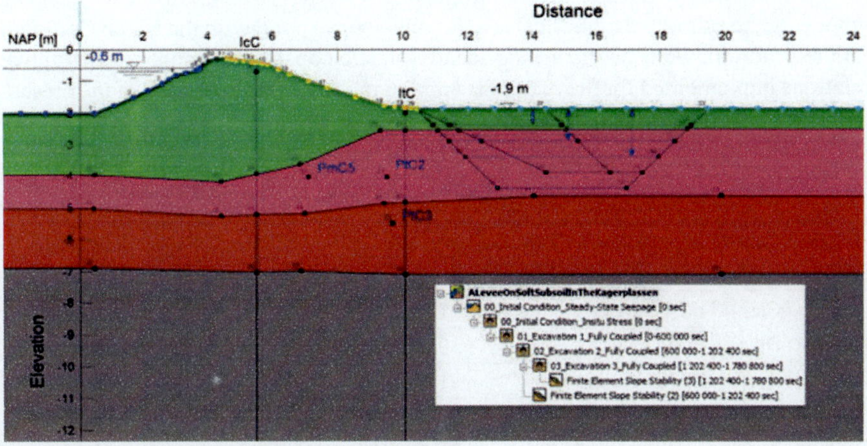

Fig. 3 FEM Model and conducted analysis types

- 4. Analysis—fully coupled hydro-mechanical
- 5. Analysis—FEM slope stability

4. Stage—3. Excavation and dewatering—(1 202 400–1 780 800 [s])

- 6. Analysis—fully coupled hydro-mechanical
- 7. Analysis—FEM slope stability

3.1 Geometry and Meshing

From the provided geometry data, the cross-section A-A with the given soil layers (Fig. 2) was modelled. The height was measured in reference to the Dutch NAP (Normal Amsterdam Peil), which will be assumed for the entire study as the reference zero for the vertical coordinate. The boundary and position of the given soil layers corresponds with the CPTu test results, were the peat layer was associated with higher friction ratio Rf [%]. Figure 3 shows the 2D numerical model used in the conducted analysis. The model is 75 m wide, 29.70 m high, and it was discretized into 4596 quad and triangle shaped finite elements with 4500 nodes. The size of the finite element's ranges from 0.3 to 3.0 m, with an average size of 0.5 m. The 2D model is considered with a thickness of 1.0 m.

3.2 Material Properties and Constitutive Models

All materials were assumed homogenous and isotropic. As a first approach, we considered all the materials with elastic-ideally plastic constitutive relation. In the next step, for soils, were the deformation was the main concern we used the modified cam clay model—MCC, with the Druger-Prager type yield criterion [5]. On the other hand, to minimize numerical issues and to correspond with the Mohr-Coulomb failure criterion, we used the elastic-ideally plastic constitutive model. Constitutive relations implemented for the materials applied in the second approach of the present numerical solution are given in Table 1.

Table 1 Constitutive models and material properties applied in the numerical model

Col.	Name	Model	γ	c'	φ'	E'	K_f	λ	κ	υ	e
[-]	[-]	[-]	[kN/m^3]	[kPa]	[^{0}]	[kPa]	[m/s]	[-]	[-]	[-]	[-]
Grey	Clay	El.-Pl.	14	8	21.7	870	1e-7			0.42	
Green	Dike	El.-Pl.	16.5	3.9	30.35	600	1e-6			0.39	
Red	Or. cl.	MCC	13.5	8	21.7		1e-7	0.214	0.016	0.42	2.8
Purple	Peat	MCC	9.7	3.7	29		1e-8	1.063	0.167	0.33	9.5
Yellow	Sand	El.-Pl.	20		35	10000	1e-4			0.25	

Besides hydraulic K-functions, material properties, used in the analyses, were derived from the provided laboratory test results [2, 3]. In all conducted analyses, effective material parameters have been used. For the "missing" hydraulic K-functions [1], an estimation has been made, based on the GEOSTUDIO built in VWC functions and saturated hydraulic conductivity. Saturated hydraulic conductivity—K_f was only measured for the peat soil layer (in terms of K_0 odometer test). For the remaining materials, Kf was assumed based on the authors experience. In terms of a back-analysis, in the last approach, the material properties, λ and κ, of the modified cam-clay constitutive model were derived from the $e = f(\ln\sigma'_v)$ relation instead of a $e = f(\log\sigma'_v)$ curve. Material properties used in the final approach are presented in Table 1.

3.3 *Time Stepping and Boundary Conditions*

As initial hydraulic condition, we assumed a fully saturated dyke body with a steady-state seepage close to the dyke surface. To model the virtually maintained water levels at the outer and inner part of the dyke, the total head, as the hydraulic boundary condition, at the upstream dyke side was kept at −0.6 m NAP and for the inner part (polder side) it was set to −2.3 m NAP. Furthermore, to keep the dyke body fully saturated for the entire numerical solution, and to ensure the shape of the seepage acquired with continues infiltrations, the pore pressure along the downstream slope surface was set to a constant value of −0.15 m. In addition, to have the initial stress condition of the dyke body and its foundation, we also performed an in situ stress analysis, were the geostatic stresses and the hydrostatic load from the canal water were of concern. The beginning (0 [s]) of the modelled period corresponds with the start of the first excavation sequence of the performed stress test (28.09.2015 09:20).

To simulate the rest of the conducted stress test (series of 3 sequenced excavations and dewatering operations) we performed a three stage, fully coupled, hydro-mechanical analysis. The three unloading stages in the transient, fully coupled, hydro-mechanical analysis are supposed to simulate the three excavations performed at the dyke toe. Each stage of the coupled analysis was divided in 20 time steps, with a duration from 578,400 to 602,400 s. Within the coupled numerical analyses, we implemented the dewatering and filling operations with a transient hydraulic boundary condition along the excavation bases, where the total head changed with time according to the conducted operations, Fig. 4.

4 Computed Results

The computed initial conditions; seepage and contour plot of total head are presented in Fig. 5.

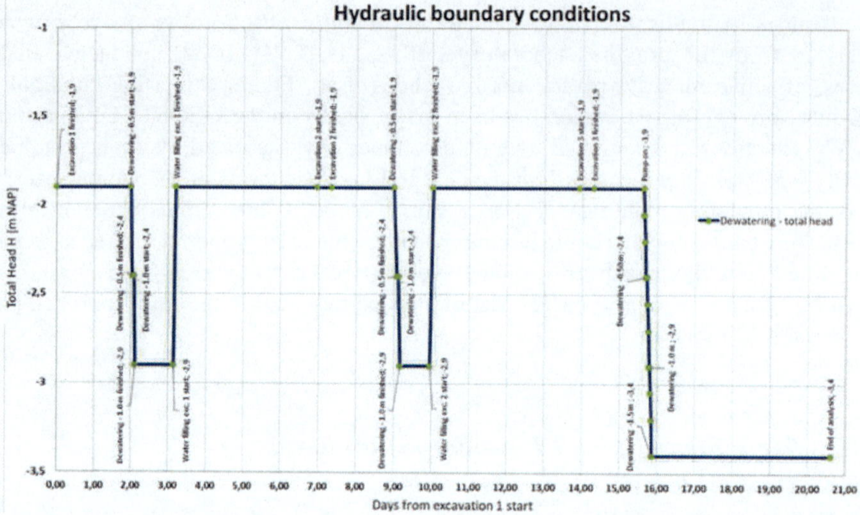

Fig. 4 Total head at the excavation bases

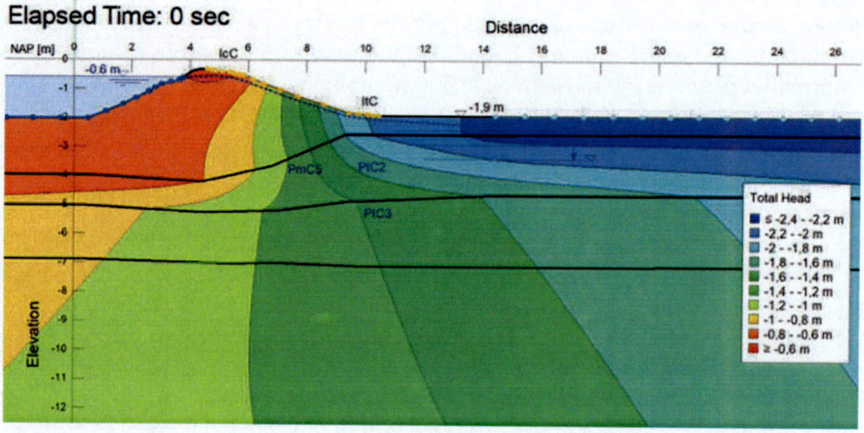

Fig. 5 Hydraulic initial conditions

The first numerical approach (elastic-ideally plastic constitutive model used only) resulted in the dyke failure right after the third dewatering sequence (−1.5 m NAP). However, after the third dewatering sequence, horizontal deformation of the peat layer at the dyke toe varied in a nonrealistic range up to 40 cm.

Resulted from the second attempt, the deformations at the dyke toe were more realistic, although the difference between the resulted and measured displacement have not been in an acceptable limit. As a comparison reference, we considered the horizontal displacements developed after the second excavation and dewatering sequence along the inclinometer ItC, located at the dyke toe. Furthermore, in this approach, the calculated factor of safety—FoS have not reached its critical value for the entire modelled period.

In the final approach we derived the material properties of the MCC constitutive model from the $e = f(\ln\sigma'_v)$ plot. Figure 6 shows the contour plot of the horizontal displacements after the first excavation and dewatering sequence, whereas Fig. 7 after the second operation sequence.

Figure 8 shows a comparison between the computed and measured displacements along the I_{tC} Inclinometer, located at the dyke toe. After the first excavation and dewatering, the difference between the computed and measured results is on the line of acceptance. On the contrary, the values computed after the second operation

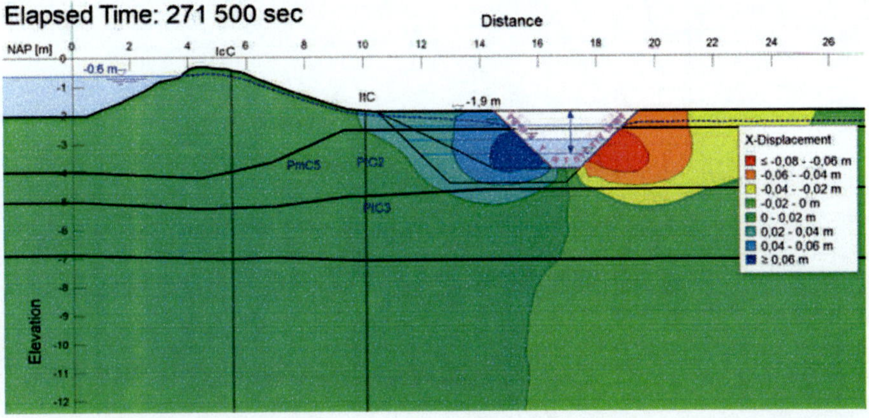

Fig. 6 Horizontal displacements after the first operation sequence

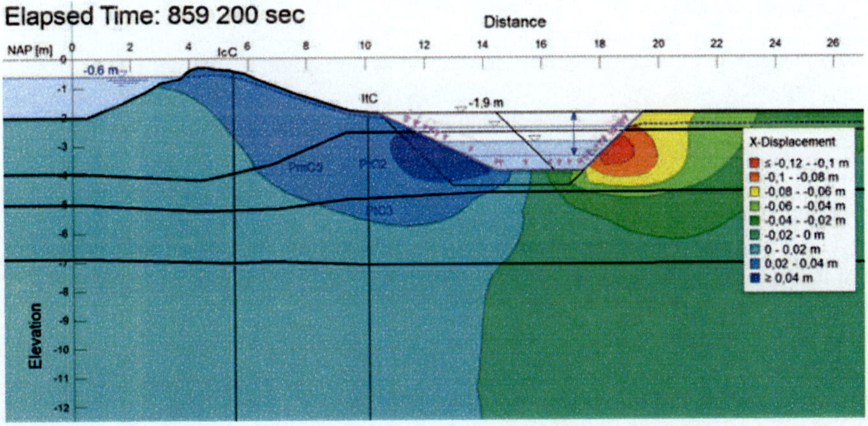

Fig. 7 Horizontal displacements after the second operation sequence

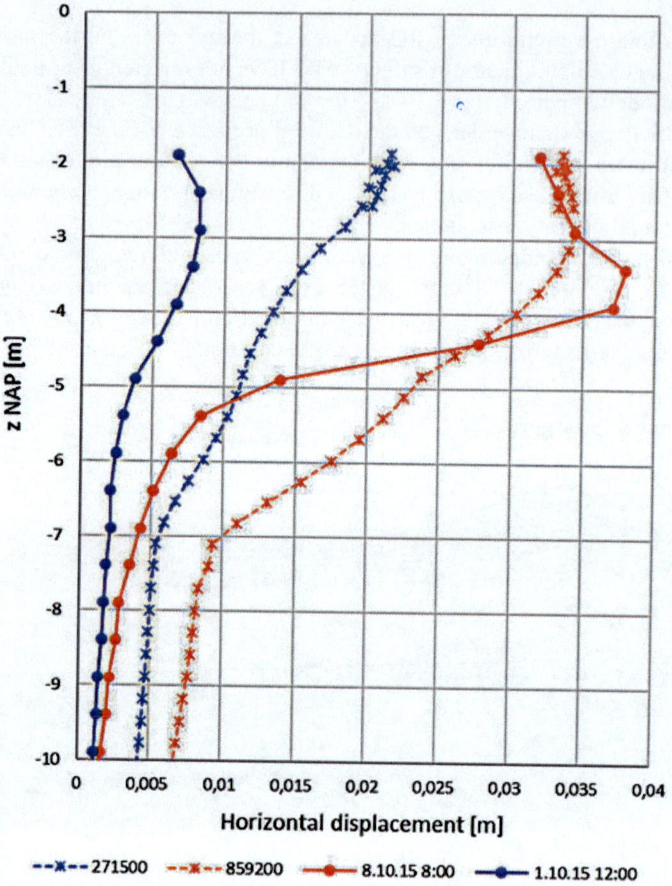

Fig. 8 Computed and measured displacement at the Inclinometer located at the dyke toe

sequence are, from an engineering point of view, in a good match with its measured counterparts.

Despite of the acceptable displacements, the computed total heads deviate from the values measured by the piezometers P_{tC2}, P_{tC3} and P_{mc5} (Fig. 9). However, the trend of the piezometric pressures during the considered period shows an acceptable correspondence.

The horizontal displacements, computed at the dyke toe after the third excavation and its dewatering to -1.5 m NAP, varied in a range from 6 to 12 cm. The results are presented in Fig. 10.

In order to predict the dyke failure, we conducted a slope stability analysis based on the normal stresses computed in the fully coupled hydro-mechanical analysis. First, to evaluate FoS of the downstream dyke slope, we considered all materials

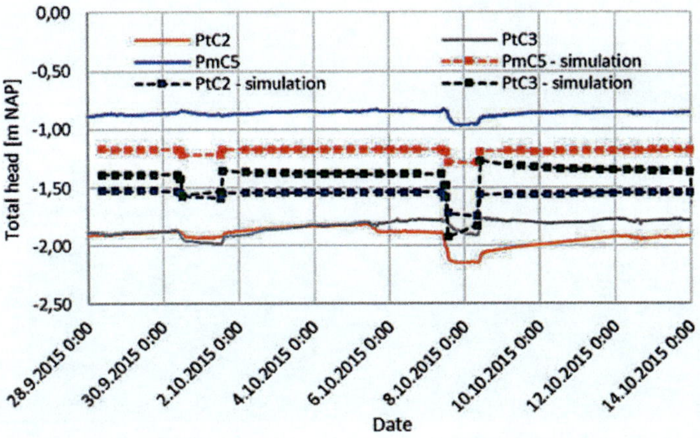

Fig. 9 Total head at the excavation bases

with the Mohr-Coulomb material model, which resulted in FoS > 1.0 for the entire modelled period (Fig. 11).

After dewatering the excavations to −1.5 m NAP the computed FoS were sorely close to the critical value 1.0 indicating the dyke failure. Therefore, instead of modelling further dewatering operations, in the slope stability analysis we modelled the peat layer with a bilinear material model as indicated in Rétháti [4] (bilinear properties: $c' = 1{,}96$ kPa, $\varphi'_1 = 37^{\circ}$, $\varphi'_2 = 29{,}8^{\circ} \rightarrow \sigma_n \geq 8.32$ kPa). This resulted in the dyke failure after the third excavation and its dewatering to −1.5 m NAP (Fig. 12); 14.10.2015 corresponding to the conducted stress test.

5 Conclusions

In this numerical solution, we assessed the pre-failure behavior of the presented structure and predicted its failure. The main findings and conclusions of the presented work are as follows:

1. Prediction of failure of the presented structure can be made sufficiently with currently available computational techniques. Although, great care must be taken to achieve a correct prediction. Based on the conducted numerical analyses we assume, that failure of the structure occurred 14.10.2015, after dewatering the excavation to −1.5 m NAP. However, after the second operation sequence FoS, of the downstream dyke slope, varied in a range close to its critical value, i.e. small change in the material properties (shear strength) of the subsoil peat layer would have resulted in a failure of the structure.
2. The dyke body and the underlying soft subsoils are made of coherent materials, this indicates, that the shape of the critical slip surface should be considered

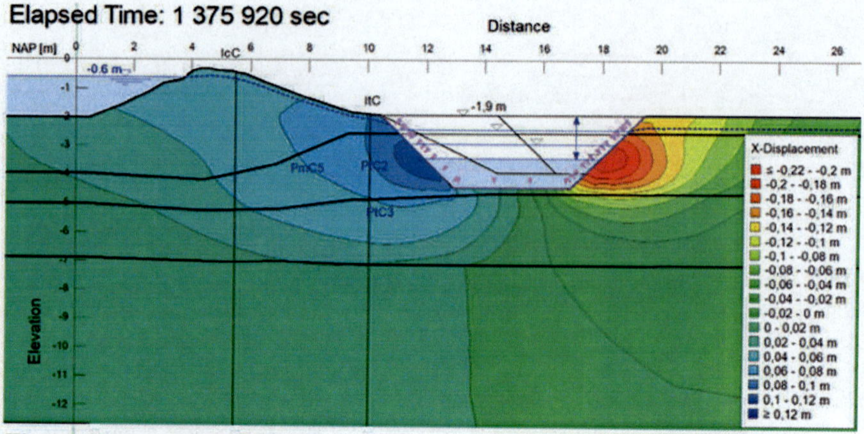

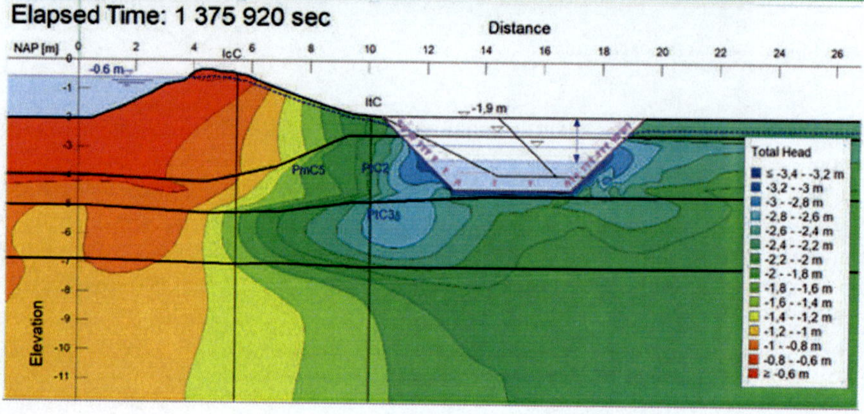

Fig. 10 Results after third operation sequence; contour plot of displacement (top); contour plot of total head (bottom)

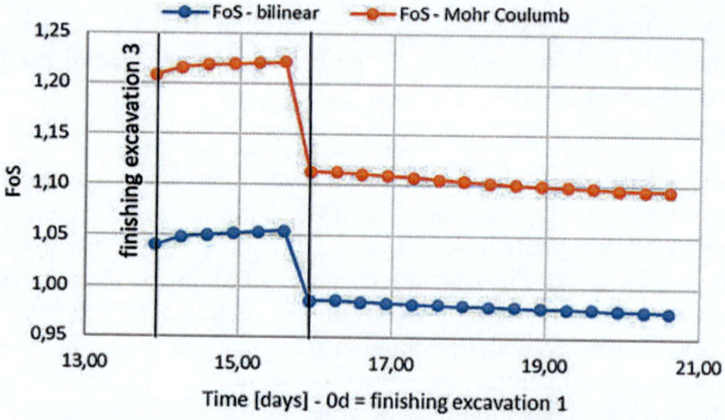

Fig. 11 FoS of the critical slip surface versus time during the modelled period

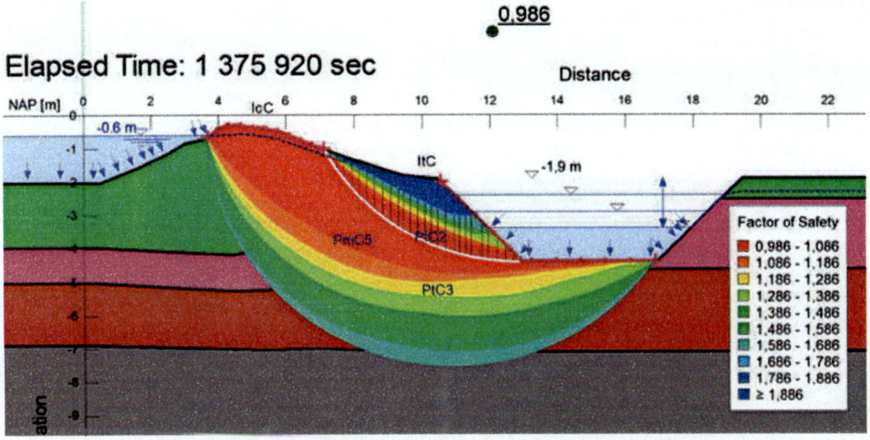

Fig. 12 FoS after third excavation sequence, with bilinear model in case of the peat layer

cylindrical or curved, as used in our solution. To support/contradict this assumption, we tried the slip surface optimization technique, which is a built-in option of the utilized software, leading to nonrealistic shapes of the failure surface.

3. Uncertainty of the solution is emphasized by the lack of a thorough investigation of the hydraulic material properties.
4. The utilized software, GEOSTUDIO 2016, is a very powerful solver for computing coupled hydro-mechanical analyses. However, for these analyses, it offers only the elastic, elastic-ideally plastic and the modified cam clay constitutive models. While all the offered models have their advantages, since not only the pre-failure behavior but also a prediction of the failure was of concern, neither of these models appear advantages enough to fully fit both requirements at the same time.
5. In the performed analyses, we have not been able to make a reasonable prediction of the pore pressure development within the modelled cross-section. Although, we achieved a good match of the computed displacements with their measured counterparts.
6. In this numerical solution, we performed a back-analysis based on the monitoring data to refine our prediction. All the performed refinements were in a framework of some reasonable assumption. In our opinion, given the complexity of the numerical solution, further refinements or parametric studies could lead to confusing results or to their misinterpretation.

Acknowledgements We gratefully acknowledge Doc. Ing. Ivan Slávik, PhD. for his valuable suggestions and discussions.

References

1. Fredlund DG, Rahardjo H, Fredlund MD (2012) Unsaturated Soil Mechanics in Engineering Practice. Wiley, Hoboken
2. Grognet M (2011) The boundary conditions in direct simple shear tests, developments for peat testing at low normal stress. M.Sc. thesis, Delft University of Technology, Delft, p 92
3. Head KH (1998) Manual of soil laboratory testing, Effective stress test, 2nd edn., vol 3. Wiley, London
4. Réthási L (1995) Alapozás kedvezőtlen talajokon, Akadémiai kiadó, Budapest
5. Stress-Deformation Modeling with SIGMA/W (2013) An engineering methodology, July 2013 Edition, GEO-SLOPE International Ltd., pp 85
6. Zienkiewicz OC, Taylor RL, Zhu JZ (2013) The finite element method: its basis and fundamentals, 7th edn. Butterworth-Heinemann, Oxford

Coupled Hydro-Mechanical Analysis of the Leendert de Boerpolder Dyke Stress-Test

C. Marulanda, J. Tello, and D. León-Vanegas

Abstract This paper summarizes the results and the main features of the fully coupled hydro-mechanical analysis of the stress test carried out in the Leendert de Boerpolder levee in the Netherlands. The analyses have been performed with software Midas GTS NX and Abaqus, using two-dimensional coupled pore-displacement elements for transient consolidation analysis. The Midas model featured the Modified Mohr-Coulomb model while the Abaqus model used Modified Cam Clay plasticity with porous elasticity. Material parameters were calibrated based on laboratory data and in situ CPTu tests. Results were compared with field measurements in order to assess the ability of current models to assess the safety characteristics of earth retaining structures subject to groundwater flow and stress. The models predicted failure during the 3rd excavation, in agreement with the planned stress test schedule.

Keywords Dyke · Coupled analysis · Failure prediction · Deformation · Drawdown · Pore pressures

1 Introduction

Theme C of the 15th Benchmark Workshop on Numerical Analysis of Dams (ICOLD) consists in the coupled hydro-mechanical analysis of the pre-failure and failure behavior of the Leendert de Boerspolder levee in the Netherlands, in which STOWA performed a stress test in 2015, consisting of a sequence of large excavations and pumping operations, until failure eventually occurred during rapid drawdown.

The objective of this exercise is to evaluate the potential of existing models to track the coupled hydro-mechanical response of earth constructions for water defense, both during service life and in the prediction of failure.

In order to simulate this benchmark, in this paper we present two bidimensional (plane-strain) fully coupled numerical models with different constitutive models;

C. Marulanda (✉) · J. Tello · D. León-Vanegas
INGETEC, Cra 6 No. 30A-30, Bogotá, Colombia
e-mail: marulanda@ingetec.com.co

G. Bolzon et al. (eds.), *Numerical Analysis of Dams*, Lecture Notes in Civil Engineering 91, https://doi.org/10.1007/978-3-030-51085-5_39

the first one developed in the software Midas GTS NX with the Modified Mohr-Coulomb model, and another one developed in Abaqus with the Modified Cam-Clay plasticity model.

2 Model Description

The geometry was defined based on the CPTu results and the schematic profile provided in the Formulation document. In order to simplify the real geometry of the problem, the following assumptions were made.

- Only four materials were considered in the model, i.e. the sand layer was not considered.
- The topsoil in the polder side was considered as the same dyke material but with a unit weight of 14 kN/m^3.
- The sole difference in the Peat material between the zone under the dyke and the polder are the OCR value, the void ratio, and the permeability. Whereas, the strength parameters remain the same.
- The permeability of the different layers was assumed based on reference values and the CPTu correlations.
- The silt material was assumed to have a linear-elastic behavior with a Young modulus of 10 MPa.

2.1 A Subsection Sample

The model in Midas GTS NX has a length of 35 m and a height of 14 m, and the different layers were represented by 1611 quadrilateral linear elements and 58 triangular linear elements. In the bottom boundary fully fixities were used, while the lateral boundaries were free to move in the vertical direction.

In the reservoir side an head of −0.6 m NAP was kept constant during the simulation, meanwhile, in the excavation walls a changing head value was considered in the dewatering and water filling steps. The lateral and bottom boundaries were considered closed. Additionally, the steps of the stress test were performed with transient fully coupled steps with the true duration register in the stress test.

The Modified Mohr-Coulomb model. The Modified Mohr Coulomb (MMC) is a double hardening model; with a shear yielding up to Mohr Coulomb failure surface and another surface represents compression yielding. This model uses nonlinear elasticity, which considers the difference stiffness of the soil under primary loading and unloading/reloading, and a stress dependent stiffness according to a power-law.

The main strength parameters of this model are the friction angle (ϕ') and the cohesion intercept (c'), which were calculated based on the relationship of Eq. (3). Furthermore, the basic parameters for soil stiffness are the secant stiffness in triaxial

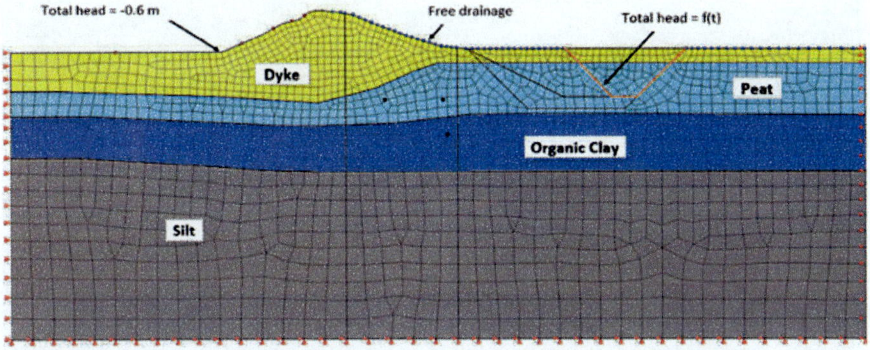

Fig. 1 Numerical model in midas GTS NX

test (E_{50}^{ref}), the tangent stiffness for primary oedometric loading (E_{oed}^{ref}), and the elastic modulus at unloading (E_{ur}^{ref}), the latter two modulus were estimated with the Eqs. (1), (2), which are valid for a power stress-level dependency of stiffness $m = 1$.

$$E_{oed}^{ref} = \frac{p_{ref}(1 + e_0)}{\lambda} \tag{1}$$

$$E_{ur}^{ref} \approx \frac{2p_{ref}(1 + e_0)}{\kappa} \tag{2}$$

2.2 *Abaqus*

The Abaqus model consists of two-dimensional, quadratic, fully coupled pore pressure-displacement elements (CPE8RP) with reduced integration. The finite element mesh can be observed in Fig. 2. The exact distribution of strata was approximated based on the available information.

The Modified Cam Clay Model. The MCC model in Abaqus, termed Clay-Plasticity, is a Critical-State plasticity model, see Fig. 3, in which:

- The yield function can depend on the three stress invariants, p, q, and r, although in the present implementation the dependence on r has been ignored by setting the flow stress ratio, $K = 1$.
- Can have an anisotropic yield surface, but has been assumed isotropic as no information on anisotropy has been provided.
- Can define the hardening by means of an exponential form characterized by the logarithmic plastic bulk modulus, λ, or by means of a tabular form by defining the relationship between p_c and ε_{vol}^{pl}.

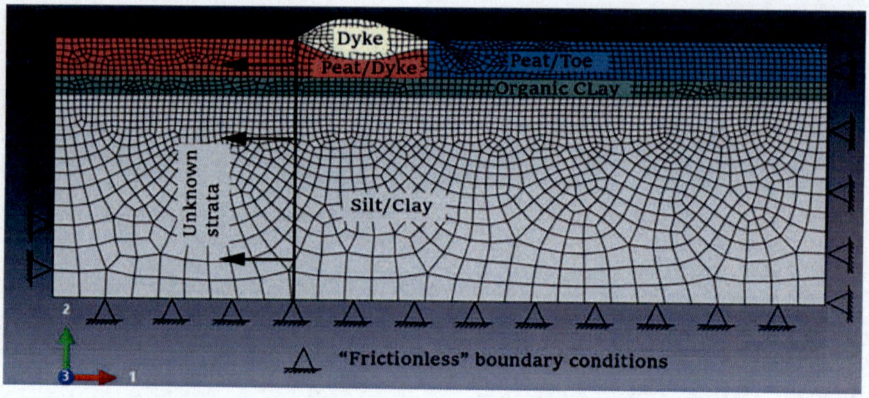

Fig. 2 Finite element mesh for Abaqus model

Fig. 3 Cam Clay yield surface in the p-q plane, with $\beta = \mathrm{K} = 1$

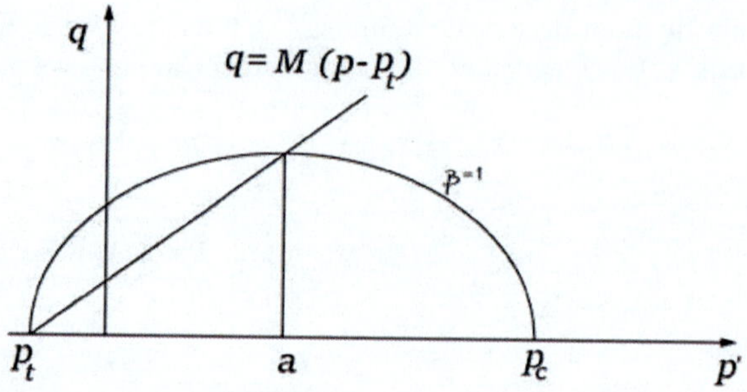

- Can include some cohesion by means of the use of the parameter $p_t < 0$ in the tabular form of hardening (see Fig. 4).
- Can be used in combination with the porous elasticity material model, or if using the tabular hardening, the linear elastic model as well. For the materials in this benchmark, the porous elastic material model has been implemented with a constant shear modulus and hence, variable Poisson ratio.

The most robust form of the Modified Cam Clay model was found to be that in which hardening is defined by means of the tabular entry of p_c and ε_{vol}^{pl}, as it allows for the definition of a tensile yield strength, p_t. This is important, firstly, because both direct simple shear tests as well as triaxial tests indicate that the materials do exhibit a tensile strength, and secondly, because certain regions in the model can experience low confinement and high deviatoric behavior, leading to critical state failure where none should occur. This form of the clay plasticity formulation avoids these difficulties.

Establishing Initial Equilibrium. Since the elastic part of the material model requires that the pressure be initially positive, an estimate of the initial stresses in the soil is required in order to start the simulation. This is usually achieved by means of

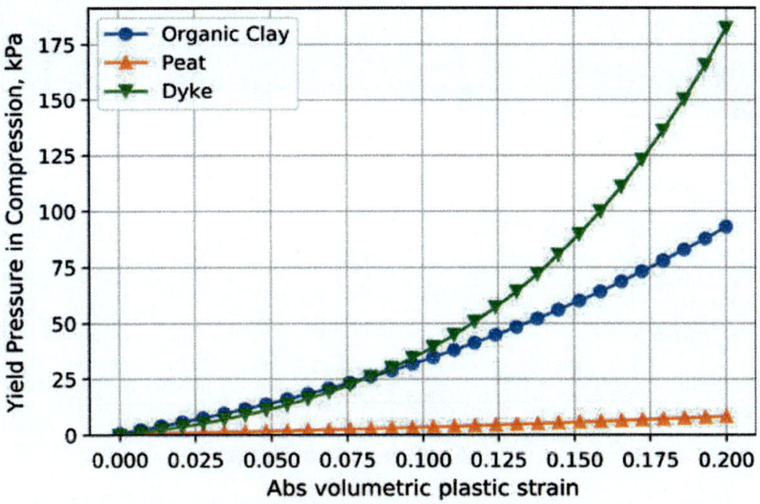

Fig. 4 Tabular hardening curves coming from the logarithmic form of the Cam Clay Model

a geostatic step in which an initial estimate of the stress distribution is provided by the user.

In order to provide a stress distribution that is nearly in equilibrium despite the strata featuring sloping interfaces, the simulation starts without the dyke material, with a horizontal free surface where the effective vertical stress is assumed to be zero. The soil is initially fully saturated, and the void ratio is defined based on the information provided in the benchmark formulation.

Assuming no variation in the horizontal direction, the equilibrium equation in terms of effective vertical stress must be integrated numerically because the density and porosity vary as a function of position. The resulting vertical stress using the parameters at the toe of the dyke can be observed in Fig. 5.

Flow Boundary Conditions. Several types of boundary conditions on the pore fluid field equations were necessary in order to faithfully model particular types of surfaces. Where a constant water level is present, such as on the upstream face of the dyke, a direct pore pressure specification suffices. However, for cases where the phreatic surface reaches a freely draining surface where water cannot flow back into the domain, a free-draining boundary condition is necessary whereby the flow velocity normal to the surface is proportional to the pore pressure for positive pore pressures, and zero for negative pore pressures. This condition will lead to the expected behavior of zero pore pressure below the phreatic surface approximately [1]. Finally, when water surface moves in time, a user subroutine was implemented to define the flow of water at the surface as a function of the difference between the current pore water pressure and a "sink" pore water pressure equal to the hydrostatic pressure in the water with a seepage coefficient as the proportionality constant as described in [2].

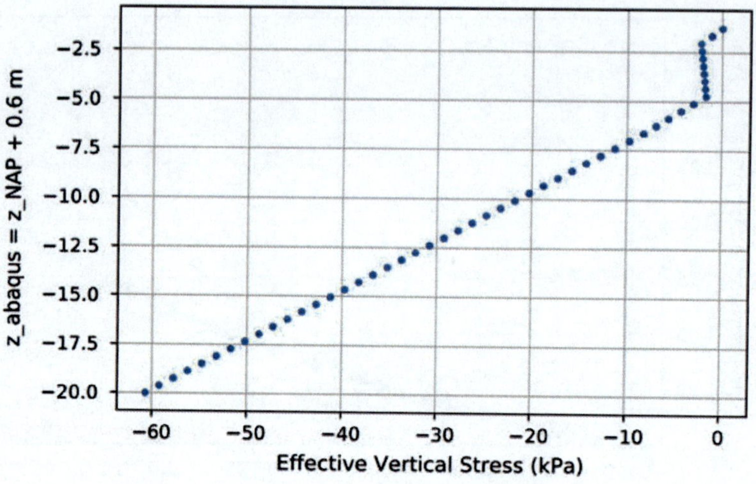

Fig. 5 Initial guess for vertical effective stress with horizontal strata

3 Geotechnical Characterization

The characterization of the different layers was made based on the different laboratory tests provided. To interpret the Triaxial CU test, it was necessary to convert normal and shear stresses ($\sigma_n - \tau$) to pressure and von Mises measures ($p - q$) as the Modified Cam Clay model are formulated in terms of these invariant quantities. This conversion often requires the adoption of certain assumptions about pore pressure and Poisson ratio during the tests, leading to possible uncertainties in the characterization of the materials.

The effective stress paths in the p-q plane at the different confinement pressures was plotted to estimate the slope (M) and the intersection with the p-axis (p_t) of the failure line, as shown in Fig. 6.

For the Direct Simple Shear test, a Poisson ratio of 0.3 was assumed in order to calculate the lateral stresses required to compute the equivalent normal and deviatoric stresses for the model calibration. This assumption is not based on actual data provided in the benchmark, and it is likely that Poisson ratio actually varies with confinement pressure, an effect that can be captured by the soft-rock constitutive model in Abaqus, but not by those forms of elasticity employed in the current models.

The Mohr Coulomb parameters were determined from the Eq. (3) which is the relationship between the slope of the failure line M and the parameters of the Mohr Coulomb (friction angle and cohesion) for a compression stress state. Besides that, in the DDS test the Mohr Coulomb parameters were determined in a $\sigma - \tau$ plot assuming that the horizontal plane is a plane of maximum stress obliquity.

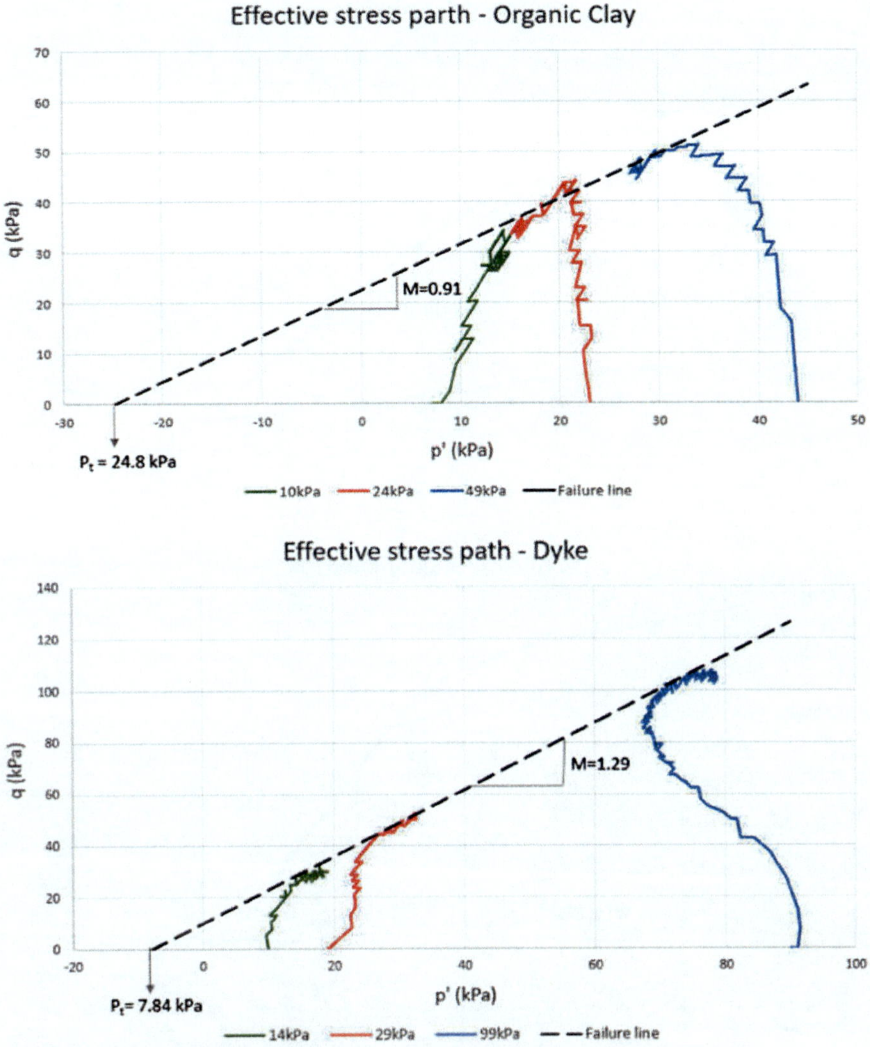

Fig. 6 Stress path: triaxial CU test of organic clay and dyke

$$\phi' = \left[\frac{3M}{6+M}\right]; c' = P_t M\left(\frac{3 - sinsin\phi'}{\phi'}\right) \tag{3}$$

In the edometric tests (Fig. 7) the compression (C_c) and swelling (C_s) indices were calculated as the slopes of the approximated straight lines, when plotting the logarithm of the vertical effective stress as a function of void ratio. Based on the parameters obtained from the one-dimensional compression test, the logarithmic plastic bulk modulus (λ) and logarithmic bulk modulus (κ) from an isotropic compression

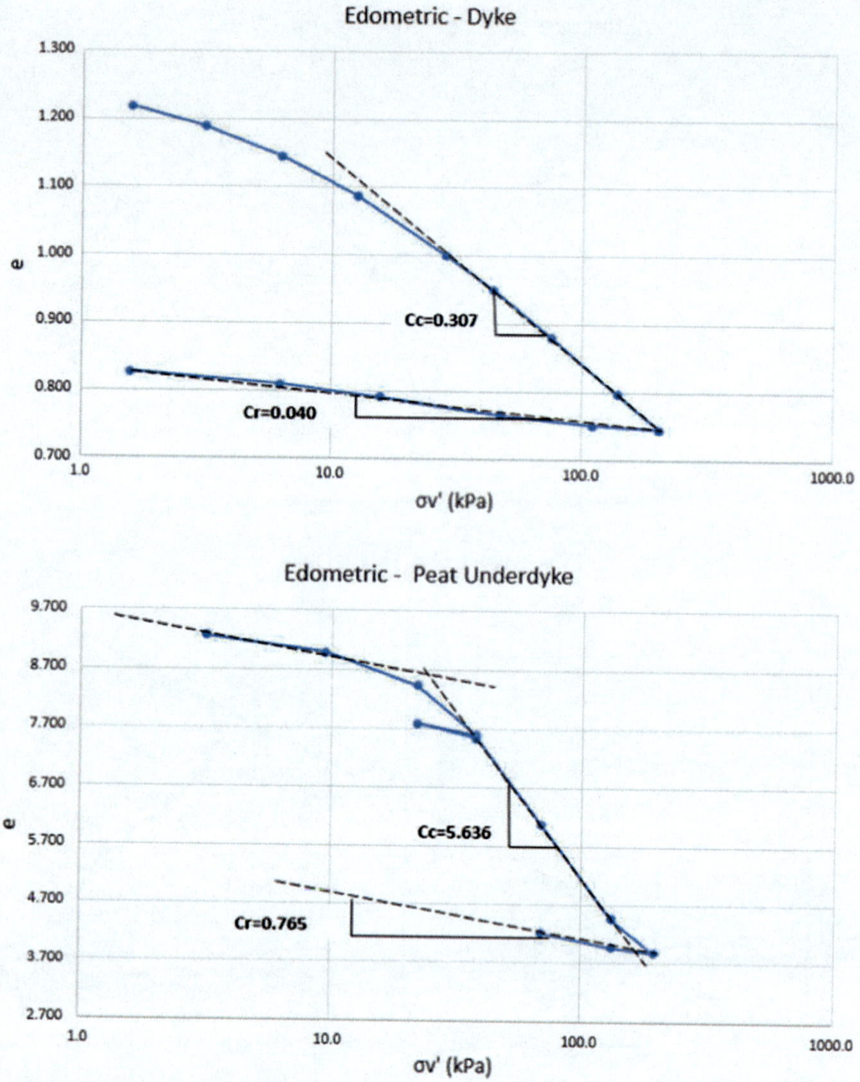

Fig. 7 Edometric test dyke and peat underdyke

were estimated with the relations of Eq. (4). It is worth noting that these equations constitute an approximation to the actual soil behavior, since during one-dimensional unloading the ratio of horizontal and vertical stresses is not constant.

$$\lambda = \frac{C_c}{2.3}; \kappa \approx \frac{2C_s}{2.3} \tag{4}$$

4 Prediction of Failure

4.1 When Did the Failure Occur?

The Finite Element model in Midas GTS NX (Modified Mohr–Coulomb model) predicted the failure during the dewatering 3, when the water level in the excavation was at −2.97 m NAP, as evidenced by a drastic increase in deformations and plastic strains and an inability to achieve convergence, implying that there is no stress distribution that can satisfy the failure criterion of the constitutive model and be in equilibrium with the applied loads [3]. According to the stress test plan, the failure happened approximately on 10/14/2015 at 5:00 AM

A time history of deformation at the toe (Fig. 8) illustrates the occurrence of failure in the Abaqus model.

4.2 Failure—Drained or Undrained Conditions

The evolution of pore pressure during the stress test shows that the maximum change of pore pressure head was 26 cm during the drawdown of 100 cm in the excavation 2 (in piezometer PtC2 located about 5 m away from the excavation).

The inclinometers also show the highest deformations in this region and at this time, suggesting that changes in pore pressure will affect the failure response.

Hence, failure will be best reproduced by drained conditions, because it is to be expected that the change in pore pressure at the same location during the 3rd

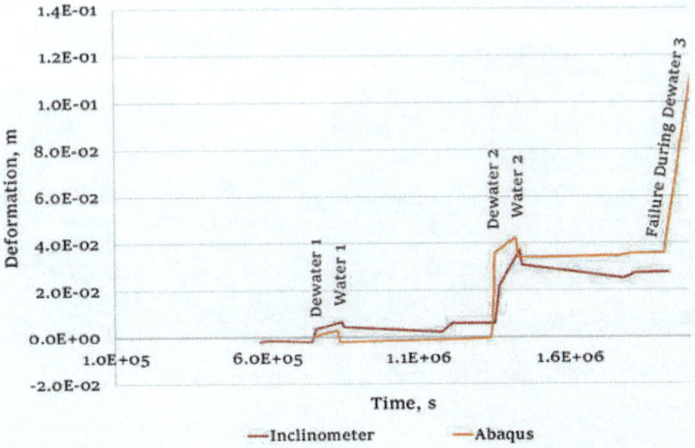

Fig. 8 Time history of horizontal deformation at a depth of -3.4 m, where the maximum deformation registered by the toe inclinometer occurs, compared with the results of the Abaqus model

dewatering would be even higher, since it is closer to the piezometer and the planned drawdown was 150 cm compared to the 100 cm of the 2nd dewatering.

4.3 Shape of the Failure Surface

The shape of the failure surface can be evidenced by the accumulation of plastic strain right before non-convergence in the Midas Model. The failure starts in the toe of the excavation and it extends along a horizontal plane in the peat and then through the dyke still the reservoir level. This kind of failure agrees with the failure in peat slope and peat dykes, which tend to fail along a horizontal plane [4].

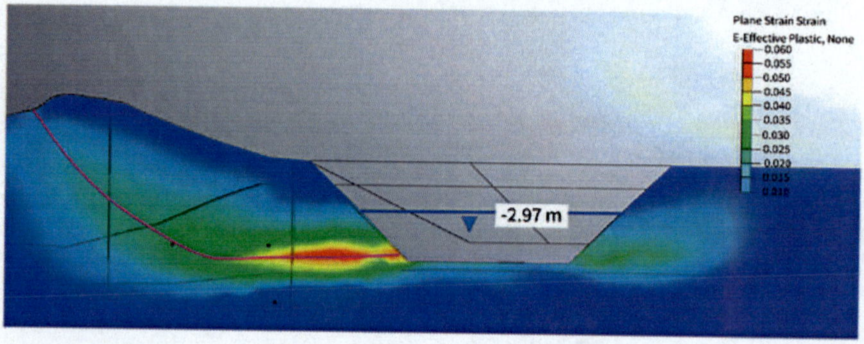

Fig. 9 Failure surface in the Midas model

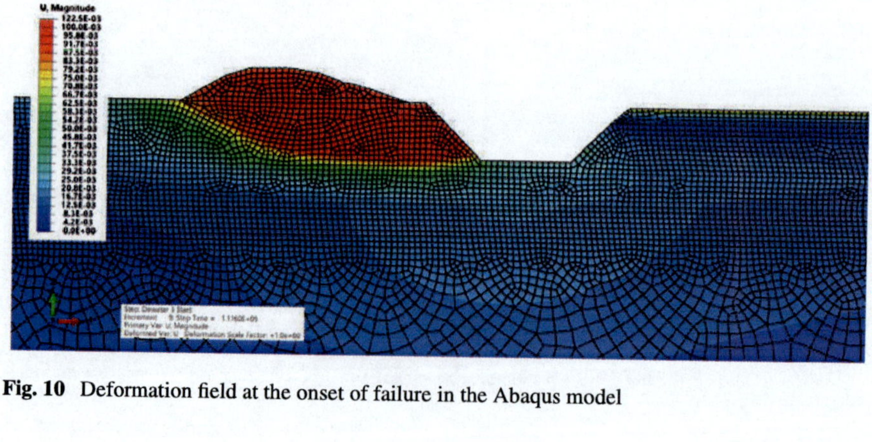

Fig. 10 Deformation field at the onset of failure in the Abaqus model

4.4 Uncertainty

According to the horizontal displacements registered by in the inclinometers, the failure surface seems to be lower than the one predicted in the FE models. Therefore, it is possible that failure may have propagated at the interface between the Peat and the Organic Clay, which made lead to complex behavior not captured by the models implemented, because no information about the mechanical behavior of this interface is forthcoming.

The failure was induced by the drawdown in the excavation and hence it is considerably affected by the permeability of the soil, since for low permeabilities the pore pressure could remain high meanwhile the stabilizing effect of the water in the excavation is lost, considerably reducing the factor of safety. In consequence, the permeability is an important parameter in the prediction of failure and at the same time is one aspect of the model most affected by uncertainty, since the soil permeability was determined by means of empirical correlations of for CPTu's and with representative values from literature.

Another aspect of the models that was found to have a significant impact on deformations was the assumed value of lateral coefficient, K_0. Since during excavations the horizontal stress of the soil is replaced by the hydrostatic pressure of water, if the horizontal stress is significantly different from this pressure, the excavation will induce proportionally large horizontal deformations, which are not observed in the instrumentation.

5 Pre-failure Response

5.1 Material Behavior Observed in the Laboratory

The accurate simulation of an excavation should consider the difference in loading/unloading stiffness of the soil. Therefore, for the simulation of this benchmark problem, the Modified Mohr-Coulomb model and the Modified Cam-Clay model were considered.

The parameter for these constitutive models were initially predicted based on the laboratory test and then calibrated with the simulation of element triaxial CU test and the simulation of the Direct Simple Shear Test (DSS) with the S*oilTest* facility in Midas GTS NX. In order to simplify the calibration procedure, the parameters of the models were adjusted to fit the stress-strain curve of the triaxial CU test and the DSS test, whereas the oedometric tests were not considered. Figure 11 shows the stress-strain curve for the experimental results and the comparison with the numerical simulation.

The parameters for the Modified Mohr-Coulomb obtained after the calibration procedure are presented in the Table 1.

Fig. 11 Numerical simulation of the CU triaxial test and DSS test

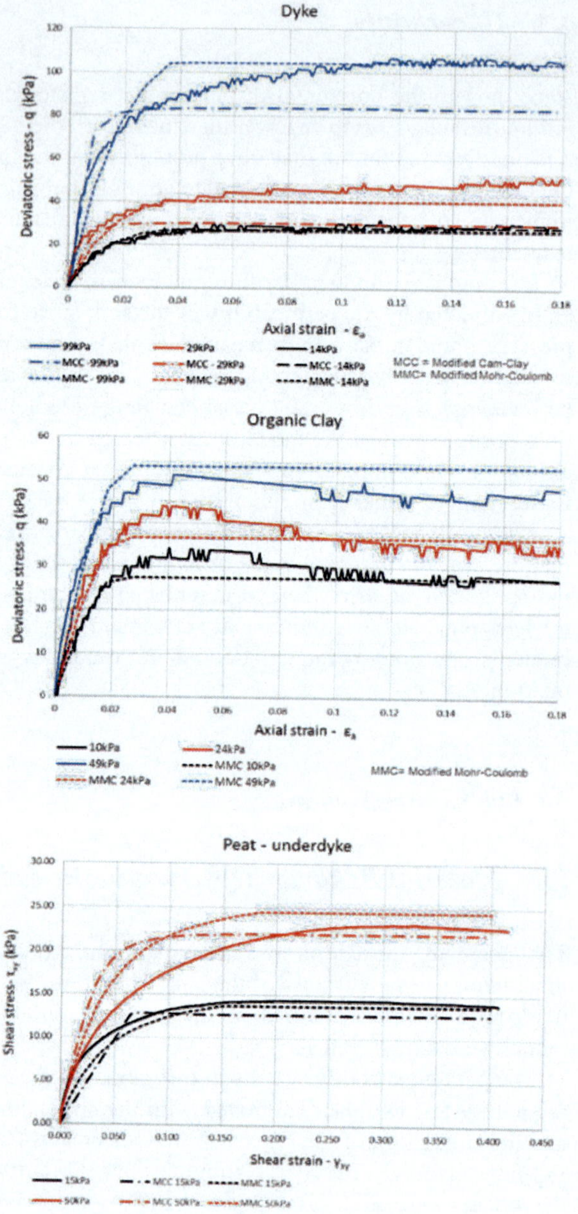

Table 1 Parameter for the modified Mohr-Coulomb model

Parameter	Dyke	Peat	Organic-Clay
$E_{50}^{Ref}\ (kPa)$	7000	2800	6000
$E_{oed}^{Ref}\ (kPa)$	3500	1400	3000
$E_{ur}^{Ref}\ (kPa)$	20000	5000	10000
m	1	0.7	1
$\phi'(°)$	33.3	25	21
$c'(kPa)$	10	6.5	21
R_f	0.5	0.8	0.3
$k(m/s)$	1×10^{-7}	1×10^{-7}	1×10^{-8}

The results presented in Fig. 11 show that the Modified Mohr-Coulomb model represent better the material behavior observed in the laboratory. This is primarily due to the hyperbolic stress-strain relationship of this model, which represents the decreasing stiffness of the soil with the increasing strain.

5.2 *Soil Response for the Stress Test*

Horizontal displacement. The horizontal displacements during the excavation and dewatering steps is mainly due to the stress release and the settlement cause by the lowering of the water table. In this case, the stress release during the excavation is very low, because in the excavation steps the water level is kept in the surface and the unit weight of the peat material is very close to the one of water. Therefore, the biggest horizontal displacements are expected to happen in the dewatering steps.

Figure 12 shows the horizontal displacement obtained with the numerical model in the dyke's toe and the comparison with values registered with inclinometers. The results agree with the instrumentation measures in the sense that they show small jumps during the excavations steps and big jumps during the dewatering steps. The water filling steps also register small jumps, reflecting that in the dewatering steps the principal movement is owing to plastic strains.

Figure 13 shows the horizontal displacement at the dyke crest. As the above results, the big displacements are presenting in the dewatering steps due to the stress release. These numerical results show a bigger difference with the inclinometer data in contrast with the results in the dike´s toe, highlighting that the elastic recoverable strains in the field is larger than the one predicted with the numerical models.

The numerical results with the Modified Mohr-Coulomb accurately represent the shape and the magnitude of the deformation register by the inclinometers. However, the results overestimated the deformation in the initial steps.

It is found that this behavior is due on one hand for the initial stress calculated with the well-known gravity loading procedure, because the horizontal effective stress in the polder side are bigger than the vertical effective stress in an average depth of

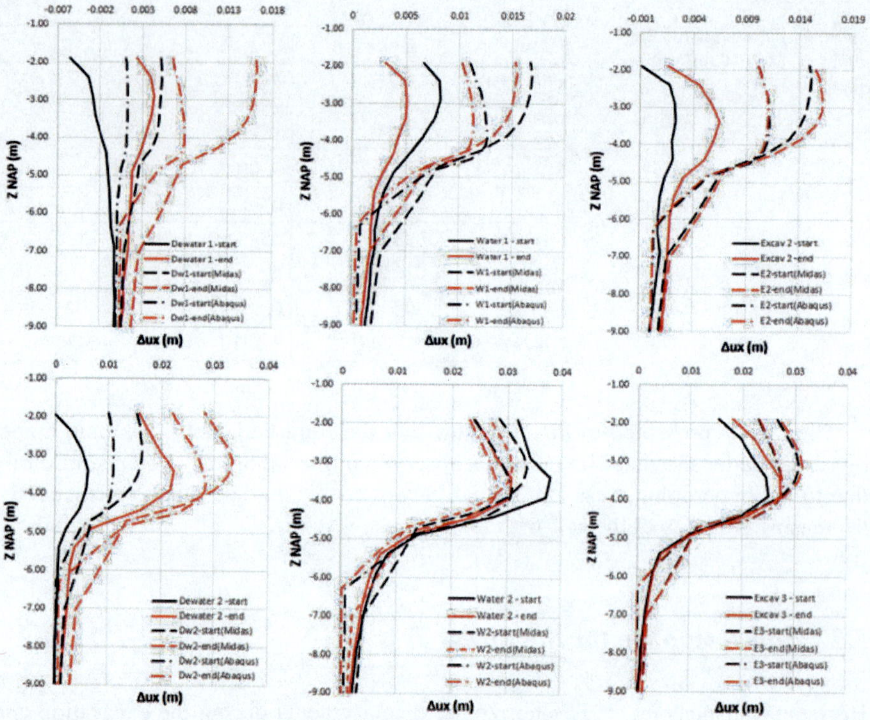

Fig. 12 Horizontal displacements in toe inclinometer

2 m, producing a big stress release in the first excavation stage. On the other hand, the truly elastic range of the Modified Mohr-Coulomb model is very small; therefore, the magnitude of plastic strains predicted in the initial steps with this model is bigger than the actual strains produced during the stress test.

Pore pressure. The changes in pore pressure during the stress test have a complex behavior, because there are changes owing to the change in the water level and changes associated with the deformation. The former has a direct relationship, while the latter is affected by pressure stresses and may depend on the complex volumetric change during shearing exhibited by many soils.

The pore pressure head registered for the three piezometers and the numerical results of pore pressure head is presented in Fig. 14. These results show that the Modified Mohr-Coulomb model accurately represents the pore pressure changes in the three piezometers, but there is an important difference between the pore pressure head values and the initial steps in the piezometers PtC3 and PmC5. This difference is likely due to not simulating in full detail the wetting step prior to excavations, as well as to likely variability and anisotropy in the permeability which were not considered in the analyses.

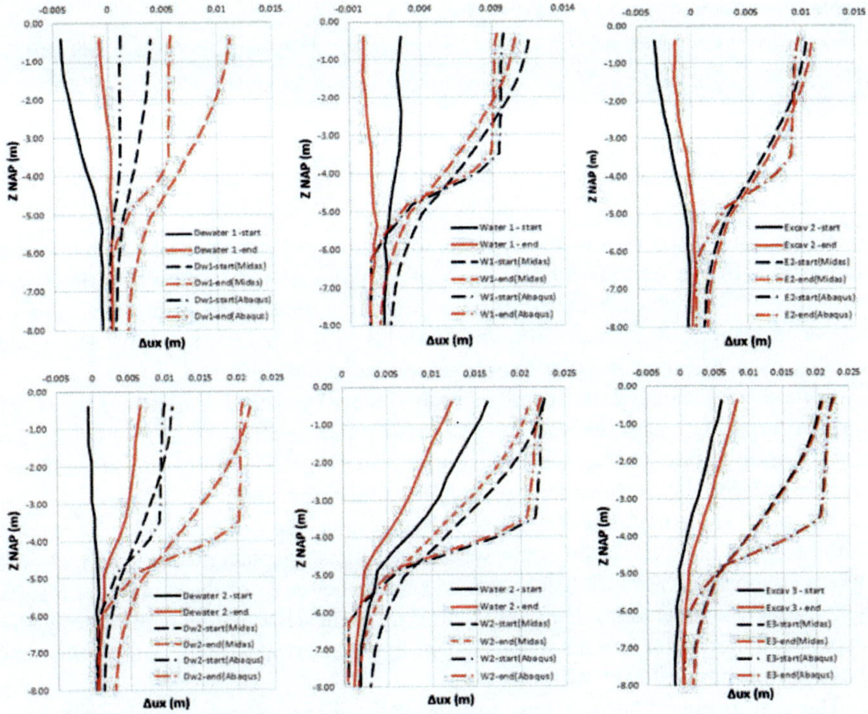

Fig. 13 Horizontal displacements in crest inclinometer

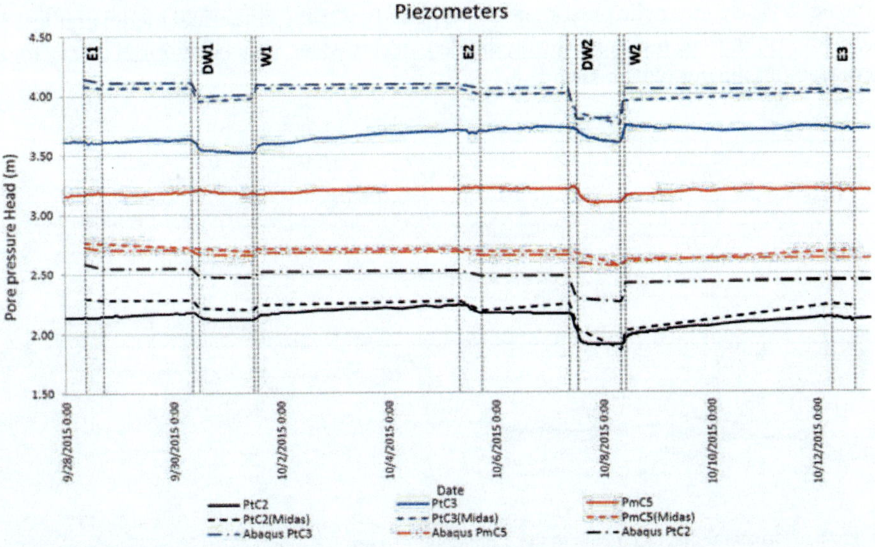

Fig. 14 Pore pressure

Table 2 Permeability found by back-analysis

Dyke (m/s)	Peat under dyke (m/s)	Peat polder (m/s)	Organic Clay (m/s)	Silt (m/s)
5×10^{-8}	7×10^{-8}	1×10^{-7}	5×10^{-9}	5×10^{-9}

5.3 *Back-Analysis*

Back-analysis procedures were implemented to improve the results of the numerical simulation. In the Abaqus model, the back analysis aimed to get an initial stress with lower horizontal effective stress in the polder side by adjusting the value of lateral coefficient, K_0. Meanwhile, in the Midas model, the permeability and initial conditions were varied in order to obtain more accurate results.

In the Midas model better results are obtained when in the initial step the water level in the polder was established at −2.1 m NAP and the effective stress is calculated using Bishop's effective stress relationship, considering the partially saturated effects, and the different layers have the permeability registered in the Table 2.

To better appreciate the difference in the results, the horizontal displacements in the toe inclinometer at the depth where the largest displacement (−3.4 m NAP) occurs is shown in Fig. 15. These results show a better fit of the numerical results with the inclinometer data, accurately representing the deformation in the dewatering steps, but underestimating the displacement in the excavation steps, suggesting that the materials have a lower stiffness.

The pore pressure head in the location of the three piezometers after the back analysis is shown in Fig. 16. From these results can be concluded that the permeability values defined with the back analysis represent better the behavior of the material in the field. However, in the piezometers PtC3 and PmC5 the numerical model predicted an increase in pore pressure during the dewatering steps, this behavior is owing to a contraction during plastic shearing.

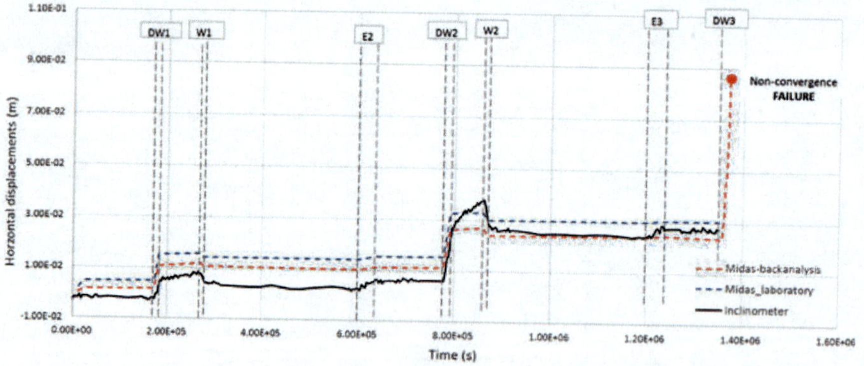

Fig. 15 Horizontal displacements in the toe inclinometer at −3.4 m NAP

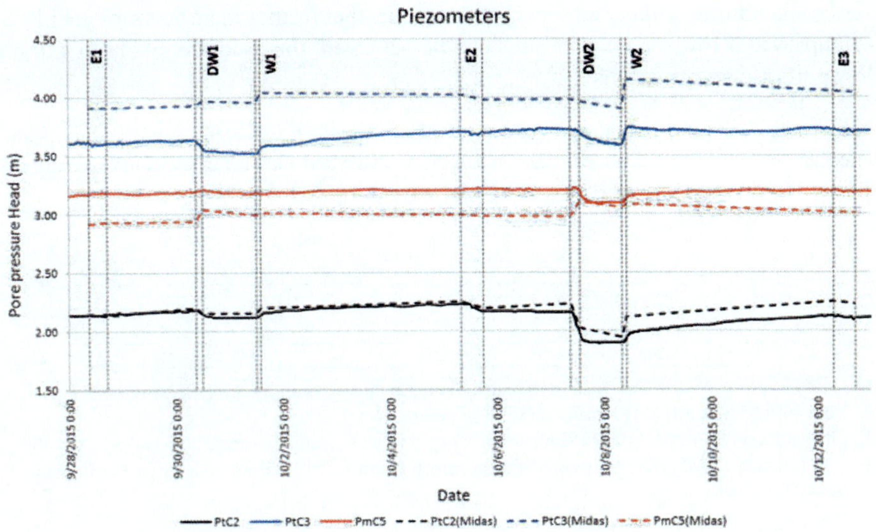

Fig. 16 Pore pressure after back-analysis

6 Conclusions

The present benchmark provided a unique opportunity to test the modeling capabilities of two popular commercial Finite Element codes, Abaqus Standard and Midas GTS NX, for evaluating the safety of earth retaining structures subject to ground water flow and large deformations.

Many aspects of the modeling process have been studied in depth, including the determination of material strength parameters based on laboratory and field investigations, the detailed establishment of initial stresses in the finite element models, as well as the implementation of special types of non-standard boundary conditions in Abaqus.

The Modified Mohr-Coulomb material model was found to present several advantages as implemented in Midas GTS NX, including the fact that no initial stresses are necessary to initiate a simulation, whereas the Modified Cam Clay model in Abaqus does require initial pressures to be present, unless a linear elastic model is combined with the tabular hardening options. However, if the initial stresses are not known exactly but can be reasonably estimated, the simulation can proceed by means of transient consolidation steps.

As can be evidenced by the deformation results in the models, the clay plasticity models in Abaqus somewhat over-estimate the deformations compared to the inclinometer readings, which may be due to the fact that the pore pressures in the model are consistently higher than those reported by the piezometers. This is true from the beginning of the simulation, where no pumping was included in the polder side, as no detailed information on how the low water levels downstream of the dyke were achieved, but this is likely an aspect of the model that further modifications are likely to improve. If lower pore pressures can be achieved, the model is likely to exhibit lower plastic deformations.

Acknowledgements The authors wish to thank the formulators for providing detailed high-quality information, which is often laborious and expensive to obtain and allowing participants to contribute and expand our experience in such important aspect of geotechnical engineering.

References

1. Calculation of phreatic surface in an earth dam. Abaqus Example Guide Section 10.1.2
2. Pore Fluid Flow. Abaqus Analysis Guide, Section 34.4.7
3. Griffiths DV, Lane PA (1999) Slope stability analysis by finite. Géotechnique 49(3):387–403
4. Van Baars S (2005) The horizontal failure mechanism of the Wilnis peat dyke. Géotechnique 55(4):319–321

Analysis of Pre-failure and Failure of a Levee on Soft Soil with Modified Cam-Clay Model in DIANA

M. V. Schwager, A. D. Tzenkov, and G. M. A. Schreppers

Abstract In order to evaluate available models for the safety assessment of levees on soft soils, a full-scale experiment was performed by Dutch authorities bringing a historical dyke to failure. In the paper, the pre-failure and the failure state of the levee are investigated by means of finite element analyses using the software code DIANA. The calculated pre-failure response is successfully validated against the displacements and pore pressures measured during the full-scale experiment. The limit state of the levee is estimated by means of strength reduction analyses using simple plasticity models as well as by more comprehensive stress-strain analyses using the Modified Cam-clay model. Based on the results of the different analyses performed, conclusions are drawn about the type of analysis recommended for the safety assessment of such levees on soft soils. Recommendations are formulated regarding the constitutive model, the hydraulic conditions and the influence of the spatial variation of strength parameters.

Keywords Levee · Soft soil · Coupled analysis · Modified Cam-Clay model

1 Introduction

In order to evaluate current models for the safety assessment of levees on soft soils, the foundation for research on regional dykes in the Netherlands STOWA performed a full-scale test on a regional dyke in the Kagerplassen, north of Leiden [1, 2]. To this aim, this historical dyke closing a polder was brought to failure. The full-scale test started with an excavation of the levee foundation on the polder side and included

M. V. Schwager (✉)
Swiss Federal Office of Energy SFOE, Bern, Switzerland
e-mail: markus.schwager@bfe.admin.ch

A. D. Tzenkov
Stucky Ltd, Renens, Switzerland

G. M. A. Schreppers
DIANA FEA BV, Delft, The Netherlands

G. Bolzon et al. (eds.), *Numerical Analysis of Dams*, Lecture Notes in Civil Engineering 91, https://doi.org/10.1007/978-3-030-51085-5_40

a sequence of excavation and dewatering phases, until failure eventually occurred during rapid drawdown at the toe of the levee. During the test, displacements and pore water pressures were measured in three sections of the levee.

Within the framework of the given numerical benchmark formulation [1], both the pre-failure and the failure state of the levee are investigated by means of finite element analyses using the software code DIANA. In a first step, the limit state is estimated in a preliminary way performing

- a total stress strength reduction analysis assuming undrained conditions (see Sect. 3.1) and
- an effective stress strength reduction analysis assuming drained conditions (see Sect. 3.2).

In a second step, the parameters of the Modified Cam-clay model are calibrated on the results of the laboratory tests (see Sect. 4) in order to be used for

- a stress-strain analysis assuming undrained conditions and a user-defined initial pore pressure field (see Sect. 5.1),
- a staggered stress-strain analysis assuming drained conditions (see Sect. 5.2) and
- a coupled hydro-mechanical stress-strain analysis allowing for partially drained conditions (see Sect. 5.3).

This paper presents a comparison of the limit state of the levee estimated by five different types of analyses performed. Furthermore, the pre-failure response obtained by three different kind of stress-strain analyses is validated against the displacements and pore pressures measured during the full-scale test. Moreover, the potential influence of the spatial variability of the shear strength is investigated (see Sect. 6). Finally, conclusions are drawn regarding the type of analysis recommended for the safety assessment of such levees on soft soils.

2 Model Set-up

2.1 Geometry

The analyses are performed on a two-dimensional finite element model. Its geometry is shown in Fig. 1. The dimensions of the model are 15 m in vertical and 35 m in horizontal direction. The middle of the crest of the levee is located at a distance of 9 m from the left and at 26 m from the right model boundary. In vertical direction, a reference coordinate system is introduced with its origin at NAP being 0.29 m above the crest level (see Fig. 2). The riverbed to the left of the levee is located at -2 m NAP; the ground surface of the polder to the right of the levee is at -1.9 m NAP. The excavation in the polder is performed close to the toe of the levee removing parts of its foundation.

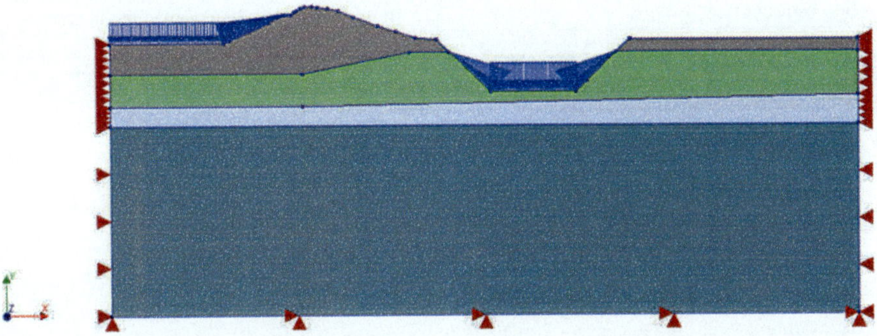

Fig. 1 Model geometry, soil strata and boundary conditions

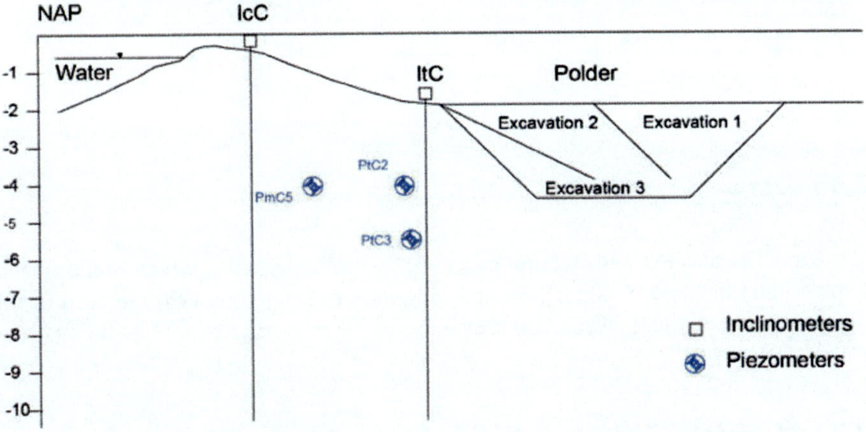

Fig. 2 Cross-section of the excavations

2.2 Boundary Conditions

At the bottom boundary, kinematic constraints in both horizontal and vertical directions are considered (see red symbols in Fig. 1). At the two side boundaries, only the horizontal displacements are constrained. In the riverbed (on the left of the levee) a hydrostatic load corresponding to the water-level at −0.6 m NAP is applied. The excavation (see Fig. 2) and dewatering phases performed in the excavation are considered by introducing the following stages performed in the model:

Description of each stage	Date when each stage is reached
Initial stage	21-9-2015
Excavation 1—removing soil—water-level at −2 m NAP	28-9-2015
Excavation 1—dewatering to −2.5 m NAP	30-9-2015
Excavation 1—dewatering to −2.5 m NAP	30-9-2015
Excavation 1—refilling water to −2 m NAP	30-9-2015
Excavation 2—removing soil – water-level at −2 m NAP	5-10/2015
Excavation 2—dewatering to −2.5 m NAP	7-10-2015
Excavation 2—dewatering to −3 m NAP	7-10-2015
Excavation 2—refilling water to −2 m NAP	8-10-2015
Excavation 3—removing soil – water-level at −2 m NAP	12-10-2015
Excavation 3—dewatering to −2.5 m NAP	14-10-2015
Excavation 3—dewatering to −3 m NAP	14-10-2015
Excavation 3—dewatering to −3.5 m NAP	14-10-2015

2.3 *Mesh*

The model is meshed with second order plane-strain elements, which mainly have a quadrilateral shape (8 nodes). Some triangular elements (6 nodes) are used to fill the gaps between the quadrilateral elements.

2.4 *Materials*

The model comprises four soil layers with different material properties (see Fig. 1). From the top to the bottom, they are named 'Dyke material', 'Peat', 'Organic Silt-clay' and 'Foundation'. The density and porosity assigned to these four materials are given in Table 1.

Table 1 Dry density and porosity

Soil layer	Dry density [kg/m^3]	Porosity [-]
Dyke material	1290	0.56
Peat	100	0.90
Organic Silt-clay	610	0.74
Foundation	900	0.50

3 Prediction of Failure

In a first step, a prediction of the limit state is made using relatively simple strength assessment techniques. For this purpose, strength reduction analyses are performed using the Newton-Raphson iteration scheme (full tangent stiffness matrix).

3.1 Total Stress Strength Reduction Analysis Assuming Undrained Conditions

A strength reduction analysis is performed in total stress formulation assuming the elasto-plastic Tresca material model for undrained conditions. Both the undrained shear strength and the Young's modulus are derived from shear tests performed in undrained conditions. Please note that the laboratory tests have been performed on normally or slightly over-consolidated samples. Therefore, the shear strength is expected to be underestimated for soil layers being over-consolidated in field conditions. The parameters obtained are listed in Table 2.

In order to mimic the undrained volumetric behavior, Poisson's ratio is set to be 0.49 which is close to incompressible behavior.

In the nonlinear analysis performed, the self-weight, the hydrostatic pressures in the riverbed and in the excavation are applied first. Afterwards, the shear strength is reduced till the mobilized undrained shear strength is reached at the limit state. The factor of safety (FS) is obtained dividing the mobilized shear strength by the shear strength of the material. Assuming different water levels in the excavation pit results in different factors of safety (see Table 3). The critical water level is found to be in between −2.1 m and −2.2 m NAP.

Table 2 Elastic and shear strength parameters for undrained conditions

Soil layer	Young's modulus [MPa]	Poisson's ratio [-]	Shear strength [kPa]
Dyke material	9	0.49	20
Peat (both sections)	3	0.49	4.5
Organic Silt-clay	6	0.49	17
Foundation	6	0.49	20

Table 3 Factor of safety for different water levels in excavation 3 assuming total stress Tresca model for undrained conditions

Water level NAP [m]	FS [-]
−2.0	1.150
−2.1	1.125
−2.2	<1.0
−2.5	<1.0

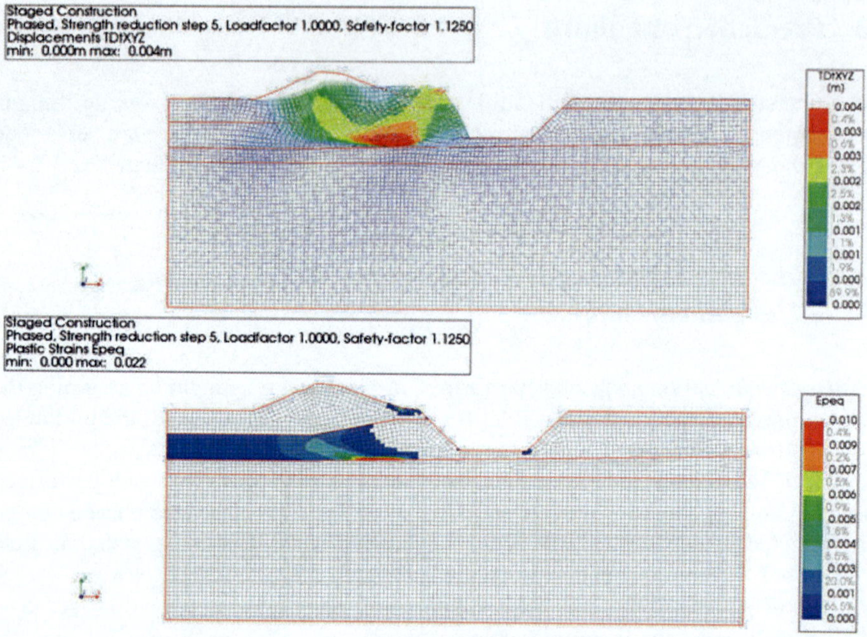

Fig. 3 Failure mechanism in the undrained strength reduction analysis (water level at −2.1 NAP): Displacement field (top) and equivalent plastic strains (bottom)

The failure mechanism obtained is shown in Fig. 3. The failure mechanism is given by a circular sliding surface starting on the bottom corner of excavation 3, touching the bottom of the Peat layer and going up to the upstream face of the levee. Plastic strains are found to start developing within the Peat layer.

3.2 *Effective Stress Strength Reduction Analysis Assuming Drained Conditions*

Another strength reduction analysis is performed in effective stresses assuming elasto-plastic Mohr-Coulomb material model for drained conditions. The effective shear strength parameters and the Young's moduli are again derived from results of the laboratory shear tests (see Table 4).

In addition, the pore pressure field is given assuming hydrostatic pressure increase with depth in combination with a user defined phreatic line connecting linearly the upstream to the downstream water level in the excavation. The assumed pore pressure field (see Fig. 4) is changing for each stage with a different water table in the excavation.

Table 4 Elastic and effective shear strength parameters

Soil layer	Young's modulus [MPa]	Poisson's ratio [-]	Cohesion [kPa]	Friction angle [°]	Dilatancy angle [°]
Dyke material	7.8	0.3	7.2	30	6
Peat (both sections)	2.6	0.3	3.6	30	6
Organic Silt-clay	5.2	0.3	7.4	28	6
Foundation	5.2	0.3	9.0	31	10

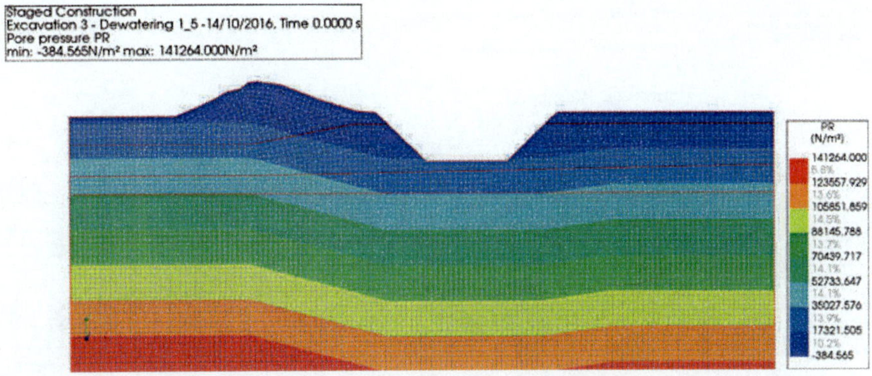

Fig. 4 Pore-pressure field assumed (water level in the excavation at −2.4 m NAP)

The nonlinear strength reduction analysis for drained conditions is performed in the same way as the previous analysis assuming undrained conditions. The factors of safety FS obtained for different water levels in the excavation pit are given in Table 5. The critical water level is found to be in between −2.5 m and −2.6 m NAP. The respective failure mechanism is shown in Fig. 5. Plastic strains are again found to start developing within the Peat layer.

Table 5 Factor of safety for different water levels in excavation 3 assuming effective stress Mohr-Coulomb model for drained conditions

Water level NAP [m]	FS [-]
−2.0	1.100
−2.1	1.075
−2.2	1.050
−2.3	1.038
−2.4	1.013
−2.5	1.013
−2.6	<1.0

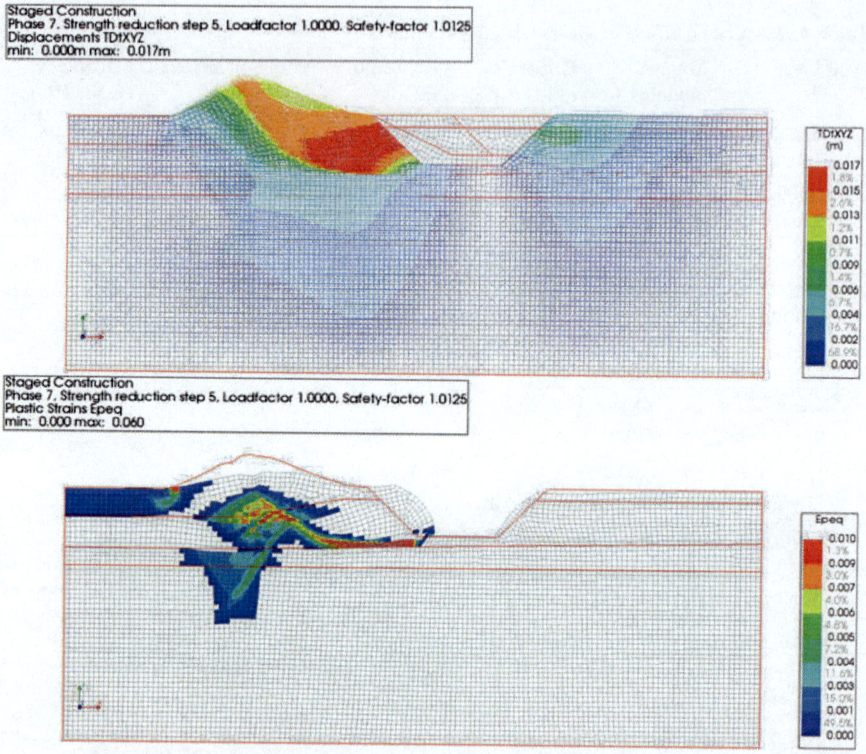

Fig. 5 Failure mechanism in the drained strength reduction analysis (water level at −2.5 NAP): Displacement field (top) and equivalent plastic strains (bottom)

As to be expected, the critical water level when failure occurs is found to be lower assuming drained conditions (critical water level at −2.6 m NAP) than assuming undrained conditions (critical water level −2.2 m NAP). Since the undrained shear strength considered in the total stress analysis is implicitly taking into account the influence of excess pore pressure due to shearing, the shear strength considered is lower and consequently resulting in a configuration being closer to limit conditions.

3.3 Limitations of the Analyses Performed so Far

The two strength reduction analyses performed so far are based on assumed fields of constant pore pressures with time and rather simple elasto-perfectly-plastic constitutive models. Therefore, the two analyses can only give an estimate of a lower and a higher value for the critical water level. The failure is supposed to happen in between these two bounds due to possibly partially drained conditions. However, these two analyses do not allow for an accurate prediction of failure in terms of deformations or in terms of the critical water level.

In order to obtain a more accurate prediction of the limit conditions, the initial pore pressure field as well as its evolution with time owing to the changing boundary conditions and owing to potential excess pore pressure should be taken into account more precisely. Therefore, a seepage analyses has to be performed providing the pore pressure field. Furthermore, a constitutive model should be used being able to capture the development of excess pore pressure due to shearing in undrained conditions. Therefore, the constitutive model is required to predict accurately the volumetric behavior of the type of soil under study. Additionally, the seepage and the mechanical analyses should be coupled in order to be able to take into account effects of partially drained conditions which may be important for the investigated boundary value problem.

4 Calibration to Laboratory Measurements

The Modified Cam-clay constitutive model [3], which is allowing both for shear and volumetric yielding, is assumed to be appropriate to capture the volumetric behavior of the soft soils under study. Its parameters are calibrated on the results of laboratory tests.

For latter purpose, the element tests performed in laboratory are simulated numerically using a finite element model consisting of one 4-nodes axial ring element. For the oedometer tests, fully drained conditions are considered neglecting the time dependency due to consolidation at each load step. For the triaxial tests, undrained conditions are applied during the shearing phase such that excess pore pressure can develop. For the non-linear analysis, an iteration scheme using always the linear stiffness matrix is applied since such iteration scheme is more suitable for the conditions expected to be strongly varying within the integration step.

4.1 Dyke Material

The parameters for the Dyke material are obtained based on the results of oedometer and undrained triaxial tests. The parameters derived are given in Table 6. Please note that the stiffness parameters λ and κ as they are implemented in DIANA do relate to the current void ratio. Thus, the void ratio which is governing together with λ and κ the incremental bulk stiffness is changing for each integration step.

For the oedometer test sample, no preconsolidation pressure is considered whereas a preconsolidation pressure of 21 kPa is considered for the triaxial test samples.

Table 6 Modified Cam-clay material parameters for the Dyke material

Parameters	Dyke material
Friction angle ϕ [°]	30
Plastic hardening κ [-]	0.15
Elastic hardening λ [-]	0.04
Cap-shape factor α [-]	1.7
Cohesion C [kPa]	7.2

Figure 6 shows the comparison between the measured and the calculated laboratory test results. The comparison contains the axial stress-strain behavior in oedometric conditions (σ_a–ε_a), the total (p, q) and effective (p', q) stress path in triaxial shearing as well as the corresponding deviatoric behavior (ε_s, q).

In general, a reasonable agreement between the measurements and the analysis results is found. However, the deviatoric stress calculated becomes constant at large shear strains, whereas the deviatoric stress measured is further increasing. This difference is supposed to be due to the phase change from contractive to dilative volumetric behavior, which cannot be captured by the Modified Cam-clay model. Therefore, the strain hardening capacity at large stains is underestimated by the constitutive model used. Nevertheless, the model used allows for an appropriate prediction of the shear strength for the range of relevant strains before failure.

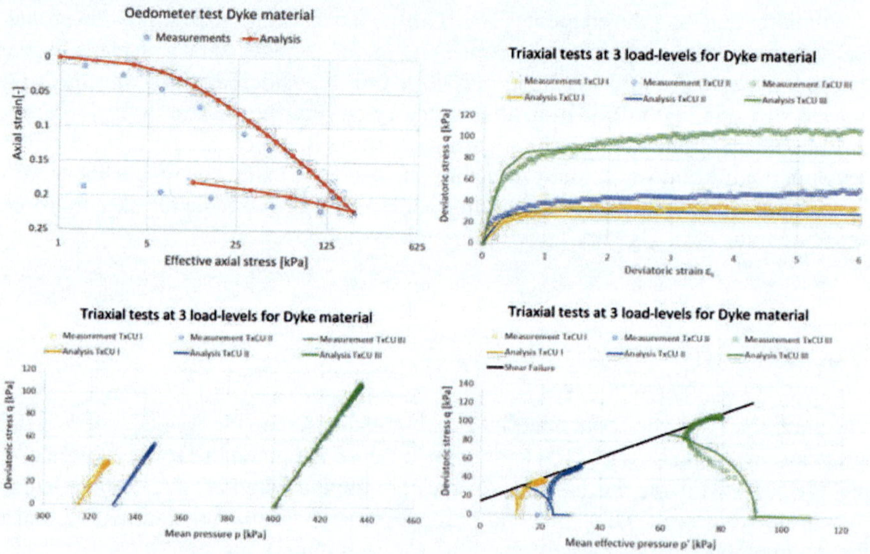

Fig. 6 Comparison of the measured and calculated oedometer and triaxial shear tests performed on the Dyke material

Table 7 Modified Cam-clay material parameters for the Peat material

Parameters	Peat material
Friction angle ϕ [°]	30
Plastic hardening κ [-]	2
Elastic hardening λ [-]	0.4
Cap-shape factor α [-]	1
Cohesion C [kPa]	3.6

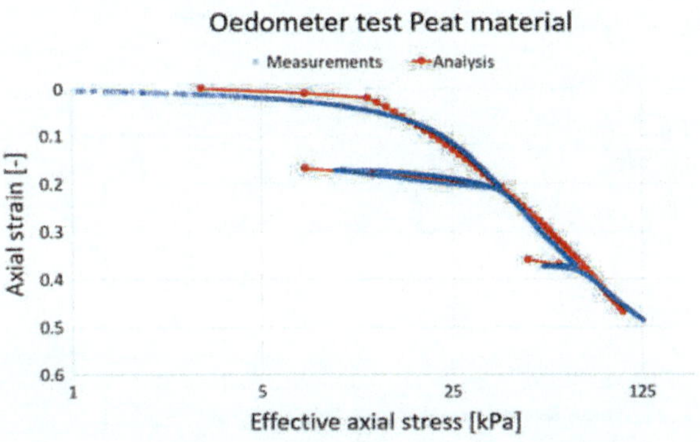

Fig. 7 Comparison of the measured and calculated oedometer test performed on the Peat material

4.2 Peat Material

For the Peat material, the shear strength parameters are obtained from the direct simple shear tests provided. The parameters governing the volumetric behavior are calibrated on the results of an oedometer test. An overview of the material parameters for the Peat material is given in Table 7.

Figure 7 shows the strains measured in the oedometer test as well as the strains calculated for calibration purpose assuming a presonsolidation pressure of 10 kPa. The Modified Cam-clay model is found to describe very well the volumetric behavior of the Peat material.

4.3 Organic Silt-Clay Material

For Organic Silt-clay material, the parameters of the Modified-Cam-clay material model are derived from oedometer and undrained triaxial shear tests. The parameters are given in Table 8. In addition, a preconsolidation pressure of 8 kPa and 24 kPa was considered for the oedometer and the triaxial test samples respectively.

Figure 8 shows the comparison between the measured and the calculated laboratory test results. The Modified Cam-clay model is found to capture reasonably well both the volumetric and the deviatoric behavior of the Organic Silt-clay material.

Table 8 Modified Cam-clay material parameters for the Organic Silt-clay material

Parameters	Organic Silt-clay material
Friction angle ϕ [°]	28
Plastic hardening κ [-]	0.38
Elastic hardening λ [-]	0.1
Cap-shape factor α [-]	2.5
Cohesion C [kPa]	7.4

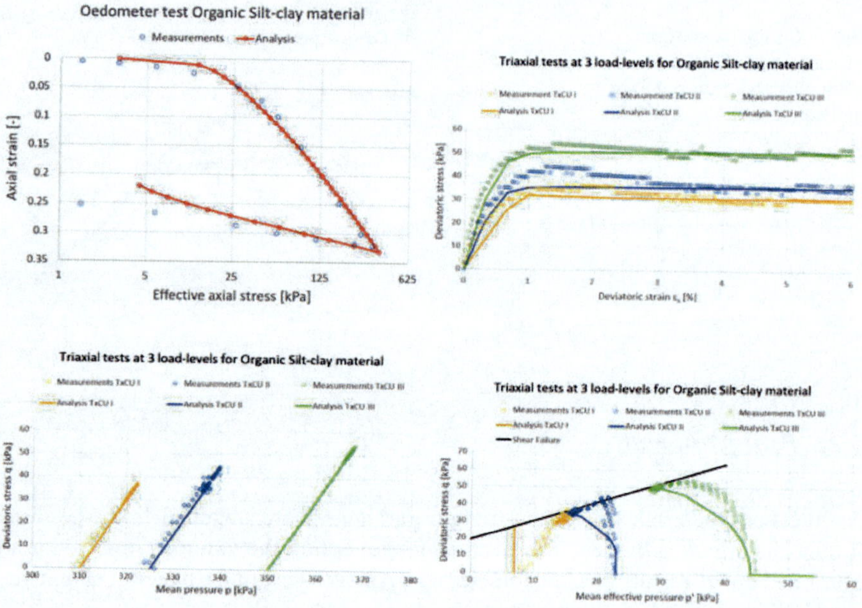

Fig. 8 Comparison of the measured and calculated oedometer and triaxial shear tests performed on the Organic Silt-clay material

4.4 *Foundation Material*

The Foundation layer is assumed to show a similar, but however a slightly stiffer and stronger, behavior than the Organic Silt-clay material. The parameters assumed for the Modified Cam-clay material are given in Table 9.

5 Pre-failure Response

In order to overcome the limitations of the analyses performed so far and presented in Sect. 3, more comprehensive modelling approaches are used. For all approaches,

Table 9 Modified Cam-clay material parameters for the Foundation material

Parameters	Foundation material
Friction angle ϕ [°]	31
Plastic hardening κ [-]	0.1
Elastic hardening λ [-]	0.01
Cap-shape factor α [-]	2.5
Cohesion C [kPa]	9.0

the analysis is performed in effective stresses assuming the Modified Cam-clay model with the parameters derived in the previous sections. Three different modelling approaches are investigated exploring different hydraulic conditions:

- In the first approach (see Sect. 5.1), the analysis is performed assuming perfectly undrained conditions. The initial pore pressure field considered is given assuming hydrostatic pressure increase with depth and a user defined phreatic line connecting linearly the upstream to the downstream water level in the polder. During the dewatering phases, the assumed pore pressure field in the levee remains unchanged.
- In the second approach (see Sect. 5.2), a staggered steady-state groundwater-flow-stress analysis is performed assuming entirely drained conditions. For each stage, a flow analysis is performed to calculate the pore pressure field corresponding to the current hydraulic boundary conditions. Each pore pressure field calculated is then used in the subsequent stress-strain analysis. For comparison purpose, the same analysis is performed again assuming undrained conditions within each stage.
- In the third approach (see Sect. 5.3), a fully coupled transient mixture analysis is performed, where displacements and pore pressure potential unknowns are calculated in the same iterative procedure. The groundwater-flow and force-equilibrium problems are solved fully coupled which allows to capture partially drained conditions. Thus, the porosity becomes dependent on the deformations. Furthermore, the fluid is assumed to be incompressible and the soil is assumed to be fully saturated.

The hydraulic conductivities assumed for the second and the third approach are given in Table 10.

The over-consolidation ratios OCR are estimated based on the laboratory tests performed. The values considered are given in Table 11.

Table 10 Hydraulic conductivities of the different soil layers

Layer	Hydraulic conductivity [m/s]
Dyke material	3e-8
Peat	1e-8
Organic Silt-clay	1e-9
Foundation	5e-9

Table 11 OCR values considered for different soil layers

Layer	OCR [-]
Dyke material	1.1
Peat below the dyke	1.1
Peat below the polder	3.0
Organic Silt-clay	2.5
Foundation	2.5

All the analyses were conducted using a Newton-Raphson iteration scheme and requiring a force convergence criterion of 0.001 within maximum 10 iterations per step. If the convergence criterion is not reached within the maximum number of iterations, the analysis is allowed to continue.

5.1 Stress-Strain Analysis Assuming Undrained Conditions

Assuming undrained conditions and a user defined initial pore pressure field, convergence was obtained for all stages except the one of lowering the water table in excavation 3 from −3.0 NAP to −3.5 NAP. This may point to failure of the levee during this stage. The critical water level is consequently expected to be in between −3.0 and −3.5 m NAP.

Figure 9 shows the corresponding equivalent plastic strain and the deformed mesh (with the deformations scaled by a factor of 10). The failure mechanism is clearly found to develop from the stage with the downstream water level at −3.0 m NAP to the consecutive stage. Furthermore, the contours of the horizontal displacements as well as their vertical profiles where the inclinometers are located are shown in the same figure.

More detailed information about the pre-failure horizontal displacements at the inclinometer located at the toe of the dam is provided in Fig. 10. The maximum horizontal displacement after dewatering of excavation 2 is found to be 4.2 cm, which is quite close to the measured maximum displacement of 3.8 cm. Also, the reversed displacement in unloading when the excavation 2 is refilled is captured very well by the model. Thus, the model is found to be able reproduce the measured maximum pre-failure displacements.

However, there are time dependent effects observed in the experiment which cannot be addressed by the model. The maximum measured horizontal displacement for example developed with time from 2.8 cm after dewatering to 3.8 cm just before refilling of excavation 2 (see Fig. 25), whereas the calculated displacement occurred almost immediately. Such time dependency may be related to effects of progressive failure, of rate dependency of the shear strength or owing to the fluid flow.

The pore-pressure values for different stages at the three locations of the piezometers (see Fig. 2) are summarized in a bar chart given in Fig. 11.

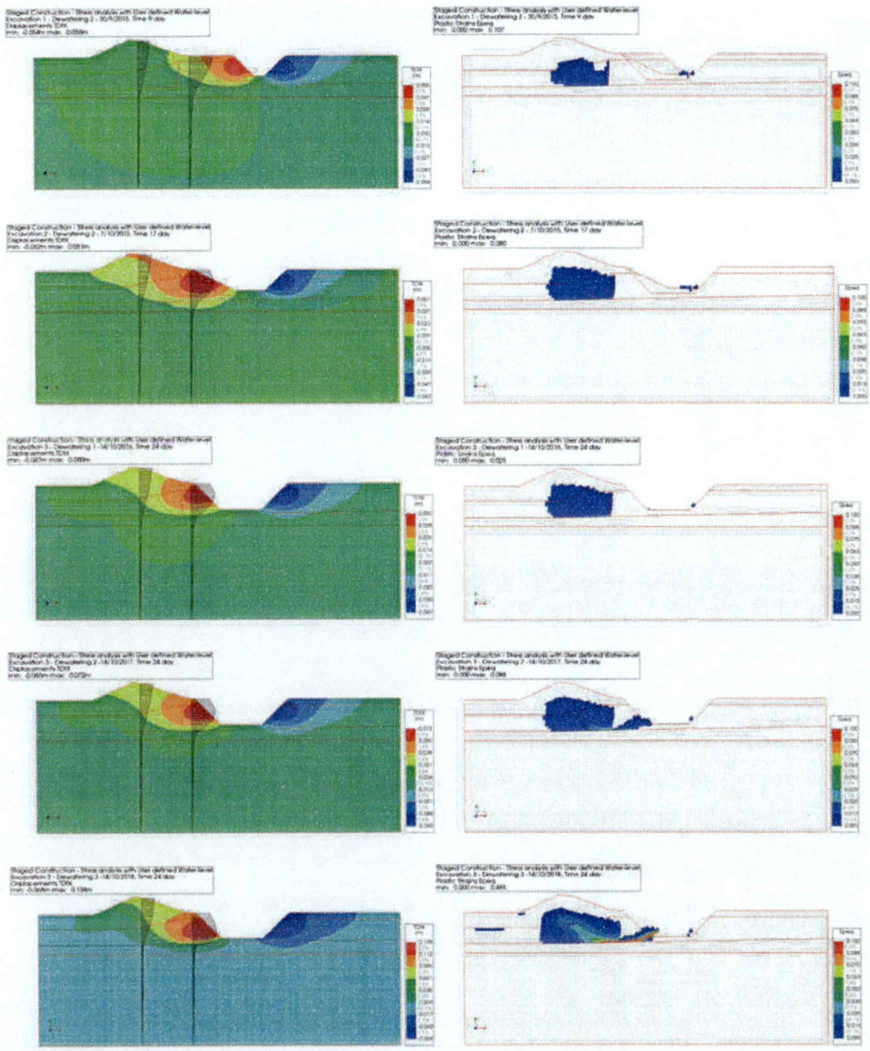

Fig. 9 Undrained stress-strain analysis: horizontal displacement and equivalent plastic strain contours at the end of different stages (excavation 1 with WL at −3.0 m NAP, excavation 2 with WL at −3.0 m NAP, excavation 3 with WL at −2.5, −3 and −3.5 m NAP)

5.2 Stress-Strain Analysis Assuming Drained Conditions

Performing a staggered analysis and assuming perfectly drained conditions, the analysis does not reach the convergence criterion in the last two stages (with the water level being at −3 m and −3.5 m NAP). The remaining unbalance is around 5% of the first estimate of the force-vector. Since numerical indications for failure are not

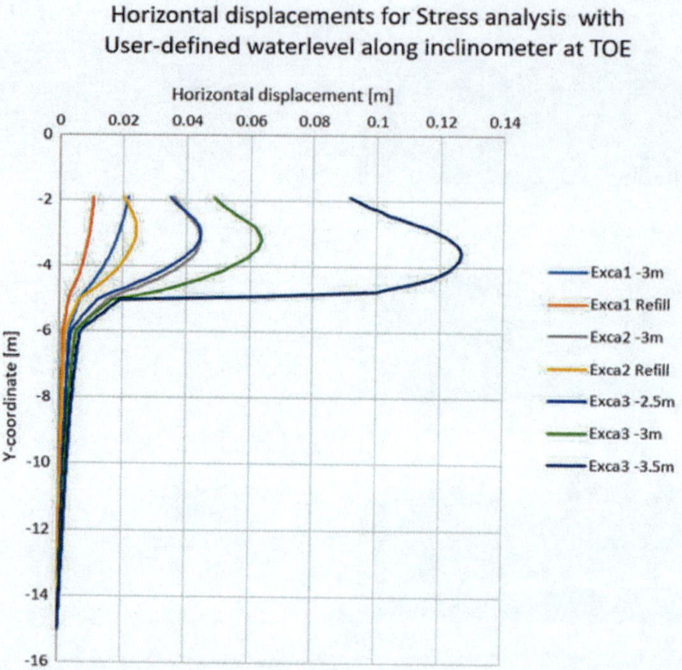

Fig. 10 Undrained stress-strain analysis: calculated horizontal displacements along the inclinometer at the toe of the dam

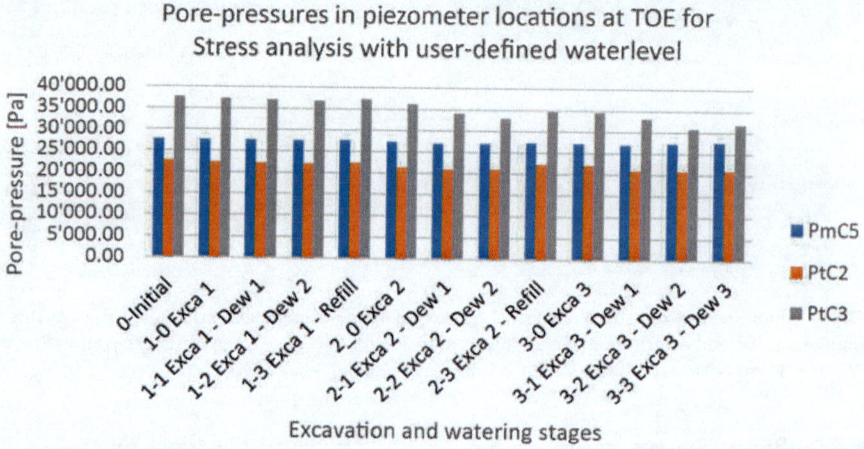

Fig. 11 Undrained stress-strain analysis: calculated pore pressures at the end of different stages at the locations of the piezometers

very strong, it is difficult to estimate the critical water level. It is expected to be considerably below −3.0 m NAP.

Figure 12 shows the calculated pore-pressure field for the initial stage and the last stage; Fig. 13 gives the pore pressures calculated at the locations of the piezometers for each stage.

The shape of the failure mechanism is found to be slightly different in drained than in undrained conditions (see Fig. 14). It is less curved and its outcrop on the upstream face of the dam is a bit lower.

In drained conditions, displacements are obtained to be larger (see Fig. 15), which is supposed to be due to the volumetric strains which can take place. Due to this softer behavior of the model, the measured displacements are over-estimated assuming drained conditions.

Since the pre-failure deformations are found to be over-estimated, another analysis is performed for comparison purpose assuming undrained conditions within each stage. Please note, that the soil is free draining at the beginning of each stage when

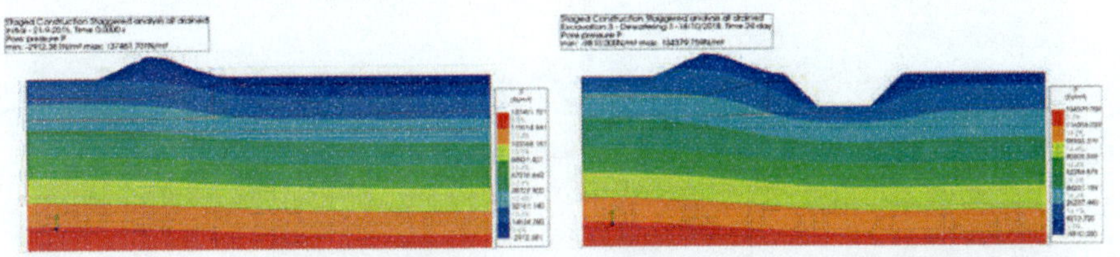

Fig. 12 Calculated pore-pressure field in the staggered analysis in the initial stage (left) and in the final stage of excavation 3 with the water table at −3.5 NAP (right)

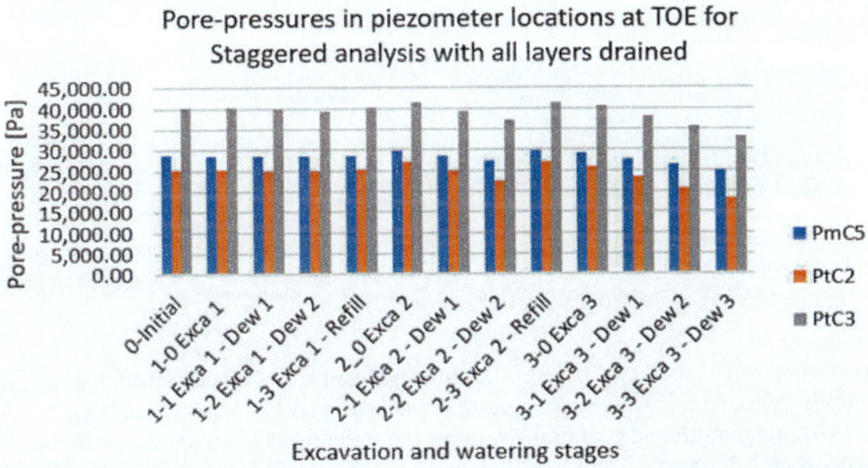

Fig. 13 Drained stress-strain analysis: calculated pore pressures at the end of different stages at the locations of the piezometers

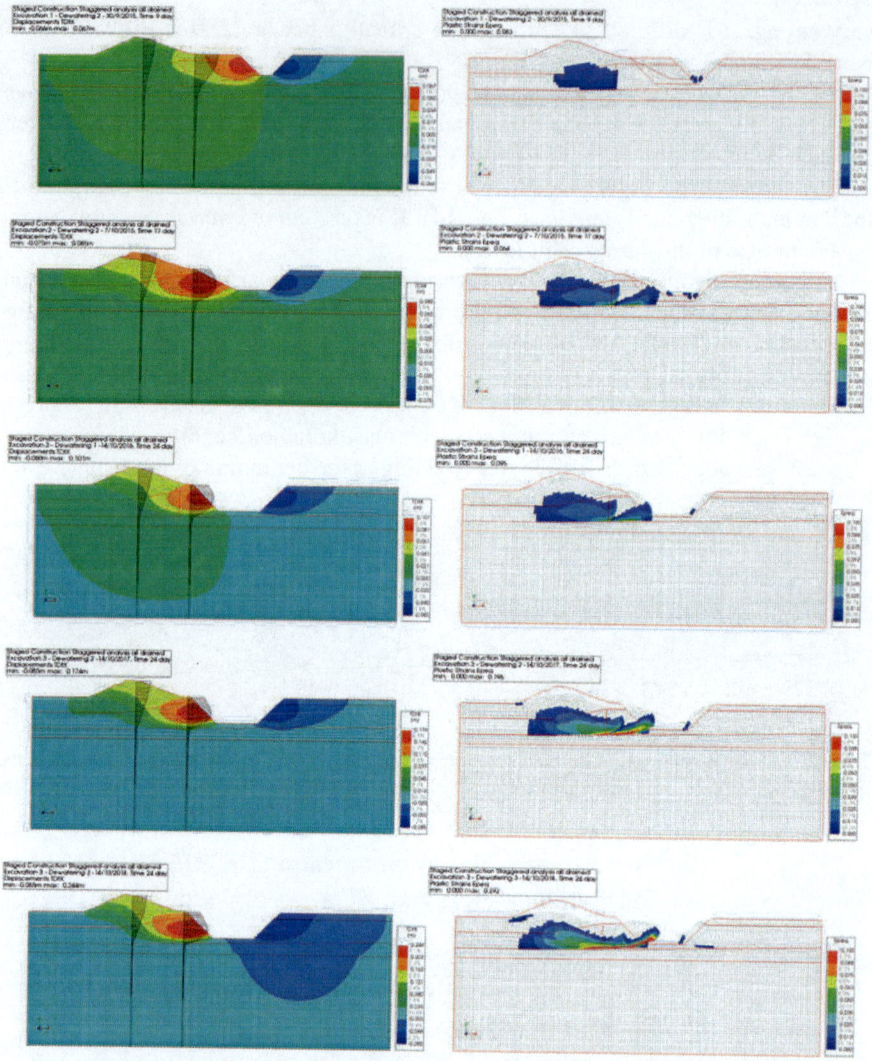

Fig. 14 Drained stress-strain analysis: horizontal displacement and equivalent plastic strain contours at the end of different stages (excavation 1 with WL at −3.0 m NAP, excavation 2 with WL at −3.0 m NAP, excavation 3 with WL at −2.5, −3 and −3.5 m NAP)

the seepage analysis is performed. The authors are aware of this approach being not entirely consistent. However, the analysis is seen as a sort of sensitivity check.

Assuming undrained conditions within each stage results in a critical water level in between −3 m and −3.5 m NAP, but closer to −3 m NAP. Furthermore, the failure is found to be much more pronounced assuming undrained conditions (see Fig. 16). As to be expected, the pre-failure deformations calculated (see Fig. 17) are smaller

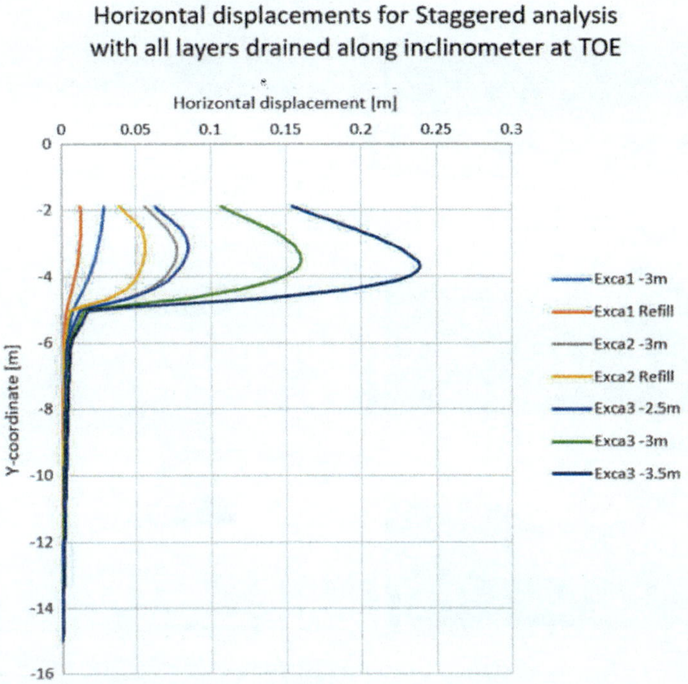

Fig. 15 Drained stress-strain analysis: calculated horizontal displacements along the inclinometer at the toe of the dam

than for fully drained conditions but larger than for fully undrained conditions (see Sect. 5.1).

5.3 *Coupled Hydro-Mechanical Stress-Strain Analysis*

Within the framework of the fully coupled mixture analysis, the change in the boundary conditions are considered with time. Whereas the preceding analyses consisted of 13 stages of one step each, the mixture analysis has got four stages. Each stage consists of one load step in which equilibrium is calculated after the respective parts of the excavation are removed while the water-level is remaining constant followed by several time-steps in which the water-level is changing and the water is flowing through the porous soil. There are 166 time-steps in total; they are varying from 15 min to 10 h per step.

Figure 18 shows the calculated total pore pressure as well as the excess pore pressure field at the initial stage and at the last stage. Positive excess pore pressure is found in the Peat below the dam. It is probably due to the volumetric behavior of soil in shearing. Negative excess pore pressure is found next the excavation. It is

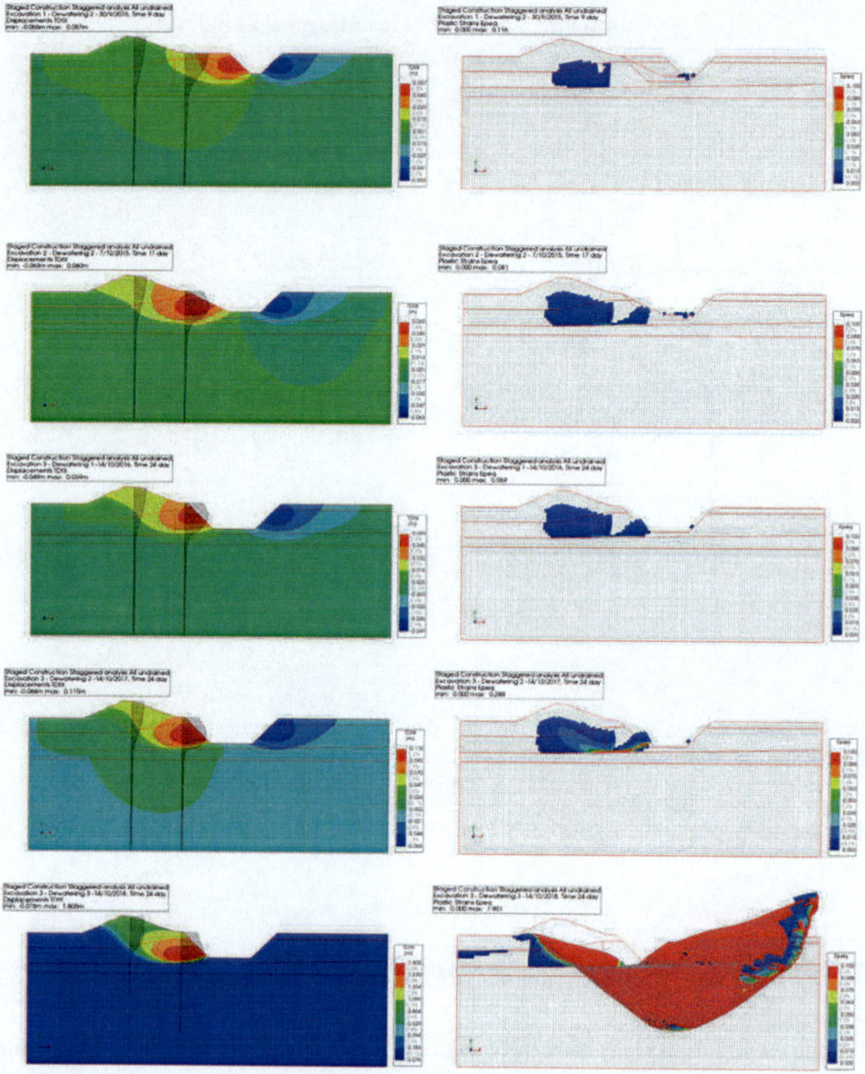

Fig. 16 Undrained staggered stress-strain analysis: horizontal displacement and equivalent plastic strain contours at the end of different stages (excavation 1 with WL at −3.0 m NAP, excavation 2 with WL at −3.0 m NAP, excavation 3 with WL at −2.5 m, −3 m and −3.5 m NAP)

supposed to be due to the transient pore pressure changes caused by the dewatering of the excavation.

During the excavation phase 3, convergence problems occurred just after reaching the water level of −3.0 m NAP (see Fig. 19). Thus, the critical water level is expected to be a bit lower than −3.0 NAP. In order to assess the critical water level and the

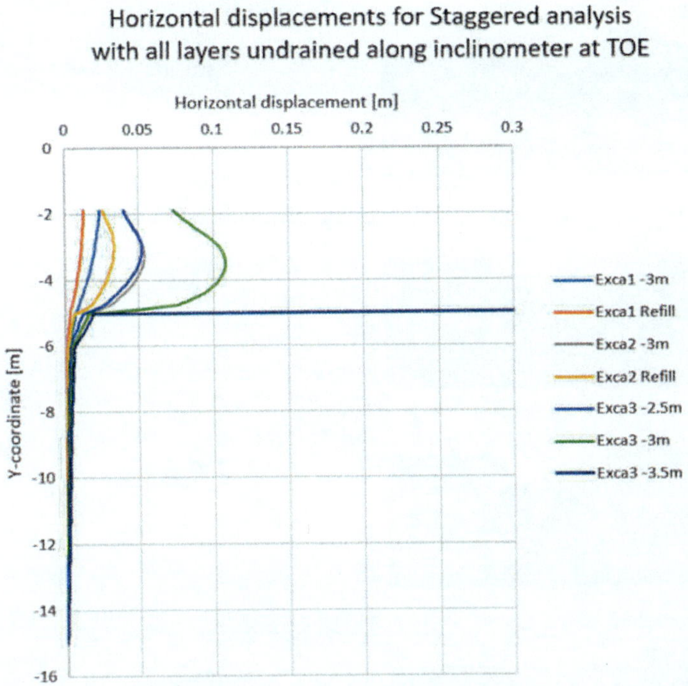

Fig. 17 Undrained staggered stress-strain analysis: calculated horizontal displacements along the inclinometer at the toe of the dam

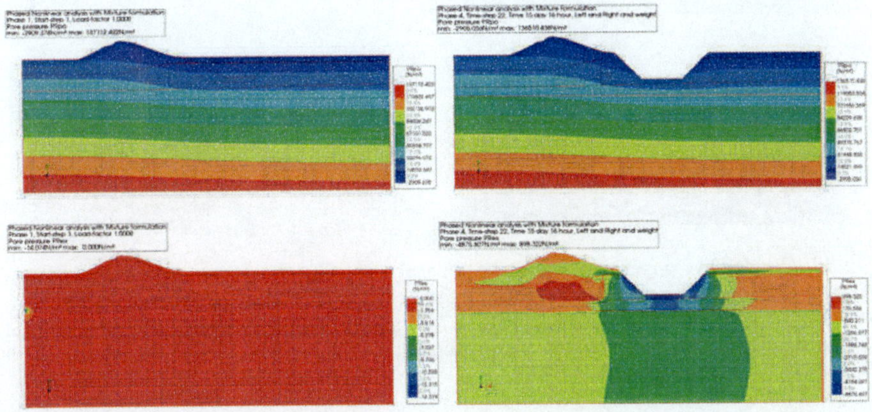

Fig. 18 Calculated total and excess pore pressure fields in the initial and the last stage

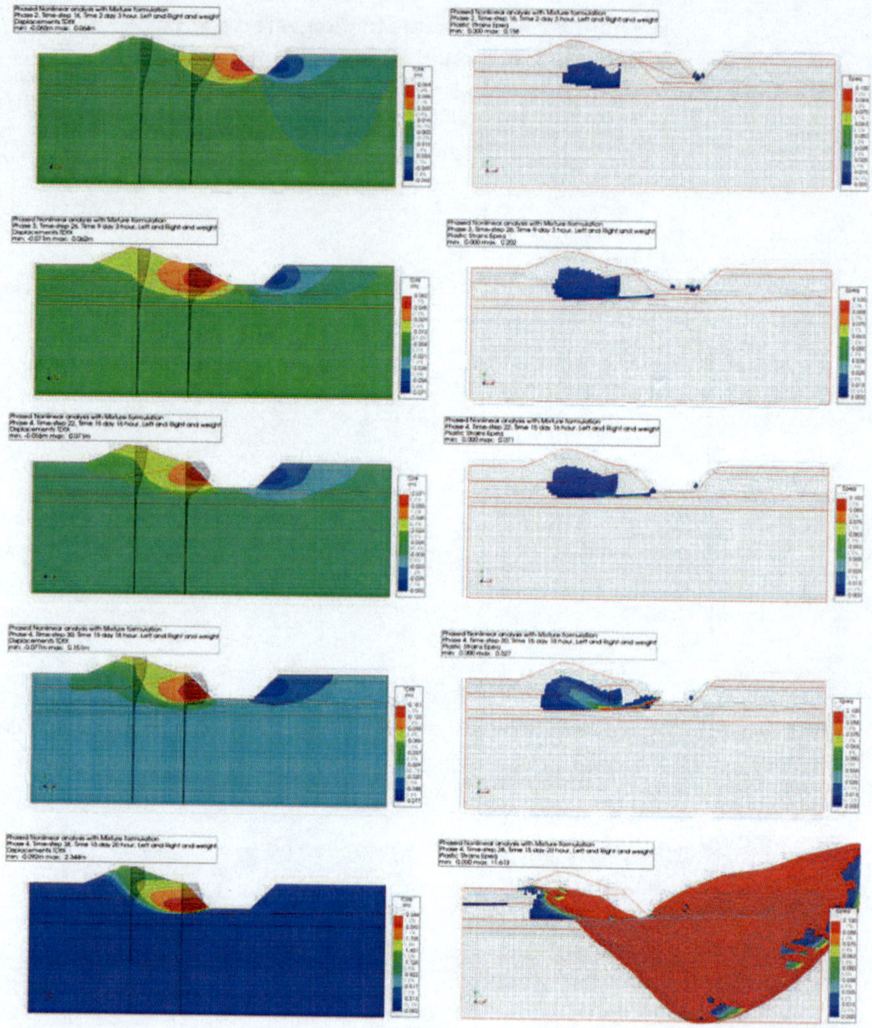

Fig. 19 Coupled hydro-mechanical stress-strain analysis: horizontal displacement and equivalent plastic strain contours at the end of different stages (excavation 1 with WL at −3.0 m NAP, excavation 2 with WL at −3.0 m NAP). Coupled hydro-mechanical stress-strain analysis: horizontal displacement and equivalent plastic strain contours at the end of different stages (excavation 3 with WL at −2.5 m, −3 m and −3.5 m NAP)

failure mechanism in a more precise way, the results of the different time steps are studied (see Fig. 20). The failure mechanism is found to form entirely when the water level is reaching −3.2 m NAP. The corresponding maximum calculated displacements is increasing from 16 cm for the water level of −3.1 m NAP to 46 cm

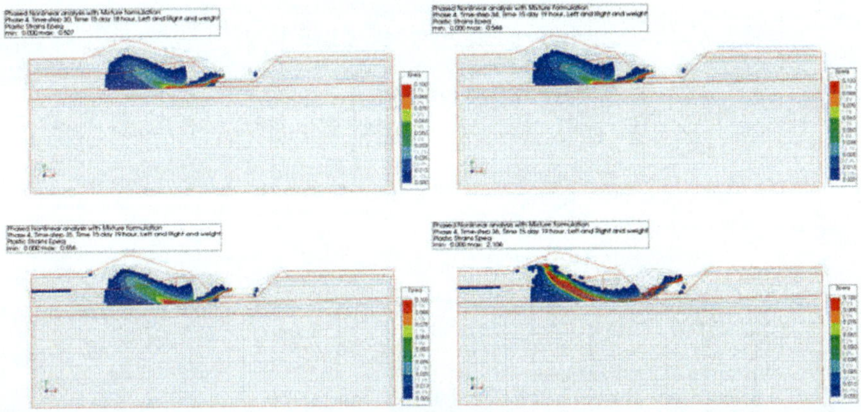

Fig. 20 Equivalent plastic strains and deformed mesh at the end of dewatering to −3.0 m NAP (top left), just at the beginning of further dewatering (top right), when the water level reaches −3.1 m (bottom left) and −3.2 m NAP (bottom right)

for the water level of −3.2 m NAP. In both time steps, the convergence after 10 iterations was around 10%, pointing to very unstable conditions.

The pre-failure horizontal displacements along the inclinometer located at the toe of the dam are shown in the Fig. 21. The pre-failure deformations are found to be a little bit larger than the ones obtained assuming undrained conditions (see Sect. 5.1). Nevertheless, the calculated deformations compare reasonably well to the measured displacements.

Figure 22 depicts the horizontal displacements of two nodes (at −3.5 m NAP below the toe of the dam where the inclinometer is located and at −2 m NAP below the crest of the dam) as a function of time. The onset of the time scale is at the beginning of excavation 1. The deformations are found to happen when the shape of the excavation or the water level in it is changing.

Figures 23 and 24 show the calculated total pore-pressures where the three piezometers are located. Also, the pore pressures are found to show an immediate reaction to changing boundary conditions. However, the immediate change in pore pressure is rather small compared to the change in the boundary conditions.

5.4 Validation Against Field Measurements

For validation purpose, the results of the stress-strain analyses are compared to the field measurements provided in the problem formulation [1] (i.e. the inclinometer and piezometers readings taken during the large-scale experiment).

The displacements calculated are compared to the displacement profile measured in inclinometer ItC located at the toe of the dam (see Fig. 25). For the analysis assuming undrained conditions (see Sect. 5.1) and for the mixture analysis (see

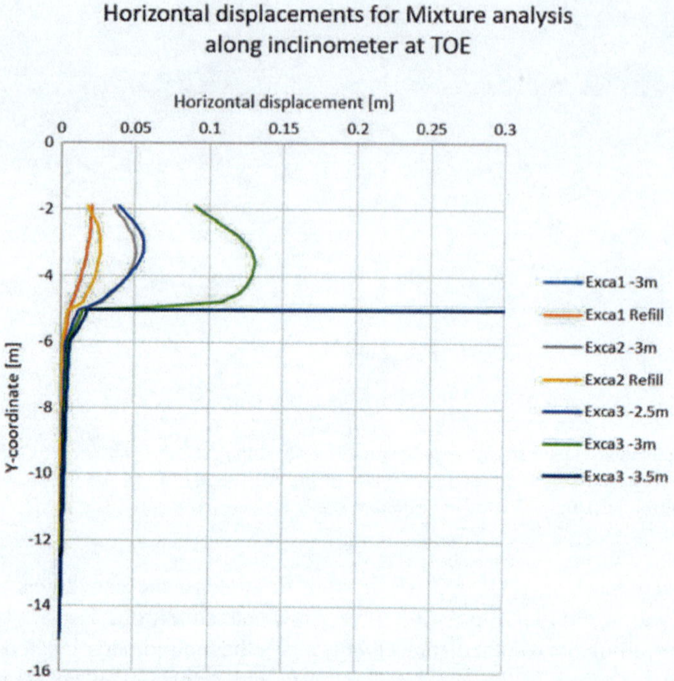

Fig. 21 Coupled hydro-mechanical stress-strain analysis: calculated horizontal displacements along the inclinometer at the toe of the dam

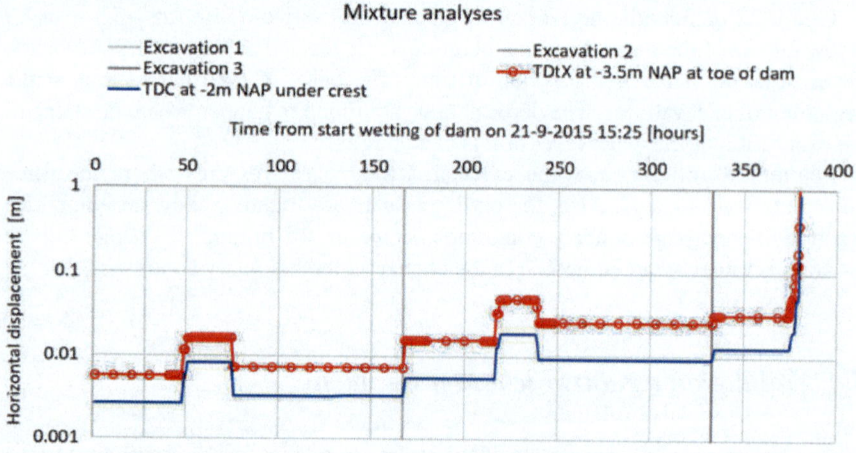

Fig. 22 Calculated horizontal displacements at −3.5 m NAP below the toe and at −2 m NAP below the crest

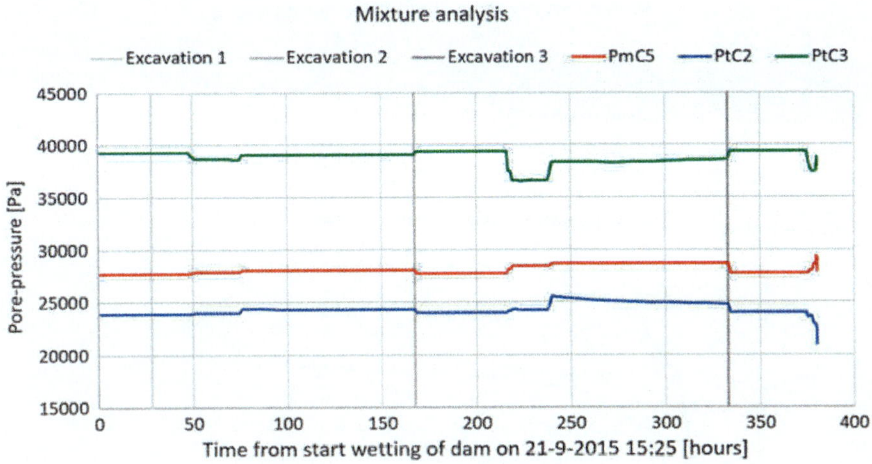

Fig. 23 Calculated pore pressures at the locations of the piezometers

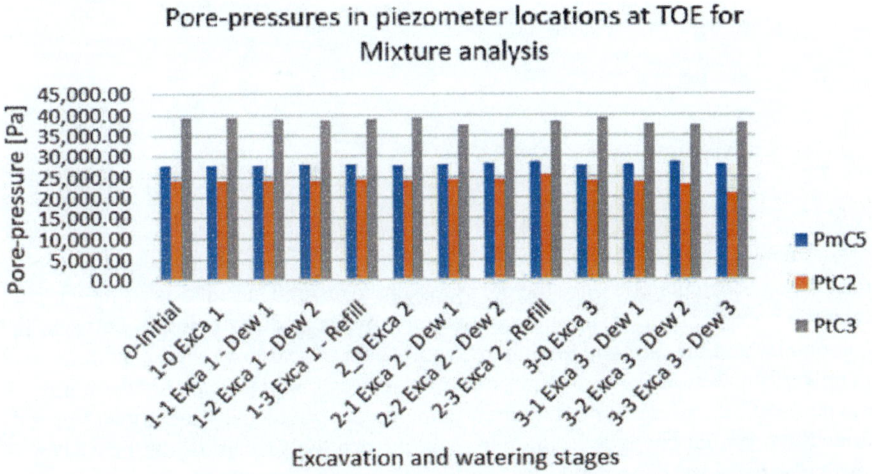

Fig. 24 Coupled hydro-mechanical stress-strain analysis: calculated pore pressures at the end of different stages at the locations of the piezometers

Sect. 5.3), a reasonable agreement is obtained in terms of pre-failure displacements. The displacements calculated assuming drained conditions (see Sect. 5.2) are considerably larger than the displacements measured. The comparison of displacements is indicating that conditions in the experiment were rather undrained than drained.

In addition, the pore pressures calculated are compared to the pressures measured by the piezometers PtC2, PtC3 and PmC5. The pore pressures measured in PtC2 and PtC3 are confirming the initial pore pressure fields calculated. However, the initial

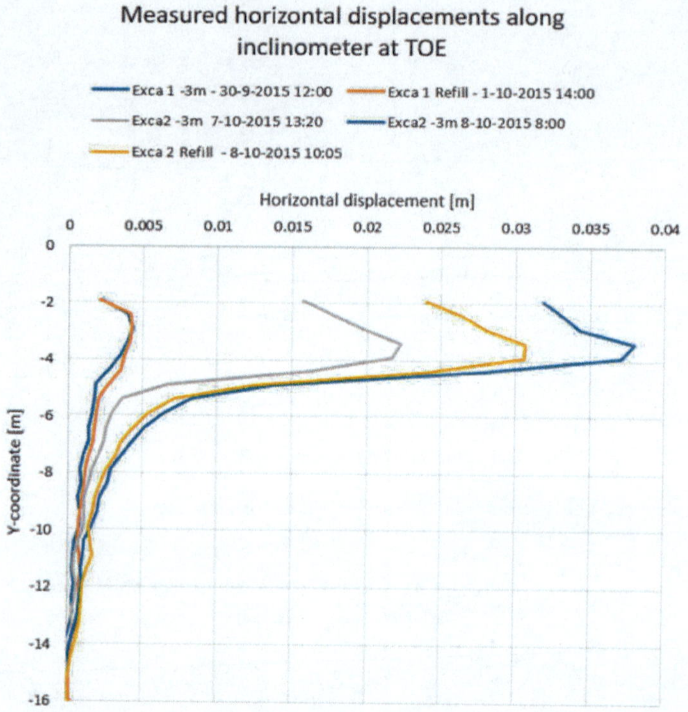

Fig. 25 Displacement profile measured in inclinometer ItC located at the toe of the dam

pressure measured in PmC5 in the Peat layer below the dam is around 7 kPa higher than the ones calculated. Since the pore pressure measured at PmC5 is slightly higher than the pore pressure assuming a hydrostatic pressure distribution in between the piezometer and the phreatic line, there seems to be a local flow having a component in upwards direction. Thus, one possible explanation could be a spatial variability of the permeability which is not captured in the model but causing locally an upwards flow. Such spatial changes in the permeability may be present in the Peat layer in particular because its permeability can be highly pressure and direction dependent. Another possible explanation for this finding could simply be an offset in the pore pressure measurements taken at PmC5.

Regarding the change in pore pressures caused by the excavating and dewatering phases, the analyses assuming undrained (see Sect. 5.1) and fully coupled conditions (see Sect. 5.3) show a response being very close to the pore pressure changes measured (see Fig. 26). In both analysis as well as in the measurements, the pore pressure changes are rather limited.

However, the pore pressure changes calculated assuming drained conditions (see Sect. 5.2) are found to be considerably larger than the ones measured. Based on the comparison of the pore pressure measurements, it can be concluded that the conditions in the experiment are supposed to be close to undrained conditions.

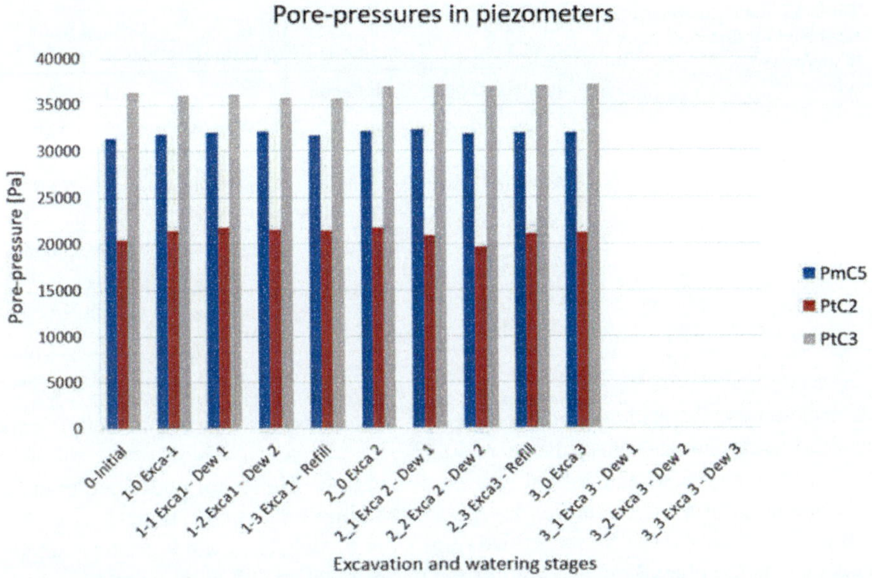

Fig. 26 Measured pore-pressures at different stages in three piezometers

Consequently, the modelling approach assuming drained conditions is found to be not entirely appropriate for the boundary value problem under study.

However, two other modelling approaches applied (i.e. the undrained analysis with a user defined pore pressure field and the fully coupled mixture analysis) are successfully validated against the over-all behavior observed in the field. Both approaches are found to be suitable to predict the deformations and pore pressures in such kind of a boundary value problem.

5.5 Discussion

The analyses performed allow for estimation of the limit state, which is given by a certain reduction in the excavation pit water level causing its failure. A summary of the critical water levels estimated is provided in Table 12.

In general, the critical water level is found to be higher assuming linear elastic/perfect plastic constitutive models than assuming the Modified Cam-clay model. Since the strength parameters used for the simple plasticity models have been calibrated on laboratory tests performed on normally consolidated soils, the shear strength assigned is rather low. However, considerable over-consolidation (particularly in the Peat layer below the polder) is considered in the analyses with the Modified Cam-clay model resulting in higher (drained peak and undrained) shear strength.

Table 12 Critical water level estimated by different types of analyses

Constitutive model	Hydr. conditions	Pore pressures	Critical water level
Lin. el., Tresca	Undrained	–	−2.2 m NAP
Lin. el., Mohr C.	Drained	User defined	−2.6 m NAP
Modified Cam-clay	Undrained	User defined	−3.0 m NAP
Modified Cam-clay	Drained	Staggered analysis	below −3.0 m NAP
Modified Cam-clay	–	Coupled analysis	−3.2 m NAP

Consequently, the critical water level is obtained to be lower when accounting for the influence of over-consolidation of soil.

As to be expected, the critical water is found to be higher for undrained than for drained conditions. This finding is true for the simple plasticity models as well as for the more comprehensive Modified Cam-Clay model. The most accurate analysis, however, is the one taking account for the hydro-mechanical coupling resulting in a critical water level of −3.2 m NAP. The latter water level is found to be relatively close to the critical water level obtained for undrained conditions (i.e. −3.0 m NAP). This leads again to the conclusion that the hydraulic conditions in the experiment have been partially drained, but relatively close to fully undrained conditions.

6 Three-Dimensional Model with Spatial Variation of Shear Strength Properties

Since the levee is not straight but slightly curved, a three-dimensional model (see Fig. 27) is set-up to investigate the influence of the geometry. The levee is assumed to form an arch with radius of 125 m in plan view. The three-dimensional model is made out of 145 000 quadratic solid elements having a size of 0.5 m.

Applying the same parameters and load-scheme as in the two-dimensional models, the levee could be expected to reach its limit state slightly earlier due to its curved shape in plan view. However, the analysis does not confirm this hypothesis; failure occurs at the same time. The reason for this finding is supposed to be the larger element size in the three-dimensional model which does not allow the deformations in failure to develop as well as in the two-dimensional model. Due to the considerable computational costs, the mesh could not be further refined within the framework of the benchmark study.

In order to investigate the influence of the spatial variation of the shear strength, a simpler three-dimensional model being equivalent to the two-dimensional model is created by extruding the two-dimensional model over a distance of 5 m (see Fig. 28).

Fig. 27 Three-dimensional model of the curved levee

The model contains 32 000 quadratic solid elements having a size of 0.25 m in the area where plastic strains develop to up to 1 m in the foundation. A staggered analysis with 13 stages is performed.

In a first step, a comparison is made in between the results of the three-dimensional and the two-dimensional model. Apart from the last stage for which no convergence is reached with both models, the maximum horizontal displacements are found to be almost identical (see Table 13). In addition, the plastic strains and the deformed mesh are compared for the pre-last stage (see Fig. 29). The shape of the contours is again found to be almost identical. However, the maximum plastic strain value in the two-dimensional model is 0.288, whereas in the three-dimensional model a value of 0.419 is calculated. This difference is explained by the different integration schemes in the elements. The 8-noded quadrilateral elements in the two-dimensional model have a 2×2 integration scheme, whereas the 20-noded brick elements in the three-dimensional model have a $3 \times 3 \times 3$ integration scheme. In the latter, the integration points are located closer to the edges of the elements and therefore can describe more pronounced local peaks.

In a second step, the spatial variation of the shear strength parameters is introduced by a random field approach. For this purpose, the spatial variation of the friction angle is derived from the CPT profiles [4, 5] assuming an exponentially decaying correlation function [6]. The stochastic properties obtained are summarized in Table 14.

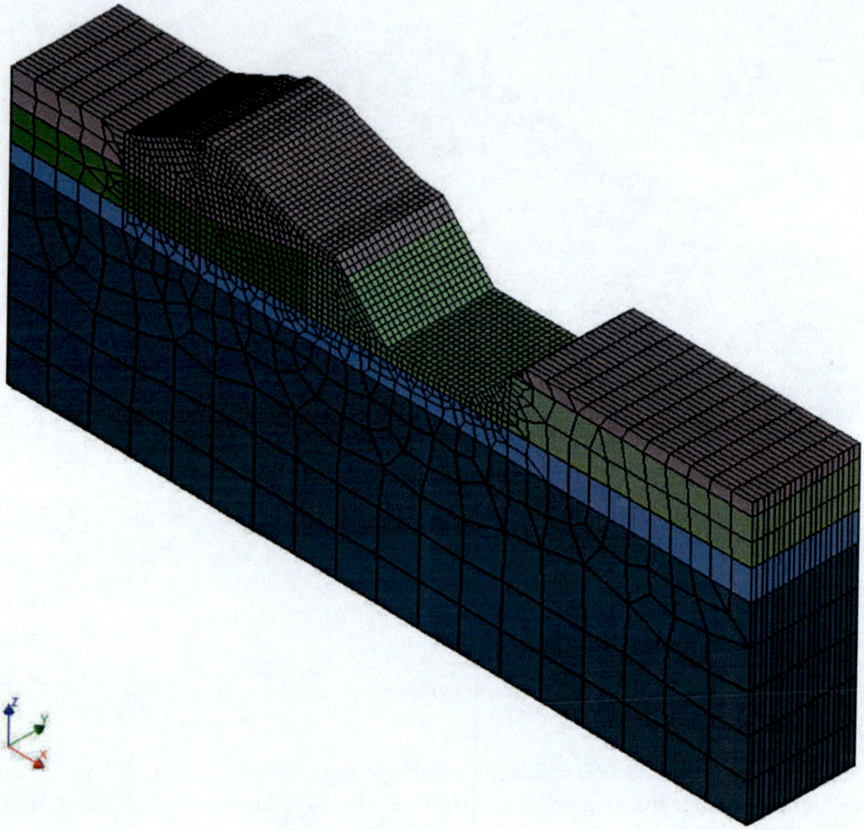

Fig. 28 Three-dimensional extruded model of the levee

Table 13 Comparison of the maximum horizontal displacements of the two-dimensional undrained staggered model and the three-dimensional extruded undrained staggered model

Stage	Max. X-displ. 2D [m]	Max X-displ. 3D [m]
1_2: Excavation 1, WL = −3 m	0.057	0.056
2_2: Excavation 2, WL = −3 m	0.060	0.058
3_1: Excavation 3, WL = −2.5 m	0.059	0.060
3_2: Excavation 3, WL = −3 m	0.115	0.115
3_3: Excavation 3, WL = −3.5 m	1.805*	0.231*

(*) no convergence reached in last analysis step

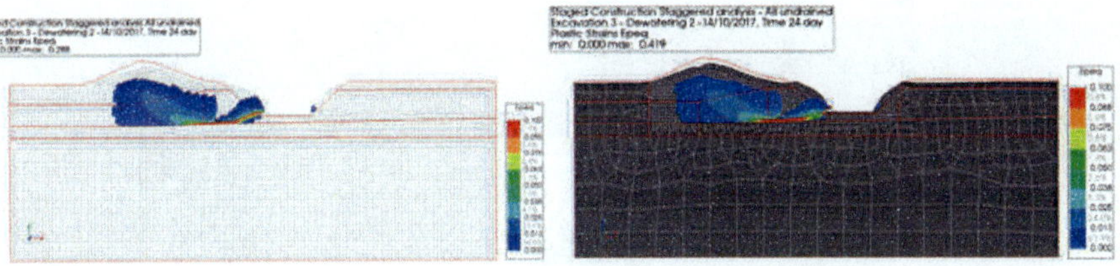

Fig. 29 Equivalent plastic strain and the deformed mesh at the end of stage 3_2: two-dimensional model (left), three-dimensional model (right)

Table 14 Stochastic properties of the friction angle in different soil-layers

Layer	Average [°]	Standard deviation [°]	Vertical correlation length [m]
Dyke	30	3.7	0.10
Peat	28	3.8	0.60
Organic Silt-clay	30	1.6	0.25
Foundation	31	0.8	0.25

The random field is defined for a regular orthogonal grid with 0.25 m spacing. A random field sample (see Fig. 30) is generated neglecting the minor spatial variation in the Foundation layer and assuming the horizontal correlation length to be 2.5 times the vertical correlation length. Table 15 gives an overview of the statistics of the three layers in the sample.

The results obtained considering the random field sample are given in Figs. 31 and 32. Due to the random field, the displacements are found to be slightly larger for the first four stages and slightly lower for the last stage. Hence, the influence of the considered random spatial variation of the friction angle on the results is found to be minor.

7 Conclusions

Based on the set of results obtained with different types of analyses, conclusions can be drawn about the type of analysis recommended for the safety assessment of such levees on soft soils. Recommendations are formulated regarding the constitutive model, the hydraulic conditions and the influence of the spatial variation of strength parameters.

In terms of constitutive models, the Modified Cam-clay model is found to be appropriate for the natural deposits investigated. Furthermore, the Modified Cam-clay model can be used as an approximation for the behavior of the dyke material,

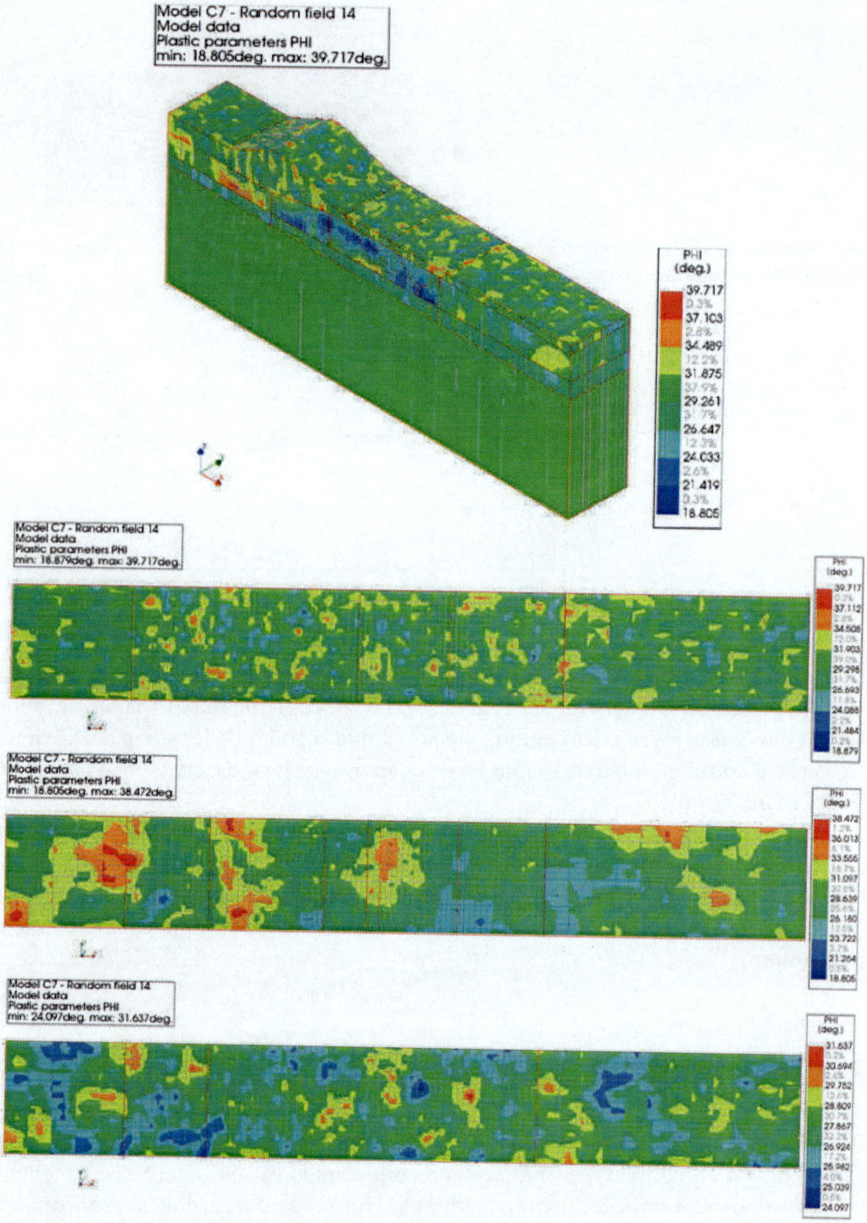

Fig. 30 Random field of the friction angle in 3D, in top view of the Dyke material (top picture), in a horizontal cross-section at top of the Peat layer (middle picture) and in a horizontal cross-section at the top of the Organic Silt-clay layer (bottom picture)

Table 15 Statistics for the friction angle in different soil-layers for the sample random field shown in Fig. 30

Layer	Minimum value [°]	Maximum value [°]	Average [°]	Standard deviation [°]
Dyke	15.0	47.0	30.0	3.72
Peat	15.4	43.4	29.5	3.75
Organic Silt-clay	21.6	35.0	27.9	1.66

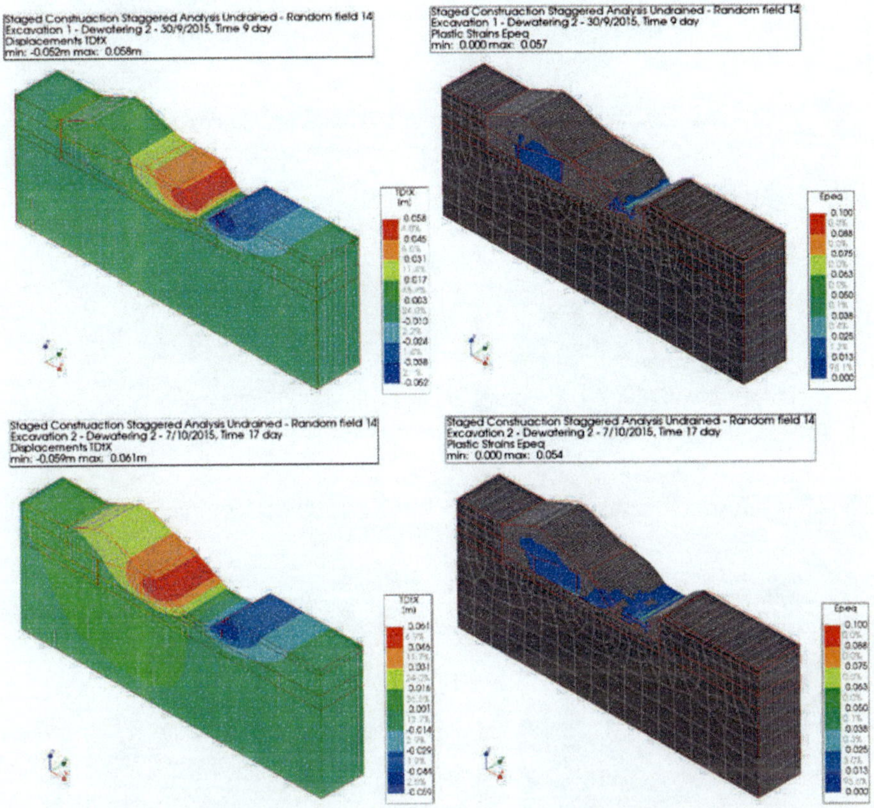

Fig. 31 Horizontal displacement and equivalent plastic strain contours at the end of different stages (excavation 1 with WL at −3.0 m NAP, excavation 2 with WL at −3.0 m NAP)

although the model cannot capture the measured phase change from contractive to dilative volumetric behavior.

Nevertheless, simple plasticity models can be used as well in order to get an estimate of the limit state. Matching the undrained shear strength on tests performed on normally consolidated soils can provide a conservative estimate of the limit state if the soil in situ is expected to be slightly over-consolidated.

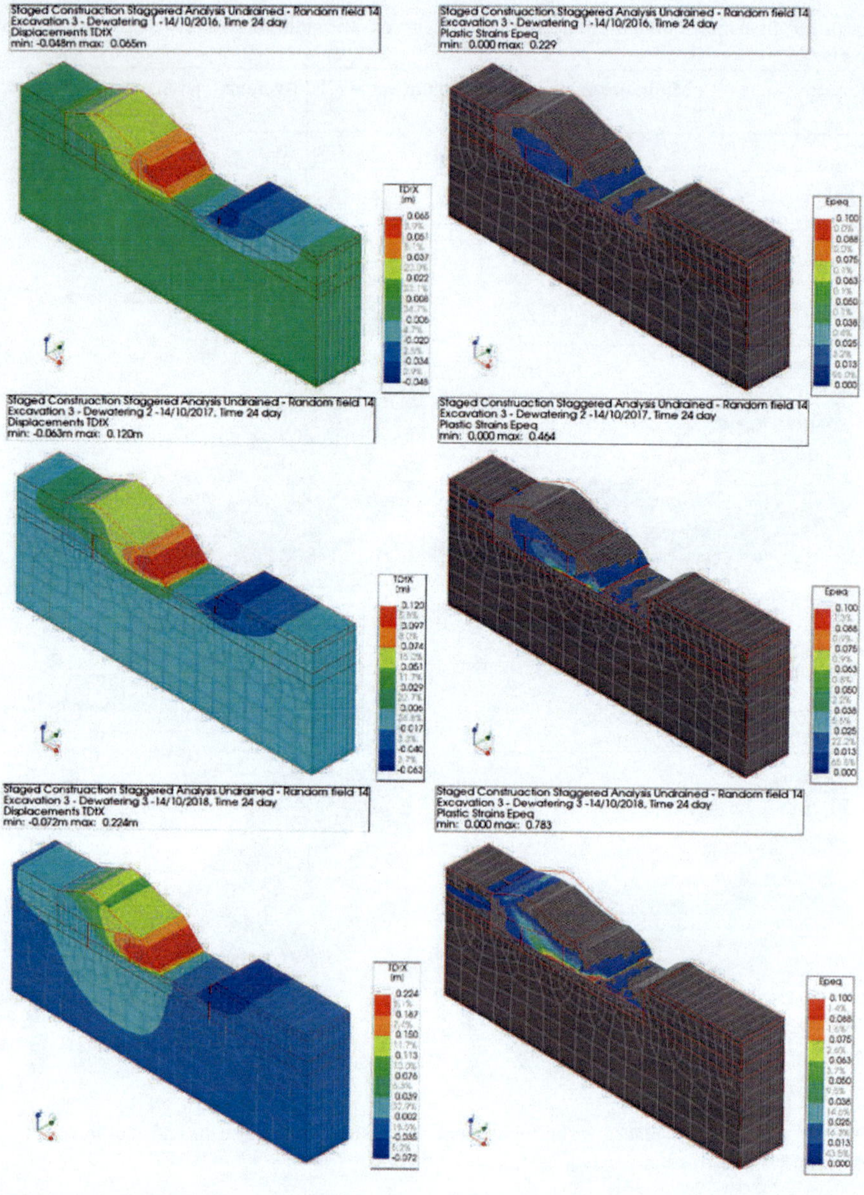

Fig. 32 Horizontal displacement and equivalent plastic strain contours at the end of different stages (excavation 3 with WL at −2.5, −3 and −3.5 m NAP)

Regarding the hydraulic conditions to be assumed, the coupled analysis is expected to be the most realistic type of analysis. A good choice of one of the two extreme conditions can also lead to a very good approximation. For the case investigated, the assumption of undrained conditions is found to be very close to what has been observed in the experiment. However, assuming the other extreme condition (i.e. drained conditions for the investigated case) may not be appropriate.

The influence of the random spatial variation of shear strength is found to be minor for the case investigated. Therefore, the focus is recommended to be rather on the systematic spatial changes of shear resistance than on its random variation. In this respect, the spatial distributions of the permeabilities (governing the pore pressures) and the stress history (resulting in over-consolidation of soils) are considered to be essential for a safety assessment of such type of levee.

References

1. Jommi C, de Gast T, Ponzoni E, Valimberti N, Muraro S, Wentholt L, van Hemert H (2019) Theme C: coupled hydro-mechanical analysis of the pre-failure and the failure behaviour of a dyke on soft subsoil. In: 15th International Benchmark workshop on numerical analyses of Dams, Milano, Italy (2019)
2. Theodoridis M (2017) Numerical simulation of the coupled hydro-mechanical response of the Leendert de Boerspolder dyke. Master thesis, TU Delft, the Netherlands
3. Roscoe K, Burland JB (1968) On the generalized stress-strain behaviour of wet clay. In: Heyman J, Leckie FA (eds) Engineering plasticity. Cambridge University Press, pp 535–610
4. Senneset K, Sandven R, Janbu N (1989) Evaluation of soil parameters from piezocone tests. Trans Res Rec 1235
5. Mayne PW (2007) Cone penetration testing. Nat Coop Highw Res Program Synth 368
6. Griffiths DV, Fenton GA (2004) Probabilistic slope stability analysis by finite elements. J Geotech Geoenvironmental Eng 130(5):507–518

Open Themes

The idea of inserting a special session dedicated to Open Themes in the Benchmark Workshops was put forward for the first time during the 9th Benchmark Workshop that took place in St. Petersburg (Russia) in 2007. Basically, two reasons supported this decision, unusual in a context dedicated to benchmarks: (i) to provide an audience to innovative methodologies for the numerical modelling of dams; (ii) to identify possible case-histories of interest to be proposed in subsequent Benchmark Workshops.

The idea was taken up a few years later, during the 13th Benchmark Workshop held in Lausanne (Switzerland) in 2015 and in the subsequent 14th Benchmark Workshop in Stockholm, in 2017, also in consideration of the interesting results of the initiative.

During the event organized in Milan in September 2019, the Open Themes session included the 11 contributions of high technical and scientific profile collected in this Section.

Among them, the case-history related to the Nam Ngum 3 dam, presented by F. Andrien, N. Ulrich and M. Monkachi (ARTELIA, Echirolles, France), has been pro-posed to be one of the Theme for the next Benchmark Workshop that will be held in Slovenia. The dam, located on a major tributary of the Mékong River and presently under construction in Laos, will be one of the highest CFRDs in the world with its maximum height above foundation of 210 m.

Numerical Analysis of the 210 m-High Nam Ngum 3 CFRD

F. Andrian, N. Ulrich, and M. Monkachi

Abstract The The Nam Ngum 3 dam, actually under construction in Lao PDR, with a maximum height above foundation of 210 m, will be one of the highest CFRDs. ARTELIA is the Owner's Engineer of Electricité du Laos (EDL) for this 480 MW HPP built on a major tributary of the Mékong. One of the main challenges in the design of the dam comes up from the tight valley. Actually, based on international feedback, extensive concrete face cracking is expected if specific construction provisions were not adopted. Moreover, the concrete face is constructed immediately upstream of the narrowest part of the valley. This results in high stiffness contrast between the banks where the concrete face lies on the bedrock and the central part where it is supported by the rockfill. These peculiar boundary conditions induce a diagonal bending which needs to be addressed with care. Consequently, ARTELIA's numerical model aims at checking that the design of the dam and its construction provisions are efficient to avoid a detrimental cracking of the concrete face. The Hardening Soil Model constitutive law is used for the rockfill. The material parameters are calibrated by means of large-scale apparatus laboratory tests. A size effect assessed by means of a rational approach is taken into account. Finally, a creep constitutive law is considered in order to anticipate the effects of delayed deformation of the rockfill on the concrete face behavior. The modeled concrete face includes the compression joints with initial opening for which the efficiency in reducing the compression stress is clearly highlighted. Based on the results of the numerical analysis, this paper describes the supplementary construction provisions proposed in order to secure a safe behavior of the dam and guarantee an acceptable watertightness of the concrete face.

Keywords Concrete face rockfill dam (CFRD) · Numerical analysis · Hardening soil model · Vertical joints · FLAC3D

F. Andrian (✉) · N. Ulrich · M. Monkachi
ARTELIA, Echirolles, France
e-mail: frederic.andrian@arteliagroup.com

G. Bolzon et al. (eds.), *Numerical Analysis of Dams*, Lecture Notes in Civil Engineering 91, https://doi.org/10.1007/978-3-030-51085-5_41

1 Introduction

The Nam Ngum 3 HPP includes the highest dam to be constructed in Lao PDR [1]. The 210 m high CFRD will actually be one of the highest of its type in the world. As per July 2019, 160 m of rockfill have been placed and the commissioning is planned in the end of 2021.

Checking and eventually validating the numerical analysis carried out by the Contractor is one of ARTELIA's assignments, being the Owner's Engineer for EDL for the project at construction stage.

The dam and the concrete face are expected to exhibit a complex 3D and non-linear behavior. Thus, it is not possible to check the results of the Contractor's numerical analysis by means of simplified analytical formulas.

Moreover, the dam has one very singularity which cannot allow a simple comparison of its behavior to that of the other CFRDs of similar height and valley shape factor. Actually, due to the choice of the location of the dam axis, the upstream face is constructed immediately upstream of the narrowest part of the valley. The concrete face is consequently expected to exhibit high horizontal tension strains at the banks due to stiffness contrast in addition to the more usual high horizontal compression strains at the center of the valley (Fig. 1).

Consequently, compared to the Constructor's numerical model, a comprehensive numerical analysis was conducted with a few major improvements to even better capture the mechanical behavior of the dam and the key aspects to be thoroughly checked.

This predictive model is being used to propose some supplementary construction provisions for securing a safe behavior of the dam. The main calculation approaches used are described in the current paper.

Moreover, as monitoring data are now available, the model is gradually updated and calibrated with the monitoring data. Some preliminary conclusions on the behavior of the rockfill are being drawn.

Fig. 1 Picture of the valley from upstream with the footprint of the concrete face (left) and horizontal cross section 120 m above the bottom of the dam (right)

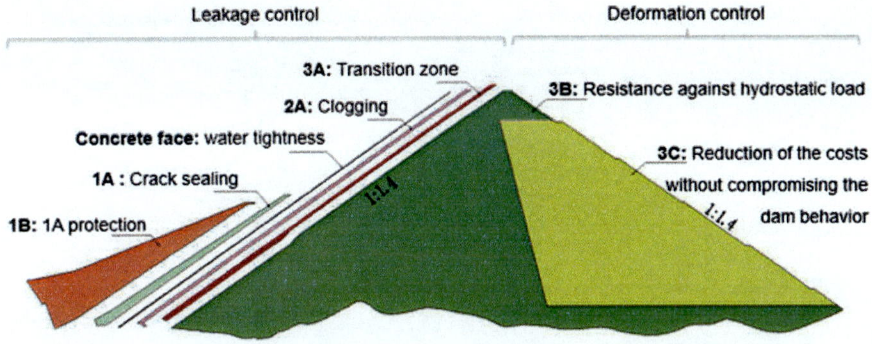

Fig. 2 Zoning of the Nam Ngum 3 dam—Valley center upstream to downstream section

2 Background

2.1 CFRDs and Concrete Face Failure

CFRDs are built based on both past experiences and continuous evolution in construction provisions and methods. Still a few numbers of high CFRDs have exhibited high leakage rate (up to 1 700 l/s) when being impounded [2–4].

These leakages are due to failure of the concrete face for which the deformation is highly prescribed by that of the rockfill. Actually, the concrete face, which unique role is to ensure the watertightness of the dam, is not able to withstand the water pressure due to its obvious slenderness (Fig. 2).

The stiffness of the rockfill and the valley shape of the dam are the main parameters governing the risks of failure of the concrete face. Several studies show that narrow valleys combined with deformable rockfill increases the concrete face failure risks.

In the case of Nam Ngum 3 dam, the expected modulus of deformation of rockfill, based on large scale laboratory tests and including a reduction factor related to size effect [5] is about 80 MPa. One may use several typical diagrams [6, 7] in order to preliminary assess the risk of concrete face failure for a dam (Fig. 3). In this purpose, the valley shape factor A/H^2 is about 3, A being the area of concrete face and H being the maximum height above the foundation. L_0 is the width of the concrete face at its lowest elevation (usually its minimum width). The Nam Ngum 3 is consequently at the threshold of damaged concrete face based on the supposed parameters if there is not any appropriate construction provisions adopted.

2.2 Mitigation Measures

As there is a size effect between rockfill of the same petrographic nature but if different gradations, the ideal solution would be to use small diameters of rockfill

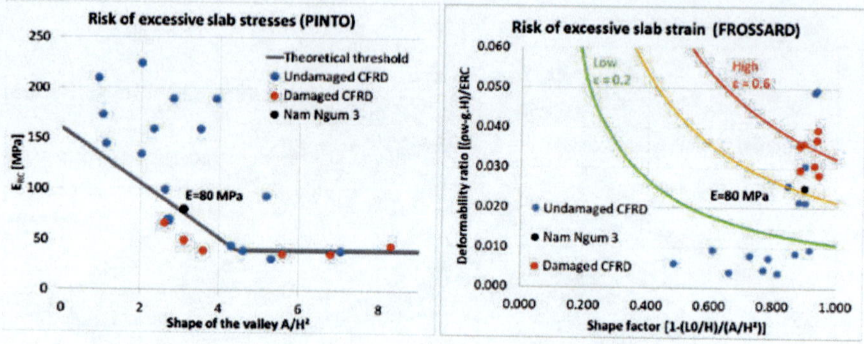

Fig. 3 Examples of diagrams for analyzing the risk of concrete face failure

for high CFRDs in order to increase their apparent Young's modulus. However, this is not realistic due to the huge increase in construction cost due to the rock crushing. Consequently, it is preferable to apply mitigation measures which allow the deformation of the rockfill to be accommodated by the concrete face without any failure.

Two means of mitigation measures are state-of-the-art nowadays.

The first type concerns the construction schedule which involves the construction of the concrete face after the completion of the rockfill. This avoids the concrete face to be prestressed by the construction settlement of the rockfill. In the case of Nam Ngum 3 dam, two stages of concrete face are planned in order to even avoid any prestress due to the early time-dependent settlement of the rockfill (Fig. 4).

The second type involves a series of vertical expansion joints on the concrete face which allows a significant decrease in the stresses due to the impounding. The Nam Ngum 3 dam includes three types of construction joints (Fig. 5):

- Tension joints without any initial opening but able to open under normal operating conditions;
- Compression joints 2 with 5 cm initial opening and expected to experience partial closure under normal operating conditions;

Fig. 4 Construction sequence of the two-staged concrete face of Nam Ngum 3 dam

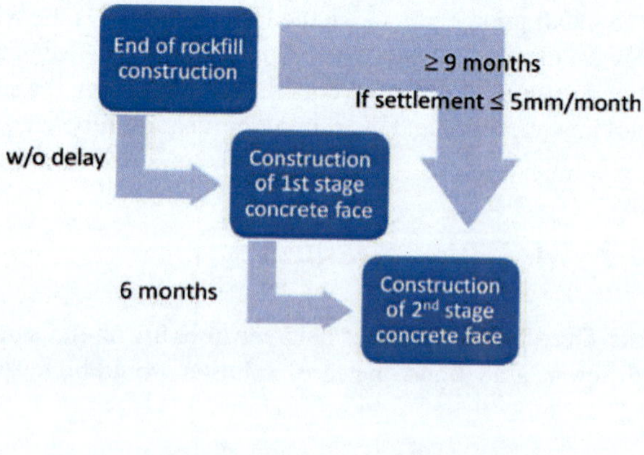

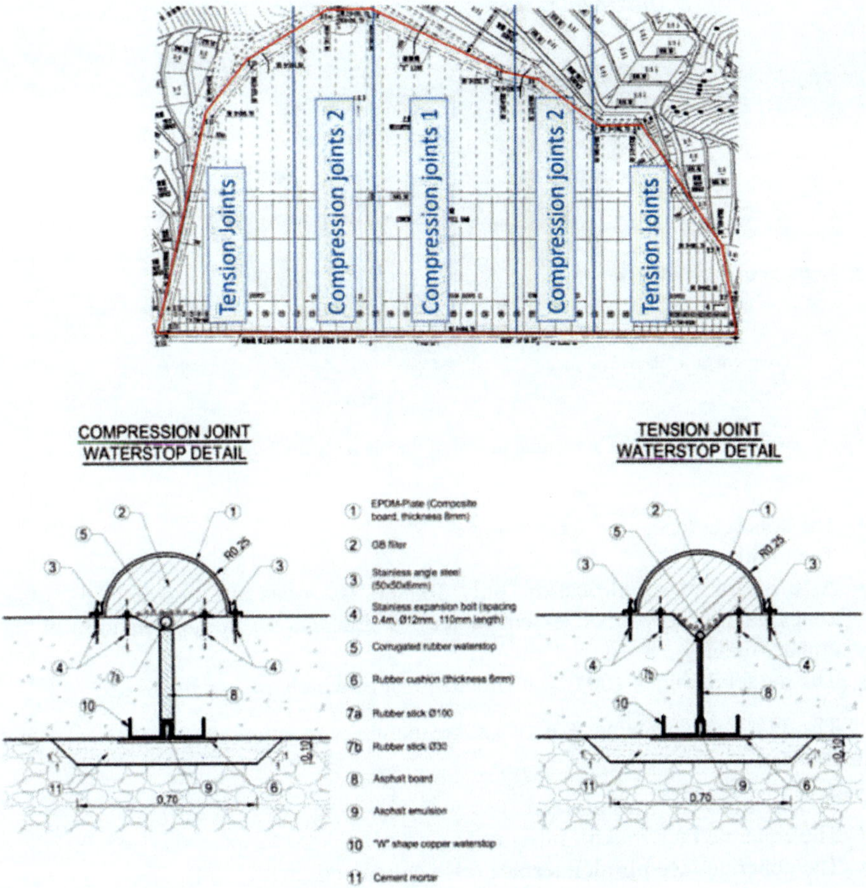

Fig. 5 Plane view of the concrete face (top), details of vertical construction joints (bottom)

- Compression joints 1 with 2.4 cm initial opening which are of intermediate type.

3 Numerical Analysis

3.1 Features of the Numerical Model

The numerical model performed by ARTELIA by means of FLAC3D [8] is rather detailed and includes (Fig. 6):

- A limited extent of the bedrock;
- The main rockfill zones 3B and 3C;

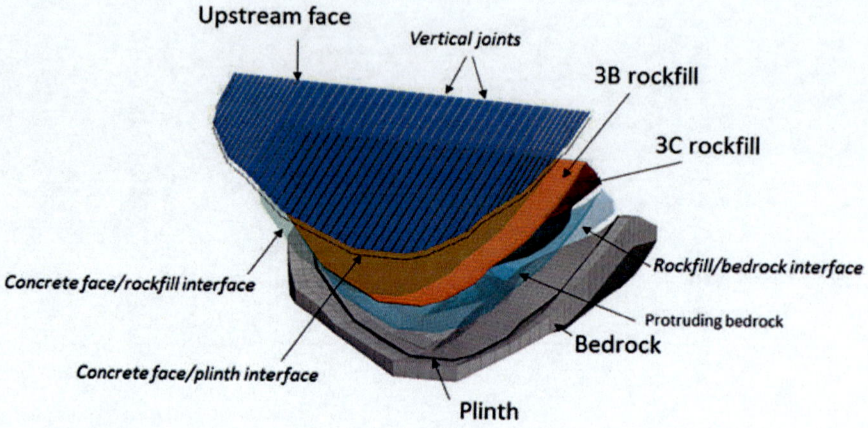

Fig. 6 Exploded view of the numerical model of the Nam Ngum 3 dam

- The concrete face;
- The plinth.
- As a means of simplification, only 3B and 3C zones are modelled, the other zones having very small extent hence few impacts on the deformation of the embankment.
- The model includes 100,000 elements most of which are hexahedra.

The relative displacements between the members are allowed by means of several interfaces:

- The rockfill/bedrock interface;
- The concrete face/rockfill interface;
- The concrete face/plinth interface (also called peripheral joint);
- The vertical expansion joints of the concrete face with the possible initial opening.

The explicit integration scheme of FLAC3D allows the highly non-linear behavior of the dam (initial opening of the vertical joints, HSM law of the rockfill) to be modeled without any convergence issues.

3.2 Constitutive Laws and Material Parameters

The concrete face, the plinth and the bedrock follow a linear elastic constitutive law.

The material parameters for these materials are given in Table 1.

The rockfill is modelled by means of the Hardening Soil Model (HSM) [9] which includes:

- A hyperbolic stress-strain relationship during axial compression with a Mohr-Coulomb failure criterion (Fig. 7a);

Table 1 Material parameters for linear elastic material

Material	Density (kg/m^3)	Young's modulus (GPa)	Poisson's ratio
Foundation	0	3	0.14
Plinth	0	30	0.20
Concrete face	2400	30	0.20

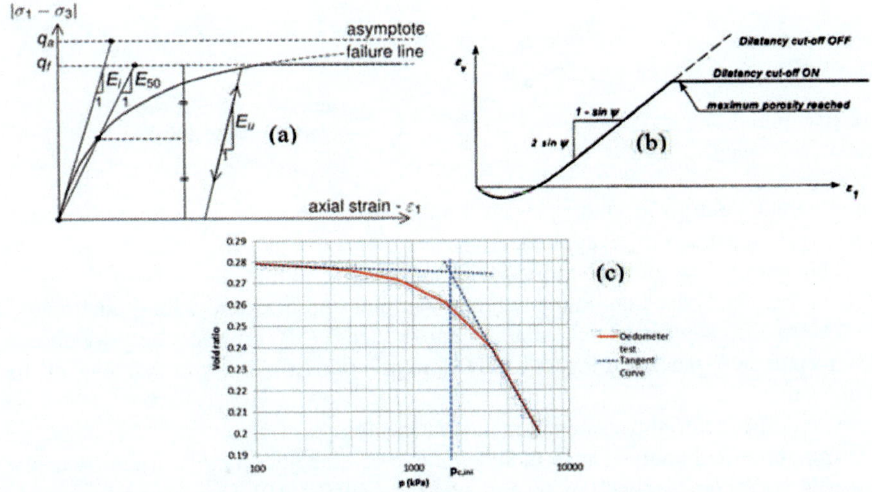

Fig. 7 Main features of the Hardening Soil Model

- A stress-dependent elastic stiffness according to a power law;
- The soil dilatancy with a dilatancy cut-off (Fig. 7b);
- A plastic straining due to primary compression with the memory of a pre-consolidation stress (Fig. 7c);
- An elastic unloading/reloading compared to virgin load (Fig. 7a).

A feature allowing the variation of friction angle as a function of the confinement stress has been added to the HSM for the rockfill and follows the following relation:

$$\Phi = \Phi_0 - \Delta\Phi \cdot \log(-\sigma_3/p_{ref}) \quad (1)$$

with:

- Φ the friction angle at the reference pressure;
- $\Delta\Phi$ the friction angle at infinite pressure.

The HSM parameters have been calibrated with data available from several large-scale laboratory tests (Fig. 8). However, a size effect taking into account the difference

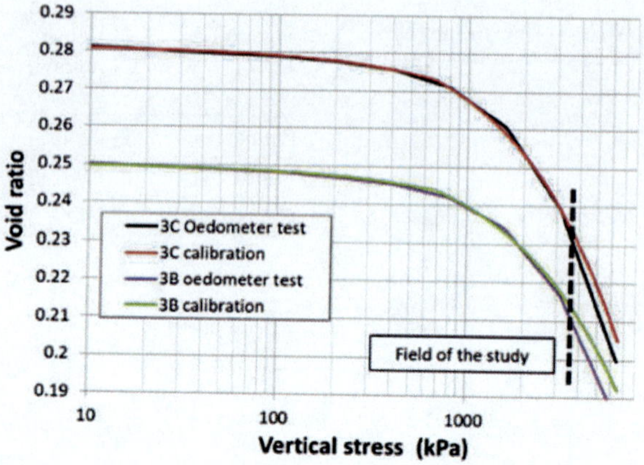

Fig. 8 Calibration of the numerical model to the laboratory tests: triaxial tests (top) and oedometer tests (bottom)

of apparent Young's moduli between the laboratory and the in situ rockfills as per Frossard's approach [5] is applied.

The modulus factor accounting for this size effect in the case of Nam Ngum 3's rockfill equals 2 based on both laboratory samples and in situ rockfill gradations.

The rockfill is planned to be mainly made of fresh (3B) or moderately weathered (3C) sandstone, placed in layers of 800 mm and with a maximum diameter of the same value.

The HSM material parameters are given in Table 2.

The numerical analysis also includes the delayed settlement/deformation of the rockfill, based on feedback of several existing CFRDs [10]. One may refer to [11] and [12] for more detailed explanations about the used approach. The behavior of the dam is predicted up to 30 years after the first impoundment of its reservoir.

4 Main Results

The settlements calculated at the end of construction and after 30 years of operation are given in Fig. 9. Their values are respectively 1.9 m and 2.4 m. The calculated value at the end of construction is in the range of values measured in several dams of similar heights.

An assessment of the apparent Young's modulus according to the ICOLD's approach [13] leads to 82 MPa. This value does not take into account any 3D effect and the resulting horizontal arching. Consequently, such modulus is rather high and cannot be used as is in the numerical model.

The stress state of the concrete face is displayed in terms of magnitude and directions of the minimum (most compressive) and the maximum (most tensile) principal stresses.

The results clearly highlight the beneficial effects of the vertical compression joints (Fig. 10). After 30 years of operation, a wide area of the concrete face would be

Table 2 Material parameters for rockfill

Material	Density (kg/m^3)	$E_{50,ref}$ (MPa)	$E_{oed,ref}$ (MPa)	$E_{ur,ref}$ (MPa)	m	R_f	Φ_0 (°)	$\Delta\Phi$ (°)	ν	K_{nc}	$P_{c,ini}$ (kPa)
3B	2200	40	27	80	0.28	0.77	50	7.5	0.2	0.34	500
3C	2170	31	22	62	0.26	0.84	48.5	6.7	0.2	0.36	500

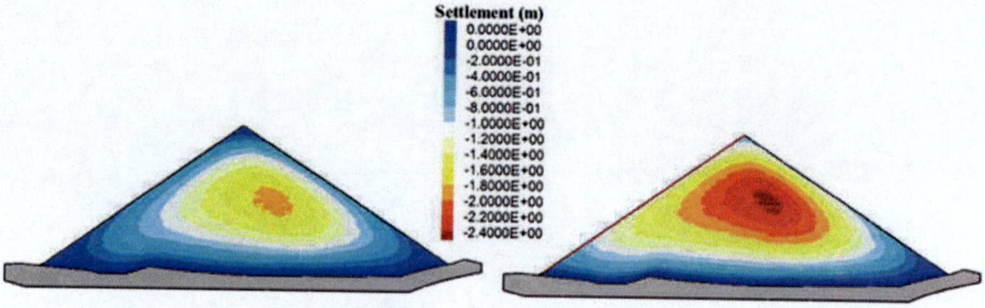

Fig. 9 Calculated settlements at the end of construction (left) and after 30 years of operation (right)

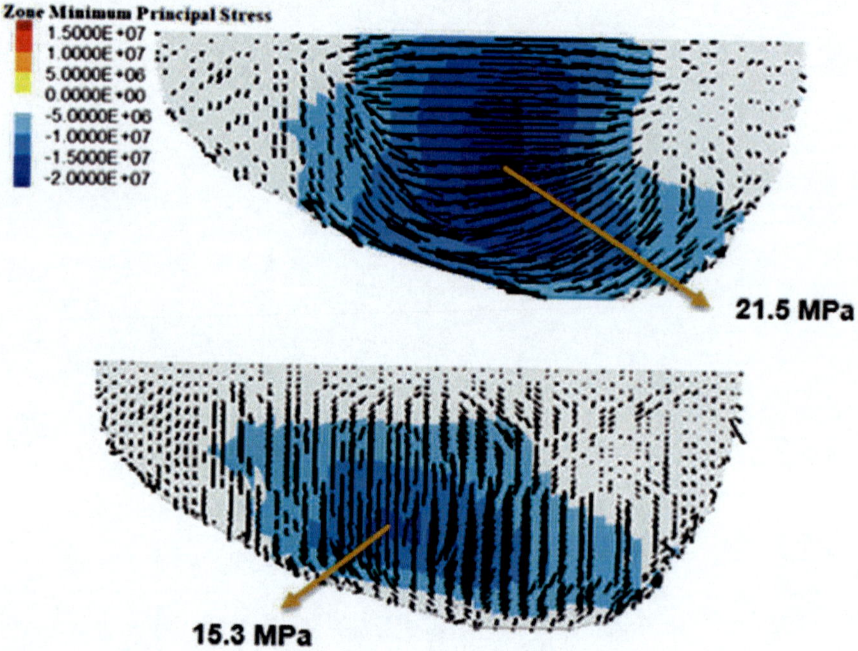

Fig. 10 Magnitude (Pa) and directions of the most compressive stresses of the concrete face without (top) and with (bottom) the vertical compression joints after 30 years of operation

subjected to horizontal compressive stresses higher than 20 MPa (specified compressive strength) without the vertical compression joints. When they are considered, the compressive stresses are in the slope direction and the maximum values are between 10 to 15 MPa for a limited area only. On the other hand, an almost total closure of the central part of the compression joints is expected after 30 years of operation.

The horizontal tensile stresses are released by the vertical joints. However, the peculiar shape of the valley generates a diagonal bending at the higher part of the abutments where the upper part of the concrete face behaves as an overhanging

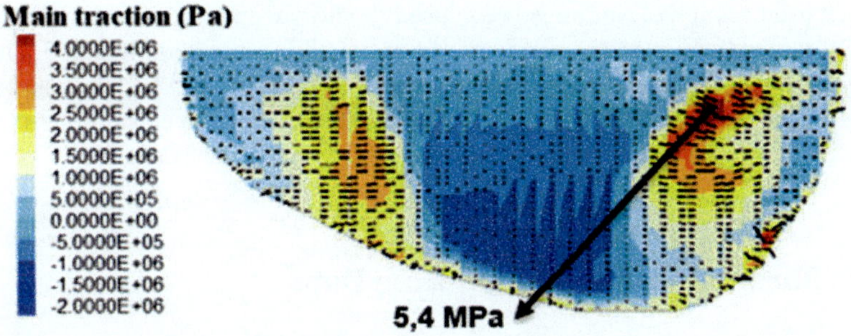

Fig. 11 Magnitude (Pa) and directions of the most tensile stresses of the concrete face after 30 years of operation

beam due to the stiffness contrast between the abutment and the rockfill (Fig. 11). A maximum tensile stress of about 5 MPa is calculated after 30 years of operation and needs to be balanced appropriately by means of steel reinforcement.

The concrete face detachment issue encountered in high CFRDs is generally solved by means of the delayed construction of the concrete face after that of the rockfill [14]. Still in the case of the Nam Ngum 3 dam, the stiffness contrast of the concrete face support generates a detachment due to hydrostatic load: an opening is generated between the concrete face and the rockfill.

The contour of the contact opening with the footprint of the abutments highlights the role played by the peculiar abutments in the calculated opening: the opening is minimal for the slab panels located in the centre (Fig. 12).

The effect of the construction of a horizontal tension joint continuous from a bank to another is analysed and proves to be efficient at reducing the concrete face detachment. Both the maximum opening and the slope direction extension are reduced of 40% or more. Moreover, a decrease of 20% in the maximum tensile stress previously analysed is achieved.

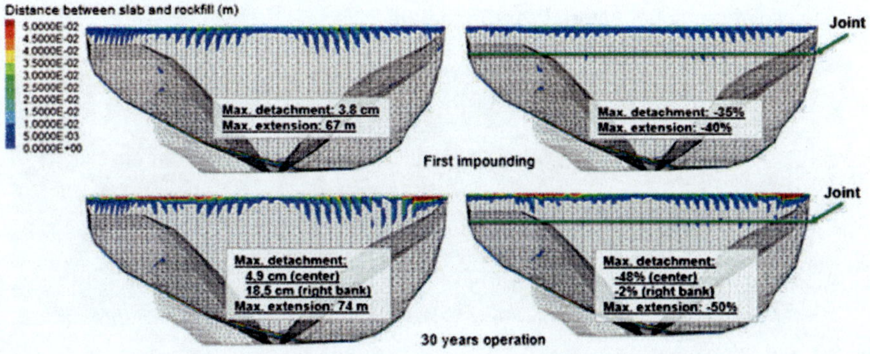

Fig. 12 Contours of concrete face detachment under different configuration and load cases

Such joint at high elevation is being used as anti-seismic provision for the recent CFRDs constructed in high seismicity regions [15]. For the Nam Ngum 3 dam, the Safety Evaluation Earthquake horizontal acceleration is 0.12 g, which is moderate for this height of dam. The construction of a horizontal joint at this elevation is expected to contribute to the reduction in the seismic tensile stresses of the slab.

5 Calibration with the Monitoring Data

The results of the calibration of the numerical with the five electromagnetic settlement gauges in May 2019 (Fig. 13) are described hereafter.

So far it is possible to simulate the settlement of the dam with one unique series of parameters as the 3C rockfill has been replaced by 3B type very early in the construction. The ability of the numerical model to simulate the measured settlement is satisfactory for different dates and different height of placed rockfill. It can be noticed that the accuracy of the calibration is less satisfactory at early age: the delayed deformation is only simulated after the construction of the concrete face. In the future, it will be possible to activate the delayed deformation constitutive law as soon as each layer is placed in order to improve the model accuracy. Only the calibration of three settlement gauges is presented in Fig. 14 for clarity.

As per May 2019, the unique $E_{50,ref}$ used is 107 MPa which is 2.7 times higher than the value expected for 3B zone. The apparent Young's modulus as per ICOLD's approach is higher than 200 MPa. The preliminary explanation to this reassuring high stiffness for the rockfill arises from the following:

- A high energy rockfill compaction compared to what is usually encountered: 10 passes of 26t vibrating roller;
- A resulting achieved low porosity: 17.7% average value compared to the target 20 or 21% for respectively 3B and 3C zones;
- A probable finer final in situ gradation compared to the design gradation, leading a new value of size effect modulus factor to be assessed in the future;
- The use of a unique material for the rockfill.

Fig. 13 Layout of the electromagnetic settlement gauges and downstream view of the construction progress in May 2019

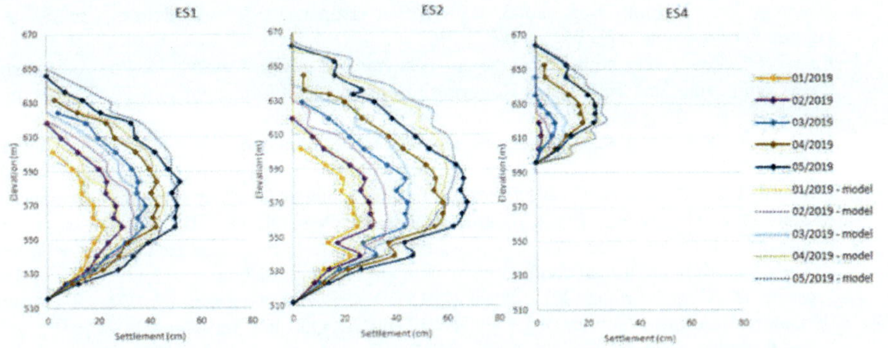

Fig. 14 Comparison between measured and computed settlements

6 Conclusions

The numerical analysis carried out by ARTELIA on Nam Ngum 3 dam has allowed most of the results of the Contractor's numerical analysis to be checked. Still some key aspects have been highlighted owing to a more refined approach. At the time being, several discussions with the Contractor are engaged in order to:

- Increase the reinforcement sections of rebars in the concrete face at specific locations due to the diagonal bending related to the peculiar shape of the abutments;
- Construct a horizontal articulated joint without any crossing rebar at higher elevation in order to reduce the detachment which may be expected between the concrete face and the rockfill;
- Saw the extruded curb behind the compression joints for which the closures are expected to be high (almost 5 cm after several years) in order to avoid hard contact zones behind the concrete face or tearing of the waterstop systems of the joints.

In May 2019, the calibration of the numerical analysis with the preliminary measured settlement gives reassuring results: the stiffness of the rockfill appears to be more than 2.5 times higher than the design value probably owing to a high energy rockfill compaction.

Both the Contractor and ARTELIA will update their numerical models and will conclude about the previously mentioned adjustments on the construction provisions.

References

1. Ducos X, Monkachi M, Bercher Q, Phet K (2018) New high CFRD dam in Lao PDR for the Nam Ngum 3 hydro plant. In: 7th international conference and exhibition on water resources and renewable energy development in Asia, Hydropower and Dams, Danang

2. Marques Filho P, Pinto NLS (2005) CFRD dam characteristics learned from experience. Hydropower & Dams 1:72–76
3. Pinto NLS (2007) Very high CFRD dams—behavior and design features. In: Proceedings of the III symposium on CFRD Dams Honoring J Barry Cooke. Florianapolis 25–27 October, pp 3–16
4. Johannesson P, Tohlang SL (2007) Lessons learned from Mohale. Int Water Power Dam Constr 59(8):16–25
5. Frossard E, Nieto-Gamboa C (2015) Scale effects in granular fill shear strength—Incidences on stability and behaviour of large rockfill dams. In: XXV ICOLD CIGB, Q98, Stavanger
6. Pinto NLS (2007) A challenge of very high CFRD dams: Very high concrete face compressive stresses. In: 5th International conference on dam engineering, Lisbon
7. Frossard E, Nieto-Gamboa C, Robles J (2015) Design and construction of Mazar CFRD (Ecuador): Behavior in service. In: XXV ICOLD CIGB, Q98-R2, Stavanger
8. Itasca Consulting Group Inc. (2018) FLAC3D—Fast Lagrangian Analysis of Continua in Three-Dimensions, Ver. 6.0. Minneapolis
9. Schanz T, Vermeer PA, Bonnier PG (1999) The hardening soil model: formulation and verification. In: Beyond 2000 in Computational Geotechnics, Rotterdam, Balkema
10. Silveira JFA, Bandeira OM (2017) Analysis of the CFRD behavior related to the rockfill deformability and its leakage consequences after reservoir filling. In: XXXI National seminar on large dams, CBDB, Brazil
11. Andrian F, Monkachi M., Ulrich N, Ducos X (2018) The Nam Ngum 3 CFRD dam: an advanced numerical analysis to prevent upstream face damages. In: HYDRO 2018, Gdansk
12. Nieto C, Ulrich N, Monkachi M (2019) Contribution of 3D non-linear modelling to the design and supervision of Concrete Face Rockfill Dams. In: Justification des Barrages: Etat de l'Art et Perspectives, Colloque CFBR, Chambéry, November 27–28
13. ICOLD: Concrete Face Rockfill Dams: Concepts for design and construction. Bulletin 141 (2010)
14. Chen Y (2012) Modélisation du comportement mécanique des grands CFRD: Identification des caractéristiques des enrochements et comportement du masque d'étanchéité amont. Ecole Centrale de Lyon
15. Hongqi M, Chi F (2016) Technical progress on researches for the safety of high Concrete-Faced Rockfill Dams. Chinese Academy of Engineering and Higher Education Press Limited Company, Elsevier LTD

Elasto-Plastic Damage Time-History Analysis of Dams: The Case of Ridracoli

G. Buffi, P. Manciola, L. De Lorenzis, V. Gusella, M. Mezzi, C. Tamagnini, A. Gambi, and G. Montanari

Abstract FEM simulations are widely recognized as essential tools in the analysis of the behaviour of dam systems. A detailed representation of the dam structure allows for a better understanding of the local response of important structural elements. The present paper intends to provide a FE modelling procedure of concrete arch-gravity dams. Case of study is the arch-gravity dam of Ridracoli. The vertical construction joints are included in the model as solid elements; their influence on the dynamic properties of the structure is investigate and, adopting the CDP—Concrete Damage Plasticity—model for the filling mortar, elasto-plastic damage time-history analyses are performed. Under MCE—Maximum Credible Earthquake—and varying the water level, the damage parameters evolution is analysed. All simulations integrate the structure-foundation and the fluid-structure interaction by means of rock mass solid and acoustic elements, respectively; moreover, the adopted damping coefficients of rock mass and structure are calibrated on linear elastic dynamic analyses. The inclusion of vertical construction joints into finite element models of dams allows us to verify the local behaviour of such real discontinuities under severe seismic events and therefore to verify the seismic vulnerability assessment of the whole structure.

Keywords Arch-Gravity dam · FE model · Vertical construction joints · CDP model · Non-Linear dynamic analysis

G. Buffi (✉) · A. Gambi · G. Montanari
Romagna Acque Società delle Fonti S.p.A, Forlì, Italy
e-mail: gbuffi@romagnacque.it

P. Manciola · V. Gusella · M. Mezzi · C. Tamagnini
Department of Civil and Environmental Engineering, University of Perugia, Perugia, Italy

L. De Lorenzis
Technische Universitat Braunschweig, Institute of Applied Mechanics, Brunswick, Germany

G. Bolzon et al. (eds.), *Numerical Analysis of Dams*, Lecture Notes in Civil Engineering 91, https://doi.org/10.1007/978-3-030-51085-5_42

1 Introduction

It is commonly accepted that the FEM—Finite Element Method—modeling of dams is paramount in the safety evaluation of dam structures [1, 2]. Their calibration against monitoring data allows a better understanding of the dam system response [3]. Indeed, FEM analyses can simulate past events or possible scenarios of the structure [3–5]. Dam model reconstruction is a fundamental aspect into static and dynamic simulations and the detailed modelling allows to quantify the influence of important structure discontinuities on the behavior, likewise joints, spillways and ancillary works [6]. For structures such as segmented arch dams built on valleys with complex topography, the detailed definition of the structure geometry—including the joints between the different blocks—and of the foundation rock mass is of primary importance in the FEM modelling of such systems. As underlined by ICOLD—International Committee on Large Dams—the effects of historical earthquakes on concrete dams are mostly related to contraction joints [7]. Indeed, in literature several models are available to simulate the opening, closing and/or shear sliding of joints under seismic events, also including non-linear effects [8, 9]. Also, experimental studies on the influence of vertical joints have been carried out [10]. According to reports in all the previous works, in arch, gravity and arch-gravity dams a crucial role is played by vertical construction joints mostly when dynamic forces such as seismic events occur. Therefore, the present work provides a FE modelling procedure of a concrete arch-gravity dam including vertical joints as solid elements in order to analyse the evolution of damage parameters under MCE—Maximum Credible Earthquake. All simulations include structure-foundation and the fluid-structure interaction and Rayleigh damping coefficients calibrated for the rock mass and the structure.

Firstly, the case study is presented, the Ridracoli arch-gravity dam, Santa Sofia (Italy). Secondly, the three-dimensional solid modelling of a whole dam system is explained. Then, elasto-plastic damage time-history analyses investigate the tensile and compression damage parameter evolution under a severe seismic event and varying the water level. At the end, conclusions and future developments are presented.

2 Case Study: Ridracoli Dam

The Ridracoli Dam is located in the village of Santa Sofia in the centre of Italy, the primary use is to supply drinking water, Fig. 1. The construction of the Ridracoli Dam started in 1975 and it was completed in 1982. After the testing phase, in 1988 the reservoir became operational. The Ridracoli Dam is an arch-gravity dam in simple concrete. The dam structure has 103.50 m maximum height and 432.00 m crest length, in the key section the top thickness is 7.00 m and the bottom thickness is 36.00 m. The dam body is a double-curved structure, symmetrical with respect to the main section,

Fig. 1 **a** The Ridracoli Dam; **b** The lake created by the presence of the dam, scale 1:500 m (INGV)

resting on a pulvino foundation base that extends all around the perimeter of the abutments. The upstream and downstream facings, following specific analytical laws, give progressively increasing thickness to the arches along the horizontal sections, from the middle to the sides and along the vertical section from the top to the bottom. Three different arches, with variable curvature centers in the three-dimensional space, characterize the downstream face and only one characterizes the upstream face. After the construction of the pulvino, which regularizes the excavation of the abutments and allows for the redistribution of stresses and strains, the dam body was built in blocks. Indeed, the construction in subsequent blocks avoids structural cracking due to concrete shrinkage and variations in hydrostatic level and temperature. The connection between the 27 cantilevers was ensured by the injection of mortar (cement plus sand) into the joints. The injections were performed through the inspection galleries midway between two blocks, in a radial direction and repeated several times in order to fill the contacts as much as possible. Moreover, during the concrete casting of adjacent blocks, plastic water-stop elements were placed along the perimeter of the inspection galleries and along the external perimeter of the joint. They have the dual function of preventing infiltration of the water and spillage of the cement grout, used for the suture of the joints. In the central part of the crest, there are eight square free overflow (ogee crest) spillways with 600 m^3/s maximum capacity.

3 FE Solid Modelling

The three-dimensional solid modelling of the whole dam system has been derived from a UAV dense point cloud [11]. It is the base for the construction of the solid model of the dam system.

Fig. 2 Solid elements. **a** a joint, a block, pulvino foundation, the left and right weight concrete blocks and the stilling basin; **b** the rock mass; **c** the water reservoir

3.1 Three-Dimensional Solid Geometry

Significant points, that can describe elements with different mechanical characteristics, interactions or boundary conditions, are extracted. The criteria are based on the knowledge of the dam system elements, RGB information associated to the points and/or on the recognizable curvature changes. Connecting the significant points, lines are drawn: these are the contours for close poly-surfaces that, including a volume, are solids by definition. The high accuracy of the survey allows for the modelling of some important details such as the vertical joints, the spillways, the stilling basin and the concrete blocks of the right and left sides, Fig. 2.

3.2 Vertical Joint Modelling

The model simulates the connections between the base of the blocks and the pulvino foundation by means of surface-based "tie" constraint, which makes the translational degrees of freedom equal for the pair of surfaces. The model simulates the joints between the blocks as independent solid elements, splitting the discontinuity surface and spacing the two resulting surfaces at 10 cm, Fig. 3. This distance is adopted intentionally larger than the mortar joint because it is also intended to represent, by simulating a heavier discontinuity, the interfaces with the adjacent blocks. These thin slices between blocks are characterized by their own mechanical properties, as Table 1 shows. Then joints and blocks are connected by means of surface-based "tie" constraint, making, like in the monolithic model, the translational degrees of freedom equal for the pair of surfaces.

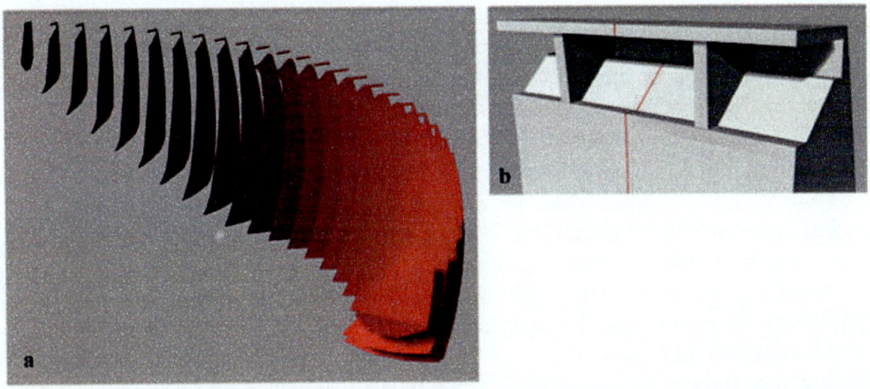

Fig. 3 **a** joints of the solid element joint representation; there is an element of discontinuity between adjacent blocks; **b** a zoom on the joint between the two blocks

Table 1 Mechanical properties of rock mass, block concrete, joint mortar and water. E_s static elastic modulus, E' drained Young's modulus, E_u undrained Young's modulus, G' drained shear modulus, G_u undrained shear modulus, ν Poisson's coefficient, ρ bulk density, B bulk modulus

	Rock Mass	Concrete of blocks	Mortar of joints	Water
E_s (kPa)	–	3.07×10^7	2.20×10^7	–
E' (kPa)	4.80×10^7	–	–	–
E_u (kPa)	2.17×10^7	–	–	–
G' (kPa)	1.92×10^7	–	–	–
G_u (kPa)	8.70×10^6	–	–	–
ν	0.25	0.20	0.20	–
ρ (t/m^3)	2.635	2.470	2.100	1.000
B (kPa)	–	–	–	2.15×10^6

3.3 General Assignments

Four-node trilinear tetrahedral elements C3D4 are used to discretize the concrete structure: pulvino foundation, blocks, joints, when they are present, as in the third model, and ancillary works such as weight blocks and stilling basin 827,431 linear tetrahedral elements characterize the whole solid element joint model with water reservoir close to the normal reservoir level (557.3 m a.s.l.).

In dynamic analysis, the hydrodynamic overpressure effect has to be added to the hydrostatic load. Different methods are available [12, 13]. The asymmetric shape of the acting water mass can influence the dynamic response of the dam structure. Consequently, a proper extension of the water mass has to be taken into account.

Adopting, from literature, 3÷4H, where H is the maximum water height [14], the condition of infinite reservoir is achieved. In the present study, an extension of 5H

(5H~500 m) is employed: a circle with its centre on the main cantilever at the crown level and with a 500 m radius isolates a cylinder of water and rock masses. Also, for the foundation-structure interaction, different methods and simplifications are available. The dimension of the rock mass is therefore related to the infinite reservoir condition and it has a minimum thickness, starting from the base structure, of 1.5 times the maximum height of the dam (1.5 H–150 m).

Four-node linear tetrahedral elements C3D4 and acoustic four-node linear tetrahedral elements AC3D4 are used to discretize respectively the rock mass and the water reservoir volume. Although a propter extension of the water mass is adopted, no reflecting surfaces are applied on the far end of the fluid. Moreover, the water reservoir is directly coupled by a surface-based tie constraint with the upstream face of the dam structure as well as with the bed rock, without any bottom absorption phenomena. Finally, the top of the water reservoir is modeled as a free surface without any sloshing mechanism. The excavation perimeter is created by splitting the base of the pulvino foundation; the rock mass and the foundation are directly coupled by means of a surface-based tie constraint. The base and the lateral surfaces of the rock mass are restrained in general by fixed constraints.

Linear elastic homogeneous isotropic materials are employed, specifying for each of them the static or dynamic Young's modulus and Poisson's coefficient, Table 1 [15–17].

Linear elastic dynamic analyses are performed on the rock mass model (without the dam structure) and on a complete model (rock mass plus dam structure) with mass less rock (only the deformability of the rock mass is taken into account) to calibrate Rayleigh damping coefficients: from the comparison with real recordings, the damping ratio results 5% for the rock mass and 2% for the structure [18].

4 Elasto-Plastic Damage Time-History Analysis

The solid element joint model, including the vertical joints as slices with their own thickness and material, allows us to investigate the damage evolution throughout such discontinuities. Indeed, the CDP—Concrete Damage Plasticity—model [19, 20] is adopted for the hydraulic mortar of the vertical joints while the other materials have linear elastic constitutive behaviour. The mechanical characteristics of the hydraulic mortar are derived from technical datasheets of the materials usually employed for the injection of joints. The adopted compression behaviour, function of cylinder concrete design compressive stress, is defined by Kent and Park [21]. This stress-strain relationship is developed for confined concrete, but the equation of the ascending branch is equal for confined and unconfined concrete, instead of it differs for the descending branch. The compressive and tensile behaviours of the hydraulic mortar of the joints are included in tabular form, as stress-strain and stress-displacement relationships respectively, Fig. 4a, c. The compressive damage C can be specified by means of a function of the inelastic strain and the tensile damage T can be specified by means of a function of the cracking displacement, Fig. 4b, d.

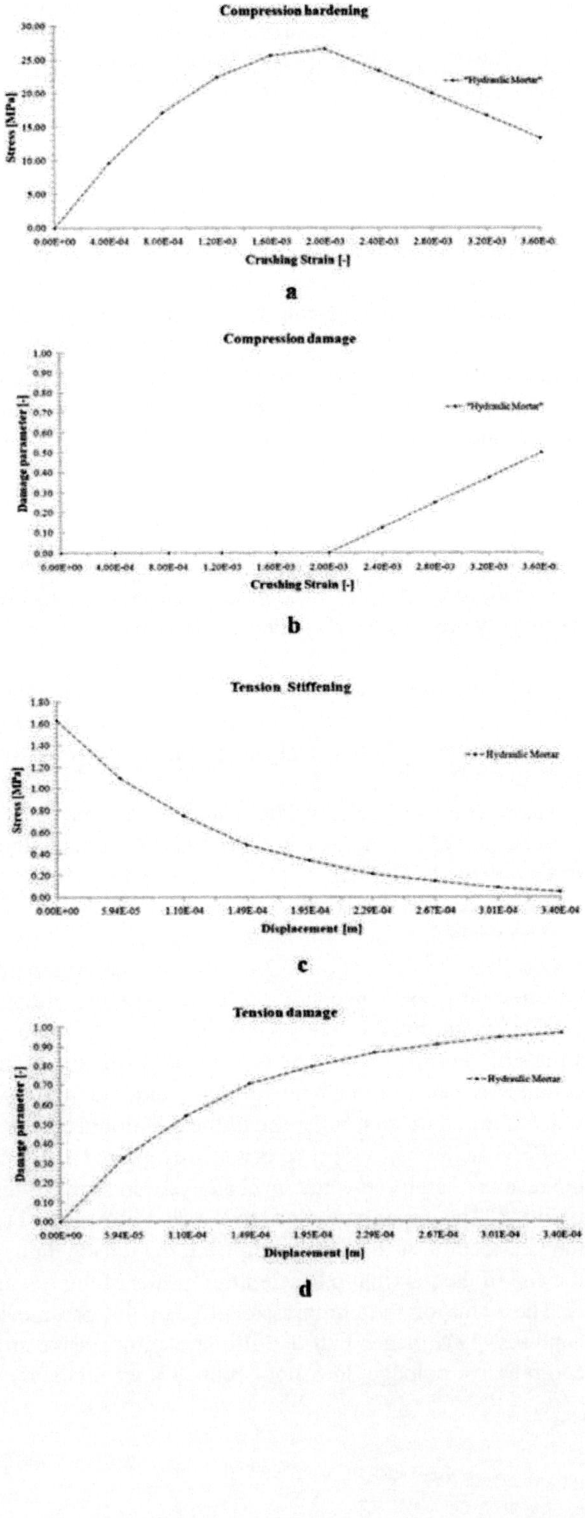

Fig. 4 Compressive and tensile behaviour of the hydraulic mortar of the joints. **a** Compression stress-strain relationship; **b** compression damage evolution; **c** Tensile stress-displacement relationship; **d** tension damage evolution

Table 2 Parameters of the CDP model. β_d is the dilatation angle, m is the eccentricity, f_y is the ratio of biaxial compressive yield stress to uniaxial compressive yield stress, K_I describes the shape of the yield function and μ_P is the viscosity parameter

CDP parameters				
β_d	m	f_y	K_I	μ_P
35	0.1	1.16	0.667	0.01

Indeed, the adoption of a stress-displacement relationship rather than a stress-strain relationship reduces the mesh sensitivity of the results in tensile behavior [22]. Until the ultimate tensile/compressive strength is reached, no damage occurs; indeed, it is defined, for the descending branch of the relationships, as the unit minus the ratio between the generic stress and the ultimate peak strength. Moreover, the CDP model requires the definition of some parameters, which identify the shape of the flow of the potential surface and the yield surface, Table 2 [23].

In order to analyse damage evolution throughout the joint elements, a strong seismic motion has to be applied. Therefore, given that, the events that occurred on the Ridracoli site were not sufficiently intense, a suitable seismic event has to be employed. The Italian and European regulations allow, in the assessment of seismic vulnerability by means of nonlinear analysis, the use of suitable real seismic time-history records, as long as their spectra fits with the design target spectrum of the site.

Adopting for the Ridracoli site a collapse limit state (SLC) target spectrum, the selection, performed by means of REXELite [24, 25], extracts seven groups of seismic events that occurred in central Italy. Between them, the Accumoli (RI) event (ML = 6.1, Max PGA = 5.47 m/s^2, duration = 60 s), which occurred on 30/10/2016 is adopted into the analysis. The accelerations time-history is uniformly applied at the base surface of the FEM model: the N-S component along the upstream-downstream direction, the E-W component along the left-right direction and the U-D component along the vertical direction.

The analysis is performed taking into account three significant water levels: the normal reservoir level of 557.3 m a.s.l., the minimum operating level of 523 m a.s.l. and the empty reservoir condition. The time step of the analysis is defined in relation to the time sampling: the inverse of the instrumental frequency is assumed as maximum values of the time period ($\Delta t = 0.005$ s). Instead, the minimum value is fixed short enough to be compatible with the computational time. The compressive C and tensile T damage parameters, at the end of the maximum acceleration peaks, for the joints of the FEM model with empty reservoir are reported in Fig. 5a, b respectively. Moreover, the results of all three-water level analyses in terms of tensile T damage are reported in Fig. 6. The upper part, close to the dam crowning, of the joint 1, located close to the key section of the dam, is showed: the tensile damage parameter is reported, at the end of the maximum acceleration peaks of the seismic even.

The value of the compression C damage parameter appears null in all three analyses. This means that the ultimate compressive strength is never reached and compressive damage does not occur in joint elements. As expected, the tensile T

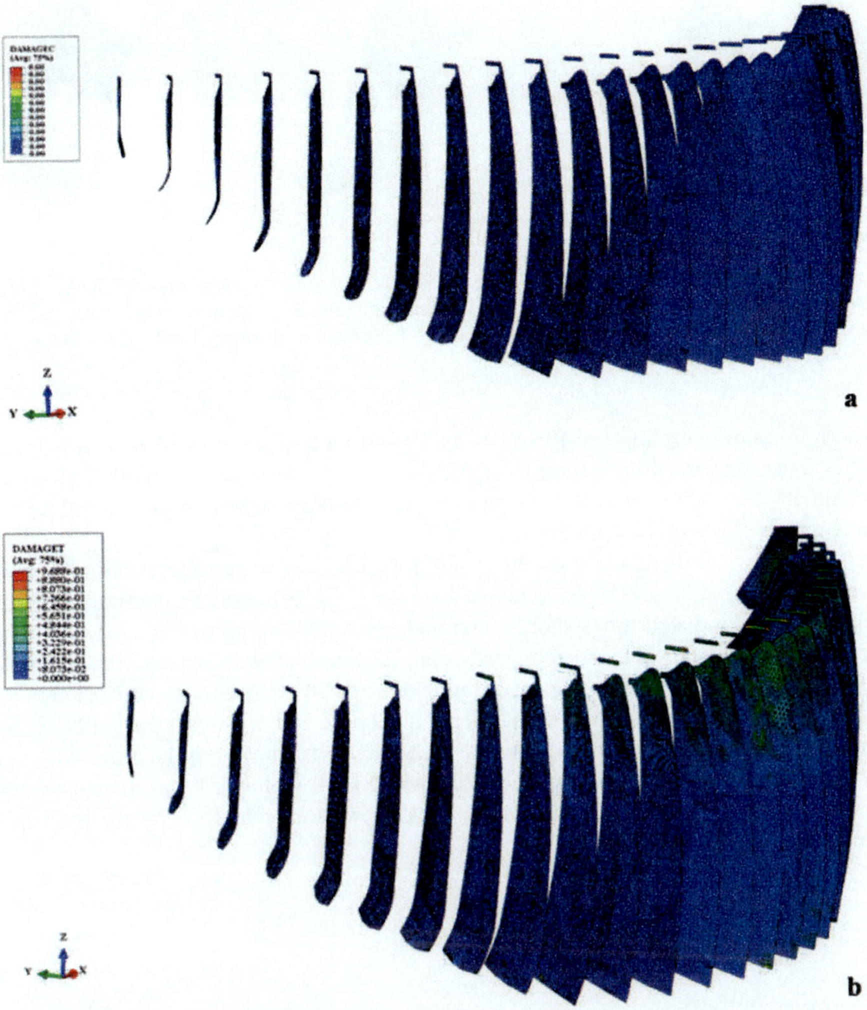

Fig. 5 **a** Compressive C damage parameter distribution of the vertical joints at the end of the maximum acceleration peaks of the Accumuli (RI) seismic event and with empty reservoir condition. No compressive damage occurs; **b** tensile T damage parameter distribution of the vertical joints at the end of the maximum acceleration peaks of the Accumoli (RI) seismic event and with empty reservoir condition. The tensile damage appears concentrated in the central joint and close to the dam crowing

damage is the most significant parameter for describing the behaviour of the hydraulic mortar. As the water level increases in the reservoir, resonant frequencies generally decrease due to the additional mass added to the dam–reservoir system [26]. Consequently, the effect of the accelerations time-history applied at the base of the model is less damaging on the structure. Passing from the empty reservoir condition, to

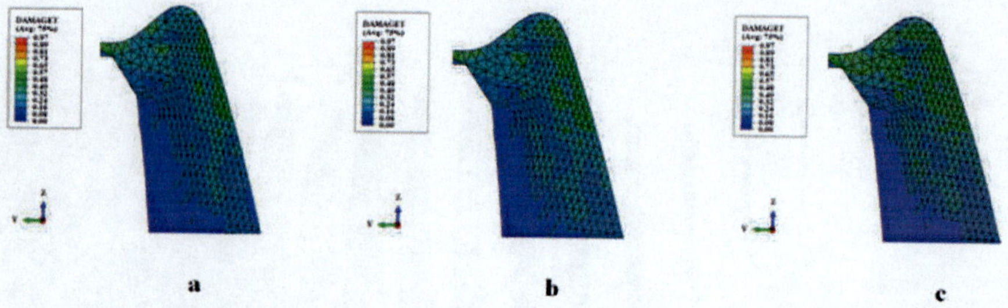

Fig. 6 The upper part of the joint 1, located close to the key section. The tensile damage, under Accumoli (RI) seismic event, progressively increases with the decrease of the water level, passing from **a** normal reservoir level (557.3 m a.s.l.), to **b** minimum operating level (523 m a.s.l.), to **c** empty reservoir condition

the minimum operating level (523 m a.s.l.) and then to the normal reservoir level (557.3 m a.s.l.), the propagation of the tensile damage decreases throughout the joint elements, Fig. 6. The tensile damage appears generally located in the central joints and mostly close to the dam crest, Fig. 5.

Moreover, all three analyses show that the damage T parameter reaches the unit value (red colour), which physically means the complete loss of material stiffness, just in really reduced areas such as the end part of the ramp of the spillway; it is probably due to a local tensile concentration. In general, the decrease of the stiffness of the hydraulic mortar assumes maximum value of 50% close to the cortical area of the joint, as suggested by the colour scale in Figs. 5 and 6, near to the spillway. In every case, such stiffness reduction does not involve fracturing of the material.

Therefore, the dam can withstand the heavier SLC—Collapse Limit State—which corresponds the applied seismic input, shaking without release of water from the reservoir.

5 Conclusions

Nowadays, FEM simulations are widely recognized as essential tools in the analysis of the behaviour of dam systems. A detailed representation of the dam structure allows for a better understanding of the local response of important structural elements. From this perspective, the detailed representation allows for the investigation of the influence of the vertical joint modelling procedures on the dynamic properties of the structure.

If the goal is the analysis of the local behaviour of the structure under seismic events, sophisticated models, such as that with solid elements and vertical joints, have to be employed. The vertical construction joints are planes of weakness of the structure, being in reality, injected with hydraulic mortar with mechanical properties that are lower than those of the concrete of the blocks. Therefore, they have to be included in FEM model for verifying, under strong seismic events, that the tensile and compressive strengths are not exceeded or, when they are, in which way the damage is distributed throughout these elements.

Acknowledgements The authors would like to express their gratitude to the Eng. Franco Farina, to the technical manager Fabrizio Cortezzi, and to all the technicians of Romagna Acque Società delle Fonti S.p.A. who cooperated on this activity.

References

1. Yochum SE, Goertz LA, Jones PH (2008) Case study of the Big Bay dam failure: accuracy and comparison of breach predictions. J Hydraul Eng 134(9):1285–1293
2. ICOLD: Bulletin 155—Guidelines for use of numerical models in dam engineering. (2013)
3. Wang J-T, Zhang C-H, Jin F (2012) Nonlinear earthquake analysis of high arch dam–water–foundation rock systems. Earthq Eng Struct Dyn 41(7):1157–1176
4. Løkke A, Chopra AK (2015) Response spectrum analysis of concrete gravity dams including dam-water-foundation interaction. J Struct Eng 141(8):0401–4202
5. MacDonald TC, Langridge-Monopolis J (1984) Breaching charateristics of dam failures. J Hydraul Eng 110(5):567–586
6. Li S, Cain S, Wosnik M, Miller C, Kocahan H, Wyckoff R (2011) numerical modeling of probable maximum flood flowing through a system of spillways. J Hydraul Eng 137(1):66–74
7. ITCOLD: Working group—Behaviour of the existing dams under seismic actions (2017)
8. Lau DT, Noruziaan B, Razaqpur AG (1998) Modelling of contraction joint and shear sliding effects on earthquake response of arch dams. Earthq Eng Struct Dyn 27(10):1013–1029
9. Ahmadi MT, Izadinia M, Bachmann H (2001) A discrete crack joint model for nonlinear dynamic analysis of concrete arch dam. Comput Struct 79(4):403–420
10. Wang SS, Zhang YF, Cao MS, Xu W (2015) Effects of contraction joints on vibrational characteristics of arch dams: experimental study. Shock Vib 2015:1–7
11. Buffi G, Manciola P, Grassi S, Barberini M, Gambi A (2017) Survey of the Ridracoli Dam: UAV–based photogrammetry and traditional topographic techniques in the inspection of vertical structures. Geomatics Nat Hazards Risk 8(2):1562–1579
12. Zee CH, Zee R (2016) Earthquake hydrodynamic pressure on dams. J Hydraul Eng 132(11):1128–1133
13. Chen BF, Yuan YS (2011) Hydrodynamic Pressures on arch dam during earthquakes. J Hydraul Eng 137(1):34–44
14. Sevim B, Altunişik AC, Bayraktar A, Akkose M, Calayir Y (2011) Water length and height effects on the earthquake behavior of arch dam-reservoir-foundation systems. J Civil Eng 15:295–303
15. Grandori R, Lembo-Fazio A, Ribacchi R (1990) Excavation of the ridracoli hydraulic tunnels using a double-shield TBM. Rock Mech Rock Eng 23(3):141–165
16. Oberti G, Bavestrello F, Rossi PP, Flamigni F (1986) Rock mechanics investigations, design and construction of the Ridracoli dam. Rock Mech Rock Eng 19(3):113–142
17. Mapei, http://www.mapei.com/it/it/prodotti-e-soluzioni/prodotti/dettaglio/mapegrout-bm, last accessed 2019/03/15
18. Buffi G, Manciola P, De Lorenzis L, Cavalagli N, Comodini F, Gambi A, Gusella V, Mezzi M, Niemeier W, Tamagnini C (2017) Calibration of finite element models of concrete arch-gravity dams using dynamical measures: the case of Ridracoli. Procedia Eng 199:110–115
19. Lee J, Fenves G (1998) Plastic-damage model for cyclic loading of concrete structures. J Eng Mech 124(8):892–900
20. Nasrin S, Ibrahim A (2018) Finite–element modeling of UHPC hybrid bridge deck connections. Int J Adv Struct Eng 10:199–210

21. Kent DC, Park R (1971) Flexual members with confined concrete. J Struct Div 97(7):1969–1990
22. Hillerborg A, Modéer M, Petersson PE (1976) Analysis of crack formation and crack growth in concrete by means of fracture mechanics and finite elements. Cement Concrete Res 6(6):773–781
23. Jankowiak T, Àodygowski T (2005) Identification of parameters of concrete damage plasticity constitutive model. Foundation Civil Environ Eng 6:53–69
24. Iervolino I, Galasso C, Cosenza E (2010) REXEL: computer aided record selection for code-based seismic structural analysis. Bull Earthq Eng 8(2):339–362
25. Iervolino I, Galasso C, Paolucci R, Pacor F (2011) Engineering ground motion record selection in the ITalian ACcelerometric Archive. Bull Earthqu Eng 9:1761–1778
26. Proulx J, Darbre GR, Kamileris N (2004) Analytical and experimental investigation of damping in archdams based on recorded earthquakes In: 13th World conference on earthquake engineering, Vancouver, B.C., Canada

Heightening of Very High Gravity Dams: The Case Study of the Grande Dixence

B. Clerc, P. Manso, and G. De Cesare

Abstract Heightening very high gravity dams is one of the solutions considered as part of the energy transition and to mitigate the effects of climate change. The paper presents a stepwise approach to assess the potential for dam heightening and its impact on the hydroelectric scheme. A first diagnosis focusses on identifying and characterising the main site and operation constraints to dam heightening. Then alternative heightening concepts are screened for feasibility, in particular heightening concepts such as gravity dam, arch dam or multi-arch dam. The selected concept(s) are then developed to create several variants with diverse geometries. Subsequently several steps are undertaken to verify internal and external stability for given selected design criteria. This approach is applied to the case study of the Grande Dixence, the highest gravity dam in the world located in Switzerland, considering heightening solutions up to a maximum of 30 m. Within this heightening window the additional loads of reservoir water and dam weight should be in principle acceptable for the already known foundation conditions. The main site constraints are in fact due to the partial submergence of the main headwater conveyance tunnel and the need to adapt the downstream surge tank of the Fionnay's power plant. Structurally, heightening the dam with a similar structural concept was preferred from inception when considering joint behaviour of the original and heightened structures. Four alternative heightening geometries were compared in terms of their overall stability for various heights of elevation. The results, obtained by analytical and computational models, showed acceptable values for all four variants. Regarding the economic analysis, a preliminary analysis of the Levelized Cost of Electricity (LCOE) computed considering the additional electricity and the construction costs is remarkably low in comparison with other projects within the Energy Transition and point out that a height increase within 10–15 m would likely be optimal. This study confirms the interest to pursue investigations and studies beyond the feasibility stage, in view of determining the optimal heightening design and further develop the business plan for high-value hydropower production.

B. Clerc (✉) · P. Manso · G. De Cesare
Department of Civil Engineering, Platform of Hydraulic Constructions (PL-LCH), Ecole Polytechnique Fédérale de Lausanne, Lausanne, Switzerland
e-mail: basile.clerc17@gmail.com

G. Bolzon et al. (eds.), *Numerical Analysis of Dams*, Lecture Notes in Civil Engineering 91, https://doi.org/10.1007/978-3-030-51085-5_43

Keywords Very high gravity dams · Dam heightening · FE method · Surge tanks · Grande dixence

1 Introduction

As part of the energy transition and to mitigate the effects of climate change, Switzerland is considering increasing its capacity to store and transfer water from wet to dry periods, mainly between seasons. Due to global warming, summer inflows have been increasing in glaciated catchments and will continue to increase in the coming decades [1]. Without additional storage, Swiss hydropower utilities will not be able to value such inflows further than run-of-the-river production (the reservoirs are already full in summer) or increase their winter supply. Dam heightening can provide large incremental positive impacts on storage with minimum negative impacts, as "[…] moderate heightening of the existing dams, which represents less than 10% of initial height, [could add] a reservoir volume of 700 hm^3 […] as sum of about 20 feasible projects" [2].

Raising dams requires deep knowledge of the structure and its foundation. Very high gravity dams are usually well studied and documented due to their importance and complexity. Such profound knowledge of the dam-reservoir-foundation system considerably reduces the uncertainty about site conditions already at an early stage of design. Furthermore, the availability of monitoring data and safety assessment tools (FE models, predictive behaviour models) strongly reduce the preparation time to reach feasible design solutions.

Switzerland aims to phase out nuclear energy until 2050. Hydropower should increase production and revenues (through flexible production) whilst facilitating the increase in wind and solar PV production. The Grande Dixence dam, located in the Canton of Valais, creates the largest reservoir in Switzerland, providing 10% of the country's storage energy. If heighten, this dam could be used to transfer a larger share of the summer inflows to produce electricity in winter, as well as increasing its ability to attenuate flash floods, with low incremental negative impacts on the natural and built environment.

A general approach is established to identify and develop solutions for the heightening of very high gravity dams, as well as performances criteria to determine the design variant to be retained and applied to the case study of the Grande Dixence dam.

2 Methods

The methodology consists of two main steps. The first step is a diagnosis of the existing hydroelectric scheme and site constrains, while the second step consists of screening possible dam heightening solutions. These two steps result in an initial catalogue of feasible dam heightening options.

Table 1 Site and scheme constraints

Technical constraints	Logistics constraints	Socio-economic and environmental constraints
• Dam geometry and particularities • Valley shape • Dam materials • Foundation and abutments • Impacts on auxiliary installations and reservoir	• Access to site • Construction site facilities • Construction machinery and methods • Building materials	• Operation non-stoppage • Nearby buildings and infrastructures • Local tourism • Wildlife

2.1 Step 1: Assessment of the Site Potential and Constraints

This step concerns the diagnosis of the present state of the main structures, of their operational principles and functions, as well as of the operational business plans, in view of identifying the untapped potential of the site and any eventual site constraints to dam heightening. The environmental, logistical and socio-economic aspects related to the operation or impacted by the heightening also need to be taken into consideration. The constraints identification and characterisation are essential to determine which ones may, depending on their severity, result in restrictions or modifications which should be anticipated and considered when assessing the feasibility of the heightening concepts. If constraints are deemed too restrictive, they may limit the height of elevation or cause certain concepts to become infeasible. The constraints are specific to each scheme but can be classified into three categories. Table 1 summarises these categories and the elements to be analysed:

Furthermore, to justify the heightening, it is essential to determine the additional water supply that can be stored by relying on actual measurements and estimations in the short, medium and long term. The volume of the lake as a function of altitude must also be computed in order to estimate the height of elevation required to store a certain volume and vice versa.

2.2 Step 2: Screening of Heightening Concepts

Generating intervention variants is a process that follows several steps investigating the different possibilities of heightening.

First, the possibilities of integrating the heightening concepts have to be evaluated. There are three main heightening concepts: gravity dam, arch dam and multiple-arch dam. This evaluation is carried out using predefined criteria and leads to a selection of feasible solutions to be developed into several variants. The evaluation criteria are partly based on the reference state established previously. The variants must then be evaluated and compared using performance and economic criteria.

Table 2 Concepts evaluation criteria

Conceptual criteria	Constructive criteria	Economic criteria
Feasibility in comparison to: • Surrounding geology • Dam geometry • Valley shape	• Flexibility • Construction methods • Additional specific works • Heightening contribution and synergy with the entire structure	• Impact on HPP operation • Quantity of materials

The analysis of heightening concepts is performed using evaluation criteria that are designed to assess the performance of the different concepts against the studied structure. The evaluation criteria can be classified into categories which Table 2 summarises:

Generation of alternative heightening concepts consists of elaborating different geometric shapes according to the requirements of the scheme or standards in use. This step can become iterative if structural concerns arise during the verifications requiring a geometry change.

The study of alternatives must highlight the advantages and disadvantages of each of them based on the overall structure stability and ultimate resistance verifications, as described in the selected standards. An analytical model is first used to obtain the results of the various verifications, allowing a preliminary comparison between alternatives. Subsequently, a computational model is developed to validate the calculation hypotheses previously made.

Finally, an economic analysis comparing the construction costs and the annual energy produced by the studied heightening should establish the cost price of each alternative.

2.3 *Case Study of the Grande Dixence Dam*

The Grande Dixence scheme, located in the southern part of the Swiss Alps, uses the waters of a 420 km^2 watershed containing 35 glaciers from Zermatt valley to Lake of Dix, where water is collected by the use of 75 water inlets connected by 100 km of galleries and 5 pumping stations. The water is then stored in the Lake of Dix, formed by the Grande Dixence dam, whose maximum volume capacity is 400 hm^3, before being turbinated by three power plants located on two distinct levels, Fionnay, Nendaz and Bieudron power plants. The Grande Dixence dam is a concrete gravity dam 285 m high. The dam crest is located at an altitude of 2365 masl, is 15 m wide and 700 m long. The reservoir's maximum operating water level is 2364 masl, which leads to a maximum hydrostatic load of 284 m of water column.

Figure 1 represents the dam central cross-section that displays the geometry used for verifications and various computations. Its main geometrical characteristics are summarized in Table 3.

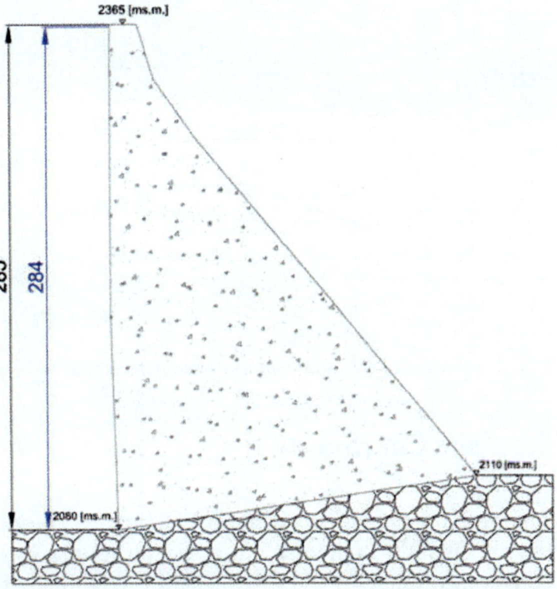

Fig. 1 Dam central cross-section

Table 3 Main geometrical characteristics of the Grande Dixence dam

Characteristics	
Height	285 m
Base width	200 m
Dam crest width	15 m
Dam crest developed length	700 m

3 Results

3.1 Additional Water Supply and Available Storage

Actual data gathered from 2011 to 2017 [6] show an average cumulated volume of 500 hm^3 of water per year. Considering the Grande Dixence dam storage is nowadays of 400 hm^3, an additional volume of 100 hm^3 could be stored by heightening. However, climate predictions indicate a water supply decrease by the end of this century. On the basis of the estimations [3], an additional volume that can be filled yearly on the long-term seems more viable and profitable. An additional volume of 60 hm^3 is selected for which the corresponding height of elevation is 13.60 m as shown in Fig. 2.

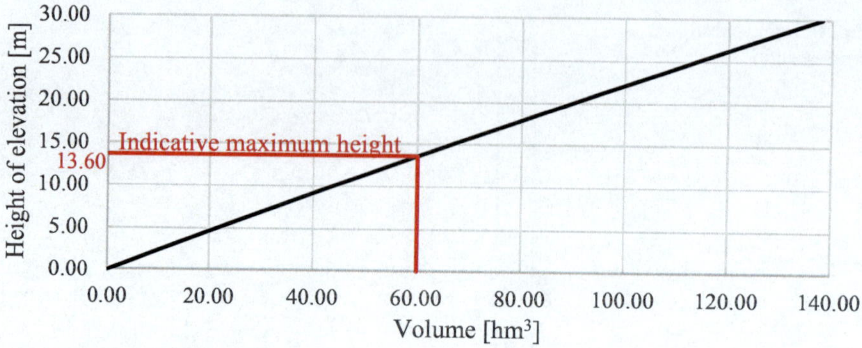

Fig. 2 Indicative additional volume that can be stored and corresponding height

3.2 *Site Constraints*

The constraints assessment determined that some of them had a greater impact on the heightening implementation and on its maximal height. The main headwater conveyance tunnel of the hydroelectric scheme and the surge tanks of Bieudron's and Fionnay's power plants are the most restrictive constraints.

Regarding the main headwater conveyance tunnel, its arrival being almost at the same level as the lake, a water elevation would result in pressurizing a certain length of the structure. The backwater curve was determined in order to evaluate the pressurized length and to verify if the section would still be pressurized where the water of Arolla is pumped into the collector. The structure would not be pressurized at the connection for a maximal heightening of 8 m.

Concerning the surge tanks, both downstream systems were simulated using the software *Hydraulic System*. The critical point to evaluate was the evolution of the pressure wave in the surge tanks created by an emergency stop of the operation. Simulations (Fig. 3) showed that Bieudron's surge tank could easily withstand an increased level in the reservoir of a height exceeding the project's scope, considering an emergency stop of the operation of 75 s.

Simulations of Fionnay's downstream system (Fig. 4) showed very different results as the surge tank summit was reached during an emergency stop of 100 s with an increase of the reservoir water level of 3 m. This value can be described as a maximum height limiting the dam heightening if no modifications are made.

The remaining technical constraints, i.e. the geometry and particularities of the dam and the shape of the valley, are to be considered when analysing the feasible concepts. However, they have, in theory, no influence on the limitation of the elevation height. Logistical, environmental and socio-economic constraints have no impact on the choice of variants. However, they are aspects to be considered when constructing the elements of the chosen heightening design.

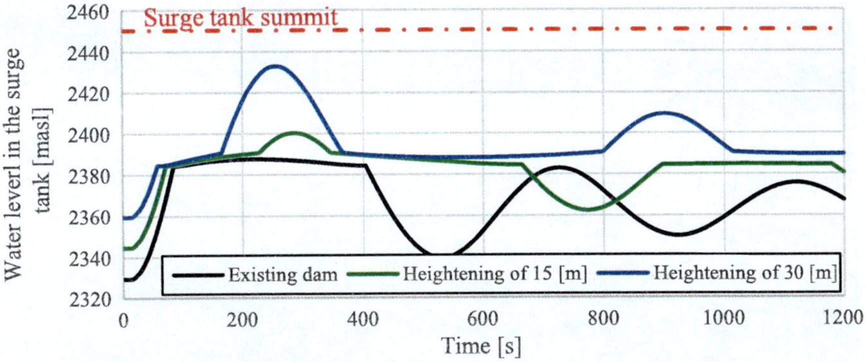

Fig. 3 Water level rise in Bieudron's surge tank during an emergency stop for the existing dam and with the heightening of 15 and 30 m

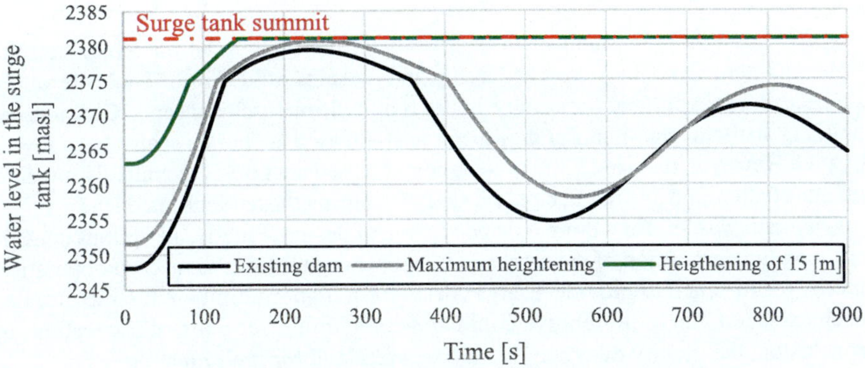

Fig. 4 Water level rise in Fionnay's surge tank during an emergency stop for the existing dam, the maximum heightening and the heightening of 15 m

3.3 Alternative Heightening Concepts

Simply extending the existing gravity dam structure is conceptually the simpler structural solution although not necessarily the one leading to the lowest effort and investment. The heightening concept may also vary with the heightening level as well, starting with large-size parapet walls for few meters which can be easily accommodated in the existing 15 m crest width. On the one hand, the plan shape of the dam crest, as can be seen on Fig. 5, strongly limits the integration of an arch as carried out recently in the Vieux-Emosson dam in Switzerland [4].

On the other hand, the presence of joints every 16 m is a strong constraint to any solution of buttress heightening concepts or multiple-arch heightening concepts. The surrounding geology is of good quality. However, topographically speaking, the integration on the abutments of an arch extension is unfeasible: indeed, although

Fig. 5 Plan view of Grande Dixence dam (*Source* Grande Dixence [7])

the left abutment is of steep good quality rock, there would be no support on the right abutment since the valley widens up and presents mild slopes. Thus, the valley shape is more favourable for gravity, buttress or multiple-arch concepts. Considering building artificial abutments and/or using post-stressed anchors has also been investigated. However, the use of 300 m long post-stressed anchors is technically unheard off and challenging and has been discarded. Anchoring the tendons inside the original gravity dam was left for a more advanced stage of design. Finally, all options considering partial demolition of the crest and temporary restrictions to reservoir operation levels, for instance to provide a larger basis for a high multiple-arch structure, has been excluded due to prohibitive conflicts with hydropower must-run operation. In conclusion, the gravity dam heightening was retained for preliminary design.

3.4 Gravity Dam Heightening Options

Four variants are generated from the chosen heightening concept. A new crest width is fixed to 5 m for all of them. The first variant (Fig. 6) replicates the existing gravity profile. The second variant (Fig. 7) consists of building a symmetrical trapezoid standing on the dam crest, where the vertical component of the additional hydrostatic

Fig. 6 Cross-section of variant 1 "Standard"

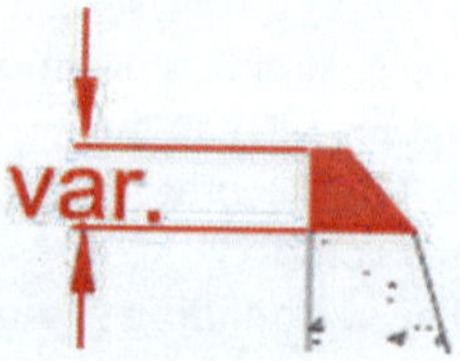

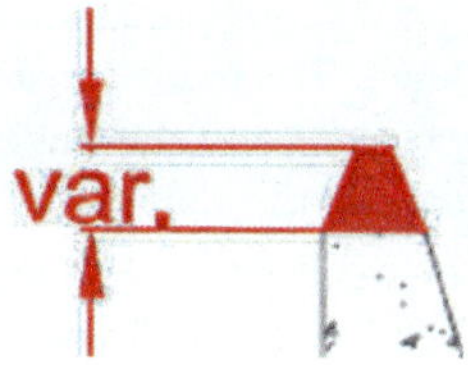

Fig. 7 Cross-section of variant 2 "Symmetrical-faced"

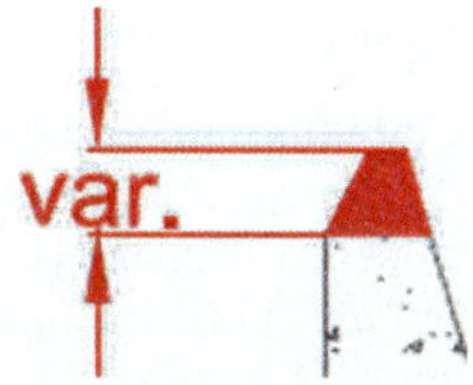

Fig. 8 Cross-section of variant 3 "Asymmetrical"

Fig. 9 Cross-section of variant 4 "Offset"

pressure acts in favour of dam stability. The third variant (Fig. 8) is built from the former, by extending the downstream face with a similar slope: the asymmetric face slopes increase the stabilising role of the additional hydrostatic pressure with regards to variant 2 but locate the heightening weight farther downstream. A fourth variant (Fig. 9) was generated after analysing the preliminary results obtained for the former three variants. In fact, the emergence of tensile stresses on the upstream face of the dam indicates a need to apply a greater load upstream of the existing dam. The symmetrical trapezoidal geometry of variant 2 is used and offset upstream. The hydrostatic pressure favourable to the stability of the structure is therefore still at play. However, the construction of a cantilever is necessary and the volume of concrete is by default larger. The cantilever's width was arbitrarily chosen as 3 m sloped at 45°.

3.5 Stability and Structural Analysis

The verifications were carried out using the "Directive on the Safety of Water Retaining Facilities" established by Swiss Federal Office of Energy SFQE [5]. To analyse the changing behaviour, all checks were performed at elevation heights of 6, 9, 12 and 15 m. These values are a consequence of the estimations of the additional volume that can be stored. The verifications recommended to be carried out by the Directive are the sliding stability, the overturning stability and the ultimate resistance. The loads considered for these verifications are the dam weight, the hydrostatic pressure and the under-pressures, which were estimated on the basis of measured data at the dam (STUCKY. 2017 [8]). The stability criteria are established by the guidelines of the Directive. The ultimate resistance criteria are the maximum stresses that the concrete can sustained. Usual recommended values of concrete tensile resistance are 0 MPa and compressive resistance are 8 MPa. However, the studied dam is an existing structure exhibiting satisfactory behaviour. Therefore, a higher tensile resistance criterion of 1 MPa is considered.

Computational model. A 2D model carried out using *ANSYS Workbench* software (Version 17.2) allows structure modelling using the finite elements method. The modelled structure is composed of three elements: the foundation, the existing dam and the studied heightening. The mesh elements, composed of quadrilaterals and triangles, have a different size depending on the element of the dam they form. Thus, the foundation has elements of 15 m, the existing dam has elements of 3 m and the studied heightening of 1 m (Fig. 10). The use of computational model allows to observe and compare variations of normal and shear stresses and displacements throughout the entire studied structure. Figure 11 shows the normal stresses for the existing dam and for variant 4 (heightening of 15 m) respectively.

Stability verifications. The stability of the heightened dam is assessed for sliding and overturning by means of rigid body analytical analysis. Sliding stability is verified for the most unfavourable potential critical surfacing, the foundation-dam interfacing, by taking into account the characteristics of the foundation and the slope of the interface. The verification is guaranteed if the global sliding security factor S_G is higher than 1.0. All results are satisfactory.

The *ANSYS* sign convention considers compressive stresses as negative and tensile stresses as positive.

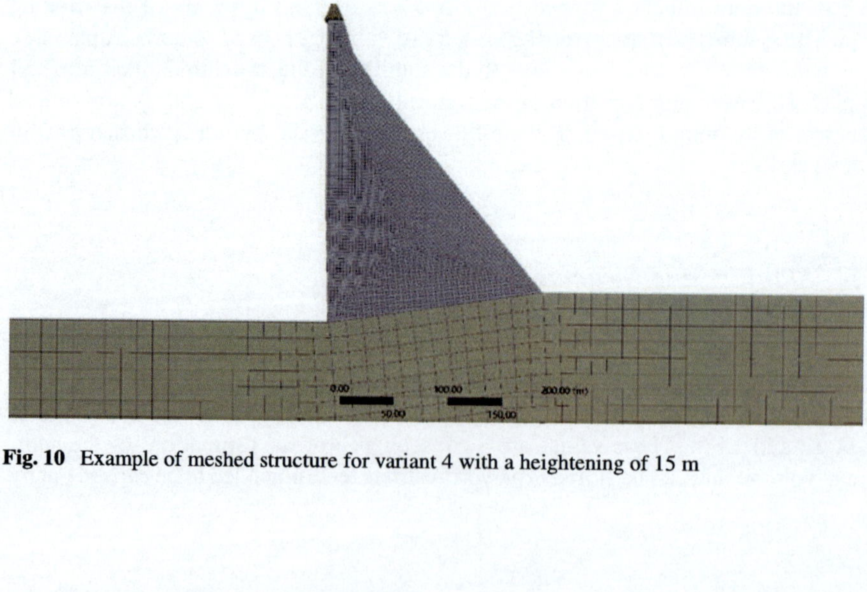

Fig. 10 Example of meshed structure for variant 4 with a heightening of 15 m

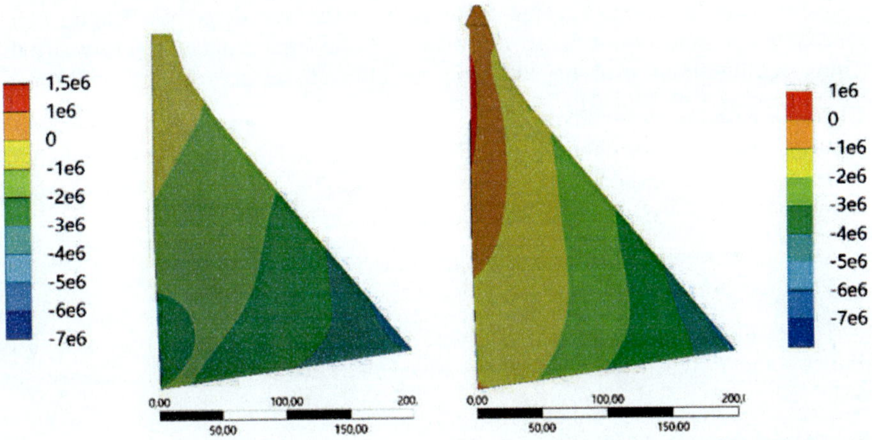

Fig. 11 Normal stresses of existing dam (left) and variant 4 with a heightening of 15 m (right)

Table 4 Sliding stability results—global sliding security factor

S_G				
Existing dam	1.46			
Height m	Variant 1	Variant 2	Variant 3	Variant 4
6	1.38	1.38	1.38	1.38
9	1.35	1.35	1.35	1.35
12	1.31	1.31	1.31	1.32
15	1.28	1.28	1.28	1.28

Table 5 Overturning stability results—dimensionless horizontal location of the resulting force (X)

X				
Existing dam	0.58			
Height m	Variant 1	Variant 2	Variant 3	Variant 4
6	0.59	0.59	0.59	0.59
9	0.60	0.60	0.60	0.60
12	0.61	0.61	0.61	0.61
15	0.62	0.62	0.62	0.62

The objective of the overturning verification is to ensure that no unacceptable tensile stresses are generated at the foundation-dam interface. For a normal combination of loads, the resulting vector of all applied forces must fit within the central third of the section (Tables 4 and 5).

Where X is the ratio of the application point on the considered base length, which means that X values must be greater than 0.33 and smaller than 0.66 to verify the tilting stability. In view of this, all results are also satisfactory.

Ultimate resistance verification. The ultimate resistance verification consists of ensuring that the stresses acting in the dam do not exceed the concrete strength limits set in the design criteria. Under full lake conditions, the critical stresses to check are as follows:

- Minimal stresses on upstream facing;
- Maximal stresses on downstream facing.

The results obtained by analytical and 2D computational model are shown in Tables 6 and 7. Tensile stressed are negative values, while compressive stresses are positive.

Economic analysis. The economic analysis is carried out considering the additional annual energy produced thanks to the heightening and the total costs of the construction work. The results are given for variants 1 to 3 and variant 4, because the first three variants have the same section.

The construction total costs are estimated using a lumped unit rate for concrete of 600 CHF/m^3 based on previous construction experience. The annual costs are

Table 6 Minimal stresses (in kPa) on upstream facing

Existing dam	Analytical model				Computational model			
	130.19				34.20			
Height m	Variant				Variant			
	1	2	3	4	1	2	3	4
6	174.09	116.36	173.06	115.62	149.00	65.40	58.30	52.70
9	190.54	164.63	196.08	168.19	73.60	111.00	68.40	81.70
12	26.01	−0.19	16.51	123.94	−128.00	−43.30	−132.00	−102.00
15	−177.39	−208.37	−212.53	−55.27	−361.00	−266.00	−364.00	−341.00

Table 7 Maximal stresses (in kPa) on downstream facing

Existing dam	Analytical model				Computational model			
	3960.36				5760.00			
Height m	Variant				Variant			
	1	2	3	4	1	2	3	4
6	4240.24	4239.68	4238.80	4236.98	6160.00	6100.00	6170.00	6080.00
9	4385.36	4384.31	4383.31	4380.20	6400.00	6300.00	6380.00	6500.00
12	4533.98	4532.44	4531.66	4526.92	6620.00	6540.00	6620.00	6730.00
15	4686.15	4684.11	4683.90	4677.17	6840.00	6770.00	6860.00	6970.00

estimated considering a project horizon of 50 years and a 5% discount rate. The values per meter of heightening are summarised in Table 8.

The additional annual energy is estimated using the highest fall of 1883 m, which is created by Bieudron downstream system. Its average value is of 960 MWh per meter of elevation.

The ratio between the annualised investment costs and the additional energy production in an average hydrological year provides an estimate of the Levelized Cost of Electricity (LCOE), known indicator for comparative feasibility analysis. The values of the LCO obtained for different heightening concept and heights of up to 25 m are shown in Table 9.

The LCOE tends to a stable value for dam height increases above 10 m for all concepts (Fig. 12).

Table 8 Total and annual costs per meter of heightening

Design Alternatives	Total construction costs MioCHF	Annuity CHF/y
1 to 3	4.2	230'000
4	6.7	368'000

Table 9 Cost prices for various height of elevation

Height of elevation m	Cost price cts. CHF/MWh	
	Variants 1 to 3	Variant 4
1	2.2578	3.6124
5	2.3691	2.9377
10	2.3827	2.8474
15	2.3866	2.8162
20	2.3879	2.7999
25	2.3883	2.7896

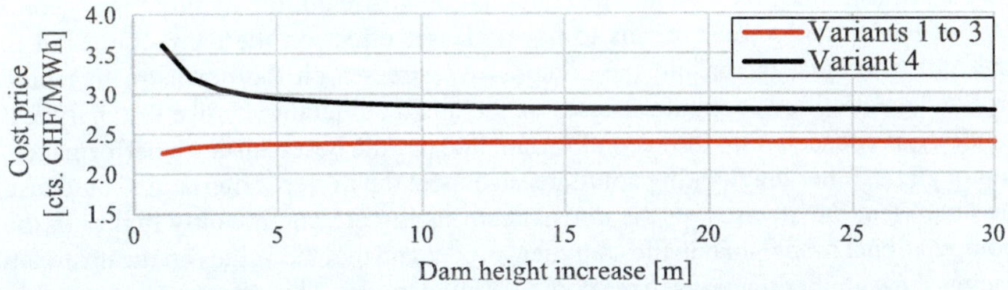

Fig. 12 LCOE of the retained alternative concepts as a function of dam height increase

4 Discussion

4.1 Comparison of Alternative Solutions

This analysis is carried out on the basis of the results obtained with both analytical and computational models and concerns the structural stability and resistance performance.

Comparison based on the analytical model results. The overall stability verifications of the structure show almost no difference between the variants. The increase of the considered loads being almost similar for each variant, to obtain such results is not surprising. In terms of ultimate resistance, the different geometries of the variants cause similar behaviour in the existing dam, but with a variable amplitude depending on the variant. On the upstream facing, while all variants have an increase in compressive stresses from a height of 6 to 9 m of elevation, these are reduced from a height of 12 m. A decrease in compression and the emergence of traction can be observed from this height up to 15 m, but the fourth variant presents with the lowest value of traction. This indicates that moving the weight of the heightening upstream improves the dam behaviour on its upstream facing. Moreover, there is no clear difference of stresses between the variants at the downstream facing. In summary, although each studied solution meets the project criteria, the fourth variant shows better ultimate resistance performances than the others.

Comparison of computational model results. The results of the computational model show, for the minimum stresses on the upstream facing, values of the same range as those obtained with the analytical model. The only notable differences are the higher minimum stress of variant 1 at a height of elevation of 6 m, as well as the higher stresses of the second variant, which also has lower tractions. Overall, variant 4 has the highest tensile stresses. On the downstream facing, the compressive stresses are very similar and it is variant 2 that has the lowest stresses, unlike variants 1 and 3. The increase in stress is almost linear with the increase in elevation height. Considering the results upstream and downstream, the symmetrical geometry of the second variant seems to have a lesser effect on the dam, compared to the other variants, increasing the compressive stresses on the downstream facing less and creating lower tensile stresses on the upstream facing. Unlike the analytical model, it is consequently the second variant that has the best resistance performance, although the other heightening solutions also meet the project criteria. It should also be noted that the stresses on the downstream facing are significantly higher in the computational model than in the analytical model and that the values on the upstream facing show greater tractions in the computational model. This difference is probably due to the analytical model which uses oversimplifying equations to represent the behaviour of such a complex structure. Although, divergences at the computational model extremities can be expected for this kind of simulation.

Comparison with existing dam. Both analytical and computational models show increased compressive stresses on the downstream facing and a loss of compression and even the emergence of tensile stresses on the upstream facing. These observations represent significant changes to the existing dam behaviour that are tolerated according to set design criteria at this stage of analysis.

4.2 Heightening Justification

All the elements and results that could justify the construction of the heightening, such as the storable volume, the structural feasibility, the economical and structural benefits of the heightening and the main obstacles to overcome, are analysed.

In terms of available and storable volume. The purpose of a heightening is to increase the volume of a reservoir. However, this is not necessary if there is no excess water supply. As explained above, this surplus was estimated at an average of 100 hm^3 between 2011 and 2017. This volume is therefore turbinated to avoid a spill over the crown and probably sold at a low price. A higher elevation would enable this additional volume to be stored, which could then be turbinated when the electricity network needs it, improving its profitability. However, due to climate change, these water supplies are most likely to change in the coming decades. Estimates indicate that water supplies will increase in the short and medium term before decreasing drastically. In order to have a long-term vision, it is preferable to carry out a heightening creating an additional volume that cannot store all the water supplies currently observed, but which ensures regular filling of the lake over a longer period. Therefore, a heightening allowing the creation of an additional volume approaching 60 hm^3 of maximum capacity seems more promising and interesting to achieve.

In terms of structural feasibility. The overall stability and ultimate resistance verifications carried out using analytical and computational model tools for a limited number of loads cases show that is structurally feasible to heighten the Grande Dixence dam using some of heightening concepts presented in this paper. Indeed, each of them meets the preliminary design criteria up to a height higher than the indicative maximum value of 13.60 m discussed previously. Figure 13 illustrates the maximum elevation heights that can meet the different project criteria.

In terms of site constraints. The assessment of the reference state identified two major constraints to the heightening of the Grande Dixence dam: the hydraulic behaviour of the main collector and of the Fionnay surge tank. Fionnay's surge tank has a quite low operation range of maximum 3 m. Increasing the tank's volume or modifying the orifice characteristics can be envisaged as adaptation measures. Restricting the operation discharge range of Fionnay HPP while the lake is close to full capacity could also be introduced. Regarding the main headwater conveyance tunnel, should the maximum water level of the reservoir be raised by more than 8 m the upstream conveyance structures become pressurized, raising possible flow

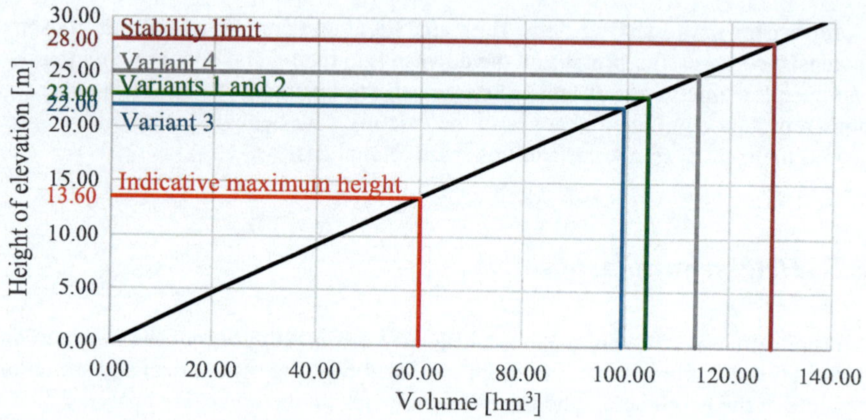

Fig. 13 Limits of design criteria for each design variant and corresponding increase in storage volume

stability issues, as well as increased water pressure at the connection with pumped waters from Arolla and a need to implement an aeration system. The development of feasible adaptation measures is left for subsequent studies.

In terms of costs and benefits. The economic analysis shows an equivalent cost price (LCOE) for alternatives 1 to 3, all being lower than alternative 4. With a cost price of less than 0.003 cts. CHF per kWh, the profitability of the heightening project is significant, considering even proxy spot market prices as low as 4 to 5 cts. CHF per kWh [9, 10]. However, the additional storage shall be used to shift electricity production to periods with higher remuneration (i.e. higher sell prices) thus increasing the generation revenues. In fact, no increase in electricity production is expected, since there is no spillage in the present system. The main benefit will be to reduce the number of production hours in summer remunerated at low prices (i.e. off-peak, because of "must run" conditions) by a larger number of peak production hours in winter (thus replacing expensive imports). The estimated cost prices showed little variation for heights of elevation higher than 10 m. At this stage of project development, there is no significant economic difference between heightening projects of 10 or 15 m. These preliminary conclusions and the associated assumptions merit further investigation. Adaptation of the auxiliary structures may increase investment costs.

4.3 Determination of Optimal Height of Elevation and Variant

Additional studies are needed to determine an optimal height of elevation, but it can already be assumed that this height is situated in a range of 10 to 15 m, considering the economic analysis and the water supplies estimations. The results obtained so far point out directions but must be extended in view of determining an optimal solution. All four alternative concepts meet the project's criteria and options 1 to 3 show the lowest cost price. Additional assessment using other evaluation criteria, such as aesthetic criteria, must be developed.

5 Conclusion

The paper presents a general approach to identify and value dam heightening solutions for very large gravity dams, illustrated herein for the case study of the Grande Dixence dam (the highest gravity dam worldwide with 285 m of height). The approach comprises two steps to obtain an initial catalogue of feasible dam heightening solutions. The first step is an assessment of the reference state of the dam and associated infrastructure. It allows identifying the main constraints limiting dam heightening. The second step is the generation and comparison of several heightening concepts using predefined design criteria, including buttress dam, multiple arch dam and gravity dam structural options.

Heightening very high gravity dams is more convenient with a similar structural concept, as shown for the case study. It facilitates structural bonding and ensemble behaviour of both the base dam and the heightening body for hydrostatic, thermal and seismic loads. Four alternative gravity heightening options were compared, considering a new crest width of 5 m and height increase options up to max. 30 m. Analytical and 2D computational models allowed assessing the changes in dam behaviour. All four concepts are structurally feasible, meeting the target design criteria. The alternatives that allow increasing the vertical hydrostatic load on the upstream dam face and heel likely improve dam behaviour. The additional water storage provided by dam heightening allows increasing the sell price of summer production that is today remunerated at below-premium rates and even increasing winter production at premium hours. The levelized cost of electricity (LCOE) is extremely low when compared to other potential hydropower projects providing also peak power and winter supply. A height increase within 10 to 15 m is likely optimal. Adaptation of headwater conveyance systems as well as of pressurized waterways and surge tanks may limit the dam heightening and increase costs and must therefore be considered from the very early stages of analysis.

There is a clear interest to heighten very gravity dams for which the incremental impacts on the natural and built environments are marginal with regards to the present footprint. The selection of the dam raise solution must however be confirmed with further investigations and studies, in parallel with further development of the business plan for high-value (flexible) hydropower production.

Acknowledements The authors would like to acknowledge Dr. Thomas Menouillard (CADFEM, Switzerland), Dr. Nicolas Bretz (HydroExploitation SA, Switzerland), Mr. Reynald Berthod (Stucky SA) and Dr. Nicolas Adam and Mr. Raphael Leroy (AlpiQ SA) for their help, as well as the financial support of the Swiss Competence Center for Energy Research—Supply of Electricity (SCCER-SoE, under contract CTI/2013/0288).

References

1. Schaefli B, Manso P, Fisher M, Huss M, Farinotti D (2019) The role of glacier retreat for Swiss hydropower production. Renew Energy 132:615–627
2. Manso P, Schleiss AJ, Stähli M, Avellan F (2016) Electricity supply and hydropower development in Switzerland. Int J Hydropower Dams 23(5):41–47
3. Manso P, Schleiss AJ, Monay B, Dujardin J, Zappa M (2017) Assessing the potential increase of seasonal energy storage to mitigate the impact of climate change in Switzerland: case study of the Grande Dixence dam. Hydro 2017, Seville
4. CSB: Dams in Switzerland—Source for Worldwide Dam Engineering (2011)
5. Office fédéral de l'énergie OFEN: Directive sur la sécurité des ouvrages d'accumulation (2015)
6. Grande Dixence: Rapports annuels d'exercice. (In French) (2011–2017)
7. Grande Dixence: Documentation technique. (In French) (2015)
8. STUCKY: Expertise quinquennale 2016 par M. Reynald Berthod, mémoire et annexes. Grande Dixence, Martigny. (In French) (2017)
9. Romande Energie: Tarifs d'électricité professionnels. (In French) (2017)
10. Groupe E: Tarifs d'électricité. (In French) (2017)

Prediction of Piezometric Levels at the Rock Concrete Interface Considering the Non-linearity of Permeability in the Foundations

M. de Granrut, B. Berthomé, and A. Simon

Abstract Being able to describe the state of a dam regarding the safety requirements is an obligation that dam owners have to fulfil. A constant surveillance is thus established, based on the measurement of structural behavioural parameters. The measurements are used to feed descriptive statistical models which are classically linear, such as the historical HST (Hydrostatic-Season-Time) model. This assumed linearity is a significant shortcoming, which only permit a poor description of the reality when it comes to analysing the piezometry in the foundations of arch dams, and more particularly at the rock-concrete interface. Indeed, the contact between rock and concrete evolves during seasons between an open and closed state, under the influence of the thermal loads, and of the hydrostatic load, but also during the years due to ageing effects. Thus, the effects of those three influencing loads on the piezometry are not merely additive. Consequently, analysing such a complex phenomenon with accuracy is only possible when considering complex interactions between the influencing quantities. This paper presents a model that describes the piezometry at the rock-concrete interface as being a fraction of the total upstream hydrostatic load. Since it has been empirically observed that the permeability variations and the piezometry are non-linearly correlated, this non-linearity is explicitly included into the model formulation. In this work, the resulting model is applied to analyse monitored piezometric levels recorded at the interface of a French arch dam. This approach greatly improves the analysis that could be made by HST, and it permits a thorough physical interpretation of the studied piezometry. Eventually, the temporal evolution of the state of aperture of the contact is assessed, which is a great improvement for dam surveillance.

Keywords Arch dams · Piezometric levels · Dam surveillance

M. de Granrut
Electricité de France, Université Grenoble Alpes, Grenoble INP 3SR Laboratory, 38000 Grenoble, France

B. Berthomé · A. Simon (✉)
Electricité de France Division Technique Générale, 38000 Grenoble, France
e-mail: alexandre-gilles.simon@edf.fr

G. Bolzon et al. (eds.), *Numerical Analysis of Dams*, Lecture Notes in Civil Engineering 91, https://doi.org/10.1007/978-3-030-51085-5_44

1 Introduction

Being able to describe the state of a dam regarding the safety requirements is an obligation that dam owners have to fulfil. A constant surveillance is thus established, based on the measurement of structural behavioural parameters, such as measurements of piezometric levels, leakages, displacements, topographical survey etc. The provided data is used to feed behavioural statistical models, which permit to describe in real-time the state of the dam and assess its safety. Classically, linear regression models are used, such as the historical HST (Hydrostatic-Season-Time) model [1], which models the effects of the external loads as being additive. This assumed linearity is a significant shortcoming, which only permits a poor description of the reality when it comes to analysing the piezometry in the foundations of arch dams, and more particularly at the rock-concrete interface.

Indeed, the contact between rock and concrete evolves during seasons between an open and closed state, under the influence of the thermal loads, and of the hydrostatic load, but also with time during the years due to ageing effects. The way those influencing loads (thermal, hydrostatic, time) affect the piezometry is not merely additive [2]. Actually, depending on the thermal state of concrete, the arch either drifts in the downstream or the upstream directions. This redistributes the mechanical strains in the foundation and modifies the state of compression of the rock-mass. The hydraulic conductivity at the interface is thus directly influenced by the thermal state of the structure, and this has a knock-on effect on the impact of the hydraulic load on the piezometry. Consequently, the thermal and the hydrostatic loads have combined effect on the structure. Interaction terms between the effect of the season and of the hydrostatic load, but even potentially of time, are thus required to describe this piezometry.

This paper presents a nonlinear regression model that describes the piezometry at the rock-concrete interface as being a fraction of the total upstream hydrostatic load. This fraction is directly linked to the permeability variations of the rock mass in the foundations, which in turn evolve with the mechanical stresses that the whole structure (foundations plus dam) is submitted to. Since it has been empirically observed that the permeability variations and the piezometry are non-linearly correlated, this non-linearity is explicitly included into the model formulation. An application of the resulting model to the analysis of monitored piezometric levels recorded at the interface of a French arch dam is then presented. This approach greatly improves the analysis that could be made by HST, and it permits a thorough physical interpretation of the studied piezometry. Eventually, the temporal evolution of the state of aperture of the contact is assessed, which is a great gain for dam surveillance.

2 Methods

2.1 State of the Art: HST (Hydrostatic Season Time)

HST is the historical linear regression model which was developed in 1967 by EDF [1], the use of which has been established by custom in the international community of dam owners.

This model is based on the principle that three loads have an effect on the behaviour of a structure. Those are:

- the thermal load, which accounts for all the thermal variations that affect the dam (the temperatures of the air, of the water, of the foundations, the solar radiation etc.). Those variations are cyclical to first order, characterised by different frequencies. Their respective effect on the structure is thus modelled as a sinusoidal function of the season S. S is an angle that varies from 0° to 360° from January 1st to December 31st, defined by $S = 2\pi(t/365.25 - floor(t/365.25))$, where t is the date of measurement, expressed in number of days since a chosen origin. The function floor stands for the integer part. The global annual thermal load that affects the dam is the sum of those thermal variations, so its effect is a sinusoidal function itself (see Eq. (1))

$$f_1(S) = a_1 cos S + a_2 sin S + a_3 cos 2S + a_4 sin 2S \tag{1}$$

- the hydrostatic load, which accounts for the water pressure on the dam. Its effect is modelled as a four-degree-polynomial function Eq. (2) of the relative trough $Z = (h_{norm} - h)/(h_{norm} - h_{emp})$ where h is the water level in the reservoir, h_{norm} is the normal water level and h_{emp} is the water level when the reservoir is empty. h, h_{norm} and h_{emp} are expressed in meters above sea level.

$$f_2(Z) = a_5\,Z + a_6\,Z^2 + a_7\,Z^3 + a_8\,Z^4 \tag{2}$$

The thermal and hydrostatic loads have instantaneous reversible effects on the structure, but some irreversible influences also have to be taken into account when modelling the behaviour of a dam. A third type of influence is thus considered:

- time, which accounts for all of the phenomena that evolve along time, such as the ageing of concrete, the swelling of concrete, plugging of drains, loss of watertightness etc. The effect of time is modelled as a linear influence of time t (see Eq. (3)).

$$f_3(t) = a_9\,t \tag{3}$$

Most of the time, the linear term is sufficient to describe the long-term temporal evolutions, but some more complex laws can be adopted. Adding for example an exponential term and/or a polynomial term can be useful when studying the early age of concrete or the first impoundment of the dam.

Consider an observed phenomenon X_i, which is measured at a given time t_i. The sample is composed of N observations collected during a given period of time, so i varies from 1 to N. HST thus assumes that X_i can be modelled as follows Eq. (4):

$$\begin{aligned} X_i = & a_0 + f_1(S_i) + f_2(Z_i) + f_3(t_i) + \varepsilon_i = a_0 + a_1\, cos S_i + a_2\, sin S_i + \\ & + a_3\, cos 2S_i + a_4\, sin 2S_i + a_5\, Z_i + a_6\, Z_i^2 + a_7\, Z_i^3 + a_8\, Z_i^4 + a_9\, t_i + \varepsilon_i \end{aligned} \tag{4}$$

$(a_0, a_1, \ldots, a_9)$ are the coefficients to be optimised by minimising the sum of squared residuals, and S_i (resp. Z_i) is the season (resp. relative trough) at time t_i. ε_i is the error term, that accounts for the imprecision of the model and any random error that are contained in the measurements (e.g. measuring errors).

Once the model is optimised, a strong assumption is made: the reversible laws are considered to remain unchanged as time passes. This permits to define the corrected measurements (CM), by removing the reversible effects f_1 and f_2 from the raw measurements, see Eq. (5).

$$CM_i = X_i - \widehat{f_1}(S_i) - \widehat{f_2}(Z_i) \tag{5}$$

This can also be estimated the following way of Eq. (6) for the period of analysis:

$$CM_i = \widehat{a_0} + \widehat{f_1}\left(S_{ref}\right) + \widehat{f_2}\left(z_{ref}\right) + f_3(t_i) + \widehat{\varepsilon_i} \tag{6}$$

with S_{ref} and Z_{ref} such that $\widehat{f_1}\left(S_{ref}\right) = 0$ and $\widehat{f_2}\left(Z_{ref}\right) = 0$. $\widehat{\varepsilon_i}$ being the sum of ε_i and the error corresponding to the estimation of $\widehat{f_1}$ and $\widehat{f_2}$. $\widehat{a_0}$ is the estimation of the coefficient a_0.

This operation amounts to calculating the measurements that would have been observed, had the season been an average season and had the water level in the reservoir been equal to h_{norm}, those conditions being called standard conditions. In this way, measurements can be compared to each other under "identical conditions", and temporal effects or abnormal evolutions can be identified. The CM are thus closely examined by the engineers in charge of the surveillance of dams.

Additionally, by summing the seasonal and hydrostatical laws, HST makes another strong assumption, which is that the hydrostatic and the thermal loads have purely independent effects. For most mechanical phenomenon (e.g. horizontal or vertical displacements), this assumption is valid, so HST achieves high prediction performances. However, as soon as the explained phenomenon can no longer be explained by additive influences, as for hydraulic phenomenon, including interaction terms is necessary.

2.2 NL HST: Non-linear HST

The statistical analysis that is classically used to analyse and explain monitored phenomenon is the HST model. HST assumes that the irreversible and the reversible evolutions of the phenomenon are independent one from another. However, the effects of the thermal and hydrostatic loads actually are interdependent. Indeed, for a given water level, the effect of the hydrostatic load differs from a thermal state to another. In winter, the low temperatures induce (for an arch dam) a thermal contraction of concrete which in turn induces a shifting of the arch in the downstream direction. This modifies the permeability at the interface by increasing it at the heel of the dam. The rock-concrete interface is said to be decompressed or open, and allows the hydrostatic load to propagate toward downstream. Uplift pressures in the rock-concrete interface thus increase. Reciprocally, in summer, the thermal expansion of concrete tends to close the interface and reduces its permeability, which limits the propagation of the hydrostatic load. Therefore, the effect of the hydrostatic load is modulated by the effect of the thermal state. Those two reversible effects are coupled. Some time-depending evolutions can also influence the permeability of the interface, such as clogging, or a loss of watertightness. From those reflections, one can infer that the effects of the different influencing loads (thermal state, hydrostatic load, time) are not merely additive, which invalidates the HST model. The model that is described in this article aims at taking into consideration those couplings, by explicitly integrating them into a non-linear formulation. This model is sought to keep a high interpretability, which is indispensable in the context of dam monitoring in order to understand the link between the external loads and the structure.

Consider (Fig. 1):

- $P(x)$ the piezometry measured by a piezometer situated at the contact, at a distance x from the upstream end of the heel. The notation is simplified to $\boldsymbol{P}$;
- $\boldsymbol{P_{do}}$ the piezometry measured at the downstream end of the toe of the dam;
- $k(x)$ a dimensionless factor, written $\boldsymbol{k}$;

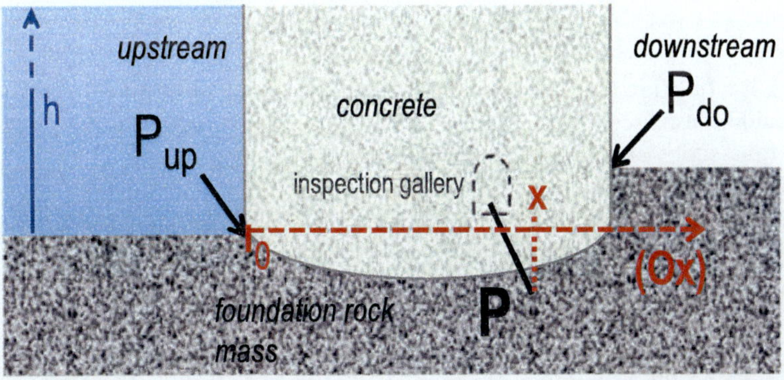

Fig. 1 Representation of the rock-concrete interface, definition of the quantities of interest

- $\boldsymbol{H_l} = h - P_{do}$ the head loss between upstream and downstream.

P is modelled as the sum of the downstream piezometry and a fraction of the hydrostatic load (see Eq. (7)).

$$P = P_{do} + k\ H_l \quad (7)$$

k expresses the link between the head loss between the upstream end and the piezometer (as a fraction of H_l), and the measured piezometric level. If the permeability of the medium was homogeneous and isotropic, this link would be a proportional and constant link (Darcy law). However, this is not always the case, and the permeability of the foundation rock mass is actually influenced by the mechanical strains of the foundation, which are themselves a function of the mechanical stresses that affect the structure (dam plus foundation).

First, the effects of those mechanical stresses on the structure are modelled linearly, as the sum of a hydrostatic function, a thermal function and a temporal function. This choice was conditioned by the following observation: the vertical displacements being an image of the mechanical strain of the structure (*meca*), a preliminary linear regression model was adjusted on measurements of vertical displacements (VD) recorded between 1994 and 2018 by a liquid level system situated at the rock-concrete interface (see Eq. (8)).

$$VD \approx meca = f_1(S) + f_2(Z) + f_3(t) + \varepsilon_i \quad (8)$$

where f_1, f_2 and f_3 are the same functions as in the HST model ((1), (2), (3)) and ε_i is the modelling error. The accuracy of fit is satisfactory: the root mean square error (RMSE) reaches 0.165 mm, which corresponds to a reduction of dispersion of 90% when compared to the initial standard deviation of 1.65 mm.

Second, the effect of the global mechanical strains on the permeability, and thus on the piezometric levels, is not linear (see Eq. (9)).

$$k = g_{non\ linear}(meca) \quad (9)$$

where *g* is a function to be determined. In order to get an idea of the type of function that could be used, complementary monitoring data was used. The same displacements time-series as the one previously used (7) permitted to draw Fig. 2, where *k* is plotted versus the vertical displacements. *k* was calculated using piezometric measurements expressed as a percentage of the total upstream load: $k = 100\ (P - P_{do})/H_l$. This plot thus permits to visualise the relation between *k* and the global mechanical strains. Said differently, it is an image of the function $g_{non\ linear}$. Since the measurements are distributed following a hyperbolic tangent type function, the function $g_{non\ linear}$ was chosen as a hyperbolic tangent function.

Considering the expression of the relative trough *Z*, H_l can be expressed as a function of the water level in the reservoir *h*, see Eq. (10)

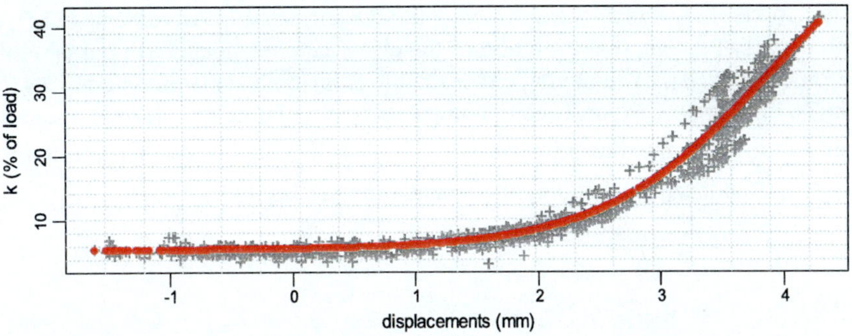

Fig. 2 Fitting of a hyperbolic tangent function (red) to the curve of the permeability versus the vertical displacements (grey)

$$H_l = h_{norm} - Z \cdot \left(h_{norm} - h_{emp}\right) - P_{do} \tag{10}$$

Finally, for a monitored piezometric time series $P_{i\epsilon\{1;N\}}$ containing N observations, the global model can be expressed as follows (see Eq. (11)):

$$\begin{aligned} P_i =& P_{do} + [b_1 + b_2\, tanh(a_0 + a_1 cos S_i + a_2 sin S_i + a_3 cos 2S_i + a_4 sin 2S_i \\ &+ a_5 Z_i + a_6 Z_i^2 + a_7 Z_i^3 + a_8 Z_i^4 + a_9 t_i)][h_{norm} - Z\,(h_{norm} - h_{emp}) \\ &- P_{do}] + \varepsilon_i i \in \{1; N\} \end{aligned} \tag{11}$$

This expression constitutes the regression model in which the coefficients b_1, b_2, a_0, ..., a_9 are the regression coefficients. They are determined by nonlinear least squares fitting, using the Levenberg Marquardt algorithm [3, 4], which iteratively minimizes the sum of squared residuals. ε_i is the residue of the model, which stands for the measurement errors and the inaccuracy of the model.

P_{do} is a constant that is not actually precisely known on dams. Consequently, the choice was made to optimize it also as a regression parameter, thus P_{do} becomes b_0. After the fitting is carried out, it is important to check the optimised value of b_0, which should be in accordance with the real limit conditions of the dam. The expression that is eventually optimised thus contains 13 regression parameters, see Eq. (12):

$$\begin{aligned} P_t =& b_0 + [b_1 + b_2\, tanh(a_0 + a_1 cos S_1 + a_2 sin S_1 + a_3 cos 2S_1 + a_4 sin 2S_1 + a_5 Z_1 \\ &+ a_6 Z_i^2 + a_7 Z_i^3 + a_8 Z_i^4 + a_9 t_i)\big]\big[h_{norm} - Z_i\left(h_{norm} - h_{emp}\right) - b_0\big] + \varepsilon_i \\ & i \in \{1; N\} \end{aligned} \tag{12}$$

Once the 13 coefficients are optimised, the reversible effects (hydrostatic and thermal effects) can be computed by simulating the piezometric levels having either Z or S vary on its range, and the remaining variables being fixed to well-chosen baseline values. With HST, it is also easy to compute the corrected measurements (CM), so as to identify the temporal evolution of the observed phenomenon (in our case, the piezometric levels), by removing the reversible part from the raw measurements (6). With NL HST, the CM are equally computed by simulating the piezometric levels thanks to the estimated coefficients $\left(\widehat{b_0}, \ldots, \widehat{a_9}\right)$, having Z and S set to reference values S_{ref} and Z_{ref}, and adding the residuals $\widehat{\varepsilon_i}$, see Eq. (13).

$$\begin{aligned} CM_i = {} & \widehat{b_0} + \Big(\widehat{b_1} + \widehat{b_2} tanh\big(\widehat{a_0} + \widehat{a_1} cos S_{ref} + \widehat{a_0} sin S_{ref} + \widehat{a_3} cos 2S_{ref} \\ & + \widehat{a_4} sin 2S_{ref} + \widehat{a_5} Z_{ref} + \widehat{a_6} Z_{ref}^2 + \widehat{a_7} Z_{ref}^3 + \widehat{a_8} Z_{ref}^4 \\ & + \widehat{a_9} t_i \big(h_{norm} - Z_{ref}\big(h_{norm} - h_{emp}\big) - b_0\big) + \widehat{\varepsilon_i}\Big) \\ & i \in \{1; N\} \end{aligned} \tag{13}$$

As the model includes interaction terms between Z, S and t; the temporal evolutions are much more accurate than with the classical additive model. In compensation, since the loads have combined effects on the output of the model, it is no longer sufficient to observe the three effects independently. On the contrary, it is now required to examine a panel of combinations in order to understand how the loads jointly influence the piezometric levels. However, only the most relevant combinations will be presented in this paper.

3 Case Study

3.1 The Studied Dam

The dam is an arch dam with a height of 73 m and with a crest length of 220 m. Its thickness varies between 5 and 13 m. It was built in the 1990s and is situated in the French Massif Central. The structure is equipped with 7 piezometers situated at the rock-concrete interface, and the proposed case study focuses on the piezometer C5 which is localised in a central cantilever of the arch (Fig. 3). Because of its localisation, the piezometric levels that C5 records are directly influenced by the state of aperture of the contact.

The time series that is studied stretches from January 2002 to February 2012. It contains 321 observations, the data acquisition frequency being in average equal to one measurement every 2 weeks. The standard deviation of the raw data is equal to 14 m.

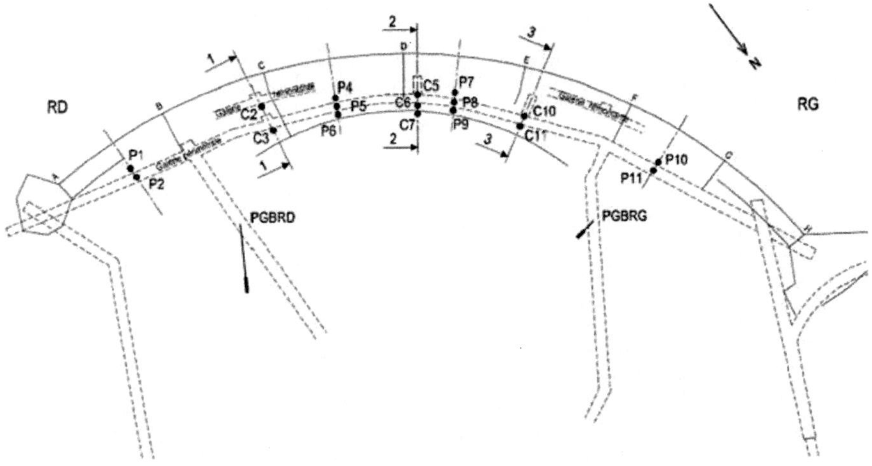

Fig. 3 Position of the studied piezometer C5 on the arch part of the dam. View from the top of the central block. Altitudes are indicated in m above the sea level

3.2 *Analysis by NL HST, Comparison to HST*

The study is encoded using the R environment. Both HST and NL HST were applied to analyse the piezometric time-series, and their prediction performances are compared. When looking at the residual scattering, HST results in a RMSE of 2 m; with NL HST, it falls to 0.8 m, which represents a reduction of 94% compared to the initial standard deviation. Including non-linearity thus improves the HST results by 60%.

When examining the response of dams to the external stresses using HST, plotting the hydrostatic law, the thermal law and the corrected measurements permits to identify the influence of each load on the dam. The principle is to look at the model's predictions having one load vary at a time, while the other two are set to fixed values. An equivalent breakdown can be carried out using the predictions from NL HST. However, with NL HST, the three loads are no longer independent, so it is of upmost interest to parameterise the varying load by one of the other two variables.

In practise, plotting the **hydrostatic effect** is made by having Z vary from its minimum to its maximum observed values. In order to see how the effect of the hydrostatic load is combined to that of the thermal state, the season is chosen as the parameter. This implies that a set of four different values is chosen for the season, corresponding to the quarters of a year centred on January, April, July and October. Finally, one date is chosen. The four corresponding predictions are plotted on the same graph (Fig. 4).

This graph reveals that including the non-linearity greatly improves the description of the effect of the hydrostatic load. Indeed, NL HST shows that when the water level rises from 910 to 940 m, the effect on the piezometric level is different, according to the season. This brings out the difference between the seasons and is related to

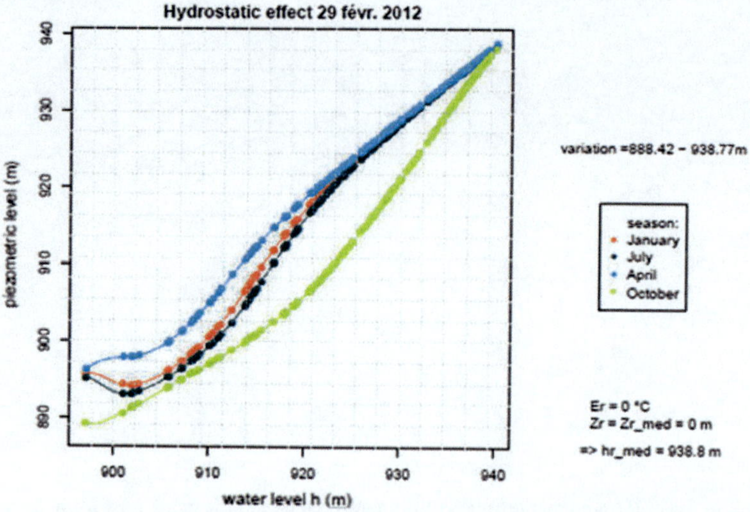

Fig. 4 The NL HST hydrostatic effect, parameterised by the season, on February 29, 2012

the thermal sensitivity of concrete: April corresponds to the colder quarter of the year (in terms of concrete temperatures), which induces the largest aperture of the interface. Conversely in October when the temperatures are the highest, the aperture is reduced by the thermal expansion of concrete, which minimises the propagation of the hydrostatic load. July and January are intermediate seasons; the corresponding hydrostatic effect is thus situated in between. As the water level rises to its maximal value, the four curves tend to converge towards a second threshold, which could correspond to a state of complete aperture of the interface.

Similarly, the **seasonal effect** is parameterized by the water level, using the same date as for the hydrostatic effect (Fig. 5).

The overall effect of the thermal state varies in a sinusoidal way, presenting a minimal effect between August and December, and a maximal effect between February and April. As mentioned previously, this is in accordance with the variation of the size of the aperture with the temperature of concrete. The threshold is also approached for water level higher than 938 m that corresponds to the point at which the interface is completely open and the thermal load no longer modulates the size of the aperture.

Finally, the **corrected measurements** are computed with NL HST, the season being the parameter, and the water level being set to its median value (Fig. 6).

As the direct consequence of the low residual modelling error obtained by NL HST compared to HST, the scattering reduction is better when using NL HST. This permits to bring out more precisely the temporal trend. Focusing on the curves corresponding to April, January and July, the carrier line shows that the piezometric levels tend to slightly increase on the period of analysis, which reveals a gradual

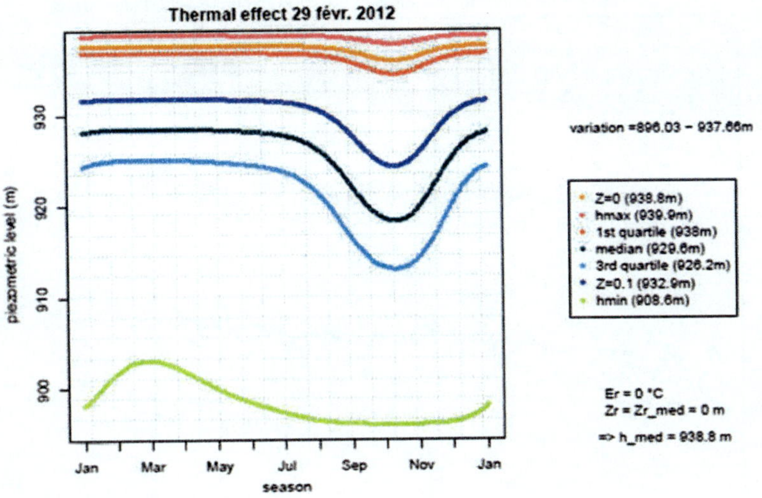

Fig. 5 The NL HST seasonal effect, parameterised by the water level (quartile values and extrema), on May 10, 2017

opening of the rock-concrete interface; this appears clearly for October, and shows a temporal evolution with a significant impact.

4 Conclusions

This comparison between HST and the NL-HST applied to analyse the state of aperture of the rock-concrete interface of an arch dam, via the piezometric measurements, highlights major improvements.

Taking into account the combined and non-additive effects of the influencing loads enables to identify clearly the contribution of each of them, and permits to overcome the difficulties that HST faced as it artificially separated the effect of each load. Additionally, it permits to take into account the thresholds that characterise the piezometry at the interface, putting forward a limit between a closed and open state of the aperture.

This model is thus an excellent alternative to HST to be used to analyse piezometric or leakages measurements at the interface. More particularly, the correction of the data proves to be a key tool for the operational monitoring, as it makes it possible to make a fine diagnosis. Conversely, it can also reveal early sign of temporal deterioration, which permits to direct servicing operations or to elaborate particular exploitation strategies.

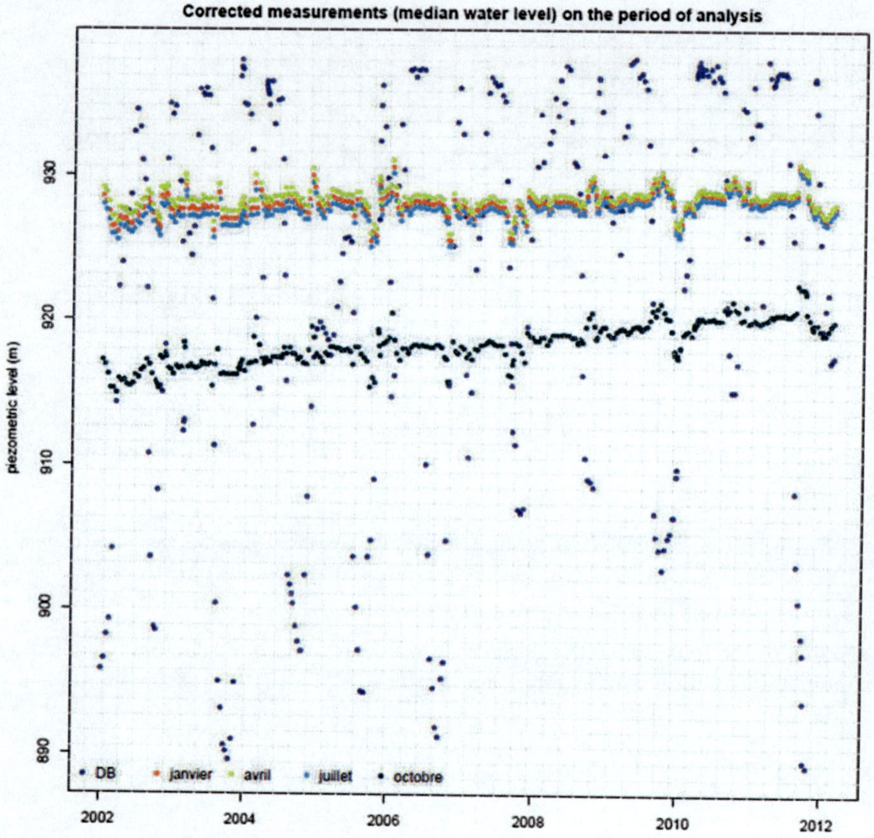

Fig. 6 The corrected measurements provided by NL HST, parameterised by the season, for a median water level; comparison to the raw data (DB, blue)

Finally, this nonlinear model is relatively easy to implement compared to more complex models such as data mining techniques, and once the model is adjusted, its interpretation is very rich.

To sum up, this comparison shows that NL HST can greatly improve the surveillance of dams and the interpretation of their behaviour, taking advantage of the rich and quality monitoring database that has been built over the past decades.

References

1. Willm G, Beaujoint N (1967) Les méthodes de surveillance des barrages au service de la production hydraulique d'Electricité de France, problèmes anciens et solutions nouvelles. In: IXth international congress on large dams, pp 529–550

2. Simon A, Royer M, Mauris F, Fabre J (2013) Analysis and interpretation of dam measurements using artificial neural networks. In: 9th ICOLD European club symposium
3. Levenberg K (1944) A method for the solution of certain problems in least squares. Q Appl Math 2:164–168
4. Marquardt DW (1963) An algorithm for least-squares estimation of nonlinear parameters. J Soc Ind Appl Math 11:431–441

Direct Method for Dynamic Soil-Structure Interaction Based on Seismic Inertia Forces

D. Froio, A. U. Bariletti, M. Eusebio, R. Previtali, and E. Rizzi

Abstract The seismic response of dams is significantly influenced by dynamic Soil-Structure Interaction (SSI) phenomena. Most of the dedicated methods to perform SSI analyses may be classified within two main approaches, namely, the so-called Direct Method and the Substructure Method. The present discussion focuses on a specific formulation of the Direct Method, valid for uniform free-field motion, and less-often adopted in the main stream practical applications, by which the seismic action is converted into inertial loads directly applied to the structure. A Substructure Method implementation is also considered, in canonical form. First, the equivalence of the Direct Method and Substructure Method approaches is theoretically discussed; then, the seismic responses of the Pine Flat dam gathered from both methods are compared, to confirm the analytical outcomes. This study demonstrates that the adopted Direct Method approach provides a wholly equivalent dynamic response, to that from the Substructure Method, with the advantages of being robust enough and rather efficient for self-implementations, possibly employing commercial FEM computer programs.

Keywords Seismic analysis of dams · Soil-Structure Interaction (SSI) · Direct method · Uniform free-field motion · Pine Flat Dam

1 Introduction

In simplified dynamic structural analyses, a structure fixed at the ground level and subjected to a base motion is normally assumed; alternatively, a massless idealization of the foundation medium, which only accounts for the foundation flexibility, is very popular in the engineering practice, thus ignoring any wave propagation phenomena within the foundation medium.

D. Froio · A. U. Bariletti · M. Eusebio (✉) · R. Previtali
Engineering & Environment (ISMES Division), CESI S.p.a, Bergamo, Italy
e-mail: marina.eusebio@cesi.it

E. Rizzi
Dipartimento di Ingegneria e Scienze Applicate, Università degli Studi di Bergamo, Bergamo, Italy

G. Bolzon et al. (eds.), *Numerical Analysis of Dams*, Lecture Notes in Civil Engineering 91, https://doi.org/10.1007/978-3-030-51085-5_45

For slender structures anchored to monolithic rock or firm ground (very stiff foundation), the stress resultants of the inertial loads acting on the structure do not lead to any additional deformation of the base, so that the dynamic response of the soil-structure interface results equal to that of the free-field. In that case, the seismic input may be directly applied to the soil-structure interface and the response of the structure only depends on its structural properties.

Nevertheless, when seismic waves impinge on an extended foundation displaying a stiffness much higher than that of the surrounding soil, the foundation cannot accommodate the imposed spatial variation of the motion; as a result, the free-field motion is distorted by the effect of the waves scattered by the foundation (kinematic interaction), as well as those generated by the structure vibration transmitted to the ground (inertial interaction). In addition, accounting for the foundation flexibility may modify to an important extent the frequency response function of the whole system, thus modifying the structural response to dynamic loads of any kind.

The need for analyzing a given structure not as if it were isolated, but rather as a part of a seismic environment and of an ensemble of soil and other structures interacting between them has made dynamic Soil-Structure Interaction (SSI) analysis imperative for an increasing range of stiff and massive structures, such as dams, especially concerning possible catastrophic consequences of structural failure in case of strong earthquakes.

The complex wave mechanism characterizing a dynamic SSI problem for earthquake excitation may be briefly described as follows. When waves propagating into the soil reach the structures, scattering occurs, causing actions on the structure. Because of such dynamic loads, the structure starts vibrating and radiating waves into the soil, thus relieving the structural stress. All these mechanical processes, even from the soil side, are fully coupled.

Energy dissipation by wave propagation away from the structure (radiation) usually constitutes the main damping effect of SSI analysis. In fact, only outgoing waves can be propagated from a structure within an unbounded medium, such that, once they leave the structure, they will not return backward. This avoids an energy build up, resulting in a global damping mechanism. Thus, models allowing for an appropriate wave radiation must be selected.

The complexity of SSI analysis, due to its multidisciplinary nature and to the need of considering both bounded and unbounded media requires sophisticated numerical treatments. Most of the dedicated techniques may be classified within two main approaches, namely, the Substructure Method (SM) and the Direct Method (DM). In the SM the soil-structure system is considered as composed by two substructures, the structure, possibly including a neighboring soil portion, and the remaining unbounded soil; once the dynamic stiffness matrix of the unbounded soil region is calculated, it is inserted into the equations of motion of the finite element model of the structure [8]. In the DM, both the structure and the soil are modelled by finite elements, and special boundaries are imposed along the boundary of the truncated soil model to simulate the wave radiation towards infinity [9].

Although the analysis of SSI problems displays a long history and appears nowadays rather well recognized, there still exists the need of a comprehensive discussion and confrontation, especially for practical applications, because of the large amount of solution strategies and computational techniques developed in the last fifty years, towards identifying a unified approach capable of encompassing the various possible formulations in a systematic manner.

In the present analysis, a specific formulation of the DM is considered, valid for linear SSI analysis and spatially uniform seismic excitations, by which the seismic action is converted into inertial loads directly applied to the structure; such a formulation, seldom adopted in the main-stream practical applications, provides equivalent results than for a well-known SM implementation, with the advantages of being robust enough and efficient for self-implementations within FEM commercial computer programs.

Hence, the aims of the present study are as follows: (i) to rigorously demonstrate the equivalence of the SM and the DM in exact terms, meaning that the same response is obtained if both methods are consistently implemented; (ii) to inspect such an equivalence, by resorting to a specific numerical example, the response of the Pine Flat Dam to the S69E component of Taft earthquake, a well-known case study in the SSI literature [5].

The present paper is organized as follows. The formulations of the SM and the DM for linear dynamic SSI analysis and their exact equivalence are the subjects of Sect. 2. A consistent validation comparison between the numerical outcomes of the autonomously reimplemented SM and DM is produced in Sect. 3. Finally, conclusive considerations are summarized in Sect. 4.

2 Formulation of Soil-Structure Interaction Problems

In the present section, the equivalence of the SM and the DM, both consistently formulated, is demonstrated in relative response components. Only the general case of flexible foundations is considered. The soil-structure is assumed to be idealized as an assembly of finite elements.

Let the soil-structure system shown in Fig. 1a be divided into three sets of nodes. The nodal displacements of the structure at the soil-structure interface with respect to a fixed Cartesian coordinate system are identified with vector $\mathbf{u}_i^t$ (total displacements); the other displacements of the structure are denoted by $\mathbf{u}_s^t$. The ground displacements are indicated by $\mathbf{u}_g^t$.

By assuming a lumped mass formulation and no external dynamical forces acting on the system, the linear equations of motion of the soil-structure system in the time domain become

$$\begin{bmatrix} \mathbf{M}_{ss}^S & \mathbf{0} & \mathbf{0} \\ \mathbf{0} & \mathbf{M}_{ii}^{S+G} & \mathbf{0} \\ \mathbf{0} & \mathbf{0} & \mathbf{M}_{\mathbf{gg}}^{\mathbf{G}} \end{bmatrix} \begin{Bmatrix} \ddot{\mathbf{u}}_s^t \\ \ddot{\mathbf{u}}_i^t \\ \ddot{\mathbf{u}}_g^t \end{Bmatrix} + \begin{bmatrix} \mathbf{K}_{ss}^S & \mathbf{K}_{si}^S & \mathbf{0} \\ \mathbf{K}_{is}^S & \mathbf{K}_{ii}^{S+G} & \mathbf{K}_{ig}^G \\ \mathbf{0} & \mathbf{K}_{gi}^G & \mathbf{K}_{gg}^G \end{bmatrix} \begin{Bmatrix} \mathbf{u}_s^t \\ \mathbf{u}_i^t \\ \mathbf{u}_g^t \end{Bmatrix} = \begin{Bmatrix} \mathbf{0} \\ \mathbf{0} \\ \mathbf{0} \end{Bmatrix}; \qquad (1)$$

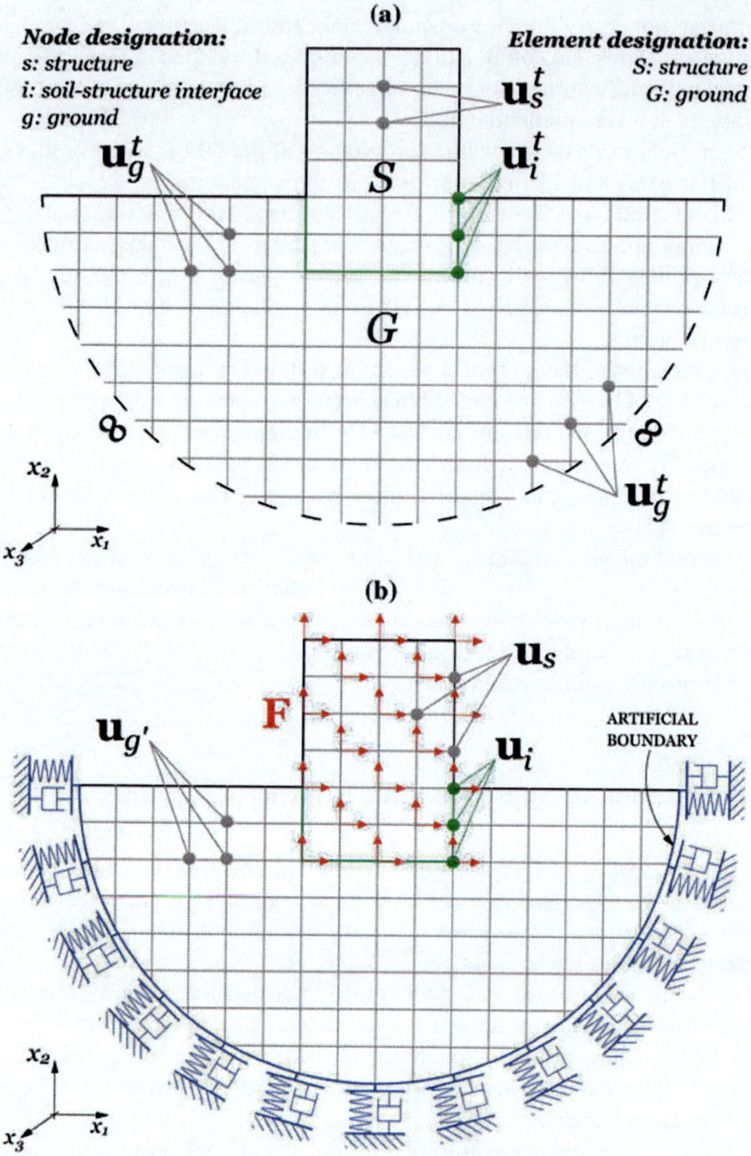

Fig. 1 Reference subsystems for SSI analysis (**a**) and schematic model for the DM (**b**)

where the mass (M) and stiffness (K) matrices have been partitioned according to the partition of vector u^t. Notice that matrices M_{ii}^{S+G} and K_{ii}^{S+G} are the sum of the corresponding contributions from both the structure (S) and the soil (G) finite elements; further, all characteristic matrices referred to the soil nodes in Eq. (1) have theoretically infinite dimensions, the soil being an unbounded medium (infinite finite element mesh).

Up to this point, everything has been developed within the time domain. A more efficient alternative is represented by formulating Eq. (1) in the frequency domain, as follows

$$\begin{bmatrix} \overline{\mathrm{K}}_{ss}^{S} & \overline{\mathrm{K}}_{si}^{S} & \mathbf{0} \\ \overline{\mathrm{K}}_{is}^{S} & \overline{\mathrm{K}}_{ii}^{S+G} & \overline{\mathrm{K}}_{ig}^{G} \\ \mathbf{0} & \overline{\mathrm{K}}_{gi}^{G} & \overline{\mathrm{K}}_{gg}^{G} \end{bmatrix} \begin{Bmatrix} \overline{\mathrm{u}}_s^t \\ \overline{\mathrm{u}}_i^t \\ \overline{\mathrm{u}}_g^t \end{Bmatrix} = \begin{Bmatrix} \mathbf{0} \\ \mathbf{0} \\ \mathbf{0} \end{Bmatrix}; \tag{2}$$

where $\overline{\mathrm{u}}$ is the Fourier transform of u, $\overline{\mathrm{K}}_{kl} = \mathrm{K}_{kl} - \omega^2 \mathrm{M}_{kl}$ is the dynamic stiffness (impedance) matrix and ω is the angular frequency.

Suppose the system be subjected to a seismic motion u^f. Let also $\mathrm{u} = \mathrm{u}^t - \mathrm{u}^f$ be the displacements of the system relative to the ground excitation input. In general, several definitions of u^f are allowed, from which the physical interpretation of u follows.

According to a well-established practice in the SSI literature, input motion u^f is usually provided by the so-called free-field ground motion, i.e. the response that would have occurred before the placing of the structure on the site. Accordingly, vector u stands for the interaction displacement, i.e. the amount of the total response due to the interaction effects induced by the actual presence of the structure.

According to the definition of the free-field motion, the governing equations of motion of soil system G subjected to u^f may be written as

$$\begin{bmatrix} \overline{\mathrm{K}}_{ii}^{G} & \overline{\mathrm{K}}_{ig}^{G} \\ \overline{\mathrm{K}}_{gi}^{G} & \overline{\mathrm{K}}_{gg}^{G} \end{bmatrix} \begin{Bmatrix} \overline{\mathrm{u}}_i^f \\ \overline{\mathrm{u}}_g^f \end{Bmatrix} = \begin{Bmatrix} \mathbf{0} \\ \mathbf{0} \end{Bmatrix}; \tag{3}$$

where $\overline{\mathrm{u}}^f$ is the Fourier transform of free-field motion u^f. Then, by combining Eqs. (2), (3) the equation of motion for relative displacement u in the frequency domain follows

$$\begin{bmatrix} \overline{\mathrm{K}}_{ss}^{S} & \overline{\mathrm{K}}_{si}^{S} & \mathbf{0} \\ \overline{\mathrm{K}}_{is}^{S} & \overline{\mathrm{K}}_{ii}^{S+G} & \overline{\mathrm{K}}_{ig}^{G} \\ \mathbf{0} & \overline{\mathrm{K}}_{gi}^{G} & \overline{\mathrm{K}}_{gg}^{G} \end{bmatrix} \begin{Bmatrix} \overline{\mathrm{u}}_s \\ \overline{\mathrm{u}}_i \\ \overline{\mathrm{u}}_g \end{Bmatrix} = - \begin{Bmatrix} \overline{\mathrm{K}}_{ss}^{S} \overline{\mathrm{u}}_s^f + \overline{\mathrm{K}}_{si}^{S} \overline{\mathrm{u}}_i^f \\ \overline{\mathrm{K}}_{is}^{S} \overline{\mathrm{u}}_s^f + \overline{\mathrm{K}}_{ii}^{S} \overline{\mathrm{u}}_i^f \\ \mathbf{0} \end{Bmatrix}. \tag{4}$$

Notice that, by writing the governing equation relative to the free-field motion, only the degrees of freedom of the structure are excited by dynamic forces.

Since the free-field motion does not account for the actual presence of the structure, components $\overline{\mathrm{u}}_s^f$ in the second member of Eq. (4), i.e. the free-field motion assigned to the nodes of the structure not belonging to the soil-structure interface, may be chosen as completely arbitrary [1]; to define $\overline{\mathrm{u}}_s^f$, the following pseudo-static transmission is usually adopted

$$\mathrm{K}_{ss}^S \overline{\mathrm{u}}_s^f + \mathrm{K}_{si}^S \overline{\mathrm{u}}_i^f = \mathbf{0}; \tag{5}$$

such that the first component in the right-hand side of Eq. (4) vanishes, resulting in the following general formulation of the SSI problem, written for interaction displacements in the frequency domain (the time domain version is analogous):

$$\begin{bmatrix} \overline{\mathrm{K}}_{ss}^S & \overline{\mathrm{K}}_{si}^S & \mathbf{0} \\ \overline{\mathrm{K}}_{is}^S & \overline{\mathrm{K}}_{ii}^{S+G} & \overline{\mathrm{K}}_{ig}^G \\ \mathbf{0} & \overline{\mathrm{K}}_{gi}^G & \overline{\mathrm{K}}_{gg}^G \end{bmatrix} \begin{Bmatrix} \overline{\mathrm{u}}_s \\ \overline{\mathrm{u}}_i \\ \overline{\mathrm{u}}_g \end{Bmatrix} = - \begin{Bmatrix} -\omega^2 \mathrm{M}_{ss}^S \overline{\mathrm{u}}_s^f \\ \overline{\mathrm{K}}_{is}^S \overline{\mathrm{u}}_s^f + \overline{\mathrm{K}}_{ii}^S \overline{\mathrm{u}}_i^f \\ \mathbf{0} \end{Bmatrix}. \tag{6}$$

Up to this point it appears fundamental to underline that dynamic stiffness submatrices $\overline{\mathrm{K}}_{ig}^G$, $\overline{\mathrm{K}}_{gi}^G$ and $\overline{\mathrm{K}}_{gg}^G$ in Eq. (6) have theoretically infinite dimensions, the soil being an unbounded medium, and, thus, Eq. (6) cannot be directly employed to solve the SSI problem. The computational strategy for handling an unbounded soil is the main and unique difference among the SM and the DM, as it will be discussed in the following.

2.1 *Substructure Method in Relative Components*

By performing a dynamic condensation of the contribution of soil degrees of freedom (g) in Eq. (6), reaction forces $\overline{\mathrm{K}}_{ig}^G \overline{\mathrm{u}}_g$ exerted on the structure by the soil may be directly correlated with transformed displacements at the soil-structure interface $\overline{\mathrm{u}}_i$:

$$\begin{bmatrix} \overline{\mathrm{K}}_{ss}^S & \overline{\mathrm{K}}_{si}^S \\ \overline{\mathrm{K}}_{is}^S & \overline{\mathrm{K}}_{ii}^S + \mathrm{S}_{ii}^G \end{bmatrix} \begin{Bmatrix} \overline{\mathrm{u}}_s \\ \overline{\mathrm{u}}_h \end{Bmatrix} = - \begin{Bmatrix} -\omega^2 \mathrm{M}_{ss}^S \overline{\mathrm{u}}_s^f \\ \overline{\mathrm{K}}_{is}^S \overline{\mathrm{u}}_s^f + \overline{\mathrm{K}}_{ii}^S \overline{\mathrm{u}}_i^f \end{Bmatrix}; \tag{7}$$

where $\mathrm{S}_{ii}^G = \overline{\mathrm{K}}_{ii}^G + \overline{\mathrm{K}}_{ig}^G {\overline{\mathrm{K}}_{gg}^G}^{-1} \overline{\mathrm{K}}_{gi}^G$ is the dynamic stiffness matrix of the unbounded soil region, identifying the action of the unbounded soil on the degrees of freedom of the soil-structure interface.

Impedance matrix S_{ii}^G may be obtained by closed-form semi-analytical solutions for regular homogeneous or horizontally-layered half-spaces supporting structures without embedment, while advanced numerical techniques are still needed in the case of more complicated geometries, due to the lack of analytical formulations.

By rewriting Eq. (7) in total displacements and accounting for Eq. (5), the usual formulation of the SM in the frequency domain may be retrieved [12]:

$$\begin{bmatrix} \overline{\mathbf{K}}^S_{ss} & \overline{\mathbf{K}}^S_{si} \\ \overline{\mathbf{K}}^S_{is} & \overline{\mathbf{K}}^S_{ii} + \mathbf{S}^G_{ii} \end{bmatrix} \begin{Bmatrix} \overline{\mathbf{u}}^t_s \\ \overline{\mathbf{u}}^t_i \end{Bmatrix} = \begin{Bmatrix} \mathbf{0} \\ \mathbf{S}^G_{ii}\overline{\mathbf{u}}^f_i \end{Bmatrix}. \tag{8}$$

Thus, Eq. (7) denotes the formulation of the SM in relative coordinates. In the following section, the equivalence of Eq. (7) with the formulation of the DM will be established.

2.2 *Direct Method in Relative Components*

In the DM, both the structure and a portion of the soil are modelled by FEM [2]; to keep the computation feasible, the unbounded soil domain must be truncated to a finite computational domain.

Let one examine the interaction problem, in which suitable transmitting boundary conditions are imposed along the artificial border of the soil model, trying to simulate the wave propagation towards infinity. For a truncated model, Eq. (6) becomes

$$\begin{bmatrix} \overline{\mathbf{K}}^S_{ss} & \overline{\mathbf{K}}^S_{si} & \mathbf{0} \\ \overline{\mathbf{K}}^S_{is} & \overline{\mathbf{K}}^{S+G}_{ii} & \overline{\mathbf{K}}'^G_{ig'} \\ \mathbf{0} & \overline{\mathbf{K}}'^G_{g'i} & \overline{\mathbf{K}}'^G_{g'g'} \end{bmatrix} \begin{Bmatrix} \overline{\mathbf{u}}_s \\ \overline{\mathbf{u}}_h \\ \overline{\mathbf{u}}_{g'} \end{Bmatrix} = - \begin{Bmatrix} -\omega^2 \mathbf{M}^S_{ss}\overline{\mathbf{u}}^f_s \\ \overline{\mathbf{K}}^S_{is}\overline{\mathbf{u}}^f_s + \overline{\mathbf{K}}^S_{ii}\overline{\mathbf{u}}^f_i \\ \mathbf{0} \end{Bmatrix}; \tag{9}$$

with apex $'$ pointing out to the difference between the unbounded model (Fig. 1a) and the truncated one with transmitting boundaries (Fig. 1b); further, by placing the artificial boundary beyond the soil-structure interface, it follows that $\overline{\mathbf{K}}'^G_{ig'} = \overline{\mathbf{K}}^G_{ig}$ and $\overline{\mathbf{K}}'^G_{g'i} = \overline{\mathbf{K}}^G_{gi}$.

Then, by again performing a dynamic condensation of the contribution of the ground degrees of freedom in Eq. (9), the final equations of motion of the structure are obtained

$$\begin{bmatrix} \overline{\mathbf{K}}^S_{ss} & \overline{\mathbf{K}}^S_{si} \\ \overline{\mathbf{K}}^S_{is} & \overline{\mathbf{K}}^S_{ii} + \overline{\mathbf{S}}'^G_{ii} \end{bmatrix} \begin{Bmatrix} \overline{\mathbf{u}}_s \\ \overline{\mathbf{u}}_i \end{Bmatrix} = - \begin{Bmatrix} -\omega^2 \mathbf{M}^S_{ss}\overline{\mathbf{u}}^f_s \\ \overline{\mathbf{K}}^S_{is}\overline{\mathbf{u}}^f_s + \overline{\mathbf{K}}^S_{ii}\overline{\mathbf{u}}^f_i \end{Bmatrix}; \tag{10}$$

where $\overline{\mathbf{S}}'^G_{ii} = \overline{\mathbf{K}}^G_{ii} + \overline{\mathbf{K}}'^G_{ig'}\overline{\mathbf{K}}'^{G\ -1}_{g'g'}\overline{\mathbf{K}}'^G_{g'i}$ is the condensed dynamic stiffness matrix at the soil-structure interface of the truncated soil model with transmitting boundaries.

Therefore, under the assumption that the transmitting boundary conditions of the truncated model can accurately reproduce the mechanical behavior of the corresponding unbounded system, i.e. $\overline{\mathbf{S}}'^G_{ii} = \overline{\mathbf{S}}^G_{ii}$, Eq. (10), derived from the application of the DM, coincides with Eq. (7) of the SM in relative coordinates. The corresponding equivalence for the total response is straightforward.

Hence, the SM and the DM are shown to be equivalent, if both methods are coherently implemented, e.g. the material damping mechanism is consistently represented, and if an effective and accurate transmitting boundary is available to reproduce the wave radiation towards infinity.

2.3 Direct Method for Uniform Free-Field Motion

Let one suppose the soil-structure system to be subjected to a uniform free-field motion. The latter condition is usually assumed in the engineering practice,[1] since very little is known about the differences between the ground motion at two adjacent points during the same earthquake, due to the lack of suitable ground motion records at several points in the neighbourhood of the structure.

Let v^f be the vector collecting the three components of the free-field motion along the coordinate axes; the 1D and 2D cases may be obtained by deleting the corresponding components of v^f. Then, the spatially uniform free-field displacement vector in the frequency domain may be written as

$$\bar{u}^f = \begin{Bmatrix} \bar{u}_s^f \\ \bar{u}_i^f \\ \bar{u}_g^f \end{Bmatrix} = \begin{bmatrix} \mathbf{I} & \mathbf{0} \\ \mathbf{I} & \mathbf{0} \\ \mathbf{0} & \mathbf{I} \end{bmatrix} \begin{bmatrix} r_i^{x_1} & r_i^{x_2} & r_i^{x_3} \\ r_g^{x_1} & r_g^{x_2} & r_g^{x_3} \end{bmatrix} \begin{Bmatrix} v_{x_1}^f \\ v_{x_2}^f \\ v_{x_3}^f \end{Bmatrix}; \tag{11}$$

where incidence matrices r_i and r_g are two matrices containing only zeros and ones and expressing the rigid-body nature of the surface ground motion. Notice that, although $\bar{u}_s^f$ has not be derived from Eq. (5), which would have provided a null vector ($\bar{u}_s^f = \mathbf{0}$), Eq. (5) still continues being satisfied, a uniform free-field being a rigid-body motion.

Substitution of Eq. (11) into the second component of the forcing term of Eq. (9) yields

$$\overline{K}_{is}^S \bar{u}_s^f + \overline{K}_{ii}^S \bar{u}_i^f = -\omega^2 M_{ii}^S; \tag{12}$$

because, again, rigid-body motions lead to vanishing internal forces. Hence, in the case of a uniform free-field, the frequency domain formulation of the DM for the relative response reads

$$\begin{bmatrix} \overline{K}_{ss}^S & \overline{K}_{si}^S & \mathbf{0} \\ \overline{K}_{is}^S & \overline{K}_{ii}^{S+G} & \overline{K'}_{ig'}^G \\ \mathbf{0} & \overline{K'}_{g'i}^G & \overline{K'}_{g'g'}^G \end{bmatrix} \begin{Bmatrix} \bar{u}_s \\ \bar{u}_i \\ \bar{u}_{g'} \end{Bmatrix} = - \begin{Bmatrix} -\omega^2 M_{ss}^S \bar{u}_s^f \\ -\omega^2 M_{ii}^S \bar{u}_i^f \\ \mathbf{0} \end{Bmatrix}. \tag{13}$$

[1] The latter assumption should instead be considered inappropriate for long base and/or deeply embedded structures.

Therefore, the relative displacements can be computed by using an interaction model, where all the exciting dynamic forces are the product of the inertial properties of the structure and of the free-field acceleration.

3 Numerical Simulations

To corroborate the equivalence between the SM and the DM, the latter formulated according to Eq. (13), a numerical demonstration is presented in this section with reference to the dynamic SSI analysis of the Pine Flat Dam (Fig. 2) on Kings' River near Fresno, California, to the S69E horizontal component (Fig. 3) of the ground acceleration recorded at the Taft Lincoln School Tunnel during the Kern County, California, earthquake of 21 July 1952.

A two-dimensional plain strain model of the tallest non-overflow monolith of Pine Flat Dam, resting on a half-plane, is considered here. Since the present analysis focuses on the modelling of dynamic SSI effects, for convenience, the impounding water has not been included in the analysis.

The dam concrete and the underlying rock (granites and basalts) are assumed to behave as homogeneous, isotropic, linear viscoelastic solid materials; their mechanical properties, listed in Table 1, correspond to those adopted in [5]; in particular,

Fig. 2 Downstream view of Pine Flat Dam

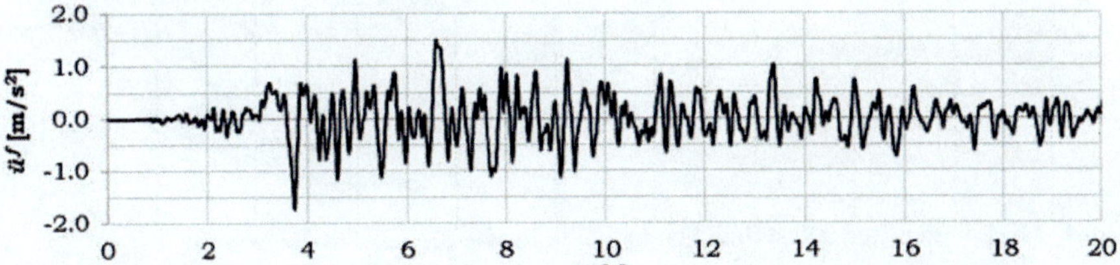

Fig. 3 S69E component of Taft ground acceleration, 21 July 1952 (first 20 s)

Table 1 Reference mechanical properties of the system

Parameter	Concrete	Rock
Dynamic Young's modulus [GPa]	22.41	68.95
Dynamic Poisson's ratio	0.20	0.33
Mass density [kg/m^3]	2483	2643
Hysteretic damping factor	0.10	0.10

the elastic modulus of concrete was determined by forced vibration tests on the dam [11].[2]

By assuming a constant hysteretic damping model, each dynamic stiffness matrix becomes

$$\overline{\mathrm{K}}_{kl} = \mathrm{K}_{kl}(1 + i\eta) - \omega^2 \mathrm{M}_{kl}; \tag{14}$$

where η is the hysteretic damping coefficient, selected equal to 0.1 for both the dam and the soil, which approximately corresponds to a 5% (viscous) damping factor for all the modes of vibration [5].

Since a hysteretic damping model has been adopted, the seismic analyses have been performed in the frequency domain, by means of a computer program autonomously implemented within a MatLab environment [10] and a FEM modelization developed within COMSOL Multiphysics [4], for the SM and the DM, respectively. Then, response time histories have been obtained by applying the inverse Fourier transform, from the frequency to the time domain.

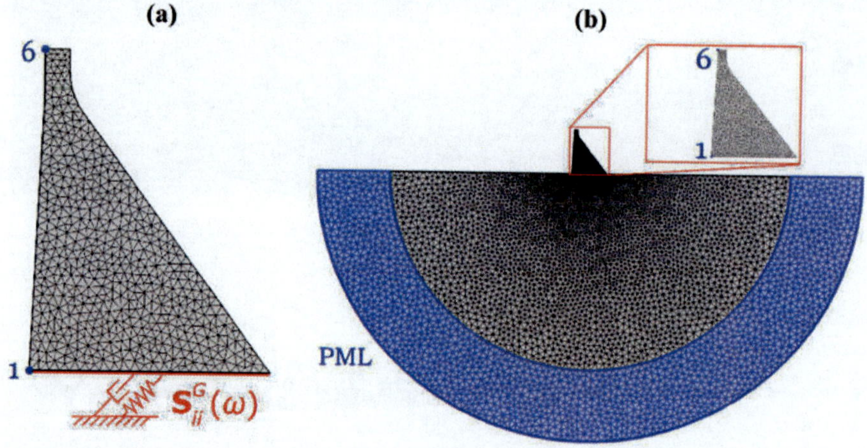

Fig. 4 FEM mesh of the tallest Pine Flat monolith for the SM (**a**) and the DM (**b**)

[2]The numerical model used to estimate the elastic modulus of concrete did not include soil-structure interaction.

The FEM idealization of the system employed in the SM (Fig. 4a), comprises 972 linear triangular elements (maximum mesh size of 4 m) and 536 nodes, corresponding to a moderately refined mesh. By maintaining the same dam mesh, the FEM model for the DM (Fig. 4b), including also a portion of soil, has been assembled; the model consists of 10275 linear triangular elements and 5284 nodes.

The dynamic response of the soil-dam system was computed for the excitation frequency range between 0 and 50 Hz, with a frequency step of 0.012 Hz, which looks adequate for the recorded Taft ground motion and for describing the most relevant wave components of the system. By taking advantage of the frequency response analysis employed in the present investigation, the wave radiation condition in the DM model has been simulated by a Perfectly Matched Layer (PML) [3] (for presentation and perspectives on PML treatment, see also [6, 7]) at the external soil domain of the COMSOL model (Fig. 4b).

Using the early mentioned numerical implementations of the SM and the DM, a complete frequency domain analysis of the dynamic interaction response (relative to the free-field) of the dam monolith to the action of the horizontal components of the Taft ground motion was carried out. The dynamic responses obtained from both methods are illustrated and compared.

For brevity, the reported results of the computer analyses consist of the horizontal displacements and accelerations at base point 1 and at crest point 6 (Fig. 4);

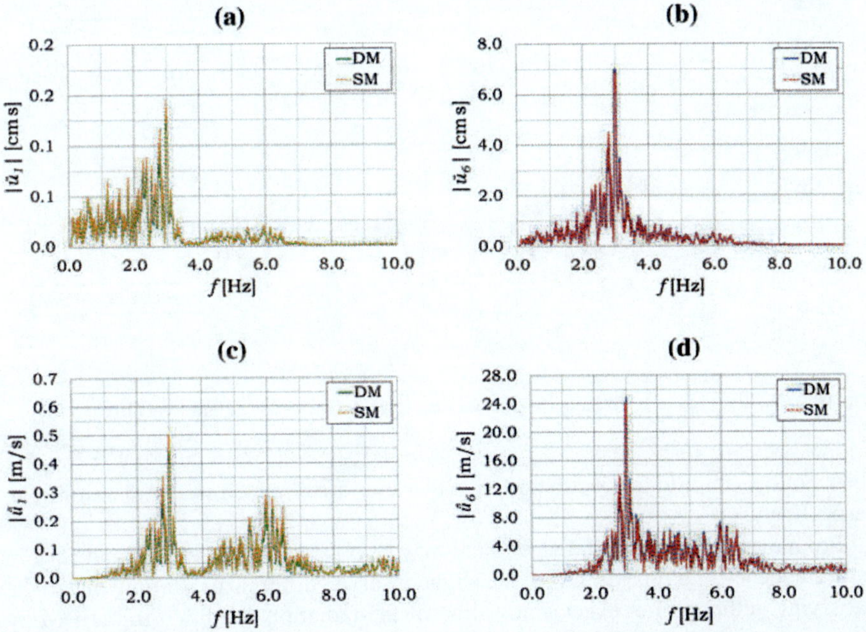

Fig. 5 Frequency response spectrum amplitude of horizontal displacements (1 (**a**), 6 (**b**)) and accelerations (1 (**c**), 6 (**d**))

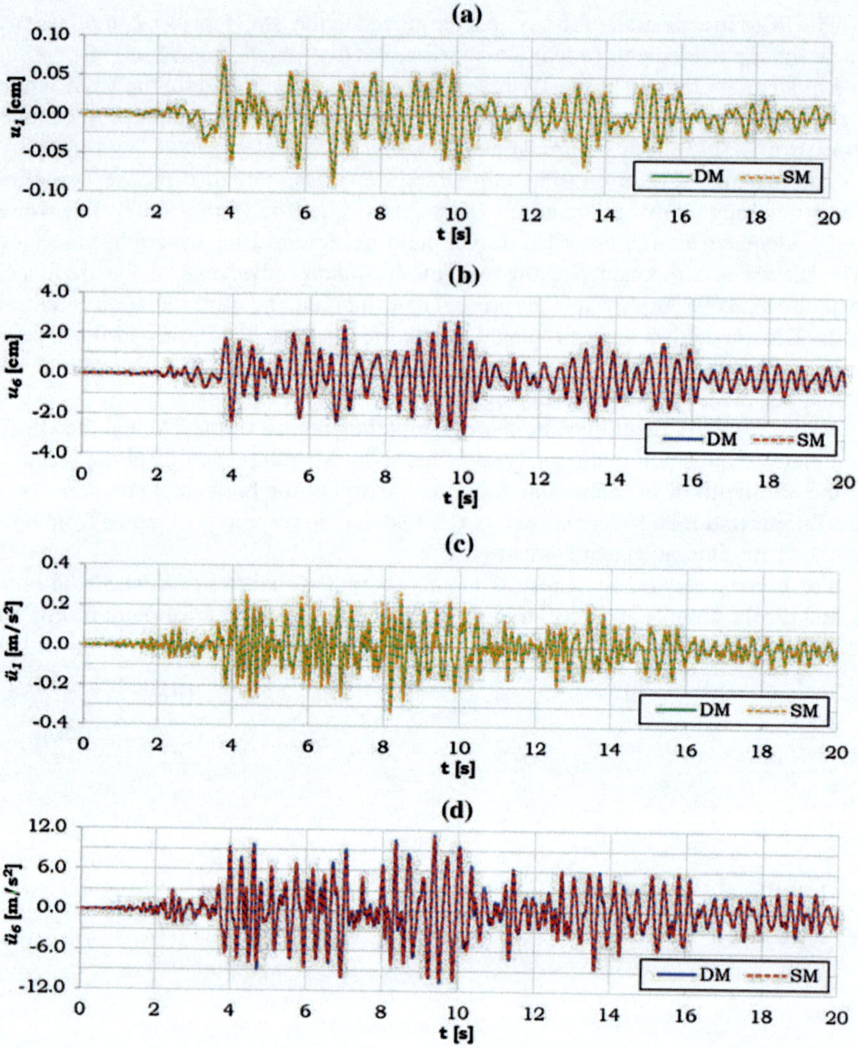

Fig. 6 Time histories of horizontal displacements (1 (**a**), 6 (**b**)) and accelerations (1 (**c**), 6 (**d**))

their Fourier amplitude spectra and time histories are depicted in Fig. 5 and Fig. 6, respectively.

From these figures, a good agreement may be observed between SM and DM, in both frequency and time domains; the slight observed differences are only due to the satisfying, although not exact, imposition of the radiation condition in the DM by the action of the PML.

The response spectrum of the interaction response is located into the main frequency band defined by the frequency content of the seismic excitation, between about 0.5 and 10 Hz, with a slight dynamic amplification at around 3 and 6 Hz, in correspondence with the lowest natural frequency of the dam on a rigid foundation, pointing out to an amplification which is mainly due to the dynamic properties of the structure.

Looking at the acceleration response of the base point (Fig. 6c), it may be observed that it represents a small fraction of the free-field motion (Fig. 3), meaning that the dam base response appears substantially equal to the undisturbed free-field motion.

This was expected, the adopted foundation being rather rigid; in fact, SSI effects are more pronounced in case of very stiff and massive structures lying on loose foundations [12]. Nevertheless, SSI effects cannot be overlooked since the induced damping mechanism by wave radiation still constitutes a key characteristic feature of the soil-structure system, which cannot be emulated by any other simplified model, such as the massless one.

4 Conclusions

In this paper, a specific version of the DM, by which the seismic action is converted into inertial body loads directly applied to the structure, is illustrated, with focus on the seismic analysis of a specific gravity dam.

The present DM approach, which applies in the recurring case of a uniform free-field motion, displays the main advantage of making the formulation of seismic SSI problems easier, especially within modern commercial FEM codes.

To verify the rigor of the method, a comparison assessment with respect to widespread SM has been carried out, allowing to affirm the following statements.

The analytical derivation of Sect. 2 has pointed out that the SM and the DM become fully equivalent, when they are consistently implemented, namely when (i) the transmitting boundary employed in the DM is really able to simulate the radiation condition and (ii) the same material damping model is employed in both methods.

The numerical results on the benchmark problem of the Pine Flat Dam gathered from both methods (Sect. 3) have shown a perfect match, for both response time histories and Fourier spectra at the dam's base and crest, confirming the exact theoretical derivations of Sect. 2.

Thus, despite a major branch of knowledge of the SM for the solution of dynamic SSI problems, the present analysis shows that the DM still constitutes a viable and effective alternative toward practical and readable simulations, as testified by the considered example and derived results.

References

1. Aydinoğlu MN (1980) Unified formulations for soil-structure interaction. In: 7th World Conference on Earthquake Engineering, vol 6. Turkish National Committee on Earthquake Engineering, Istanbul, Turkey, pp 121–128

2. Basu U, Chopra AK (2003) Perfectly matched layers for time-harmonic elastodynamics of unbounded domains: theory and finite-element implementation. Comput Methods Appl Mech Eng 192(11–12):1337–1375
3. Bayo E, Wilson EL (1983) Numerical techniques for the evaluation of soil structure interaction effects in the time domain. Technical Report 83-4, Earthquake Engineering Research Center (EERC), Berkeley, California
4. COMSOL Multiphysics (2018) Version 5.4. Stockholm, Sweden
5. Fenves G, Chopra AK (1984) Earthquake analysis of concrete gravity dams including reservoir bottom absorption and dam-water-foundation rock interaction. Earthq Eng Struct Dyn 12(5):663–680
6. Froio D, Rizzi E, Simões FMF, Pinto da Costa A (2020) DLSFEM–PML formulation for the steady-state response of a taut string on visco-elastic support under moving load. Meccanica 55(4):765–790
7. Froio D, Rizzi E, Simões FMF, Pinto da Costa A (2020) True PML for elastic-supported beam steady-state vibration under moving load by DLSFEM formulation, Computers & Structures. https://doi.org/10.1016/j.compstruc.2020.106295
8. Gutierrez JA, Chopra AK (1984) A substructure method for earthquake analysis of structures including structure-soil interaction. Earthq Eng Struct Dyn 6(1):51–69
9. Kausel E (1988) Local transmitting boundaries. J Eng Mech (ASCE) 114(6):1011–1027
10. MatLab (2018) Version R2018b. Natick, Massachusetts
11. Rea D, Liaw C-Y, Chopra AK (1975) Mathematical models for the dynamic analysis of concrete gravity dams. Earthq Eng Struct Dyn 3(3):249–258
12. Wolf JP (1985) Dynamic soil-structure interaction, 1st edn. Prentice-Hall, Englewood Cliffs, New Jersey

The Influence of Microscopic Parameters on Deformation Properties of Rockfill Materials

C. Ma, G. Zenz, E. J. Staudacher, and J. Yang

Abstract To analyze the influence of microscopic parameters on deformation properties of rockfill materials in triaxial test simulation with discrete element method (DEM), a control variable method was adopted to determine the influence on stress-strain and volumetric strain curves. In addition, the influence mechanism was explained. Firstly, the characteristics and development trend of triaxial test simulations of rockfill materials were summarized in detail, and it is determined that the contact bond model is the best choice for simulating triaxial tests of rockfill materials. Subsequently, a detailed DEM model of the triaxial test of rockfill materials was established to study the influences and mechanism of friction coefficient, normal and shear stiffness, normal and shear bond strength. Finally, the relationship between deformation properties and broken bond of rockfill materials was investigated. It is considered that the broken bonds of rockfill materials were obviously influenced by microscopic parameters, and the deformation properties were significantly affected by broken bonds. Therefore, in this paper, the influence of microscopic parameters on the deformation properties of rockfill materials was explored, providing a reference for the calibration of microscopic parameters and laying a foundation for the characteristics research of rockfill materials.

Keywords Rockfill materials · Discrete element method · Triaxial test · Microscopic parameters · Deformation properties

---

C. Ma (✉) · J. Yang
Institute of Water Resources and Hydro-Electric Engineering, Xi'an University of Technology, Xi'an, China
e-mail: shanximachunhui@stu.xaut.edu.cn

C. Ma · G. Zenz · E. J. Staudacher
Institute of Hydraulic Engineering and Water Resources Management, Graz University of Technology, Graz, Austria

G. Bolzon et al. (eds.), *Numerical Analysis of Dams*, Lecture Notes in Civil Engineering 91, https://doi.org/10.1007/978-3-030-51085-5_46

1 Introduction

As a common building material, the rockfill material is widely adopted in infrastructure construction, such as water conservancy, hydraulic engineering, roads, railways, airports and so on, and it is also an important scientific research topic in landslides, blasting and river channel evolution. With the development of hydraulic engineering throughout the world, a rockfill dam has become one of the most preferred dam types in many projects. Now, rockfill dam heights are approaching 300 m [1], which requires a further understanding of the deformation mechanism of rockfill materials. Due to the constraints of test principles, test environment, test cost and so on, laboratory triaxial tests are difficult to comprehensively, accurately and truly reflect the mechanical properties of rockfill materials, which is one of the reasons for large settlements of rockfill dams during operation. In recent years, the discrete element method (DEM) [2] has gained wide attention in the numerical simulation of geotechnical experiments to explore microscopic phenomena because of its clear relationship between physical and mechanical properties.

The contact model and its microscopic parameters of DEM have an obvious influence on simulation results of geotechnical tests. Due to the complexity of microstructures and the limitations of current research means, a detailed description of the relationship between microscopic parameters and macroscopic behavior is still a research hotspot [3]. The calibration of microscopic parameters is a challenging task due to the limited knowledge of the rockfill's behavior. Current studies of DEM focus on the interaction between microscopic and macroscopic parameters from qualitative and quantitative perspectives: Zhou et al. [4] studied the size effects on the behavior of crushable granular materials, concluding that the size effect under high confining pressure is significant. Zhu et al. [5] considered the influence of the relative density and gradations of soil on the triaxial shear properties and proposed that the porosity simulated by the relative density is feasible. Zhou et al. [6] studied the true triaxial test simulation of crushable granular materials and analyzed the correlation between microscopic parameters. Yang et al. [7] researched quantitative relationships between microscopic parameters, Young's modulus and Poisson's ratio of the rock parallel bonded model. Yoon [8] employed experimental design and optimization in uniaxial compression simulation for calibrating contact-bonded particle models. Xu et al. [9] presented a set of empirical formulas to describe the correlation between macroscopic and microscopic elastic constants of sand material. Zhao et al. [10] investigated systematically the influence of microscopic parameters on macroscopic properties of the parallel bonded model, and the formulas between microscopic and macroscopic properties are proposed qualitatively and quantitatively. About rockfill materials, Xu et al. [11] studied the the creep and stress relaxation behavior of crushable granular material, finding that the deviatoric stress level has a significant influence on the evolution of the agglomerate breakage number and mechanical coordination number. Li et al. [12] adapted the response surface function to achieve the high precision nonlinear mapping between the deformation data and the microscopic parameters of rockfill materials, then microscopic parameters were back calculated.

The abovementioned researchers reveal the influence of microscopic parameters on the stress-strain laws to a certain degree, but they are hard to apply to the microscopic simulation of rockfill materials directly. On the one hand, the objects in abovementioned research are mostly brittle materials such as rock and concrete, but the rockfill material as a nonlinear material has no specific damage criterion with a large strength after the stress peak. On the other hand, most of abovementioned research study the macroscopic behavior by using the Mohr Coulomb strength criterion, and there are few studies that focus on the deformation property curves that are regularly used in engineering. Since material samples of rockfill dams mainly contain a large number of particles of random accumulation, the triaxial tests have complex and irregular force chain. Thus, there is a complex nonlinear and uncertain relationship between the microscopic parameters and the macroscopic behavior, which is challenging to describe accurately and quantitatively. Therefore, the influence of microscopic parameters on deformation properties of rockfill materials is investigated in this paper.

2 Contact Model of Rockfill Materials

Currently, the DEM can be divided into discontinuous deformation analysis [13] and particle flow method [2]. The discontinuous deformation analysis is mostly used for the stability analysis of rock mass, and the particle flow method is often applied in geotechnical experiment simulations. The contact model is the core principle in particle flow method, describing the basic law between the particle deformation, force, and motion. The frequent contact models divide into three parts: stiffness model, slip model and bond model. The stiffness model establishes the relationship between the contact force and displacement. Whether the particles have relative motion is determined by the slip model. The maximum joint strength of the normal and shear force is described by the bonding model. The bonding models can be divided into a contact bond model and a parallel bond model. The contact bond model connects balls at a point, and the normal and shear bond strength can be set. The principle is shown in Fig. 1. The bonds break if the force is larger than the bond strength, and then the balls will follow the slip model.

Through the summary of the contact models of triaxial tests of rockfill materials with DEM in recent years, it can be known that the linear stiffness model was commonly used in rockfill simulations in the early stage [14, 15]. As a result of further research, the contact bond model was introduced to simulate breakage of rockfill particles [12, 16, 17]. Compared with the linear stiffness model alone, due to the addition of bonded, the stiffness parameters greatly reduce in the contact bond model. Additionally, the early two-dimensional model gradually turned over to the three-dimensional model in triaxial test simulations. Compared with natural rock or soil, the rockfill material is composed of irregularly shaped and multi-angled particles, of which the characteristics are large bite force and shear strength. If the rockfill simulation is carried out by using spherical particles only, it will inevitably result in a

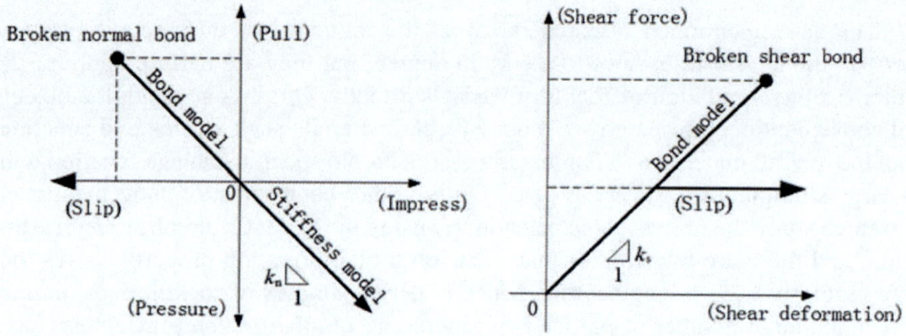

(a) Normal force versus surface gap (b) shear force versus surface gap

Fig. 1 Force-displacement law of the linear contact bond model

single arrangement of particles and a weak bite force. Therefore, a three-dimensional model with the contact bond model and breakable clusters is established to improve the accuracy of the triaxial test simulation.

3 Influence of Microscopic Parameters

3.1 Triaxial Sample of DEM

To compare with other research, the particle size gradation and its laboratory triaxial test results of rockfill material from the Sujiahekou dam is adapted from Shao Lei [18]. According to the particle size gradation, the DEM sample of the triaxial test was generated by the extrusion repulsion method, and particles with a larger diameter were simulated as breakage clusters. The linear stiffness model is used between the balls with small diameter, and the contact bond model for the clusters.

The engineering, laboratory and design gradations of the rockfill are shown in Fig. 2, and the triaxial sample of DEM is shown in Fig. 3. The size of the DEM sample is $\Phi 300 \times 650$ mm, and the initial porosity ratio is 0.40. In total 5026 balls, together with 508 balls which belong to the clusters were discretized in the model. The isostatic consolidation and loading are controlled by a servo control program in the DEM, and the confining pressure is 800 kPa. The movement speed of the upper and bottom loading platens was set to 0.15 m/s to simulate the static loading.

Through the trial calculation, the deformation curves are obtained with high fitting precision as shown in Fig. 4. The microscopic parameter values are normal stiffness $k_n = 3.5$ MN/m, shear stiffness $k_s = 2.6$ MN/m, friction coefficient $\mu = 0.09$, normal bond strength $b_n = 2.2$ kN, and shear bond strength $b_s = 1.6$ kN. Compared with the results in Shao Lei [18], the DEM simulation in this paper has good fitting precision, with increasing deviation of volumetric strain at higher axial strains. The possible reason is that the simulation of breakage particle still has a narrow gap in the actual situation.

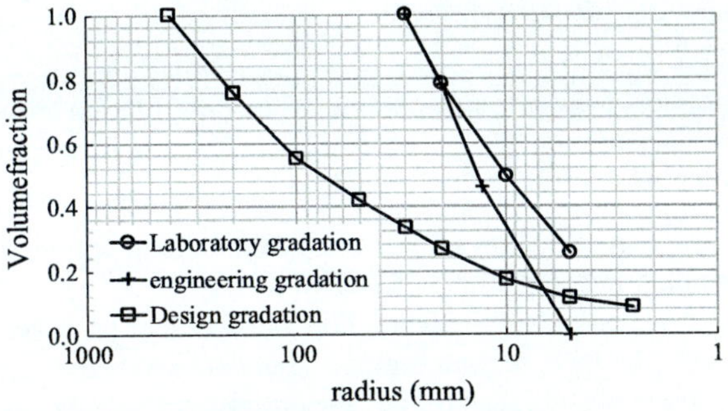

Fig. 2 Gradation curves of rockfill materials

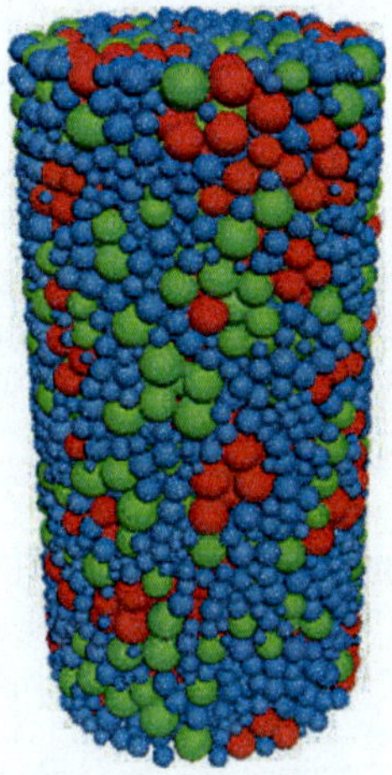

Fig. 3 The DEM sample of triaxial test

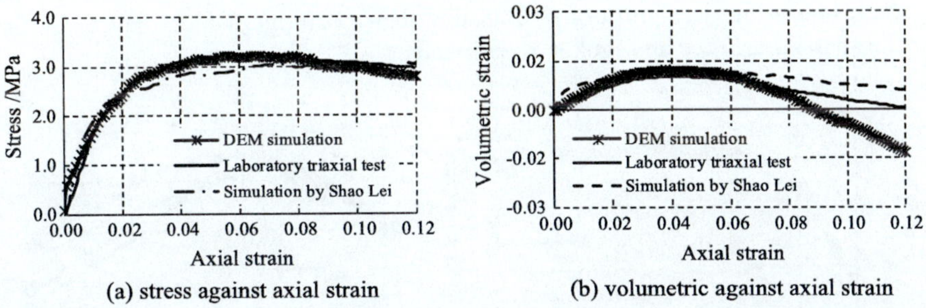

Fig. 4 Comparison of deformation curves between the DEM simulation and laboratory triaxial test

3.2 Microscopic Parameters

To analyze the effect of microscopic parameters on deformation properties of rockfill materials in triaxial test simulation, a control variable method is adapted to determine the influence on stress-strain and volumetric strain curves.

Normal stiffness. Figure 5 shows the stress-strain and the volumetric strain curves of the DEM simulation for normal stiffness values given as 3.8 MN/m, 3.5 MN/m, 3.2 MN/m and 2.9 MN/m, respectively. For the stress-strain curve, as the normal stiffness increases, the balls require higher stress to produce the same strain, which leads to an increase in the initial slope, i.e., the initial macroscopic modulus of rockfill materials. If the bond strength remains unchanged, the stress-strain curve will reach the peak earlier as the normal stiffness increases, indicating the axial strain of the peak will decrease. Meantime, due to the mutual constraints between the particles, the resisting force of the particles increases, so the stress-strain curve increases slightly as the normal stiffness increases. For the volumetric strain curve, the peak value of the volumetric strain curve decreases as the normal stiffness increases, and the axial strain of dilatancy advances. With the increase of the normal stiffness, the axial strain of the triaxial sample will decrease under the same loading, so the peak value of the volumetric strain decreases according to the following equation:

$$\varepsilon_{\mathrm{v}} = \varepsilon_{\mathrm{a}} - 2 \cdot \varepsilon_{\mathrm{r}} \tag{1}$$

where ε_{v} is volumetric strain, and ε_{a} and ε_{r} are axial and radial strain, respectively.

In addition, the increase of the normal stiffness accelerates the change rate of the volumetric strain, resulting that the volumetric strain curve is compressed in the horizontal axis.

If the normal bond strength remains the same value, the number of broken normal bonds will increase as the normal stiffness increases. At the same time, even if the shear stiffness remains unchanged, due to the increase of normal stiffness, the particles have the possibility of re-adjusting their positions, which may cause the increase in shear force and broken shear bonds.

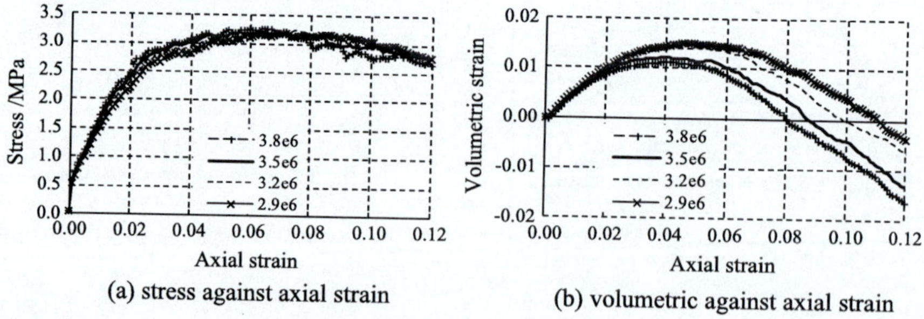

Fig. 5 The influence of normal stiffness on simulation curves

Shear stiffness. When the values for the shear stiffness are set to 2.9 MN/m, 2.6 MN/m, 2.3 MN/m and 2.0 MN/m, respectively, the resulting stress-strain and volumetric strain curves of the DEM simulation have the shapes as shown in Fig. 6. Compared with the normal stiffness, the shear stiffness is very limited on the deformation properties curve, and there is a certain fluctuation effect after the dilatancy in the volumetric strain curve. If the shear bond strength remains the same value, the number of broken shear bonds will have a small increase as the normal stiffness increases.

Frictional coefficient. In Fig. 7, The influence of friction coefficient on the stress-strain as well as the volumetric strain curves of the DEM simulation are depicted. For the stress-strain curve, with the increase of the friction coefficient, the initial slope and stress peak raise significantly, and the stress peak is more prominent. An increment in the friction coefficient means that the shear force increases which is required because the ball slips. Thus, the slip of the particles and the radial deformation of the triaxal test samples are limited, and therefore the resistance against deformation has increased. For the volumetric strain curve, with the increase of the friction coefficient, the radial deformation of the triaixal sample is reduced, so the peak value increases according to Eq. (1). In addition, the axial strains of dilatancy at different friction coefficients are similar.

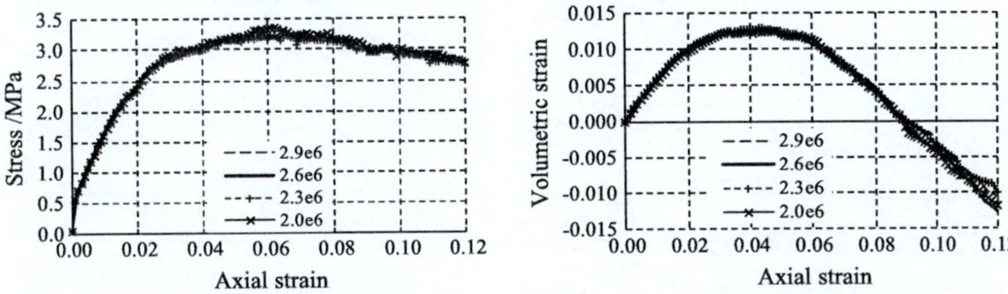

(a) stress against axial strain (b) volumetric against axial strain

Fig. 6 The influence of shear stiffness on simulation curves

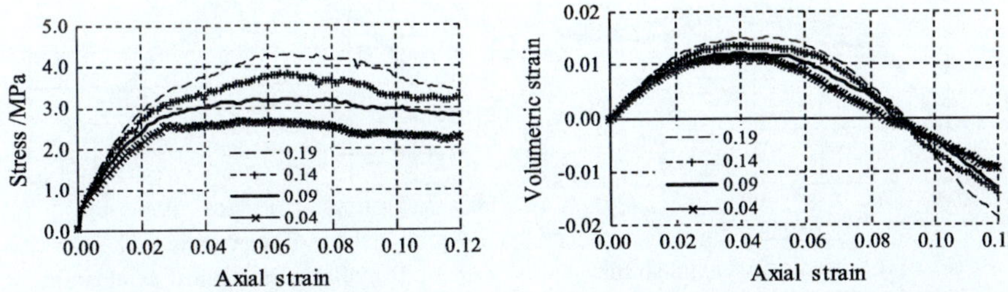

(a) stress against axial strain (b) volumetric against axial strain

Fig. 7 The influence of frictional coefficient on simulation curves

Normal bond strength. The results of the DEM simulations are shown in Fig. 8 considering a variation of the normal bond strength in the range from 2.8 kN, to 1.9 kN. For the stress-strain curve, the initial slope remains unchanged because most of the normal bonds don't break in this stage. When the curve reaches the peak, the normal bond gradually begins to break. As the normal bond strength increases, the peak value and its axial strain increases. For the volumetric strain curve, the axial deformation capacity expands with the increase in the normal bond strength. Therefore, this behavior leads to higher peak values of the volumetric strain curves, and the delay after peak decreases according to Eq. (1).

Shear bond strength. The results for a variation of the shear bond strength (values between 2.2 kN and 1.3 kN) are shown in Fig. 9. The initial slope of the stress-strain curve remains unchanged because most of shear bonds do not break at this stage. Compared with the normal bond strength, the influence of shear bond strength appeared earlier, indicating that the damage of the shear bond was initiated at lower strains. Besides, as the shear bond strength increases, the peak value increases, and further on the axial strain decreases. With the increase in the shear bond strength, the radial deformation capacity increases. Thus, the peak value of the body strain curve decreases, and the peak decreases after the peak according to Eq. (1).

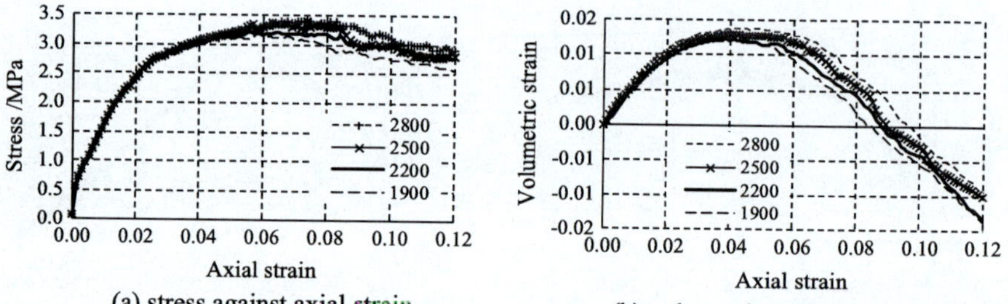

(a) stress against axial strain (b) volumetric against axial strain

Fig. 8 The influence of normal bond strength on simulation curves

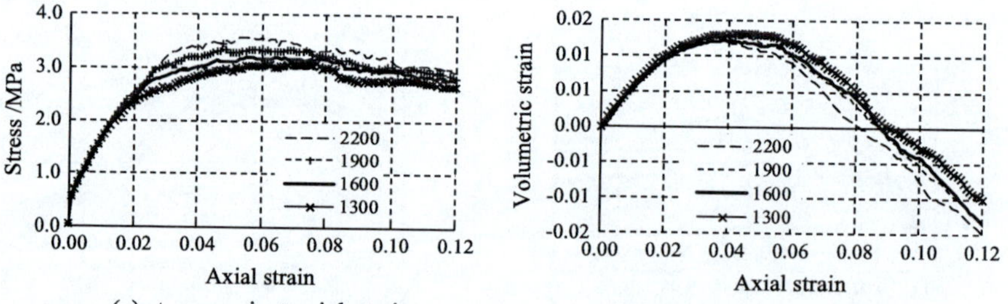

(a) stress against axial strain (b) volumetric against axial strain

Fig. 9 The influence of shear bond strength on simulation curves

According to the abovementioned calculation and analysis, the influence laws of microscopic parameters of rockfill materials on deformation properties are shown in Table 1. It can be seen from Table 1 that there is a complex nonlinear relationship between microscopic parameters and macroscopic performance of rockfill materials. It is shown that the microscopic parameters are interdependent, hence it is difficult to express this behaviour with a simple formula. The triaxial sample consists of a large number of different sized particles, therefore contact surfaces and forces as well as deformation of each particle are different. Generally speaking, the mechanical properties of most particles in the triaxial sample can be expected, while the small particles are not affected by force, bond, and other factors. The deformation properties of the triaxial sample are determined mainly by the particles, so the influence of the microscopic parameters on the macroscopic behavior can be qualitatively analyzed and predicted. However, the macroscopic behavior of the triaxial sample has certain inconstancy or randomness that is affected by a small number of unexpected particles.

There are also mutual constraints between microscopic parameters of rockfill materials, so the triaxial sample will balance with the appropriate combination of microscopic parameters. If the microscopic parameters change within a certain range, the abovementioned laws are correct. There is a possibility that the abovementioned laws are not work, if one of the microscopic parameters exceeds its range too much. Additionally, due to the complexity of rockfill materials and the current limitation of computational resources, the effects of particle shape, particle size, loading speed

Table 1 The influence of microscopic parameters on deformation properties of rockfill materials

			Stiffness model		Slip model	Bond model	
			k_n	k_s	μ	b_n	b_s
Stress-strain curve	Initial slope		↑↑		↑		
	Peak	Axial stress	↑	↑	↑↑	↑	↑
		Axial strain	↓			↑	↓
	Curve fluctuation after peak				↑		
Volumetric strain curve	Initial slope						
	Peak	Volumetric strain	↓		↑↑	↑	↓
		Axial strain	↓				
	Curve fluctuation after peak			↑	↑	↑	↓
Broke bonds	Broke normal bonds		↑↑			↓	
	Broke shear bonds		↑	↑			↓

Note ↓ means decrease as the variable increases. With the increase in the variable, ↑ means increase, and ↑↑ means increases significantly and other factors is not be considered in this paper. To get a better understanding of the sophisticated behavior of rock fill material at macroscopic and microscopic level, further research is needed.

4 Influence of Broke Bond

The broken rate of rockfill materials has a significant influence on its macroscopic mechanical properties. According to the abovementioned calculation and analysis, the macroscopic mechanical behaviour of rockfill materials is analyzed in the perspective of breakage particles. The macroscopic performance of the rockfill material and the statistics of broken bonds are shown in Fig. 10. All curves divide into three stages: (1) Elastic deformation stage. In this stage, the particles in the triaxial sample are in the compression and deformation stage with a small amount of slip and broken bonds. The macroscopic behavior is mainly affected by the stiffness of particles. (2) Transitional stage. The slip and broken rate of the particles increases, and the triaxial sample passes from the elastic to the plastic state. In this stage, the macroscopic behavior is mainly affected by the friction coefficient. (3) In the plastic deformation stage, a large number of stable particles will break and slip, which is mainly affected by the bond strength of the particles. In this stage, the stress-strain and volumetric strain curves reach their peaks, and the volumetric strain curve begins to decrease. It is worth mentioning that only the volumetric strain curve begins to fall (the radial deformation is greater than 0.5 times the axial deformation), then the stress-strain curve will remain or fall. Therefore, there is a close relationship between

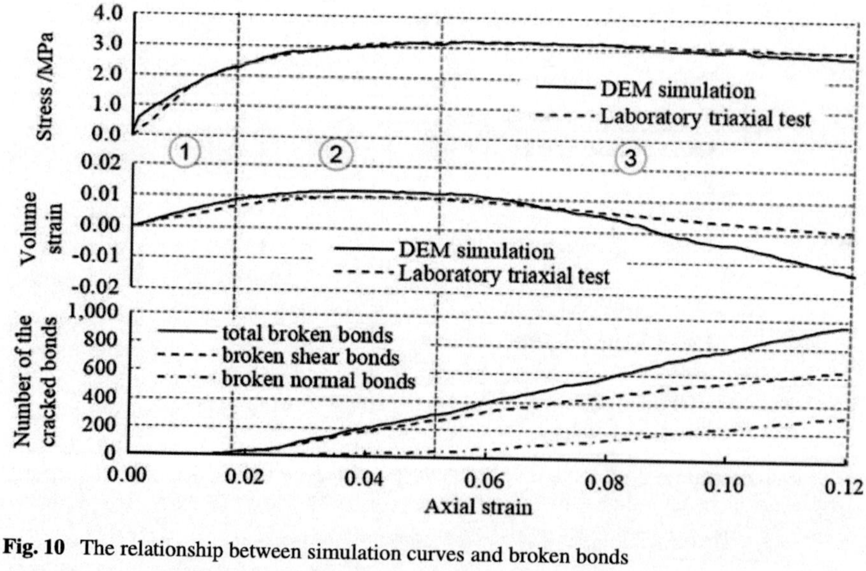

Fig. 10 The relationship between simulation curves and broken bonds

the breakage particle and the macroscopic mechanical behavior of rockfill materials. The stiffness, friction coefficient, and bond strength have a noticeable influence on the break of the particles and the mechanical behavior after break.

5 Conclusions

The control variable method was employed to analyze the influence of microscopic parameters on deformation properties in the triaxial test simulation of rockfill materials, and the influence mechanism was explained. Summarizing the current research results, the contact bond model and the three-dimensional model are the development directions of the triaxial test simulation of rockfill materials. In addition, a DEM model with the contact bond model and breakable clusters has been established, and the influence of several microscopic parameters (normal and shear stiffness, friction coefficient as well as normal and shear bond strength) has been analyzed. And the influence mechanism on the stress-strain and volumetric strain curve has been investigated. Finally, the relationship between the stress-strain curve, the volumetric strain curve and the broken bonds curve of the rockfill material has been explored. It is notable that there is a close relationship between the breakage particle and the macroscopic mechanical behavior of rockfill materials. Therefore, the influence of microscopic parameters on the deformation properties of rockfill materials which were investigated in this paper provides a possible guidance for the calibration of microscopic parameters and set a foundation for further research in this topic.

Acknowledgements This research was supported by the Key projects of natural science basic research program of Shaanxi province (Grant No. 2018JZ5010), science and technology projects of the state grid Xinyuan company LTD. (Grant No. SGXYZA00GCJS1900034) and the Water Science Plan Project of Shaanxi Province (Grant No. 2018SLKJ-5). The authors are grateful to Shao Lei for the data.

References

1. Wen L, Chai J, Xu Z, Qin Y, Li Y (2018) A statistical review of the behavior of concrete face rockfill dams based on case histories. Géotechnique 68(9):749–771
2. Cundall PA, Strack OD (1979) A discrete numerical model for granular assemblies. Geotechnique 29(1):47–65
3. Shi C, Xu WY (2018) Numerical simulation technology and application with particle flow code(5.0). China Architecture and Building Press, Beijing
4. Zhou W, Yang L, Ma G, Chang X, Lai Z, Xu K (2016) DEM analysis of the size effects on the behavior of crushable granular materials. Granular Matter 18(3):64
5. Zhu J, Guo WL, Xu JC, Chu FY (2017) DEM analysis on impact of gradation and compactness on coarse-grained soil in tri-axial test. J Chongqing Jiaotong Univ (Nat Sci) 36(6):70–74
6. Zhou W, Yang L, Ma G, Chang X, Cheng Y, Li D (2015) Macro–micro responses of crushable granular materials in simulated true triaxial tests. Granular Matter 17(4):497–509

7. Yang B, Jiao Y, Lei S (2006) A study on the effects of microparameters on macroproperties for specimens created by bonded particles. Eng Comput 23(6):607–631
8. Yoon J (2007) Application of experimental design and optimization to PFC model calibration in uniaxial compression simulation. Int J Rock Mech Min Sci 44(6):871–889
9. Xu XM, Ling DS, Chen YM, Huang B (2010) Correlation of microscopic and macroscopic elastic constants of granular materials based on linear contact model. Chinese J Geotechn Eng 32(7):991–998
10. Zhao GY, Dai B, Ma C (2012) Study of effects of microparameters on macroproperties for parallel bonded model. Chin J Rock Mechan Eng 31(7):1491–1498
11. Xu M, Hong J, Song E (2018) DEM study on the macro-and micro-responses of granular materials subjected to creep and stress relaxation. Comput Geotech 102:111–124
12. Shouju L, De L, Shen Y (2015) Meso-parameter inversion of constitutive model for rockfill materials based on macro experimental data. J Shandong Univ Sci Technol 5:20–26
13. Shi G (2015) Contact theory. Sci China Technol Sci 58(9):1450–1496
14. Yang G, Xiao Y, Gao DQ (2010) Study of three-dimensional particle flow numerical simulation and failure criterion for coarse-grained soils. Rock and Soil Mech 31:402–406
15. Alaei E, Mahboubi A (2012) A discrete model for simulating shear strength and deformation behavior of rockfill material, considering the particle breakage phenomenon. Granular Matter 14(6):707–717
16. Han HX, Chen W, Qiu ZF, Fu XD (2016) Numerical simulation of two-dimensional particle flow in broken rockfill materials. Chinese J Geotechn Eng 38(s2):234–239
17. Han H, Chen W, Huang B, Fu X (2017) Numerical simulation of the influence of particle shape on the mechanical properties of rockfill materials. Eng Comput 34(7):2228–2241
18. Shao L, Chi SC, Zhou LJ, Wang YZ (2013) Discrete element simulation of crushable rockfill materials. Water Sci Eng 6(2):215–229

Cost Risk Assessment of 13 km Long Headrace Tunnel in the Himalayas

F. Raggi, A. Tamburini, and L. Altarejos-Garcia

Abstract ELC Electroconsult S.p.A. (Italy), in association with NEWJEC Inc. (Japan), is carrying out the updating of a feasibility study and the detailed design of a storage hydropower project in the Himalayas, under a grant from the Asian Development Bank. The project layout includes two powerhouses. The main powerhouse, with a capacity of 600 MW, is located at some 14 km distance from the reservoir. The headrace tunnel to the main powerhouse has a length of 13.3 km. It is located in a region of high seismicity, with lack of access along the tunnel alignment, which it makes it difficult to get comprehensive geotechnical information along the tunnel route. Despite the uncertainties present, it is necessary to provide and assessment of the headrace tunnel cost and schedule to determine the project feasibility. This paper shows the procedure followed by the Consultant to assess a risk-based contingency cost for the tunnel accounting for the uncertainties present. The uncertainties have been grouped in three types: (1) extent of the geological formations, (2) geomechanical properties, and (3) the occurrence of adverse events such as collapses, rockfalls, squeezing and tunnel flooding due to groundwater. The objective is to assess a level of cost contingency associated with a certain probability of not being exceeded, to inform the decision on the amount of the contingency cost that should be considered at this stage of the project.

Keywords Risk assessment · Numerical modeling · Uncertainty · Cost

F. Raggi · A. Tamburini
ELC-Electroconsult, Milan, Italy

L. Altarejos-Garcia (✉)
Technical University of Cartagena, Cartagena, Spain
e-mail: luis.altarejos@upct.es

G. Bolzon et al. (eds.), *Numerical Analysis of Dams*, Lecture Notes in Civil Engineering 91, https://doi.org/10.1007/978-3-030-51085-5_47

1 Introduction

The hydropower project is located in the Himalayas. Reservoir total storage volume is 1,581 Mm^3 and will generate some 3,400 GWh per year, with an average energy production of 1,300 GWh in dry season—from December to May—and 2,100 GWh in wet season—from June to November. Main dam is an approximately 220 m high Concrete Face Rockfill Dam (CFRD). The project includes two powerhouses. Main Powerhouse has a capacity of 600 MW and is located 14 km south from the reservoir. Water is conveyed from the reservoir to Main Powerhouse through a 13.3 km long headrace tunnel. The second powerhouse, located at the dam toe, has a capacity of 35 MW and its purpose is to turbine the environmental flow releases required downstream of the dam.

The design and construction of a 13.3 km long headrace tunnel in the Himalayas, in a high seismic region, is a challenging task. There is the possibility of having significant cost increases and delays during the construction of the tunnel which in turn may have strong implications on the feasibility of the project. The way to consider these uncertainties is to include a contingency cost in the project budget, as it is common practice in any hydropower project. Considering a prudent, high contingency cost will likely cover the cost increase due to uncertainties, but a too high contingency cost may render the project unfeasible from the economic and/or financial point of view. Therefore, it is an interesting exercise to try to say something about the probability of exceedance associated with a certain contingency cost amount. An estimate of the order of magnitude of the exceedance probability associated with a contingency cost figure is a piece of useful information when making a decision on the level of contingency cost that should be included in the project's budget.

The purpose of the cost risk analysis presented in this paper is to assess a contingency cost to account for the uncertainties present in the design. Thesc uncertainties have bcen grouped in three types: (1) extent of the geological formations, (2) geomechanical properties, and (3) the occurrence of adverse events such as collapses, rockfalls, squeezing and tunnel flooding due to groundwater. The objective is to assess a level of cost contingency associated with a certain probability of not being exceeded, according with available information.

The paper is structured as follows. Section 2 includes tunnel description and main geological features. Section 3 describes the approach followed and Sect. 4 describes uncertainties included and how they have been treated in the simulation model. Section 5 includes main results obtained and Sect. 6 summarizes the main conclusions drawn from the study.

2 Tunnel Description

2.1 Layout

Headrace tunnel inner diameter is 8.3 m, and its alignment is some 13.3 km long. It runs according to an almost N–NE to S–SW direction, starting from the left abutment, upstream of dam site. Tunnel route crosses a mountain range characterized by high peaks and deep valleys. Rock cover reaches and exceeds some 1,100 m within a phyllite rock mass type, whereas minimum rock cover is some 150 m in the vicinity of a creek valley, not far from the tunnel intake. Tunnel diameter has been optimized through a classical exercise that considers estimates of Tunnel Diameter versus Construction Cost and Tunnel Diameter versus Energy Losses Cost relationships.

2.2 Geological and Geotechnical Features

From a regional point of view, the tunnel is expected to cross major and minor regional tectonic lineaments mostly E–W oriented. Main tectonic feature is the E–W thrust crossing the tunnel at a distance of about 7.5 km from the intake. According to literature [1], this thrust is probably associated to the MCTZ (Main Central Thrust Zone), which separates the Lesser Himalayan units outcropping to the north from the Higher Himalayan units thrusted over the Lesser Himalayan ones. The schematic geological cross section (Fig. 1) provides a simplified interpretation of the regional geological framework of the project area.

According to literature [1], MCTZ in the project area is not active, because it has been cut off from the active MCTZ part. Results of preliminary seismic hazard analysis carried out within the framework of the project confirm this hypothesis. Geo-mechanical profile of the tunnel is shown for the northern part (Fig. 2) and southern part (Fig. 3).

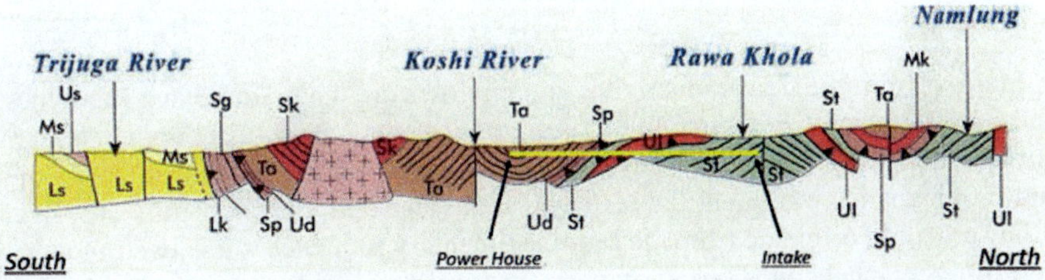

Fig. 1 A Regional S–N geological cross section oriented along the tunnel, showing the structural framework of this section of the Himalayan Range. Yellow line represents the position of the tunnel

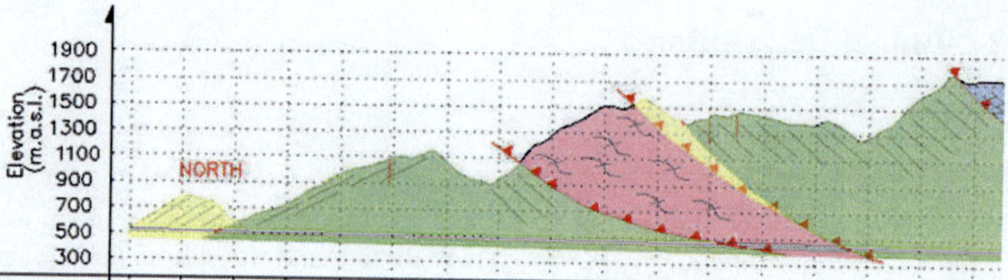

Extension (m)	611	132	1200	60	2204	593	491	309
Max Overb. (m)	260	227	570	570	1097	925	981	860
Geology (–)	Qz	Qz/Ph	Ph	Fault	Ph	Thrust	Gn	Thrust
UCS (MPa)	200	80	60	60	60	100	175	100
GSI (–)	55	35	50	20	50	25	45	25
mi (–)	20	15	7	7	7	15	25	15
D (m)	0.2	0.1	0.2	0.1	0.2	0.1	0.2	0.1

Fig. 2 Geo-mechanical profile of the tunnel. Northern part

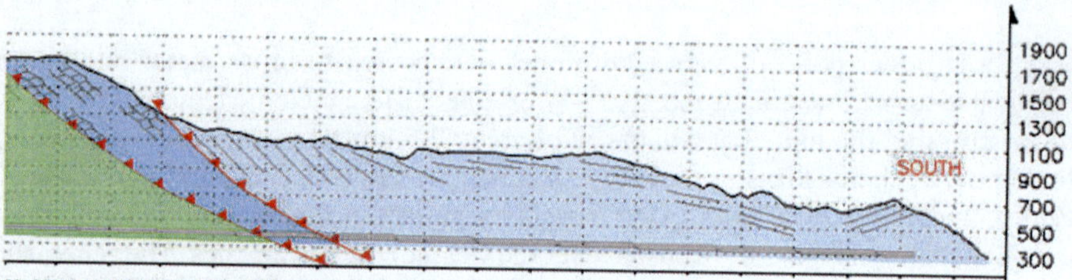

2680	100	407	100	4413
1300	740	747	748	736
Ph	Thrust	Mb–D	Thrust	Msh – Gn
60	65	75	65	150
50	25	45	25	50
7	15	9	15	25
0.2	0.1	0.2	0.1	0.2

Fig. 3 Geo-mechanical profile of the tunnel. Southern part

Starting from the intake, quartzites and phyllites are the predominant lithotypes. The contact between these two lithotypes is not a sedimentary boundary but a set of surfaces along which there may have been thrusts and/or intercalations, therefore this limit is not always well identifiable in a clear passage and it is not continuous. Even if they probably originate from the same sedimentary sequence, these two lithotypes have different mechanical behavior: quartzites are more rigid and tend to break while phyllites are more ductile and tend to fold, therefore the contact between these units is complex and variable.

Moving to the south, at some 3 km from the intake, a tectonic contact between phyllite and gneiss rocks was identified by surface mapping. Inclination of the thrusting surfaces, as well as the other ones described later, was measured only on outcrop. According to recent literature [2], these thrusting surfaces are not planar but probably tend to become less inclined with depth. Furthermore, these thrust zones do not necessarily represent sharp boundaries between geological units but transitional zones with thickness ranging from few to thousands of meters, as reported in literature [3, 4]. After another major thrusting surface, phyllites and quartzites were found again on outcrop. According to the previous observation made about the geometry of the thrusting surfaces, the gneiss zone could not necessarily be present at the tunnel's depth and may close before intersecting the tunnel axis.

Going southern, at about 6.2–6.5 km from the tunnel's northern intake, along the road near the tunnel layout, a tectonic contact between phyllites and marble-calcareous rocks belonging to the Halesi Dolomite was found. According to literature [1], the Halesi Dolomite composes the footwall of the MCTZ. These rocks are bounded by thrusts and divide the Lesser Himalayan Sequence from the Higher Himalaya Crystalline rocks. Karst phenomena are quite common in Halesi Dolomite and may have implications for the project.

At some 7.5 km from the intake, another lithological change between marbles and dolomites and gneissic-micaschists rocks was mapped. The limit between those units was not clearly visible on site but has clearly a tectonic origin and is related to the MCTZ.

Remaining part of the tunnel is expected to cross this gneiss and micaschists unit.

According to what has been previously stated, and due to the large depth of the tunnel it is not possible to predict with precision the position of the boundaries between geological units that will be intersected by the tunnel axis, as well as the thickness of such units.

As far as the rock cover over the tunnel and related initial stress conditions are concerned, maximum rock cover is about 1,360 m in the middle part of the tunnel, in chainage 6 + 700, at the interface between Phyllite and Dolomite. This is a contact zone with expected low-quality conditions of the rock mass. Another critical point in terms of rock cover is in Dolomite, not far from the next contact zone. At foreseen tunnel depths, squeezing of the rock mass is expected to be a critical issue.

Criticality of the rock mass response to excavation for any deeply seated tunnel, especially in some kind of medium/weak rock mass, like Phyllite, and associated with severe in situ stress conditions due to high depths, has been anticipated through examination of specific diagrams suited for this purpose. In fact, stability of tunnels in relatively weak rock is controlled by the ratio of uniaxial compressive strength of the rock mass, or the Global Rock Mass Strength, to maximum in situ stress. This ratio provides a guide for a first estimate of support requirements [5]. Results of latest studies [6] on this topic are summarized in Fig. 4.

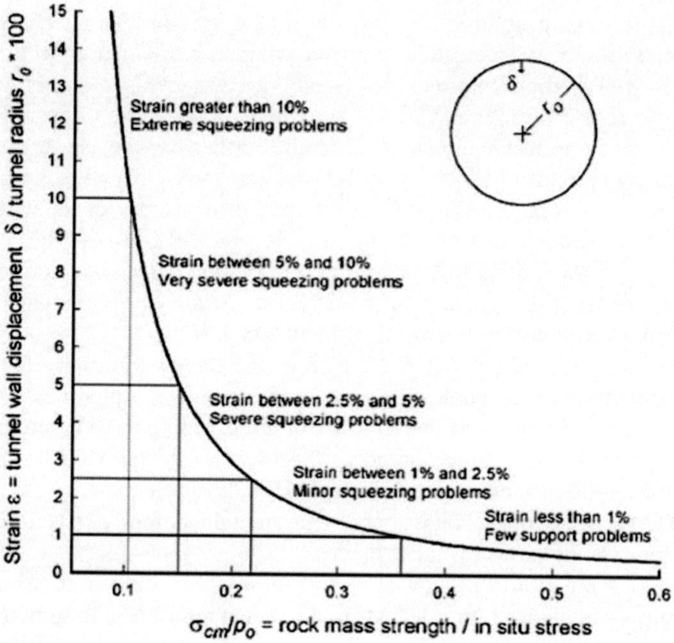

Fig. 4 Classification of plastic potential

2.3 *Tunnel Sections*

From a regional point of view, the tunnel is expected to cross major and minor regional tectonic lineaments. In order to take into account expected geotechnical conditions of the rock mass, six different sections: A, B, C, D, E and F, with several support and concrete lining design features have been designed to be used along the tunnel (Fig. 5).

3 Proposed Approach

Methodology adopted to estimate a probability of exceedance of the contingency cost is the Monte Carlo simulation method. A cost model for the tunnel that considers key variables is prepared. Some of the variables have been considered as deterministic, with fixed values, while others are considered as random variables, with their own probability distribution functions. Therefore, tunnel cost becomes a random variable itself.

Random variables considered are: (1) length of each of the 14 geological stretches; (2) % of length of each geological stretch affected by adverse events like rockfall,

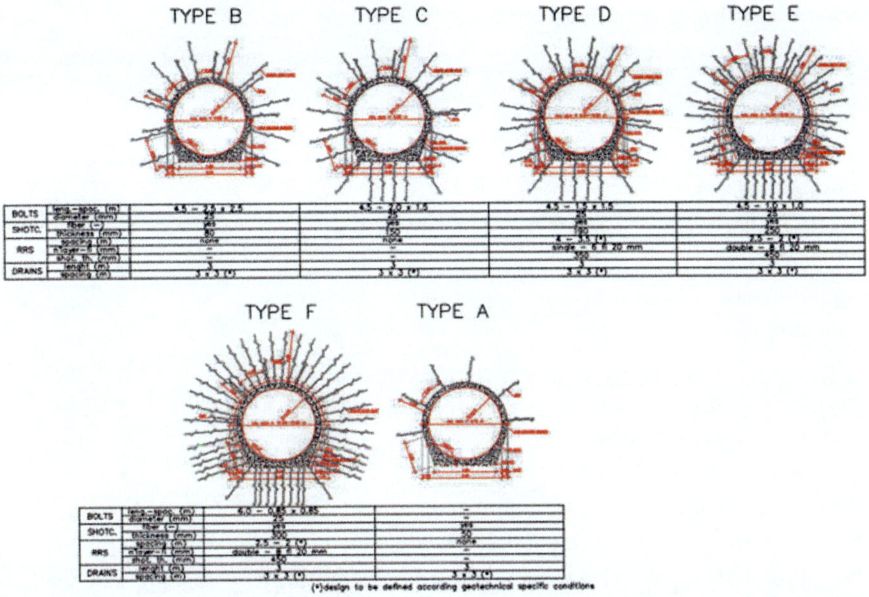

Fig. 5 Tunnel cross sections

squeezing, flooding, etc.; and (3) cost increase per unit length of the different cross sections when affected by adverse events.

Deterministic variables considered are: (1) % of use of each cross section, A to F, on each of the 14 geological stretches, previously determined by specific Monte Carlo simulations; (2) cost per unit length of the different tunnel cross sections, A to F.

Monte Carlo simulation method samples N times the random variables and calculates N times the tunnel cost, making it possible to derive information on how the tunnel cost is distributed in a probabilistic way.

4 Analysis of Uncertainties

4.1 Uncertainty in Extent of Geological Formations

The tunnel has been divided into 14 zones. According to information available, initial estimate of each zone's length, or Base Case, is shown in Table 1.

Position of boundaries between different zones is not known with precision. Different location of boundaries implies different zone lengths. To deal with this uncertainty, position of each boundary, defined by the chainage or progressive value,

Table 1 Length of geological zones in tunnel for Base Case

Geological zones along tunnel	Mean length (m)	Chainage
Initial point		0
Quartzites—Contact	611	611
Contact—Phyllite	132	743
Phyllite—Fault	1,200	1 + 943
Fault—Phyllite	60	2 + 003
Phyllite—Thrust	2,204	4 + 207
Thrust—Gneiss	593	4 + 800
Gneiss—Thrust	491	5 + 291
Thrust—Phyllite	309	5 + 600
Phyllite—Thrust	2,680	8 + 280
Thrust—Marbles and Dolomites	100	8 + 380
Marbles and Dolomites—Thrust	407	8 + 787
Thrust—Micaschists and Gneiss 1	100	8 + 887
Micaschists and Gneiss 1—Micaschist and Gneiss 2	2,463	11 + 350
Micaschist and Gneiss 2—Final point	1,950	13 + 300
Final Point		13 + 300

is considered a random variable and may vary over a certain range. Uniform probability distribution functions are used to sample the chainage position over a certain variation range. Ranges of variation for boundaries are defined to allow maximum variability within each zone, subjected to the following restrictions: (a) zone lengths should be positive (> 0) for all random simulations; (b) chainage at both ends of the tunnel, 0 + 000 and 13 + 300, are fixed values, therefore total length of the tunnel should be always 13,300 m.

Under these conditions, ranges of variation are shown in Table 2.

Let the boundary between two adjacent zones, I and II, be defined by the Base Case progressive chainage, Ch_i^*, and let R_i^* be the length of the range of variation for the position of the boundary. For each simulation, k, a random value, $R_i(k)$ is sampled from the uniform probability distribution defined as $U[-R_i^*/2; +R_i^*/2]$.

Chainage of boundary for that simulation is then set to $Ch_i(k) = Chi_i^* + R_i(k)$.

Same procedure is applied to the next boundary between zones II and III, which is defined by Base Case chainage Ch_j^*. Let R_j^* be the length of the range of variation for the position of this boundary. For each simulation, k, a random value, $R_j(k)$ is sampled from the uniform probability distribution defined as $U[-R_j^*/2; +R_j^*/2]$. For the k simulation a value of $Ch_j(k)$ is obtained.

Length of zone II for the k simulation is then estimated as $L_{II}(k) = Ch_j(k) - Ch_i(k)$.

Same process is applied for all boundaries through simulations.

Table 2 Ranges of variation of extent of geological zones

Geological zones along tunnel	Mean length (m)	Base Case Chainage	Uncertainty range, R, for Chainage (m)
Initial point		0	0
Quartzites—Contact	611	611	130
Contact—Phyllite	132	743	130
Phyllite—Fault	1,200	1 + 943	60
Fault—Phyllite	60	2 + 003	60
Phyllite—Thrust	2,204	4 + 207	500
Thrust—Gneiss	593	4 + 800	480
Gneiss—Thrust	491	5 + 291	300
Thrust—Phyllite	309	5 + 600	300
Phyllite—Thrust	2,680	8 + 280	100
Thrust—Marbles and Dolomites	100	8 + 380	100
Marbles and Dolomites—Thrust	407	8 + 787	100
Thrust—Micaschists and Gneiss 1	100	8 + 887	100
Micaschists and Gneiss 1—Micaschist and Gneiss 2	2,463	11 + 350	1,000
Micaschist and Gneiss 2—Final point	1,950	13 + 300	0

4.2 *Uncertainty in Geo-Mechanical Properties*

In addition to the deterministic design of the tunnel, to cope with the existing geo-mechanical uncertainties, a previous Monte Carlo Analysis has been carried out for each relevant geological formation type.

In particular, the most important aspect faced with this type of analyses is the identification of the rock mass quality. Geo-mechanical data collected mainly come from the surficial geo-mechanical stations and from exploratory adits, so there is no direct information obtained at tunnel elevations. Thus, Geological Strength Index, GSI, has been assumed as a random variable uniformly distributed in the range highlighted in Fig. 6 for Phyllite and in Fig. 7 for Quartzite. Considering that data have been obtained from outcrops and exploratory adits, lower and higher values have not been considered. It is expected an average rock mass quality characterized by a GSI greater than 50 at tunnel elevations.

Another aspect taken into account in Monte Carlo Analysis is variability of in situ stresses. For each homogeneous stretch with same rock mass type, the whole variation range of in situ stresses has been considered, from minimum to maximum,

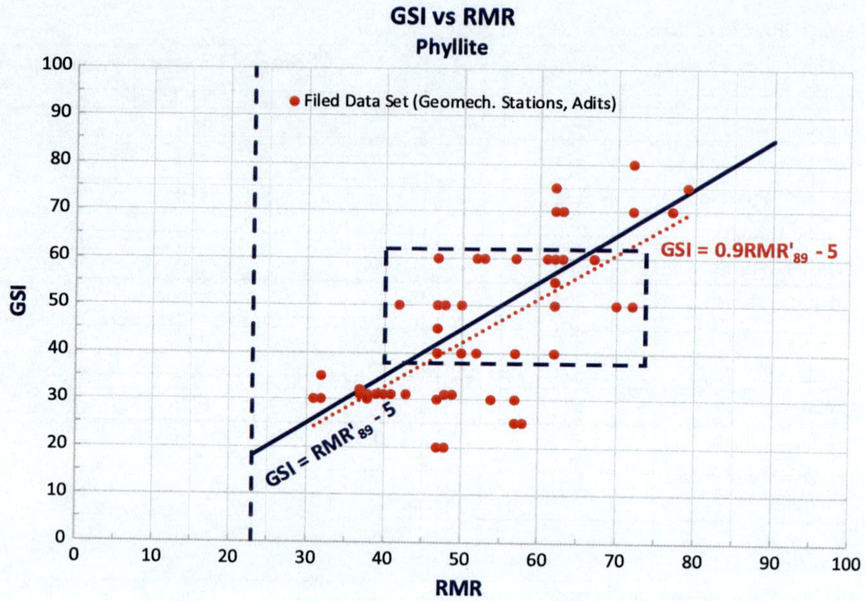

Fig. 6 GSI versus RMR'$_{89}$ for Phyllite (blue rectangle shows range of variation for Monte Carlo analysis)

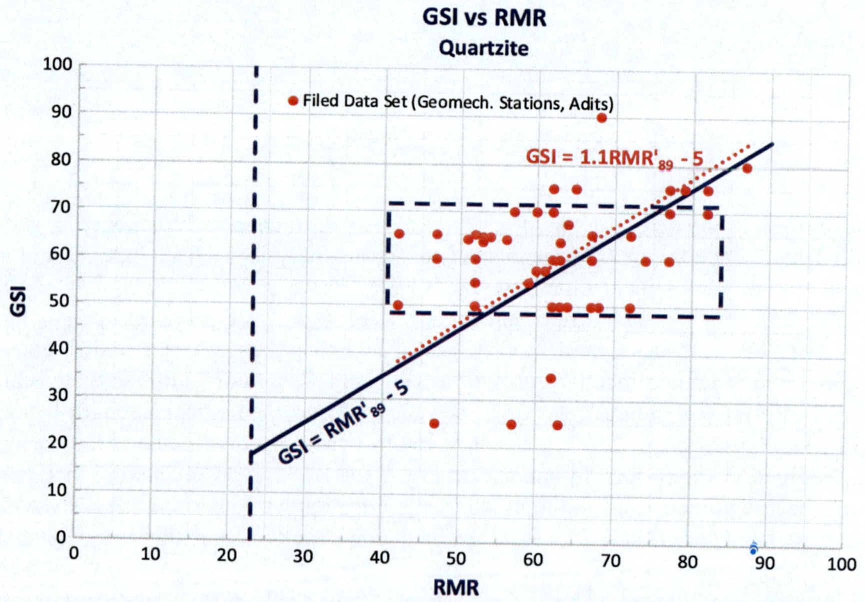

Fig. 7 GSI versus RMR'$_{89}$ for Quartzite (blue rectangle shows range of variation for Monte Carlo analysis)

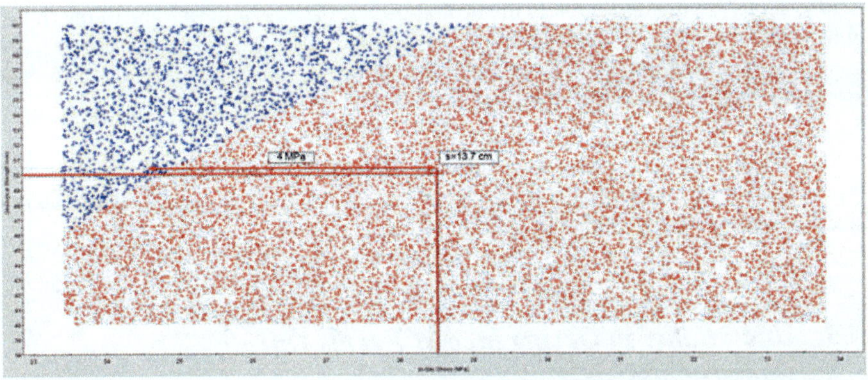

Fig. 8 Safety factor for section C in phyllite (stretch 2,680 m long)

accordingly to longitudinal profile of the tunnel. The adopted method, Ground Reaction Curve-Longitudinal Displacement Profile (GRC-LDP), considers stress ratios of horizontal stress to vertical stress of 1.0 and considers only circular tunnels. However, for quantifying uncertainties related to geo-mechanical aspects, the method can be considered reliable.

For each stretch with same rock mass type, the output of the Monte Carlo Analyses in terms of Probability of Failure of a specific excavation section has been considered to calculate the probability of the use of that section. Consistently with the GRC-LDP method, Safety Factor adopted to detect failure has been calculated as the ratio of the support strength to the support stress. For each couple *quality*-in situ *stress* i.e. (GSI, p_0), a value of Safety Factor, SF, has been obtained and depicted, as shown in Fig. 8, where red is for SF < 1 and blue is for SF ≥ 1. In this 2D graph the x-axis is p_0 and the y-axis is GSI.

Probability of failure has been calculated as the ratio between area with SF < 1 (red) and total area of the graph. Probability of Failure of the *i*th section is calculated using Eq. (1).

$$p_{f,i} = \frac{A_{SF<1,i}}{A_{tot,i}} = \frac{A_{SF<1,i}}{A_{SF<1,i} + A_{SF\geq 1,i}} = \frac{A_{red,i}}{A_{red,i} + A_{blue,i}} \tag{1}$$

Probability of use a certain section has been obtained using Eq. 2.

$$P_{u,n} = 100 - P_{f,n} - \sum_{i=1}^{n-1} P_{u,i} = P_{f,n-1} - P_{f,n} \tag{2}$$

where:

$P_{u,n}$ is the probability of the use of the *n*th section, starting from the lightest to the heaviest

$P_{f,n}$ and $P_{f,n-1}$ are respectively the Probability of Failure of the nth and of the $(n-1)^{th}$, which is the section lighter than the nth section

$P_{u,i}$ is the probability of use of the ith section.

Table 3 shows expected distribution of use of different cross sections along the tunnel.

For shorter stretches, such as Fault and Thrust stretches, only deterministic analyses have been carried out. In these stretches, considered overburden is the average overburden measured along the stretch.

4.3 Uncertainty in Occurrence of Events During Construction

Events during construction. Tunnel cost may be affected by the occurrence of certain undesirable events during construction [7]. The following events are frequent: (1) collapse/rockfall; (2) excessive deformation; (3) flooding; (4) rock burst/spalling; (5) portal failure.

Collapse of heading and/or failure of heading/lining implies fall of rocks of major dimensions. Different mechanisms involved are wedge or planar failure. Excessive deformation can occur due to construction defects and/or presence of swelling and squeezing ground, which had not been predicted. Flooding events comprises cases where the tunnel is invaded by large quantities of underground water. Rock burst and spalling means overstressing of massive or intact brittle rock, i.e. stresses developed in the ground exceed local strength of material. It can cause spalling or in the worst cases sudden and violent failure of the rock mass. Portal failure may happen at certain locations of a tunnel, where there is a lower resistance of ground mass and/or concentration of stresses.

Collapse/rockfall are the most reported events in tunnel construction (70%) while excessive deformation (11%) and flooding (9%) follow in frequency. These types of events cover 90% of reported incidents during tunnel construction [8, 9]. Therefore, in the cost model prepared, these 3 groups of events have been considered.

Main reported causes for occurrence of these events include unpredicted geology, presence of water with high pressure and earthquakes.

Consequences derived from these events may appear in the form of stoppage of work, equipment damage, injuries to workers and reconstruction of affected section, adding significant costs to the project. To account for the effect of these uncertain events on the tunnel cost it is needed to assess 2 aspects: % of tunnel length affected by the event and % of cost increase in the length affected by the event.

Tunnel length affected by events. Tunnel length affected by events is uncertain, so it has been considered as a random variable. The tunnel is divided in 14 zones. For each zone, 3 possible types of events are considered: (1) collapse/rockfall, (2) excessive deformation and (3) flooding. For each zone and each type of event, the ratio of length affected per total length will be sampled from a certain range, defined

Table 3 Ranges of variation of extent of geological zones

Start chainage (km)	End chainage (km)	Length (m)	A (%)	B (%)	C (%)	D (%)	E (%)	F (%)		A (m)	B (m)	C (m)	D (m)	E (m)	F (m)
0 + 000	0 + 611	611	100.0	0.0%	0.0	0.0	0.0	0.0		611	0	0	0	0	0
0 + 611	0 + 743	132	100.0	0.0	0.0	0.0	0.0	0.0		132	0	0	0	0	0
0 + 743	1 + 943	1200	69.7	26.2	4.1	0.0	0.0	0.0		836	314	49	0	0	0
1 + 943	2 + 003	60	0.0	0.0	0.0	0.0	0.0	100.0		0	0	0	0	0	60
2 + 003	4 + 207	2204	5.7	32.4	39.7	22.2	0.0	0.0		126	714	875	489	0	0
4 + 207	4 + 800	593	0.0	0.0	0.0	100.0	0.0	0.0		0	0	0	593	0	0
4 + 800	5 + 291	491	61.0	38.6	0.4	0.0	0.0	0.0		300	190	2	0	0	0
5 + 291	5 + 600	309	0.0	0.0	100.0	0.0	0.0	0.0		0	0	309	0	0	0
5 + 600	8 + 280	2680	0.0	0.0	18.9	66.5	14.6	0.0		0	0	507	1782	391	0
8 + 280	8 + 380	100	0.0	0.0	100.0	0.0	0.0	0.0		0	0	100	0	0	0
8 + 380	8 + 787	407	0.0	61.5	38.5	0.0	0.0	0.0		0	250	157	0	0	0
8 + 787	8 + 887	100	0.0	0.0	100.0	0.0	0.0	0.0		0	0	100	0	0	0
8 + 887	11 + 350	2463	93.0	7.0	0.0	0.0	0.0	0.0		2291	172	0	0	0	0
11 + 350	13 + 300	1950	93.0	7.0	0.0	0.0	0.0	0.0		1814	137	0	0	0	0
4413									Total length (m)	6109	1777	2098	2864	391	60
									Total (%)	45.9	13.4	15.8	21.5	2.9	0.5

Table 4 Ranges of variation of length affected by events

Geological zones along tunnel	Rockfall		Excessive deformation		Flooding	
	Min	Max	Min	Max	Min	Max
Quartzites	0.00	0.10	0.00	0.05	0.00	0.20
Contact	0.00	0.20	0.00	0.10	0.00	0.20
Phyllite	0.00	0.15	0.00	0.20	0.00	0.10
Fault	0.00	0.20	0.00	0.20	0.00	0.20
Phyllite	0.00	0.15	0.00	0.20	0.00	0.10
Thrust	0.00	0.20	0.00	0.20	0.00	0.20
Gneiss	0.00	0.05	0.00	0.05	0.00	0.10
Thrust	0.00	0.20	0.00	0.20	0.00	0.20
Phyllite	0.00	0.15	0.00	0.20	0.00	0.10
Thrust	0.00	0.20	0.00	0.20	0.00	0.20
Marbles and Dolomites	0.00	0.05	0.00	0.05	0.00	0.20
Thrust	0.00	0.20	0.00	0.20	0.00	0.20
Micaschists and Gneiss 1	0.00	0.10	0.00	0.05	0.00	0.10
Micaschists and Gneiss 2	0.00	0.10	0.00	0.05	0.00	0.10

by a minimum and a maximum number. These ranges are shown in Table 4. Values are based on engineering judgment, taking into account geo-mechanical features of different zones (phyllites, gneiss, fault, contact zones, thrust zones, etc.) in the light of information available. For each simulation, k, a random value of the ratio of affected length is sampled from the defined range, assuming a Uniform probability distribution, and it is applied to the corresponding simulated length.

Cost increase in the length affected by events. Economic impact of events depends on their magnitude, which is uncertain. Therefore, magnitude of impacts has been considered as a random value. It is expressed in terms of the ratio between cost increase per unit length of tunnel affected respect the base case or 'no impact' cost per unit length. For each zone and type of event, the ratio of cost increase will be sampled from a certain range, defined by a minimum and a maximum ratio. These ranges are shown in Table 5. For each simulation, k, a random value of the ratio of cost increase is sampled for each zone and for each event considered from the defined

Table 5 Ranges of variation of cost increase of section affected by events

Event	Ratio of cost increase respect the cost of 'no event' case	
	Min	Max
Collapse/rockfall	0.00	1.00
Excessive deformation	0.00	0.50
Flooding	0.00	0.20

Table 6 Baseline Cost of tunnel

Tunnel Section Type	Length (m)	Baseline Unit Cost (USD/m)	Cost (MUSD)
A	6,110	11,322	69.18
B	1,777	12,255	21.78
C	2,098	13,616	28.57
D	2,864	15,213	43.57
E	391	18,510	7.24
F	60	21,440	1.29
Total	13,300		171.63

range, assuming a Uniform probability distribution. The cost increase is calculated multiplying the length affected by the cost increase per unit length.

5 Results: Cost Simulations with Monte Carlo

5.1 Baseline Cost

Baseline Cost of tunnel is obtained multiplying the assumed values of length in the Base Case for each typical tunnel section: A, B, C, D, E and F, by the baseline or 'no adverse event impact' unit cost per m of each section. Result for Baseline Cost is 171.63 M USD, as shown in Table 6.

5.2 Cost Simulations

A total of 1,000 simulations of the cost model have been performed. Histogram illustrates how cost is distributed in terms of probability is shown in Fig. 9. Cumulative probability distribution is shown in Fig. 10. Summary of results obtained is shown in Table 7.

Results of the analysis show that contingency cost is expected to vary in a range from 4 to 20 M USD, approximately. Contingency cost with a 50% probability of being exceeded is 9.7 M USD and represents the 5.6% of the Baseline cost. Contingency cost with a 10% probability of being exceeded is 13 M USD and represents the 7.6% of the Baseline cost. Contingency cost with a 5% probability of being exceeded is 14.3 M USD and represents the 8.3% of the Baseline cost. Contingency cost with a 1% probability of being exceeded is 16 M USD and represents the 9.3% of the Baseline cost.

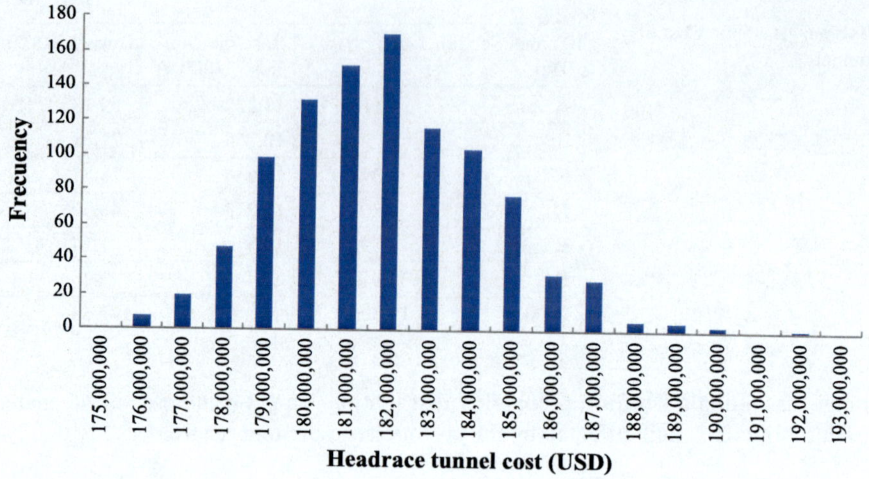

Fig. 9 Histogram of tunnel cost after 1,000 simulations

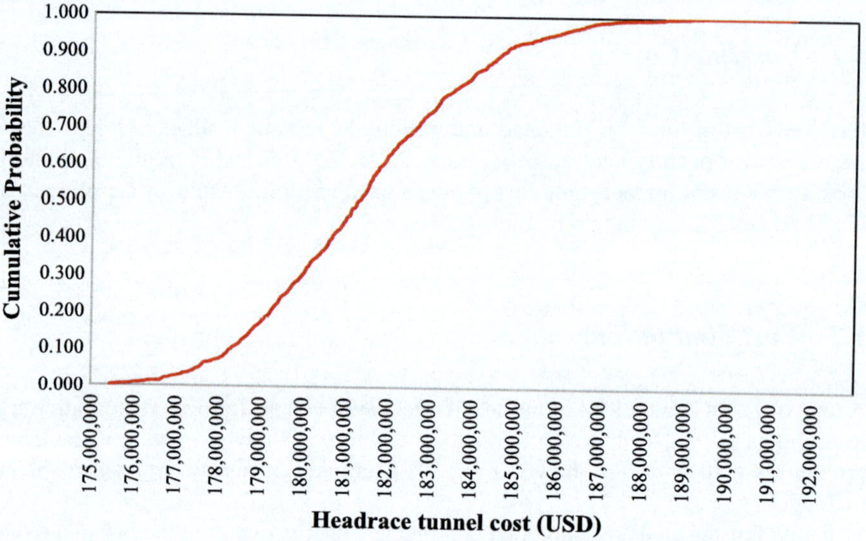

Fig. 10 Cumulative probability distribution of tunnel cost after 1,000 simulations

6 Conclusions

Total cost of tunnel is formed by the summation of two terms: (1) Baseline cost, and (2) Contingency cost.

Table 7 Summary of tunnel cost with contingency after 1,000 simulations

	Cost (M USD)
Minimum Cost	175.44
Maximum Cost	191.52
Mean Cost	181.39
Median Cost	181.26
25% Exceedance probability cost	183.07
10% Exceedance probability cost	184.64
5% Exceedance probability cost	185.86
1% Exceedance probability cost	187.56

Baseline cost is estimated in 171.6 M USD. It is a deterministic value, based on the following assumptions: length of each geological zone is known; geotechnical properties will vary along each zone in a certain, known range, without spatial correlation; distribution percentages of each typical cross section (A to F) along each geological zone is known; undesirable events, such as rockfall/collapse, squeezing and flooding will not occur, as the design matches the foreseen geological conditions.

Contingency cost is a random value, associated with the following uncertainties: length of each geological zone is not known with precision therefore it is considered as a random value; undesirable events such as rockfall/collapse, squeezing and flooding may occur due to the aleatoric, natural variability of the geological properties that may be present along the tunnel the 13.3 km long tunnel; length of tunnel potentially affected by these events is not known with precision therefore it is considered as a random value; cost increase associated with the occurrence of the events is not known with precision therefore it is considered a random value.

The analysis performed helps to assess the amount of cost contingency to be allocated in the budget and the expected associated exceedance probability.

For instance, if the target is to have a 10% Exceedance Probability cost (a cost with a 90% probability not being exceeded) then a cost contingency of 184.6—171.6 = 13 M USD should be considered.

It can be concluded that the analysis performed is helpful to inform the decision on what should be the most appropriate level of contingency cost to be considered in the project, taking into account existing uncertainties arising from natural variability of geological rock masses, their geo-mechanical properties and level of information available. It is also useful to communicate and made explicit the criteria used to cope with uncertainty, so it can be revised and enriched as more information becomes available during the progress of the project.

References

1. Dhital MR (2015) Geology of the Nepal Himalaya: regional perspective of the classic collided orogen. In:Regional geology review. Springer, Heidelberg

2. Carosi R, Montomoli C, Iaccarino S (2018) 20 years of geological mapping of metamorphic core across Central and Eastern Himalayas. Earth Sci Rev 177:124–138
3. Groppo C, Rolfo F, Indares A (2012) Partial melting in the Higher Himalayan Crystallines of Eastern Nepal: the effect of decompression and implications for the "channel flow" model. J Petrol 53:1057–1088
4. Mosca P, Groppo C, Rolfo F (2014) The geology between Khimti Khola and Likhu Khola valleys: a field trip along the Numbur Cheese Circuit (central-eastern Nepal Himalaya). J Virtual Explor (Electronic Edition 47, paper 4)
5. Hoek E Practical rock engineering. https://www.rocscience.com/learning/hoeks-corner. Accessed 05 June 2019
6. Marinos P, Hoek E (2000) GSI: a geologically friendly tool for rock mass strength estimation. In: Proceedings of GeoEng 2000 at the International conference on geotechnical and geological engineering
7. Sousa RL (2010) Risk analysis for tunneling projects. Doctoral Thesis, Massachusetts Institute of Technology, USA
8. Eskesen SD, Tengborg P, Kampmann J, Veicherts TH (2004) Guidelines for tunnelling risk management: International Tunnlling Association, Working Group No 2. Tunn Undergr Space Technol 19:217–237
9. Isakson T (2002) Model for estimation of time and cost based on risk evaluation applied on tunnel projects. Doctoral Thesis, Royal Institute of Technology, Sweden

Identification of Dam Behavior by Means of Machine Learning Classification Models

F. Salazar, A. Conde, and D. J. Vicente

Abstract The improvements in monitoring devices result in databases of increasing size showing dam behaviour. Advanced tools are required to extract useful information from such large amounts of data. Machine learning is increasingly used for that purpose worldwide: data-based models are built to estimate the dam response in front of a given combination of loads. The results of the comparison between model predictions and actual measurements can be used for decision support in dam safety evaluations. However, most of the works to date consider each device separately. A different approach is used in this contribution: a set of displacement records are jointly considered to identify patterns using a classification model. First, potential anomaly scenarios are defined and the response of the dam for each of them is obtained with numerical models under a realistic load combination. Then, the resulting displacements are used to generate a machine learning classifier. This model is later used to predict the most probable class of dam behavior corresponding to a new set of records. The methodology is applied to a double-curvature arch dam, showing great potential for anomaly detection.

Keywords Machine learning · Random forest · Arch dam · Anomaly detection

1 Introduction

The tendency towards the installation of automatic data acquisition systems in dam monitoring results in an increasing amount of available data. This has motivated researchers and practitioners to use machine-learning-based predictive models for dam safety assessment, as shown by the number of scientific publications in the field [1].

Most of the published works share the same structure: some period of data measurements is taken, for which both the loads (mainly the reservoir level and the temperature) and the response are known. The displacements in concrete dams

F. Salazar (✉) · A. Conde · D. J. Vicente
International Center for Numerical Methods in Engineering (CIMNE), Madrid, Spain
e-mail: fsalazar@cimne.upc.edu

G. Bolzon et al. (eds.), *Numerical Analysis of Dams*, Lecture Notes in Civil Engineering 91, https://doi.org/10.1007/978-3-030-51085-5_48

are more frequently analyzed, though other variables have also been addressed (e.g. leakage [2]). Some data-based predictive model is fitted to part of the available data (i.e. training set), then the model is applied to predict the dam response for the remaining period (i.e. test or validation set). Prediction accuracy is measured by comparing the model predictions with the actual readings.

The main goal of these approaches is early detection of anomalies, for which some threshold is typically set so that if the deviation of the actual reading from the model prediction is greater than the threshold, some warning is issued.

This approach provides advantages over conventional statistical models such as HST [3], including more flexibility and accuracy [4], therefore, it allows setting more constrained safety thresholds and better control of dam response.

Other works deal with the interpretation of the dam response by analyzing the model. Although machine learning (ML) algorithms are often considered as 'black box' models, some tools are available for their analysis, which have been shown to be useful for understanding dam behavior [5–8].

However, these approaches also have an important limitation, namely that each monitoring device is analyzed separately. Thus, the implementation of these models for anomaly detection requires fitting and analyzing as many models as relevant monitoring devices: their predictions need to be compared to the readings, then the results interpreted as regards potential anomalies or failure modes. This interpretation must be based on knowledge of the historical behavior of the dam in different situations.

The simultaneous analysis of a set of monitoring devices by means of an expert system would improve the process of anomaly detection. For example, in the event of some reading error in one device, a prediction model could detect it more or less quickly, but a subsequent analysis is required to determine whether the difference between the prediction and the measurement corresponds to a certain anomaly.

However, the joint analysis of the readings of a set of devices has been much less explored. Mata et al. [9] proposed a methodology based on Principal Component Analysis (PCA) and presented an example application with one potential failure scenario under constant hydrostatic load. The idea is to identify patterns among a set of readings to elucidate whether they correspond to a safe state or to some anomaly. For such purpose, the anomaly considered (an unacceptable relative sliding between dam and foundation in the right bank) was simulated with a numerical model.

In this work, we followed a similar approach with the following novelties:

1. Several anomaly scenarios were considered.
2. Low-relevant anomalies are analyzed, to verify the potential for early detection.
3. The thermal load was considered in addition to the hydrostatic load.
4. The method was verified for different load combinations.
5. The chosen algorithm is capable of dealing with a high amount of input variables (load and monitoring devices), without the need for doing variable selection.

In addition, the proposed methodology can be useful to support the design of the monitoring system: the algorithm automatically computes the relative influence of the inputs (here, the monitoring devices) as regards their usefulness to identify response scenarios, so the most relevant ones can be selected for being automated.

2 Methodology

The proposed methodology includes the following steps:

1. Determination of the potential anomalies or failure modes to consider. They need to be susceptible of being reproduced with a numerical model with enough accuracy. In the pilot test, an arch dam was selected and anomalies were defined that reproduce imposed displacements on the abutments and on the foundation. These conditions have been introduced into the model by modifying the corresponding boundary condition.
2. Computation of the dam response with the Finite Element Method (FEM). In our implementation, we use an in-house developed code [10] implemented in the Kratos environment [11] and coupled with the pre and post processor GiD [12]. Both the normal and the anomalous scenarios are computed and the predicted response in the monitoring devices is stored. In the case study presented we considered the displacements measured by pendulums. We used real load combinations, according to the actual evolution of the reservoir level and air temperature. If applied before construction, realistic combinations of environmental conditions should be generated.
3. Creation of a database including the variable loads (reservoir level and temperature in our case), the dam response at the location of the monitoring devices and the identifier of the corresponding scenario (normal or anomalous).
4. Generation of a machine learning classifier, fitted to the data. This model takes the loads and displacements as inputs, together with the identifier of the corresponding scenario, then 'learns' patterns associated to each of the simulated states. Once fitted, the model can be used to compute the most probable scenario for a set of readings, given some load combination.

Once the classifier has been trained and verified, it can be used to predict the most probable class of dam behavior corresponding to a new set of records.

3 Case Study

3.1 Description of the Dam

The methodology has been applied to a Spanish double curvature arch dam with 80 m height over foundation and 20 cantilevers. The monitoring data for the period 1999–2006 were considered, including the reservoir level and the air temperature, as well as the displacements in 28 stations of 7 pendulums. Figure 1 shows the time series of the reservoir level and air temperature in the period analyzed. We considered both the tangential and radial displacements at all the available locations for the analysis. Figure 2 shows a scheme of the dam and the location of the monitoring devices.

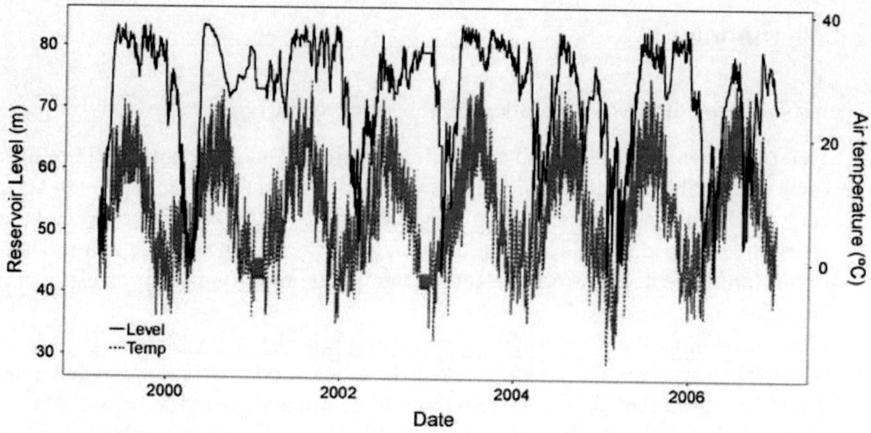

Fig. 1 Evolution of the reservoir level and air temperature in the period considered

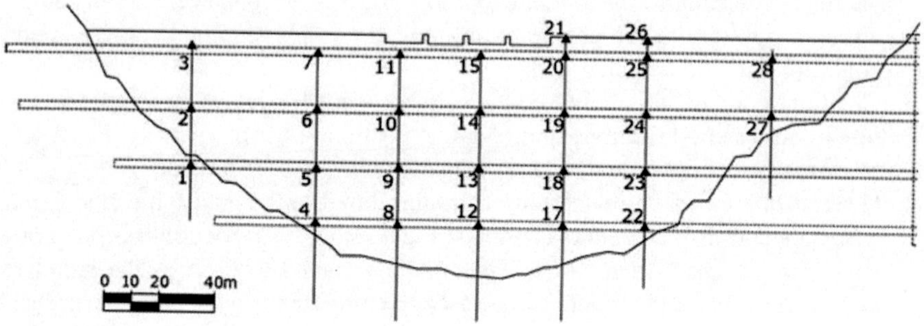

Fig. 2 Dam body and position of the pendulums. View from downstream

Since both the reservoir level and the concrete temperature are influential in the dam displacements, and the latter is dependent on the initial temperature considered, we run a preliminary analysis to obtain a realistic thermal field to be used as initial temperature in the dam body. For that purpose, a transient analysis was run over the period analyzed (1999–2006) with a constant value of the initial temperature (8 °C) and a time step of 12 h. The resulting thermal field at the end of this preliminary computation was taken as the initial temperature for all the scenarios considered. A similar approach was used by Santillán et al. [13].

Material properties are included in Table 1. The mesh is formed by tetrahedral linear elements of variable size: a finer mesh was used in the dam body, enough to ensure at least three elements along the radial direction, while increasing size was chosen for the foundation, up to 25 m. This resulted in 33,000 nodes forming 173,000 tetrahedrons.

Table 1 Material properties

Material properties	Concrete	Foundation	Units
Young modulus	3e10	4.9e10	Pa
Poisson	0.2	0.25	[–]
Density	2400	3000	Kg/m^3

3.2 *Scenarios Considered*

First, we run a transient analysis representing the actual behavior of the dam, to be taken as a reference of the normal or safe state (Scenario 0). We verified that this model sensibly represents the observed dam response by comparing the evolution of displacements in the model with the measurements recorded. Figure 3 includes this comparison for one of the locations.

Then we defined modifications with respect to Scenario 0, representing potential anomalies (Table 2 and Fig. 4).

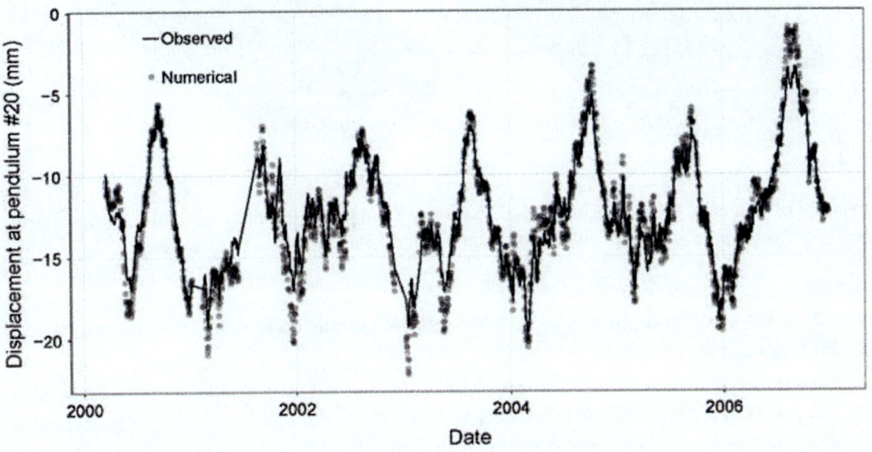

Fig. 3 Comparison between the radial displacement measured on pendulum 20 and the results of the numerical model for Scenario 0

Table 2 Anomaly scenarios considered

Scenario	Description	Magnitude
1	Imposed displacement in the left abutment	1 mm
2		0.5 mm
3	Imposed displacement in the right abutment	1 mm
4		0.5 mm
5	Imposed displacement in the riverbed	1 mm

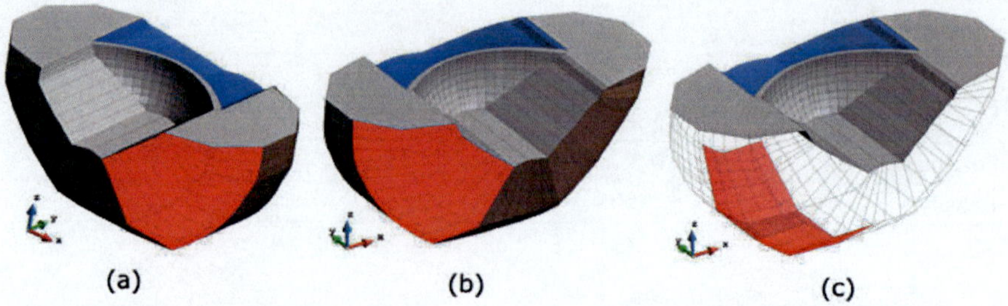

Fig. 4 Areas with modified boundary conditions (in red) for Scenarios 1–2 (**a**), 3–4 (**b**) and 5 (**c**)

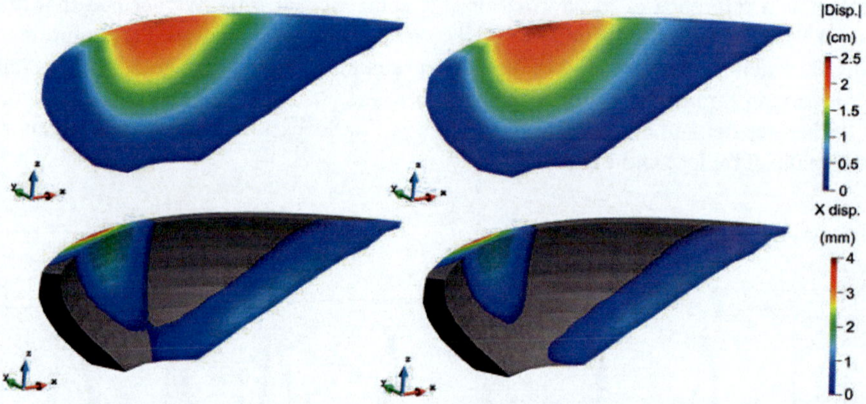

Fig. 5 Difference in the displacement field between Scenario 0 (left column) and 3 (right column). First row: total displacements. Bottom row: positive displacements in X direction

The difference in displacement fields among scenarios is small in general, as seen in the example in Fig. 5.

As a result of these calculations, we obtained a database including 34,152 records for the period 1999–2006 for 6 Scenarios. Each record includes reservoir level and air temperature, as well as tangential and radial displacements in 28 locations. The last column contains the identifier of the scenario corresponding to each set of records.

Although time is not explicitly considered, we divided the dataset into a training period corresponding to years 1999–2002, and left the remaining data for validation. This approach represents a realistic application, in which past performance can be used to build a model that can be later applied to real-time safety assessment. In practice, the model could be updated with certain frequency to enlarge the training set and thus the predictive accuracy. The effect of the training set size on the performance of ML models was assessed in a previous work, including criteria for updating the model [4]. In this case, we tested three different periods for training, namely 1999–2000, 1999–2001, and 1999–2002. For validation, we used the data for 2003–2006 for all models.

The data used, generated by numerical models, include no measurement errors. We modified them by adding a random variable with zero mean and a standard deviation of 0.10 mm to simulate errors due to measurement accuracy.

4 Classification Task

Among the machine learning algorithms available for classification, we used random forest in this work [14], since it is acknowledged to be appropriate in settings with many highly correlated input variables [15]. The algorithm automatically selects the more relevant variables and discards those with low influence in the results, which in our case are those with low usefulness to identify the response scenario.

The same algorithm was previously employed in regression problems in different applications, e.g. to build dam predictive models [2], to interpret dam response to seismic loads [16] and to better understand the behavior of labyrinth spillways [17].

First, we took all the available inputs. Then, the process was repeated taking only those inputs which showed to have highest relevance for the classification. We used this approach to build models with the 10 and 20 topmost relevant inputs. The availability of an accurate model using a reduced number of inputs can be useful to choose which devices should be automated in an existing dam.

We used the library random forest [18] and the R software [19]. All the models were run with default training parameters.

5 Results and Discussion

The raw outcome of the model is the probability of belonging to each of the defined classes. Then, the predicted scenario is that with the highest probability. Table 3 shows a summary of the results obtained. They correspond to the classification of the validation data, i.e. the period 2003–2006. There is a clear increase in predictive accuracy when 20 inputs are used instead of 10, and a small benefit when all 64 variables are considered.

The size of the training set has a relevant influence in the classification accuracy. The general improvement is low when year 2002 is added, but including data for 2001 results in a decrease of around 30% in average misclassification error. It should be kept in mind that we considered 730 records per year, e.g. two measurements per day. On a different note, the prediction task is challenging, since the validation data includes 17,532 sets of measurements to be classified among the scenarios considered. In such setting, the resulting accuracy can be considered as a useful result.

The misclassification rate for Scenario 0 (last column in Table 3) is useful result for practical purposes, since it represents the percentage of normal records that were wrongly classified as potentially anomalous (e.g. false positive rate). The results show

Table 3 Results of the classification task

Model Id	Inputs	Training set	Average classification error (%)	Scenario with highest error	Classification error for scenario 0 (%)
A	All (64)	1999–2000	4.79	2	2.09
B		1999–2001	3.48	4	1.71
C		1999–2002	3.04	3	0.86
D	More relevant (10)	1999–2000	13.93	5	17.86
E		1999–2001	10.71	5	15.40
F		1999–2002	9.29	5	13.42
G	More relevant (20)	1999–2000	6.21	2	8.35
H		1999–2001	4.50	0	6.37
I		1999–2002	3.83	0	5.30

that model performance improves as more information (e.g. number of devices) is included.

Table 4 shows the confusion matrix, i.e., the predicted versus the actual class for each sample, for the case with all inputs and training period 1999–2002. As expected, Scenario 0 is misclassified with Scenarios 2 and 4, which feature the lowest magnitude of the imposed displacement (0.5 mm). All situations for Scenarios 1 and 3 are correctly identified as not pertaining to Scenario 0.

As mentioned before, the default predicted class is the one that obtains the highest probability. However, the results can be analyzed in more detail by observing the probabilities assigned by the model to all classes. As an example, errors in the classification of Scenario 0 of the complete model have been investigated. Figure 6 shows the probabilities assigned to Scenario 0 (colored circles) and those corresponding to the other scenarios (black squares). Although Scenario 0 does not have the highest probability, the predicted values are clearly different from zero, often close to the highest among the remaining classes, and never the minimum.

Table 4 Example of confusion matrix. Model C (all inputs and training period 1999–2002)

		Actual scenario					
Predicted scenario		0	1	2	3	4	5
	0	2898	0	42	0	97	67
	1	0	2781	8	0	0	0
	2	8	141	2872	0	0	0
	3	0	0	0	2789	4	0
	4	14	0	0	133	2818	20
	5	2	0	0	0	3	2835

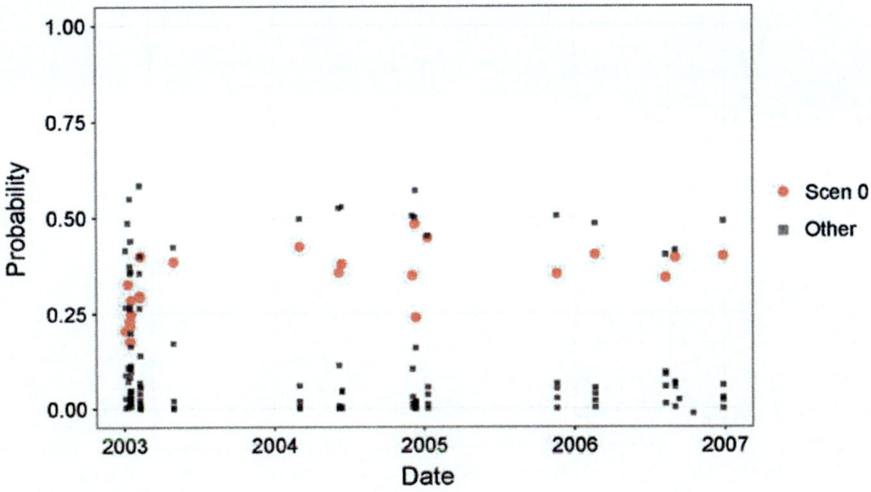

Fig. 6 Probability of Scenario 0 (colored circles) and that for Scenarios 1–5 (grey squares) for the false positive cases

The average probability of Scenario 0 in cases erroneously classified as anomalous is 0.33. This value is greater than the average probability assigned to Scenario 0 in truly anomalous scenarios, which are respectively 0.008, 0.13, 0.01, 0.15 and 0.12.

Another aspect that can be considered in practice is the temporal evolution of the prediction: in the example considered, every misclassification of Scenario 0 was followed by a correct prediction as normal behavior. Therefore, the reliability of the prediction can be associated to the number of samples consecutively predicted with the same class. From a practical viewpoint, the occurrence of a set of consecutive anomaly predictions can be established as a requirement for the issuance of safety warnings. Similar results were obtained for false negatives, i.e. anomalous scenarios wrongly classified as safe.

Classification models can be further analyzed to extract useful information. A measure of variable importance is computed for each input during model fitting [14]. The result for the model with all available variables is shown in Fig. 7. It is based on the average result of all scenarios considered.

Figure 8 shows that the most influential devices are located at the bottom part of the dam body. These results are reasonable, since the modifications to the reference case include imposed displacements on the boundary of the foundation, therefore their effect is higher in that area, and tend to be compensated by the monolithic response of the structure. This is in contrast to the conventional practice in dam safety: the displacements in the upper area of the higher cantilevers are more frequently analyzed, because they typically result in higher range of variation.

These results depend on the nature of the anomaly to detect, but show that when all devices are jointly considered, deviations with respect to normal behavior are more easily detected in areas with lower range of variation in normal operation conditions.

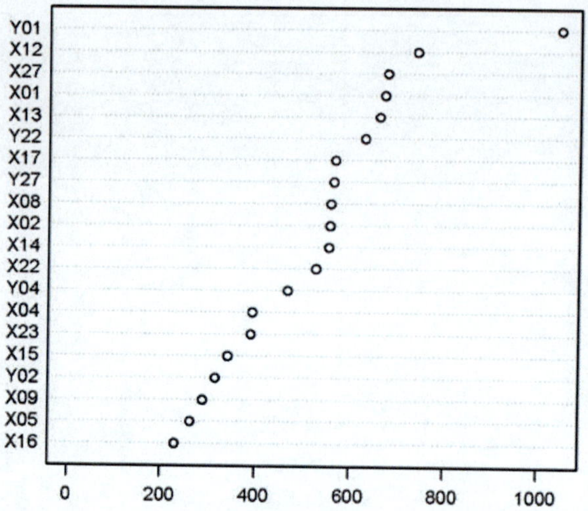

Fig. 7 Relative influence of the 20 more important inputs in the full model

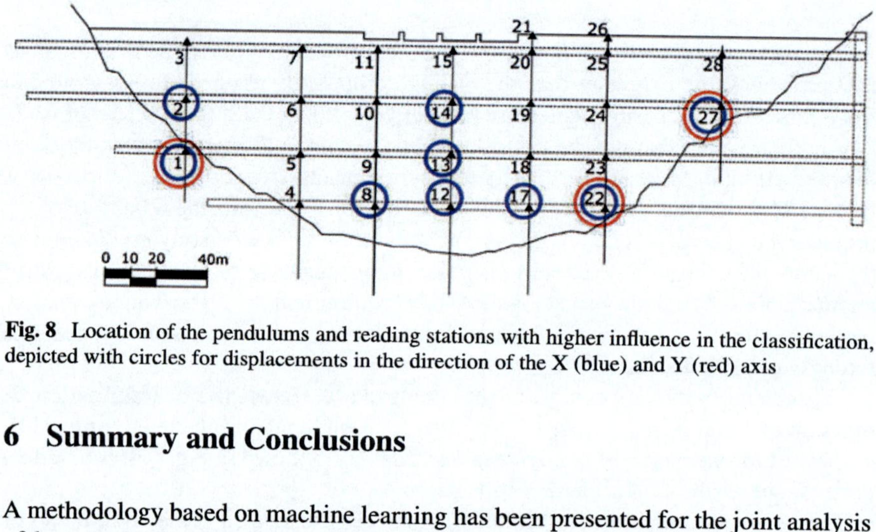

Fig. 8 Location of the pendulums and reading stations with higher influence in the classification, depicted with circles for displacements in the direction of the X (blue) and Y (red) axis

6 Summary and Conclusions

A methodology based on machine learning has been presented for the joint analysis of dam monitoring data, which allows classifying the response of the structure among a series of previously defined possible states. The results show that the method can be useful as a support for dam safety analysis, thus allowing the identification of even small deviations from normal behavior.

The main limitation of the approach presented is that only those scenarios that can be numerically modeled with sufficient precision can be considered. This limits its application to certain situations. However, it could be applied to others not considered in this work, such as the opening of the dam-foundation contact in concrete dams, or the appearance of preferential seepage zones in earth and rock-filldams. The latter would be reflected in certain reading patterns at the piezometers. This line of work is currently underway.

Furthermore, a possible anomaly cannot, in principle, be directly identified if it has not been defined and reproduced beforehand. In such situation, the result of the model could be anomalous in terms of the probability of belonging to the considered scenarios, which could also be useful for identification. This is also a research under development: these situations could be considered by adding an 'unknown' scenario to the potential anomalies.

Acknowledgements The authors acknowledge the financial support to CIMNE via the CERCA Programme/Generalitat the Catalunya. This work was also partially funded by the Spanish Ministry of Science, Innovation and Universities (*Ministerio de Ciencia, Innovación y Universidades*) through the projects NUMA (RTC-2016-4859-5) and TRISTAN (RTI2018-094785-B-I00).

References

1. Salazar F, Morán R, Toledo MÁ, Oñate E (2017) Data-based models for the prediction of dam behaviour: a review and some methodological considerations. Arch Comput Methods Eng 24(1):1–21
2. Salazar F, Toledo MA, Oñate E, Morán R (2015) An empirical comparison of machine learning techniques for dam behaviour modelling. Struct Saf 56:9–17
3. Willm G, Beaujoint N (1967) Les méthodes de surveillance des barrages au service de la production hydraulique d'Electricité de France, problems anciens et solutions nouvelles. In: IXth International Congress on Large Dams, Istanbul, pp 529–550 (In French)
4. Salazar F, Toledo MÁ, González JM, Oñate E (2017) Early detection of anomalies in dam performance: a methodology based on boosted regression trees. Struct Control Health Monit 24(11):e2012
5. Mata J (2011) Interpretation of concrete dam behaviour with artificial neural network and multiple linear regression models. Eng Struct 33(3):903–910
6. De Granrut M, Simon A, Dias D (2019) Artificial neural networks for the interpretation of piezometric levels at the rock-concrete interface of arch dams. Eng Struct 178:616–634
7. Tinoco JAB, Granrut MD, Dias D, Miranda TF, Simon AG (2018) Using soft computing tools for piezometric level prediction. In: Third International Dam World Conference, pp 1–10
8. Salazar F, Toledo MÁ, Oñate E, Suárez B (2016) Interpretation of dam deformation and leakage with boosted regression trees. Eng Struct 119:230–251
9. Mata J, Leitão NS, De Castro AT, Da Costa JS (2014) Construction of decision rules for early detection of a developing concrete arch dam failure scenario. A discriminant approach. Comput Struct 142:45–53
10. Vicente DJ, San Mauro J, Salazar F, Baena CM (2017) An interactive tool for automatic predimensioning and numerical modeling of Arch Dams. Math Probl Eng. article ID 9856938
11. Dadvand P, Rossi R, Oñate E (2010) An object-oriented environment for developing finite element codes for multi-disciplinary applications. Archives Comput Methods Eng 17(3):253–297
12. Ribó R, Pasenau M, Escolano E, Pérez J, Coll A, Melendo A, González S (2008) GiD The Personal Pre and Postprocessor. Reference Manual, version, 9
13. Santillán D, Salete E, Vicente DJ, Toledo MÁ (2014) Treatment of solar radiation by spatial and temporal discretization for modeling the thermal response of arch dams. J Eng Mech 140(11):05014001

14. Breiman L (2001) Random forests. Mach Learn 45(1):5–32
15. Díaz-Uriarte R, De Andres SA (2006) Gene selection and classification of microarray data using random forest. BMC Bioinform 7(1):3
16. Salazar F, Hariri-Ardebili MA (2009) Machine learning based seismic stability assessment of dams with heterogeneous concrete. In: 3rd Meeting of EWG Dams and Earthquakes, Lisbon, Portugal, 06–09 May 2019
17. Salazar F, Crookston BM (2019) A performance comparison of machine learning algorithms for arced labyrinth spillways. Water 11(3):544
18. Liaw A, Wiener M (2002) Classification and regression by random forest. R News 2(3):18–22
19. R Core Team (2019) A language and environment for statistical computing. R Found Stat Comput. Vienna, Austria. https://www.R-project.org/

Interpretation of Dam Monitoring Data Combining Visualisation Tools and Machine Learning. Eberlaste Dam Case Study

F. Salazar, R. Kohler, A. Conde, and F. Landstorfer

Abstract The installation of automatic data acquisition systems, together with the use of machine learning, allow obtaining useful information on the behaviour of dams. In this contribution, an example of application for a machine learning based predictive model is presented. Specifically, the level in a piezometer and its association with the reservoir level is studied for an embankment dam. The results show the model's ability to identify changes in dam response by taking full advantage of the available monitoring data. The flexibility of the algorithm allows different types of variables to be analysed without the need to determine a priori which are the most influential loads or how they affect the target value. The model has been implemented in a software tool that includes additional functionalities, specific for the treatment and exploration of dam monitoring data. It can be applied to different dam types and response variables.

Keywords Machine learning · Boosted regression trees · Data analysis · Data exploration · Embankment dam

1 Introduction

With the implementation of automatic monitoring and data acquisition systems, databases of increasing size are generated for dam behaviour, which can be very useful in the analysis of the response as well as in safety management. However, this increase in the volume of available data also presents challenges in terms of data analysis and processing tools, as well as methods for generating predictive models.

F. Salazar (✉) · A. Conde
International Center for Numerical Methods in Engineering (CIMNE), Madrid, Spain
e-mail: fsalazar@cimne.upc.edu

R. Kohler
VERBUND, Villach, Austria

F. Landstorfer
VERBUND, Schwarzach, Austria

G. Bolzon et al. (eds.), *Numerical Analysis of Dams*, Lecture Notes in Civil Engineering 91, https://doi.org/10.1007/978-3-030-51085-5_49

Traditionally, the representation of dam monitoring data has been limited to conventional spreadsheets, which was sufficient when manual data with low reading frequency were handled. The safety evaluation was based fundamentally on the expert analysis of graphs showing the evolution of the most relevant variables, as well as their relationship to the main loads, which in most of the dams are limited to the reservoir level and temperature.

In other fields of science and engineering, where large databases are available, tools are being developed to extract information from the data. These tools include highly customizable and interactive visualisation environments, which allow data to be presented in different formats, so that certain patterns and erroneous data can be identified. On the other hand, it is possible to generate data-based prediction models using machine learning algorithms that offer relevant advantages over traditional statistical methods.

VERBUND, in its digitalization process, is implementing tools of this type to extract information from the behaviour of its main dams, which already have automatic data acquisition systems. This contribution presents the result of the collaboration with CIMNE in the development of a software tool for this purpose, as well as an example of application to the Eberlaste embankment dam.

2 Case Study

2.1 *Eberlaste Dam*

Eberlaste is an embankment dam with an asphalt concrete core and a cut-off wall as underground sealing that was completed in 1968, with a maximum height of 28 m and a crest length of 480 m. The foundation is a deep, heterogeneous, alluvial deposit. Although bedrock was found close to the surface at the abutments, the upper surface of such competent foundation features high slope towards the valley, where only river-deposited gravel and sands were found in a 125 m deep drill [1]. As a result, large settlements and relevant seepage were expected already in the design stage. A 50 m long stabilising berm was also built in the dam foreland to improve safety against ground failure [2]. The monitoring system was also designed according to these unconventional features. As such, 15 relief wells of 25 cm pipe diameter were installed, 60 m deep, at the downstream dam toe. All of them are artesian wells and the discharge is monitored continuously. 14 automated monitored and 13 manually measured piezometers allow measuring the pore water pressure in various depths. A plan view of the dam is included in Fig. 1, with the location of the monitoring devices considered.

In this work the data from the automated devices, which include discharge at 14 relief wells, the total seepage flow, hydraulic head at 7 of them, 14 piezometers, suspended solids (mg/l), reservoir level, air temperature, snowfall and three records of rainfall (at three different stations) are used.

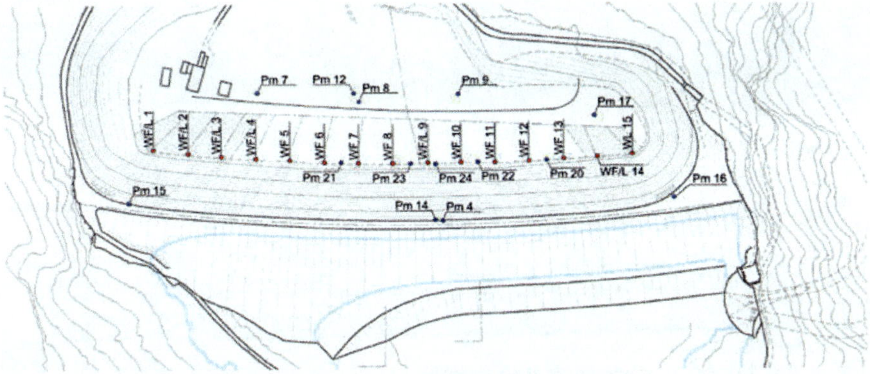

Fig. 1 Eberlaste Dam. Plan view. WF/L = relief well; Pm = Piezometer

With regard to the location within the dam body, the devices can be classified in three groups:

1. At the dam crest, right behind the asphalt core and the cut-off: piezometers 1, 2, 4–6, 13–16, 18, 19.
2. At the downstream dam toe: Piezometers 20–24, relief wells (all).
3. At the downstream end of the berm: Piezometers 7, 8, 9, 12, 17.
4. The remaining devices are sparsely located: Piezometers 10 and 11 are 100 and 200 m away from the berm toe, respectively.

2.2 Data Analysis Process

The analysis of the monitoring data was performed with a methodology including the following steps for each output variable:

1. Exploration of the time evolution and its relation to reservoir level with the time series plots and scatterplots.
2. Fitting of a predictive model based on boosted regression trees (BRT).
3. Interpretation of the model in terms of the most influential inputs and the nature of their effect on the variable under consideration.

The process is illustrated with the description of its application to Pm14 (level at piezometer #14).

Exploration. The time series plot of the output variable, together with the reservoir level, is shown in Fig. 2. Pm14 is displayed on the left vertical axis and the reservoir level (RL) on the right vertical axis. A possible change in dam response can be seen, since Pm14 appears above RL before 2010, then both series roughly overlap until 2014, and finally RL is above from that year on.

From the exploration of the time series, it is possible to identify the existence of reading errors or missing data periods. A specific functionality was developed

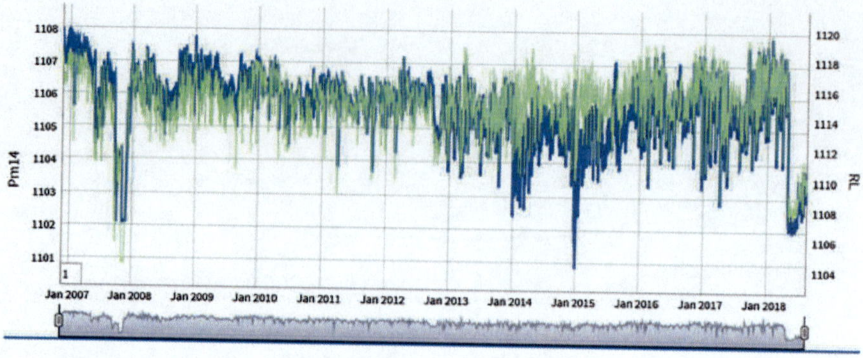

Fig. 2 Pm14 (blue) as a function of the reservoir level (green)

to interactively correct these imperfections by selecting erroneous data or periods without data and replace them with interpolated values.

The relation between RL and Pm14, as well as its evolution over time, can be best observed in a scatterplot. This graph has functionalities to rotate the axes and zoom, for a better analysis. Figure 3 shows two viewpoints of the data recorded for Pm14, as a function of the date and the reservoir level.

The application allows tuning the colour of the points as a function of some of the variables in the data set. Here, we used "Year" for the colours. It should be noted that "Year" variable is generated as numeric (with two decimal positions), not as integer.

This visualisation suggests that Pm14 responds linearly to RL for the most recent period (red points). The dots show higher scatter for the initial years (in blue) and appear above the straight alignment of red points. This means that Pm14 was more closely connected to the reservoir in 2007. Green points (around 2012) are closer to the more recent records.

Of course, although RL is expected to be the most influential input, Pm14 can in principle be dependent on other input variables. If that is the case, this interpretation of the scatterplot regarding the evolution over time might be wrong. This can be investigated by machine learning and the application developed.

Model fitting. Data exploration can provide interesting information, especially if the right tools are available and the task is carried out by dam engineering technicians with high knowledge on the dam under study. However, dam behaviour is often complex, and the acting loads are correlated, which makes it nearly impossible to identify certain effects by simply observing the data.

It is common to use statistical models to generate predictions of certain dam response variables and to analyse the contribution of each of the acting loads. The most commonly used example is the HST model, developed for the study of displacements in concrete dams [3]. It is based on the hypothesis that these displacements are the result of the combination of the effect of hydrostatic load, temperature and time (which includes irreversible deformations of different origin). This method has

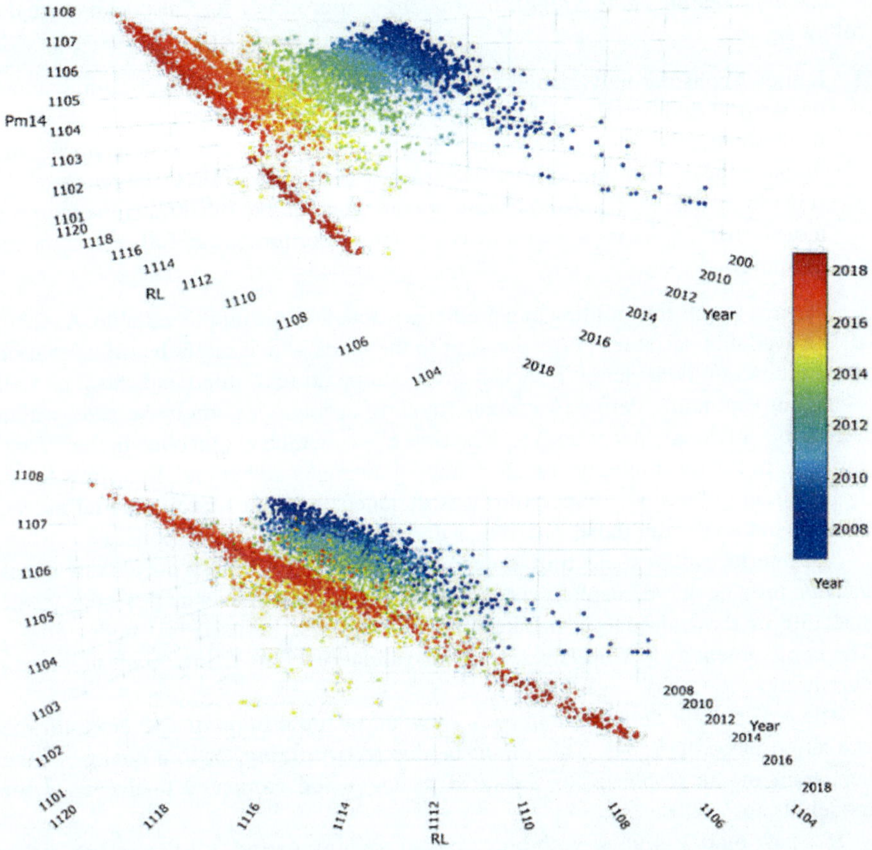

Fig. 3 Different views of the 3D scatterplot of Pm14 as a function of reservoir level (RL) and date (Year)

often been used both in professional practice and at the research level, although its limitations have also been identified [4].

A good number of alternatives have been proposed to overcome these limitations of HST: It does not consider the actual temperature, does not adapt well to variables other than the displacements of concrete dams, and does not allow consideration of the correlation between loads (as is the case with temperature, which is affected by the level of the reservoir). These methodologies range from advanced statistical models (e.g. HST-Grad [5], hybrid [6]) to machine learning models, which are constructed exclusively from the data (e.g. Neural Networks [7, 8]).

In this work, we make use of Boosted Regression Trees (BRT), a machine learning algorithm that was the most advantageous in a comparative study [9] and which was already used for the analysis of the behaviour of an arch dam [10].

The main features of this algorithm making it appropriate for this problem are the following:

1. It allows considering variables of different nature and range of variation without the need of additional transformations.
2. It automatically selects the most relevant variables in the dam response and discards those with little influence. Hence, variable selection is not necessary.
3. It is robust when it comes to training parameters, unlike other machine learning models that require in-depth knowledge of the algorithm to carefully select model fit options.

The next step is the building of a predictive model for the output variable. As input all the available variables corresponding to the loads which might be influential on the response are considered: reservoir level, temperature, rainfall and snow, as well as the corresponding derived variables (moving average or cumulative sum, which need to be previously generated) and the time effect, which we encoded in the "Year" variable. In this example, we created moving averages and cumulative sums for 2, 7, 15, 30 and 60 days. Temperature was included to verify that the model indeed discards automatically those variables without influence on the response.

The interface allows selecting some training parameters, though the default values usually provide good results. The main decision to be made in this step is the selection of the training period, i.e. the data that will be used for model fitting. The complementary period is reserved for validation. This is important to control overfitting.

The results with default parameters show an increase in predictive error for the validation period (Fig. 4). This might be due to overfitting, or to a change in the behaviour of the response for the most recent period compared to that used for model fitting.

If a new model is fitted with an extended training period, till the end of 2017, testing accuracy increases (Fig. 5). This confirms that some change occurred in the response of Pm14 to the variation of the hydraulic load between 2015 and 2017. Additional verification can be obtained from the model interpretation as described in the next subsection.

Model interpretation. The first result of model interpretation is the identification of the most influential variables, i.e. those inputs with stronger association with the analysed response. Figure 6 shows that the moving average of two days of the reservoir level is the most influential input, followed by the time.

As expected, reservoir level is the most influential variable in the piezometric level considered. However, the two-days moving average shows higher importance than the raw (actual) level. This result can be verified going back to the scatterplot. Figure 7 shows the relation between Pm14 and both RL and RL_02. A polygon was overdrawn to better observe that indeed there is less scatter of Pm14 in the RL_02 plot. This means that the association is stronger, as identified by the BRT model. This can reveal some inertia in the response of Pm14 to changes in the reservoir level.

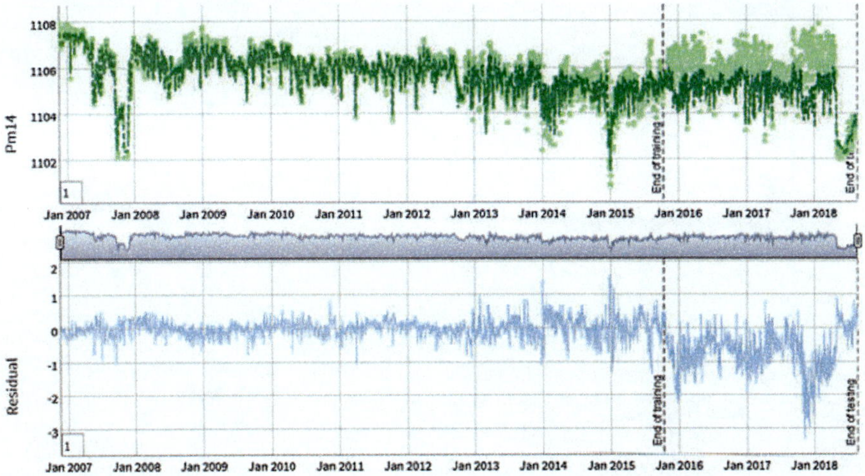

Fig. 4 Results of model fitting for Pm14 and default parameters. Mean absolute error (MAE) increases from the training set (0.2) to the testing set (0.77). This and the evolution of the residual (bottom) depict changes in the response variable for the test period

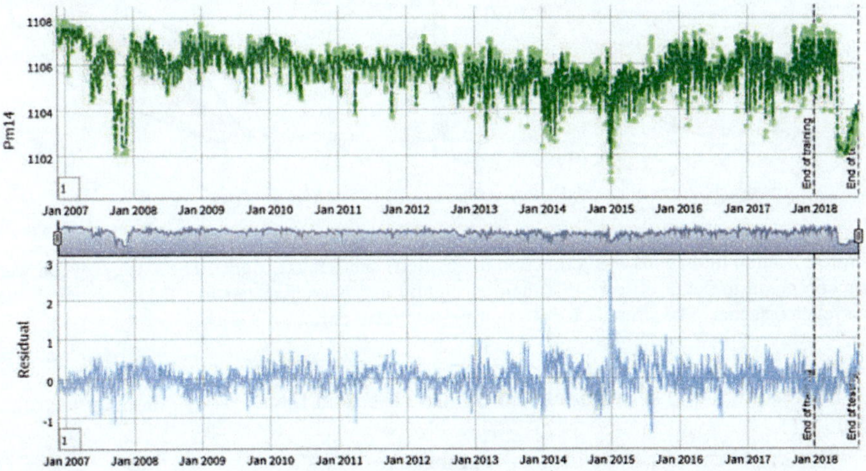

Fig. 5 Results of model fitting for Pm14 with extended training period. Accuracy is higher for the testing period (MAE = 0.23), and residual more stable

Figure 7 also shows that there seems to be higher influence of RL for the most recent period (red dots). To verify if the algorithm is capable of identifying this effect, a new model was fitted taking the period 2015/01/01–2017/12/31 for training. The results of the interpretation of this model for the relative influence (Fig. 8) are in agreement with the visualisation of the scatterplot: The reservoir level at the day

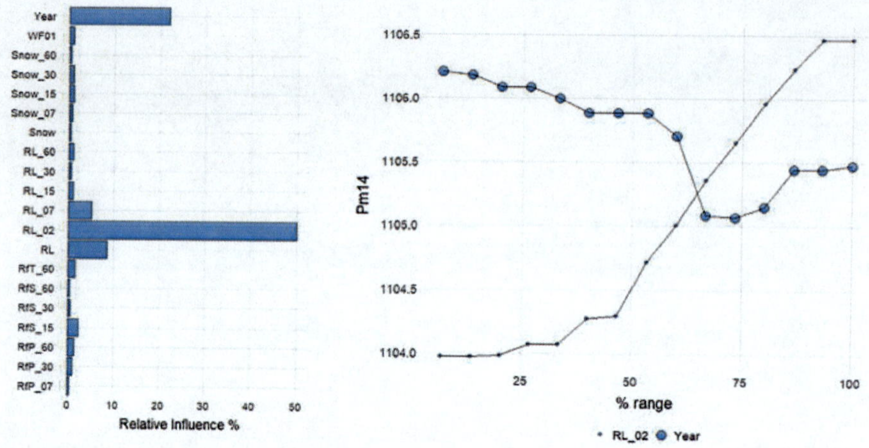

Fig. 6 Model interpretation. Left: Relative influence of the inputs considered. Right: Partial dependence of Pm14 on the two most influential variables

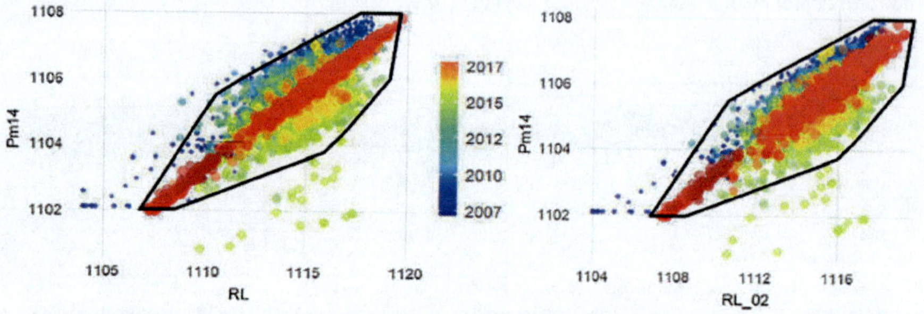

Fig. 7 Scatterplot of Pm14 as a function of RL (left) and RL_02 (right). The same polygon was superimposed to better observe that there is slightly less overall scatter for RL_02, although the correlation seems to be stronger for RL in the most recent period (red dots)

of the reading (RL) is more relevant in this case than the two-days moving average (RL_02).

Two conclusions can be drawn from these results:

1. The relative influence computed by the software accounts for the interaction between inputs: a small difference between RL and RL_02 results in a high difference in importance, as computed by the model. This shows the capability of the algorithm for dealing with highly correlated inputs.
2. The interpretation of the BRT model allows identifying effects which are hard to detect by data exploration, even if advanced visualization tools are employed.

As for the time effect, the partial dependence plot shows sharp changes in the response along time (Fig. 8 right). This result is obtained for all the values of the

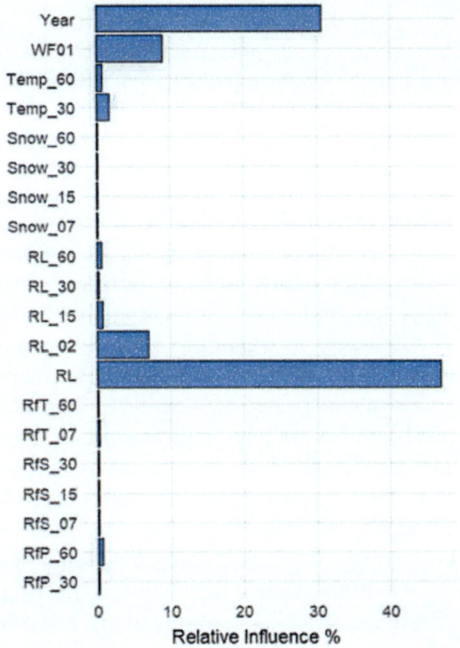

Fig. 8 Relative influence of input variables for a BRT model fitted with 2015–2017 as the training period. RL is the most influential variable, in accordance with the visual impression from the scatterplot (Fig. 7)

reservoir level, as observed in Fig. 9, where the combined average effect of reservoir level and time is depicted.

The effect of the hydraulic load is conventional: Positive correlation is observed, with a slope change around 1110 m.a.s.l. As regards time, there is a decrease until 2014, followed by a two-year stable period and a sharp increase around the end of 2015. These sudden changes are consistent with the operation report of the

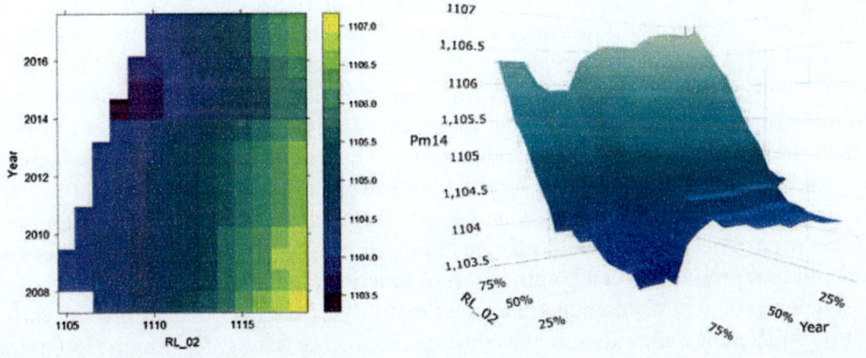

Fig. 9 Combined partial dependence of Pm14 on RL_02 and time. The influence of time is nearly stable for all values of the hydrostatic load

piezometer. The second was probably due to the piezometer flush performed in November 2015.

3 Discussion

Prediction models based on BRT can be useful for interpreting the behaviour of dams by analysing the monitoring data. They have the advantage of being very flexible in terms of the nature of the dam response to be analysed, as well as in terms of the amount of input data. Previous studies resulted in a good prediction capacity, as well as robustness to training parameters, which means that it is not necessary to have in-depth knowledge of the method for its practical use.

The model has a fundamental limitation that must always be taken into account: Its results, when making predictions outside the range of training data, are not reliable. This implies, for example, that the model prediction will be poor for a higher reservoir level than the historical maximum recorded in the dam. In order to control this extrapolation effect, the software developed verifies whether the test data are within the training range, and issues a warning in case of extrapolation.

The implication of this feature to the time effect has another interpretation: In this type of analysis, it is common to fit a model to the historical behaviour of the dam and apply it to analyse the current or future response of the structure. This necessarily implies extrapolating in terms of the time variable. Therefore, according to the above, the prediction would not be reliable. However, the BRT model has the particularity that in these cases it applies a constant time effect, equal to that identified in the most recent period. Therefore, a sudden increase in the discrepancy between prediction and observation reflects a change in behaviour of the variable under consideration with respect to that at the end of the training period. This can be useful for anomaly detection.

As a result, BRT models can be used for both, to identify changes in the behaviour of output variables, when analysing past behaviour (back analysis) and to detect deviations from normal performance, when applied to real-time data. In this sense, the flexibility of the algorithm allows capturing any kind of time effect without the need to make a trial and error process with functions of different shape, as is the case with conventional polynomial fitting. In the case study presented, the effect of time shows the changes in the association between the hydraulic load and the piezometric level which indeed occurs when some major action is performed on the well, such as cleaning or readjustment (as was the case in the analysed device, according to the reports). This implies sharp changes in the time effect which would have been difficult to reproduce with a combination of functions to be defined a priori.

The example shown demonstrates the flexibility of BRT models to predict variables of different nature than displacements in concrete dams, for which HST was developed and is often used. In this regard, BRT also provided useful results in terms of prediction and interpretation for other kind of variables such as leakage flow in arch dams [11], or seepage and flow at relief wells in embankment dams.

4 Summary and Conclusions

A software tool for dam monitoring data analysis and interpretation was presented, with an example of application. It includes functionalities for data exploration, with high degree of interactivity. Different kinds of scatterplots and time series plots can be generated and analysed. Also, wrong data can be fixed on the fly, in the same environment.

In addition, it includes an implementation of BRT models which can be used for predicting response variables of different nature, thus extracting high degree of information from the available data. These models can be analysed to explore the degree of association between the inputs considered and the output under analysis, as well as the shape of each partial dependence. The flexibility of the algorithm results in the automatic exclusion of low relevant inputs without the need for specific variable selection. Furthermore, input effects of irregular shape can be captured by the model, as those due to modifications in the recording device. Again, this can be obtained without any specific preliminary operation or decision by the user.

The software tool can be used for different dam types and output variables, as far as adequate monitoring data are available and the results are interpreted by experienced dam engineers.

Acknowledgements The authors acknowledge the financial support to CIMNE via the CERCA Programme/Generalitat the Catalunya. This work was also partially funded by the Spanish Ministry of Science, Innovation and Universities (Ministerio de Ciencia, Innovación y Universidades) through the project TRISTAN (RTI2018-094785-B-I00).

References

1. Hoeg K, Valstad T, Hansteen OE (1995) Transverse cracking in embankment dams. A literature review and finite element study. Norwegian Geotechnical Institute
2. ATCOLD-Austrian National Committee on Large Dams (2018) Pumped storage hydropower in Austria
3. Willm G, Beaujoint N. (1967) Les méthodes de surveillance des barrages au service de la production hydraulique d'Electricité de France, problèmes anciens et solutions nouvelles. In: IXth International Congress on Large Dams, Istanbul, pp 529–550 (In French)
4. Salazar F, Morán R, Toledo MÁ, Oñate E (2017) Data-based models for the prediction of dam behaviour: a review and some methodological considerations. Arch Comput Methods Eng 24(1):1–21
5. Tatin M, Briffaut M, Dufour F, Simon A, Fabre JP (2015) Thermal displacements of concrete dams: accounting for water temperature in statistical models. Eng Struct 91:26–39
6. Perner F, Obernuber P (2009) Analysis of arch dam deformations. In: 2nd International Conference for Long Term Behaviour of Dams, Graz
7. De Granrut M, Simon A, Dias D (2019) Artificial neural networks for the interpretation of piezometric levels at the rock-concrete interface of arch dams. Eng Struct 178:616–634
8. Mata J (2011) Interpretation of concrete dam behaviour with artificial neural network and multiple linear regression models. Eng Struct 33(3):903–910

9. Salazar F, Toledo MA, Oñate E, Morán R (2015) An empirical comparison of machine learning techniques for dam behaviour modelling. Struct Saf 56:9–17
10. Salazar F, Toledo MÁ, González JM, Oñate E (2017) Early detection of anomalies in dam performance: a methodology based on boosted regression trees. Struct Control Health Monitor 24:e2012
11. Salazar F, Toledo MÁ, Oñate E, Suárez B (2016) Interpretation of dam deformation and leakage with boosted regression trees. Eng Struct 119:230–251

Uncertainty Quantification and Reduction in the Structural Analysis of Existing Concrete Gravity Dams

G. Sevieri, A. De Falco, and G. Marmo

Abstract The failure of a large gravity dam might have catastrophic effects putting at risk human lives, not counting the considerable economic consequences. Most of dams are located in natural hazard prone areas so the structural control and the evaluation of the dam fragility (in particular against to flood and earthquake) assume great importance both to apply early warning procedures and to define resilience-enhancing strategies. Numerical models assume great importance to predict the seismic behaviour of the complex dam-soil-reservoir interacting system, nevertheless they are affected by different uncertainties. The effects of uncertainties can be reduced by calibrating finite element models with all available data about the structure. Measurements recorded by monitoring systems and in situ test results take on a major role as important sources of information. This paper investigates the effect of the uncertainties in the static and dynamic analysis of existing concrete gravity dams by means of two case studies. The general Polynomial Chaos Expansion technique is used to propagate the uncertainties through the numerical models of the case studies even without High Performance Computing. The effects of the uncertainties are thus quantified in terms of model output variation. General Polynomial Chaos Expansion-based predictive models are then used for the solution of the inverse problem thus reducing the computational burden.

Keywords Concrete dams · Uncertainty quantification · General polynomial chaos expansion · Bayesian inference · Model parameter calibration

G. Sevieri (✉)
University College London, London, UK
e-mail: g.sevieri@ucl.ac.uk

A. De Falco
University of Pisa, Pisa, Italy

G. Marmo
Italian Ministry of Infrastructure and Transport, Rome, Italy

G. Bolzon et al. (eds.), *Numerical Analysis of Dams*, Lecture Notes in Civil Engineering 91, https://doi.org/10.1007/978-3-030-51085-5_50

1 Introduction

Concrete gravity dams are fundamental infrastructures because of their use for energy production, flood control and industrial supply. The partial or total collapse of a dam might have catastrophic consequences in terms of both economic losses and human lives. In addition to having high exposures, existing concrete dams are usually located in natural hazard prone areas, so the quantification of their risk against flood and earthquake [1], as well as the development of reliable control systems, are paramount for our society. In addition, a better understanding of the risk would enable portfolio prioritizations and resilience-enhancing strategies to be developed [2].

The prediction of the static and dynamic behaviour of dams is a challenging task because of the complexity of the physical phenomena behind the interacting system dam-soil-basin. Refined numerical models are then fundamental to quantify the structural fragility and to perform dam control. However, numerical models are massively affected by uncertainties mainly related to the material mechanical characteristics and the interaction between monoliths [1]. All available information about the structure must then be used to reduce the uncertainties, thus obtaining a reliable prediction of the dam behaviour.

According to De Falco et al. [2], the main sources of uncertainty in this context can be divided in two groups. One group is related to the definition of the deterministic model. The choice of the geometrical model is crucial, as huge differences can be found between 2D or 3D models [1]. The Soil-Structure Interaction (SSI) [3] and Fluid-Structure Interaction (FSI) [4] modelling approaches strongly affect the seismic analysis of concrete gravity dams. In addition, the constitutive models adopted for dam concrete and soil affect the way the collapse is simulated [5].

The second group of uncertainties is related to model parameters: epistemic uncertainties arise once a deterministic model is defined. As discussed by De Falco et al. [5] and Sevieri et al. [6], uncertainties related to elastic parameters strongly affect numerical results.

The effect of epistemic uncertainties can be reduced through model calibration procedures [7, 8]. Such procedures aim to identify the model parameters set which lead to the best fitting between recorded and simulated behaviour of the structure. From a mathematical perspective, model calibration procedures are ill-posed inverse problems in the Hadamard's sense [9]. Regularisation approaches, defined in a deterministic or probabilistic setting [8], are then needed to solve the inverse problem. Deterministic approaches are based on the definition of objective functions, which are then minimised to achieve a good matching between recorded and simulated behaviours. Probabilistic approaches (e.g., the Bayesian inference) regularise the inverse problem by introducing additional levels of information, namely the prior distributions of the uncertain parameters. Pragmatically, probabilistic approaches are more computationally expensive than deterministic ones, but they provide parameter probability density functions (pdf) as solution. Working with parameter pdfs rather than single values is a clear advantage with the aim to perform a reliable seismic fragility analysis [8].

In this paper, the effects of the uncertainties related to the mechanical parameters of the materials on the static and dynamic behaviour of concrete dams are investigated and discussed. The general Polynomial Chaos Expansion (gPCE) technique [10] is used to propagate the uncertainties through the numerical models, thus speeding up the procedures and making possible the solution of forward and inverse problems even without High Performance Computing (HPC). The Bayesian procedures proposed by Sevieri et al. [6, 11], are applied in order to reduce the effects of the uncertainties both in static and dynamic case. Two case studies of large Italian dams show how uncertainties affect the numerical results as well as the importance of reducing them when monitoring data are available.

2 Uncertainty Quantification and Reduction

2.1 Introduction

Uncertainty Quantification (UQ) is the science of quantitative characterisation of uncertainties in both computational and real world. This is a fundamental topic in engineering field because it can be used to assess how the output of a numerical model changes as model parameters vary. The practical implication in dam engineering ranges from the possibility to account for the material uncertainties in the calculation of the structural fragility to the opportunity to calibrate a proper threshold of the measured quantity in the dam control. Several different approaches can be used to propagate uncertainties through a deterministic model. The most used is the Monte Carlo simulation [12], which allows propagating uncertainties by just running simulations until the convergence of the probabilistic analysis is achieved. These techniques rely on the available computational capability so, when it is limited, only simplified models or small number of samples can be considered. Alternatively, less computationally expensive approaches such as spectral methods [13] can be used, to reduce the computational burden. The more complex mathematical framework of these methods allows strongly reducing the number of samples needed to achieve the convergence. Among the spectral approaches the gPCE is a particularly interesting method which allows both solving the forward problem with a relatively small number of samples and defining a surrogate model of the output.

Let us consider the uncertain model output $u(\mathbf{x}, \boldsymbol{\theta}_{\text{model}})$ which is a function of deterministic parameters $\mathbf{x}$ of the model and random variables $\boldsymbol{\theta}_{\text{model}}$. The uncertain model output is then approximated through a polynomial expansion (i.e., the response surface $\hat{u}(\mathbf{x}, \boldsymbol{\theta}_{\text{model}})$ with orthogonal basis functions $\hat{u}(\mathbf{x}, \boldsymbol{\theta}_{\text{model}})$ whose form is chosen with regard to the distributions of the random parameters $\boldsymbol{\theta}_{\text{model}}$. Assuming that $\boldsymbol{u}^{(\alpha)}(\mathbf{x})$ is the matrix of the combination coefficients and $\mathbf{I}$ the so-called multi-index, the functional approximation can be written as,

$$u(\mathbf{x}, \boldsymbol{\theta}_{\text{model}}) \approx \hat{u}(\mathbf{x}, \boldsymbol{\theta}_{\text{model}}) = \sum_{\alpha \in \mathbf{I}} \boldsymbol{u}^{(\alpha)}(\mathbf{x})\, \boldsymbol{\Psi}_{(\alpha)}(\boldsymbol{\theta}_{\text{model}}). \tag{1}$$

The combination coefficients are determined with regard to some reference solutions of the deterministic model whose parameter sets are sampled from the prior distributions of $\boldsymbol{\theta}_{\text{model}}$. Different computational strategies can be adopted to calculate the coefficients (e.g., regression, Bayesian inference) [10]. Once the coefficients are calibrated, the forward problem can be easily solved by exploiting the orthogonal property of the basis functions, and $\hat{u}(\mathbf{x}, \boldsymbol{\theta}_{\text{model}})$ can be used to surrogate the numerical model output within the inverse analysis algorithm. This clearly leads to an additional error related to the discrepancy between model output and functional approximation. However, if the gPCE is properly built such an error is negligible.

The gPCE-based functional approximations are then used to create predictive models of dam displacements and modal characteristics [8, 12] needed for the solution of the inverse problem. In fact, predictive models describe the probabilistic relationship between observations, predictions (i.e., numerical model, functional approximation), random variables and error terms.

The inverse problem can thus be easily solved in the Bayesian framework by using the results of material tests for the definition of prior distributions of the model parameters. Whereas, the observations recorded by the monitoring system are used to update the state of knowledge. The epistemic uncertainties within the model are thus reduced. This procedure also enables the estimation of error term distributions [8, 12], a fundamental aspect for the comparison of different deterministic models and for the selection of the more accurate one.

In the next sections two case studies are presented in order to show how epistemic uncertainties related to the material parameters affect the result of static and dynamic analysis (i.e., UQ) and how they can be reduced through the Bayesian inference. Bulk and shear modulus, K and G respectively, of concrete and soil are treated as random variables in the following applications. In this context, the measurable deterministic parameters collected in $\mathbf{x}$ are the environmental conditions, such as basin level, air and water temperatures.

2.2 *Static Monitoring System*

Several studies have been carried out focusing on the use of the static monitoring system for model parameter calibration in dam engineering [14]. In fact, static monitoring systems, which record displacements in upstream-downstream direction and the environmental conditions (e.g., basin level, temperatures of air and water), are commonly installed on concrete dams. Most of the available studies are defined in a deterministic setting. In this case, predictive models are based on functional approximation.

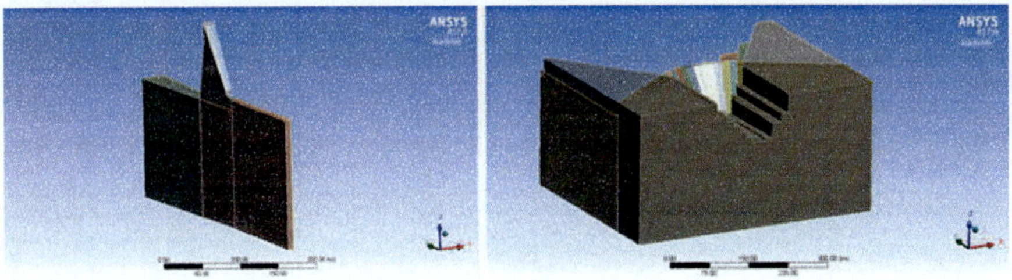

Fig. 1 Single monolith model and 3D complete model

In this section, the procedure proposed by Sevieri et al. [6] is applied for the study of a real large Italian dam. This method makes use of the gPCE (Eq. 1) and Fourier series for the definition of predictive models of the dam displacements $\delta(\mathbf{x}, \boldsymbol{\theta}_{\text{model}})$. In particular the gPCE is adopted to simulate the component of deformation related to the basin level variation $\delta^H(\mathbf{x}, \boldsymbol{\theta}_{\text{model}})$, while the Fourier series are used for the thermal part $\delta^T(\mathbf{x}, \boldsymbol{\theta}_{\text{model}})$. A linear function of time is finally used to describe the drift term $\delta^K(\text{t})$. The predictive model of the dam displacements is completed by adding the error term and a correction factor for the further reduction of the bias. The parameters collected in $\boldsymbol{\theta}_{\text{model}}$, the Fourier coefficients and the error term are treated as random variables and then updated in the solution of the inverse problem.

The analysed gravity dam has curved shape in plant and is composed of 11 monoliths separated by vertical contraction joints. The maximum height is 108 m, while the crest length is 234.25 m. Two models are built in order to simulate two different vertical joint conditions: a complete 3D model, in which monoliths are bonded, and a single monolith model representing the central spillway one. The two models (Fig. 1), are analysed by means of ANSYS r17. The complete 3D model has 29,981 Hex-dominant elements, while the single monolith one has 3587 Hex-dominant elements and free lateral displacements.

The reference displacement point is in the upper part of the central spillway monolith, where the displacements are recorded by an inverse pendulum. Figure 2 shows the displacements and the basin levels during the time period considered for the updating in this application. The elastic parameters of the concrete K_{cls} and G_{cls} as well as those of the foundation soil K_{soil} and G_{soil}, are treated as random variables and then objectives of the updating process. The prior distributions of the model parameters are defined based on the results of the in situ tests. The prior distribution statistics of the mechanical parameters are shown in Table 1.

The first step of the procedure is the construction of the gPCE. In particular, 3th order Hermitian polynomials are used as basis functions and trained on 1024 analysis results for each model. The results in terms of mean and variance are reported in Tables 2 and 3. Figure 3 shows the comparisons between prior and posterior distributions.

The posterior distributions derived for the single monolith model have mean values higher than those of the prior distributions and those of the complete model. This is due to the lack of stiffness of the single monolith model related to the absence

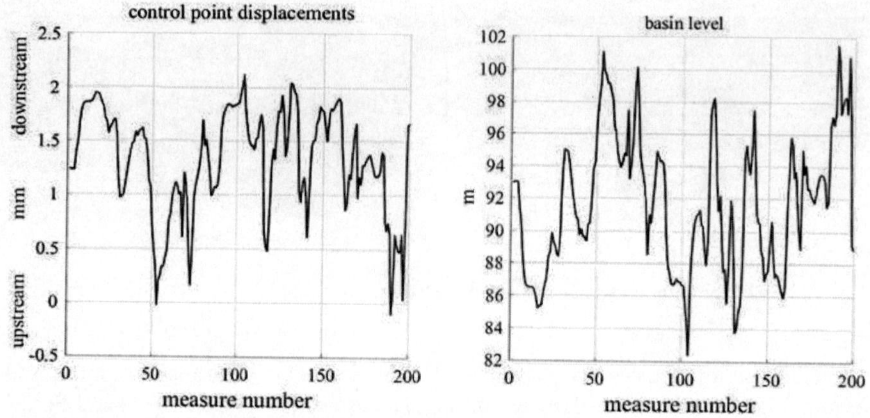

Fig. 2 Control point displacements (left) and basin level (right)

Table 1 Static case: prior distributions of the mechanical parameters

	K_{cls} (MPa)	G_{cls} (MPa)	K_{soil} (MPa)	G_{soil} (MPa)
Distributions	*LN*	*LN*	*LN*	*LN*
Mean values	12,148	8364.2	41,000	23,428
Standard deviations	4915.1	3384.2	9393.1	5367.4

Table 2 Posterior distributions of the 3D complete model parameters

	K_{cls} (MPa)	G_{cls} (MPa)	K_{soil} (MPa)	G_{soil} (MPa)	σ
Mean values	12,149	8653.9	40,769	22,517	0.0332
Standard deviations	220.17	152.05	553.10	169.53	0.0016

Table 3 Posterior distributions of the single monolith model parameters

	K_{cls} (MPa)	G_{cls} (MPa)	K_{soil} (MPa)	G_{soil} (MPa)	σ
Mean values	36,264	23,545	40,224	22,571	0.0661
Standard deviations	8436.5	5152.5	3032.3	958.29	0.0984

of interaction with other monoliths. In addition, the single monolith model shows a higher variability of the result. In fact, the posterior distributions of the single monolith model parameters have high values of standard deviations.

Even though a solution of the inverse problem can be derived for every kind of problem, the excessively high values of the concrete mechanical parameters obtained in the case of the single monolith model clearly show that the interaction between monolith must be considered in the analysis.

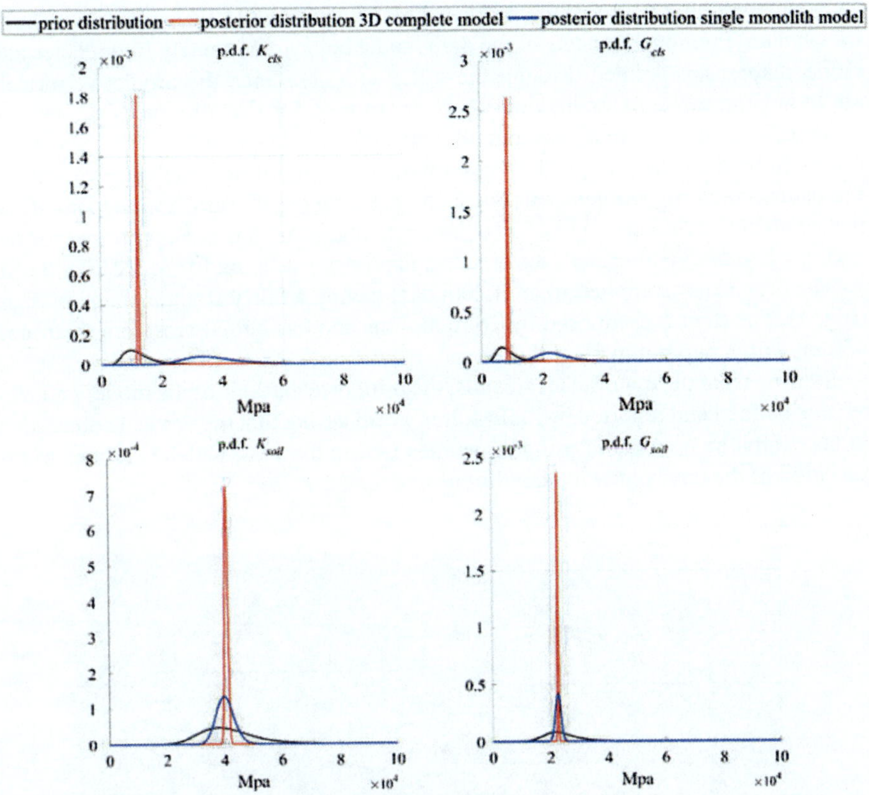

Fig. 3 Comparison between prior and posterior distributions of *K* and *G* of soil and concrete for the single monolith model and for the complete 3D model

2.3 Dynamic Monitoring System

Dynamic monitoring systems are more appropriate than static ones to control the structural health state and to calibrate dynamic model parameters. The observations recorded by a dynamic monitoring system based on ambient vibrations [15] can be either used in the updating process or elaborated through Operational Modal Analysis (OMA) in order to derive the modal characteristics of the system, i.e. frequencies, mode shapes and damping. Modal characteristics are commonly adopted as reference information in civil engineering, because their use simplifies the resolution of the inverse problem. In fact, modal analyses are thus used within the updating process.

Dynamic Structural Health Monitoring systems (SHM) are very rare in dam engineering, nevertheless some applications are available in the literature [14]. Most of them aim to verify the feasibility of the installation of dynamic monitoring systems, rather than to discuss the implementation of the structural control or the model parameter calibration.

In this section, the procedure proposed by Sevieri et al. [12] is applied to a large Italian dam. Predictive models of the dam modal behaviour, namely frequencies and mode shapes, are defined through the gPCE (Eq. 1). Once the predictive models are trained on the numerical solutions, they are used for the solution of the inverse problem instead of the model output, thus speeding up the procedure. The procedure proposed by Sevieri et al. [12] defines a numerical algorithm which allows solving the *mode matching problem* without using the concept of *system mode shapes* or that of *objective functions* [16]. This numerical algorithm has been proposed for the specific case of dynamic analyses of dams, to properly account for both SSI and FSI.

The procedure is applied to an Italian dam having a fairly straight shape in plant (Fig. 4). The dam is composed by 26 monoliths and has a total crest length around 450 m, with a maximum height of 65 m.

Starting from the material test results, the prior distributions of the model parameters are defined and reported in Table 4. It is worth noting that the vertical contraction joints exhibit an opening–closing movement during the year, which is related to the variation of the environmental conditions and temperature.

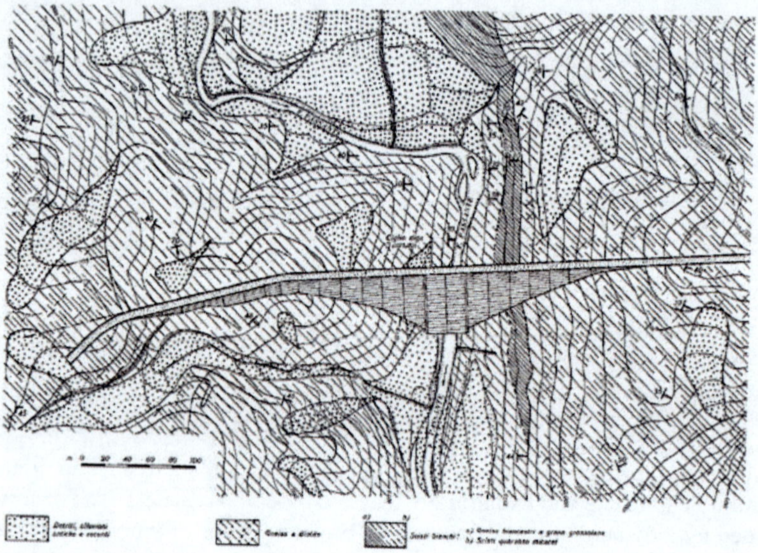

Fig. 4 Dam technical drawing and FEM

Table 4 Dynamic case: prior distributions of the mechanical parameters

	K_{cls} (MPa)	G_{cls} (MPa)	K_{soil} (MPa)	G_{soil} (MPa)
Distribution	LN	LN	LN	LN
Mean	14880.0	10424.0	19210.0	10446.0
S. D.	5824.3	2520.5	19590.0	5203.0

This observation, that cannot be directly used for the updating of the dynamic properties, must be considered by defining two limit cases:

- vertical joints are in contact, so monoliths are modeled as “bonded”;
- adjacent monoliths can have relative displacements, the contacts are modelled as “frictionless”.

In both cases the contacts are defined as relationship between two surfaces. A 3D FE model of the system dam-soil-basin is created in ABAQUS® v 6.14 (Fig. 5). It is composed of 40,638 quadratic tetrahedral mechanical elements C3D10 for the soil and 14,397 for the dam body, 28,707 linear tetrahedral acoustic elements AC3D4 for the basin and 1550 linear hexahedral one-way infinite elements as boundary condition for the soil domain. Acoustic impedances at the end of the reservoir avoid incident waves reflection.

The outputs of 350 analyses for each model are used to determine the gPCE coefficients (Eq. 1), while the polynomial expansion degrees are selected in order to minimize the errors in terms of mean and variance.

Only the modes related to the dam body are significant for the updating, while those related to the soil are usually not recorded. Three first modes that mobilize the largest amount of dam mass (Fig. 6) are chosen as reference for the solution of

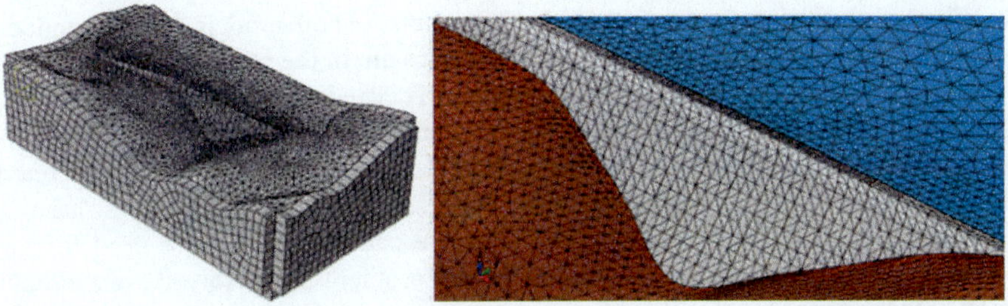

Fig. 5 Dynamic FE model

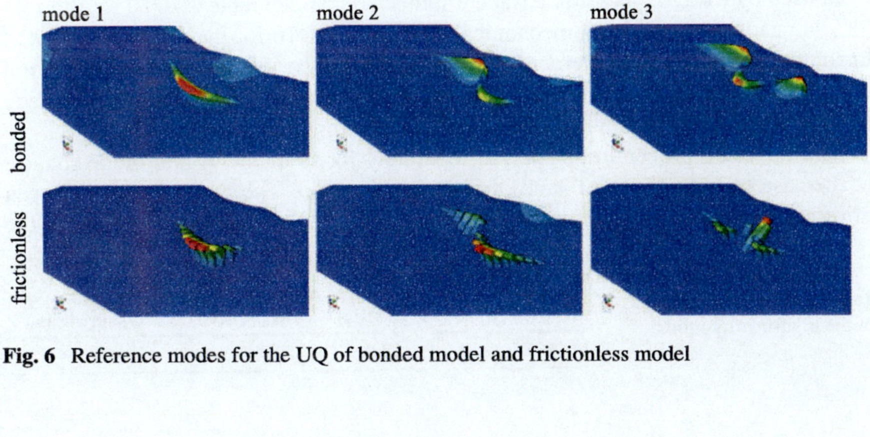

Fig. 6 Reference modes for the UQ of bonded model and frictionless model

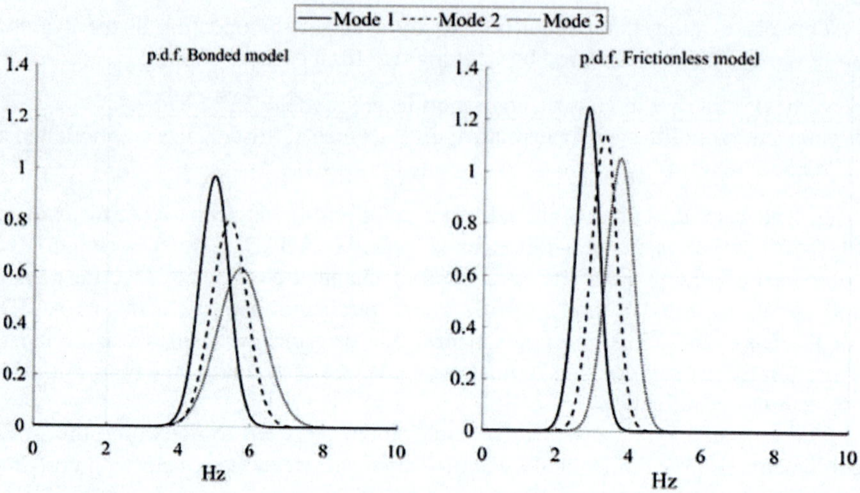

Fig. 7 Distributions of the first three frequencies

the forward problem. In particular, 5th-order Hermite polynomials are used for the definition of the basis functions for both frequencies and mode shapes.

Figure 7 shows the distributions of the frequencies of both models. In the "bonded" case the mean values of the first three frequencies are in the range 5–6 Hz, while in the "frictionless" one they are in the range 3–4 Hz. The lack of stiffness of the "frictionless" model, due to the absence of the interaction among monoliths in upstream–downstream direction, leads to lower frequencies. The standard deviations increase toward higher frequency and they are relatively smaller in the "frictionless" case.

Once the gPCE based predictive models are calibrated, they are used for the resolution of the inverse problem. In this way the problem can be solved even without HPC. Since no records of ambient vibrations are available in this study, a high-fidelity model is used to simulate the experimental behaviour of the dam. The high-fidelity model is a more refined version of the "bonded" model previously presented. It is composed by 81,276 quadratic tetrahedral mechanical elements C3D10 for the soil, 28,794 quadratic tetrahedral mechanical elements C3D10 for the dam body, 57,414 linear tetrahedral acoustic elements AC3D4 for the basin and 3100 linear hexahedral one-way infinite elements as boundary conditions for the soil domain. The elastic parameters of the high-fidelity model are reported in Table 5.

The Bayesian procedure is applied separately for frequencies and mode shapes, in the case of both "bonded" and "frictionless" model. The comparisons between prior and posterior distributions are shown in Fig. 8. For the "bonded" model the

Table 5 Elastic parameters of the high-fidelity model

K_{cls} (MPa)	G_{cls} (MPa)	K_{soil} (MPa)	G_{soil} (MPa)
16667.0	15217.0	21930.0	22321.0

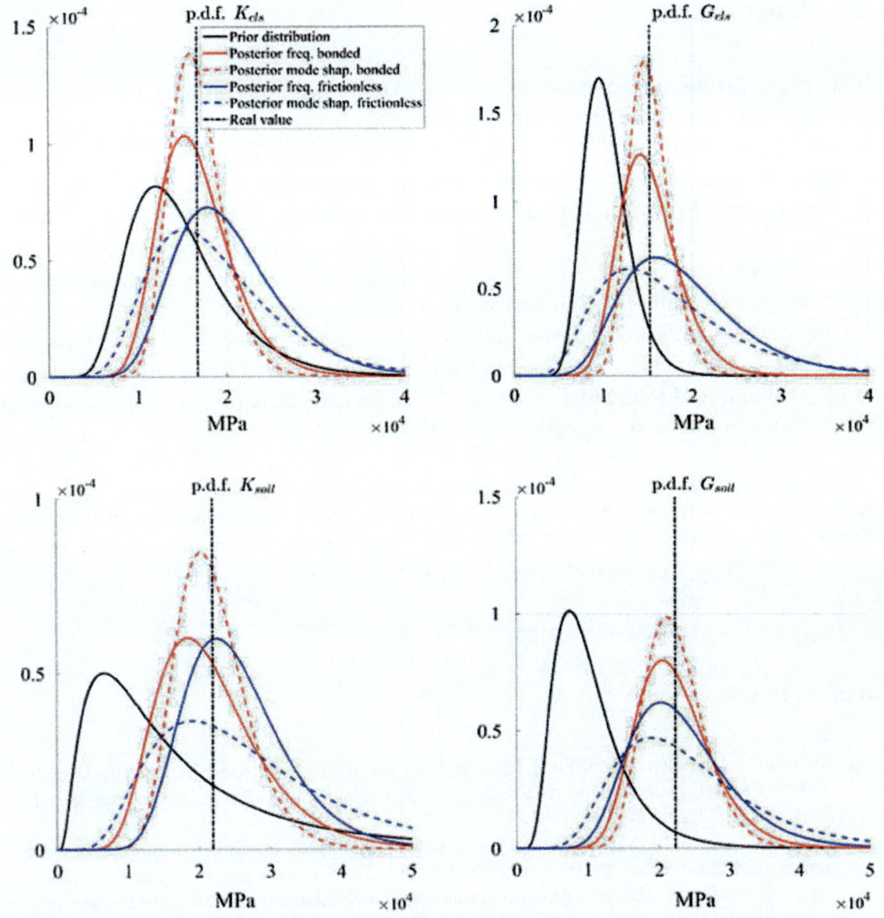

Fig. 8 Comparison between prior and posterior distributions of the elastic parameters

updating provides the same values assumed for the high-fidelity model, either using frequencies or mode shapes predictive models. In the "frictionless" case the use of different predictive models leads to different results which are in contrast with the assumed values, because of the lack of stiffness due to the absence of interaction between adjacent monoliths. These results highlight the need to perform predictive models that consider the state of the vertical contraction joints, i.e. close or open. This issue may be addressed by modelling the state of each vertical joint as random variable, and to correlate these states with the environmental conditions, i.e. reservoir level and temperatures.

3 Conclusion

This paper discusses the effect of the model parameter uncertainties on the static and dynamic analysis of concrete gravity dams. In particular two case studies are used to show the effects of the epistemic uncertainties and how to reduce them when observations recorded by a static or a dynamic monitoring system are available.

The general Polynomial Chaos Expansion technique for the definition of the predictive models of the dam behaviour allows drastically reducing the computational burden thus making possible the use of probabilistic procedures even in presence of complex models without High Performance Computing.

The Bayesian inference allows calibrating the model parameters in order to reduce the effects of the uncertainties. One of the main advantages of the Bayesian inference is the estimation of the model error which can be used to compare different kind of deterministic models and to set up a proper threshold for dam control.

Acknowledgements This research work has been supported by the Directorate of Dams of the Italian Ministry of Infrastructures and Transport and by the research programme "Multi-scale modelling in structural engineering" within the "Progetti di Ricerca di Ateneo 2018" of the University of Pisa.

References

1. Andreini M, De Falco A, Marmo G, Mori M, Sevieri G (2017) Modelling issues in the structural analysis of existing concrete gravity dams. In: Proceedings of the 85th ICOLD annual meeting, Prague, Czech Republic, pp 363–383
2. De Falco A, Mori M, Sevieri G (2018) Bayesian updating of existing concrete gravity dams model parameters using static measurements. In: Owen R, de Borst R, Reese J, Pearce C (eds) 6th European conference on computational mechanics & 7th European conference on computational fluid dynamics, ECCM-ECFD 2018, vol 1. International Center for Numerical Methods in Engineering (CIMNE), pp 2245–2256
3. De Falco A, Mori M, Sevieri G (2019) Soil-structure Interaction modeling for the dynamic analysis of concrete gravity dams. In: Papadrakakis M, Fragiadakis M (eds) 7th International conference on computational methods in structural dynamics and earthquake engineering, COMPDYN, vol 3. Institute of Structural Analysis and Antiseismic Research, School of Civil Engineering, National Technical University of Athens (NTUA), Greece, pp 5662–5673
4. De Falco A, Mori M, Sevieri G (2018) FE models for the evaluation of hydrodynamic pressure on concrete gravity dams during earthquakes. In: Owen R, de Borst R, Reese J, Pearce C (eds) 6th European conference on computational mechanics & 7th European conference on computational fluid dynamics, ECCM-ECFD, vol 1. International Center for Numerical Methods in Engineering (CIMNE), pp 1731–1742
5. De Falco A, Mori M, Sevieri G, Zani N (2017) Simulation of concrete crack development in seismic assessment of existing gravity dams. In: Braga F, Salvatore W, Vignoli A (eds) 17th Convegno ANIDIS—L'ingegneria Sismica in Italia, vol 1. Pisa University Press, pp 1–2
6. Sevieri G, Andreini M, De Falco A, Matthies HG (2019) Concrete gravity dams model parameters updating using static measurements. Eng Struct 196, article 109231
7. International Commission on Large Dams (2000) Bullettin 118—automated dam monitoring systems guidelines and case histories. ICOLD, Paris

8. De Falco A, Girardi M, Pellegrini D, Robol L, Sevieri G (2018) Model parameter estimation using Bayesian and deterministic approaches: the case study of the Maddalena Bridge. Procedia Struct Integr 11:210–217
9. Hadamard J (1923) Lectures on Cauchy's problem in linear partial differential equations. New Haven Yale University Press
10. Xiu D (2010) Numerical methods for stochastic computations, 1st edn. Princeton University Press
11. Sevieri G, De Falco A (2010) Concrete gravity dams FE model parameters updating using ambient vibrations. In: Papadrakakis M, Papadopoulos V, Stefanou G (eds) 3rd international conference on uncertainty quantification in computational sciences and engineering, vol 1. UNCECOMP, Institute of Structural Analysis and Antiseismic Research, School of Civil Engineering, National Technical University of Athens (NTUA), Greece, pp 286–297
12. Kottegoda NT, Rosso R (2008) Applied statistics for civil and environmental engineers, 2nd edn. Blackwell Publishing
13. Ghanem RG, Spanos PD (1991) Stochastic finite elements: a spectral approach, 1st edn. Springer, New York
14. Bukenya P, Moyo P, Beushausen H, Oosthuizen C (2014) Health monitoring of concrete dams: a literature review. J Civil Struct Health Monitor 4:235–244
15. Sevieri G, De Falco A (2020) Dynamic structural health monitoring for concrete gravity dams based on the Bayesian inference. Journal of Civil Structural Health Monitoring 10:235–250
16. Huang Y, Shao C, Wu B, Beck JL, Li H (2018) State-of-the-art review on Bayesian inference in structural system identification and damage assessment. Adv Struct Eng 22(6):1329–1351

A Non-destructive Parameter Identification for an Embankment Dam

J. Toromanovic, H. Mattsson, J. Laue, and S. Knutsson

Abstract When performing predictions of future behaviour and assessments of the safety for embankment dams, numerical modelling if often needed as support. For embankment dams a usual issue is to obtain values of the material parameters for the material in the dam. Sampling is not easily performed, and it could also negatively affect the performance and safety of the dam. One way to determine values of the material parameters is to create a digital copy of the dam by utilizing so called inverse analysis. The created numerical model is calibrated towards field measurements. To make the reality and the numerical model correspond to each other, the values of material parameters for constitutive models are calibrated. The calibration is done by an automatized process. Further, the calibrated model could be used in predictions of future dam behaviour. The methodology has been shown to work for field data containing errors of a usual magnitude.

Keywords Embankment dam · Inverse analysis · Digital twin · Predictions · Noise

1 Digital Twins and Embankment Dams

When assessing deformations and stability for earth- and rockfill dams, numerical modelling is often used as support. This has e.g. been common in situations when traditional methods based on limit equilibrium are regarded as not sufficient [1].

In order to perform reliable numerical analyses, values of material parameters defining the stress-strain relationships for the different materials in the dam are needed. The materials of an embankment dam vary, as the different zones have different performance. For already existing dams, it is often a challenge to obtain information about the mechanical behaviour of the materials. Many dams are old and a limited amount of information may be available about the materials and the method of construction. In geotechnics, usually, sampling is performed and followed by laboratory testing in order to determine the mechanical behaviour of the materials.

J. Toromanovic (✉) · H. Mattsson · J. Laue · S. Knutsson
Luleå University of Technology, Luleå, Sweden
e-mail: jasmina.toromanovic@ltu.se

G. Bolzon et al. (eds.), *Numerical Analysis of Dams*, Lecture Notes in Civil Engineering 91, https://doi.org/10.1007/978-3-030-51085-5_51

Traditional sampling may affect the dam body negatively considering the performance and safety, especially if the sampling is performed in the central impervious part of the dam. Non-destructive methods would therefore be suitable to apply when there are negative consequences of utilising traditional sampling.

Today methods based on artificial intelligence are used to make replicas of real objects. The air plane industry have come far with their concept of "digital twins" [2]. The concept is based on that a model of the real object is updated towards real measurements, where the method can identify critical components and give suggestions for amendments. For embankment dams this method has not been extensively tested. Measurements are performed on and inside an embankment dam, these measurements could be utilised to create a digital twin through inverse analysis. If this is done, the need of destructive sampling decreases.

In this paper, the methodology of inverse analysis and its application on an embankment dam is presented. This is followed by predictions of the dam behaviour when dam safety measures are performed, where the modelling is based on values of parameters obtained from inverse analysis. Finally, effects of measurement errors on inverse analysis are discussed.

2 Case Study

The studied dam is a traditional embankment dam, built in the 1960s to a height of 45 m. The dam is constructed with a central core consisting of till. Fine and coarse filters are placed on each side of the core. The shoulders consist of rockfill. Dam safety measures, in form of stabilising berms, have been undertaken on the downstream side of the dam.

A section of the dam is presented in Fig. 1. The sketch is somewhat simplified, since there are horizontal filters in the dam also. There are different types of instrumentation in the dam for measurement of deformations, pore pressure and seepage. In Fig. 1, the position of an inclinometer is shown.

The numerical modelling of the dam was performed with the finite element program PLAXIS 2D. The modelling was done in the following steps:

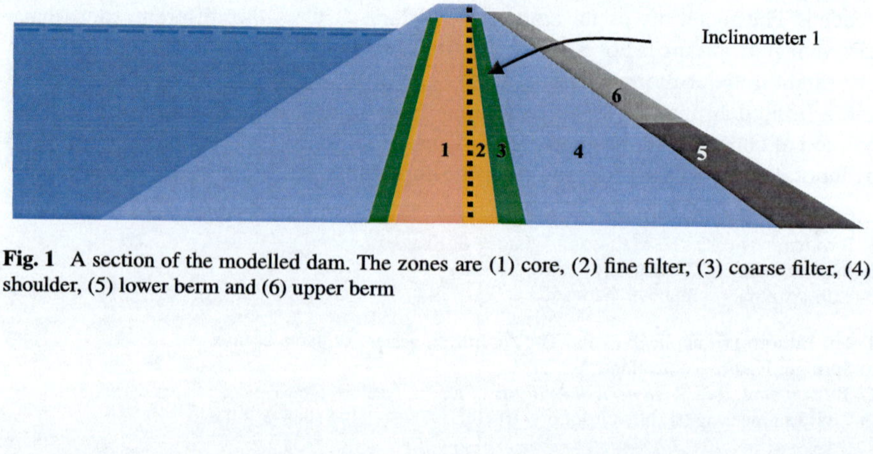

Fig. 1 A section of the modelled dam. The zones are (1) core, (2) fine filter, (3) coarse filter, (4) shoulder, (5) lower berm and (6) upper berm

- **Construction of the dam**
  - Modelled step-wise with horizontal layers
- **Impoundment of the reservoir**
- **Construction of the downstream berms**
  - Lower part of berm
    Upper part of berm

In order to perform the modelling, knowledge about the stress-strain behaviour in the dam is needed. Here the problems arise with inadequate information of mechanical properties and difficulties to retrieve samples. Therefore, it would be beneficial to obtain values of material parameters through inverse analysis.

3 Inverse Analysis—A Non-destructive Method for Retrieving Values of Material Parameters

3.1 Why Inverse Analysis?

Inverse analysis is a non-destructive method for identification of values for material parameters. The method is based on calibrating numerical models towards field measurements. In the calibration process the values of the material parameters are changed, the input data for the chosen stress-strain relationship (constitutive model), until the difference between the result from the chosen field measurement and the associated response from the numerical model is minimised. In order to automatize the process, a search algorithm is utilised. The calibration of the numerical model is done when the mentioned discrepancy is minimised. The result of the inverse analysis are ranges of appropriate values of the material parameters for the materials in the dam, that were retrieved without any physical damage on the dam structure. A sketch of the process is shown in Fig. 2.

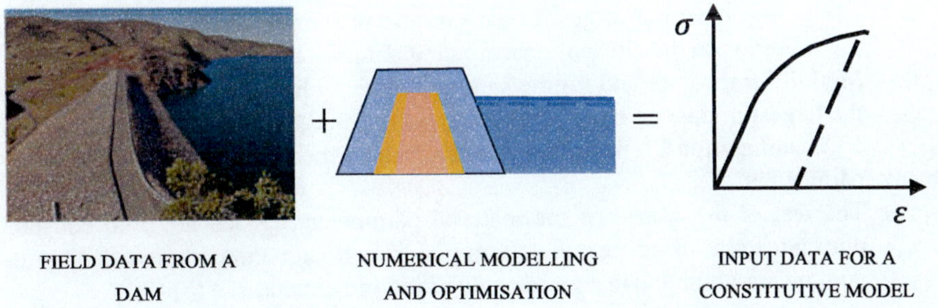

Fig. 2 From field data to input data

Based on ideas by [3], inverse analysis has been applied in geomechanics by e.g. [4–6]. Different search strategies have been applied in inverse problems. A genetic algorithm, a form of artificial intelligence, was firstly introduced in a geomechanical inverse problem by [7]. At Luleå University of Technology this research has been applied to embankment dams [8, 9]. Considering that it is difficult to take up samples from an existing embankment dam, inverse analysis is a suitable tool to determine mechanical behaviour instead of traditional destructive sampling.

3.2 *Calibration Process—How is the Inverse Analysis Performed for the Embankment Dam?*

Inverse analysis has been performed for the dam in Fig. 1. In order to carry out the inverse analysis, some preparations are needed:

- **Measurement data**
  - Choice of what monitoring data from field to use.
  - Horizontal deformations from "Inclinometer 1" have been used.
- **Numerical model**
  - A numerical model of the dam is created in PLAXIS 2D.
  - A section of the dam is modelled, suitable for assuming plane strain conditions, making two dimensional modelling reasonable.
  - The case of adding the upper part of the downstream berm has been used for the inverse analysis. Deformations have been reset before the construction of the lower part of the berm.
- **Constitutive model**
  - The constitutive model Hardening soil [10] has been utilised.
- **What parameters to optimise?**
  - A sensitivity study is performed to assess which parameters in the constitutive model that have the largest impact on the response in the numerical model associated with the chosen measurement data.
  - Moduli for the core and for the shoulder material have been identified to have the largest impact on horizontal deformations in the position of "Inclinometer 1" in simulations. These two quantities have been chosen as optimisation parameters.
  - The rest of the values of the material parameters needed are held constant during the optimisation, and determined from documentation provided by the dam owner and by using some engineering judgement, see [8].

When the preparations for the inverse analysis have been done, values of the chosen material parameters are going to be optimised. The inverse analysis is performed by the finite element program communicating with the optimisation program. The optimisation program consists of:

- **Objective function**
  - A function that calculates the difference between the measured values in field and the associated numerical response. The goal is to minimise the difference during the optimisation.
- **Search strategy**
  - Searches for solutions, values of the optimisation parameters. It has a strategy to identify the best solutions.
  - A genetic algorithm was utilised, that is a form of artificial intelligence, based on [11–13].
- **Interaction with a numerical model**
  - A numerical model is updated until the objective function is minimised.

A loop in Fig. 3 corresponds to testing of a value of the modulus for the core and a value of the modulus for the shoulder material by a computation in PLAXIS. Initially, the algorithm is searching for solutions randomly within the search domain. Thereafter, focus is on improving the results through so-called "generations". The algorithm is based on evolutionary principles where the solutions with least objective function value are kept. Even better solutions are created in coming generations using the present solutions as a base—"survival of the fittest". The solutions are gradually focused to a smaller area in the search domain when the algorithm works. The fact that the solutions are improving is also seen in the decreasing values of the objective function, which means that the difference between simulated values and real measurement values are decreasing. In the end, the best solutions are presented in form of appropriate intervals for the chosen optimisation parameters. In Fig. 4 it is shown how the algorithm works.

4 Predictions with Values of Material Parameters from Inverse Analysis

A big advantage of performing inverse analysis, is that this calibration process results in values of material parameters that can be further used in modelling of future measures.

When a strengthening measure for the dam from Fig. 1 was planned, the dam owner wanted a prediction of the deformations that were expected due to the new construction. The dam was to be strengthened with a new berm on the downstream

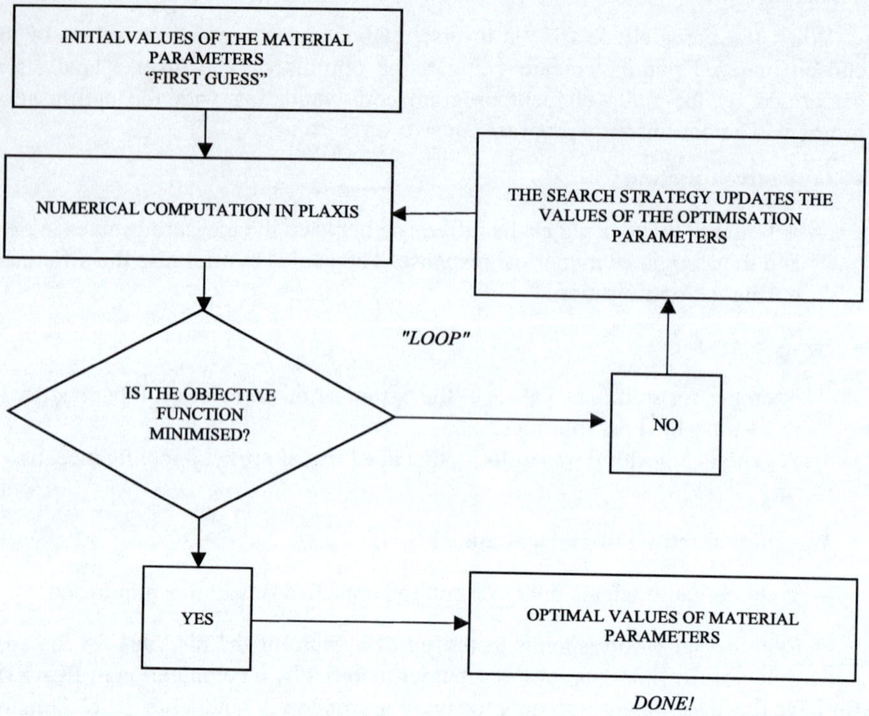

Fig. 3 Calibration process

side, see Fig. 5. Based on values of material parameters from inverse analysis, for a geometry without the new berm, a numerical analysis of the deformations caused by construction of the new berm was done. When the strengthening action was performed, a new inclinometer was installed—"Inclinometer 2" as shown in Fig. 5. From "Inclinometer 2" measurement data became available for comparison with the computed deformations.

In Fig. 6 inclinometer deformations from field and modelled deformations in the corresponding position in the numerical model are compared. From the calibration process several good solutions, values of the optimisation parameters, are obtained. The values of the optimisation parameters can be interpreted as intervals. If new simulations are done based on results from an optimisation performed, with minimum and maximum values of the optimisation parameters, the grey area in Fig. 6 is obtained as the model response. If the modelling is based on average values, the dashed line is obtained. A reasonable agreement between modelled and observed deformations were obtained. The study indicated that dam behaviour can be predicted based on values obtained through inverse analysis.

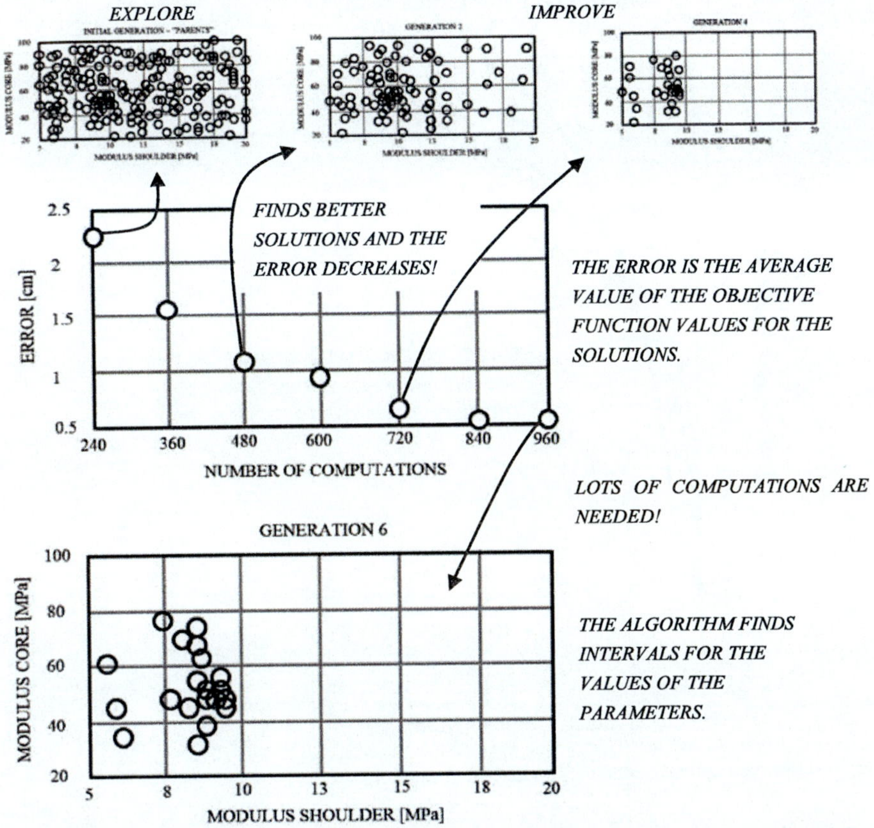

Fig. 4 The amount of solutions kept is gradually decreasing in the search of new and better solutions

Fig. 5 Modelled dam with the new berm

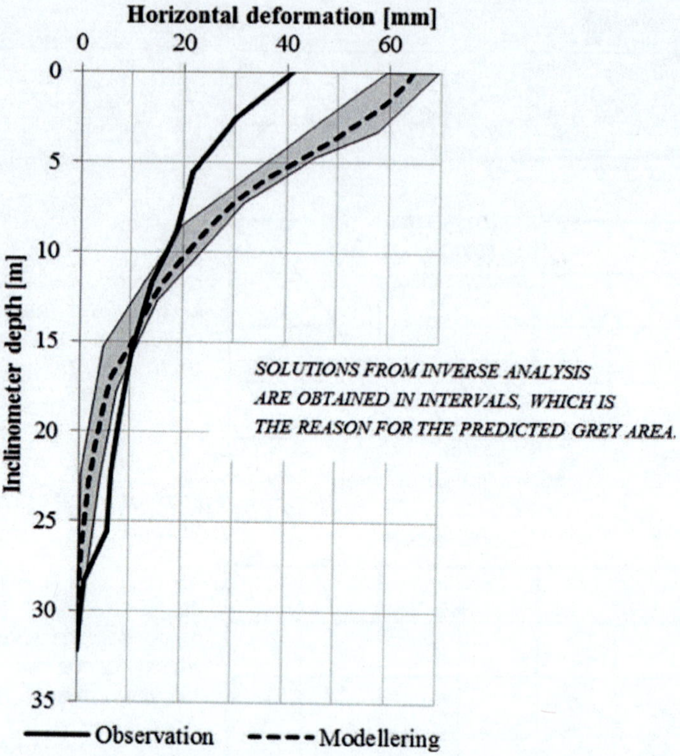

Fig. 6 Observed and modelled inclinometer deformations

5 Can Errors in Measurement Data Be Handled by Inverse Analysis?

Errors are usually occurring in field measurements, related to e.g. handling, installation or even the precision of the equipment. In order to assess how the chosen genetic algorithm handles errors, inverse analyses towards data containing errors were performed. In the study, data including measurement errors has been numerically generated.

In Fig. 7a, a sketch of the deformations from "Inclinometer 1" is shown, after construction of the upper part of the berm in Fig. 1. The horizontal deformations are represented by ΔU. The deformations in Fig. 7a have been numerically simulated from an optimal solution, named "OPT" in Fig. 8. The solution "OPT" was obtained by inverse analysis towards real measurements. In Fig. 7b) errors have been added to the deformations.

In order to assess if the applied genetic algorithm can find good solutions when the inclinometer deformations are containing errors, inverse analyses have been done towards inclinometer data containing errors. The proximity of the found solution for

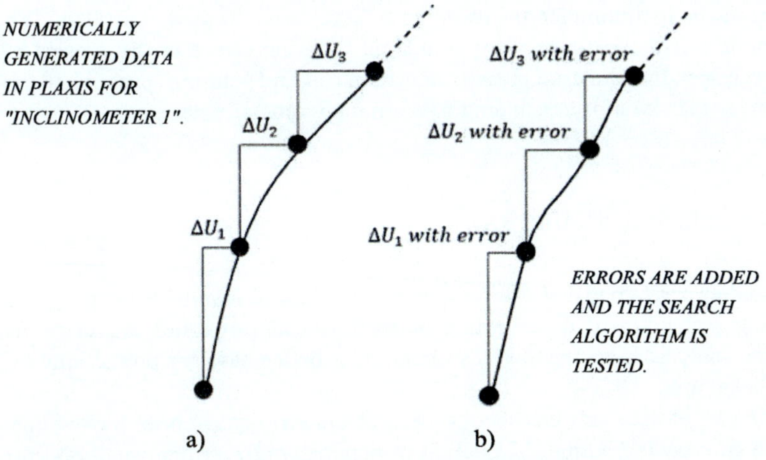

Fig. 7 Sketch of how the errors are added to the inclinometer data

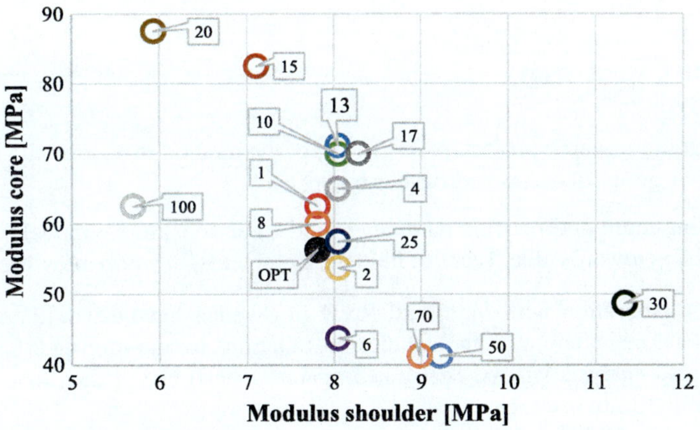

Fig. 8 The best solutions for inclinometer data with errors

data containing errors to the optimal solution "OPT" has been examined, see Fig. 8. The number next to the solutions is presenting a percentage, this percentage is the "amount of error" added to the incremental deformation.

Solutions for many of the lower errors are gathering around "OPT" in Fig. 8. Even if the inclinometer data contains errors, reasonable solutions are still found by the utilised genetic algorithm for the lower errors which are of a magnitude considered as normal in inclinometer measurements. The utilised algorithm continues to find solutions, without any numerical problems when errors are added. This ability is regarded as positive, since errors are likely to occur in field measurements.

Errors in inclinometer measurements have been studied, see [14]. When the magnitude of these errors were compared with the errors in the present study, it has been seen that the used genetic algorithm can find solutions even when the errors are larger than what is considered usual in inclinometer data.

6 Concluding Remarks

The studies performed have shown that inverse analysis is suitable to apply on already existing embankment dams, when the mechanical properties are to be decided. Inverse analysis is a step towards obtaining a better base for predictions of future dam behaviour.

In a part of the study effects of measurement errors have been looked upon. The search strategy is finding good values of parameters for errors usually occurring in inclinometer measurements. The measurement data utilised in inverse analysis does not need to be flawless in order to be useful.

7 Future Research

In the ongoing research project, practical use of the method inverse analysis on dam structures is going to be studied further by:

- including more optimisation parameters and other constitutive models
- optimising towards other types of field measurements, e.g. pore pressure, seepage
  - A collaboration with Vattenfall R&D in Sweden has been initiated. A test embankment dam with geotechnical installations for monitoring is being built. Inverse analysis will be performed against measurement data from this test dam in future research
- applying other search algorithms
- comparing results from inverse analysis to laboratory test results
  - A collaboration with LKAB in Sweden has been started, where the results from inverse analysis of a tailings dam is going to be compared to laboratory test results of the material in the dam.

References

1. Clough RW, Woodward RJ (1967) Analysis of embankment stresses and deformations. J Soil Mech Found Div 96(SM4):1629–1653
2. General Electric GE. The Digital Twin. https://www.ge.com/digital/lp/digital-twin-compressing-time-value-digital-industrial-companies. Accessed 14 Aug 2019
3. Tarantola A (1987) Inverse problem theory. Elsevier, Amsterdam, The Netherlands
4. Gioda G, Maier G (1980) Direct search solution of an inverse problem in elastoplasticity: identification of cohesion, friction angle and in situ stress by pressure tunnel tests. Int J Numer Methods Eng 15(12):1823–1848
5. Swoboda G, Ichikawa Y, Dong Q, Zaki M (1999) Back analysis of large geotechnical models. Int J Numer Anal Methods Geomech 23(13):1455–1472
6. Gens A, Ledesma A, Alonso EE (1996) Estimation of parameters in geotechnical backanalysis—II. Application to a tunnel excavation problem. Comput Geotech 18(1):29–46
7. Levasseur S (2007) Inverse analysis in geotechnics: development of a method based on genetic algorithms. University Joseph Fourier, Grenoble, France
8. Toromanovic J (2018) On parameter identification for better predictions of dam behaviour. Luleå University of Technology, Luleå, Sweden
9. Vahdati P (2014) Identification of soil parameters in an embankment dam by mathematical optimization. Luleå University of Technology, Luleå, Sweden
10. Schanz T, Vermeer A, Bonnier P (1999) The hardening soil model: formulation and verification. In: Proceedings of beyond 2000 computational geotechnics 10 years PLAXIS, Rotterdam, The Netherlands, pp 281–296
11. Holland JH (1975) Adaptation in natural and artificial systems: an introductory analysis with applications to biology, control, and artificial intelligence. MIT Press, Cambridge, MA, USA
12. Goldberg D (1989) Genetic algorithms in search, optimization, and machine learning. Addison-Wesley, Reading, MA, USA
13. Levasseur S, Malécot Y, Boulon M, Flavigny E (2008) Soil parameter identification using a genetic algorithm. Int J Numer Anal Methods Geomech 32(2):189–213
14. Mikkelsen P (2003) Advances in inclinometer data analysis. In: Proceedings of FMGM—symposium on field measurements in geomechanics, Oslo, Norway